W9-AJQ-830

Concepts in

ELEVENTH EDITION

Biology

Eldon D. Enger
Frederick C. Ross
David B. Bailey

Delta College

Mc Graw Hill **Higher Education**

Boston   Burr Ridge, IL   Dubuque, IA   Madison, WI   New York   San Francisco   St. Louis
Bangkok   Bogotá   Caracas   Kuala Lumpur   Lisbon   London   Madrid   Mexico City
Milan   Montreal   New Delhi   Santiago   Seoul   Singapore   Sydney   Taipei   Toronto

Higher Education

CONCEPTS IN BIOLOGY, ELEVENTH EDITION

Published by McGraw-Hill, a business unit of The McGraw-Hill Companies, Inc., 1221 Avenue of the Americas, New York, NY 10020. Copyright © 2005, 2003, 2000, 1997 by The McGraw-Hill Companies, Inc. All rights reserved. No part of this publication may be reproduced or distributed in any form or by any means, or stored in a database or retrieval system, without the prior written consent of The McGraw-Hill Companies, Inc., including, but not limited to, in any network or other electronic storage or transmission, or broadcast for distance learning.

Some ancillaries, including electronic and print components, may not be available to customers outside the United States.

This book is printed on recycled, acid-free paper containing 10% postconsumer waste.

International   1 2 3 4 5 6 7 8 9 0 QPD/QPD 0 9 8 7 6 5 4 3
Domestic   1 2 3 4 5 6 7 8 9 0 QPD/QPD 0 9 8 7 6 5 4 3

ISBN 0–07–255289–1
ISBN 0–07–111121–2 (ISE)

Publisher: *Martin J. Lange*
Sponsoring editor: *Thomas C. Lyon*
Senior developmental editor: *Deborah Allen*
Director of development: *Kristine Tibbetts*
Marketing manager: *Tamara Maury*
Lead project manager: *Peggy J. Selle*
Senior production supervisor: *Laura Fuller*
Senior media project manager: *Jodi K. Banowetz*
Senior media technology producer: *John J. Theobald*
Designer: *Rick D. Noel*
Cover designer: *Mary Sailer, Spring Valley Studio*
Cover image: *Getty Images, Inc., CA21916 Dragonfly, Larry West*
Senior photo research coordinator: *Lori Hancock*
Photo research: *LouAnn K. Wilson*
Supplement producer: *Brenda A. Ernzen*
Compositor: *Shepherd, Inc.*
Typeface: *10/12 Sabon*
Printer: *Quebecor World Dubuque, IA*

The credits section for this book begins on page 560 and is considered an extension of the copyright page.

Library of Congress Cataloging-in-Publication Data

Enger, Eldon D.
  Concepts in biology / Eldon D. Enger, Frederick C. Ross, David B. Bailey. — 11th ed.
    p.   cm.
  ISBN 0–07–255289–1 (hard : alk. paper)
  1. Biology. I. Ross, Frederick C. II. Bailey, David B., 1967–. III. Title.

QH308.2.E54   2005
570—dc22
                                        2003020497
                                        CIP

INTERNATIONAL EDITION ISBN 0–07–111121–2
Copyright © 2005. Exclusive rights by The McGraw-Hill Companies, Inc., for manufacture and export. This book cannot be re-exported from the country to which it is sold by McGraw-Hill. The International Edition is not available in North America.

www.mhhe.com

Brief Contents

Contents

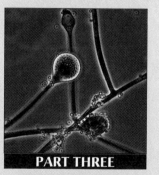

PART THREE

Cell Division and Heredity   157

8 Mitosis: The Cell-Copying Process   157

PART FOUR

Evolution and Ecology 205

PART FIVE

Physiological Processes   339

PART SIX

The Origin and Classification of Life   435

Preface

There are several things about this eleventh edition of *Concepts in Biology* that we find exciting. A major change is that David Bailey has joined the author team. David, a gifted teacher and valued colleague, was previously involved in many behind-the-scenes activities related to the Online Learning Center. With this edition he has taken on additional responsibilities for revising a portion of the text. This revision, as with previous editions, is very much a collaborative effort. When we approach a revision, we carefully consider comments and criticisms of reviewers and discuss how to address their suggestions and concerns. As we proceed through the revision process, we solicit input from one another and we critique each other's work. We feel the result is a readable, accurate, current, and interesting introduction to the science of biology.

The Eleventh Edition

Now in the eleventh edition, the origin of this book remains deeply rooted in our concern for the education of college students in the field of biology. We believe that large, thick books intimidate introductory-level students who are already anxious about taking science courses. With each edition, we have worked hard to provide a book that is useful, interesting, and engaging to students while introducing them to the core concepts and current state of the science.

There are several structural changes in this eleventh edition:

There are **more headings** to show relationships among topics and to divide material into easily digested sections.

The former Key Concepts and Applications have been reorganized into **Goals and Objectives,** as suggested by several reviewers. Each objective is a measurable activity that allows students to assess their mastery of the material as they progress through a chapter.

The **Learning Connections** at the end of each chapter have been reorganized and updated, and correspond precisely to the online study material. In addition, over 100 **new questions** have been written and many others were revised.

Cross-references have been added throughout the text that call attention to related or background material located elsewhere in the text.

A list of **prefixes** and **suffixes** has been added to the inside covers to serve as a ready reference and to help students master new vocabulary and appreciate the origin of many scientific words.

Three chapters were completely rewritten in this edition: **Chapter 2, *The Basics of Life: Chemistry,*** provides a wider survey of chemistry topics. The chapter presents the basic concepts of energy and matter upon which later chapters will build as they explain biological structures and processes from the nature of cells to the complexity of ecosystems. To make these topics easier to follow, all topics dealing with chemistry or physics have been correlated with biological examples that should be familiar to most students.

Chapter 6, *Biochemical Pathways,* was reorganized to discuss processes common to both aerobic cellular respiration and photosynthesis. An overview of both respiration and photosynthesis is presented before each is discussed in more detail. Because of the differing desires of instructors, this material is organized into two different levels: fundamental and detailed. Care was taken to make sure that figures and their descriptions appear close to one another so that students can easily follow processes as they are described. New boxed readings discuss the evolution of photosynthesis and how knowledge of metabolic processes can be applied in developing herbicides and treating disease.

Chapter 7, *DNA and RNA: The Molecular Basis of Heredity,* has been rewritten to provide a better transition from enzymes and metabolism to cell division and inheritance. This new format stresses connections between the cell's need to reliably produce enzymes and the processes that regulate protein production. New terms and explanations have been added to help students identify with news items in the emerging fields of genomics and proteomics. As a result of the reorganization of these chapters, many new illustrations have been substituted for previous illustrations. Most of the remaining illustrations were modified as well.

There are many other changes throughout the text that strengthen the content and deal with emerging topics. Some of these include:

Part 2, *Cells: Anatomy and Action,* has additional material on the glycemic index, signal transduction in cells, the function of immune system cells, metabolic disorders and enzyme activity, and the nature of vitamins, cofactors, and coenzymes.

Part 3, *Cell Division and Heredity,* has new material on cell organelles, stem cells, cloning, the human genome project, and cystic fibrosis.

Part 4, *Evolution and Ecology,* was modified so that material on changes in beak sizes of Darwin's finches replaces the section on peppered moths. New material was added about the evolution of diseases such as SARS, monkey pox, and new strains of influenza. A new section on stabilizing, directional, and disruptive selection was added. The title of chapter 17 was changed to *Evolutionary and Ecological Aspects of Behavior* to more accurately reflect the content of the chapter. Also, changes were made in the headings and order of material to make it easier for the reader to follow.

Part 5, *Physiological Processes,* has new material on the molecular basis of obesity and measuring the caloric value of foods.

Part 6, *The Origin and Classification of Life,* has greatly modified sections on geologic time, AIDS, and cladistics. Additions include new material on microbes and biological warfare, sudden oak death, emerald ash borer, how antidiarrheal preparations work, and products obtained from plants. In addition, there have been changes in headings and organization to make the flow of material easier to follow.

Features

Each chapter in this text contains a number of features that actively involve students in the learning process:

Chapter Outline
At the opening of each chapter, the outline lists the major headings in the chapter as well as the boxed readings.

Goals and Objectives
The goals provide a general statement about the significance and importance of the material. The objectives give more precise directions to readers about what they should be able to do after studying the chapter.

Topical Headings
Throughout the chapter, headings subdivide the material into meaningful sections that help the reader recognize and organize information.

Full-Color Graphics
The line drawings and photographs illustrate concepts or associate new concepts with previously mastered information. Every illustration emphasizes a point or helps teach a concept.

How Science Works and Outlooks
Each of these boxed readings was designed to catch the interest of the reader by providing alternative views, historical perspectives, or interesting snippets of information related to the content of the chapter.

Cross References
Cross references have been added throughout that call attention to related or background material located elsewhere in the text. These guide students to review basic coverage as they study new topics and to see connections between concepts.

Chapter Summary
The summary at the end of each chapter clearly reviews the concepts presented.

Thinking Critically
This feature gives students an opportunity to think through problems logically and arrive at conclusions based on the concepts presented in the chapters. Guidelines to assist the student in thinking about these questions are found on the *Online Learning Center.*

Concept Maps
Constructing concept maps provides students with an opportunity to strengthen their understanding of the chapter by organizing terms or ideas from the chapter into a logical relationship. Sample solutions and additional interactive concept maps for each chapter are found on the *Online Learning Center.*

Key Terms
A list of key terms in each chapter helps students identify the vocabulary they need to understand the concepts and ideas presented in the chapter. Definitions are found in the glossary at the end of the text. Students can practice learning Key Terms with interactive flashcards on the *Online Learning Center.*

Learning Connections
Each chapter ends with Learning Connections that (1) provide review questions keyed to each major heading and (2) identify additional resources available on the *Online Learning Center.*

Student Support Materials
Online Learning Center

This online resource offers an extensive array of resources to support the learning experience, including the **Essential Study Partner,** a collection of interactive study modules. Other online resources include:

- Chapter multiple-choice quizzes with immediate feedback
- Animations and simulations to help students understand difficult concepts

- PowerWeb: Biology—A great resource for research papers; with weekly updates
- Interactive flashcards to help students learn key vocabulary terms
- Answers to questions and problems in the text
- Sample solutions for concept maps in the text, plus additional concept maps
- BioLabs to help students master skills vital to success in the laboratory with dynamically generated data, a thorough review, and a printable lab report
- Labeling Exercises, Case Studies, Key Points, Quick Overviews, and more!

www.mhhe.com/enger11

Student Study Guide ISBN 0-07-284303-9

This student study guide provides an overview for each chapter as well as a set of learning activities. There is a focus on vocabulary terms and three types of questions to help students measure their mastery of the learning objectives for each chapter. These include short answer questions, label/diagram/explain activities, and multiple choice questions. Answers to all questions are provided in the Study Guide.

Laboratory Manual ISBN 0-07-255290-5

The Laboratory Manual features 29 carefully designed, class-tested exploratory investigations. Each exercise contains safety messages, an introduction to the material, step-by-step procedures, ample space to record and graph data, and review questions. The laboratory gives students an opportunity to "get their hands into biology," going beyond reading and studying to actually enter into the process of doing science.

Instructor Support Materials

Digital Content Manager

This easy-to-use instructor's CD-ROM contains every piece of line art and many of the photographs and tables from the text in both a .jpg format and a PowerPoint file that allows you to easily integrate full-color images into your classroom presentations. In addition, you will find a library of 700 photos, arranged by topic. Finally, 90 pieces of McGraw-Hill's Active Art provided in a PowerPoint format allow you to visually build complex topics step-by-step in your classroom. Your Digital Content Manager puts you in control of a powerful set of image resources to engage students throughout your lectures.

Instructor's Testing and Resource CD

This CD-ROM provides powerful tools for creating exams utilizing Brownstone Diploma® testing software for Windows or Mac. The CD also contains Word files of the test bank that you can edit or print. In addition, the CD contains the Instructor's Manual and the Laboratory Resource Guide.

Instructor's Manual

For each chapter, the Instructor's Manual provides:

1. A brief narrative summarizing the chapter;
2. A list of the chapter goals and objectives that can be printed or used to create student handouts; and
3. A list of the overhead transparencies and DCM resources supporting that chapter.

The Instructor's Manual can be found on the Instructor's Testing and Resource CD-ROM and online in the Instructor's Edition of the Online Learning Center at www.mhhe.com/enger11.

Overhead Transparencies

A set of 150 full-color **transparencies** is available free to adopters of the eleventh edition of *Concepts in Biology*. This set includes tables and figures from the text.

Laboratory Resource Guide

The Laboratory Resource Guide provides information on acquiring, organizing, and preparing laboratory equipment and supplies. Estimates of the time required for students to complete individual laboratory experiences are provided, as well as answers to questions in the laboratory manual. The Laboratory Resource Guide can be found on the Instructor's Testing and Resource CD-ROM and online in the Instructor's Edition of the Online Learning Center at www.mhhe.com/enger11.

Online Learning Center

Instructors have access to all student materials available through the Online Learning Center. In addition, the Instructor's Edition of the OLC provides access to the Instructor's Manual, the Laboratory Resource Guide, concept map solutions, and links to other online professional resources.

Acknowledgments

A large number of people have helped us write this text. Our families continued to give understanding and support as we worked on this revision. We acknowledge the thousands of students in our classes who have given us feedback over the years concerning the material and its relevancy. They were the best possible sources of criticism.

We gratefully acknowledge the invaluable assistance of the following reviewers throughout the development of the manuscript:

Radman Ali, Morris College
Casey Basset, University of Montevallo
Barbra Burdett, Lincoln College
Nira Clark, Southwestern College
Manhosha Dasanayaka, Mott Community College
Jason Fitzgerald, Southeastern Illinois College
Edward Fliss, St. Louis Community College
Jonathan Frye, McPherson College
Andrew Goliszek, North Carolina A & T University
Herb Grover, Hardin-Simmons University
Danny Herman, Montcalm Community College

Cris Hochwender, University of Evansville
Juanita Hopper, Benedict College
Shawn Lester, Montgomery College
Michelle Mater, Montcalm Community College
Jayne McClain, Greenville Technical College
John Minichiello, Middlesex Community College
James Moran, Suffolk County Community College
Margaret Nichols, Central Ohio Technical College
Margaret Nsofor, Southern Illinois University at Carbondale
Leba Sarkis, Aims Community College
John Steiner, College of Alameda
C. Michael Stinson, Southside Virginia Community College
Mark Storey, Texarkana College
Bradley Swanson, Central Michigan University
Jacob Van Houten, Delta College
Robert Williams, Southeast Community College

Eldon D. Enger
Frederick C. Ross
David B. Bailey

Guided Tour

Chapter Outline A chapter outline appears at the beginning of each chapter as a quick guide.

Goals and Objectives A list of goals and objectives organizes concepts and directs the study of each chapter.

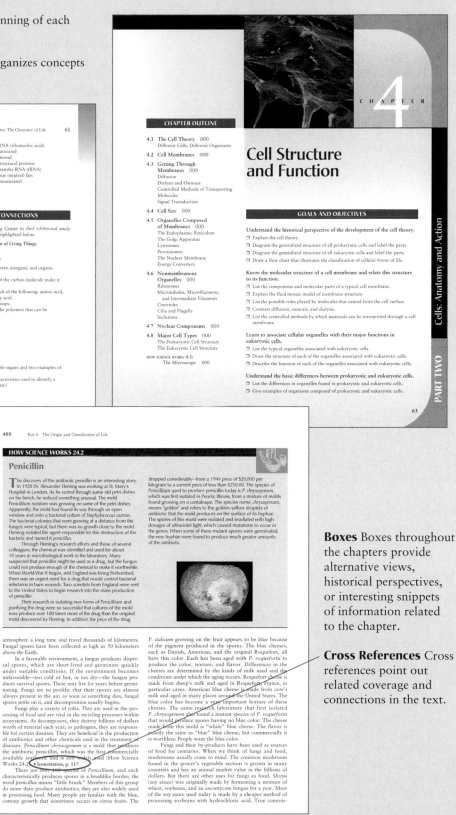

Summary An end-of-chapter summary clearly reviews the concepts presented in each chapter.

Thinking Critically This section provides an opportunity to think through problems and apply concepts presented in the chapter.

Concept Maps Helps strengthen your understanding by providing opportunities to organize ideas into logical relationships.

Key Terms A list of key terms reinforces learning of new vocabulary needed to understand chapter concepts.

Learning Connections Helps organize study resources by major section and are repeated and expanded on the Online Learning Center. At the Online Learning Connection, review questions are answered and highlighted resources can be accessed with a click of the mouse.

Boxes Boxes throughout the chapters provide alternative views, historical perspectives, or interesting snippets of information related to the chapter.

Cross References Cross references point out related coverage and connections in the text.

Table of Boxes

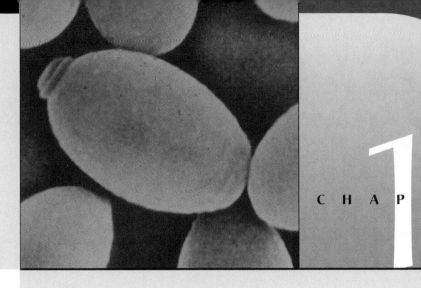

What Is Biology?

GOALS AND OBJECTIVES

Understand the process of science.

❏ State two characteristics of a good hypothesis.

❏ Explain why events that are correlated may not show a cause-and-effect relationship.

❏ List and describe five elements of the scientific method.

❏ Differentiate among a hypothesis, a theory, and a scientific law.

Recognize the difference between scientific and nonscientific areas of study.

❏ List characteristics that differentiate science from nonscience.

❏ Define and give an example of pseudoscience.

❏ Differentiate between applied and theoretical science.

Recognize that science has limitations.

❏ Give examples of problems caused by unwise use of biological information.

❏ Identify questions that science is not able to answer.

Understand how living and nonliving things differ.

❏ List and give an example of five characteristics typical of living things.

❏ State the differences among a cell, an organ, and an organism.

❏ Explain why a cell can be an organism.

Appreciate the significance of the science of biology in modern society.

❏ List three ways in which the science of biology has improved your quality of life.

❏ Describe two historical examples that illustrate how a lack of understanding of biology resulted in major problems.

Introduction

PART ONE

1.1 The Significance of Biology in Your Life

Many college students question the need for science courses such as biology in their curriculum, especially when their course of study is not science related. However, it is becoming increasingly important that all citizens be able to recognize the power and limitations of science, understand how scientists think, and appreciate how the actions of societies change the world in which we and other organisms live. Consider how your future will be influenced by how the following questions are ultimately answered:

> Does electromagnetic radiation from electric power lines, computer monitors, cell phones, and microwave ovens affect living things?
>
> Is DNA testing reliable enough to be admitted as evidence in court cases?
>
> Is there a pill that can be used to control a person's weight?
>
> Can physicians and scientists manipulate our genes in order to control certain disease conditions we have inherited?
>
> Will the thinning of the ozone layer of the upper atmosphere result in increased incidence of skin cancer?
>
> Will a vaccine for AIDS be developed in the next 10 years?
>
> Will new, inexpensive, socially acceptable methods of birth control be developed that can slow world population growth?
>
> Are human activities really causing the world to get warmer?
>
> How does extinction of a species change the ecological situation where it once lived?

As an informed citizen in a democracy, you can have a great deal to say about how these problems are analyzed and what actions provide appropriate solutions. In a democracy it is assumed that the public has gathered enough information to make intelligent decisions (figure 1.1). This is why an understanding of the nature of science and fundamental biological concepts is so important for any person, regardless of his or her vocation. *Concepts in Biology* was written with this philosophy in mind. The concepts covered in this book are core concepts selected to help you become more aware of how biology influences nearly every aspect of your life.

Most of the important questions of today can be considered from philosophical, social, and scientific standpoints. None of these approaches individually presents a solution to most problems. For example, it is a fact that the human population of the world is growing very rapidly. Philosophically, we may all agree that the rate of population growth should be slowed. Science can provide information about why populations grow and which actions will be the most effective in slowing population growth. Science can also develop meth-

Figure 1.1

Biology in Everyday Life
These news headlines reflect a few of the biologically based issues that face us every day. Although articles such as these seldom propose solutions, they do inform the general public so that people can begin to explore possibilities and make intelligent decisions leading to solutions.

ods of conception control that would limit a person's ability to reproduce. Killing infants and forced sterilization are both methods that have been tried in some parts of the world within the past century. However, most would contend that these "solutions" are philosophically or socially unacceptable. Science can provide information about the reproductive process and how it can be controlled, but society must answer the more fundamental social and philosophical questions about reproductive rights and the morality of controls. It is important to recognize that science has a role to play but that it does not have the answers to all our problems.

1.2 Science and the Scientific Method

You already know that biology is a scientific discipline and that it has something to do with living things such as microorganisms, plants, and animals. Most textbooks define **biology** as the science that deals with life. This basic definition seems clear until you begin to think about what the words *science* and *life* mean.

The word *science* is a noun derived from a Latin term (*scientia*) meaning *knowledge* or *knowing*. Humans have accumulated a vast amount of "knowledge" using a variety of methods, some by scientific methods and some by other methods.

Science is distinguished from other fields of study by *how* knowledge is acquired, rather than by the act of accumulating facts. **Science** is actually a process used to solve problems or develop an understanding of natural events that involves testing possible answers. The process has become known as the *scientific method*. The **scientific method** is a way of gaining information (facts) about the world by forming possible solutions to questions followed by rigorous testing to determine if the proposed solutions are valid (*valid* = meaningful, convincing, sound, satisfactory, confirmed by others).

When using the scientific method, scientists make several fundamental assumptions. There is a presumption that:

1. There are specific causes for events observed in the natural world,
2. The causes can be identified,
3. There are general rules or patterns that can be used to describe what happens in nature,
4. An event that occurs repeatedly probably has the same cause,
5. What one person perceives can be perceived by others, and
6. The same fundamental rules of nature apply regardless of where and when they occur.

For example, we have all observed lightning associated with thunderstorms. According to the assumptions that have just been stated, we should expect that there is an explanation that would explain all cases of lightning regardless of where or when they occur and that all people could make the same observations. We know from scientific observations and experiments that lightning is caused by a difference in electrical charge, that the behavior of lightning follows general rules that are the same as that seen with static electricity, and that all lightning that has been measured has the same cause wherever and whenever it occurred.

Scientists are involved in distinguishing between situations that are merely correlated (happen together) and those that are correlated and show *cause-and-effect relationships*. When an event occurs as a direct result of a previous event, a cause-and-effect relationship exists. Many events are correlated, but not all correlations show a cause-and-effect relationship. For example, lightning and thunder are correlated and have a cause-and-effect relationship. However, the relationship between autumn and trees dropping their leaves is more difficult to sort out. Because autumn brings colder temperatures many people assume that the cold temperature is the cause of the leaves turning color and falling. The two events are correlated. However there is no cause-and-effect relationship. The cause of the change in trees is the shortening of days that occurs in the autumn. Experiments have shown that artificially shortening the length of days in a greenhouse will cause the trees to drop their leaves even though there is no change in temperature. Knowing that a cause-and-effect relationship exists enables us to make predictions about what will happen should that same set of circumstances occur in the future.

This approach can be used by scientists to solve particular practical problems, such as how to improve milk production in cows or to advance understanding of important concepts such as evolution that may have little immediate practical value. Yet an understanding of the process of evolution is important in understanding genetic engineering, the causes of extinction, or human physiology—all of which have practical applications. The scientific method requires a systematic search for information and a continual checking and rechecking to see if previous ideas are still supported by new information. If the new evidence is not supportive, scientists discard or change their original ideas. Scientific ideas undergo constant reevaluation, criticism, and modification.

The scientific method involves several important identifiable components, including careful observation, the construction and testing of hypotheses, an openness to new information and ideas, and a willingness to submit one's ideas to the scrutiny of others. However, it is not an inflexible series of steps that must be followed in a specific order. Figure 1.2 shows how these steps may be linked and table 1.1 gives an example of how scientific investigation proceeds from an initial question to the development of theories and laws.

Observation

Scientific inquiry often begins with an observation that an event has occurred repeatedly. An **observation** occurs when we use our senses (smell, sight, hearing, taste, touch) or an extension of our senses (microscope, tape recorder, X-ray machine, thermometer) to record an event. Observation is more than a casual awareness. You may hear a sound or see an image without really observing it. Do you know what music was being played in the shopping mall? You certainly heard it but if you are unable to tell someone else what it was, you didn't "observe" it. If you had prepared yourself to observe the music being played, you would be able to identify it. When scientists talk about their observations, they are referring to careful, thoughtful recognition of an event—not just casual notice. Scientists train themselves to improve their observational skills since careful observation is important in all parts of the scientific method.

The information gained by direct observation of the event is called **empirical evidence** (*empiric* = based on experience; from the Greek *empirikos* = experience). Empirical evidence is capable of being verified or disproved by further observation. If the event occurs only once or cannot be repeated in an artificial situation, it is impossible to use the scientific method to gain further information about the event and explain it.

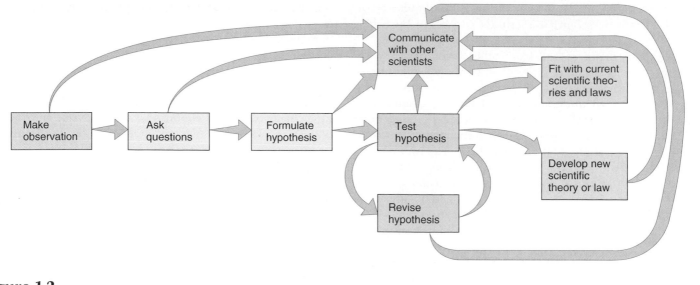

Figure 1.2

The Scientific Method

The scientific method is a way of thinking that involves making hypotheses about observations and testing the validity of the hypotheses. When hypotheses are disproved, they can be revised and tested in their new form. Throughout the scientific process, people communicate about their ideas. Scientific theories and laws develop as a result of people recognizing broad areas of agreement about how the world works. These laws and theories help people formulate their approaches to scientific questions.

Questioning and Exploration

As scientists gain more empirical evidence about an event they begin to develop *questions* about it. How does this happen? What causes it to occur? When will it take place again? Can I control the event to my benefit? The formation of the questions is not as simple as it might seem because the way the questions are asked will determine how you go about answering them. A question that is too broad or too complex may be impossible to answer; therefore a great deal of effort is put into asking the question in the right way. In some situations, this can be the most time-consuming part of the scientific method; asking the right question is critical to how you look for answers.

Let's say, for example, that you observed a cat catch, kill, and eat a mouse. You could ask several kinds of questions:

1a. Does the cat like the taste of the mouse?
1b. If given a choice between mice and canned cat food, which would a cat choose?
2a. What motivates a cat to hunt?
2b. Do cats hunt only when they are hungry?

Obviously, 1b and 2b are much easier to answer than 1a and 2a even though the two sets of questions are attempting to obtain similar information.

Once a decision has been made about what question to ask, scientists *explore other sources of knowledge* to gain more information. Perhaps the question has already been answered by someone else or several possible answers have already been rejected. Knowing what others have already done

allows one to save time and energy. This process usually involves reading appropriate science publications, exploring information on the Internet, or contacting fellow scientists interested in the same field of study. Even if the particular question has not been answered already, scientific literature and other scientists can provide insights that may lead toward a solution. After exploring the appropriate literature, a decision is made about whether to continue to explore the question. If the scientist is still intrigued by the question, a formal hypothesis is constructed and the process of inquiry continues at a different level.

Constructing Hypotheses

A **hypothesis** is a statement that provides a possible answer to a question or an explanation for an observation that can be tested. A good hypothesis must be logical, account for all the relevant information currently available, allow one to predict future events relating to the question being asked, and be testable. Furthermore, if one has the choice of several competing hypotheses, one should use the simplest hypothesis with the fewest assumptions. Just as deciding which questions to ask is often difficult, the formation of a hypothesis requires much critical thought and mental exploration. If the hypothesis does not account for all the observed facts in the situation, doubt will be cast on the work and may eventually cast doubt on the validity of the scientist's work. If a hypothesis is not testable or is not supported by the evidence, the explanation will be only hearsay and no more useful than mere speculation.

Table 1.1

THE NATURE OF THE SCIENTIFIC METHOD

Component of Science Process	Description of Process	Example of the Process in Action
Observation	Recognize something has happened and that it occurs repeatedly. (Empirical evidence is gained from experience or observation.)	Doctors observe that many of their patients, who are suffering from tuberculosis, fail to be cured by the use of the medicines (antibiotics) traditionally used to treat the disease.
Question formulation	Ask questions about the observation, evaluate the questions, and keep the ones that will be answerable.	Have the drug companies modified the antibiotics? Are the patients failing to take the antibiotics as prescribed? Has the bacterium that causes tuberculosis changed?
Exploration of alternative resources	Go to the library to obtain information about this observation. Talk to others who are interested in the same problem. Visit other researchers or communicate via letter, fax, or computer to help determine if your question is a good one or if others have already explored the topic.	Read medical journals. Contact the Centers for Disease Control and Prevention. Consult experts in tuberculosis. Attend medical conventions. Contact drug companies and ask if their antibiotic formulation has been changed.
Hypothesis formation	Pose a possible answer to your question. Be sure that it is testable and that it accounts for all the known information. Recognize that your hypothesis may be wrong.	Tuberculosis patients who fail to be cured by standard antibiotics have tuberculosis caused by antibiotic resistant populations of the bacterium *Mycobacterium tuberculosis.*
Test hypothesis (Experimentation)	Set up an experiment that will allow you to test your hypothesis using a control group and an experimental group. Be sure to collect and analyze the data carefully.	Set up an experiment in which samples of tuberculosis bacteria are collected from two groups of patients; those who are responding to antibiotic therapy but still have bacteria and those who are not responding to antibiotic therapy. Grow the bacteria in the lab and subject them to the antibiotics normally used. Use a large number of samples. The bacteria from the patients who are responding positively to the antibiotics are the control. The samples from those that are not responding constitute the experimental group. Experiments consistently show those patients who are not recovering have strains of bacteria that are resistant to the antibiotic being used.
Agreement with existing scientific laws and theories or New laws or theories are constructed	If your findings are seen to fit with other major blocks of information that tie together many different kinds of scientific information, they will be recognized by the scientific community as being consistent with current scientific laws and theories. In rare instances, a new theory or law may develop as a result of research.	Your results are consistent with the following laws and theories: Mendel's laws of heredity state that characteristics are passed from parent to offspring during reproduction. The theory of natural selection predicts that when populations of organisms like *Mycobacterium tuberculosis* are subjected to something that kills many individuals in the population, those individuals that survive and reproduce will pass on the characteristics that allowed them to survive to the next generation and that the next generation will have a higher incidence of the characteristics. The discovery of the structure of DNA and subsequent research has led to the development of a major new theory and has led to a much more clear understanding of how changes (mutations) occur to genes.
Conclusion and communication	You arrive at a conclusion. Throughout the process, communicate with other scientists both by informal conversation and formal publications.	You conclude that the antibiotics are ineffective because the bacteria are resistant to the antibiotics. This could be because some of the individual bacteria contained altered DNA (mutation) that allowed them to survive in the presence of the antibiotic. They survived and reproduced passing their resistance to their offspring and building a population of antibiotic resistant tuberculosis bacteria. A scientific article is written describing the experiment and your conclusions.

Keep in mind that a hypothesis is based on observations and information gained from other knowledgeable sources and predicts how an event will occur under specific circumstances. Scientists test the predictive ability of a hypothesis to see if the hypothesis is supported or is disproved. If you disprove the hypothesis, it is rejected and a new hypothesis must be constructed. However, if you cannot disprove a hypothesis, it increases your confidence in the hypothesis, but it does not prove it to be true in all cases and for all time. Science always allows for the questioning of ideas and the substitution of new ones that more completely describe what is known at a particular point in time. It could be that an alternative hypothesis you haven't thought of explains the situation or you have not made the appropriate observations to indicate that your hypothesis is wrong.

Testing Hypotheses

The test of a hypothesis can take several forms. It may simply involve the collection of pertinent information that already exists from a variety of sources. For example, if you visited a cemetery and observed from reading the tombstones that an unusually large number of people of different ages died in the same year, you could hypothesize that there was an epidemic of disease or a natural disaster that caused the deaths. Consulting historical newspaper accounts would be a good way to test this hypothesis.

In other cases a hypothesis may be tested by simply making additional observations. For example, if you hypothesized that a certain species of bird used cavities in trees as places to build nests, you could observe several birds of the species and record the kinds of nests they built and where they built them.

Another common method for testing a hypothesis involves devising an experiment. An **experiment** is a re-creation of an event or occurrence in a way that enables a scientist to support or disprove a hypothesis. This can be difficult because a particular event may involve a great many separate happenings called **variables.** For example, the production of songs by birds involves many activities of the nervous system and the muscular system and is stimulated by a wide variety of environmental factors. It might seem that developing an understanding of the factors involved in birdsong production is an impossible task. To help unclutter such situations, scientists use what is known as a *controlled experiment.*

A **controlled experiment** allows scientists to construct a situation so that only one variable is present. Furthermore, the variable can be manipulated or changed. A typical controlled experiment includes two groups: one in which the variable is manipulated in a particular way and another in which there is no manipulation. The situation in which there is no manipulation of the variable is called the **control group;** the other situation is called the **experimental group.**

The situation involving birdsong production would have to be broken down into a large number of simple questions, such as: Do both males and females sing? Do they sing during all parts of the year? Is the song the same in all cases? Do some individuals sing more than others? What anatomical structures are used in singing? What situations cause birds to start or stop singing? Each question would provide the basis for the construction of a hypothesis that could be tested by an experiment. Each experiment would provide information about a small part of the total process of birdsong production. For example, in order to test the hypothesis that male sex hormones produced by the testes are involved in stimulating male birds to sing, an experiment could be performed in which one group of male birds had their testes removed (the experimental group), whereas the control group was allowed to develop normally. The presence or absence of testes is manipulated by the scientist in the experiment and is known as the **independent variable.** The singing behavior of the males is called the **dependent variable** because if sex hormones are important, the singing behavior observed will change depending on whether the males have testes or not (the independent variable). In an experiment there should only be one independent variable and the dependent variable is expected to change as a direct result of manipulation of the independent variable. After the experiment, the new data (facts) gathered would be analyzed. If there were no differences in singing between the two groups, scientists could conclude that the independent variable evidently did not have a cause-and-effect relationship with the dependent variable (singing). However, if there was a difference, it would be likely that the independent variable was responsible for the difference between the control and experimental groups. In the case of songbirds, removal of the testes does change their singing behavior.

Scientists are not likely to accept the results of a single experiment because it is possible a random event that had nothing to do with the experiment could have affected the results and caused people to think there was a cause-and-effect relationship when none existed. For example, the operation necessary to remove the testes of male birds might cause illness or discomfort in some birds, resulting in less singing. A way to overcome this difficulty would be to subject all birds to the same surgery but to remove the testes of only half of them. (The control birds would still have their testes.) Only when there is just one variable, many replicates (copies) of the same experiment are conducted, and the results are consistently the same, are the results of the experiment considered convincing.

Furthermore, scientists often apply statistical tests to the results to help decide in an impartial manner if the results obtained are **valid** (meaningful, fit with other knowledge) and **reliable** (give the same results repeatedly) and show cause and effect, or if they are just the result of random events.

During experimentation, scientists learn new information and formulate new questions that can lead to even more experiments. One good experiment can result in 100 new questions and experiments. The discovery of the structure of

Figure 1.3

The Growth of Knowledge
James D. Watson and Francis W. Crick are theoretical scientists who, in 1953, determined the structure of the DNA molecule, which contains the genetic information of a cell. This photograph shows the model of DNA they constructed. The discovery of the structure of the DNA molecule was followed by much research into how the molecule codes information, how it makes copies of itself, and how the information is put into action.

the DNA molecule by Watson and Crick resulted in thousands of experiments and stimulated the development of the entire field of molecular biology (figure 1.3). Similarly, the discovery of molecules that regulate the growth of plants resulted in much research about how the molecules work and which molecules might be used for agricultural purposes.

If the processes of questioning and experimentation continue, and evidence continually and consistently supports the original hypothesis and other closely related hypotheses, the scientific community will begin to see how these hypotheses and facts fit together into a broad pattern. When this happens, a theory has come into existence.

The Development of Theories and Laws

A **theory** is a widely accepted, plausible generalization about fundamental concepts in science that explain *why* things happen. An example of a biological theory is the germ theory of disease. This theory states that certain diseases, called *infectious* diseases, are caused by living microorganisms that are capable of being transmitted from one individual to another. When these microorganisms reproduce within a person and their populations rise, they cause disease.

As you can see, this is a very broad statement that is the result of years of observation, questioning, experimentation, and data analysis. The germ theory of disease provides a broad overview of the nature of infectious diseases and methods for their control. However, we also recognize that each kind of microorganism has particular characteristics that determine the kind of disease condition it causes and the methods of treatment that are appropriate. Furthermore, we recognize that there are many diseases that are not caused by microorganisms.

Because we are so confident that the theory explains why some kinds of diseases spread from one person to another, we use extreme care to protect people from infectious microorganisms by treating drinking water, maintaining sterile surroundings when doing surgery, and protecting persons with weakened immune systems from sources of infection.

Theories and hypotheses are different. A hypothesis provides a possible explanation for a specific question; a theory is a broad concept that shapes how scientists look at the world and how they frame their hypotheses. For example, when a new disease is encountered, one of the first questions asked would be, "What causes this disease?" A hypothesis could be constructed that states, "The disease is caused by a microorganism." This would be a logical hypothesis because it is consistent with the general theory that many kinds of diseases are caused by microorganisms (germ theory of disease). Because they are broad, unifying statements, there are few theories. However, just because a theory exists does not mean that testing stops. As scientists continue to gain new information they may find exceptions to a theory or, even in rare cases, disprove a theory.

A **scientific law** is a uniform or constant fact of nature that describes *what* happens in nature. An example of a biological law is the biogenetic law, which states that all living things come from preexisting living things. While laws describe what happens and theories describe why things happen, in one way laws and theories are similar. They have both been examined repeatedly and are regarded as excellent predictors of how nature behaves.

In the process of sorting out the way the world works, scientists use generalizations to help them organize information. However, the generalizations must be backed up with facts. The relationship between facts and generalizations is a two-way street. Often as observations are made and hypotheses are tested, a pattern emerges, which leads to a general conclusion, principle, or theory. This process of developing general principles from the examination of many sets of specific facts is called **induction** or **inductive reasoning.** For example, when people examine hundreds of species of birds, they observe that all kinds lay eggs. From these observations, they may develop the principle that egg laying is a fundamental characteristic of birds, without examining every single species of bird.

Once a rule, principle, or theory is established, it can be used to predict additional observations in nature. When general principles are used to predict the specific facts of a situation, the process is called **deduction** or **deductive reasoning.** For example, after the general principle that birds lay eggs is established, one could deduce that a newly discovered species of bird would also lay eggs. In the process of science, both induction and deduction are important thinking processes used to increase our understanding of the nature of our world.

Communication

One central characteristic of the scientific method is the importance of communication. For the most part, science is conducted out in the open under the critical eyes of others who are interested in the same kinds of questions. An important part of the communication process involves the publication of articles in scientific journals about one's research, thoughts, and opinions. The communication can occur at any point during the process of scientific discovery.

People may ask questions about unusual observations. They may publish preliminary results of incomplete experiments. They may publish reports that summarize large bodies of material. And they often publish strongly held opinions that may not always be supportable with current data. This provides other scientists with an opportunity to criticize, make suggestions, or agree. Scientists attend conferences where they can engage in dialog with colleagues. They also interact in informal ways by phone, e-mail, and the Internet.

The result is that most of science is subjected to examination by many minds as it is discovered, discussed, and refined.

1.3 Science, Nonscience, and Pseudoscience

Fundamental Attitudes in Science

As you can see from this discussion of the scientific method, a scientific approach to the world requires a certain way of thinking. There is an insistence on ample supporting evidence by numerous studies rather than easy acceptance of strongly stated opinions. Scientists must separate opinions from statements of fact. A scientist is a healthy skeptic.

Careful attention to detail is also important. Because scientists publish their findings and their colleagues examine their work, they have a strong desire to produce careful work that can be easily defended. This does not mean that scientists do not speculate and state opinions. When they do, however, they take great care to clearly distinguish fact from opinion.

There is also a strong ethic of honesty. Scientists are not saints, but the fact that science is conducted out in the open in front of one's peers tends to reduce the incidence of dishonesty. In addition, the scientific community strongly condemns and severely penalizes those who steal the ideas of others, perform shoddy science, or falsify data. Any of these infractions could lead to the loss of one's job and reputation.

From Discovery to Application

The scientific method has helped us understand and control many aspects of our natural world. Some information is extremely important in understanding the structure and functioning of things in our world but at first glance appears to have little practical value. For example, understanding the life cycle of a star or how meteors travel through the universe may be important for people who are trying to answer questions about how the universe was formed, but it seems of little value to the average citizen. However, as our knowledge has increased, the time between first discovery to practical application has decreased significantly.

For example, scientists known as *genetic engineers* have altered the chemical code system of microorganisms so that they may produce many new drugs such as antibiotics, hormones, and enzymes. The ease with which these complex chemicals are produced would not have been possible had it not been for the information gained from the basic, theoretical sciences of microbiology, molecular biology, and genetics (figure 1.4). Our understanding of how organisms genetically control the manufacture of proteins has led to the large-scale production of enzymes. Some of these chemicals can remove stains from clothing, deodorize, clean contact lenses, remove damaged skin from burn patients, and "stone wash" denim for clothing.

Another example that illustrates how fundamental research can lead to practical application is the work of Louis Pasteur, a French chemist and microbiologist. Pasteur was interested in the theoretical problem of whether life could be generated from nonliving material. Much of his theoretical work led to practical applications in disease control. His theory that there are microorganisms that cause diseases and decay led to the development of vaccinations against rabies and the development of pasteurization for the preservation of foods (figure 1.5).

Science and Nonscience

Both scientists and nonscientists seek to gain information and improve understanding of their fields of study. The differences between science and nonscience are based on the assumptions and methods used to gather and organize information and, most important, the way the assumptions are tested. The difference between a scientist and a nonscientist is that a scientist continually challenges and tests principles and assumptions to determine a cause-and-effect relationship, whereas a nonscientist may not be able to do so or may not believe that this is important. For example, a historian may have the opinion that if President Lincoln had not appointed Ulysses S. Grant to be a General in the Union Army, the Confederate States of America would have won the Civil War. Although there can be considerable argument about the topic, there is no way that it can be tested. There-

(a)

(b)

Figure 1.4

Genetic Engineering
Genetic engineers have modified the genetic code of bacteria, like *Escherichia coli,* commonly found in the colon (*a*) to produce useful products such as vitamins, protein, and antibiotics. The bacteria can be cultured in special vats where the genetically modified bacteria manufacture their products (*b*). The products can be extracted from the mixture in the vat.

fore, it is not scientific. This does not mean that history is not a respectable field of study. It is just not science.

Once you understand the scientific method, you won't have any trouble identifying astronomy, chemistry, physics, and biology as sciences. But what about economics, sociology, anthropology, history, philosophy, and literature? All of these fields may make use of certain central ideas that are derived in a logical way, but they are also nonscientific in some

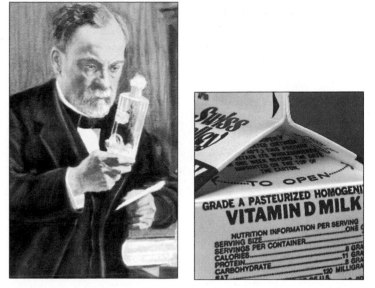

Figure 1.5

Louis Pasteur and Pasteurized Milk
Louis Pasteur (1822–1895) performed many experiments while he studied the question of the origin of life, one of which led directly to the food-preservation method now known as pasteurization.

ways. Some things are beyond science and cannot be approached using the scientific method. Art, literature, theology, and philosophy are rarely thought of as sciences. They are concerned with beauty, human emotion, and speculative thought rather than with facts and verifiable laws. On the other hand, physics, chemistry, geology, and biology are almost always considered sciences.

Many fields of study have both scientific and nonscientific aspects. The style of clothing worn is often shaped by the artistic creativity of designers and shrewd marketing by retailers. Originally, animal hides, wool, cotton, and flax were the only materials available and the choice of color was limited to the natural color of the material or dyes extracted from nature. The development of synthetic fabrics and dyes, machines to construct clothing, and new kinds of fasteners allowed for new styles and colors (figure 1.6). Similarly, economists use mathematical models and established economic laws to make predictions about future economic conditions. However, the reliability of predictions is a central criterion of science, so the regular occurrence of unpredicted economic changes indicates that economics is far from scientific. Many aspects of anthropology and sociology are scientific in nature but they cannot be considered true sciences because many of the generalizations in these fields cannot be tested by repeated experimentation. They also do not show a significantly high degree of cause and effect, or they have poor predictive value.

(a)

(b)

Figure 1.6

Science and Culture
While the design of clothing is not a scientific enterprise, scientific discoveries have altered the possible choices available. (*a*) Originally, clothing could only be made from natural materials with simple construction methods. (*b*) The discovery of synthetic fabrics and dyes and the invention of specialized fasteners resulted in increased variety and specialization of clothing.

Pseudoscience

Pseudoscience (*pseudo* = false) is not science but uses the appearance or language of science to convince, confuse, or mislead people into thinking that something has scientific validity. When pseudoscientific claims are closely examined, it is found that they are not supportable as valid or reliable. The area of nutrition is a respectable scientific field, however, there are many individuals and organizations that make unfounded claims about their products and diets (figure 1.7).

We all know that we must obtain certain nutrients like amino acids, vitamins, and minerals from the food we eat or we may become ill. Many scientific experiments have been performed that reliably demonstrate the validity of this information. However, in most cases, it has not been demonstrated that the nutritional supplements so vigorously advertised are as useful or desirable as advertised. Rather, selected bits of scientific information (amino acids, vitamins, and minerals are essential to good health) have been used to create the feeling that additional amounts of these nutritional supplements are necessary or that they can improve your health. In reality, the average person eating a varied diet will obtain all of these nutrients in adequate amounts and nutritional supplements are not required.

In addition, many of these products are labeled as organic or natural, with the implication that they have greater

Figure 1.7

"Nine out of Ten Doctors Surveyed Recommend Brand X"
It is obvious that there are many things wrong with this statement. First of all, is the person in the white coat a physician? Second, if only 10 doctors were asked, the sample size is too small. Third, only selected doctors might have been asked to participate. Finally, the question could have been asked in such a way as to obtain the desired answer: "Would you recommend brand X over Dr. Pete's snake oil?"

nutritive value because they are organically grown (grown without pesticides or synthetic fertilizers) or because they come from nature. The poisons curare, strychnine, and nicotine are all organic molecules that are produced in nature by plants that can be grown organically, but we wouldn't want to include them in our diet.

Limitations of Science

By definition, science is a way of thinking and seeking information to solve problems. Therefore the scientific method can be applied only to questions that have factual bases. Questions concerning morals, value judgments, social issues, and attitudes cannot be answered using the scientific method. What makes a painting great? What is the best type of music? Which wine is best? What color should I paint my house? These questions are related to values, beliefs, and tastes; therefore, the scientific method cannot be used to answer them.

Science is also limited by the ability of people to pry understanding from the natural world. People are fallible and do not always come to the right conclusions because information is lacking or misinterpreted, but science is self-correcting. As new information is gathered, old incorrect ways of thinking must be changed or discarded. For example, at one time scientists were sure that the Sun went around the Earth. They observed that the Sun rose in the east and traveled across the sky to set in the west. Because scientists could not feel the Earth moving it seemed perfectly logical that the Sun traveled around the Earth. Once they understood that the Earth rotated on its axis, they began to re-

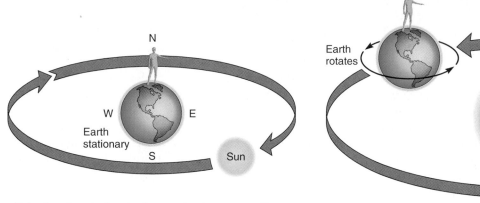

Scientists thought that the Sun revolved around the Earth.

We now know that the Earth rotates on its axis and also revolves around the Sun.

Figure 1.8

Science Must Be Willing to Challenge Previous Beliefs
Science always must be aware that new discoveries may force a reinterpretation of previously held beliefs. Early scientists thought that the Sun revolved around the Earth in a clockwise direction. This was certainly a reasonable theory at the time. Subsequently, we have learned that the Earth revolves around the Sun in a counterclockwise direction, at the same time rotating on its axis in a counterclockwise direction. This rotation of the Earth on its axis gives us the impression that the Sun is moving.

alize that the rising and setting of the Sun could be explained in other ways. A completely new concept of the relationship between the Sun and the Earth developed (figure 1.8).

Although this kind of study seems rather primitive to us today, this change in thinking about the Sun and the Earth was a very important step toward understanding the universe and how the various parts are related to one another. This background information was built upon by many generations of astronomers and space scientists, and finally led to space exploration.

People need to understand that science cannot answer all the problems of our time. Although science is a powerful tool there are many questions it cannot answer and many problems it cannot solve. Most of the problems societies face are generated by the behavior and desires of people. Famine, drug abuse, and pollution are human-caused and must be resolved by humans. Science may provide some tools for social planners, politicians, and ethical thinkers, but science does not have, nor does it attempt to provide, all the answers to the problems of the human race. Science is merely one of the tools at our disposal.

1.4 The Science of Biology

The science of biology is, broadly speaking, the study of living things. It draws on chemistry and physics for its foundation and applies these basic physical laws to living things. Because there are many kinds of living things, there are many special areas of study in biology. Practical biology—such as medicine, crop science, plant breeding, and wildlife management—is balanced by more theoretical biology—such as medical microbiological physiology, photosynthetic biochemistry, plant tax-

onomy, and animal behavior (ethology). There is also just plain fun biology like insect collecting and bird watching. Specifically, biology is a science that deals with living things and how they interact with their surroundings.

At the beginning of the chapter, biology was defined as the science that deals with living things. But what does it mean to be alive? You would think that a biology textbook could answer this question easily. However, this is more than just a theoretical question because in recent years it has been necessary to construct legal definitions of what life is and especially of when it begins and ends. The legal definition of death is important because it may determine whether a person will receive life insurance benefits or if body parts may be used in transplants. In the case of heart transplants, the person donating the heart may be legally "dead," but the heart certainly isn't. It is removed while it still has "life," even though the person is not "alive." In other words, there are different kinds of death. There is the death of the whole living unit and the death of each cell within the living unit. A person actually "dies" before every cell has died. Death, then, is the absence of life, but that still doesn't tell us what life is. At this point, we won't try to define life but we will describe some of the basic characteristics of living things.

Characteristics of Life

Living things have special abilities and structures not typically found in things that were never living. The ability to manipulate energy and matter is unique to living things. Energy is the ability to do work or cause things to move. Matter is anything that has mass and takes up space. Developing an understanding of how living things modify matter and use energy will help you appreciate how living things differ from nonliving objects. Living things show five characteristics that the nonliving do not display: (1) metabolic processes, (2) generative processes, (3) responsive processes, (4) control processes, and (5) a unique structural organization. It is important to recognize that while these characteristics are typical of all living things, they may not necessarily all be present in each organism at every point in time. For example, some individuals may reproduce or grow only at certain times. This section gives a brief introduction to the basic characteristics of living things that will be expanded upon in the rest of the text.

(1) **Metabolic processes** involve the total of all chemical reactions and associated energy changes that take place within an organism. This set of reactions is often simply referred to as **metabolism** (*metabolism* = Greek *metaballein*, to turn about, change, alter). Energy is necessary for movement, growth, and many other activities. The energy that organisms use is stored in the chemical bonds of complex molecules. The chemical reactions used to provide energy and raw materials to organisms are controlled and sequenced. There are three essential aspects of metabolism: *nutrient uptake, nutrient processing,* and *waste elimination.* All living things expend energy to take in nutrients (raw materials) from their environment. Many animals take in these materials by eating or swallowing other organisms. Microorganisms and plants absorb raw materials into their cells to maintain their lives. Once inside, nutrients enter a network of chemical reactions. These reactions manipulate nutrients in order to manufacture new parts, make repairs, reproduce, and provide energy for essential activities. However, not all materials entering a living thing are valuable to it. There may be portions of nutrients that are useless or even harmful. Organisms eliminate these portions as waste. These metabolic processes also produce unusable heat energy, which may be considered a waste product.

(2) **Generative processes** are activities that result in an increase in the size of an individual organism—*growth*—or an increase in the number of individuals in a population of organisms—*reproduction.* During growth, living things add to their structure, repair parts, and store nutrients for later use. Growth and reproduction are directly related to metabolism because neither can occur without gaining and processing nutrients. Since all organisms eventually die, life would cease to exist without reproduction. There are a number of different ways that various kinds of organisms reproduce and guarantee their continued existence. Some kinds of living things reproduce by *sexual reproduction* in which two individuals contribute to the creation of a unique, new organism. *Asexual reproduction* occurs when an individual organism makes identical copies of itself.

(3) **Responsive processes** allow organisms to respond to changes within their bodies and in their surroundings in a meaningful way. Responsive processes have been organized into three categories: *irritability, individual adaptation,* and *adaptation of populations,* which is also known as *evolution.*

Irritability is an individual's ability to recognize a stimulus and rapidly respond to it, such as your response to a loud noise, beautiful sunset, or noxious odor. The response occurs only in the individual receiving the stimulus and the reaction is rapid because the structures and processes that cause the response to occur (i.e., muscles, bones, and nerves) are already in place.

Individual adaptation also results from an individual's reaction to a stimulus but is slower because it requires growth or some other fundamental change in an organism. For example, when the days are getting shorter, a varying hare responds such that its fur color will change from its brown summer coat to its white winter coat—genes responsible for the production of brown pigment are "turned off" and new white hair grows. Similarly, the response of our body to disease organisms requires a change in the way cells work to attack and eventually destroy the disease-causing organism. Or the body responds to lower oxygen levels by producing more red blood cells, which carry oxygen. This is why athletes like to train at high elevations. Their ability to transport oxygen to muscles is improved by the increased number of red blood cells.

Evolution involves changes in the kinds of characteristics displayed by individuals within the population. It is a slow change in the genetic makeup of a *population* of organisms over generations. This process occurs over long periods of time and enables a species (population of a specific kind of organism) to adapt and better survive long-term changes in its environment over many generations. For example, the development of structures that enable birds to fly long distances, allow them to respond to a world in which the winter season presents severe conditions that would threaten survival. Similarly, the development of the human brain and the ability to reason allowed our ancestors to craft and use tools. The use of tools allowed them to survive and be successful in a great variety of environmental conditions.

(4) **Control processes** are mechanisms that ensure an organism will carry out all metabolic activities in the proper sequence (*coordination*) and at the proper rate (*regulation*). All the chemical reactions of an organism are coordinated and linked together in specific pathways. The orchestration of all the reactions ensures that there will be specific stepwise handling of the nutrients needed to maintain life. The molecules responsible for coordinating these reactions are known as *enzymes.* **Enzymes** are molecules, produced by organisms, that are able to increase and control the rate at which life's chemical reactions occur. Enzymes also regulate the amount of nutrients processed into other forms. The physical activities of organisms are coordinated also. When an insect walks, the activities of the muscles of its six legs are coordinated so that an orderly movement results.

Many of the internal activities of organisms are interrelated and coordinated so that a constant internal environment is maintained. The process of maintaining a constant internal environment is called **homeostasis.** For example, when we begin to exercise we use up oxygen more rapidly so the amount of oxygen in the blood falls. In order to maintain a "constant internal environment" the body must obtain more oxygen. This involves more rapid contractions of the muscles that cause breathing and a more rapid and forceful pumping of the heart to get blood to the lungs. These activities must occur together at the right time and at the correct rate, and when they do, the level of oxygen in the blood will remain normal while supporting the additional muscular activity.

(5) The **unique structural organization** of living things can be seen at both the organism and cellular levels. All living things are made up of complex, structural units called **cells.** Cells have an outer limiting membrane and several kinds of internal structures. Each structure has specific functions. Some living things, like you, consist of trillions of cells while others, such as bacteria or yeasts, consist of only one cell. Any unit that is capable of functioning independently is called an **organism,** whether it consists of a single cell or complex groups of interacting cells (figure 1.9). Nonliving materials, such as rocks, water, or gases, do not share a structurally complex common subunit. Figure 1.10 summarizes the characteristics of living things.

Levels of Biological Organization

Biologists and other scientists like to organize vast amounts of information into conceptual chunks that are easier to relate to one another. One important concept in biology is that all living things share the structural and functional characteristics we have just discussed. Another important organizing concept is that organisms are special kinds of matter that interact with their surroundings at several different levels (table 1.2). When biologists seek answers to a particular problem they may attack it at several different levels simultaneously. They must understand the molecules that make up living things, how the molecules are incorporated into cells, how tissues, organs, or systems within an organism function, and how populations and ecosystems are affected by changes in individual organisms.

For example, in the 1950s people began to notice a decline in the populations of certain kinds of birds. In 1962 Rachel Carson wrote a book entitled *Silent Spring* in which she linked the use of certain kinds of persistent pesticides with the changes in populations of animals. This controversial book launched the modern environmental movement and led to a great deal of research on the impact of persistent organic molecules on living things. The pesticide, DDT, which has been banned from use in much of the world because of its effects on populations of animals, presents a good case study to illustrate how biologists must be aware of the different levels of organization when studying a particular problem. DDT is an organic molecule that dissolves readily in fats and oils. It is also a molecule that does not break down very quickly. Therefore, once it is present it will continue to have its effects

Euplotes

Yeast

Orchid

Humans

Figure 1.9

An Organism Can Be Simple or Complex
Each individual organism, whether it is simple or complex, is able to independently carry on metabolic, generative, responsive, and control processes. Some organisms, like yeast or the protozoan *Euplotes*, consist of single cells while others, like orchids and humans, consist of many cells organized into complex structures.

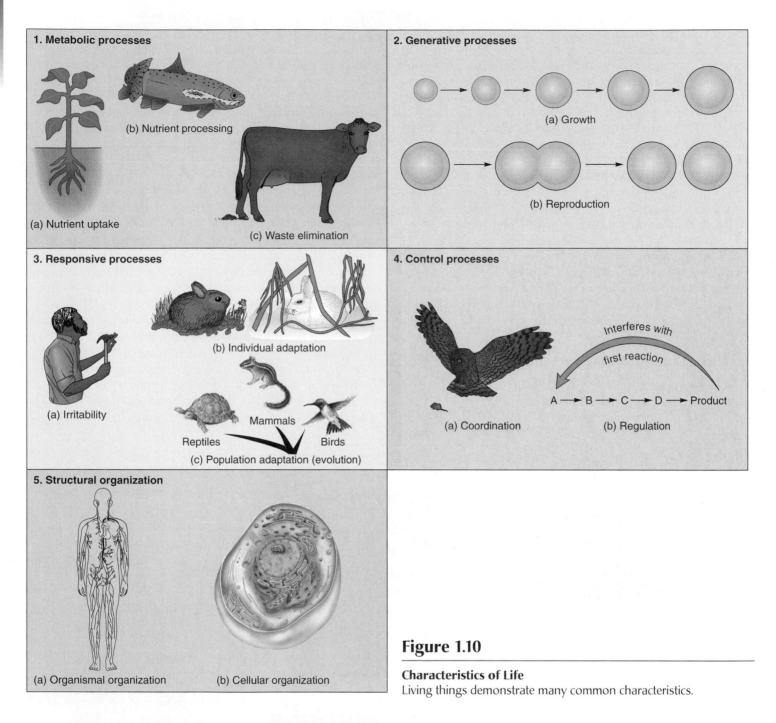

Figure 1.10

Characteristics of Life
Living things demonstrate many common characteristics.

for years. Since DDT dissolves in oils, it is often concentrated in the fatty portions of animals, and when a carnivore eats an animal with DDT in its fat, the carnivore receives an increased dose of the toxin. Birds are particularly affected by DDT, since it interferes with the ability of many kinds of birds to synthesize egg shells. Carnivorous birds like eagles are particularly vulnerable to increased levels of DDT in their bodies, because carnivores consume fats from their prey. Fragile shells are easily broken and the ability of the birds to reproduce falls sharply and their populations fall. • **biomagnification, p. 297**

Thus, determining why the populations of certain birds were declining, involved: (1) knowledge of the nature of the molecules involved, (2) how the affected animals interacted in a community of organisms, (3) where DDT was found in the bodies of animals, (4) what organ systems it affected, and ultimately (5) how it affected the ability of specialized cells to produce egg shells.

Table 1.2

LEVELS OF ORGANIZATION FOR LIVING THINGS

Category	Characteristics/Explanation	Example/Application
Biosphere	The worldwide ecosystem	Human activity affects the climate of the Earth. Global climate change, hole in ozone layer
Ecosystem	Communities (groups of populations) that interact with the physical world in a particular place	The Everglades ecosystem involves many kinds of organisms, the climate, and the flow of water to south Florida.
Community	Populations of different kinds of organisms that interact with one another in a particular place	The populations of trees, insects, birds, mammals, fungi, bacteria, and many other organisms that interact in any location
Population	A group of individual organisms of a particular kind	The human population currently consists of over 6 billion individual organisms. The current population of the California condor is about 220 individuals.
Individual organism	An independent living unit	A single organism Some organisms consist of many cells—you, a morel mushroom, a rose bush. Others are single cells—yeast, pneumonia bacterium, *Amoeba*.
Organ system	Groups of organs that perform particular functions	The circulatory system consists of a heart, arteries, veins, and capillaries, all of which are involved in moving blood from place to place.
Organ	Groups of tissues that perform particular functions	An eye contains nervous tissue, connective tissue, blood vessels, and pigmented tissues, all of which are involved in sight.
Tissue	Groups of cells that perform particular functions	Blood, groups of muscle cells, and the layers of the skin are all groups of cells that perform a particular function.
Cell	The smallest unit that displays the characteristics of life	Some organisms are single cells. Within multicellular organisms there are several kinds of cells—heart muscle cells, nerve cells, white blood cells.
Molecules	Specific arrangements of atoms	Living things consist of special kinds of molecules, such as proteins, carbohydrates, and DNA.
Atoms	The fundamental units of matter	Hydrogen, oxygen, nitrogen and about 100 others

The Significance of Biology

To a great extent, we owe our current high standard of living to biological advances in two areas: food production and disease control. Plant and animal breeders have developed organisms that provide better sources of food than the original varieties. One of the best examples of this is the changes that have occurred in corn. Corn is a grass that produces its seed on a cob. The original corn plant had very small ears that were perhaps only three or four centimeters long. Through selective breeding, varieties of corn with much larger ears and more seeds per cob have been produced. This has increased the yield greatly. In addition, the corn plant has been adapted to produce other kinds of corn, such as sweet corn and popcorn.

Figure 1.11

Biological Research Contributes to Increased Food Production
This graph illustrates a steady increase in milk yield, largely because of changing farming practices and selective breeding programs.

Data from the U.S. Department of Agriculture, *National Agricultural Statistics.*

Corn is not an isolated example. Improvements in yield have been brought about in wheat, rice, oats, and other cereal grains. The improvements in the plants, along with changed farming practices (also brought about through biological experimentation), have led to greatly increased production of food.

Animal breeders also have had great successes. The pig, chicken, and cow of today are much different animals from those available even 100 years ago. Chickens lay more eggs, dairy cows give more milk, and beef cattle grow faster (figure 1.11). All of these improvements raise our standard of living. One interesting example is the change in the kinds of hogs that are raised. At one time, farmers wanted pigs that were fatty. The fat could be made into lard, soap, and a variety of other useful products. As the demand for the fat products of pigs declined, animal breeders developed pigs that gave a high yield of meat and relatively little fat. Today, plant and animal breeders can produce plants and animals almost to specifications.

Much of the improvement in food production has resulted from the control of plants and animals that compete with or eat the organisms we use as food. Control of insects and fungi that weaken plants and reduce yields is as important as the development of new varieties of plants. Because these are "living" pests, biologists have been involved in the study of them also.

There has been fantastic progress in the area of health and disease control. Many diseases, such as polio, whooping cough, measles, and mumps, can be easily controlled by vaccinations or "shots" (How Science Works 1.1). Unfortunately, the vaccines have worked so well that some people no longer worry about getting the shots, and some of these diseases, such as diphtheria, are reappearing. These diseases have not been eliminated, and people who are not protected by vaccinations are still susceptible to them.

The understanding of how the human body works has led to treatments that can control such diseases as diabetes, high blood pressure, and even some kinds of cancer. Paradoxically, these advances contribute to a major biological problem: the increasing size of the human population.

Consequences of Not Understanding Biological Principles

Now we will look at some of the problems that have been created by well-intentioned individuals who inadequately understood or inappropriately applied biological principles.

As European settlers spread over North America in the eighteenth and nineteenth centuries, they utilized natural resources such as timber, coal, game, oil, and soil. As long as the human population remained small and dispersed, many of these resources could be sustained by regrowth or reproduction—thus they are called renewable resources (e.g., timber, game, soil). The supply of nonrenewable resources such as oil and coal appeared to be large enough to last for centuries. However, as the population increased and demands for these resources grew, a need to conserve our resources for future generations became clear. Restricting the rate of use of renewable resources would allow for their regrowth and reproduction and preserve them for future generations. To this end, the first national park (Yellowstone) was established in 1872. At the time, people thought the idea of "setting aside" a piece of the landscape in this fashion was a great way to solve the problem of scarce resources. Since that time millions of acres of deserts, forests, mountain ranges, and prairies have been designated as preserves, monuments, parks, and national forests. It was believed that by compartmentalizing our country we could keep harmful influences away from these areas and preserve dwindling resources for the future.

With the passage of time, scientists have recognized that compartmentalizing our land does not keep harmful things from happening inside the parks. Damage resulting from human activities outside these "preserves" has crept across our artificial boundaries. Some of the damage has been severe. For example, although Everglades National Park in Florida has been well managed by the National Park Service, this ecosystem is experiencing significant destruction. Commercial and agricultural development adjacent to the park have caused groundwater levels in the Everglades to drop so low that the very existence of the park is threatened. In addition, fertilizer has entered the park from surrounding farmland and encouraged the growth of plants that change the nature of the ecosystem. In 2000, Congress authorized the expenditure of $1.4 billion to begin to implement a plan that will address the problems of water flow and pollution.

Figure 1.12

Effects of Clear-Cutting on Forests

This satellite photo of Washington's Olympic Peninsula shows the extent of deforestation resulting from commercial timber harvesting. The darker shades of red indicate forested regions, lighter shades show recent growth, and the light blue highlights deforested areas. The photo inset is a typical clear-cut area and corresponds to the light blue in the satellite photo.

The historic emphasis on managing forests for timber production has also caused concerns about the degradation of ecosystems. The Pacific Northwest (Washington, Oregon, British Columbia, and northern California) presents an example. The practice of clear-cutting (stripping the forest of all trees) large regions of forest for lumber and paper pulp appears to have negatively affected many people as well as the animal and plant life in the region. Clear-cutting to the edge of streams has resulted in decreases in the populations of salmon and other important organisms. Satellite photos as well as photos taken from aircraft reveal extensive ecosystem destruction (figure 1.12). Scientists working in conjunction with the federal government have now proposed a long-term, regional approach they hope will bring the ecosystems of the region back into balance. This approach takes into consideration all species, including humans, and the needs of each to utilize the natural resources of the region.

Another problem has been caused by the introduction of exotic (foreign) species of plants and animals. In North America, this has had disastrous consequences in a number of cases. Both the American chestnut and the American elm have been nearly eliminated by diseases that were introduced by accident. Other organisms have been introduced on purpose because of shortsightedness or a total lack of understanding about biology. The starling and the English (house) sparrow were both introduced into this country by people who thought that they were doing good. Both of these birds have multiplied greatly and have displaced some native birds. The gypsy moth is also an introduced species; the moths were brought to the United States by silk manufacturers in hopes of interbreeding the

HOW SCIENCE WORKS 1.1

Edward Jenner and the Control of Smallpox

Edward Jenner (1749–1823) was born in Berkeley in Gloucestershire in the west of England. As was typical at the time, he became an apprentice to a local doctor and then eventually went to London as a pupil of an eminent surgeon. In 1773, he returned to Berkeley and practiced medicine there for the rest of his life.

At this time in history in Europe and Asia, smallpox was a common disease that nearly everyone developed usually early in life. This resulted in large numbers of deaths, particularly in children. It was known that after infection the person was protected from future smallpox infection. Various cultures had developed ways of reducing the number of deaths caused by smallpox by deliberately infecting people with the smallpox virus. If deliberate infections were given when the patient was otherwise healthy, it was likely that a mild form of the disease would develop and the person would survive and be protected from the disease in the future. In the Middle East, material from the pocks was scratched into the skin. This practice of deliberately infecting people with smallpox was introduced into England in 1717 by Lady Mary Wortley Montagu, the wife of the ambassador to Turkey. She had observed the practice of deliberate infection in Turkey and had her own children inoculated. This practice was common in England in the early 1700s, and Jenner carried out such deliberate inoculations of smallpox as part of his practice. He also frequently came in contact with individuals who had smallpox as well as individuals who were infected with cowpox—a mild disease similar to smallpox.

In 1796, Jenner introduced a safer way to protect against smallpox, which was the result of his 26-year study of these two diseases, cowpox and smallpox. Jenner made two important *observations*. Milkmaids and others who had direct contact with infected cows often developed a mild illness with pocklike sores after milking cows with cowpox sores on their teats. In addition those who had been infected with cowpox rarely became sick with smallpox. He asked the *question*, "Why don't people who have had cowpox get smallpox?" He developed the *hypothesis* that the mild disease caused by cowpox somehow protected them from the often fatal smallpox. This led him to perform an *experiment*. In his first experiment, he took puslike material from a sore on the hand of a milkmaid named Sarah Nelmes and rubbed it into small cuts on the arm of an 8-year-old boy named James Phipps. James developed the normal mild infection typical of cowpox and completely recovered. Subsequently, Jenner inoculated Phipps with material from a person suffering from smallpox. (Recall that this was a normal practice at the time.) James Phipps did not develop any disease. He was protected from smallpox by being purposely exposed to cowpox. The word that was used to describe the process was vaccination. The Latin word for cow is *vacca* and the cowpox disease was known as *vaccinae*.

When these results became known, public reaction was mixed. Some people thought that vaccination was the work of the devil. However, many European rulers supported Jenner by encouraging their subjects to be vaccinated. Napoleon and the Empress of Russia were very influential and, in the United States, Thomas Jefferson had some members of his family vaccinated. Many years later, following the development of the *germ theory of disease*, it was discovered that cowpox and smallpox are caused by viruses that are very similar in structure. Exposure to the cowpox virus allows the body to develop immunity against the cowpox virus and the smallpox virus at the same time. Subsequently, a slightly different virus was used to develop a vaccine against smallpox, which was used worldwide. In 1979, almost 200 years after Jenner developed his vaccination, the Centers for Disease Control and Prevention (CDC) in the United States and the World Health Organization (WHO) of the United Nations declared that smallpox had been eradicated.

The advent of bioterrorism raises awareness about the value of vaccinations. There is a vaccine against anthrax; however, since anthrax is not a communicable disease it is not likely to cause an epidemic. Even though smallpox was eliminated as a disease, the United States and Russia retained samples of smallpox. If terrorists were to obtain samples of the smallpox virus, the virus could be used with deadly effect, because it is contagious. It could easily spread among people of the world, especially those who have not recently been vaccinated.

Today, vaccinations (immunizations) are used to control many diseases that were common during the 1900s. Many of these diseases were known as childhood diseases because essentially all children got them. Today, they are rare in populations that are vaccinated. The following chart shows the schedule of immunizations recommended by the Advisory Committee on Immunization Practices of the American Academy of Pediatrics, and the American Academy of Family Physicians.

gypsy moth with the silkworm moth to increase silk production. When the scheme fell short of its goal and moths were accidentally set free, the moths quickly took advantage of their new environment by feeding on native forest trees.

Many human diseases have also found their way into the country, with devastating results. The smallpox virus arrived in America with explorers and spread through the susceptible Native American population, killing hundreds of thousands. Syphilis bacteria did the same. Dangerous microbes have also found their way into the country on imported research animals. Infected monkeys carried a strain of Ebola virus into the United States. Yet, with these examples to instruct us, there are still people who try to sneak exotic plants and animals into the country without thinking about the possible consequences.

Technological advances and advances in our understanding of human biology have presented us with a series of ethical situations that we have not been able to resolve satis-

HOW SCIENCE WORKS 1.1 *(continued)*

Recommended Childhood and Adolescent Immunization Schedule United States, 2003

AGE / VACCINE	Birth	1 month	2 months	4 months	6 months	12 months	15 months	18 months	24 months	4–6 years	11–12 years	13–18 years
Hepatitis B	First		Second			Third				Hep B series if missed		
DPT: diphtheria, tetanus, pertussis (whooping cough)			First	Second	Third		Fourth			Fifth	Tetanus and diphtheria	
Haemophilus influenzae type B influenza			First	Second	Third	Fourth						
Injectable inactivated polio			First	Second		Third				Fourth		
Pneumococcal conjugate (pneumonia)			First	Second	Third	Fourth		Additional vaccinations for high-risk groups				
MMR: measles, mumps, rubella (German measles)						First				Second	If any doses missed	
Varicella (chickenpox)						First		If first dose missed				2 doses if never had by age 13
Hepatitis A										Children in certain parts of country		
Influenza						Yearly for high-risk groups						

Source: Advisory Committee on Immunization Practices, American Academy of Pediatrics and American Academy of Family Physicians.

factorily. Major advances in health care in this generation have prolonged the lives of people who would have died a generation earlier. Many of the techniques and machines that allow us to preserve and extend life are extremely expensive and are therefore unavailable to most citizens of the world. Furthermore, many people in the world lack even the most basic health care, while the rich nations of the world spend money on cosmetic surgery and keep comatose patients alive with the assistance of machines.

Future Directions in Biology

Where do we go from here? Although the science of biology has made major advances, many problems remain to be solved. For example, scientists are seeking major advances in the control of the human population and there is a continued interest in the development of more efficient methods of producing food.

One area that will receive much attention in the next few years is the relationship between genetic information and such diseases as Alzheimer's disease, stroke, arthritis, and cancer. These and many other diseases are caused by abnormal body chemistry, which is the result of hereditary characteristics. Curing certain hereditary diseases is a big job. It requires a thorough understanding of genetics and the manipulation of hereditary information in all of the trillions of cells of the organism.

Another area that will receive much attention in the next few years is ecology. Climate change, destruction of natural ecosystems to feed a rapidly increasing human population, and pollution are all still severe problems. Most people need to learn that some environmental changes may be acceptable and that other changes will ultimately lead to our destruction. We have two tasks. The first is to improve technology and increase our understanding about how things work in our biological world. The second, and probably the more difficult, is to educate, pressure, and remind people that their actions determine the kind of world in which the next generation will live.

It is the intent of science to learn what is going on in these situations by gathering the facts in an objective manner. It is also the role of science to identify cause-and-effect relationships and note their predictive value in ways that will improve the environment for all forms of life. Scientists should also make suggestions to politicians and other policymakers about which courses of action are the most logical from a scientific point of view.

SUMMARY

The science of biology is the study of living things and how they interact with their surroundings. Science and nonscience can be distinguished by the kinds of laws and rules that are constructed to unify the body of knowledge. Science involves the continuous testing of rules and principles by the collection of new facts. In science, these rules are usually arrived at by using the scientific method—observation, questioning, exploring resources, hypothesis formation, and the testing of hypotheses. When general patterns are recognized, theories and laws are formulated. If a rule is not testable, or if no rule is used, it is not science. Pseudoscience uses scientific appearances to mislead.

Living things show the characteristics of (1) metabolic processes, (2) generative processes, (3) responsive processes, (4) control processes, and (5) a unique structural organization. Biology has been responsible for major advances in the areas of food production and health. The incorrect application of biological principles has sometimes led to the elimination of useful organisms and to the destruction of organisms we wish to preserve. Many biological advances have led to ethical dilemmas that have not been resolved. In the future, biologists will study many things. Two areas that are certain to receive attention are the relationship between heredity and disease, and ecology.

THINKING CRITICALLY

For guidelines to answer this question, visit the Online Learning Center.

The scientific method is central to all work that a scientist does. Can this method be used in the ordinary activities of life? How might a scientific approach to life change how you choose your clothing or your recreational activities, or which kind of car you buy? Can these choices be analyzed scientifically? Should they be analyzed scientifically? Is there anything wrong with looking at these matters from a scientific point of view?

CONCEPT MAP TERMINOLOGY

The construction of a concept map is a technique that helps students recognize how separate concepts are related to one another. Some concept maps may be simple orderly lists. Others may form networks of connections that help to show how ideas are linked. It is important to understand that there is not just one way that things can be put together. The examples show two different ways of looking at the same concepts and organizing them in a meaningful way. (Take another look at figure 1.2. It is a variety of concept map.)

Construct a concept map using these terms.

biology	observation	scientific method
experiment	science	theory
hypothesis		

Visit the Online Learning Center to use an interactive format to construct another concept map.

KEY TERMS

Use interactive flash cards, on the Online Learning Center, to help you learn the meanings of these terms.

atom
biology
biosphere
cell
community
control group
control processes
controlled experiment
deductive reasoning (deduction)
dependent variable
ecosystem
empirical evidence
energy
enzymes
experiment
experimental group
generative processes
homeostasis
hypothesis
independent variable

inductive reasoning (induction)
matter
metabolic processes
metabolism
molecule
observation
organ
organ system
organism
population
pseudoscience
reliable
responsive processes
science
scientific law
scientific method
theory
tissue
valid
variable

LEARNING CONNECTIONS

Visit the Online Learning Center to find additional study resources including those highlighted below.

1.1   The Significance of Biology in Your Life

Experience This
• Finding Biology in the News

1. List five issues that biological research may help us solve in the near future.

1.2   Science and the Scientific Method

2. What is the difference between simple correlation and a cause-and-effect relationship?
3. How does a hypothesis differ from a scientific theory or a scientific law?
4. List three objects or processes you use daily that are the result of scientific investigation.
5. The scientific method cannot be used to deny or prove the existence of God. Why?
6. What are controlled experiments? Why are they necessary to support a hypothesis?
7. List the parts of the scientific method

1.3   Science, Nonscience, and Pseudoscience

Experience This!
• Science or pseudoscience in advertisements

Interactive Concept Map
• Different ways of knowing

8. What is the difference between science and nonscience?
9. How can you identify pseudoscience?
10. Why is political science not a science?

1.4   The Science of Biology

Animations and Review
• Life characteristics

Labeling Exercises
• The characteristics of life
• Levels of biological organization part I
• Levels of biological organization part II
• Levels of biological organization part III

Interactive Concept Map
• Characteristics of life

11. List three advances that have occurred as a result of biology.
12. List three mistakes that could have been avoided had we known more about living things.
13. What is biology?
14. List five characteristics of living things.
15. What is the difference between regulation and coordination?

Test yourself with multiple choice questions that provide immediate feedback on the Online Learning Center with PowerWeb at

www.mhhe.com/enger11

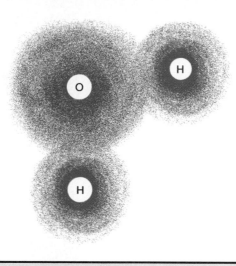

The Basics of Life: Chemistry

GOALS AND OBJECTIVES

Understand that all substances arc comprised of matter and energy.

❏ Contrast matter and energy.

❏ Summarize the difference between potential and kinetic energy.

❏ State why energy is necessary for life.

Recognize the basic structure of atoms.

❏ Discriminate between atoms and elements; molecules and compounds.

❏ Describe the information found in the periodic table of the elements.

❏ Describe how isotopes differ from one another.

❏ Differentiate among atomic weight, atomic number, atomic mass, and mass number.

❏ Explain how atoms stick together to form compounds.

❏ Contrast chemical bonds and hydrogen bonds.

❏ Describe the differences among liquids, solids, and gases.

Describe the chemical nature of water.

❏ Understand the roles water plays in maintaining life.

❏ Differentiate among solution, solvent, and solute.

**Recognize how compounds may be broken down
and reconnected in different ways.**

❏ List the five types of chemical reactions associated with all living things.

❏ Explain the difference between reactants and products.

❏ Identify compounds that are acids, bases, or salts.

❏ Work with the pH scale.

Cells: Anatomy and Action

PART TWO

2.1 The Structure and Function of Living Things

All life forms display highly specialized structures and functions. Since the structure of all matter involves chemical substances, a basic understanding of *chemistry* is important to an understanding of living things. **Chemistry** is the science concerned with the study of the composition, structure, and properties of substances and the transformations they undergo. Many people are familiar with the biological terms *anatomy* and *physiology*. **Anatomy** is the study of the structure of living things. An organism's structure is composed of **matter,** which is anything that has mass and takes up space (volume). Don't confuse mass and the concept of weight. Mass refers to an amount of matter, while weight refers to the amount of *force* (a push or pull capable of changing the position of an object) with which that object is attracted by gravity. Since weight is determined by gravity, a person's weight will be different on the Moon and on Earth, but their mass will be the same. **Physiology** is the scientific study of living processes, activities, and functions. Physiological processes are all the chemical activities of an organism and involve the handling of *energy* (figure 2.1). **Energy** is the ability to do work and *work* is done when an object is moved from one place to another. This means that when physiological processes occur, some part of the organism or its environment is moved from one place to another. This could be the movement of a flower bud opening or the blinking of an eyelid. • **muscle contraction, p. 403**

2.2 Energy and Life

Where do living things get their energy? Most people would answer that many organisms get their energy from "food" or nutrients while other organisms receive theirs from sunlight. However, the answer is more complex. Actually there are two general types of energy: *potential* and *kinetic*. **Kinetic energy** is energy of motion. For example, a fish swimming through a pond displays kinetic energy as does a molecule of table sugar being broken down to smaller parts. Energy that is not yet doing work is *potential energy.* You might also think of **potential energy** as stored energy. When we talk of *chemical energy,* we are really talking about the potential energy in matter. This energy has the potential to be released as kinetic energy to do work, such as causing an organism to move faster, digest food, or build muscle.

An object that appears to be motionless does not necessarily lack energy. An object on top of a mountain may be motionless, but still may contain significant amounts of potential energy, or "energy of position," that can be released as kinetic energy if the object rolls down the mountain. Keep in mind that potential energy also increases whenever things experiencing a repelling force are pushed together. You experience this every time you use kinetic energy to "click" your ballpoint pen, which compresses the spring inside. This gives the spring more potential energy that is converted back

Figure 2.1

Anatomy and Physiology
In order to understand living things researchers must investigate both their structure and function. This team in the Animal Health Diagnostic Laboratory at Michigan State University is gathering information about the heart and breathing rates of this horse.

into kinetic energy when the spring expands as the ink cartridge is retracted into the case. The kinetic energy for compressing the spring came from your biological ability to harvest energy from food molecules. Potential energy also increases whenever things that attract each other are pulled apart. An example of this occurs when you stretch a rubber band. That increased potential energy is converted to the "snapping" back of the band when you let go. Again, the kinetic energy needed to stretch the rubber band came from you. It is important to understand that energy is not consumed or used up in these processes. The same amount of energy is released when the rubber band is snapped back as was stored when you stretched the band. One of the important scientific laws, the **law of conservation of energy,** or the *first law of thermodynamics,* states that energy is never created or destroyed. Energy can be converted from one form to another but the *total* energy remains constant. All living things obey this law. • **cell metabolism, p. 104**

There are five forms of energy and each can be either kinetic or potential: (1) *mechanical,* (2) *nuclear,* (3) *electrical,* (4) *radiant,* and (5) *chemical.* All organisms in some way interact with these forms. Mechanical energy is the energy most people associate with machines or things in motion. A track athlete displays potential mechanical energy at the start line; the energy becomes kinetic mechanical energy when the athlete is running (figure 2.2). Nuclear energy is the form of energy from reactions involving the innermost part of matter, the nucleus. In a nuclear power plant, nuclear energy is used to generate electrical energy and if not controlled properly can have profound effects on all life. Electrical energy or electricity is the flow of charged particles. All organisms use

(a)

(b)

Figure 2.2

Potential Mechanical Energy
The pole vaulter in (a) has the potential energy needed to make the jump. Converting this form of energy to kinetic energy takes place during this running approach to the bar (b).

the flow of charged particles as a part of their metabolism. This is how nervous systems operate. Radiant energy is most familiar as heat and visible light but there are other forms such as X-radiation and microwaves. Chemical energy is a kind of internal potential energy that is stored in matter and can be released as kinetic energy when chemicals are changed from one form to another. For example, chemical bond energy is released when wood is burned. The burning of wood generates heat and light. A slower, controlled burning process called *cellular respiration* releases energy from food molecules in living systems.

2.3 What Is the Nature of Matter?

The idea that substances are composed of very small particles can be traced back to certain early Greek philosophers. Democritus, during the fifth century B.C., wrote that matter was empty space filled with tremendous numbers of tiny, indivisible particles called *atoms*. (The word *atom* is from the Greek word meaning "uncuttable.") Aristotle, another Greek philosopher, rejected the idea of atoms. He believed that matter was continuous and made up of only four parts: earth, air, fire, and water. Aristotle's belief about matter predominated

through the 1600s. Galileo and Newton believed the ideas about matter being composed of tiny particles, or atoms, since this theory seemed to explain the behavior of matter. Widespread acceptance of the particle model did not occur, however, until strong evidence was developed through the science of chemistry in the late 1700s and early 1800s. The experiments finally led to a collection of assumptions about the small particles of matter and the space around them. Collectively, the assumptions were called the *kinetic molecular theory*. The **kinetic molecular theory** states that all matter is made up of tiny particles that are in constant motion.

Atoms and Molecules

Atoms are the smallest units of matter that can exist alone. There are over 100 different kinds of atoms, the fundamental units of which all matter is made. **Elements** are fundamental chemical substances that are made up of only one kind of atom. The **periodic table of the elements** lists the elements (figure 2.3). Traditionally, elements are represented in a shorthand form by letters called *chemical symbols*. The table is called *periodic* because the properties of elements recur periodically (at regular intervals) when the elements are listed in order of their size. The table has horizontal rows of elements called *periods* and vertical columns of elements called *families*. It is made up of rows and columns of squares, with each element having its own square in a specific location. The arrangement has a meaning, both about atomic structure and about chemical functions. The periods are numbered from 1 to 7 on the left side. Period 1, for example, has only two elements: H (hydrogen) and He (helium). Period 2 starts with Li (lithium) and ends with Ne (neon). The two rows at the bottom of the table are actually part of periods 6 and 7 (between atomic numbers 57 and 72, 89 and 104). They are moved so that the table is not so wide.

Families (or group) are identified with Roman numerals and letters at the top of each column. Group IIA, for example, begins with Be (beryllium) at the top and has Ra (radium) at the bottom. The A families are in sequence from left to right. The B families are not in sequence, and one group contains more elements than the others. Elements in vertical columns have the same configurations, and that structure is responsible for the chemical properties of an element.

Don't worry. You will not have to know the entire table. The 11 main elements comprising living things have the *chemical symbols* C, H, O, P, K, I, N, S, Ca, Fe, and Mg. (A mnemonic trick to help remember these is: CHOPKINS CaFé, Mighty good!). • diet and nutrition, p. 367

Elements are composed of atoms, which move about and may interact with other atoms to form *compounds*. A **compound** is a chemical substance made up of atoms of two or more elements combined in a specific ratio and arrangement. One unit of a compound is called a *molecule*. A **molecule** is a single particle of a substance composed of two or more atoms attached to one another by attractive forces called *chemical bonds*. A molecule still retains the characteristics of that substance.

Figure 2.3

Periodic Table of the Elements
The table provides information about all the known elements. Notice that the atomic weights of the elements increase as you read left to right along the periods. Reading top to bottom in a family gives you a glimpse of a group that has similar chemical properties.

For example, pure water is composed of 2 atoms of hydrogen and 1 atom of oxygen. It is represented by a chemical *formula*, H_2O. The **formula** of a compound describes what elements it contains (as indicated by a chemical symbol) and in what proportions they occur (as indicated by the subscript number). In this case, the subscript "2" indicates 2 atoms of the element hydrogen, and the symbol for oxygen without a subscript indicates that there is only 1 atom of oxygen present in this molecule.

The terms *atom* and *molecule* are usually used with reference to quantities of substance: for example, 16 atoms of

helium and 5 molecules of sugar. The element helium is a gas and the compound table sugar is a solid.

Some atoms, such as oxygen, hydrogen, and nitrogen, bond together to form *diatomic* (*di-* = two) molecules (Outlooks 2.1). In our atmosphere, these elements are found as the gases H_2, O_2, and N_2. The subscript indicates the number of atoms of that element in a single molecule of that substance. Other elements are not normally diatomic but exist as single or *monatomic* (*mon-* = one) units; for example, the gases helium (He) and neon (Ne). These chemical symbols or initials indicate a single atom of that element.

OUTLOOKS 2.1

Sticking Together

Some molecules are strongly attracted to one another, forming solid material—for example, sand—while the attraction between others is weaker and results in the formation of liquid—for example, water. The force that holds the *same kind of molecules* together is called a *cohesive force*. (A group of people who stick together on an issue is called a cohesive group.) The force that holds *different kinds of molecules* together is called an *adhesive force*. Adhesive tape bonds the bandage to your skin.

States of Matter

The strength of the attractive forces between molecules and the kind of arrangements they form results in three **states of matter:** *solid, liquid,* and *gas.* A **solid** (i.e., bone) contains molecules packed tightly together. With the least amount of kinetic energy, these molecules vibrate in place and are at fixed distances from one another. Strong forces bind them together. Solids have definite shapes and volumes under ordinary temperature and pressure conditions. The hardness of a solid is the resistance of a solid to forces that tend to push its molecules farther apart. The amount of kinetic energy in a solid is less than that in a liquid of the same material. A **liquid** (i.e., the watery component of blood and lymph) has molecules attracted to each other but slightly farther apart than in a solid. Because they are moving more rapidly, they sometimes slide past each other as they move. While liquids can change their shape under ordinary conditions, they maintain a fixed volume under ordinary temperature and pressure conditions—that is, a liquid of a certain volume will take the shape of the container into which it is poured but it will take up the same amount of space regardless of its shape. This gives liquids the ability to *flow*, so we call them *fluids.* Still more energetic are the molecules of a **gas** (i.e., air and the components of air that are present in the blood). The attraction the gas (air) molecules have for each other is overcome by the speed with which the individual molecules move. Because they are moving the fastest, their collisions tend to push them farther apart, and so a gas expands to fill its container. The shape of the container and pressure determine the shape and volume of gases. The term *vapor* is used to describe the gaseous form of a substance that is usually in the liquid state. Water vapor, for example, is the gaseous form of liquid water. • **nitrogen cycle, p. 291**

2.4 Molecules and Kinetic Energy

Common experiences let you know that all matter has a certain amount of kinetic energy. If you were to open a bottle of perfume in a closed room with no air movement, it wouldn't take long for the aroma to move throughout the room. The kinetic molecular theory explains this by saying that the molecules diffuse, or spread, throughout the room because they are in constant, random motion. This theory also predicts that the rate at which they diffuse depends on the temperature of the room—the higher the air temperature, the greater the kinetic energy of the molecules, the more rapid the diffusion of the perfume.

Temperature is a measure of the average kinetic energy of the molecules making up a substance. The two most common numerical scales used to measure temperature are the Fahrenheit and Celsius scales. If people comment on the temperature of something they usually are making a comparison. For example, the air temperature today is "colder" or "hotter" than it was yesterday. They may also refer to a scale for comparison (i.e., the temperature is 20°C [68°F]).

The concept of *heat* is not the same as the concept of temperature. Temperature deals with the comparative "hotness" or "coldness" of different things. **Heat** is the total internal kinetic energy of molecules. The heat, or internal kinetic energy, of molecules can change as a result of interactions with the environment. This is what happens when you rub your hands together. It is thought that friction results in increased temperatures because molecules on one moving surface will catch on another surface, stretching the molecular forces that are holding them. They are pulled back to their original position with a "snap," resulting in an increase of vibrational kinetic energy.

Heat is measured in units called *calories*. A **calorie** is the amount of heat necessary to raise the temperature of 1 gram of water 1 degree Celsius (°C). *Note that heat (measured in calories) and temperature (measured in Celsius) are not the same thing but are related to one another.* The heat that an object possesses cannot be measured with a thermometer. What a thermometer measures is the temperature of an object. The temperature is really a measure of how fast the molecules of the substance are moving and how often they bump into other molecules. This is a measure of the kinetic energy of the molecules. If heat energy is added to an object, the molecules will vibrate faster. Consequently, the temperature will rise because the added heat energy will result in a speeding up of the movement of the molecules. So, although there is a relationship between heat and temperature, the amount of heat in calories that an object has depends on the size of the object and its particular properties such as density, volume, and pressure. For example, if you measure the temperature as you heat liquid water, the temperature rises steadily until it reaches 100°C (212°F). This is followed by a period when the temperature does not rise even though you continue to add heat. The additional heat energy goes into the molecules and overcomes the cohesive forces between water molecules. When this is accomplished, the water boils. The amount of heat required to cause this state change from liquid water to a water vapor is 540 calories of heat energy for each gram of water that evaporates (540 cal/g). • **temperature regulation in animals, p. 516**

Why do we take a person's temperature? Since the body's size and composition usually do not change in a short time, any change in temperature means that the body has either gained or lost heat. If the temperature is high, the body has usually increased its heat content as a result of increased metabolism. This increase in temperature is a symptom of abnormality. A low body temperature can also indicate abnormal body functioning.

2.5 Structure of the Atom

An atom is constructed of three major particles; two of them are in a central region or core called the **atomic nucleus.** The third type of particle is in the region surrounding the nucleus (figure 2.4). The weight, or mass, of the atom is concentrated in the nucleus. One type of particle located in the nucleus is the *neutron*. **Neutrons** were named to reflect their lack of electrical charge. They are neutral. **Protons,** the second type of particle in the nucleus, have a positive electrical charge. Moving around the nucleus in certain areas called **energy levels** are the **electrons.** Each electron has a negative electrical charge. The number of electrons determines the space that an atom occupies.

An atom's charge is neutral if the number of positively charged protons equals the number of negatively charged electrons. For instance, hydrogen, with 1 proton, would have 1 electron; carbon, with 6 protons, would have 6 electrons. You can determine the number of either of these two particles in a neutral atom if you know the number of the other particle.

All the atoms of the same element have the same number of protons. The number of protons determines the identity of the element. For example, carbon always has 6 protons and no other element has that number. Oxygen always has 8 protons. The **atomic number** of an element is the number of protons in an atom of that element; therefore, each element has a unique atomic number. Because oxygen has 8 protons, its atomic number is 8. The mass of a proton is 1.67×10^{-24} grams. Because this is an extremely small mass and is awkward to express, 1 proton is said to have a mass of 1 **atomic mass unit** (abbreviated as AMUμ) (table 2.1).

Although all neutral atoms of the same element have the same number of protons and electrons, they do not always have the same number of neutrons. In the case of oxygen, over 99% of the atoms have 8 neutrons, but there are others with more or fewer neutrons. Each atom of the same element with a different number of neutrons is called an **isotope** of that element. Since neutrons have a mass very similar to that of a proton, isotopes that have more neutrons will have a greater mass than those that have fewer neutrons.

Elements occur in nature as a mixture of isotopes. The **atomic weight** of an element is an average of all the isotopes present in their normal proportions. For example, of all the hydrogen isotopes on Earth, 99.985% occur as an isotope without a neutron and 0.015% as the isotope with one neutron. There is a third isotope with two neutrons but it is not considered because it is highly unstable. When the math is done to

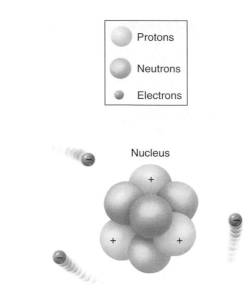

Nucleus

Figure 2.4

Atomic Structure
The nucleus of the atom contains the protons and the neutrons, which are the massive particles of the atom. The electrons, much less massive, are in constant motion about the nucleus. Therefore, the neutrons and protons give an atom its mass (weight) and the volume of an atom is determined by how many and how far out the electrons encircle the neutrons and protons.

Table 2.1			
COMPARISON OF ATOMIC PARTICLES			
	Protons	**Electrons**	**Neutrons**
Location	Nucleus	Outside nucleus	Nucleus
Charge	Positive (+)	Negative (–)	None (neutral)
Number present	Identical to the atomic number	Equal to number of protons	Atomic weight minus atomic number
Mass	1 AMU(μ)	1/1,836 AMU(μ)	1 AMU(μ)

account for the relative amounts of the various isotopes of hydrogen, the atomic weight turns out to be 1.0079 μ.

The sum of the number of protons and neutrons in the nucleus of an atom is called the **mass number.** Mass numbers are used to identify isotopes. A hydrogen atom with 1 proton and 1 neutron has a mass number of 1 + 1, or 2, and is referred to as hydrogen-2 (also called deuterium). A hydrogen atom with 1 proton and 2 neutrons has a mass number of 1 + 2, or 3, and is referred to as hydrogen-3 (also called tritium) (figure 2.5).

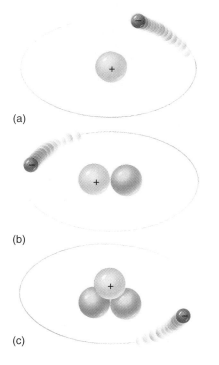

(a)

(b)

(c)

Figure 2.5

Isotopes of Hydrogen

(a) The most common form of hydrogen is the isotope that is 1 μ (AMU). It is composed of one proton and no neutrons. (b) The isotope *deuterium* is 2 μ and has one proton and one neutron. (c) *Tritium*, 3 μ, has two neutrons and one proton. Each of these isotopes of hydrogen also has one electron but, because the mass of an electron is so small, the electrons do not contribute significantly to the mass as measured in μ. All three isotopes of hydrogen are found on Earth, but the most frequently occurring has 1 μ and is commonly called *hydrogen*. Most scientists use the term "hydrogen" in a generic sense (i.e., the term is not specific but might refer to any or all of these isotopes).

Locating Electrons

Our understanding of the structure of the atom has changed significantly since the concept of the atom was first introduced. At one time, people thought of atoms as miniature solar systems with the nucleus in the center and electrons in orbits around the nucleus. However, as more experimental data were gathered and interpreted, a new model was formulated. In this model, electrons occupy certain regions of space called *energy levels*. Each energy level contains electrons moving at approximately the same speed and, therefore, electrons of a given level have about the same amount of kinetic energy. Each energy level is numbered in increasing order with energy level 1 containing electrons with the lowest amount of energy; energy level 2 having electrons with more energy than those found in energy level 1; energy level 3 having electrons with even more energy than those in level 2, and so forth. In addition, electrons do not encircle the atomic nucleus in flat, two-dimensional paths. Some move around the atomic nucleus in a three-dimensional region that is spherical, forming cloudlike or fuzzy layers

about the nucleus. Others move in a manner that resembles the figure eight (8), forming fuzzy regions that look like dumbbells or hourglasses (figure 2.6).

The first energy level can contain 2 electrons; the second energy level can contain a total of 8 electrons; the third energy level, 8; and so forth (table 2.2). Also note in table 2.2 that for some of the atoms (He, Ne, Ar), the outermost energy level contains the maximum number of electrons it can hold. Elements such as He and Ne, with filled outer energy levels, are particularly stable.

All atoms have a tendency to seek such a stable, filled outer energy level arrangement, a tendency referred to as the *octet (8) rule*. The rule states that atoms attempt to acquire an outermost energy level with 8 electrons through chemical reactions. Since elements like He and Ne have full outermost energy levels under ordinary circumstances, they do not normally undergo chemical reactions and are therefore referred to as *noble* (implying that they are too special to interact with other elements), or *inert*. Atoms of other elements have outer energy levels that are not full; for example, H, C, Mg, and will undergo reactions to fill their outermost energy level in order to become stable. (An important exception to the octet rule is hydrogen. It becomes stable with 2 electrons.) It is important for chemists and biologists to focus on electrons in the outermost energy level since it is these electrons that are involved in the chemical changes of all physiological processes.

2.6 Chemical Change

Atoms gain or lose electrons through a chemical reaction to become more stable. Such a reaction results in the formation of a *chemical bond*. A **chemical bond** is an attractive force that holds atoms together as a molecule. The two types of bonds discussed are: (1) ionic bonds and (2) covalent bonds.

Ions and Ionic Bonds

Ionic bonds are formed after atoms *transfer* electrons to achieve a full outermost energy level. Electrons are donated or received in the transfer, forming a positive and a negative *ion*, a process called *ionization*. Any positively or negatively charged atom or molecule is called an **ion**. The force of attraction between oppositely charged ions forms **ionic bonds**, and *ionic compounds* are the result. Ionic compounds are formed when an element from the left side of the periodic table reacts with an element from the right side. This results in the formation of a stable group that has an orderly arrangement and is a crystalline solid. Ions and ionic compounds are very important in living systems. For example, sodium chloride is a crystal solid known as table salt. A positively charged sodium ion is formed when a sodium atom loses one electron. This results in a stable, outermost energy level with 8 electrons. When an atom of chlorine receives an electron to stabilize its outermost energy level, it becomes a negative ion. All positively charged ions are called *cations*

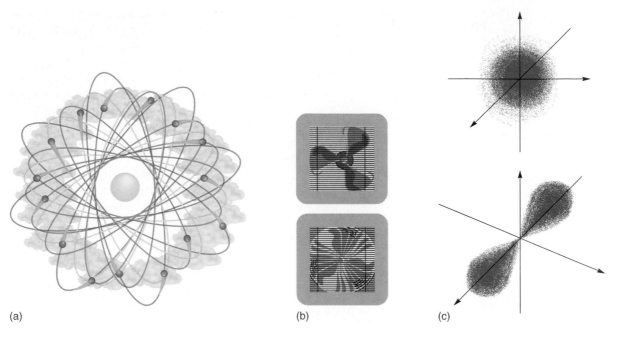

(a)   (b)   (c)

Figure 2.6

The Electron Cloud

So fast are the electrons moving around the nucleus that they can be thought of as forming a cloud around it, rather than an orbit or single track. (*a*) You might think of the electron cloud as hundreds of photographs of an atom. Each photograph shows where an electron was at the time the picture was taken. But when the next picture is taken, the electron is somewhere else. In effect, an electron is everyplace in its energy level at the same time just as the fan blade of a window fan is everywhere at once when it is running. (*b*) No matter where you stick your finger in the fan, you will be touched (ouch!) by the moving blade. Although we are able to determine where an electron is at a given time, we do not know the exact path it uses to go from one place to another. (*c*) A better way to represent the spherical or hourglass routes traveled by electrons.

Table 2.2

THE NUMBER OF ELECTRONS POSSIBLE IN ENERGY LEVELS

Element	Symbol	Atomic Number	Number of Electrons Required to Fill Each Energy Level			
			Energy Level 1	Energy Level 2	Energy Level 3	Energy Level 4
Hydrogen	H	1	1			
Helium	He	2	2			
Carbon	C	6	2	4		
Nitrogen	N	7	2	5		
Oxygen	O	8	2	6		
Neon	Ne	10	2	8		
Sodium	Na	11	2	8	1	
Magnesium	Mg	12	2	8	2	
Phosphorus	P	15	2	8	5	
Sulfur	S	16	2	8	6	
Chlorine	Cl	17	2	8	7	
Argon	Ar	18	2	8	8	
Potassium	K	19	2	8	8	1
Calcium	Ca	20	2	8	8	2

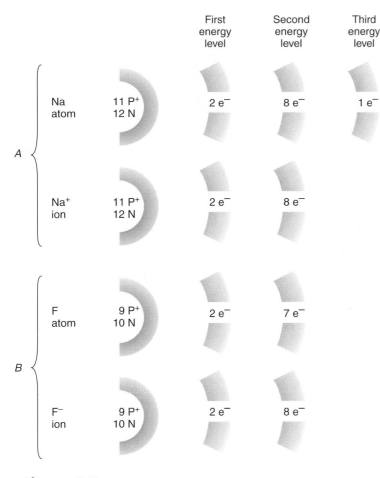

		First energy level	Second energy level	Third energy level
A	Na atom 11 P⁺ 12 N	2 e⁻	8 e⁻	1 e⁻
	Na⁺ ion 11 P⁺ 12 N	2 e⁻	8 e⁻	
B	F atom 9 P⁺ 10 N	2 e⁻	7 e⁻	
	F⁻ ion 9 P⁺ 10 N	2 e⁻	8 e⁻	

Figure 2.7

Ion Formation

A sodium atom (*A*) has two electrons in the first energy level, eight in the second energy level, and one in the third level. When it loses its one outer electron, it becomes a sodium cation. An atom of fluorine (*B*) has two electrons in the first energy level and only seven in the second energy level. To become stable, fluorine picks up an extra electron from an electron donor to fill its outermost energy level, thus satisfying the octet rule. It becomes a fluorine anion.

and all negative charged ions are called *anions* (figure 2.7). When these oppositely charged ions are close to one another, the attractive force between them forms an ionic bond. The dots in the following diagram represent the electrons in the outermost energy levels of each atom. This kind of diagram is called an electron dot formula. • **nerve impulse, p. 389**

$$\text{Na·} + \text{·}\ddot{\underset{..}{\text{Cl}}}\text{:} \rightarrow \text{Na}^+\text{Cl}^-$$

When many ionic compounds are dissolved in water, the ionic bonds are broken and the ions separate, or *dissociate*, from one another. For example, solid sodium chloride dissociates in water to become ions in solution:

$$\text{NaCl} \rightarrow \text{Na}^+ + \text{Cl}^-$$

Any substance that dissociates into ions in water and allows the conduction of electric current is called an **electrolyte**.

Covalent Bonds

Most substances do not have the properties of ionic compounds since they are not composed of ions. Consider what would happen if the majority of bonds holding an organism together were ionic and it rained. Most substances are molecular, composed of electrically neutral groups of atoms that are tightly bound together. As noted earlier, many gases are diatomic, occurring naturally as 2 atoms bound together as an electrically neutral molecule. Hydrogen, for example, occurs as molecules of H_2 and no ions are involved. The hydrogen atoms are held together by a *covalent bond*. A **covalent bond** is a chemical bond formed by the sharing of a pair of electrons. In the diatomic hydrogen molecule, each hydrogen atom contributes a single electron to the shared pair. Hydrogen atoms both share one pair of electrons, but other elements might share more than one pair. • **structure of carbohydrates, p. 47**

Consider how the covalent bond forms between two hydrogen atoms by imagining two hydrogen atoms moving toward one another. Each atom has a single electron. As the atoms move closer and closer together, their outer energy levels begin to overlap. Each electron is attracted to the oppositely charged nucleus of the other atom and the overlap tightens. Then the repulsive forces from the like-charged nuclei will halt the merger. A state of stability is reached between the 2 nuclei and 2 electrons because the outermost energy level is full, and an H_2 molecule has been formed. The electron pair is now shared by both atoms and the attraction of each nucleus for the electron of the other holds the atoms together (figure 2.8).

Two dots can represent a lone pair of outer orbital electrons, or they can represent a bonding pair, a single pair of electrons being shared by 2 atoms. Bonding pairs of electrons are often represented by a simple line between 2 atoms as in the example:

H : H   is shown as   H − H

and

A covalent bond in which a single pair of electrons is shared by 2 atoms is called a *single covalent bond* or simply a single bond. Some atoms have 2 unpaired electrons and can share more than 1 electron pair. A double bond is a covalent bond formed when *2 pairs* of electrons are shared by 2 atoms. This happens mostly in compounds involving atoms of the elements C, N, O, and S. For example, ethylene, a gas given off from ripening fruit, has a double bond

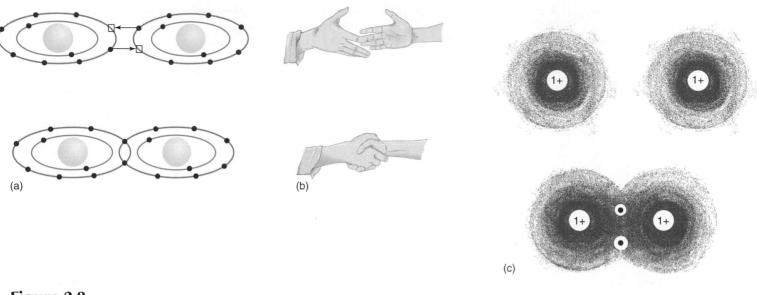

Figure 2.8

Covalent Bond Between Hydrogen Atoms
(*a*) When two atoms come sufficiently close to each other that the locations of the outermost electrons overlap, an electron from each one can be shared to "fill" that outermost energy-level area. (*b*) When two people shake hands, they need to be close enough to each other so that their hands can overlap to form a handshake. (*c*) When the hydrogen atoms bond, a new electron distribution pattern forms around the entire molecule, and both electrons share the outermost molecular energy level.

between the two carbons. The electron dot formula for ethylene is

A triple bond is a covalent bond formed when *three pairs* of electrons are shared by two atoms. Triple bonds occur mostly in compounds with atoms of the elements C and N. Atmospheric nitrogen gas, for example, forms a triple covalent bond

$$\ddot{N} \,\vdots\!\vdots\!\vdots\, \ddot{N}$$

2.7 Water: The Essence of Life

Water seems to be a simple molecule, but it has several special properties that make it particularly important for living things. A water molecule is composed of 2 atoms of hydrogen and 1 atom of oxygen joined by covalent bonds. However, electrons in these covalent bonds are not shared equally. Oxygen, with 8 positive protons, has a greater attraction for the shared electron than hydrogen does with its single proton. Therefore, the shared electrons spend more time around the oxygen part of the molecule than they do

around the hydrogen. This results in the oxygen end of the molecule being more negative than the hydrogen end. When electrons in a covalent bond are not equally shared, the molecule is said to be *polar* and the covalent bonds are given the name *polar covalent bonds*.

When the negative pole of a polar molecule is attracted to the positive pole of another similar polar molecule, the hydrogen will usually be located between the 2 molecules. Since in polar molecules the positive hydrogen end of one molecule is attracted to the negative end of a similar molecule, these attractive forces are often called *hydrogen bonds*. **Hydrogen bonds** are intermolecular (i.e., between molecules) forces of attraction and occur only between hydrogen and oxygen or hydrogen and nitrogen. *Hydrogen bonds hold molecules together; they do not bond atoms together; they are not considered true chemical bonds.* We usually represent this attraction as three dots between the attracted regions. This weak bond is not responsible for forming molecules, but it is important in determining how groups of molecules are arranged. In addition, when a very large molecule, such as a protein or DNA (genetic material), has parts of its structure slightly positive and other parts slightly negative, these two areas will attract each other and result in coiling or folding of these threadlike molecules (figure 2.9). Because water is a polar covalent compound (has slightly "+" and "−" ends), it has several significant physical and biological properties (Outlooks 2.2).

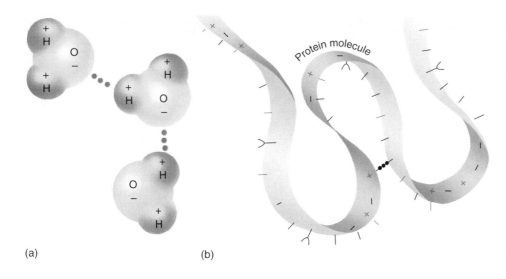

(a)   (b)

Protein molecule

Figure 2.9

Hydrogen Bonds

(*a*) Water molecules arrange themselves so that their positive portions are near the negative portions of other water molecules. The attraction force of a single hydrogen bond is indicated as three dots in a row. It is this kind of intermolecular bonding that accounts for water's unique chemical and physical properties. Without such bonds, life as we know it on Earth would be impossible. (*b*) The large protein molecule here also has polar areas. When the molecule is folded so that the partially positive areas are near the partially negative areas, a slight attraction forms that tends to keep it folded.

Mixtures and Solutions

A **mixture** is matter that contains two or more substances that are not in set proportions (figure 2.10). For example, salt water can be composed of varying amounts of NaCl and H_2O. If the components of the mixture are distributed equally throughout, it is called a homogeneous *solution*. A **solution** is a homogeneous mixture of ions or molecules of two or more substances. The process of making a solution is called *dissolving*. The amounts of the component parts of a solution are identified by the terms *solvent* and *solute*. The **solvent** is the component present in the larger amount. The **solute** is the component that dissolves in the solvent. If one of the components of a solution is a liquid, it is usually identified as the solvent. An *aqueous solution* is a solution of a solid, a liquid, or a gas in water. When sugar dissolves in water, sugar molecules separate from one another. The molecules become uniformly dispersed throughout the molecules of water. In an aqueous salt solution, however, the salt dissociates into sodium and chlorine ions.

The relative amounts of solute and solvent are described by the *concentration* of a solution. In general, a solution with a large amount of solute is "concentrated" and a solution with much less solute is "dilute." The terms *dilute* and *concentrated* are somewhat arbitrary.

2.8 Chemical Reactions

When compounds are formed or broken into other substances, new materials with new properties are produced. This kind of a change in matter is called a *chemical change*, and the process is called a *chemical reaction*. A **chemical reaction** is defined as a change in matter in which different chemical substances are created by forming or breaking chemical bonds. The elements stay the same but the compounds and their properties change when the elements are bonded in new combinations. All living things have the abil-

Figure 2.10

How Do Mixtures Compare?

Matter can be a pure substance or a mixture. The term homogeneous means "the same throughout." Homogenized milk has the same composition throughout the container. Before milk was homogenized (i.e., vigorously shaken to break fat into small globules), it would "separate." The cream (which floats to the top) could be "skimmed" off the milk below, leaving skimmed milk. A heterogeneous mixture does not have the same composition throughout.

ity to manipulate energy and matter. In other words, they are constantly performing chemical reactions.

Chemical reactions produce new chemical substances with greater or smaller amounts of potential energy. Energy is absorbed to produce new chemical substances with more potential energy. Energy is released when new chemical substances are produced with less potential energy. For example, new chemical substances are produced in green plants through the process called *photosynthesis*. A green plant uses radiant energy (sunlight), carbon dioxide, and water to produce new chemical materials and oxygen. These new chemical materials, the stuff that makes up leaves, roots, and wood, contain more chemical energy than the carbon dioxide and water from which they were formed.

OUTLOOKS 2.2

Water and Life

1. **Water has a high surface tension.** Because water molecules are polar, hydrogen bonds form between water molecules and they tend to stick more to one another (are *cohesive*) than to air molecules. Thus, water tends to pull together to form a smooth surface where water meets air. This layer can be strong enough for certain insects to walk upon. For instance, certain insects can be seen to run along the surface of a pond with the same appearance as a person walking over a pool covered with a strong layer of plastic film. Because water molecules stick to each other and to some other things (are *adhesive*) water easily climbs through narrow tubes called capillary tubes. The fact that water molecules can stick to other things explains why water can make things wet. This *capillary action* also helps water to move through soil and up the vessels in plants and through capillaries to animals.
   • hydrologic cycle p. 290

2. **Water has unusually high heats of vaporization and fusion.** Because polar water molecules hydrogen bond to one another, it requires an unusually large amount of heat energy to separate them from one another. Water resists changes in temperature. It takes 540 calories of heat energy to convert 1 gram of liquid water to its gaseous state, water vapor. More practically, this means that large bodies of water such as lakes and rivers must absorb enormous amounts of energy before they will evaporate, leaving the life within them high and dry. This also means that humans can get rid of excess body heat by sweating since when the water evaporates it removes heat from the skin. On the opposite side, a high heat of fusion means that this large amount of heat energy must be removed from liquid water before it changes from a liquid to its solid state, ice. Therefore, water can remain liquid and a suitable home for countless organisms long after the atmospheric temperature has reached the freezing point.

3. **Water has unusual density characteristics.** As heat energy is lost from a lake or other body of water and its temperature decreases, it reaches its greatest density at 4°C. This means that at 4°C it sinks, leaving the less dense, colder water on top. As the surface water changes from its liquid to solid state, the molecules form new arrangements that resemble a honeycomb. The spaces between the water molecules make the solid state, ice, less dense than the water beneath and the ice floats. It is the surface water that freezes to a solid, covering the denser, liquid water and living things that live in it.

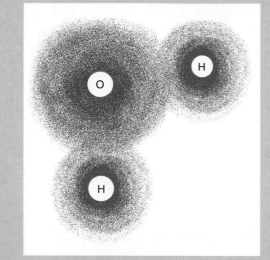

4. **Water's specific gravity is an important property.** One cubic centimeter of water weighs 1 gram. Water has a density of 1 gram/cubic centimeter. Anything with a higher density will sink in water, and anything with a lower density floats. Any substance with a specific gravity less than 1.00 floats. If you mix water and gasoline, the gasoline (specific gravity of 0.75) floats to the top. People also vary in the specific gravity of their bodies. Some persons find it very easy to float in water while others find it impossible. This is directly related to their individual specific gravity, which in turn is a measure of their ratio of body fat to muscle and bone.

5. **Water is considered to be the *"universal solvent,"*** because most other chemicals can be dissolved in water. This means that wherever water goes, either through the ground, in the air, or an organism, it carries chemicals. Most recently, it has been discovered that water in its most pure form is even capable of acting as a solvent for oils.

6. **Water comprises 80–90% of the bodies of most living things.** This is important since the chemical reactions of all living things occur in water.

7. **Water vapor in the atmosphere is known as *humidity*,** which changes with environmental conditions. The ratio of how much water vapor *is* in the air compared to how much water vapor *could be* in the air at a certain temperature is called *relative humidity*. Relative humidity is closely associated with your comfort. When the relative humidity and temperature are high, it is difficult to evaporate water from your skin, so it is more difficult to cool yourself and you are uncomfortably warm.

A **chemical equation** is a way of describing what happens in a chemical reaction. The chemical reaction of photosynthesis is described by the equation:

$$\underset{\substack{\text{energy}\\\text{(sunlight)}}}{} + \underset{\substack{\text{carbon}\\\text{dioxide}\\\text{molecules}}}{} + \underset{\substack{\text{water}\\\text{molecules}}}{} \rightarrow \underset{\substack{\text{plant}\\\text{material}\\\text{molecules}\\\text{(ex., sugar)}}}{} + \underset{\substack{\text{oxygen}\\\text{molecules}}}{}$$

$$6\,CO_2 \;+\; 6\,H_2O \;\rightarrow\; C_6H_{12}O_6 \;+\; 6\,O_2$$

The substances that are changed, carbon dioxide molecules and water molecules, are on the left side equation and are called **reactants.** The equation also indicates that energy is absorbed, since the term "energy" appears on the left side. The arrow indicates that a chemical reaction is occurring or means "yields." The new chemical substances are on the right side and are called **products.** Reading the photosynthesis reaction as a sentence you would say, "Carbon dioxide and water use energy to react, yielding plant materials and oxygen."

Notice in the photosynthesis reaction that there are numbers (coefficients) preceding each of the chemical formulas and subscripts within each chemical formula. The coefficients indicate the number of each kind of molecule involved in the reaction. The subscripts indicate the number of each kind of element in a single molecule of that compound. Chemical reactions always take place in whole number ratios. That is, only whole molecules can be involved in a chemical reaction. It is not possible to have half a molecule of water serve as a reactant or become a product. Half a molecule of water is not water. Furthermore, the numbers of atoms of each element on the reactant side must equal the numbers on the product side. Because the preceding equation has equal numbers of each element (C, H, O) on both sides, the equation is said to be "balanced."

Five of the most important chemical reactions that occur in organisms are (1) oxidation-reduction, (2) dehydration synthesis, (3) hydrolysis, (4) phosphorylation, and (5) acid-base.

Oxidation-Reduction Reactions

The following reaction between carbohydrate (ex., sugar) and oxygen is an example of an important group of chemical reactions called **oxidation-reduction.** When the term "oxidation" was first used, it specifically meant reactions involving the combination of oxygen with other atoms. But fluorine, chlorine, and other elements were soon recognized to participate in similar reactions, so the definition was changed to one describing the shifts of electrons in the reaction.

An oxidation-reduction reaction is defined as a reaction in which electrons are transferred from one atom to another, and with it, the energy contained in those electrons. As implied by the name, such a reaction has two parts and each part tells you what happens to the electrons. *Oxidation* is the part of an oxidation-reduction reaction in which an atom loses an electron. *Reduction* is the part of an oxidation-reduction reaction in which an atom gains an electron. The name also implies that in any reaction in which oxida-

tion occurs, reduction must also take place. One cannot take place without the other. Cellular respiration is an oxidation-reduction reaction that occurs in all cells:

$$\underset{\text{sugar}}{C_6H_{12}O_6} + \underset{\text{oxygen}}{6\,O_2} \rightarrow \underset{\text{water}}{6\,H_2O} + \underset{\substack{\text{carbon}\\\text{dioxide}}}{6\,CO_2} + \text{energy}$$

In this cellular respiration reaction, sugar is being oxidized (losing its electrons) and oxygen is being reduced (gaining the electrons from sugar). The high chemical potential energy in the sugar molecule is released and some of this energy is used by the organism to perform work. In the previous photosynthesis reaction, water is being oxidized (losing its electrons) and carbon dioxide is being reduced (gaining the electrons from water). The energy required to carry out this reaction comes from the sunlight and becomes stored in the product, sugar.

Dehydration Synthesis Reactions

Dehydration synthesis reactions, as the name implies, involve the loss of water as a result of the chemical change. The water is a product formed from its component parts (H and OH) that are removed from the reactants. Proteins, for example, consist of a large number of amino acid subunits joined together by dehydration synthesis:

$$\underset{\text{amino acid 1}}{H{-}NH_3C_2O{-}OH} + \underset{\text{amino acid 2}}{H{-}NH_3C_2O{-}OH} \rightarrow$$

$$\rightarrow$$

$$\underset{\text{protein}}{H{-}NH_3C_2O{-}\,NH_3C_2O{-}OH} + \underset{\substack{\text{water}\\(H_2O)}}{H{-}OH}$$

The building blocks of protein are bonded to one another to synthesize (produce) larger, more complex product molecules (i.e., protein). In dehydration synthesis reactions, water is produced as smaller reactants become chemically bonded to one another forming fewer but larger product molecules.

Hydrolysis Reactions

Hydrolysis reactions are the opposite of dehydration synthesis reactions. In a hydrolysis reaction, water is used to break one of the reactants into smaller products:

$$\underset{\text{protein}}{H{-}NH_3C_2O{-}NH_3C_2O{-}OH} + \underset{\substack{\text{water}\\(H_2O)}}{H{-}OH} \rightarrow$$

$$\rightarrow$$

$$\underset{\text{amino acid 1}}{H{-}NH_3C_2O{-}OH} + \underset{\text{amino acid 2}}{H{-}NH_3C_2O{-}OH}$$

A more familiar name for this chemical reaction is *digestion.* This is the kind of chemical reaction that occurs when a protein food, such as meat, is digested. Notice in the previous example that the H and OH component parts of the reactant water become parts of the building block products. In hydrolysis reactions, water is used as a reactant and larger molecules are broken down into smaller units.

Phosphorylation Reactions

A **phosphorylation reaction** takes place when a cluster of atoms known as a *phosphate group*

$$(O—P=O)$$

with O^- above, O^- below

is added to another molecule. This cluster is abbreviated in many chemical formulas in a shorthand form as **P** is transferred from one molecule to another. This is a very important reaction since the bond between a phosphate group and another atom contains potential energy that is used by all cells to power numerous physiological activities. Phosphorylation reactions result in the transferring of their potential energy to other molecules to power activities such as muscle contraction transport of nutrients and sperm swimming:

adenosine triphosphate   +   sugar   →
higher   lower
energy   energy

adenosine diphosphate   +   sugar monophosphate
lower   higher
energy   energy

This type of reaction is commonly involved in providing the kinetic energy needed by all organisms. It can also take place in reverse. When this occurs, energy must be added from the environment (sunlight or another phosphorylated molecule) and is stored in the phosphorylated molecule.

energy   adenosine   sugar
added   +   diphosphate   +   monophosphate   →
lower   higher
energy   energy

adenosine
triphosphate   +   sugar
higher   lower
energy   energy

The red color indicates higher energy; green, moderate energy; and blue, lower energy content. This kind of phosphorylation occurs during photosynthesis (storing sunlight energy in molecules) and cellular respiration (releasing energy).

Acid–Base Reactions

Also known as ion exchange reactions, **acid-base reactions** take place when the ions of one compound (acid) interact with the ions of another compound (base), forming a salt and water (see Section 2.9). An aqueous solution containing dissolved acid is a solution of hydrogen ions. If a solution containing a second ionic basic compound is added, a mixture of ions results. While mixed together, an ionic exchange reaction can take place. For example,

$H^+ Cl^-$ + $Na^+ OH^-$ → $Na^+ Cl^-$ +   HOH
hydrochloric   +   sodium   →   sodium   +   water
acid   hydroxide   chloride   (H_2O)

In an acid-base reaction, the H from the acid becomes chemical bonded to the OH of the base. This type of reaction frequently occurs in organisms or their environment. Since the reactants can be very harmful, such a reaction can be protective.

2.9 Acids, Bases, and Salts

Acids, bases, and salts are three classes of biologically important compounds (table 2.3). Their characteristics are determined by the nature of their chemical bonds. An **acid** is any ionic compound that releases a hydrogen ion in solution. You can think of an acid, then, as a substance able to donate a proton to a solution. A hydrogen atom without its electron is a proton. Acids have a sour taste, such as the taste of citrus fruits. However, tasting chemicals to see if they are acids can be very hazardous since many are highly corrosive. An example of a common acid is the phosphoric acid—H_3PO_4— in your cola soft drink. It is a dilute solution of this acid that gives cola drinks their typical flavor. Hydrochloric acid is another example:

Acids are ionically bonded molecules which when placed in water dissociate into hydrogen (H^+) ions.

Bases, or *alkaline* substances, have a slippery feel on the skin. They have a caustic action on living tissue by converting the fats in living tissue into a water-soluble substance. A similar reaction is used to make soap by mixing a strong base with fat. This chemical reaction results in soap having its slippery feeling. Bases are also used in certain kinds of batteries—that is, alkaline batteries. Weak bases have a bitter taste; for example, the taste of coffee. A base is the opposite of an acid in that it is an ionic compound that, when dissolved in water, removes hydrogen ions from solution. Many kinds of bases release a group known as a **hydroxide ion,** or OH^- group. This group is composed of an oxygen atom and a hydrogen atom bonded together, but with an additional electron. The hydroxide ion is negatively

Table 2.3

SOME COMMON ACIDS, BASES, AND SALTS

Acids

Acetic acid	CH_3COOH	A weak acid found in vinegar
Carbonic acid	H_2CO_3	The weak acid of carbonated beverages that provides bubbles or fizz
Lactic acid	$CH_3CHOHCOOH$	Weak acid found in sour milk, sauerkraut, and pickles
Phosphoric acid	H_3PO_4	Weak acid used in cleaning solutions, added to carbonated cola beverages for taste
Sulfuric acid	H_2SO_4	Strong acid used in batteries

Bases

Sodium hydroxide	$NaOH$	Strong base also called lye or caustic soda; used in oven cleaners
Potassium hydroxide	KOH	Also known as caustic potash; strong base used in drain cleaners
Magnesium hydroxide	$Mg(OH)_2$	Also known as milk of magnesia; a weak base used in antacids and laxatives

Salts

Alum	$Al(SO4)_2$	Found in medicine, canning, and baking powder
Baking soda	$NaHCO_3$	Used in fire extinguishers, antacids, and baking powder
Chalk	$CaCO_3$	Used in antacid tablets
Epsom salts	$MgSO_4 \cdot H_2O$	Used in laxatives and skin care
Trisodium phosphate (TSP)	Na_3PO_4	Used in water softeners, fertilizers, and cleaning agents

charged. Because it is negatively charged it will remove positively charged hydrogen ions from solution. A very strong base used in oven cleaners is sodium hydroxide, $NaOH$. Notice that free ions are always written with the type and number of their electrical charge as a superscript.

Basic (alkaline) substances are ionically bonded molecules which when placed in water dissociate into hydroxide (OH^-) ions.

Acids and bases are also spoken of as being strong or weak. Strong acids are those that dissociate nearly all of their hydrogens when in solution (ex., hydrochloric acid). Weak acids dissociate only a small percentage of their hydrogens, (ex., phosphoric acid). Strong bases dissociate nearly all of their hydroxides ($NaOH$), and weak bases, only a small percentage. The weak base sodium bicarbonate, $NaHCO_3$, will react with acids in the following manner:

$$NaHCO_3 + HCl \rightarrow NaCl + CO_2 + H_2O$$

Notice that sodium bicarbonate does not contain a hydroxide ion but it is still a base since it removes hydrogen ions from solution.

The degree to which a solution is acidic or basic is represented by a quantity known as **pH.** The pH scale is a measure of hydrogen ion concentration (figure 2.11). A pH of 7 indicates that the solution is neutral and has an equal number of H^+ ions and OH^- ions to balance each other. As the pH number gets smaller, the number of hydrogen ions in the solution increases. A number higher than 7 indicates that the solution has more OH^- than H^+. Pure water has a pH of 7. As the pH number gets larger, the number of hydroxide ions increases. It is important to note that the pH scale is logarithmic—that is, a change in one pH number is actually a 10-fold change in real numbers of OH^- or H^+. For example, there is 10 times more H^+ in a solution of pH 5 than in a solution of pH 6. • enzymes and acid, p. 96

When water dissociates it releases both hydrogen (H^+) and hydroxide (OH^-) ions. It is neither a base nor an acid. Its pH is 7, neutral.

Salts are compounds that do not release either H^+ or OH^- when dissolved in water; thus, they are neither acids nor bases. However, they are generally the result of the reaction

Figure 2.11

The pH Scale

The concentration of acid (proton donor or electron acceptor) is greatest when the pH number is lowest. As the pH number increases, the concentration of base (proton acceptor or electron donor) increases. At a pH of 7.0, the concentrations of H^+ and OH^- are equal. We usually say, as the pH number gets smaller the solution becomes more acidic. As the pH number gets larger the solution becomes more basic or alkaline.

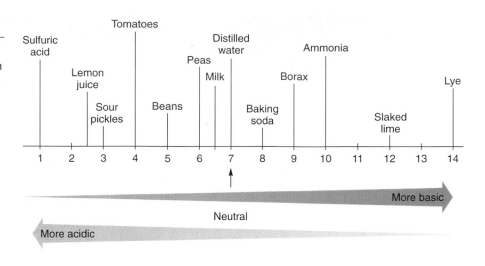

between an acid and a base in a solution. For example, when an acid such as HCl is mixed with NaOH in water, the H^+ and the OH^- combine with each other to form pure water, H_2O. The remaining ions (Na^+ and Cl^-) join to form the salt NaCl:

$$HCl + NaOH \rightarrow (Na^+ + Cl^- + H^+ + OH^-) \rightarrow NaCl + H_2O$$

Salts are ionically bonded molecules which when placed in water dissociate into ions that are neither hydrogen (H^+) nor hydroxide (OH^-).

The chemical reaction that occurs when acids and bases react with each other is called *neutralization*. The acid no longer acts as an acid (it has been neutralized) and the base no longer acts as a base.

As you can see from figure 2.11, not all acids or bases produce the same pH. Some compounds release hydrogen ions very easily, cause low pHs, and are called *strong acids*. Hydrochloric acid (HCl) and sulfuric acid (H_2SO_4) are examples of strong acids (figure 2.12a). Many other compounds give up their hydrogen ions grudgingly and therefore do not change the pH very much. They are known as *weak acids*. Carbonic acid (H_2CO_3) and many organic acids found in living things are weak acids. Similarly, there are strong bases like sodium hydroxide (NaOH) and weak bases like sodium bicarbonate—$Na^+(HCO_3)^-$ (figure 2.12 b). • evolution of life, p. 440

SUMMARY

The study of life involves learning about anatomy and physiology and the structure and function of organisms. All organisms display the chemical and physical properties typical of all matter and energy. The two kinds of energy used by organisms are potential and kinetic. The kinetic molecular theory states that all matter is made up of tiny particles that are in constant motion.

All matter is composed of atoms, which are composed of an atomic nucleus and electrons. The atomic nucleus can contain protons and neutrons while the electrons encircle the nucleus at different energy levels. Atoms tend to seek their most stable configuration and follow the octet rule, which states that they all seek a filled outermost energy level. How they achieve this stability is the nature of chemical reactions. Five of the most important chemical reactions that occur in organisms are (1) oxidation-reduction, (2) dehydration synthesis, (3) hydrolysis, (4) acid-base, and (5) phosphorylation.

There are many kinds of atoms whose symbols and traits are described by the periodic table of the elements. These atoms differ from one another by the number of protons and electrons they contain. Each is given an atomic number, based on the number of protons in the nucleus, and an atomic weight, an average of all the isotopes of a particular element. The mass number is the sum of the number of protons and neutrons in the nucleus of an atom.

Atoms may be combined by chemical reactions into larger units called molecules. There are many kinds of molecules. Two kinds of chemical bonds allow molecules to form—ionic bonds and covalent bonds. A third bond, the hydrogen bond, is a weaker bond that holds molecules together and may also help large molecules maintain a specific shape. Molecules are described by their chemical formulas, which state the number and kinds of components of which they are composed.

Energy can neither be created nor destroyed, but it can be converted from one form to another. Potential energy and kinetic energy can be interconverted. The amount of kinetic energy that the molecules of various substances contain determines whether they are solids, liquids, or gases. Temperature is a measure of the average kinetic energy of the molecules making up a substance. Heat is the

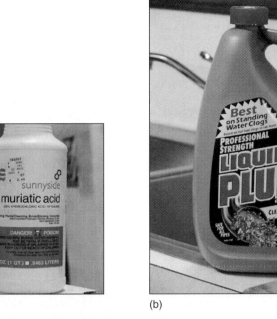

(a)   (b)

Figure 2.12

Strong Acid and Strong Base

(a) Hydrochloric acid (HCl) has the common name of muriatic acid. It is a strong acid used in low concentrations to clean swimming pools and brick surfaces. It is important that you wear protective equipment when working with a solution of muriatic acid. (b) Liquid Plumr® is a good example of a drain cleaner with a strong base. The active ingredient is NaOH.

total internal kinetic energy of molecules. The random motion of molecules, which is due to their kinetic energy, results in their being distributed throughout available space forming mixtures.

An ion is an atom that is electrically unbalanced. Ions interact to form ionic compounds, such as acids, bases, and salts. Compounds that release hydrogen ions when mixed in water are called acids; those that remove hydrogen ions are called bases. A measure of the hydrogen ions present in a solution is known as the pH of the solution.

THINKING CRITICALLY

For guidelines to answer this question, visit the Online Learning Center.

Sodium bicarbonate ($NaHCO_3$) is a common household chemical known as baking soda, bicarbonate of soda, or bicarb. It has many uses other than baking. It is a component of many products including toothpaste and antacids, swimming pool chemicals, and headache remedies. When baking soda comes in contact with hydrochloric acid, the following reaction occurs:

$$HCl + NaHCO_3 \rightarrow NaCl + CO_2 + H_2O$$

Can you describe what happens to the atoms in this reaction? In your description, include changes in chemical bonds, pH, and kinetic energy. Can you describe why the baking soda is such an effective chemical in the previously mentioned products? You might try this at home: Place a pinch of sodium bicarbonate ($NaHCO_3$) on a plate. Add a couple of drops of vinegar. Observe the reaction. Based on the previous reaction, can you explain chemically what has happened?

CONCEPT MAP TERMINOLOGY

Construct a concept map using these terms.

anion	molecule
cation	neutron
electron	proton
ion	salt
ionic bond	

Visit the Online Learning Center to use an interactive format to construct another concept map.

KEY TERMS

Use interactive flash cards, on the Online Learning Center, to help you learn the meanings of these terms.

acid	chemical reaction
acid-base reactions	chemistry
anatomy	compound
atomic mass unit (μ)	covalent bond
atomic nucleus	dehydration synthesis
atomic number	electrolyte
atomic weight	electron
atoms	elements
base	energy
calorie	energy level
chemical bond	formula
chemical equation	gas

heat
hydrogen bond
hydroxide ion
ion
ionic bond
isotope
kinetic energy
kinetic molecular theory
law of conservation of energy
liquid
mass number
matter
mixture
molecule
neutron
oxidation-reduction reaction

periodic table of the elements
pH
phosphorylation reaction
physiology
potential energy
product
proton
reactant
salt
solid
solute
solution
solvent
states of matter
temperature

LEARNING CONNECTIONS

Visit the Online Learning Center to find additional study resources including those highlighted below.

2.1   The Structure and Function of Living Things

1. What is chemistry?
2. What does anatomy have to do with the study of chemistry?
3. What does physiology have to do with the study of chemistry?

2.2   Energy and Life

Animations and Review
• Basic chemistry

Interactive Concept Maps
• Ways of looking at matter

4. What is the difference between potential and kinetic energy?
5. Define the term *work*.

2.3   What Is the Nature of Matter?

Interactive Concept Maps
• Information from the periodic table

6. What is the difference between an atom and an element?
7. What is the difference between a molecule and a compound?
8. On what bases are solids, liquids, and gases differentiated?
9. What relationship does kinetic energy have to the three states of matter?

2.4   Molecules and Kinetic Energy

10. What is the difference between temperature and heat?
11. What is a calorie?

2.5   Structure of the Atom

Animations and Review
• Atoms

12. How many protons, electrons, and neutrons are in a neutral atom of potassium having an atomic weight of 39?
13. How would two isotopes of oxygen differ?
14. Define the terms *AMU* and *atomic number*.

2.6   Chemical Change

Animations and Review
• Bonds

15. Why are the outermost electrons of an atom important?
16. Name two kinds of chemical bonds that hold atoms together. How do these bonds differ from one another?

2.7   Water: The Essence of Life

Animations and Review
• Water

Experience This!
• Hydrogen bonds and surface tension

Review
• The Special Properties of Water

17. What is the difference between a polar and a nonpolar molecule?
18. What is different about a hydrogen bond in comparison to covalent and ionic bonds?
19. What is the difference between a solute and a solvent?
20. What relationship does kinetic energy have to homogeneous solutions?

2.8   Chemical Reactions

21. Give an example of an ion exchange reaction.
22. What happens during an oxidation-reduction reaction?
23. Describe the difference between a reactant and a product.

2.9   Acids, Bases, and Salts

Animations and Review
• pH

24. What does it mean if a solution has a pH number of 3, 12, 2, 7, or 9?
25. If the pH of a certain solution changes from 8 to 9, what happens to the hydroxide ion concentration?

Test yourself with multiple choice questions that provide immediate feedback on the Online Learning Center with PowerWeb at

www.mhhe.com/enger11

Organic Chemistry
The Chemistry of Life

GOALS AND OBJECTIVES

Understand carbon atoms and their chemical nature.

❏ Distinguish between molecules that are organic and inorganic.

❏ State the features of a carbon atom that make it able to bond in chains and rings.

❏ Describe how single and double covalent bonds are formed in organic molecules.

❏ Draw an example of a carbon skeleton.

❏ Recognize the main functional groups.

Recognize different molecular structures common to organic molecules.

❏ List the major group of organic molecules associated with living things.

❏ Recognize the basic subunit of each of the major groups of organic molecules.

❏ Describe the function played by each of the major groups of organic molecules.

❏ Give examples of each of the major groups of organic molecules.

❏ Explain why certain organic molecules are considered monomers and others polymers.

❏ Diagram a simple structure of the monomers of monosaccharide, amino acid, and nucleotide.

Know the ways that molecules can react with one another.

❏ Explain the difference between dehydration synthesis and hydrolysis.

❏ Diagram simple dehydration synthesis and hydrolysis reactions.

❏ Describe how organic molecules such as proteins can have primary, secondary, tertiary, and quaternary structures.

Cells: Anatomy and Action

PART TWO

Figure 3.1

Some Common Synthetic Organic Materials

These items are examples of useful organic compounds invented by chemists. The words *organism, organ, organize,* and *organic* are all related. Organized objects have parts that fit together in a meaningful way. Organisms are separate living things that are organized. Animals, for example, have within their bodys' organ systems comprised of *organs,* and their organs are composed of unique kinds of molecules that are *organic.*

3.1 Molecules Containing Carbon

The principles and concepts discussed in chapter 2 apply to all types of matter—nonliving as well as living. Living systems are composed of various types of molecules. Most of the things we described in the previous chapter did not contain carbon atoms and so are classified as **inorganic molecules.** This chapter is mainly concerned with more complex structures, **organic molecules,** which contain carbon atoms arranged in rings or chains. • origin of life, p. 440

The original meanings of the terms *inorganic* and *organic* came from the fact that organic materials were thought to be either alive or produced only by living things. A very strong link exists between organic chemistry and the chemistry of living things, which is called **biochemistry,** or biological chemistry. Modern chemistry has considerably altered the original meanings of the terms organic and inorganic, because it is now possible to manufacture unique organic molecules that cannot be produced by living things. Many of the materials we use daily are the result of the organic chemist's art. Nylon, aspirin, polyurethane varnish, silicones, Plexiglas, food wrap, Teflon, and insecticides are just a few of the unique synthetic molecules that have been invented by organic chemists (figure 3.1).

In many instances, organic chemists have taken their lead from living organisms and have been able to produce organic molecules more efficiently, or in forms that are slightly different from the original natural molecule. Some examples of these are rubber, penicillin, certain vitamins, insulin, and alcohol (figure 3.2). Another example is the insecticide Pyethrin, which is based on a natural insecticide that is widely used in agriculture and for domestic purposes, and is from the chrysanthemum plant *Pyrethrum cinerariaefolium.*

Carbon: The Central Atom

All organic molecules, whether they are natural or synthetic, have certain common characteristics. The carbon atom, which is the central atom in all organic molecules, has some unusual properties. Carbon is unique in that it can combine with other carbon atoms to form long chains. In many cases the ends of these chains may join together to form ring structures (fig-

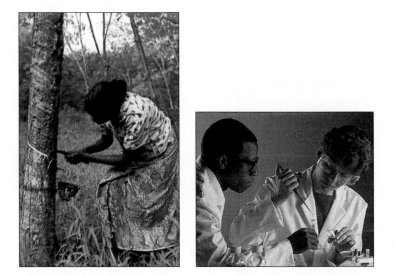

Figure 3.2

Natural and Synthetic Organic Compounds

Some organic materials, such as rubber, were originally produced by plants but are now synthesized in industry. The photograph on the left shows the collection of latex from the rubber tree. After processing, this naturally occurring organic material will be converted into products such as gloves, condoms, and tubing. The other photograph shows organic chemists testing one of the chemicals in the rubber manufacturing process.

ure 3.3). Only a few other atoms have this ability. Also unusual is that these bonding sites are all located at equal distances from one another. If you were to take a rubber ball and stick four nails into it so that they were equally distributed around the ball, you would have a good idea of the arrangement of the bonds (figure 3.4). These bonding sites are arranged this way because in the carbon atom there are 4 electrons in the second energy level. These 4 electrons do not stay in the standard positions described in chapter 2. They distribute themselves differently, that is, into four propeller-shaped orbitals. This allows them to be as far away from each other as possible. Carbon atoms are usually involved in covalent bonds. Because carbon has four places it can bond, the carbon

Figure 3.3

A Chain or Ring Structure
The ring structure shown on the bottom is formed by joining the two ends of a chain of carbon atoms.

Figure 3.4

Bonding Sites of a Carbon Atom
The arrangement of bonding sites around the carbon is similar to a ball with four equally spaced nails in it. Each of the four bondable electrons inhabits an area as far away from the other three as possible. Each can share its electron with another. Carbon can share with four other atoms at each of these sites.

Figure 3.5

A Methane Molecule
A methane molecule is composed of one carbon atom bonded with four hydrogen atoms. (*a*) These bonds are formed at the four bonding sites of the carbon. For the sake of simplicity, all future diagrams of molecules will be two-dimensional drawings, although in reality they are three-dimensional molecules. (*b*) Each line in the diagram represents a covalent bond between the two atoms where a pair of electrons is being shared. This is a shorthand way of drawing the pair of shared electrons.

Figure 3.6

Double Bonds
These diagrams show several molecules that contain double bonds. A double bond is formed when two atoms share two pairs of electrons with each other.

atom can combine with 4 other atoms by forming four separate, *single* covalent bonds with other atoms. This is the case with the methane molecule, which has 4 hydrogen atoms attached to a single carbon atom. Pure methane is a colorless and odorless gas that makes up 95% of natural gas (figure 3.5). The aroma of natural gas is the result of mercaptan (and trimethyl disulfide) added to let consumers know when a leak occurs. Outlooks 3.1 explains how chemists and biologists diagram the kinds of bonds formed in organic molecules.

Some atoms may be bonded to a single atom more than once. This results in a slightly different arrangement of bonds around the carbon atom. An example of this type of bonding occurs when oxygen is attracted to a carbon. Oxygen has 2 bondable electrons. If it shares one of these with a carbon and then shares the other with the same carbon, it forms a *double bond*. A **double bond** is two covalent bonds formed between 2 atoms that share two pairs of electrons. Oxygen is not the only atom that can form double bonds, but double bonds are common between it and carbon. The double bond is denoted by two lines between the two atoms:

$$-C=O$$

Two carbon atoms might form double bonds between each other and then bond to other atoms at the remaining bonding sites. Figure 3.6 shows several compounds that contain double bonds. Some organic molecules contain *triple covalent bonds;* the flammable gas acetylene, HC ≡ CH, is one example. Others, like hydrogen cyanide, HC ≡ N, have biological significance. This molecule inhibits the production of energy and results in death.

Although most atoms can be involved in the structure of an organic molecule, only a few are commonly found. Hydrogen (H) and oxygen (O) are almost always present. Nitrogen (N), sulfur (S), and phosphorus (P) are also very important in specific types of organic molecules.

An enormous variety of organic molecules is possible because carbon is able to (1) bond at four different sites, (2) form long chains, and (3) combine with many other kinds of atoms. The types of atoms in the molecule are important in determining the properties of the molecule. The three-dimensional arrangement of the atoms within the molecule is also important. Because most inorganic molecules are small and involve few atoms, a group of atoms can be usually arranged in only one way to form a molecule. There is only one arrangement for a single oxygen atom and 2 hydrogen atoms in a molecule of water. In a molecule of sulfuric acid, there is only one arrangement for the sulfur atom, the 2 hydrogen atoms, and the 4 oxygen atoms.

Sulfuric (battery) acid

However, consider these two organic molecules:

Dimethyl ether

Ethyl alcohol
(as found in alcoholic beverages)

OUTLOOKS 3.1

Chemical Shorthand

You have probably noticed that sketching the entire structural formula of a large organic molecule takes a great deal of time. If you know the structure of the major functional groups, you can use several shortcuts to more quickly describe chemical structures. When multiple carbons with 2 hydrogens are bonded to each other in a chain, we sometimes write it as follows:

Or we might write it this way:

$$-CH_2-CH_2-CH_2-CH_2-CH_2-CH_2-CH_2-CH_2-CH_2-CH_2-CH_2-CH_2-$$

or more simply, we may write it as follows $(-CH_2-)_{12}$. If the 12 carbons were in a pair of two rings, we probably would not label the carbons or hydrogens unless we wished to focus on a

particular group or point. We would probably draw the two 6-carbon rings with only hydrogen attached as follows:

Don't let these shortcuts throw you. You will soon find that you will be putting a –H group onto a carbon skeleton and neglecting to show the bond between the oxygen and hydrogen, just like a professional. Structural formulas are regularly included in the package insert information of most medications.

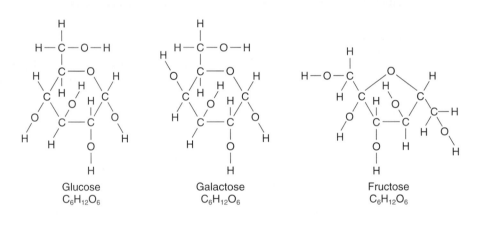

Glucose
$C_6H_{12}O_6$

Galactose
$C_6H_{12}O_6$

Fructose
$C_6H_{12}O_6$

Figure 3.7

Structural Formulas for Several Hexoses
Several 6-carbon sugars—hexoses (*hex* = 6; *-ose* = sugar)—are represented here. All have the same empirical formula, but each has a different structural formula called a *structural isomer*. Structural isomers may act quite differently from each other.

Both the dimethyl ether and the ethyl alcohol contain 2 carbon atoms, 6 hydrogen atoms, and 1 oxygen atom, but they are quite different in their arrangement of atoms and in the chemical properties of the molecules. The first is an ether; the second is an alcohol. Because the ether and the alcohol have the same number and kinds of atoms, they are said to have the same empirical formula, which in this case is written C_2H_6O. An empirical formula simply indicates the number of each kind of atom within the molecule. When the arrangement of the atoms and their bonding within the molecule is indicated, we call this a structural formula. Figure 3.7 shows several structural formulas for the empirical formula $C_6H_{12}O_6$. Molecules that have the same empirical formula but different structural formulas are called **isomers** (How Science Works 3.1).

The Carbon Skeleton and Functional Groups

To help us understand organic molecules a little better, let's consider some of their similarities. All organic molecules have a **carbon skeleton,** which is composed of rings or chains of carbons. It is this carbon skeleton that determines the overall shape of the molecule. The differences between various organic molecules depend on the length and arrangement of the carbon skeleton. In addition, the kinds of atoms that are bonded to this carbon skeleton determine the way the organic compound acts. Attached to the carbon skeleton are specific combinations of atoms called **functional groups.** Functional groups determine specific chemical properties. By learning to recognize some of the functional groups, it is possible to identify an organic molecule and to predict something about its activity. Figure 3.8

HOW SCIENCE WORKS 3.1

Generic Drugs and Mirror Image Isomers

Isomers that are mirror images of each other are called *mirror image isomers, stereo isomers, enantomers,* or *chiral compounds.* The difference among stereo isomers is demonstrated by shining polarized light through the two types of sugar. The light coming out the other side of a test tube containing D-glucose will be turned to the right—that is, *dextro*rotated. When a solution of L-glucose has polarized light shown through it, the light coming out the other side will be rotated to the left—that is, *levo*rotated.

The results of this basic research has been utilized in the pharmaceutical and health care industries. When drugs are synthesized in large batches in the lab many contain 50% "D" and 50% "L" enantomers. Various so-called "generic drugs" are less expensive because they are a mixture of the two enantomers and have not undergone the more thorough and expensive chemical processes involved in isolating only the "D" or "L" form of the drug. Generic drugs that are mixtures may be (1) just as effective as the pure form, (2) slower acting, (3) or not effective if prescribed to an individual who is a "nonresponder" to that medication. It is for these last two reasons that physicians must be consulted when there is a choice between a generic and a nongeneric medication.

Classification of Small Molecules by Functional Groups

Example of Functional Group in Biologically Important Molecule	Name of Compound

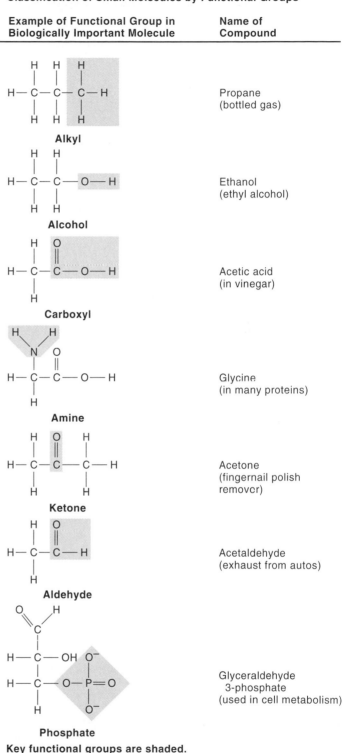

Key functional groups are shaded.

Figure 3.8

Functional Groups

These are some of the groups of atoms that frequently attach to a carbon skeleton. The nature of the organic compound changes as the nature of the functional group changes from one molecule to another.

Figure 3.9

Reactions Involving Carbohydrates
(a) In the dehydration synthesis reaction illustrated here, the two –OH groups line up next to each other so that the –OH groups can be broken from the molecules to form water, and the oxygen that remains acts as an attachment site between the two sugar molecules. (b) A hydrolysis reaction is the opposite of a dehydration synthesis reaction. Carefully compare the two. Many structural formulas appear to be complex at first glance, but if you look for the points where subunits are attached and dissect each subunit, they become much simpler to deal with.

shows some of the functional groups that are important in biological activity. Remember that a functional group does not exist by itself; it must be a part of an organic molecule.

Common Organic Molecules

One way to make organic chemistry more manageable is to organize different kinds of compounds into groups on the basis of their similarity of structure or the chemical properties of the molecules. Frequently you will find that organic molecules are composed of subunits that are attached to each other in sequence. If you recognize the subunit, then the whole organic molecule is much easier to identify. It is similar to distinguishing among the units of a pearl necklace, boat's anchor chain, and beaded key chain. They are all constructed of individual pieces hooked together. These **macromolecules** (*macro-* = large) are very large molecules that are combinations of many smaller, similar molecules joined together in a chainlike structure. Each is distinctly different because the repeating units are not the same—that is, pearls are not anchor links. In all these examples, individual pieces are called *monomers* (*mono* = single; *mer* = segment or piece). The entire finished piece composed of all the units hooked together is called a *polymer* (*poly* = many; *mer* = segments). Of the four groups of organic compounds described, those

that are constructed as polymers are the carbohydrates, proteins, and nucleic acids. In addition, there is the category called the *lipids* that are not true polymers but are macromolecules.

The monomers in a polymer are usually combined by a **dehydration synthesis reaction** (*de* = remove; *hydro* = water; *synthesis* = combine). This reaction results in the synthesis or formation of a macromolecule when water is removed from between the two smaller component parts. For example, when a monomer with an –OH group attached to its carbon skeleton approaches another monomer with an available hydrogen, dehydration synthesis can occur. Figure 3.9a shows the removal of water from between two such subunits. Notice that in this case, the structural formulas are used to help identify just what is occurring. However, the chemical equation also indicates the removal of the water. You can easily recognize a dehydration synthesis reaction because the reactant side of the equation shows numerous small molecules, whereas the product side lists fewer, larger products and water.

The reverse of a dehydration synthesis reaction is known as *hydrolysis* (*hydro* = water; *lyse* = to split or break). **Hydrolysis** is the process of splitting a larger organic molecule into two or more component parts by the addition of water (figure 3.9b). Digestion of food molecules in the stomach is an important example of hydrolysis.

Table 3.1	
THE RELATIVE SWEETNESS OF VARIOUS SUGARS AND SUGAR SUBSTITUTES	
Type of Sugar or Artificial Sweetener	**Relative Sweetness**
Lactose (milk sugar)	0.16
Galactose	0.30
Maltose (malt sugar)	0.33
Glucose	0.75
Sucrose (table sugar)	**1.00**
Fructose (fruit sugar)	1.75
Cyclamate	30.00
Aspartame	150.00
Saccharin	350.00

Figure 3.10

Molecular Formulas of Glucose and Fructose
Glucose (blood sugar) is an aldehyde, and fructose (fruit sugar) is a ketone. Both have a molecular formula of $C_6H_{12}O_6$ (see figure 3.8).

3.2 Carbohydrates

One class of organic molecules, **carbohydrates,** is composed of carbon, hydrogen, and oxygen atoms linked together to form monomers called *simple sugars* or *monosaccharides* (*mono* = single; *saccharine* = sweet, sugar) (table 3.1). Carbohydrates play a number of roles in living things. They serve as an immediate source of energy (sugars), provide shape to certain cells (cellulose in plant cell walls), are components of many antibiotics and coenzymes, and are an essential part of genes. The empirical formula for a simple sugar is easy to recognize because there are equal numbers of carbons and oxygens and twice as many hydrogens—for example, $C_3H_6O_3$ or $C_5H_{10}O_5$. We usually describe simple sugars by the number of carbons in the molecule. The ending *-ose* indicates that you are dealing with a carbohydrate. A tri*ose* has 3 carbons, a pent*ose* has 5, and a hex*ose* has 6. If you remember that the number of carbons equals the number of oxygen atoms and that the number of hydrogens is double that number, these names tell you the empirical formula for the simple sugar.

Simple sugars, such as glucose, fructose, and galactose, provide the chemical energy necessary to keep organisms alive (Outlooks 3.2). *Glucose,* $C_6H_{12}O_6$, is the most abundant carbohydrate and serves as a food and a basic building block for other carbohydrates. Glucose (also called *dextrose*) is found in the sap of plants, and in the human bloodstream it is called *blood sugar.* Corn syrup, which is often used as a sweetener, is mostly glucose. Fructose, as its name implies, is the sugar that occurs in fruits, and it is sometimes called *fruit sugar.* Both glucose and fructose have the same molecular formula, but glucose is an aldehyde sugar and fructose is a ketone sugar (figure 3.10). A mixture of glucose and fructose is found in honey. This mixture also is formed when table sugar (sucrose) is reacted with water in the presence of an acid, a reaction that takes place in the preparation of canned fruit and candies. The mixture of glucose and fructose is called *invert sugar.* Thanks to fructose, invert sugar is about twice as sweet to the taste as the same amount of sucrose.

Simple sugars can be combined with each other to form **complex carbohydrates** (figure 3.11). When two simple sugars bond to each other, a *disaccharide* (*di-* = two) is formed; when three bond together, a *trisaccharide* (*tri-* = three) is formed. Generally, we call a complex carbohydrate that is larger than this a *polysaccharide* (many sugar units). For example, when glucose and fructose are joined together they form a *disaccharide* with the loss of a water molecule:

$$C_6H_{12}O_6 + C_6H_{12}O_6 \rightarrow C_{12}H_{22}O_{11} + H_2O$$

glucose   fructose   sucrose   water

The most common disaccharide is *sucrose,* or ordinary table sugar. Sucrose occurs in high concentrations in sugarcane and sugar beets. It is extracted by crushing the plant materials, then dissolving the sucrose from the materials with water. The water is evaporated and the crystallized sugar is decolorized with charcoal to produce white sugar. Other common disaccharides include *lactose* (milk sugar) and *maltose* (malt sugar). All three disaccharides have similar properties, but maltose tastes only about one-third as sweet as sucrose. Lactose tastes only about one-sixth as sweet as sucrose. No matter which disaccharide sugar is consumed (sucrose, lactose, or maltose), it is converted into glucose and transported by the bloodstream for use by the body.

In all cases, the complex carbohydrates are formed by the removal of water from between the sugars. Some common examples of polysaccharides are starch and glycogen. Cellulose is an important polysaccharide used in constructing the cell walls of plant cells. Humans cannot digest (*hydrolyze*) this complex carbohydrate, so we are not able to use it as an energy source. On the other hand, animals known as ruminants (e.g., cows and sheep) and termites have microorganisms

OUTLOOKS 3.2

What You Become Depends on What You Eat—The Glycemic Index

It is known that the consumption of large amounts of certain *refined* carbohydrates can be detrimental. Refined carbohydrates have been milled or processed to produce a starch that is easily absorbed by the body and also removes many vitamins and minerals. *Whole grains,* on the other hand, have not been milled into fine flour. Carbohydrates are ranked on the *glycemic index* (GI). The GI is a scale that relates to how fast carbohydrates can cause an increase in blood glucose levels. Glucose, also known as blood sugar, is at the top of the GI. Certain complex carbohydrates (ex., potatoes) are also high on the GI since they are rapidly broken down into glucose by cooking or digestion. Such foods stimulate the production of insulin. With an increase in insulin, there is a greater potential for the development of problems such as insulin resistance, type II diabetes, and obesity. *Insulin resistance* is a reduced sensitivity to the action of insulin. (i.e., insulin causes glucose to move from the blood into the cells where it can be metabolized). When this occurs, the body attempts to overcome this resistance by secreting more insulin from the pancreas. The symptoms include abnormal cholesterol and/or triglyceride levels, high blood pressure, and upper body obesity—all risk factors for cardiac disease.

The development of type II diabetes occurs when the pancreas fails to sustain this increased insulin secretion. Along with high blood levels of insulin, there is an increase in appetite. Increasing your intake of low-GI foods (below 55) would be more beneficial. Stay away from high-GI foods (i.e., food with a GI of more than 70). Because of this new knowledge, it is becoming clear that our ideas of a balanced diet may need to be reexamined.

Estimated GI of Some Foods[1]

Tofu frozen dessert, nondairy	164
Dates	146
Potato—baked	121
French fries	107
Pumpkin	107
Graham crackers	106
Potato—boiled/mashed	104
Glucose	***100***
Parsnips	97
Taco shell	97
Cantaloupe	93
High-fructose corn syrup	89
Ice cream	87
Kellogg's corn flakes	84
Pita bread	82
Popcorn	79
Ice cream, low-fat	71
Chocolate	70
Baked beans	69
Oat bran bread	68
Oatmeal	55
Sweet corn	55
Apple	54
Milk—skimmed	46
Converted rice	44
Snickers candy	41
Pasta—spaghetti	38
Milk—whole	39
Yogurt, low-fat	33
Soy beans	18
Peanuts	14

• nutrition, p. 373

[1] Many of these values are for the raw, unprepared food item. Realize that pumpkin pie will not have the same GI as raw pumpkin.

(Modified from *Integrated Science,* 2nd ed., Tillery/Enger, and Ross, McGraw-Hill Publishing.)

within their digestive tracts that do digest cellulose, making it an energy source for them. Plant cell walls add bulk or fiber to our diet, but no calories. Fiber is an important addition to the diet because it helps control weight, reduce the risk of colon cancer, and control constipation and diarrhea.

Simple sugars can be used by the cell as components in other, more complex molecules. Sugar molecules are a part of other, larger molecules such as DNA, RNA, or ATP. The ATP molecule is important in energy transfer. It has a simple sugar (ribose) as part of its structural makeup.

3.3 Proteins

There are thousands of kinds of **proteins** in living things and they can be placed into three categories. Some proteins are important for maintaining the shape of cells and organ-isms—they are usually referred to as **structural proteins.** The proteins that make up the cell membrane, muscle cells, tendons, and blood cells are examples of structural proteins. The protein collagen is found throughout the human body and gives tissues shape, support, and strength. The second category of proteins, **regulator proteins,** helps determine what activities will occur in the organism. These regulator proteins include enzymes and some hormones. These molecules help control the chemical activities of cells and organisms. Enzymes are important, and they are dealt with in detail in chapter 5. Some examples of enzymes are the digestive enzymes in the intestinal tract. Two hormones that are regulator proteins are insulin and oxytocin. Insulin is produced by the pancreas and controls the amount of glucose in the blood. If insulin production is too low, or if the molecule is improperly constructed, glucose molecules are not removed from the bloodstream at a fast enough rate. The ex-

(a) Cellulose

(b) Plant starches

Amylopectin Amylose

(c) Glycogen

Figure 3.11

Complex Carbohydrates

Simple sugars are attached to each other by the removal of water from between them. Three common complex carbohydrates are (a) cellulose (wood fibers), (b) plant starch (amylose and amylopectin), and (c) glycogen (sometimes called animal starch). Glycogen is found in muscle cells. Notice how all are similar in that they are all polymers of simple sugars, but they differ from one another in how they are joined together. While many organisms are capable of digesting (hydrolyzing) the bonds that are found in glycogen and plant starch molecules, few are able to break those that link the monosaccharides of cellulose together.

cess sugar is then eliminated in the urine. Other symptoms of excess "sugar" in the blood include excessive thirst and even loss of consciousness. The disease caused by improperly functioning insulin is known as *diabetes*. Oxytocin, a second protein hormone, stimulates the contraction of the uterus during childbirth. It is also an example of an organic molecule that has been produced artificially (e.g., pitocin) and is used by physicians to induce labor. The third category of proteins is **carrier proteins.** Proteins in this category pick up and deliver molecules at one place and transport them to another. For example, proteins regularly attach to cholesterol entering the system from the diet-forming molecules called lipoproteins, which are transported through the circulatory system. The cholesterol is released at a distance from the digestive tract and the proteins return to pick up more entering dietary cholesterol.

Proteins are polymers made up of monomers known as **amino acids.** An amino acid is a short carbon skeleton that contains an amino group (nitrogen and 2 hydrogens) on one end of the skeleton and a carboxylic acid group at the other end (figure 3.12). In addition, the carbon skeleton may have one of several different side chains on it. There are about 20 common amino acids (Outlooks 3.3).

The amino acids can bond together by dehydration synthesis reactions. When two amino acids form a bond by removal of water, the nitrogen of the amino group of one is bonded, or linked, to the carbon of the acid group

Amino group   Acid group

Side chain
"R-group"

Figure 3.12

The Structure of an Amino Acid

An amino acid is composed of a short carbon skeleton with three functional groups attached: an amino group, a carboxylic acid group (acid group), and an additional variable group (R-group). It is the variable group that determines which specific amino acid is constructed.

of another. This covalent bond is termed a **peptide bond** (figure 3.13).

Any amino acid can form a peptide bond with any other amino acid. The two amino acids fit together in a specific way, with the amino group of one bonding to the acid group of the next. You can imagine that by using 20 different amino acids as building blocks, you can construct millions of

OUTLOOKS 3.3

Interesting Amino Acids

Humans require nine amino acids in their diet. These are called *essential amino acids* because the body is not able to manufacture them.

threonine	tryptophan	methionine
lysine	phenylalanine	isoleucine
valine	histidine	leucine

These essential amino acids are used by the body in the synthesis of proteins required for good health. For example, the sulfur-containing amino acid methionine is essential for the absorption and transportation of the elements selenium and potassium. It also prevents excess fat buildup in the liver, and it traps heavy metals such as lead, cadmium, and mercury, bonding with them so that they can be excreted from the body. Since they are not readily available in most plant proteins, they are most easily acquired through meat, fish, and dairy products.

Other nonessential amino acids also play interesting roles in human metabolism, for example:

- Lysine is found in such foods as yogurt, fish, chicken, brewer's yeast, cheese, wheat germ, pork, and other meats. It improves calcium uptake; in concentrations higher than the amino acid arginine, it helps to control cold sores (herpes virus infection).

- Tryptophan is found in turkey, dairy products, eggs, fish, and nuts; it is required for the manufacture of controlling depression, PMS (premenstrual syndrome), insomnia, migraine headaches, and immune function disorders.
- Glutamic acid is found in animal and vegetable proteins; it is used in monosodium glutamate (MSG), a flavor-enhancing salt, and is required for the synthesis of folic acid. It is found to accumulate in and damage brain cells following stroke, and is the only amino acid metabolized in the brain.
- Asparagine is found in asparagus. Some individuals excrete this amino acid in an aromatically noticeable amount after eating asparagus.

polypeptide constitutes its *primary structure*. The specific sequence of amino acids in a polypeptide is controlled by the genetic information of an organism. Genes are specific messages that tell the cell to link particular amino acids in a specific order; that is, they determine a polypeptide's primary structure. The kinds of side chains on these amino acids influence the shape that the polypeptide forms. Many polypeptides fold into globular shapes after they have been made as the molecule bends. Some of the amino acids in the chain can form bonds with their neighbors.

The string of amino acids in a polypeptide is likely to twist into particular shapes (a coil or a pleated sheet), whereas other portions remain straight. These twisted forms are referred to as the *secondary structure* of polypeptides. For example, at this secondary level some proteins (e.g., hair) take the form of an *alpha helix:* a shape like that of a coiled telephone cord. The helical shape is maintained by hydrogen bonds formed between different amino acid side chains at different locations in the polypeptide. Remember from chapter 2 that these forces of attraction do not form molecules but result in the orientation of one part of a molecule to another part within the same molecule. Other polypeptides form hydrogen bonds that cause them to make several flat folds that resemble a pleated skirt. This is called a *beta pleated sheet*. The way a particular protein folds is important to its function. In Alzheimer's and prion infections such as bovine spongiform encephalitis (mad cow disease) and Creutzfeldt-Jakob's disease, protein structures are not

Figure 3.13

A Peptide Bond

Part of this protein polypeptide is made up of the amino acids cysteine, valine, and lysine. The bond that results from a dehydration synthesis reaction between two amino acids is called a peptide bond. This bond forms as a result of the removal of the hydrogen and hydroxide groups. In the formation of this bond, the nitrogen is bonded directly to the carbon.

different combinations. Each of these combinations is termed a **polypeptide chain.** A specific polypeptide is composed of a specific sequence of amino acids bonded end to end. There are four levels, or degrees, of protein structure. A listing of the amino acids in their proper order within a particular

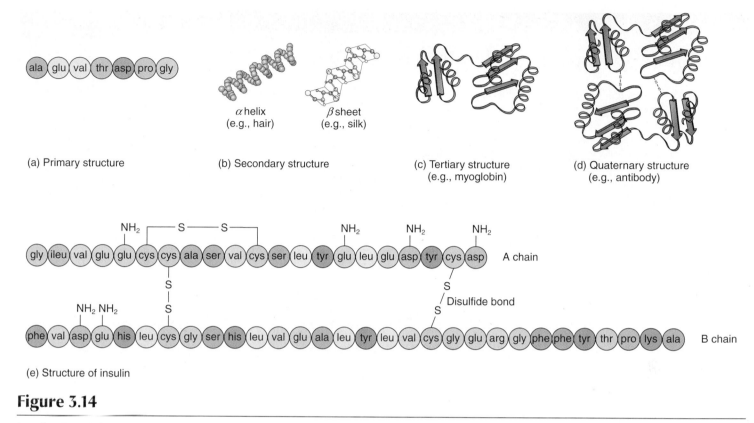

(a) Primary structure

α helix
(e.g., hair)

β sheet
(e.g., silk)

(b) Secondary structure

(c) Tertiary structure
(e.g., myoglobin)

(d) Quaternary structure
(e.g., antibody)

(e) Structure of insulin

Figure 3.14

Levels of Protein Structure

(a) The primary structure of a molecule is simply a list of its component amino acids in the order in which they occur. (b) This figure illustrates the secondary structure of protein molecules or how one part of the molecule is initially attached to another part of the same molecule. (c) If already folded parts of a single molecule attach at other places, the molecule is said to display tertiary (third-degree) structure. (d) Quaternary (fourth-degree) structure is displayed by molecules that are the result of two separate molecules (each with its own tertiary structure) combining into one large macromolecule. (e) The protein insulin is composed of two polypeptide chains bonded together at specific points by reactions between the side chains of particular sulfur-containing amino acids. The side chains of one interact with the side chains of the other and form a particular three-dimensional shape. The bonds that form between the polypeptide chains are called disulfide bonds.

formed correctly, resulting in characteristic nervous system symptoms. Prions have an amino acid sequence identical to a normal brain protein but are folded differently. The normal protein contains helical segments, whereas the corresponding segments of the prion protein are pleated sheets.

It is also possible for a single polypeptide to contain one or more coils and pleated sheets along its length. As a result, these different portions of the molecule can interact to form an even more complex globular structure. This occurs when the coils and pleated sheets twist and combine with each other. The complex, three-dimensional structure formed in this manner is the polypeptide's *tertiary* (third-degree) *structure*. A good example of tertiary structure can be seen when a coiled phone cord becomes so twisted that it folds around and back on itself in several places. The oxygen-holding protein found in muscle cells, myoglobin, displays tertiary structure: It is composed of a single (153 amino acids) helical molecule folded back and bonded to itself in several places.

Frequently several different polypeptides, each with its own tertiary structure, twist around each other and chemically combine. The larger, globular structure formed by these interacting polypeptides is referred to as the protein's *quaternary* (fourth-degree) *structure*. The individual polypeptide chains are bonded to each other by the interactions of certain side chains, which can form disulfide covalent bonds (figure 3.14). Quaternary structure is displayed by the protein molecules called *immunoglobulins* or *antibodies*, which are involved in fighting diseases such as mumps and chicken pox (Outlooks 3.4). The protein portion of the hemoglobin molecule (globin is globular in shape) also demonstrates quaternary structure. • immunity, p. 343

Individual polypeptide chains or groups of chains forming a particular configuration are proteins. The structure of a protein is closely related to its function. Any changes in the arrangement of amino acids within a protein can have far-reaching effects on its function. For example, normal hemoglobin found in red blood cells consists of two kinds of polypeptide chains called the alpha and beta chains. The beta chain is 146 amino acids long. If just one of these amino acids is replaced by a different one, the hemoglobin molecule may not function properly. A classic example of this results in a condition known as *sickle-cell anemia*. In this case, the sixth amino acid in the beta chain, which is

OUTLOOKS 3.4

Antibody Molecules: Defenders of the Body

Globular proteins that function as antibodies are known as immunoglobulins or antibodies. These proteins are manufactured by cells known as B-cells and plasma cells in response to the presence of dangerous molecules known as immunogens or antigens. Antigens may be such things as bacteria, viruses, pollen, plant oils, insect venoms, or toxic molecules. They are also protein-containing molecules. The basic structure of most antibodies is that of a slingshot, either a **Y** or a **T** configuration. The antibody takes the T shape when not combined with an antigen. After combining, the antibody changes its shape and assumes a shape that better resembles the letter Y. Antibodies are composed of two long, "heavy" polypeptide chains and two short, "light" chains attached to one another by covalent bonds giving it quarternary structure. The unit is able to spread apart, or flex, at the "hinge," making the space between the top of the Y larger or smaller. The portion of the antibody that combines with the antigen is located at the tips of the Y arms. The bonds that hold the two together are hydrogen forces. When persons are *vaccinated*, they are given a solution of disease-causing organisms, or their products that have been specially treated so that they will not cause harm. However, the *vaccine* does cause the body's immune system to produce antibody proteins that protect them against that danger.

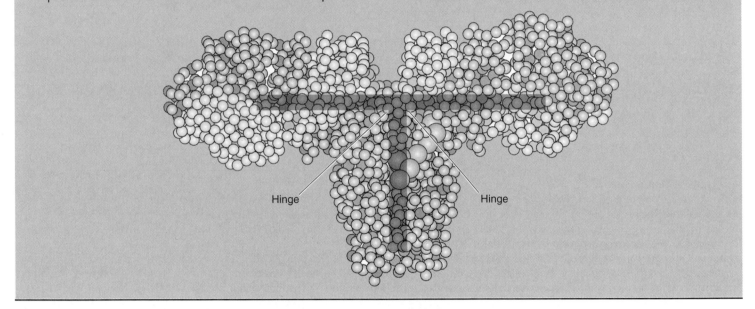

Hinge   Hinge

normally glutamic acid, is replaced by valine. This minor change causes the hemoglobin to fold differently, and the red blood cells that contain this altered hemoglobin assume a sickle shape when the body is deprived of an adequate supply of oxygen. • sickle-cell gene frequency, p. 144

When a particular sequence of amino acids forms a polypeptide, the stage is set for that particular arrangement to bond with another polypeptide in a certain way. Think of a telephone cord that has curled up and formed a helix (its secondary structure). Now imagine that you have attached magnets at several irregular intervals along that cord. You can see that the magnets at the various points along the cord will attract each other, and the curled cord will form a particular three-dimensional shape. You can more closely approximate the complex structure of a protein (its tertiary structure) if you imagine several curled cords, each with magnets attached at several points. Now imagine these magnets as bonding the individual cords together. The globs or ropes of telephone cords approximate the quaternary structure of a protein. This shape can be compared to the shape of a key. In order for a key to do its job effectively, it has to have particular bumps and grooves on its surface. Similarly, if a particular protein is to do its job effectively, it must have a particular shape. The protein's shape can be altered by changing the order of the amino acids, which causes different cross-linkages to form. Changing environmental conditions also influence the shape of the protein. Figure 3.15 shows the importance of the three-dimensional shape of the protein.

Energy in the form of heat or light may break the hydrogen bonds within protein molecules. When this occurs, the chemical and physical properties of the protein are changed and the protein is said to be **denatured.** (Keep in mind, a protein is a molecule, not a living thing, and therefore cannot be "killed.") A common example of this occurs when the gelatinous, clear portion of an egg is cooked and the protein changes to a white solid. Some medications are

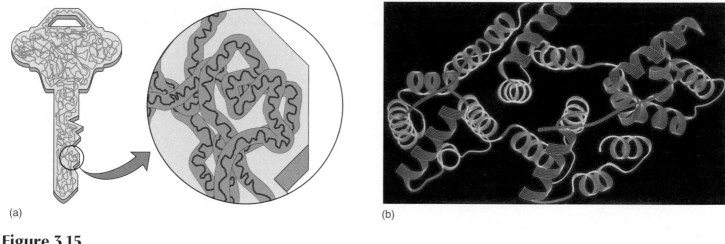

(a)

(b)

Figure 3.15

The Three-Dimensional Shape of Proteins
(a) The specific arrangement of amino acids results in side chains that are available to bond with other side chains. The results are specific three-dimensional proteins that have a specific surface geometry. We frequently compare this three-dimensional shape to the three-dimensional shape of a specific key. (b) This is the structure of the protein annexin as determined by X-ray diffraction. This molecule is located just inside cell membranes and is involved in the transport of materials through the membrane. The surface of this molecule has the shape required to attach to the molecule being transported.

proteins and must be protected from denaturation so as not to lose their effectiveness. Insulin is an example. For protection, such medications may be stored in brown-colored bottles or kept under refrigeration.

3.4 Nucleic Acids

The last group of polymers we will consider are the *nucleic acids*. **Nucleic acids** are complex organic molecules that store and transfer information within a cell. There are two types of nucleic acids: DNA and RNA. DNA serves as genetic material while RNA plays a vital role in the manufacture of proteins. All nucleic acids are constructed of fundamental monomers known as **nucleotides.** Each nucleotide is composed of three parts: (1) a 5-carbon simple sugar molecule that may be ribose or deoxyribose, (2) a phosphate group, and (3) a nitrogenous base. The nitrogenous bases may be one of five types. Two of the bases are the larger, double-ring molecules *Adenine* and *Guanine*. The smaller bases are the single ring bases *Thymine*, *Cytosine*, and *Uracil* (figure 3.16). Nucleotides (monomers) are linked together in long sequences (polymers) so that the sugar and phosphate sequence forms a "backbone" and the nitrogenous bases stick out to the side. DNA has deoxyribose sugar and the bases A, T, G, and C, while RNA has ribose sugar and the bases A, U, G, and C (figure 3.17). (Nucleotides are also components of molecules used to transfer chemical-bond energy. One, ATP and its role in metabolism, will be discussed in chapter 6.)

DNA (*deoxyribonucleic acid*) is composed of two strands that form a ladderlike structure thousands of bases long. The two strands are attached between their protruding bases according to *the base pair rule*—that is, *a*denine protruding from one strand always pairs with *t*hymine protruding from the other strand (in the case of RNA, *a*denine always pairs with *u*racil). Guanine always pairs with *c*ytosine.

A T (or A U) and G C

A meaningful message, a **gene,** is written using the nitrogenous bases as letters along a segment of a strand of DNA. (e.g., the base sequence CATTAGACT) (figure 3.18). The strand that contains this message is called the *coding strand,* from which we get the term *genetic code.* If these bases are read in groups of three, they make sense to us (i.e., "cat," "tag," and "act"). This is the basis of the genetic code for all organisms. Directly opposite the coding strand is a sequence of nitrogenous bases that are called *non-coding* since it makes no "sense" but protects the coding strand from chemical and physical damage. Both strands are twisted into a helix—that is, a molecule turned around a tubular space. Strands of helical DNA may contain tens or thousands of base pairs (AT and GC combinations) that an organism reads as a sequence of chapters in a book. Each chapter is a gene. Just as chapters in a book are identified by beginning and ending statements, different genes along a DNA strand have beginning and ending signals. They tell when to start and when to stop reading a particular gene. Human body cells contain 46 strands (books) of helical DNA, each containing thousands of genes (chapters). These strands are called **chromosomes** when they become supercoiled in preparation for cellular reproduction. Before cell reproduction, the DNA makes copies of the coding

(1)   (a) Ribose   (b) Deoxyribose

Figure 3.16

Nucleic Acids

Each nucleotide is composed of a (*1*) sugar, deoxyribose (D) or ribose (R); (*2*) one of five nitrogen-containing bases: adenine (A), guanine (G), cytosine (C), uracil (U), or thymine (T); and (*3*) an acid phosphate group (P).

(2)   (a) Adenine   (b) Guanine   (c) Cytosine   (d) Uracil   (e) Thymine

(3)   Phosphate group

Figure 3.17

DNA and RNA

(*a*)A single strand of DNA is a polymer composed of nucleotides. Each nucleotide consists of deoxyribose sugar, phosphate, and one of four nitrogenous bases: A, T, G, or C. Notice the backbone of sugar and phosphate. (*b*) RNA is also a polymer but each nucleotide is composed of ribose sugar, phosphate, and one of four nitrogenous bases: A, U, G, or C.

(a) DNA single strand   (b) RNA

A single nucleotide

A nucleotide

Backbone

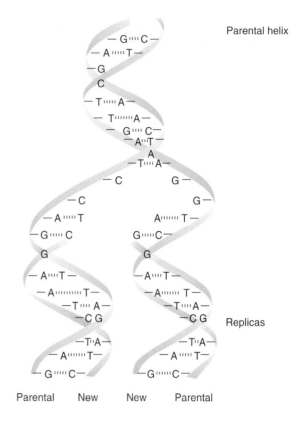

Figure 3.19

Passing the Information on to the Next Generation
This is a generalized illustration of DNA replication and does not show how the DNA supercoils into chromosomes. Each daughter cell receives a copy of the double helix. The helices are identical to each other and identical to the original double strands of the parent cell.

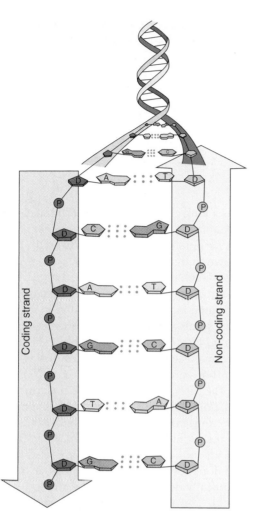

Figure 3.18

DNA
The genetic material is really double-stranded DNA molecules comprised of sequences of nucleotides that spell out an organism's genetic code. The coding strand of the double molecule is the side that can be translated by the cell into meaningful information. The genetic code has the information for telling the cell what proteins to make, which in turn become the major structural and functional components of the cell. The non-coding strand is unable to code for such proteins.

and non-coding strands, ensuring that the offspring, or *daughter* cells, will each receive a full complement of the genes required for their survival (figure 3.19). A gene is a segment of DNA that is able to (1) replicate by directing the manufacture of copies of itself; (2) mutate, or chemically change, and transmit these changes to future generations; (3) store information that determines the characteristics of cells and organisms; and (4) use this information to direct the synthesis of structural and regulatory proteins.

RNA (*ribo*nucleic *a*cid) is found in three forms. **Messenger RNA (mRNA)** is a single-strand copy of a portion of the coding strand of DNA for a specific gene. When mRNA is formed on the surface of the DNA, the base pair rule (A pairs with U and G pairs with C) applies. After mRNA is formed and peeled off, it moves to a cellular structure called the *ribosome* where the genetic message can be translated into a protein molecule. Ribosomes contain another type of RNA, **ribosomal RNA (rRNA)**. rRNA is also an RNA copy of DNA, but after being formed it becomes twisted and covered in protein to form a ribosome. The third form of RNA, **transfer RNA (tRNA)**, is also copies of different segments of DNA, but when peeled off the surface, each takes the form of a cloverleaf. tRNA molecules are responsible for transferring or carrying specific amino acids to the ribosome where all three forms of RNA come together and cooperate in the manufacture of protein molecules (figure 3.20).

DNA

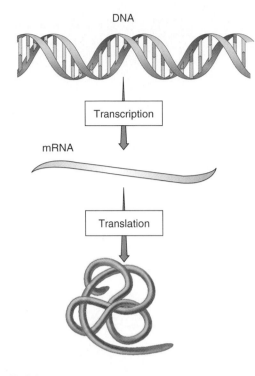

Figure 3.20

The Role of RNA
The entire process of protein synthesis begins with DNA. All forms of RNA (messenger, transfer, and ribosomal) are copies of different sequences of coding strand DNA and each plays a different role in protein synthesis. When the protein synthesis process is complete, the RNA can be reused to make more of the same protein coded for by the mRNA. This is similar to replaying a video tape in a VHS machine. The tape is like the mRNA and the tape machine is like the ribosome. Eventually the tape and machine will wear out and must be replaced. In a cell, this involves the synthesis of new RNA molecules from food.

3.5 Lipids

We generally call molecules in this group **fats.** However, there are three different types of **lipids:** *true fats* (pork chop fat or olive oil), *phospholipids* (the primary component of cell membranes), and *steroids* (most hormones). In general, lipids are large, nonpolar, organic molecules that do not easily dissolve in polar solvents such as water. They are soluble in nonpolar substances such as ether or acetone. Just like carbohydrates, the lipids are composed of carbon, hydrogen, and oxygen. They do not, however, have the same ratio of carbon, hydrogen, and oxygen in their empirical formulas. Lipids generally have very small amounts of oxygen in comparison to the amounts of carbon and hydrogen. *Simple lipids* such as steroids and prostaglandins are not able to be hydrolyzed into smaller, similar subunits. *Complex lipids* such as true fats and phospholipids can be hydrolyzed into smaller, similar units.

True (Neutral) Fats

True (neutral) fats are important, complex organic molecules that are used to provide energy, among other things. The building blocks of a fat are a glycerol molecule and fatty

acids. The **glycerol** is a carbon skeleton that has three alcohol groups attached to it. Its chemical formula is $C_3H_5(OH)_3$. At room temperature, glycerol looks like clear, lightweight oil. It is used under the name *glycerin* as an additive to many cosmetics to make them smooth and easy to spread.

Glycerol

A **fatty acid** is a long-chain carbon skeleton that has a carboxyl functional group. If the carbon skeleton has as much hydrogen bonded to it as possible, we call it **saturated.** The saturated fatty acid shown in figure 3.21*a* is stearic acid, a component of solid meat fats such as mutton tallow. Notice that at every point in this structure the carbon has as much hydrogen as it can hold. Saturated fats are generally found in animal tissues—they tend to be solids at room temperatures. Some other examples of saturated fats are butter, whale blubber, suet, lard, and fats associated with such meats as steak or pork chops.

If the carbons are double-bonded to each other at one or more points, the fatty acid is said to be **unsaturated.** The occurrence of a double bond in a fatty acid is indicated by the Greek letter ω (omega) followed by a number indicating the location of the first double bond in the molecule. Oleic acid, one of the fatty acids found in olive oil, is comprised of 18 carbons with a single double bond between carbons 9 and 10. Therefore, it is chemically designated C18:Iω9 and is a monounsaturated fatty acid. This fatty acid is commonly referred to as an omega-9 fatty acid. The unsaturated fatty acid in figure 3.21*b* is linoleic acid, a component of sunflower and safflower oils. Notice that there are two double bonds between the carbons and fewer hydrogens than in the saturated fatty acid. Linolenic acid is chemically a polyunsaturated fatty acid with two double bonds and is designated C18:2ω6, an omega-6 fatty acid. This indicates that the first double bond of this 18-carbon molecule is between carbons 6 and 7. Since the human body cannot make this fatty acid, it is called an *essential fatty acid* and must be taken in as a part of the diet. The other essential fatty acid, linoleic acid, is C18:3ω3 and has three double bonds. These two fatty acids are commonly referred to as omega-3 fatty acids. One key function of these essential fatty acids is the synthesis of prostaglandin hormones that are necessary in controlling cell growth and specialization.

Sources of Omega-3 Fatty Acids	Sources of Omega-6 Fatty Acids
Certain fish oils (salmon, sardines, herring)	Corn oil
	Peanut oil
Flaxseed oil	Cottonseed oil
Soybeans	Soybean oil
Soybean oil	Sesame oil
Walnuts	Sunflower oil
Walnut oil	Safflower oil
Canola oil	
Green leafy vegetables	

(a)   Stearic acid

(b)   Linoleic acid

Figure 3.21

Structure of Saturated and Unsaturated Fatty Acids
(a)Stearic acid is an example of a saturated fatty acid. (b)Linoleic acid is an example of an unsaturated fatty acid.

Unsaturated fats are frequently plant fats or oils—they are usually liquids at room temperature. Peanut, corn, and olive oils are mixtures of different triglycerides and are considered unsaturated because they have double bonds between the carbons of the carbon skeleton. A polyunsaturated fatty acid is one that has a great number of double bonds in the carbon skeleton. When glycerol and 3 fatty acids are combined by three dehydration synthesis reactions, a fat is formed. That dehydration synthesis is almost exactly the same as the reaction that causes simple sugars to bond together.

In nature, most unsaturated fatty acids have hydrogen atoms that are on the same side of the double bonded carbons. These are called *cis* fatty acids. If the hydrogens are on opposite sides of the double bonds, they are called *trans* fatty acids.

cis double bond: oleic acid

trans double bond: elaidic acid

Trans fatty acids also occur and are found in grazing animals such as cattle, sheep, and horses. Therefore, humans acquire them in their diets in the form of meat and dairy products. French fries, donuts, cookies, and crackers are examples of foods high in *trans* fatty acids. *Trans* fatty acids are also formed during the *hydrogenation* of either vegetable or fish oils. The hydrogenation process breaks the double bonds in the fatty acid chain and adds more hydrogen atoms. This can change the liquid to a solid. Many products display this term on their labels (i.e., "made from hydrogenated corn oil"). This process extends the products' shelf life and allows producers to convert oils to other solids such as margarine.

It has been shown in clinical studies that *trans* fatty acids tend to raise total blood cholesterol levels, but less than more saturated fatty acids. *Trans* fatty acids also tend to raise the so-called "bad fats" (low-density lipoproteins, LDLs) and lower the so-called "good fats" (high-density lipoproteins, HDLs) when used instead of *cis* fatty acid. It is suspected that this may increase the risk of heart disease. Because of their importance to cardiovascular health, the United States Department of Health and Human Services has changed food labeling regulations with regard to *trans* fatty acids. It now requires listing the amount of *trans* fatty acids under the amount of saturated fat. HHS suggests a person eat no more than 20 grams a day (about 10% of total calories), including the *trans* fatty acids. *Cis* fatty acids are found in higher concentrations in products such as canola or olive oil.

Fats are important molecules for storing energy. There is more than twice as much energy in a gram of fat as in a gram of sugar—9 calories versus 4 calories. This is important to an organism because fats can be stored in a relatively small space and still yield a high amount of energy. Fats in animals also provide protection from heat loss. Some animals have a layer of fat under the skin that serves as an insulating layer. The thick layer of blubber in whales, walruses, and seals prevents the loss of internal body heat to the cold, watery environment in which they live. This same layer of fat, together with the fat deposits around some internal organs—such as the kidneys and heart—serve as a cushion that protects these organs from physical damage. If a fat is formed from a glycerol molecule and 3 attached fatty acids, it is called a *triglyceride*; if 2, a *diglyceride*; and

Figure 3.22

A Fat Molecule

The arrangement of the three fatty acids (yellow) attached to a glycerol molecule (red) is typical of the formation of a fat. The structural formula of the fat appears to be very cluttered until you dissect the fatty acids from the glycerol; then it becomes much more manageable. This example of a triglyceride contains a glycerol molecule, two unsaturated fatty acids (linoleic acid), and a third saturated fatty acid (stearic acid).

if 1, a *monoglyceride* (figure 3.22). Triglycerides account for about 95% of the fat stored in human tissue.

Phospholipids

Phospholipids are a class of complex, water-insoluble organic molecules that resemble fats but contain a phosphate group (PO_4) in their structure (figure 3.23). One of the reasons phospholipids are important is that they are a major component of membranes in cells. Without these lipids in our membranes, the cell contents would not be separated from the exterior environment. Some of the phospholipids are better known as the *lecithins*. Lecithins are found in cell membranes and also help in the emulsification of fats. They help separate large portions of fat into smaller units. This allows the fat to mix with other materials. Lecithins are added to many types of food for this purpose (chocolate bars, for example). Some people take lecithin as nutritional supplements because they believe it leads to healthier hair and better reasoning ability. But once inside your intestines, lecithins are destroyed by enzymes, just like any other phospholipid (Outlooks 3.5). Phospholipids are essential components of the membranes of all cells and will be described again in chapter 4.

Steroids

Steroids, another group of lipid molecules, are characterized by their arrangement of interlocking rings of carbon. They often serve as hormones that aid in regulating body processes. We have already mentioned one steroid molecule that you are probably familiar with: cholesterol. Although serum cholesterol (the kind found in your blood, associated with lipoproteins) has been implicated in many cases of atherosclerosis, this steroid is made by your body for use as a component of cell membranes. It is also used by your body to make bile acids. These products of your liver are channeled into your intestine to emulsify fats. Cholesterol is also necessary for the manufacture of vitamin D. Cholesterol molecules in the skin react with ultraviolet light to produce vitamin D, which assists in the proper development of bones and teeth. Figure 3.24 illustrates some of the steroid compounds, such as testosterone and estrogen, that are typically manufactured by organisms. • conception control, p. 426

A large number of steroid molecules are hormones. Some of them regulate reproductive processes such as egg and sperm production (see chapter 21); others regulate such things as salt concentration in the blood.

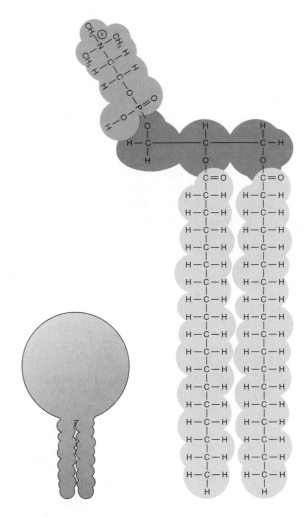

Figure 3.23

A Phospholipid Molecule

This molecule is similar to a fat but has a phosphate group (purple) in its structure. The phosphate group and two fatty acids are bonded to the glycerol by a dehydration synthesis reaction. Molecules like these are also known as lecithins. Phospholipid molecules may be shown as a balloon with two strings. The balloon portion is the glycerol and phosphate group, which are soluble in water. The strings are the fatty acid segments of the molecule, which are not water-soluble.

Fat and Your Diet

When triglycerides are eaten in fat-containing foods, digestive enzymes hydrolyze them into glycerol and fatty acids. These molecules are absorbed by the intestinal tract and coated with protein to form lipoprotein, as shown in the accompanying diagram. The combination allows the fat to better dissolve in the blood so that it can move throughout the body in the circulatory system.

The five types of lipoproteins found in the body are: (1) chylomicrons, (2) very-low-density lipoproteins (VLDL), (3) low-density lipoproteins (LDL), (4) high-density lipoproteins (HDL), and (5) lipoprotein a-Lp(a). Chylomicrons are very large particles formed in the intestine and are between 80–95% triglycerides. As the chylomicrons circulate through the body, cells remove the triglycerides in order to make hormones, store energy, and build new cell parts. When most of the triglycerides have been removed, the remaining portions of the chylomicrons are harmlessly destroyed.

The VLDLs and LDLs are formed in the liver. VLDLs contain all types of lipid, protein, and 10%–15% cholesterol, while the LDLs are about 50% cholesterol. As with the chylomicrons, the body uses these molecules for fats they contain. However, in some people, high levels of LDLs and Lp(a) in the blood are associated with the diseases atherosclerosis, stroke, and heart attack. It appears that saturated fat disrupts the clearance of the LDLs from the bloodstream. Thus, while in the blood, LDLs may stick to the insides of the vessels, forming deposits that restrict blood flow and contribute to high blood pressure, strokes, and heart attacks. Even though they are 30% cholesterol, a high level of HDLs (made in the intestine) in comparison to LDLs and Lp(a) is associated with a lower risk of

atherosclerosis. One way to reduce the risk of this disease is to lower your intake of LDLs and Lp(a). This can be done by reducing your consumption of saturated fats. An easy way to remember the association between LDLs and HDLs is "L = Lethal" and "H = Healthy" or "Low = Bad" and "High = Good."

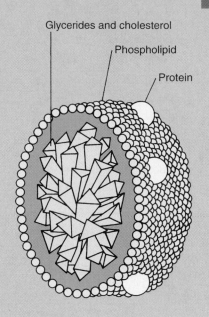

Glycerides and cholesterol
Phospholipid
Protein

Diagram of a Lipoprotein

Reprinted with permission, Best Foods, a division of CPC International, Inc.

Normal HDL Values	**Normal LDL Values**	**Normal VLDL Values**
Men: 40–70 mg/dL	Men: 91–100 mg/dL	Men: 0–40 mg/dL
Women: 40–85 mg/dL	Women: 69–100 mg/dL	Women: 0–40 mg/dL
Children: 30–65 mg/dL		
Minimum desirable: 40 mg/dL	High risk: no higher than 100 mg/dL	Desirable: below 40 mg/dL
	Moderate risk: no higher than 130 mg/dL	
	Low risk: at or below 160 mg/dL	
For Total Cholesterol Levels		
Desirable: Below 200 mg/dL	Borderline: 200–240	Undesirable: Above 240

(a) Cholesterol   (b) Testosterone   (c) Progesterone

Figure 3.24

Steroids

(*a*) Cholesterol is produced by the human body and is found in your cells' membranes. (*b*)Testosterone increases during puberty, causing the male sex organs to mature. (*c*) Progesterone is a female sex hormone produced by the ovaries and placenta.

Table 3.2			

TYPES OF ORGANIC MOLECULES FOUND IN LIVING THINGS

Type of Organic Molecule	Basic Subunit	Function	Examples
Carbohydrates	Simple sugar/monosaccharides	Provide energy	Glucose, fructose
	Complex carbohydrates/ polysaccharides	Provide support	Cellulose, starch, glycogen
Proteins	Amino acid	Maintain the shape of cells and parts of organisms	Cell membrane Hair Antibodies Clotting factors Enzymes Muscle
		As enzymes, regulate the rate of cell reactions	Ptyalin in the mouth
		As hormones, effect physiological activity, such as growth or metabolism	Insulin
		Many proteins serve as molecules that carry other molecules to distant places in the body	Lipoproteins, hemoglobin
Nucleic acids	Nucleotide	Store and transfer genetic information that controls the cell	DNA RNA
		Involved in protein synthesis	
Lipids 1. Fats	Glycerol and fatty acids	Provide energy Provide insulation Serve as shock absorber	Lard Olive oil Linseed oil Tallow
2. Phospholipids	Glycerol, fatty acids, and phosphorus compounds	Form a major component of the structure of the cell membrane	Cell membrane
3. Steroids and prostaglandins	Structure of interlocking carbon rings	Often serve as hormones that control the body processes	Testosterone Vitamin D Cholesterol

SUMMARY

The chemistry of living things involves a variety of large and complex molecules. This chemistry is based on the carbon atom and the fact that carbon atoms can connect to form long chains or rings. This results in a vast array of molecules. The structure of each molecule is related to its function. Changes in the structure may result in abnormal functions, which we call disease. Some of the most common types of organic molecules found in living things are carbohydrates, lipids, proteins, and nucleic acids. Table 3.2 summarizes the major types of biologically important organic molecules and how they function in living things.

THINKING CRITICALLY

For guidelines to answer this question, visit the Online Learning Center.

Both amino acids and fatty acids are organic acids. What property must they have in common with inorganic acids such as sulfuric acid? How do they differ? Consider such aspects as structure of molecules, size, bonding, and pH.

CONCEPT MAP TERMINOLOGY

Construct a concept map using these terms.

amino acid	polymer
dehydration synthesis	polypeptide
denature	primary structure
hydrolysis	side chain
monomer	

Visit the Online Learning Center to use an interactive format to construct another concept map.

KEY TERMS

Use interactive flash cards, on the Online Learning Center, to help you learn the meanings of these terms.

amino acid	fatty acid
biochemistry	functional groups
carbohydrate	gene
carbon skeleton	glycerol
carrier proteins	hydrolysis
chromosomes	inorganic molecules
complex carbohydrates	isomers
dehydration synthesis reaction	lipids
denature	macromolecules
DNA (deoxyribonucleic acid)	messenger RNA (mRNA)
double bond	nucleic acids
fat	nucleotide
organic molecules	RNA (ribonucleic acid)
peptide bond	saturated
phospholipid	steroid
polypeptide	structural proteins
protein	transfer RNA (tRNA)
regulator proteins	true (neutral) fats
ribosomal RNA (rRNA)	unsaturated

LEARNING CONNECTIONS

Visit the Online Learning Center to find additional study resources including those highlighted below.

3.1   Molecules Containing Carbon

Interactive Concept Maps
• Characteristics of carbon

1. What is the difference between inorganic and organic molecules?
2. What two characteristics of the carbon molecule make it unique?
3. Diagram an example of each of the following: amino acid, simple sugar, glycerol, fatty acid.
4. Describe five functional groups.
5. List three monomers and the polymers that can be constructed from them.

3.2.   Carbohydrates

Animations and Review
• Organic chemistry
• Carbohydrates

Experience This!
• Polymers and monomers

6. Give two examples of simple sugars and two examples of complex sugars.
7. What are the primary characteristics used to identify a compound as a carbohydrate?

3.3   Proteins

Animations and Review
• Proteins

8. How do the primary, secondary, tertiary, and quaternary structures of proteins differ?
9. List the three categories of proteins and describe their functions.

3.4.   Nucleic Acids

Animations and Review
• Nucleic acids

10. Describe how DNA differs from and is similar to RNA both structurally and functionally.
11. List the nitrogenous bases that base pair in DNA; in RNA.

3.5. **Lipids**

Interactive Concept Maps
• Text concept map

Animations and Review
• Lipids

12. Describe three different kinds of lipids.
13. What is meant by HDL, LDL, and VLDL? Where are they found? How do they relate to disease?

Test yourself with multiple choice questions that provide immediate feedback on the Online Learning Center with PowerWeb at

www.mhhe.com/enger11

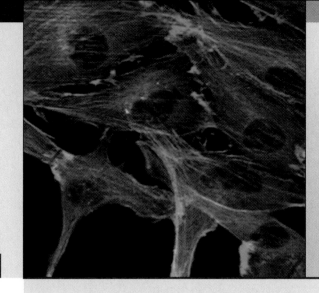

CHAPTER OUTLINE

Cell Structure and Function

GOALS AND OBJECTIVES

Understand the historical perspective of the development of the cell theory.

❑ Explain the cell theory.

❑ Diagram the generalized structure of all prokaryotic cells and label the parts.

❑ Diagram the generalized structure of all eukaryotic cells and label the parts.

❑ Draw a flow chart that illustrates the classification of cellular forms of life.

Know the molecular structure of a cell membrane and relate this structure to its function.

❑ List the components and molecular parts of a typical cell membrane.

❑ Explain the fluid mosaic model of membrane structure.

❑ List the possible roles played by molecules that extend from the cell surface.

❑ Contrast diffusion, osmosis, and dialysis.

❑ List the controlled methods by which materials can be transported through a cell membrane.

Learn to associate cellular organelles with their major functions in eukaryotic cells.

❑ List the typical organelles associated with eukaryotic cells.

❑ Draw the structure of each of the organelles associated with eukaryotic cells.

❑ Describe the function of each of the organelles associated with eukaryotic cells.

Understand the basic differences between prokaryotic and eukaryotic cells.

❑ List the differences in organelles found in prokaryotic and eukaryotic cells.

❑ Give examples of organisms composed of prokaryotic and eukaryotic cells.

Cells: Anatomy and Action

PART TWO

4.1 The Cell Theory

The concept of a cell is one of the most important ideas in biology because it applies to all living things. The idea of a cell as the smallest unit of life did not emerge all at once, but has been developed and modified over many years. It is still being modified today.

Several individuals made key contributions to the cell concept. Anton van Leeuwenhoek (1632–1723) was one of the first to make use of a **microscope** to examine biological specimens (How Science Works 4.1). When van Leeuwenhoek discovered that he could see things moving in pond water using his microscope, his curiosity stimulated him to look at a variety of other things. He studied blood, semen, feces, pepper, and tartar, for example. He was the first to see individual cells and recognize them as living units, but he did not call them cells. The name he gave to these "little animals" that he saw moving around in the pond water was *animalcules.*

The first person to use the term **cell** was Robert Hooke (1635–1703) of England, who was also interested in how things looked when magnified. He chose to study thin slices of cork from the bark of a cork oak tree. He saw a mass of cubicles fitting neatly together, which reminded him of the barren rooms in a monastery. Hence, he called them *cells.* As it is currently used, the term *cell* refers to the basic structural unit that makes up all living things. When Hooke looked at cork, the tiny boxes he saw were, in fact, only the cell walls that surrounded the living portions of plant cells. We now know that the **cell wall** is composed of the complex carbohydrate cellulose, which provides strength and protection to the living contents of the cell. The cell wall appears to be a rigid, solid layer of material, but in reality it is composed of many interwoven strands of cellulose molecules. Its structure allows certain very large molecules to pass through it readily, but it acts as a screen to other molecules.

Hooke's use of the term *cell* in 1666 in his publication *Micrographia* was only the beginning, for nearly 200 years passed before it was generally recognized that all living things are made of cells and that these cells can reproduce themselves. In 1838, Mathias Jakob Schleiden stated that all plants are made up of smaller cellular units. In 1839, Theodor Schwann published the idea that all animals are composed of cells.

Soon after the term *cell* caught on, it was recognized that the cell's vitally important portion is inside the cell wall. This living material was termed **protoplasm,** which means *first-formed substance.* The term *protoplasm* allowed scientists to distinguish between the living portion of the cell and the nonliving cell wall. Very soon microscopists were able to distinguish two different regions of protoplasm. One type of protoplasm was more viscous and darker than the other. This region, called the **nucleus** or core, appeared as a central body within a more fluid material surrounding it. **Cytoplasm** (*cyto* = cell; *plasm* = first-formed substance) is the name given to the colloidal fluid portion of the protoplasm (figure 4.1). Although the term *protoplasm* is seldom used today, the term *cytoplasm* is still very common in the vocabulary of cell biologists.

(a)

(b)

Nucleus

Cytoplasm

Protoplasm

Cell wall

(c)

Figure 4.1

Cells—Basic Structure of Life

The cell concept has changed considerably over the last 300 years. Robert Hooke's idea of a cell (*a*) was based on his observation of slices of cork (cell walls of the bark of the cork oak tree). Hooke invented the compound microscope and illumination system shown here (*b*), one of the best such microscopes of his time. One of the first subcellular differentiations was to divide the protoplasm into cytoplasm and nucleus as shown in this plant cell (*c*). We now know that cells are much more complex than this; they are composed of many kinds of subcellular structures, some components numbering in the thousands.

HOW SCIENCE WORKS 4.1

The Microscope

To view very small objects we use a magnifying glass as a way of extending our observational powers. A magnifying glass is a lens that bends light in such a way that the object appears larger than it really is. Such a lens might magnify objects 10 or even 50 times. Anton van Leeuwenhoek (1632–1723), a Dutch drape and clothing maker, was one of the first individuals to carefully study magnified cells (figure *a*). He made very detailed sketches of the things he viewed with his simple microscopes

(a) Replica of van Leeuwenhoek microscope

and communicated his findings to Robert Hooke and the Royal Society of London. His work stimulated further investigation of magnification techniques and descriptions of cell structures. These first microscopes were developed in the early 1600s.

Compound microscopes (figure *b*), developed soon after the simple microscopes, are able to increase magnification by bending light through a series of lenses. One lens, the *objective lens*, magnifies a specimen that is further magnified by the second lens, known as the *ocular lens*. With the modern technology of producing lenses, the use of specific light waves, and the immersion of the objective lens in oil to collect more of the available light, objects can be magnified 100 to 1,500 times. Microscopes typically available for student use are compound light microscopes. The major restriction of magnification with a light microscope is the limited ability of the viewer to distinguish two very close objects as two distinct things. The ability to separate two objects is termed *resolution* or *resolving power*. Some people have extremely good eyesight and are able to look at letters on a page and recognize that they are separate objects; other persons see the individual letters as "blurred together." Their eyes have different resolving powers. We can enhance the resolving power of the human eye by using lenses as in eyeglasses or microscopes. All lens systems, whether in the eye or in microscopes, have a limited resolving power.

If two structures in a cell are very close to each other, you may not be able to determine that there are actually two structures rather than one. The limits of resolution of a light microscope are related to the wavelengths of the light being transmitted through the specimen. If you could see ultraviolet light waves, which have shorter wavelengths, it would be possible to resolve more individual structures.

An *electron microscope* (figure *c*) makes use of this principle: the moving electrons have much shorter wavelengths than visible light. Thus, they are able to magnify 200,000 times and still resolve individual structures. The difficulty is, of course, that you are unable to see electrons with your eyes. Therefore, in an electron microscope, the electrons strike a photographic film or television monitor, and this "picture" shows the individual structures. Heavy metals scattered on the structures to be viewed increase the contrast between areas where there are structures that interfere with the transmission of the electrons and areas where the electrons are transmitted easily. The techniques for preparing the material to be viewed—slicing the specimen very thinly and focusing the electron beam on the specimen—make electron microscopy an art as well as a science.

Most recently the *laser feedback* and *tunneling microscopes* and new techniques enable researchers to visualize previously unseen molecules and even the surface of atoms such as chlorine and sodium.

- Objective lenses
- Binocular eye piece lenses
- Revolving objective lens control
- Stage
- Condenser lens adjustment
- Iris diaphragm control
- Mechanical stage controls
- Light
- Fine adjustment control
- Course adjustment control

(b) Compound microscope

(c) Electron microscope

The development of better light microscopes and, ultimately even more powerful microscopes and staining techniques, revealed that protoplasm contains many structures called **organelles** (*elle* = little). It has been determined that certain functions are performed in certain organelles. The essential job an organelle does is related to its structure. Each organelle is dynamic in its operation, changing shape and size as it works. Organelles move throughout the cell, and some even self-duplicate. Different cell types contain different combinations and numbers of these organelles. In multicellular organisms, tissues are composed of cells that are similar in organelle composition, allowing them to work together as a unit.

Different Cells; Different Organisms

All living things are cells or composed of cells. To date, most biologists recognize two major cell types, *prokaryotes* and *eukaryotes*. Whether they are **prokaryotic cells** or **eukaryotic cells,** they have certain things in common: (1) cell membranes, (2) cytoplasm, (3) genetic material, (4) energy currency, (5) enzymes and coenzymes. These are all necessary in order to carry out life's functions mentioned in chapter 1. Should any of these not function properly, a cell would die.

The differences among cell types are found in the details of their structure. While prokaryotic cells lack most of the complex internal organelles typical of eukaryotes, they are cells and can carry out life's functions. To better understand and focus on the nature and differences among cell types, biologists have further classified organisms into large categories called **domains.** The following diagram illustrates scientists' current hypothesis on this level of organization: • **classification of organisms, p. 456**

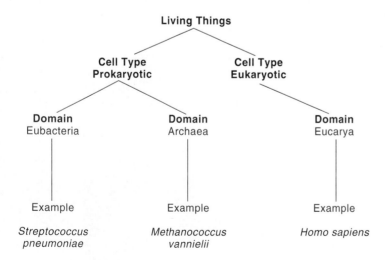

Most single-celled organisms that we commonly refer to as *bacteria* are prokaryotic cells and are classified in the Domain Eubacteria. Other less-well-known prokaryotes display significantly different traits that have caused biologists to create a second category of prokaryotes, the Domain Archaea, or the Archaebacteria. All other living things are based on the eukaryotic cell plan. Members of the kingdoms Protista (algae and protozoa), Fungi, Plantae (plants), and Animalia (animals) are all comprised of eukaryotic cells (figure 4.2).

Notice that viruses are not included in this classification system. That is because they are not cellular in nature. Viruses are not composed of the basic cellular structural components. They are composed of a core of nucleic acid (DNA or RNA, never both) and a surrounding coat or capsid composed of protein. For this reason, the viruses are called acellular or noncellular.

4.2 Cell Membranes

One feature common to all cells and many of the organelles they contain is a thin layer of material called *membrane.* Membrane can be folded and twisted into many different structures, shapes, and forms. The particular arrangement of membrane of an organelle is related to the functions the organelle is capable of performing. This is similar to the way a piece of fabric can be fashioned to function as a pair of pants, a shirt, sheets, pillowcases, or a rag doll. All cellular membranes have a fundamental molecular structure that allows them to be fashioned into a variety of different organelles.

Cellular membranes are thin sheets composed primarily of phospholipids and proteins. The current model of how membranes are constructed is known as the **fluid-mosaic model,** which proposes that the various molecules of the membrane are able to flow and move about. The membrane maintains its form because of the physical interaction of its molecules with its surroundings. The phospholipid molecules of the membrane have one end (the glycerol portion) that is soluble in water and is therefore called **hydrophilic** (*hydro* = water; *phile* = loving). The other end that is not water soluble, called **hydrophobic** (*phobia* = fear), is comprised of fatty acids. We commonly represent this molecule as a balloon with two strings (see figure 3.23). The inflated balloon represents the glycerol and negatively charged phosphate; the two strings represent the uncharged fatty acids. Consequently, when phospholipid molecules are placed in water, they form a double-layered sheet, with the water-soluble (hydrophilic) portions of the molecules facing away from each other. This is commonly referred to as a *phospholipid bilayer.* If phospholipid molecules are shaken in a glass of water, the molecules will automatically form double-layered membranes. It is important to understand that the membrane formed is not rigid or stiff but resembles a heavy olive oil in consistency. The component phospholipids are in constant motion as they move with the surrounding water molecules and slide past one another. Other molecules found in cell membranes are cholesterol and carbohydrates. Cholesterol is found in the middle of the membrane, in the hydrophobic region, because cholesterol is not water soluble. It appears to play a role in stabilizing the membrane and keeping it flexible.

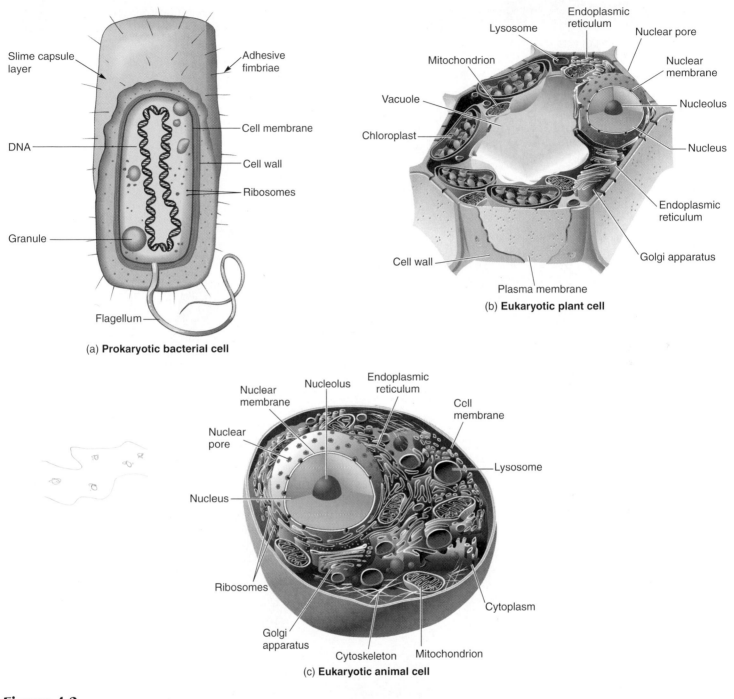

(a) **Prokaryotic bacterial cell**

(b) **Eukaryotic plant cell**

(c) **Eukaryotic animal cell**

Figure 4.2

Major Cell Types

There are two major types of cells, the prokaryotes and the eukaryotes. Prokaryotic cells are represented by the (*a*) bacteria, and eukaryotic cells by (*b*) plant and (*c*) animal cells.

The protein component of cellular membranes can be found on either surface of the membrane, or within the membrane among the phospholipid molecules. Many of the protein molecules are combined with lipids (lipoproteins) or carbohydrate molecules (glycoproteins). Some are capable of moving from one side to the other. Some of these proteins help with the chemical activities of the cell. Others aid in the movement of molecules across the membrane by forming membrane channels through which substances may travel or by acting as transport molecules (figure 4.3). Channels operate like gates in a fence (i.e., they can open and close when circumstances dictate). Cells use channels to control the movement of materials such as ions (ex., potassium ions) and protons (H^+) through the membrane.

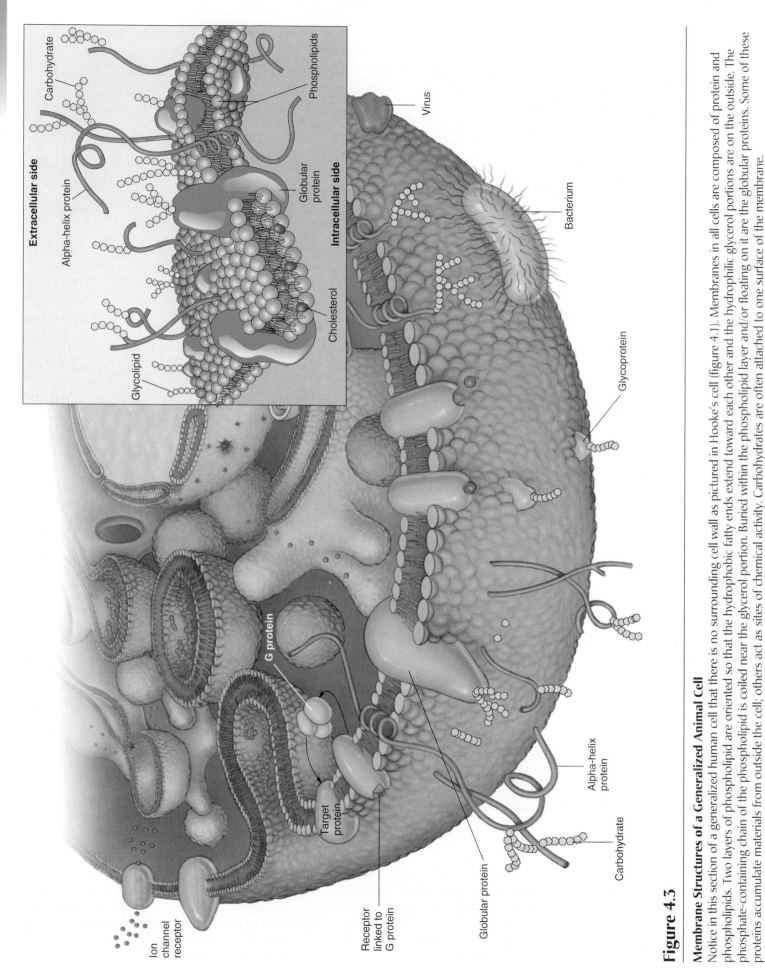

Extracellular side

Carbohydrate

Alpha-helix protein

Phospholipids

Globular protein

Intracellular side

Cholesterol

Glycolipid

Virus

Bacterium

Glycoprotein

Ion channel receptor

G protein

Target protein

Receptor linked to G protein

Globular protein

Alpha-helix protein

Carbohydrate

Figure 4.3

Membrane Structures of a Generalized Animal Cell

Notice in this section of a generalized human cell that there is no surrounding cell wall as pictured in Hooke's cell (figure 4.1). Membranes in all cells are composed of protein and phospholipids. Two layers of phospholipid are oriented so that the hydrophobic fatty ends extend toward each other and the hydrophilic glycerol portions are on the outside. The phosphate-containing chain of the phospholipid is coiled near the glycerol portion. Buried within the phospholipid layer and/or floating on it are the globular proteins. Some of these proteins accumulate materials from outside the cell; others act as sites of chemical activity. Carbohydrates are often attached to one surface of the membrane.

Cells also have proteins on their surface that act as recognition molecules. The presence of these molecules enables one cell or one organism to recognize other cells or organisms. One group of such molecules was discovered just after World War II and is known as *histocompatibility antigens* (*histo-* = tissue; compatible). It is the presence of these antigens that is responsible for the rejection of transplanted tissues or organs from donors that are "incompatible." On the white blood cells (leukocytes; *leuko* = white) of humans, the histocompatiblity antigens are called *human leukocyte antigens* (HLAs). If the donated tissue HLAs are seen as being dangerous and foreign to the recipient, they will be rejected, that is, killed. To avoid tissue transplant rejection, HLA tissue matching is done to determine the degree of tissue similarity between donor and recipient. In the case of identical twins, the immune system would see the HLAs of the donor twin to be the same as those on the cell surfaces of the recipient twin. If the donor is not closely related, the HLA type may be too different or foreign to allow that person to become a donor of such tissues or organs as blood, bone marrow, or a kidney.

In addition to phospholipids and proteins, some protein molecules found on the outside surfaces of cellular membranes have carbohydrates or fats attached to them. These combination molecules are important in determining the "sidedness" (inside–outside) of the membrane and also help organisms recognize differences among types of cells. Carbohydrates are usually found on the outside of the membrane, where they are bound to proteins or lipids. They appear to play a role in cell-to-cell interactions and are involved in binding with regulatory molecules. Your body can recognize disease-causing organisms because their surface proteins are different from those of its own cellular membranes. Some of these molecules also serve as attachment sites for specific chemicals, bacteria, protozoa, white blood cells, and viruses. Many dangerous agents cannot stick to the surface of cells and therefore do not cause harm. For this reason cell biologists explore the exact structure and function of these cell surface molecules. They are also attempting to identify molecules that can interfere with the binding of such agents as viruses and bacteria in the hope of controlling infections such as AIDS.

4.3 Getting Through Membranes

If a cell is to stay alive it must meet the characteristics of life outlined in chapter 1. This includes taking nutrients in and eliminating wastes and other by-products of metabolism. Several mechanisms allow cells to carry out the processes characteristic of life. They include diffusion, osmosis, dialysis, facilitated diffusion, active transport, and phagocytosis.

In addition, it is necessary for cells to respond to changes in their environment. For example, when an electrical or chemical signal is sent to a cell, the cell must in some way recognize the signal and respond to it. This is what occurs when light falls on your eye cells. If the cells do not "realize" that there has been an environmental change, they may respond inappropriately or not at all. This could cost them their lives.

Diffusion

There is a natural tendency in gases and liquids for molecules of different types to completely mix with each other. This is because they are moving constantly. Their movement is random and is due to the kinetic energy found in the individual molecules. Consider two types of molecules. As the molecules of one type move about, they tend to scatter from a central location. The other type of molecule also tends to disperse. The result of this random motion is that the two types of molecules are eventually mixed.

Remember that the motion of the molecules is completely random. They do not move because of conscious thought—they move because of their kinetic energy. If you follow the paths of molecules from a sugar cube placed in a glass of water, you will find that some of the sugar molecules move away from the cube, whereas others move in the opposite direction. However, more sugar molecules would move away from the original cube because there are more molecules there to start with.

We generally are not interested in the individual movement but rather in the overall movement. This overall movement is termed *net movement*. **Net movement** is the movement in one direction minus the movement in the opposite direction. The direction in which the greatest number of molecules moves (net movement) is determined by the relative concentration of the molecules. **Diffusion** is the resultant movement; it is defined as the net movement of a kind of molecule from a place where that molecule is in higher concentration to a place where that molecule is more scarce. When a kind of molecule is completely dispersed, and movement is equal in all directions, we say that the system has reached a state of **dynamic equilibrium.** There is no longer a net movement because movement in one direction equals movement in the other. It is dynamic, however, because the system still has energy, and the molecules are still moving.

Because the cell membrane is composed of phospholipid and protein molecules that are in constant motion, temporary openings are formed that allow small molecules to cross from one side of the membrane to the other. Molecules close to the membrane are in constant motion as well. They are able to move into and out of a cell by passing through these openings in the membrane.

The rate of diffusion is related to the kinetic energy and size of the molecules. Because diffusion only occurs when molecules are unevenly distributed, the relative concentration of the molecules is important in determining how fast diffusion

Figure 4.4

The Concentration Gradient

Gradual changes in concentrations of molecules over distance are called concentration gradients. This bar shows a color gradient of molecules with full color (concentrated molecules) at one end and no color (few molecules) at the other end. A concentration gradient is necessary for diffusion to occur. Diffusion results in net movement of molecules from an area of higher concentration to an area of lower concentration.

occurs. The difference in concentration of the molecules over a distance is known as a **concentration gradient** or **diffusion gradient.** When the molecules are equally distributed, no such gradient exists (figure 4.4). • gas exchange p. 348

Diffusion can take place only as long as there are no barriers to the free movement of molecules. In the case of a cell, the membrane permits some molecules to pass through, whereas others are not allowed to pass or are allowed to pass more slowly. Whether a molecule is able to pass through the membrane also depends on its size, electric charge, and solubility in the phospholipid membrane. The membrane does not, however, distinguish direction of movement of molecules; therefore, the membrane does not influence the direction of diffusion. The direction of diffusion is determined by the relative concentration of specific molecules on the two sides of the membrane, and the energy that causes diffusion to occur is supplied by the kinetic energy of the molecules themselves (figure 4.5).

Diffusion is an important means by which materials are exchanged between a cell and its environment. Because the movement of the molecules is random, the cell has little control over the process; thus, diffusion is considered a passive process—that is, chemical-bond energy does not have to be expended. For example, cells are constantly using oxygen in various chemical reactions. Consequently, the oxygen concentration in cells always remains low. The cells, then, contain a lower concentration of oxygen than the oxygen level outside the cells. This creates a diffusion gradient, and the oxygen molecules diffuse from the outside of the cell to the inside of the cell.

In large animals, many cells are buried deep within the body; if it were not for the animals' circulatory systems, cells would have little opportunity to exchange gases directly with their surroundings. The circulatory system is a transportation system within a body composed of blood vessels of various sizes. These vessels carry many different molecules from one place to another. Oxygen may diffuse into blood

(a)   (b)

Figure 4.5

Diffusion

As a result of molecular motion, molecules move from areas where they are concentrated to areas where they are less concentrated. These figures show molecules leaving and entering animal cells by diffusion. The net direction of movement is controlled by concentration (always high-to-low concentration), and the energy necessary is supplied by the kinetic energy of the molecules themselves. (a) shows net diffusion out of the cell while (b) shows net diffusion into the cell.

through the membranes of the lungs, gills, or other moist surfaces of the animal's body. The circulatory system then transports the oxygen-rich blood throughout the body. The oxygen automatically diffuses into cells. This occurs because the insides of cells are always low in oxygen inasmuch as the oxygen combines with other molecules as soon as it enters. The opposite is true of carbon dioxide. Animal cells constantly produce carbon dioxide as a waste product and so there is always a high concentration of it within the cells. These molecules diffuse from the cells into the blood, where the concentration of carbon dioxide is kept constantly low because the blood is pumped to the moist surface (gills, lungs, etc.) and the carbon dioxide again diffuses into the surrounding environment. In a similar manner, many other types of molecules constantly enter and leave cells.

Dialysis and Osmosis

Another characteristic of all membranes is that they are selectively permeable. **Selectively permeable** means that a membrane will allow certain molecules to pass across it and will prevent others from doing so. Molecules that are able to dissolve in phospholipids, such as vitamins A and D, can pass through the membrane rather easily; however, many molecules cannot pass through at all. In certain cases, the membrane differentiates on the basis of molecular size; that is, the membrane allows small molecules, such as water, to pass through and prevents the passage of larger molecules. The membrane may also regulate the passage of ions. If a particular portion of the membrane has a large number of positive ions on its surface, positively charged ions in the environment will be repelled and prevented from crossing the membrane.

Table 4.1

THE EFFECTS OF OSMOSIS ON DIFFERENT CELL TYPES

Cell Type	What Happens When Cell Is Placed in Hypotonic Solution	What Happens When Cell Is Placed in Hypertonic Solution
With cell wall (e.g., bacteria, fungi, plants)	Swells; does not burst because of the presence of protective cell wall. Cells will become swollen (*turgid*) under these conditions.	Shrinks; cell membrane pulls away from inside of cell wall and forms compressed mass of protoplasm, a process known as *plasmolysis*. Cells will shrink under these conditions. Placing cells in salt water causes certain types of bacterial cells to tear their cell membranes away from the cell wall and results in their death.
Without cell wall (e.g., human red blood cells)	Swells and may burst, a process called *hemolysis*. Red blood cells will *hemolyze* under these conditions.	Shrinks into compact mass, a process known as *crenation*.

We make use of diffusion across a selectively permeable membrane when we use a dialysis machine to remove wastes from the blood. **Dialysis** is the net movement (diffusion) of a solute through a selectively permeable membrane. If a kidney is unable to function normally, blood from a patient is diverted to a series of tubes composed of selectively permeable membranes. The toxins that have concentrated in the blood diffuse into the surrounding fluids in the dialysis machine, and the cleansed blood is returned to the patient. Thus the machine functions in place of the kidney.

Water molecules easily diffuse through cell membranes. The net movement (diffusion) of water molecules through a selectively permeable membrane is known as **osmosis.** In any osmotic situation, there must be a selectively permeable membrane separating two solutions. For example, a solution of 90% water and 10% sugar separated by a selectively permeable membrane from a sugar solution of 80% water and 20% sugar, demonstrates osmosis. The membrane allows water molecules to pass freely but prevents the larger sugar molecules from crossing. There is a higher concentration of water molecules in one solution compared to the concentration of water molecules in the other, so more of the water molecules move from the solution with 90% water to the other solution with 80% water. Be sure that you recognize that (1) osmosis is really diffusion in which the diffusing substance is water, and (2) that the regions of different concentrations are separated by a membrane that is more permeable to water than the substance dissolved in the water.

A proper amount of water is required if a cell is to function efficiently. Too much water in a cell may dilute the cell contents and interfere with the chemical reactions necessary to keep the cell alive. Too little water in the cell may result in a buildup of poisonous waste products. As with the diffusion of other molecules, osmosis is a passive process because the cell has no control over the diffusion of water molecules. This means that the cell can remain in balance with an environment only if that environment does not cause the cell to lose or gain too much water.

If cells contain a concentration of water and dissolved materials equal to that of their surroundings, the cells are said to be **isotonic** to their surroundings. For example, the ocean contains many kinds of dissolved salts. Organisms such as sponges, jellyfishes, and protozoa are isotonic because the amount of material dissolved in their cellular water is equal to the amount of salt dissolved in the ocean's water.• kidney function, p. 357

If an organism is going to survive in an environment that has a different concentration of water than does its cells, it must expend energy to maintain this difference. Organisms that live in freshwater have a lower concentration of water (higher concentration of dissolved materials) than their surroundings and tend to gain water by osmosis very rapidly. They are said to be **hypertonic** to their surroundings, and the surroundings are **hypotonic** compared to the cells. These two terms are always used to compare two different solutions. The hypertonic solution is the one with more dissolved material and less water; the hypotonic solution has less dissolved material and more water. It may help to remember that the water goes where the salt is (table 4.1). Organisms whose cells gain water by osmosis must expend energy to eliminate any excess if they are to keep from swelling and bursting (figure 4.6).

Under normal conditions, when we drink small amounts of water the cells of the brain swell a little, and signals are sent to the kidneys to rid the body of excess water. By contrast, persons who are dehydrated, such as marathon runners, may drink large quantities of water in a very short time following a race. This rapid addition of water to the body may cause abnormal swelling of brain cells because the excess water cannot be gotten rid of rapidly enough. If this happens, the person may lose consciousness or even die because the brain cells have swollen too much.

(a) Isotonic

(b) Cell in hypertonic solution

Figure 4.6

Osmotic Influences on Cells

The cells in these three photographs were subjected to three different environments. (*a*) The cell is isotonic to its surroundings. The water concentration inside the red blood cell and the water concentration in the environment are in balance with each other, so movement of water into the cell equals movement of water out of the cell, and the cell has its normal shape. (*b*) The cell is in a hypertonic solution. Water has diffused from the cell to the environment because a high concentration of water was in the cell and the cell has shrunk. (*c*) A cell has accumulated water from the environment because a higher concentration of water was outside the cell than in its protoplasm. The cell is in a hypotonic solution so it is swollen.

(c) Cell in hypotonic solution

Plant cells also experience osmosis. If the water concentration outside the plant cell is higher than the water concentration inside, more water molecules enter the cell than leave. This creates internal pressure within the cell. But plant cells do not burst because they are surrounded by a rigid cell wall. Lettuce cells that are crisp are ones that have gained water so that there is high internal pressure. Wilted lettuce has lost some of its water to its surroundings so that it has only slight internal cellular water pressure. Osmosis occurs when you put salad dressing on a salad. Because the dressing has a very low water concentration, water from the lettuce diffuses from the cells into the surroundings. Salad that has been "dressed" too long becomes limp and unappetizing.

Controlled Methods of Transporting Molecules

So far, we have considered only situations in which cells have no control over the movement of molecules. Cells cannot rely solely on diffusion and osmosis, however, because many of the molecules they require either cannot pass through the cell membranes or occur in relatively low concentrations in the cells' surroundings.

Some molecules move across the membrane by combining with specific carrier proteins. When the rate of diffusion of a substance is increased in the presence of a carrier, we call this **facilitated diffusion.** Because this movement is still diffusion, the net direction of movement is in accordance with the concentration gradient. Therefore, this is considered a *passive transport* method, although it can only occur in living organisms with the necessary carrier proteins. One example of facilitated diffusion is the movement of glucose molecules across the membranes of certain cells. In order for the glucose molecules to pass into these cells, specific proteins are required to carry them across the membrane. The action of the carrier does not require an input of energy other than the kinetic energy of the molecules (figure 4.7).

When molecules are moved across the membrane from an area of *low* concentration to an area of *high* concentration,

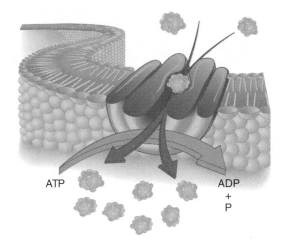

Figure 4.8

Active Transport

One possible method whereby active transport could cause materials to accumulate in a cell is illustrated here. Notice that the concentration gradient is such that if simple diffusion were operating, the molecules would leave the cell. The action of the carrier protein requires an active input of energy (the compound ATP) other than the kinetic energy of the molecules; therefore, this process is termed *active* transport.

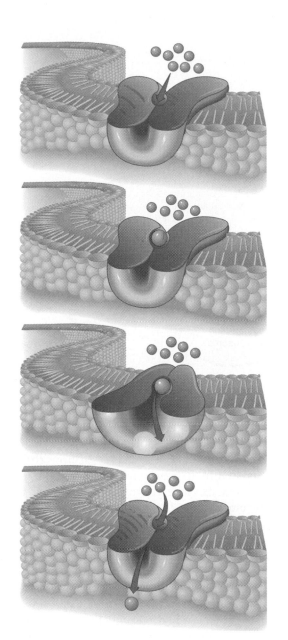

Figure 4.7

Facilitated Diffusion

This method of transporting materials across membranes is a diffusion process (i.e., a movement of molecules from a high to a low concentration). However the process is helped (facilitated) by a particular membrane protein. No chemical-bond energy is required for this process. The molecules being moved through the membrane attach to a specific transport carrier protein in the membrane. This causes a change in its shape that propels the molecule or ion from inside to outside or outside to inside.

the cell must expend energy. This is the opposite direction molecules move in osmosis and diffusion. The process of using a carrier protein to move molecules up a concentration gradient is called **active transport** (figure 4.8). Active transport is very specific: Only certain molecules or ions are able to be moved in this way, and they must be carried by specific proteins in the

membrane. The action of the carrier requires an input of energy other than the kinetic energy of the molecules; therefore, this process is termed *active* transport. For example, some ions, such as sodium and potassium, are actively pumped across cell membranes. Sodium ions are pumped out of cells up a concentration gradient. Potassium ions are pumped into cells up a concentration gradient. •nervous system function, p. 389

In addition to active transport, materials can be transported into a cell by **endocytosis** and out by **exocytosis**. **Phagocytosis** is another name for one kind of endocytosis that is the process cells use to wrap membrane around a particle (usually food) and engulf it (figure 4.9). This is the process leukocytes (white blood cells) use to surround invading bacteria, viruses, and other foreign materials. Because of this, these kinds of cells are called *phagocytes*. When phagocytosis occurs, the material to be engulfed touches the surface of the phagocyte and causes a portion of the outer cell membrane to be indented. The indented cell membrane is pinched off inside the cell to form a sac containing the engulfed material. This sac, composed of a single membrane, is called a **vacuole** or *phagosome* (*phago* = to eat; *-some* = body). Once inside the cell, the membrane of the vacuole is broken down, releasing its contents inside the cell, or it may combine with another vacuole containing destructive enzymes.

Many types of cells use phagocytosis to acquire large amounts of material from their environments. If a cell is not surrounding a large quantity of material but is merely engulfing some molecules dissolved in water, the process is termed **pinocytosis**. In this form of endocytosis, the sacs that are formed are very small in comparison to those formed during

Figure 4.9

Endocytosis and Exocytosis
The sequence illustrates a cell engulfing a microbe and surrounding it with a membrane via endocytosis. Once encased in a portion of the cell membrane (now called a phagocytic vacuole), a lysosome adds its digestive enzymes to it, which speeds the breakdown of the dangerous microbe. Finally, the hydrolyzed (digested) material moves from the vacuole to the inner surface of the cell membrane where the contents are discharged by exocytosis.

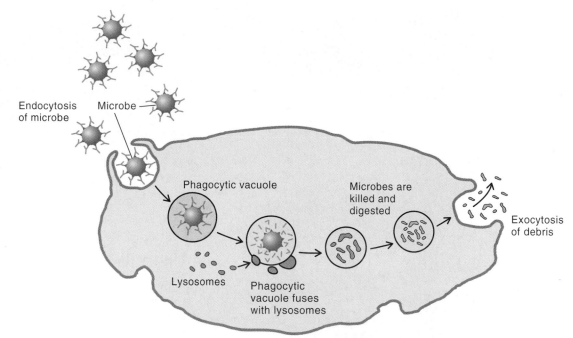

phagocytosis. Because of this size difference, they are called **vesicles.** In fact, an electron microscope is needed in order to see vesicles. The processes of phagocytosis and pinocytosis differ from active transport in that the cell surrounds large amounts of material with a membrane rather than taking the material in molecule by molecule through the membrane.

Signal Transduction

Signal transduction is the term used to describe how cells detect specific signals and how they transmit these signals to the cell's interior. These signals can be physical (electrical or heat) or chemical. Because some chemicals are capable of passing directly through the membrane, they easily enter the cell. Once inside, they can pass on their message to regulatory proteins. These proteins then enter into chemical reactions that result in a change in the cell's behavior. For example, estrogen produced in one part of the body travels through the bloodstream and passes through the tissue to make direct contact with target cells. Once the hormone passes through the target cells, the message is communicated to begin the process of female sex organ development. This is like a person smelling the cologne of his or her date through a curtain. The aroma molecules pass through the curtain to the person's nose and stimulate a response.

However, most signal molecules are not capable of entering cells in such a direct manner. Most signal molecules or *ligands* (primary messengers), remain outside their target cells. When they arrive at the cell they attach to a receptor site molecule embedded in the membrane. The receptor/ligand combination initiates a sequence of events within the membrane that transmits information through the membrane to the interior, generating internal signal molecules called *second messengers*. The secondary messengers are molecules or ions that begin a cascade of chemical reactions that enables the target cell to respond to the stimulus by making a change in how the cell functions. This would be comparable to having your little brother carry a message to you in your room from your date, who remains outside your front door. In a cell, such signal transduction results in a change in chemical activity in the cytoplasm, or genes being turned on or turned off. For example, when a ligand called *epidermal growth factor* (EGF) attaches to the membrane receptor protein called *tyrosine kinase*, it triggers a chain of events inside the membrane. Other EGF molecules migrate to the area and bind together. When enough EGF molecules cluster together, they stimulate a chemical reaction that ultimately leads to gene action, which in turn causes cell growth and division.

4.4 Cell Size

Cells vary greatly in size (figure 4.10). The size of a cell is directly related to its level of activity and the rate that molecules move across its membranes. In order to stay alive, a cell must have a constant supply of nutrients, oxygen, and other molecules. It must also be able to get rid of carbon dioxide and other waste products that are harmful to it. The larger a cell becomes, the more difficult it is to satisfy these requirements; consequently, most cells are very small. There are a few exceptions to this general rule, but they are easily explained. Egg cells, like the yolk of a hen's egg, are very large cells. However, the only part of an egg cell that is metabolically active is a small spot on its surface. The central portion of the egg is simply inactive stored food called

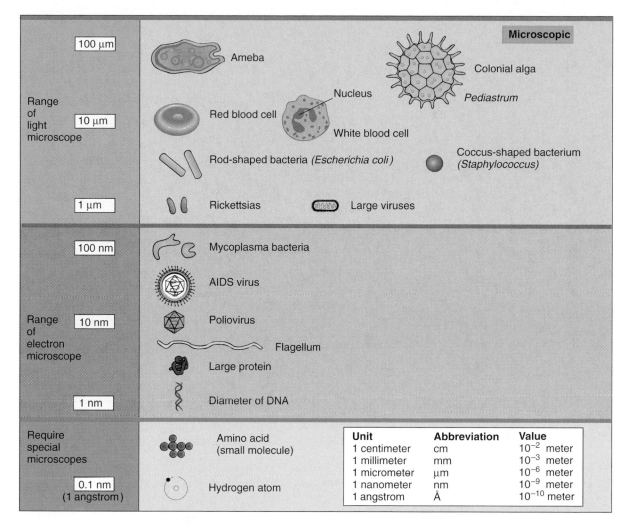

Unit	Abbreviation	Value
1 centimeter	cm	10^{-2} meter
1 millimeter	mm	10^{-3} meter
1 micrometer	μm	10^{-6} meter
1 nanometer	nm	10^{-9} meter
1 angstrom	Å	10^{-10} meter

Figure 4.10

Comparative Sizes

Most cells are too small to be seen with the naked eye. Some type of mechanism is required to magnify them in order to make them visible. Notice how much smaller, in general, prokaryotic cells are than eukaryotic cells. Also notice that most viruses are even smaller and require the use of electron microscopes for viewing.

yolk while the "white" surrounds the actual cell. Similarly, some plant cells are very large but consist of a large, centrally located region filled with water. Again, the metabolically active portion of the cell is at the surface (outer face), where exchange by diffusion or active transport is possible.

There is a mathematical relationship between the surface area and volume of a cell referred to as the *surface area-to-volume ratio*. As cells grow, the amount of surface area increases by the square (X^2) but volume increases by the cube (X^3). They do not increase at the same rate. The surface area increases at a slower rate than the volume. Thus, the surface area-to-volume ratio changes as the cell grows. As a cell gets larger, cells have a problem with transporting materials across the plasma membrane. For example, diffusion of molecules is quite rapid over a short distance, but becomes slower over a longer distance. If a cell were to get too large, the center of the cell would die because transport mechanisms such

as diffusion would not be rapid enough to allow for the exchange of materials. When the surface area is not large enough to permit sufficient exchange between the cell volume and the outside environment, cell growth stops. For example, the endoplasmic reticulum of eukaryotic cells provides an increase in surface area for taking up or releasing molecules. Cells lining the intestinal tract of humans have fingerlike extensions that also help in solving this problem. • alveoli, p. 352

4.5 Organelles Composed of Membranes

Now that you have some background concerning the structure and the function of membranes, let's turn our attention to the way cells use membranes to build the structural components of their protoplasm. The outer boundary of the eukaryotic cell is

Lysosomes

Endoplasmic reticulum

Vacuoles

Golgi body

Figure 4.11

Membranous Cytoplasmic Organelles

Certain structures in the cytoplasm are constructed of membranes. Membranes are composed of protein and phospholipids. The four structures here—vacuoles, Golgi body, endoplasmic reticulum, and lysosomes—are constructed of single-layer membranes.

termed the **cell membrane** or **plasma membrane.** It is associated with a great variety of metabolic activities including taking up and releasing molecules, sensing stimuli in the environment, recognizing other cell types, and attaching to other cells and nonliving objects. In addition to the cell membrane, many other organelles are composed of membranes. Each of these membranous organelles has a unique shape or structure that is associated with particular functions. One of the most common organelles found in cells is the *endoplasmic reticulum.*

The Endoplasmic Reticulum

The **endoplasmic reticulum,** or **ER,** is a set of folded membranes and tubes throughout the cell. This system of membranes provides a large surface upon which chemical

activities take place (figure 4.11). Because the ER has an enormous surface area, many chemical reactions can be carried out in an extremely small space. Picture the vast surface area of a piece of newspaper crumpled into a tight little ball. The surface contains hundreds of thousands of tidbits of information in an orderly arrangement, yet it is packed into a very small volume.

Proteins on the surface of the ER are actively involved in controlling and encouraging chemical activities—whether they are reactions involving growth and development of the cell or those resulting in the accumulation of molecules from the environment. The arrangement of the proteins allows them to control the sequences of metabolic activities so that chemical reactions can be carried out very rapidly and accurately.

On close examination with an electron microscope, it becomes apparent that there are two different types of ER—rough and smooth. The rough ER appears rough because it has ribosomes attached to its surface. *Ribosomes* are nonmembranous organelles that are associated with the synthesis of proteins from amino acids. They are "protein-manufacturing machines." Therefore, cells with an extensive amount of rough ER—for example, your pancreas cells—are capable of synthesizing large quantities of proteins. As members of the immune system, lymphocytes known as *B-cells* are responsible for the production of proteins called *antibodies* or *immunoglobulins*. When B-cells become specialized in the production of one particular kind of antibody, B-cells change into cells known as *plasma cells*. As cells specialized in the production of antibody proteins, plasma cells grow in size and develop extensive amounts of rough ER. Antibodies secreted from plasma cells move throughout the body, combine with potentially dangerous toxic molecules or microbes, and are involved in their destruction. Smooth ER lacks attached ribosomes but is the site of many other important cellular chemical activities, including fat metabolism and detoxification reactions that are involved in the destruction of toxic substances such as alcohol and drugs. Your liver cells contain extensive smooth ER. • immunity, p. 343

In addition, the spaces between the folded membranes serve as canals for the movement of molecules within the cell. This system of membranes allows for rapid distribution of molecules within a cell. The rough and smooth ER may also be connected to one another and to the nuclear membrane.

The Golgi Apparatus

Another organelle composed of membrane is the **Golgi apparatus.** Even though this organelle is also composed of membrane, the way in which it is structured enables it to perform jobs that are different from those performed by the ER. The typical Golgi is composed of from 5 to 20 flattened, smooth, membranous sacs, which resemble a stack of pancakes. The Golgi apparatus is the site of the synthesis and packaging of certain molecules produced in the cell. It is also the place where particular chemicals are concentrated prior to their release from the cell or distribution within the cell. Some Golgi vesicles are used to transport such molecules as mucus, carbohydrates, glycoproteins, insulin, and enzymes to the outside of the cell. The molecules are concentrated inside the Golgi, and tiny vesicles are pinched or budded off the outside surfaces of the Golgi sacs. The vesicles move to and merge with the endoplasmic reticulum or cell membrane. In so doing, the contents are placed in the ER where they can be utilized or transported from the cell. The Golgi is also responsible for preparing individual molecules for transport to the cell membrane so that they can be secreted from the cell. This process is so important that if the Golgi is damaged, new Golgi will be made from ER to accomplish this task.

Figure 4.12

Syndactylism
This person displays the trait known as syndactylism (*syn* = connected; *dactyl* = finger or toe). In most people, enzymes break down the connecting tissue, allowing the toes to separate. In this genetic abnormality, these enzymes failed to do their job.

Lysosomes

An important group of molecules necessary to the cell includes the hydrolytic enzymes. This combination of enzymes is capable of destroying carbohydrates, nucleic acids, proteins, and lipids. Because cells contain large amounts of these molecules, these enzymes must be controlled in order to prevent the destruction of the cell. The Golgi apparatus is the site where these enzymes are converted from their inactive to their active forms and packaged in membranous sacs. These vesicles are pinched off from the outside surfaces of the Golgi sacs and given the special name **lysosomes,** or "bursting body." The lysosomes are used by cells in four major ways:

1. When a cell is damaged, the membranes of the lysosomes break and the enzymes are released. These enzymes then begin to break down the contents of the damaged cell so that the component parts can be used by surrounding cells.
2. Lysosomes also play a part in the normal development of an organism. For example, as a tadpole slowly changes into a frog, the cells of the tail are destroyed by the action of lysosomes. In humans, the developing embryo has paddle-shaped hands and feet. At a prescribed point in development, the cells between the bones of the fingers and toes release the enzymes that had been stored in the lysosomes. As these cells begin to disintegrate, individual fingers or toes begin to take shape. Occasionally this process does not take place, and infants are born with "webbed" fingers or toes (figure 4.12). This developmental defect, called *syndactylism,* may be surgically corrected soon after birth.
3. In many kinds of cells, the lysosomes are known to combine with food vacuoles. When this occurs, the

enzymes of the lysosome break down the food particles into smaller and smaller molecular units. This process is common in one-celled organisms such as *Paramecium.*

4. Lysosomes are also used in the destruction of engulfed, disease-causing microorganisms such as bacteria, viruses, and fungi. As these invaders are taken into the cell by phagocytosis, lysosomes fuse with the phagocytic vacuole. When this occurs, the hydrolytic enzymes and proteins called *defensins* move from the lysosome into the vacuole to destroy the microorganisms.

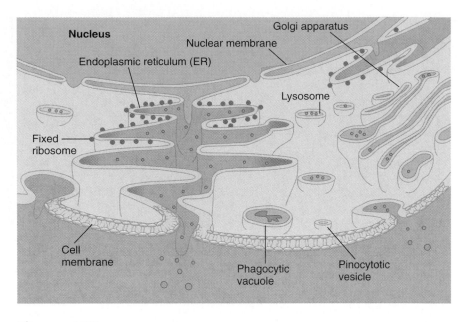

Figure 4.13

Interconversion of Membranous Organelles

Eukaryotic cells contain a variety of organelles composed of phospholipids and proteins. Each has a unique shape and function. Many of these organelles are interconverted from one to another as they perform their essential functions. Cell membranes can become vacuolar membrane or endoplasmic reticulum, which can become vesicular membrane, which in turn can become Golgi or nuclear membrane. However, mitochondria cannot exchange membrane parts with other membranous organelles.

Peroxisomes

Another submicroscopic vesicle is the **peroxisome.** In human cells, peroxisomes are responsible for producing hydrogen peroxide, H_2O_2. The peroxisome enzymes are able to manufacture H_2O_2 that is used in destroying invading microbes. The activity of H_2O_2 is easily demonstrated by mixing the enzyme catalase with H_2O_2. The enzyme converts the hydrogen peroxide to water and oxygen, which forms bubbles. It is the O_2 that is responsible for oxidizing potentially harmful microbes and other dangerous materials. Peroxisomes are also important because they contain enzymes that are responsible for the breakdown of long-chain fatty acids and the synthesis of cholesterol.

The many kinds of vacuoles and vesicles contained in cells are frequently described by their function. Thus phagosomes contain phagocytosed microbes, food vacuoles hold food, and water vacuoles store water. Specialized water vacuoles called *contractile vacuoles* are able to forcefully expel excess water that has accumulated in the cytoplasm as a result of osmosis. The contractile vacuole is a necessary organelle in cells that live in freshwater. The water constantly diffuses into the cell because the environment contains a higher concentration than that inside the cell and therefore water must be actively pumped out. The special containers that hold the contents resulting from pinocytosis are called *pinocytic vesicles.* In all cases, these simple containers are constructed of a surrounding membrane. In most plants, there is one huge, centrally located vacuole in which water, food, wastes, and minerals are stored.

The Nuclear Membrane

A nucleus is a place in a cell—not a solid mass. Just as a room is a place created by walls, a floor, and a ceiling, the nucleus is a place in the cell created by the **nuclear membrane.** This membrane separates the *nucleoplasm,* liquid material in the nucleus, from the cytoplasm. Because they are separated, the cytoplasm and nucleoplasm can maintain different chemical compositions. If the membrane was not formed around the genetic material, the organelle we call the nucleus would not exist. The nuclear membrane is formed from many flattened sacs fashioned into a hollow sphere around the genetic material, DNA. It also has large openings, called *nuclear pore complexes,* which allow thousands of relatively large molecules such as RNA to pass into and out of the nucleus each minute. These pores are held open by donut-shaped molecules that resemble the "eyes" in shoes through which the shoelace is strung. It has been determined that thousands of molecules move in and out through these pores each second.

All of the membranous organelles just described can be converted from one form to another (figure 4.13). For example, phagocytosis results in the formation of vacuolar membrane from cell membrane that fuses with lysosomal membrane, which in turn came from Golgi membrane.

Energy Converters

Two other organelles composed of membranes are incapable of interconversion. Both types of organelles are associated with energy conversion reactions in the cell. These organelles are the *mitochondrion* and the *chloroplast* (figure 4.14).

The **mitochondrion** is an organelle resembling a small bag with a larger bag inside that is folded back on itself. These inner folded surfaces are known as the **cristae.** Located on the surface of the cristae are particular proteins and enzymes in-

Mitochondrion

Inner membrane

Cristae

(b)

Outer membrane

Chloroplast

Stroma

(a)

Thylakoid membrane

Granum

Figure 4.14

Energy-Converting Organelles

(a) The chloroplast, containing the pigment chlorophyll, is the site of photosynthesis. The chlorophyll, located in the grana, captures light energy that is used to construct organic, sugarlike molecules in the stroma. (b) The mitochondria with their inner folds, called cristae, are the site of aerobic cellular respiration, where food energy is converted to usable cellular energy. Both organelles are composed of phospholipid and protein membranes.

volved in *aerobic cellular respiration.* **Aerobic cellular respiration** is the series of reactions involved in the release of usable energy from food molecules, which requires the participation of oxygen molecules. Enzymes that speed the breakdown of simple nutrients are arranged in a sequence on the mitochondrial membrane. The average human cell contains upwards of 10,000 mitochondria. Cells involved in activities that require large amounts of energy, such as muscle cells, contain many more mitochondria. When properly stained, they can be seen with a compound light microscope. When cells are functioning aerobically, the mitochondria swell with activity. But when this activity diminishes, they shrink and appear as threadlike structures.

A second energy-converting organelle is the **chloroplast.** This membranous, saclike organelle contains the green pigment **chlorophyll** and is only found in plants and other eukaryotic organisms that carry out photosynthesis. Some cells contain only one large chloroplast; others contain hundreds of smaller chloroplasts. In this organelle light energy is converted to chemical-bond energy in a process known as **photosynthesis.** Chemical-bond energy is found in food molecules. A study of the ultrastructure—that is, the structures seen with an electron microscope—of a chloroplast shows that the entire organelle is enclosed by a membrane, whereas other membranes are folded and interwoven throughout. As shown in figure 4.14a, in some areas concentrations of these membranes are stacked up or folded back on themselves. Chlorophyll molecules are attached to these membranes. These areas of concentrated chlorophyll are called **thylakoid** membranes stacked up to form the **grana** of the chloroplast. The space between the grana, which has no chlorophyll, is known as the **stroma.**

Mitochondria and chloroplasts are different from other kinds of membranous structures in several ways. First, their membranes are chemically different from those of other

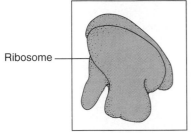

Ribosome

Figure 4.15

Ribosomes

Each ribosome is constructed of two subunits of protein and ribonucleic acid. These globular organelles are associated with the construction of protein molecules from individual amino acids. They are sometimes located individually in the cytoplasm where protein is being assembled, or they may be attached to endoplasmic reticulum (ER). They are so obvious on the ER when using electron micrograph techniques that, when they are present, we label this ER rough ER.

membranous organelles; second, they are composed of double layers of membrane—an inner and an outer membrane; third, both of these structures have ribosomes and DNA that are similar to those of bacteria; and fourth, these two structures have a certain degree of independence from the rest of the cell—they have a limited ability to reproduce themselves but must rely on nuclear DNA for assistance. The functions of these two organelles are discussed in chapter 6.

All of the organelles just described are composed of membranes. Many of these membranes are modified for particular functions. Each membrane is composed of the double phospholipid layer with protein molecules associated with it.

4.6 Nonmembranous Organelles

Suspended in the cytoplasm and associated with the membranous organelles are various kinds of structures that are not composed of phospholipids and proteins arranged in sheets.

Ribosomes

In the cytoplasm are many very small structures called **ribosomes** that are composed of ribonucleic acid (RNA) and 34 proteins. Ribosomes function in the manufacture of protein. Each ribosome is composed of two oddly shaped subunits—a

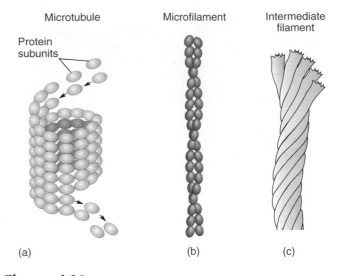

Microtubule   Microfilament   Intermediate filament

Protein subunits

(a)   (b)   (c)

Figure 4.16

Microtubules, Microfilaments, and Intermediate Filaments
(a) Microtubules are hollow tubes constructed of protein spheres called tubulin. The dynamic nature of the microtubule is useful in the construction of certain organelles, such as centrioles, spindle fibers, and cilia or flagella. (b) Microfilaments are composed of the contractile protein, actin. This is the same contractile protein found in human muscle cells. (c) Intermediate filaments are also protein but resemble a multistranded wire cable. These link microtubules and microfilaments.

large one and a small one. The larger of the two subunits is composed of a specific type of RNA associated with several kinds of protein molecules. The smaller is composed of RNA with fewer protein molecules than the large one. These globular organelles are involved in the assembly of proteins from amino acids—they are frequently associated with the endoplasmic reticulum to form rough ER. Areas of rough ER have been demonstrated to be active sites of protein production. Cells actively producing nonprotein materials, such as lipids, are likely to contain more smooth ER than rough ER. Many ribosomes are also found floating freely in the cytoplasm (figure 4.15) wherever proteins are being assembled. Cells that are actively producing protein (e.g., liver cells) have great numbers of free and attached ribosomes. The details of how ribosomes function in protein synthesis are discussed in chapter 7.

Microtubules, Microfilaments, and Intermediate Filaments

The interior of a cell is not simply filled with liquid cytoplasm. Among the many types of nonmembranous organelles found there are elongated protein structures know as **microtubules, microfilaments,** and **intermediate filaments** (figure 4.16). Their various functions are as complex as those provided by the structural framework and cables of a high-rise office building, geodesic dome (e.g., as seen at Epcot Center, FL), or skeletal and muscular systems of a large animal. All three types of organelles interconnect and some are attached to the inside of the cell membrane forming what is known as the

(a)

(b)

Cell membrane · Ribosome · Endoplasmic reticulum · **Intermediate filament** · Mitochondrion · Ribosomes · **Microfilament** · **Microtubule**

Figure 4.17

The Cytoskeleton

A complex array of microfilaments, microtubules, and intermediate filaments provides an internal framework structure for the cell. The cellular skeleton is not a rigid, fixed-in-place structure, but is dynamic and changes as the microfilament and microtubule component parts are assembled and disassembled. (a) Elements of the cytoskeleton have been labeled with a fluorescent dye to make them visible. The microtubules have fluorescent red dye and actin filaments are green. Part (b) shows how the various parts of the cytoskeleton are interconnected.

cytoskeleton of the cell (figure 4.17). These cellular components provide the cell with shape, support, and the ability to move about the environment.

The cytoskeleton also serves to transport materials from place to place within the cytoplasm. Think of the cytoskeleton components as the internal supports and cables required to construct a circus tent. The shape of the flexible canvas cover (i.e., cell membrane) is determined by the location of internal tent poles (i.e., microtubules) and the tension placed on them by attached wire or rope cables (i.e., contractile microfilaments). The poles are made light and strong by being tubular (*tubulin* protein) and are attached to the inner surface of the canvas cover at specific points. The comparable cell membrane attachment points for microtubules are cell membrane molecules known as *integrins*. The reason the poles stay in place is because of their attachment to the canvas and the tension placed on them by the cables. As the cables are adjusted, the shape of the canvas (i.e., cell) changes. The intermediate filaments serve as cables that connect microfilaments and microtubules, thus providing additional strength and support. Just as in the tent analogy, when one of the microfilaments or intermediate filaments is adjusted, the shape of the entire cell changes. For example, when a cell is placed on a surface to which it cannot stick, the internal tensions created by the cytoskeleton components can pull together and cause the cell to form a sphere. A cell's cytoskeleton also changes shape dramatically when a cell divides. It constricts, pulling the cell together in the middle and allowing the membrane to be sealed between the two new daughter cells.

Just as internal changes in tension can cause change, changes in the external environment can cause the cell to change. When forces are exerted on the outside of the cell, internal tensions shift causing physical and biochemical activity. For example, as cell tension changes, some microtubules begin to elongate and others begin to shorten. This can result in overall movement of the cell. Cells that remain flat appear to divide more frequently, cells prevented from flattening commit suicide more often, and those that are neither too flat nor spherical neither divide nor die. Enzymes attached to the cytoskeleton are activated when the cell is touched. Some of these events even affect gene activity.

Centrioles

An arrangement of two sets of microtubules at right angles to each other makes up a structure known as the **centriole.** The centrioles of many cells are located in a region known as the *centrosome*. The centrosome is usually located close to the nuclear membrane. Centrioles operate by organizing microtubules into a complex of strings called *spindle fibers*. The *spindle* is the structure upon which chromosomes are attached so that they may be properly separated during cell division. Each set is composed of nine groups of short microtubules arranged in a cylinder (figure 4.18). The functions of centrioles and spindle fibers in cell division are referred to again in chapter 8. One curious fact about centrioles is that they are present in most animal cells but not in many types of plant cells. Other structures called *basal*

Cell membrane

Centriole

Figure 4.18

The Centriole

These two sets of short microtubules are located just outside the nuclear membrane in many types of cells. The micrograph shows an end view of one of these sets. Magnification is about 160,000 times.

bodies resemble centrioles and are located at the base of cilia and flagella. • mitosis, p. 158

Cilia and Flagella

Many cells have microscopic, hairlike structures projecting from their surfaces; these are **cilia** or **flagella** (figure 4.19). In general, we call them flagella if they are long and few in number, and cilia if they are short and more numerous. They are similar in structure, and each functions to move the cell through its environment or to move the environment past the cell. They are constructed of a cylinder of nine sets of microtubules similar to those in the centriole, but they have an additional two microtubules in the center. These long strands of microtubules project from the cell surface and are covered by cell membrane. When cilia and flagella are sliced crosswise, their cut ends show what is referred to as the 9 + 2 *arrangement* of microtubules. The cell has the ability to control the action of these microtubular struc-

tures, enabling them to be moved in a variety of different ways. Their coordinated actions either propel the cell through the environment or the environment past the cell surface. The protozoan *Paramecium* is covered with thousands of cilia that actively beat a rhythmic motion to move the cell through the water. The cilia on the cells that line your trachea move mucous-containing particles from deep within your lungs.

Inclusions

Inclusions are collections of materials that do not have as well-defined a structure as the organelles we have discussed so far. They might be concentrations of stored materials, such as starch grains, sulfur, or oil droplets, or they might be a collection of miscellaneous materials known as **granules**. Unlike organelles, which are essential to the survival of a cell, the inclusions are generally only temporary sites for the storage of nutrients and wastes.

Eukaryotic Cilia and Flagella

(a)

Inner microtubules

Outer microtubules

Cell membrane

Cilium

Flagella

Arms

Cilia on surface

(b)

Filament

Hook

Prokaryotic Flagella

P ring

L ring

S ring

Rod

M ring

Figure 4.19

Eukaryotic and Prokaryotic Cilia and Flagella

(*a*) Cilia and flagella function like oars or propellers that move the cell through its environment or move the environment past the cell. Cilia and flagella are constructed of groups of microtubules as in the ciliated protozoan shown on the left and the flagellated alga on the right. Flagella are usually less numerous and longer than cilia. (*b*) The flagella of prokaryotes are composed of a single type of protein arranged in a fiber that is anchored into the cell wall and membrane. Bacterial flagella move the cell by rotating.

Some inclusion materials may be harmful to other cells. For example, rhubarb leaf cells contain an inclusion composed of oxalic acid, an organic acid. Needle-shaped crystals of calcium oxalate can cause injury to the kidneys of an organism that eats rhubarb leaves. The sour taste of this particular compound aids in the survival of the rhubarb plant by discouraging animals from eating it. Similarly, certain bacteria store in their inclusions crystals of a substance known to be harmful to insects. Spraying plants with these bacteria is a biological method of controlling the population of the insect pests, while not interfering with the plant or with humans.

In the past, cell structures such as ribosomes, mitochondria, and chloroplasts were also called *granules* because their structure and function were not clearly known. As scientists learn more about inclusions and other unidentified particles in the cells, they too will be named and more fully described.

4.7 Nuclear Components

As stated at the beginning of this chapter, one of the first structures to be identified in cells was the nucleus. The nucleus was referred to as the cell center. If the nucleus is removed from a cell, the cell can live only a short time. For example, human red blood cells begin life in bone marrow, where they have nuclei. Before they are released into the bloodstream to serve as oxygen and carbon dioxide carriers, they lose their nuclei. As a consequence, red blood cells are able to function only for about 120 days before they disintegrate.

When nuclear structures were first identified, it was noted that certain dyes stained some parts more than others. The parts that stained more heavily were called **chromatin,** which means colored material. Chromatin is composed of long molecules of deoxyribonucleic acid (DNA) in association with proteins. These DNA molecules contain the genetic information for the cell, the blueprints for its construction and maintenance. Chromatin is loosely organized DNA in the nucleus. When the chromatin is tightly coiled into shorter, denser structures, we call them **chromosomes** (*chromo* = color; *some* = body). Chromatin and chromosomes are really the same molecules but differ in structural arrangement. In addition to chromosomes, the nucleus may also contain one, two, or several *nucleoli*. A **nucleolus** is the site of ribosome manufacture. Nucleoli are composed of specific granules and fibers in association with the cell's DNA used in the manufacture of ribosomes. These regions, together with the completed or partially completed ribosomes, are called nucleoli.

The final component of the nucleus is its liquid matrix called the **nucleoplasm.** It is a colloidal mixture composed of water and the molecules used in the construction of ribosomes, nucleic acids, and other nuclear material (figure 4.20).

Figure 4.20

The Nucleus
One of the two major regions of protoplasm, the nucleus has its own complex structure. It is bounded by two layers of membrane that separate it from the cytoplasm. Inside the nucleus are the nucleoli, chromosomes or chromatin material composed of DNA and protein, and the liquid matrix (nucleoplasm). Magnification is about 20,000 times.

4.8 Major Cell Types

Not all of the cellular organelles we have just described are located in every cell. Some cells typically have combinations of organelles that differ from others. For example, some cells have nuclear membrane, mitochondria, chloroplasts, ER, and Golgi; others have mitochondria, centrioles, Golgi, ER, and nuclear membrane. Other cells are even more simple and lack the complex membranous organelles described in this chapter. Because of this fact, biologists have been able to classify cells into two major types: *prokaryotic* and *eukaryotic* (figure 4.21).

The Prokaryotic Cell Structure

Prokaryotic cells, the *Eubacteria* and *Archaea,* do not have a typical nucleus bound by a nuclear membrane, nor do they contain mitochondria, chloroplasts, Golgi, or extensive networks of ER. However, prokaryotic cells contain DNA and enzymes and are able to reproduce and engage in metabolism. They perform all of the basic functions of living things with fewer and more simple organelles. As yet, members of the Archaea are of little concern to the medical profession because none have been identified as disease-causing. They are typically found growing in extreme environments where the pH, salt concentration, or temperatures make it impossible for most other organisms to survive. The other prokaryotic cells are called *bacteria* and about 5% cause diseases such as tuberculosis, strep throat, gonorrhea, and acne. Other prokaryotic cells are responsible for the decay and decomposition of dead organisms. Although some bacteria have a type of green photosynthetic pigment and carry on photosynthesis, they do so without chloroplasts and use different chemical reactions. • microbes, p. 476

One significant difference between prokaryotic and eukaryotic cells is in the chemical makeup of their ribosomes.

Prokaryotic Cells

Characterized by few membranous organelles; DNA not separated from the cytoplasm by a membrane

Eukaryotic Cells

Cells larger than prokaryotic cells; DNA found within nucleus with a membrane separating it from the cytoplasm; many complex organelles composed of many structures including phospholipid bilayer membranes

Domain Eubacteria	Domain Archaea	Domain Eucarya			
		Kingdom Protista	Kingdom Fungi	Kingdom Plantae	Kingdom Animalia
Kingdoms not specified	Kingdoms Euryarchaeota, Korarchaeota, Krenarchaeota				
Unicellular microbes; typically associated with bacterial "diseases," but 90%–95% are ecologically important and not pathogens	Unicellular microbes; typically associated with extreme environments including low pH, high salinity, and extreme temperatures	Unicellular microbes; some in colonies; both photosynthetic and heterotropic nutrition; a few are parasites	Multicellular organisms or loose colonial arrangements of cells; organism is a row or filament of cells; decay fungi and parasites	Multicellular organisms; cells supported by a rigid cell wall of cellulose; some cells have chloroplasts; complex arrangement into tissues	Multicellular organisms with division of labor into complex tissues; no cell wall present; acquire food from the environment; some are parasites
Examples: Gram-positive bacteria such as *Steptococcus pneumonia* and Gram-negative bacteria such as *E. coli*	Examples: *Methanococcus*, halophiles, and *Thermococcus*	Examples: protozoans such as *Amoeba* and *Paramecium* and algae such as *Chlamydomonas* and *Euglena*	Examples: yeast such as bakers yeast, molds such as *Penicillium*; morels, mushrooms, and rusts	Examples: moss, ferns, cone-bearing trees, and flowering plants	Examples: worms, insects, starfish, frogs, reptiles, birds, and mammals

Figure 4.21

Cell Types and the Major Groups of Organisms
The two types of cells (prokaryotic and eukaryotic) are described in relationship to the major patterns found in all living things, the kingdoms of life. Note the similarities of all kingdoms and the subtle differences among them.

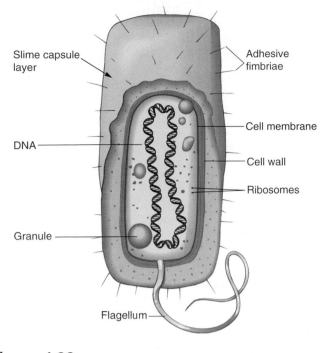

Figure 4.22

Prokaryotic Cell
All bacteria are prokaryotic cells. While smaller and less complex, each is capable of surviving on its own. Most bacteria are involved with decay and decomposition, but a small percentage are pathogens responsible for disease such as strep throat, TB, syphilis, anthrax, and gas gangrene. The cell illustrated here is a bacillus because it has a rod shape.

Table 4.2

COMPARISON OF GENERAL PLANT AND ANIMAL CELL STRUCTURE

Plant Cells	Animal Cells
CELL WALL	
Cell membrane	Cell membrane
Cytoplasm	Cytoplasm
Nucleus	Nucleus
Mitochondria	Mitochondria
CHLOROPLASTS	**CENTRIOLE**
Golgi apparatus	Golgi apparatus
Endoplasmic reticulum	Endoplasmic reticulum
Lysosomes	Lysosomes
Vacuoles/vesicles	Vacuoles/vesicles
Ribosomes	Ribosomes
Nucleolus	Nucleolus
Inclusions	Inclusions
Cytoskeleton	Cytoskeleton

The ribosomes of prokaryotic cells contain different proteins than those found in eukaryotic cells. Prokaryotic ribosomes are also smaller. This discovery was important to medicine because many cellular forms of life that cause common diseases are bacterial. As soon as differences in the ribosomes were noted, researchers began to look for ways in which to interfere with the prokaryotic ribosome's function but *not* interfere with the ribosomes of eukaryotic cells. **Antibiotics,** such as streptomycin, are the result of this research. This drug combines with prokaryotic ribosomes and causes the death of the prokaryote by preventing the production of proteins essential to its survival. Because eukaryotic ribosomes differ from prokaryotic ribosomes, streptomycin does not interfere with the normal function of ribosomes in human cells.

Most prokaryotic cells are surrounded by a *capsule* or slime layer that can be composed of a variety of compounds (figure 4.22). In certain bacteria this layer is responsible for their ability to stick to surfaces (including host cells) and to resist phagocytosis. Many bacteria also have fimbriae, hair-like protein structures, which help the cell stick to objects. Those with flagella are capable of propelling themselves through the environment. Below the capsule is the rigid cell wall comprised of a unique protein/carbohydrate complex called peptidoglycan. This provides the cell with the strength to resist osmotic pressure changes and gives the cell shape. Just beneath the wall is the cell membrane. Thinner and with a slightly different chemical composition from eukaryotes, it carries out the same functions as the cell membranes in eukaryotes. Most bacteria are either rods (bacilli), spherical (cocci), or curved (spirilla). The genetic material within the cytoplasm is DNA in the form of a loop.

The Eukaryotic Cell Structure

Eukaryotic cells contain a true nucleus and most of the membranous organelles described earlier. Eukaryotic organisms can be further divided into several categories or domains based on the specific combination of organelles they contain. The cells of plants, fungi, protozoa and algae, and animals are all eukaryotic. The most obvious characteristic that sets the plants and algae apart from other organisms is their green color, which indicates that the cells contain chlorophyll. Chlorophyll is necessary for the process of photosynthesis—the conversion of light energy into chemical-bond energy in food molecules. These cells, then, are different from the other cells in that they contain chloroplasts in their cytoplasm. Another distinguishing characteristic of plants and algae is the presence of cellulose in their cell walls (table 4.2).

The group of organisms that has a cell wall but lacks chlorophyll in chloroplasts is collectively known as *fungi*. They were previously thought to be either plants that had lost their ability to make their own food or animals that had developed cell walls. Organisms that belong in this category of eukaryotic cells include yeasts, molds, mushrooms, and the fungi that cause such human diseases as athlete's foot, jungle rot, and ringworm. Now we have come to recognize this group as different enough from plants and animals to place them in a separate kingdom.

Eukaryotic organisms that lack cell walls and cannot photosynthesize are placed in separate groups. Organisms that consist of only one cell are called protozoans—examples are *Amoeba* and *Paramecium*. They have all the cellular organelles described in this chapter except the chloroplast; therefore, protozoans must consume food as do the fungi and the multicellular animals.

Although the differences in these groups of organisms may seem to set them worlds apart, their similarity in cellular structure is one of the central themes unifying the field of biology. One can obtain a better understanding of how cells operate in general by studying specific examples. Because the organelles have the same general structure and function regardless of the kind of cell in which they are found, we can learn more about how mitochondria function in plants by studying how mitochondria function in animals. There is a commonality among all living things with regard to their cellular structure and function.

X	Y	Z
Inorganic	Organic	Organic
High concentration outside cell	High concentration inside cell	High concentration inside cell
Essential to life of cell	Essential to life of cell	Poisonous to the cell
Small and can pass through the membrane	Large and cannot pass through the membrane	Small and can pass through the membrane

With this information and your background in cell structure and function, osmosis, diffusion, and active transport, decide whether this kind of cell will continue to live in this sea, and explain why or why not.

CONCEPT MAP TERMINOLOGY

Construct a concept map using these terms.

aerobic cellular respiration	osmosis
carbon dioxide	oxygen
chloroplast	sugar
facilitated diffusion	water
mitochondrion	

Visit the Online Learning Center to use an interactive format to construct another concept map.

SUMMARY

The concept of the cell has developed over a number of years. Initially, only two regions, the cytoplasm and the nucleus, could be identified. At present, numerous organelles are recognized as essential components of both prokaryotic and eukaryotic cell types. The structure and function of some of these organelles are compared in table 4.3. This table also indicates whether the organelle is unique to prokaryotic or eukaryotic cells or found in both.

The cell is the common unit of life. We study individual cells and their structures to understand how they function as individual living organisms and as parts of many-celled beings. Knowing how prokaryotic and eukaryotic cell types resemble each other or differ from each other helps physicians control some organisms dangerous to humans.

KEY TERMS

Use interactive flash cards, on the Online Learning Center, to help you learn the meanings of these terms.

active transport	domain
aerobic cellular respiration	dynamic equilibrium
antibiotics	endocytosis
cell	endoplasmic reticulum (ER)
cell membrane	eukaryotic cells
cell wall	exocytosis
cellular membranes	facilitated diffusion
centriole	flagella
chlorophyll	fluid-mosaic model
chloroplast	Golgi apparatus
chromatin	grana
chromosomes	granules
cilia	hydrophilic
concentration gradient	hydrophobic
cristae	hypertonic
cytoplasm	hypotonic
cytoskeleton	inclusions
dialysis	intermediate filaments
diffusion	isotonic
diffusion gradient	lysosome

THINKING CRITICALLY

For guidelines to answer this question, visit the Online Learning Center.

A primitive type of cell consists of a membrane and a few other cell organelles. This first kind of cell lives in a sea that contains three major kinds of molecules with the following characteristics:

Table 4.3

SUMMARY OF THE STRUCTURE AND FUNCTION OF THE CELLULAR ORGANELLES

Organelle	Type of Cell in Which Located	Structure	Function
Plasma membrane	Prokaryotic and eukaryotic	Membranous; typical membrane structure; phospholipid and protein present	Controls passage of some materials to and from the environment of the cell
Inclusions (granules)	Prokaryotic and eukaryotic	Nonmembranous; variable	May have a variety of functions
Chromatin material	Prokaryotic and eukaryotic	Nonmembranous; composed of DNA and proteins (histones in eukaryotes and HU proteins in prokaryotes)	Contain the hereditary information that the cell uses in its day-to-day life and pass it on to the next generation of cells
Ribosomes	Prokaryotic and eukaryotic	Nonmembranous; protein and RNA structure	Site of protein synthesis
Microtubules, microfilaments, and intermediate filaments	Eukaryotic	Nonmembranous; strands composed of protein	Provide structural support and allow for movement
Nuclear membrane	Eukaryotic	Membranous; double membrane formed into a single container of nucleoplasm and nucleic acids	Separates the nucleus from the cytoplasm
Nuceolus	Eukaryotic	Nonmembranous; group of RNA molecules and DNA located in the nucleus	Site of ribosome manufacture and storage
Endoplasmic reticulum	Eukaryotic	Membranous; folds of membrane forming sheets and canals	Surface for chemical reactions and intracellular transport system
Golgi apparatus	Eukaryotic	Membranous; stack of single membrane sacs	Associated with the production of secretions and enzyme activation
Vacuoles and vesicles	Eukaryotic	Membranous; microscopic single membranous sacs	Containers of materials
Peroxisomes	Eukaryotic	Membranous; submicroscopic membrane-enclosed vesicle	Release enzymes to break down hydrogen peroxide
Lysosomes	Eukaryotic	Membranous; submicroscopic membrane-enclosed vesicle	Isolate very strong enzymes from the rest of the cell
Mitochondria	Eukaryotic	Membranous; double membranous organelle: large membrane folded inside a smaller membrane	Associated with the release of energy from food; site of aerobic cellular respiration
Chloroplasts	Eukaryotic	Membranous; double membranous organelle: large membrane folded inside a smaller membrane (grana)	Associated with the capture of light of energy and synthesis of carbohydrate molecules: site of photosynthesis
Centriole	Eukaryotic	Two clusters of 9 microtubules	Associated with cell division
Contractile vacuole	Eukaryotic	Membranous; single-membrane container	Expels excess water
Cilia and flagella	Eukaryotic and prokaryotic	Nonmembranous; prokaryotes composed of single type of protein arranged in a fiber that is anchored into the cell wall and membrane; 9 + 2 tubulin protein in eukaryotes	Flagellar movement in prokaryotic type rotate; ciliary and flagellar movement in eukaryotic type seen as waving or twisting

microfilaments
microscope
microtubules
mitochondrion
net movement
nuclear membrane
nucleoli (singular, nucleolus)
nucleoplasm
nucleus
organelles
osmosis
peroxisome
phagocytosis

photosynthesis
pinocytosis
plasma membrane
prokaryotic cells
protoplasm
ribosomes
selectively permeable
signal transduction
stroma
thylakoid
vacuole
vesicles

LEARNING CONNECTIONS

Visit the Online Learning Center to find additional study resources including those highlighted below.

4.1 The Cell Theory

1. Describe how the concept of the cell has changed over the past 200 years.
2. Define *cytoplasm*.

4.2 Cell Membranes

Labeling Exercises
- Fluid mosaic model (1)
- Fluid mosaic model (2)
- Membrane proteins

Animations and Review
- Membrane structure

3. What are the differences between the cell and the cell membrane?
4. Describe the role played by HLA cell membrane recognition molecules.

4.3 Getting Through Membranes

Interactive Concept Map
- Moving over membranes

Animations and Review
- Diffusion
- Osmosis
- Facilitated diffusion
- Active transport

Experience This!
- Diffusion, osmosis, or active transport?

5. What three methods allow the exchange of molecules between cells and their surroundings?
6. How do diffusion, facilitated diffusion, osmosis, and active transport differ?
7. Why does putting salt on meat preserve it from spoilage by bacteria?

4.4 Cell Size

Animations and Review
- Surface to volume

Human Explorations
- Cell size

8. On the basis of surface area-to-volume ratio, why do cells tend to remain small?
9. What is the advantage of having numerous folds in a cell's plasma membrane?

4.5 Organelles Composed of Membranes

Animations and Review
- Endomembrane
- Energy organelles

Labeling Exercises
- Golgi apparatus
- Mitochondria
- Chloroplast

Interactive Concept Maps
- Text concept map

10. Make a list of the membranous organelles of a eukaryotic cell and describe the function of each.
11. Define the following terms: stroma, grana, and cristae.

4.6 Nonmembranous Organelles

Animations and Review
- Cytoskeleton

Labeling Exercises
- Cytoskeleton

12. Make a list of the nonmembranous organelles of the cell and describe their functions.
13. Give three examples of cellular inclusions.

4.7 Nuclear Components

Labeling Exercises
- Anatomy of the nucleus

14. Define the following terms: chromosome and chromatin.
15. What is a nucleolus?

4.8 Major Cell Types

Animations and Review
- Prokaryotes
- Eukaryotes
- Summary

Labeling Exercises
- Animal cell (1)
- Animal cell (2)
- Plant cell (1)
- Plant cell (2)
- Bacterium (1)
- Cyanobacterium

16. Diagram a eukaryotic and a prokaryotic cell and show where proteins, nucleic acids, carbohydrates, and lipids are located.

17. What two types of organisms have prokaryotic cell structure?

Test yourself with multiple choice questions that provide immediate feedback on the Online Learning Center with PowerWeb at

www.mhhe.com/enger11

CHAPTER OUTLINE

Enzymes

GOALS AND OBJECTIVES

Understand the nature of enzyme structure and how they work.

❑ Explain why enzymes are so important to all organisms.

❑ Describe to which group of organic molecules enzymes belong.

❑ Contrast active site and binding site.

❑ Relate the shape of an enzyme to its ability to help in a chemical reaction.

❑ Define the term *activation energy*.

❑ Describe what happens when an enzyme and a substrate combine.

Understand how other molecules can assist in enzyme activity.

❑ Contract cofactors, vitamins, and coenzymes.

❑ Define the term *denature*.

Know how the environment can affect enzyme action.

❑ Describe why enzymes work in some situations and not in others.

❑ Define the term *turnover number*.

❑ List what environmental factors are able to alter enzyme activity.

Understand how enzymes can be controlled by the cell.

❑ Define the term *negative feedback inhibition*.

❑ Describe enzymatic competition.

❑ Explain the role played by gene-regulator proteins in enzyme action.

Cells: Anatomy and Action

PART TWO

5.1 How Cells Use Enzymes

All living things require energy and building materials in order to grow and reproduce. Energy may be in the form of visible light, or it may be in energy-containing covalent bonds found in nutrients. **Nutrients** are molecules required by organisms for growth, reproduction, or repair—they are a source of energy and molecular building materials. The formation, breakdown, and rearrangement of molecules to provide organisms with essential energy and building blocks are known as *biochemical reactions.* These reactions occur when atoms or molecules come together and form new, more stable relationships. This results in the formation of new molecules and a change in the energy distribution among the reactants and end products. Most chemical reactions require an input of energy to get them started. This is referred to as **activation energy.** This energy is used to make the reactants unstable and more likely to react (figure 5.1).

If organisms are to survive, they must obtain sizable amounts of energy and building materials in a very short time. Experience tells us that the sucrose in candy bars contains the potential energy needed to keep us active, as well as building materials to help us grow (sometimes to excess!). Yet, random chemical processes could take millions of years to break down a candy bar, releasing its energy and building materials. Of course, living things cannot wait that long. To sustain life, biochemical reactions must occur at extremely rapid rates. One way to increase the rate of any chemical reaction and make its energy and component parts available to a cell is to increase the temperature of the reactants. In general, the hotter the reactants, the faster they will react. However, this method of increasing reaction rates has a major drawback when it comes to living things: organisms die because cellular proteins are denatured before the temperature reaches the point required to sustain the biochemical reactions necessary for life. This is of practical concern to people who are experiencing a fever. Should the fever stay too high for too long, major disruptions of cellular biochemical processes could be fatal.

There is a way of increasing the rate of chemical reactions without increasing the temperature. This involves using substances called *catalysts.* A **catalyst** is a chemical that speeds the reaction but is not used up in the reaction. It can

Figure 5.1

The Lowering of Activation Energy
Enzymes operate by lowering the amount of energy needed to get a reaction going—the activation energy. When this energy is lowered, the nature of the bonds is changed so they are more easily broken. Whereas the cartoon shows the breakdown of a single reactant into many end products (as in a hydrolysis reaction), the lowering of activation energy can also result in bonds being broken so that new bonds may be formed in the construction of a single, larger end product from several reactants (as in a synthesis reaction).

be recovered unchanged when the reaction is complete. Catalysts function by lowering the amount of activation energy needed to start the reaction. A cell manufactures specific proteins that act as catalysts. A protein molecule that acts as a catalyst to speed the rate of a reaction is called an **enzyme.** Enzymes can be used over and over again until they are worn out or broken. The production of these protein catalysts is under the direct control of an organism's genetic material (DNA). The instructions for the manufacture of all enzymes are found in the genes of the cell. Organisms make their own enzymes. How the genetic information is used to direct the synthesis of these specific protein molecules is discussed in chapter 7.

5.2 How Enzymes Speed Chemical Reaction Rates

As the instructions for the production of an enzyme are read from the genetic material, a specific sequence of amino acids is linked together at the ribosomes. Once bonded, the chain of amino acids folds and twists to form a globular molecule. It is the nature of its three-dimensional shape that allows this

Figure 5.2

Enzyme-Substrate Complex Formation
During an enzyme-controlled reaction, the enzyme and substrate come together to form a new molecule—the enzyme-substrate complex molecule. This molecule exists for only a very short time. During that time, activation energy is lowered and bonds are changed. The result is the formation of a new molecule or molecules called the end products of the reaction. Notice that the enzyme comes out of the reaction intact and ready to be used again.

OUTLOOKS 5.1

Enzymes and Stonewashed "Genes"

The popularity of stonewashed jeans grew dramatically in the late 1960s. To get the stonewashed effect, the denim was actually washed in machines along with stones. The stones rubbed against the denim, wearing the blue dye off the surface of the material. But the stones also damaged the cotton fibers. The fiber damage shortened the life of the fabric, a feature that many consumers found unacceptable. Now, to create the stonewashed look and still maintain strong cotton fibers, enzymes are used that "digest" or hydrolyze the blue dye on the surface of the fabric. Because the enzyme is substrate or dye specific, the cotton fibers are not harmed.

enzyme to combine with a reactant and lower the activation energy. Each enzyme has a specific three-dimensional shape that, in turn, is specific to the kind of reactant with which it can combine. The enzyme physically fits with the reactant. The molecule to which the enzyme attaches itself (the reactant) is known as the **substrate.** When the enzyme attaches itself to the substrate molecule, a new, temporary molecule—the **enzyme-substrate complex**—is formed (figure 5.2). When the substrate is combined with the enzyme, its bonds are less stable and more likely to be altered and form new bonds. The enzyme is specific because it has a particular shape that can combine only with specific parts of certain substrate molecules (Outlooks 5.1).

You might think of an enzyme as a tool that makes a job easier and faster. For example, the use of an open-end crescent wrench can make the job of removing or attaching a nut and bolt go much faster than doing that same job by hand. In order to accomplish this job, the proper wrench must be used. Just any old tool (screwdriver or hammer) won't work! The enzyme must also physically attach itself to the substrate; therefore, there is a specific **binding site** or **attachment site** on the enzyme surface. Figure 5.3 illustrates the specificity of both wrench and enzyme. Note that the wrench and enzyme are recovered unchanged after they have been used. This means that the enzyme and wrench can be

used again. Eventually, like wrenches, enzymes wear out and have to be replaced by synthesizing new ones using the instructions provided by the cell's genes. Generally, only very small quantities of enzymes are necessary because they work so fast and can be reused.

Both enzymes and wrenches are specific in that they have a particular surface geometry or shape that matches the geometry of their respective substrates. Note that both the enzyme and wrench are flexible. The enzyme can bend or fold to fit the substrate just as the wrench can be adjusted to fit the nut. This is called the *induced fit hypothesis.* The fit is induced because the presence of the substrate causes the enzyme to "mold" or "adjust" itself to the substrate as the two come together.

The place on the enzyme that causes a specific part of the substrate to change is called the **active site** of the enzyme, or the place on the enzyme surface where chemical bonds are formed or broken. (Note in the case illustrated in figure 5.3 that the "active site" is the same as the "binding site." This is typical of many enzymes.) This site is the place where the activation energy is lowered and the electrons are shifted to change the bonds. The active site may enable a positively charged surface to combine with the negative portion of a reactant. Although the active site does mold itself to a substrate, enzymes do not have the ability to fit all substrates. Enzymes are specific to a certain substrate or group of very similar substrate molecules. One enzyme cannot speed the rate of all types of biochemical reactions. Rather, a special enzyme is required to control the rate of each type of chemical reaction occurring in an organism.

Because the enzyme is specific to both the substrate to which it can attach and the reaction that it can encourage, a unique name can be given to each enzyme. The first part of an enzyme's name is the name of the molecule to which it can become attached. The second part of the name indicates the type of reaction it facilitates. The third part of the name is "-ase," the ending that tells you it is an enzyme. For example, *DNA polymerase* is the name of the enzyme that attaches to the molecule DNA and is responsible for increasing its length through a polymerization reaction. A few enzymes (e.g., pepsin and trypsin) are still referred to by

(a)

(b)

Figure 5.3

It Fits, It's Fast, and It Works
(*a*) Although it could be done by hand, an open-end crescent wrench can be used to remove the wheel from this bicycle more efficiently. The wrench is adjusted and attached, temporarily forming a nut-bolt-wrench complex. Turning the wrench loosens the bonds holding the nut to the bolt and the two are separated. The use of the wrench makes the task much easier. Keep in mind that the same wrench that is used to disassemble the bicycle can be used to reassemble it. Enzymes function in the same way. (*b*) An enzyme will "adjust itself" as it attaches to its substrate, forming a temporary enzyme-substrate complex. The presence and position of the enzyme in relation to the substrate lowers the activation energy required to alter the bonds. Depending on the circumstances (what job needs to be done), the enzyme might be involved in synthesis (constructive; i.e., anabolic) or hydrolysis (destructive; i.e., catabolic) reactions.

their original names. The enzyme responsible for the dehydration synthesis reactions among several glucose molecules to form glycogen is known as *glycogen synthetase*. The enzyme responsible for breaking the bond that attaches the amino group to the amino acid arginine is known as *arginine aminase*. When an enzyme is very common, we often shorten its formal name. The salivary enzyme involved in the digestion of starch is *amylose* (starch) *hydrolase;* it is generally known as *amylase*. Other enzymes associated with the human digestive system are noted in table 18.2.

The Role of Enzyme Helpers

Certain enzymes need an additional molecule, a *cofactor,* to enable them to function. Cofactors may be certain elements or complex organic molecules. Cofactors temporarily attach to the enzyme and work with the protein catalyst to speed up a reaction. If the cofactor is not protein but another kind of organic molecule, it is called a *coenzyme*. A **coenzyme** aids a

reaction by removing one of the end products or by bringing in part of the substrate. Many coenzymes cannot be manufactured by organisms and must be obtained from their foods. In addition, coenzymes are frequently constructed from minerals (zinc, magnesium, or iron), vitamins, and nucleotides. You know that a constant small supply of vitamins in your diet is necessary for good health. The reason your cells require vitamins is to serve in the manufacture of certain coenzymes. A coenzyme can work with a variety of enzymes; therefore, you need extremely small quantities of vitamins. An example of enzyme–coenzyme cooperation is shown in figure 5.4. The metabolism of alcohol consists of a series of reactions resulting in its breakdown to carbon dioxide (CO_2), water (H_2O), and energy. During one of the reactions in this sequence, the enzyme alcohol dehydrogenase picks up hydrogen from alcohol and attaches it to NAD. In this reaction, NAD (*n*icotinamide *a*denine *d*inucleotide, manufactured from the vitamin niacin) acts as a coenzyme because NAD carries the hydrogen away from the reaction as the alcohol is broken down. The presence of the coenzyme NAD is necessary for the enzyme to function properly (Outlooks 5.2).

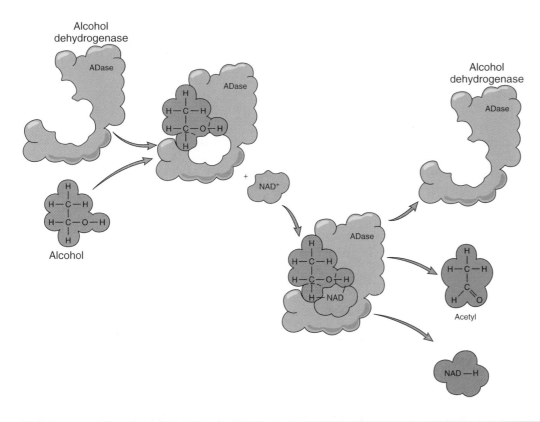

Alcohol dehydrogenase

Alcohol

Alcohol dehydrogenase

Acetyl

NAD—H

Figure 5.4

The Role of Coenzymes
NAD is a coenzyme that works with the enzyme alcohol dehydrogenase (ADase) during the decomposition of alcohol. The coenzyme carries the hydrogen from the alcohol molecule after it is removed by the enzyme. Notice that the hydrogen on the alcohol is picked up by the NAD. The use of the coenzyme NAD makes the enzyme function more efficiently because one of the end products of this reaction (hydrogen) is removed from the reaction site. Because the hydrogen is no longer close to the reacting molecules, the overall direction of the reaction is toward the formation of acetyl. This encourages more alcohol to be broken down.

OUTLOOKS 5.2

Cofactors, Vitamins, and Coenzymes

The term *cofactor* is used to describe inorganic ions or organic molecules. Ions such as zinc, iron, and magnesium assist enzymes in their performance as catalysts. These ions chemically combine with the enzyme. If an organic molecule functions as cofactor it is called a *coenzyme*. Organic cofactors are made from molecules such as certain amino acids, purines, pyrimidines, and vitamins. They are required since the cells are not able to manufacture these molecules. *Vitamins* are a group of unrelated organic molecules—that is, they are quite different from one another. Vitamins are either water-soluble (ex., vitamin B complex) or fat-soluble (ex., vitamin A). Many vitamins are used in the manufacture of final products that act as coenzymes. The vitamin riboflavin (B_2) is metabolized by cells and becomes flavin adenine dinucleotide (FAD). The vitamin niacin is metabolized by cells and becomes nicotinamide adenine dinucleotide (NAD^+). Pantothenic acid becomes coenzyme A (CoA), a molecule used to carry a specific, 2-carbon molecule, acetyl, generated in one reaction to another reaction. Like enzymes, the cell uses inorganic cofactors and coenzymes repeatedly, as shown in box figure 5.1.

Coenzymes such as NAD^+, and FAD are used to carry electrons to and from many kinds of oxidation–reduction reactions. NAD^+, FAD, and other coenzymes are only bonded temporarily to their enzymes and therefore can assist different enzymes in their reaction. Coenzymes play vital roles in metabolism. Without them, most cellular reactions would come to an end and the cell would die. • electron transport, p. 107

Box Figure 5.1

5.3 How the Environment Affects Enzyme Action

An enzyme forms a complex with one substrate molecule, encourages a reaction to occur, detaches itself, and then forms a complex with another molecule of the same substrate. The number of molecules of substrate that a single enzyme molecule can react with in a given time (e.g., reactions per minute) is called the **turnover number.**

Sometimes the number of jobs an enzyme can perform during a particular time period is incredibly large—ranging between a thousand (10^3) and 10 thousand trillion (10^{16}) times faster per minute than uncatalyzed reactions! Without the enzyme, perhaps only 50 or 100 substrate molecules might be altered in the same time. With this in mind, let's identify the ideal conditions for an enzyme and consider how these conditions influence the turnover number.

Temperature

An important environmental condition affecting enzyme-controlled reactions is temperature (figure 5.5), which has two effects on enzymes: (1) It can change the rate of molecular motion, and (2) it can cause changes in the shape of an enzyme. As the temperature of an enzyme-substrate system increases, you would expect an increase in the amount of product molecules formed. This is true up to a point. The temperature at which the rate of formation of enzyme-substrate complex is fastest is termed the *optimum temperature. Optimum* means the best or most productive quantity or condition. In this case, the optimum temperature is the temperature at which the product is formed most rapidly.

As one lowers the temperature below the optimum, molecular motion slows, and the rate at which the enzyme-substrate complexes form decreases. Even though the enzyme is still able to operate, it does so very slowly. That is why foods can be preserved for long periods by storing them in freezers or refrigerators.

When the temperature is raised above the optimum, some of the molecules of enzyme are changed in such a way that they can no longer form the enzyme-substrate complex; thus, the reaction is not encouraged. If the temperature continues to increase, more and more of the enzyme molecules will become inactive. When heat is applied to an enzyme, it causes permanent changes in the three-dimensional shape of the molecule. The surface geometry of the enzyme molecule will not be recovered, even when the temperature is reduced. We can again use the wrench analogy. When a wrench is heated above a certain temperature, the metal begins to change shape. The shape of the wrench is changed permanently so that even if the temperature is reduced, the surface geometry of the end of the wrench is permanently lost. When this happens to an enzyme, we say that it has been *denatured.* A **denatured** enzyme is one whose protein structure has been permanently changed so that it has lost its original

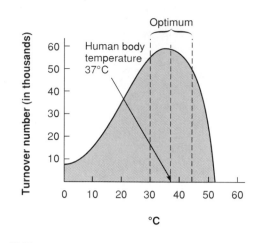

Figure 5.5

The Effect of Temperature on the Turnover Number

As the temperature increases, the rate of an enzymatic reaction increases. The increasing temperature increases molecular motion and may increase the number of times an enzyme contacts and combines with a substrate molecule. Temperature may also influence the shape of the enzyme molecule, making it fit better with the substrate. At high temperatures, the enzyme molecule is irreversibly changed so that it can no longer function as an enzyme. At that point, it has been denatured. Notice that the enzyme represented in this graph has an optimum (best) temperature range of between 30°C and 45°C.

biochemical properties. Because enzymes are molecules and are not alive, they are not "killed," but denatured. Although egg white is not an enzyme, it is a protein and provides a common example of what happens when denaturation occurs as a result of heating. As heat is applied to the egg white, it is permanently changed (denatured).

pH

Another environmental condition that influences enzyme action is pH. The three-dimensional structure of a protein leaves certain side chains exposed. These side chains may attract ions from the environment. Under the right conditions, a group of positively charged hydrogen ions may accumulate on certain parts of an enzyme. In an environment that lacks these hydrogen ions, this would not happen. Thus, variation in the most effective shape of the enzyme could be caused by a change in the number of hydrogen ions present in the solution. Because the environmental pH is so important in determining the shapes of protein molecules, there is an optimum pH for each specific enzyme. The enzyme will fit with the substrate only when it is at the proper pH. Many enzymes function best at a pH close to neutral (7.0). However, a number of enzymes perform best at pHs quite different from 7. Pepsin, an enzyme found in the stomach, works well at an acid pH of 1.5 to 2.2, whereas arginase, an enzyme in the liver, works well at a basic pH of 9.5 to 9.9 (figure 5.6).

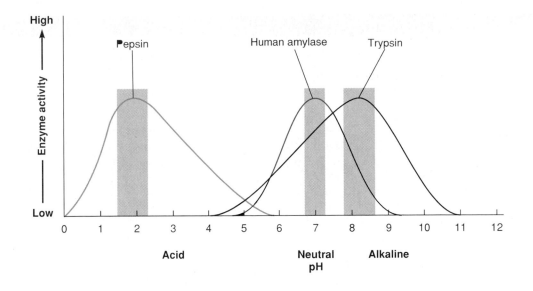

Figure 5.6

The Effect of pH on the Turnover Number

As the pH changes, the rate of the enzymatic reaction changes. The ions in solution alter the environment of the enzyme's active site and the overall shape of the enzyme. The enzymes illustrated here are human amylase, pepsin, and trypsin. Amylase is found in saliva and is responsible for hydrolyzing starch to glucose. Pepsin is found in the stomach and hydrolyzes protein. Trypsin is produced in the pancreas and enters the small intestine where it also hydrolyzes protein. Notice that each enzyme has its own pH range of activity, the optimum (shown in the color bars) being different for each.

Enzyme–Substrate Concentration

In addition to temperature and pH, the concentration of enzymes, substrates, and products influences the rates of enzymatic reactions. Although the enzyme and the substrate are in contact with one another for only a short period of time, when there are huge numbers of substrate molecules it may happen that all the enzymes present are always occupied by substrate molecules. When this occurs, the rate of product formation cannot be increased unless the number of enzymes is increased. Cells can actually do this by synthesizing more enzymes. However, just because there are more enzyme molecules does not mean that any one enzyme molecule will be working any faster. The turnover number for each enzyme stays the same. As the enzyme concentration increases, the amount of product formed increases in a specified time. A greater number of enzymes are turning over substrates; they are not turning over substrates faster. Similarly, if enzyme numbers are decreased, the amount of product formed declines.

We can also look at this from the point of view of the substrate. If substrate is in short supply, enzymes may have to wait for a substrate molecule to become available. Under these conditions, as the amount of substrate increases, the amount of product formed increases. The increase in product is the result of more substrates available to be changed. If there is a very large amount of substrate, even a small amount of enzyme can eventually change all the substrate to product; it will just take longer. Decreasing the amount of substrate results in reduced product formation because some enzymes will go for long periods without coming in contact with a substrate mole-cule. To see how abnormal enzyme activity may result in a metabolic disorder, see How Science Works 5.1.

5.4 Cellular-Controlling Processes and Enzymes

In any cell there are thousands of kinds of enzymes. Each controls specific chemical reactions and is sensitive to changing environmental conditions such as pH and temperature. In order for a cell to stay alive in an ever-changing environment, its innumerable biochemical reactions must be controlled. **Control processes** are mechanisms that ensure that an organism will carry out all metabolic activities in the proper sequence (coordination) and at the proper rate (regulation). *Coordination* of enzymatic activities in a cell results when specific reactions occur in a given sequence; for example, $A \rightarrow B \rightarrow C \rightarrow D \rightarrow E$. This ensures that a particular nutrient will be converted to a particular end product necessary to the survival of the cell. Should a cell not be able to coordinate its reactions, essential products might be produced at the wrong time or never be produced at all, and the cell would die. *Regulation* of biochemical reactions refers to how a cell controls the amount of chemical product produced. The old expression "having too much of a good thing" applies to this situation. For example, if a cell manufactures too much lipid, the presence of those molecules could interfere with other life-sustaining reactions, resulting in the death of the cell. On the other hand, if a cell does not

HOW SCIENCE WORKS 5.1

Metabolic Disorders Related to Enzyme Activity— Fabray's Disease and Cushing's Syndrome

Fabray's disease is a fat-storage disorder caused by a deficiency of an enzyme known as ceramidetrihexosidase, also called alpha-galactosidase A. This enzyme is involved in the breakdown of lipids. The gene for the production of this enzyme is located on the X chromosome. Normally a woman has 2 X chromosomes, and if one of these chromosomes contains this abnormal form of the gene, she is considered to be a "carrier" of this trait. Some carriers show cloudiness of the cornea of their eyes. Normally, males have 1 X and 1 Y chromosome. Therefore, if their mother is a carrier, they will have a 50:50 chance of inheriting this trait from their mother. Males with this abnormality have burning sensations in their hands and feet that become worse when they exercise and in hot weather. Most have small, raised, reddish-purple blemishes on their skin. As they grow older, they are at risk for strokes, heart attacks, and kidney damage. Some affected people develop gastrointestinal problems. They have frequent bowel movements shortly after eating. It is hoped that enzyme replacement and eventually gene therapy will allow patients to control, if not eliminate, the symptoms of Fabray's disease.

Recent evidence suggests that the "pot belly" or "apple belly" of obese men is the result of the overproduction of an enzyme called 11β hydroxysteroid dehydrogenase type 1. High levels of this enzyme have been found in their belly fat. This enzyme regenerates active cortisol from an inactive form, cortisone, thus increasing the effects of this hormone. This has led some to believe that men with apple bellies are displaying a mild, localized form of Cushing's syndrome.

A clinical case of Cushing's syndrome, or hypercortisolism, is a disorder of the endocrine system. The endocrine system is composed of a number of glands that communicate with one another and other tissues through chemical messengers transported throughout the body by the circulatory system. Cushing's syndrome may be caused by either prolonged exposure of the body's tissues to high levels of the hormone cortisol or by the overproduction of cortisol in the adrenal gland. Common traits seen in a person with Cushing's syndrome may include:

Severe obesity
Diabetes
Severe fatigue and muscle weakness
High blood pressure
Backache
Easy bruising
Bluish-red stretch marks on the skin

In addition, women may show increased growth of facial and body hair, and menstrual periods may become irregular or stop.

---

produce enough of an essential molecule, such as a digestive enzyme, it might also die. The cellular control process involves both enzymes and genes.

Keep in mind that any one substrate may be acted upon by several different enzymes. Although all these different enzymes may combine with the same substrate, they do not have the same chemical effect on the substrate because each converts the substrate to different end products. For example, acetyl is a substrate that can be acted upon by three different enzymes: citrate synthetase, fatty acid synthetase, and malate synthetase (figure 5.7). Which enzyme has the greatest success depends on the number of each type of enzyme available and the suitability of the environment for the enzyme's operation. The enzyme that is present in the greatest number or is best suited to the job in the environment of the cell wins, and the amount of its end product becomes greatest.

Gene Regulation

Whenever there are several different enzymes available to combine with a given substrate, **enzymatic competition** results. For example, the use a cell makes of the substrate molecule acetyl is directly controlled by the amount and kinds of enzymes it produces. The number and kind of enzymes produced are regulated by the cell's genes. It is the job of chemi-

cal messengers to inform the genes as to whether specific enzyme-producing genes should be turned on or off or whether they should have their protein-producing activities increased or decreased. Such chemical messengers are called **gene-regulator proteins.** Gene-regulator proteins that decrease protein production are called *gene-repressor proteins*, whereas those that increase protein production are gene-activator proteins. Returning to our example, if the cell is in need of protein, the acetyl could be metabolized to provide one of the building blocks for the construction of protein by turning up the production of the enzyme malate synthetase. If the cell requires energy to move or grow, more acetyl can be metabolized to release this energy by producing more citrate synthetase. When the enzyme fatty acid synthetase outcompetes the other two, the acetyl is used in fat production and storage.

Negative Feedback

Another method of controlling the synthesis of many molecules within a cell is called **negative-feedback inhibition.** This control process occurs within an enzyme-controlled reaction sequence. As the number of end products increases, some product molecules *feed back* to one of the previous reactions and have a *negative effect* on the enzyme controlling that re-

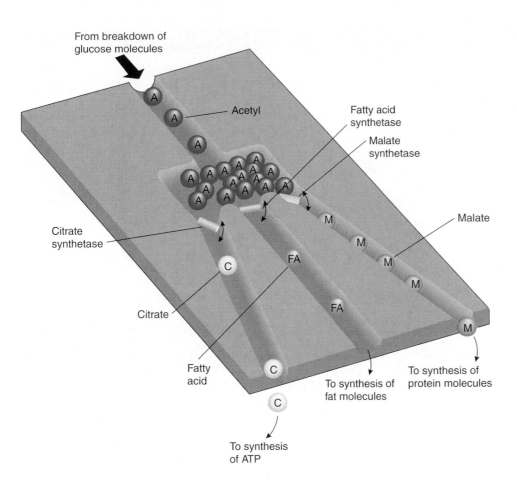

From breakdown of glucose molecules

Acetyl

Fatty acid synthetase

Malate synthetase

Citrate synthetase

Malate

Citrate

Fatty acid

To synthesis of fat molecules

To synthesis of protein molecules

To synthesis of ATP

Figure 5.7

Enzymatic Competition

Acetyl can serve as a substrate for a number of different reactions. Three such reactions are shown here. Whether it becomes a fatty acid, malate, or citrate is determined by the enzymes present. Each of the three enzymes may be thought of as being in competition for the same substrate—the acetyl molecule. The cell can partially control which end product will be produced in the greatest quantity by producing greater numbers of one kind of enzyme and fewer of the other kinds. If citrate synthetase is present in the highest quantity, more of the acetyl substrate will be acted upon by that enzyme and converted to citrate rather than to the other two end products, malate and fatty acids. The illustration represents the action of each enzyme as an "enzyme gate."

action; that is, they *inhibit* or prevent that enzyme from performing at its best.

Because the end product can no longer be produced at the same rapid rate, its concentration falls. When there are too few end product molecules to feed back they no longer cause inhibition. The enzyme resumes its previous optimum rate of operation, and the end product concentration begins to increase. This also helps regulate the number of end products formed but does not involve the genes (figure 5.8).

Inhibition

In addition, the operation of enzymes can be influenced by the presence of other molecules. An **inhibitor** is a molecule that attaches itself to an enzyme and interferes with its ability to form an enzyme-substrate complex (figure 5.9). One of the early kinds of pesticides used to spray fruit trees contained arsenic. The arsenic attached itself to insect enzymes and inhibited the normal growth and reproduction of insects. Organophosphates are pesticides that inhibit several enzymes necessary for the operation of the nervous system. When they are incorporated into nerve cells, they disrupt normal nerve transmission and cause the death of the affected organisms. In humans, death that is due to pesticides is usually caused by uncontrolled muscle contractions, resulting in breathing failure.

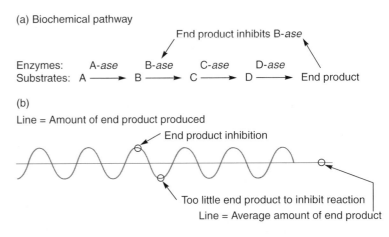

(a) Biochemical pathway

End product inhibits B-*ase*

Enzymes: A-*ase*   B-*ase*   C-*ase*   D-*ase*
Substrates: A ⟶ B ⟶ C ⟶ D ⟶ End product

(b)
Line = Amount of end product produced

End product inhibition

Too little end product to inhibit reaction
Line = Average amount of end product

Figure 5.8

Negative Feedback Inhibition

(*a*) In an enzyme-controlled biochemical pathway, the end product may inhibit one step of its formation when its concentration becomes high enough. (*b*) In negative feedback inhibition, when the amount of end product reaches a high enough concentration, the molecules significantly interfere with enzyme action. When product concentration falls below the inhibiting level, the enzyme is able to again manufacture product molecules, and the product molecule concentration rises. Notice how the amount of end product increases above the average and then decreases below the average amount of end product.

Figure 5.9

Enzymatic Inhibition

The left-hand side of the illustration shows the normal functioning of the enzyme. On the right-hand side, the enzyme is unable to function. This is because an inhibitor, malonic acid, is attached to the enzyme and prevents the enzyme from forming the normal complex with succinic acid. As long as malonic acid is present, the enzyme will be unable to function. If the malonic acid is removed, the enzyme will begin to function normally again. Its attachment to the inhibitor in this case is not permanent but has the effect of reducing the number of product molecules formed per unit of time.

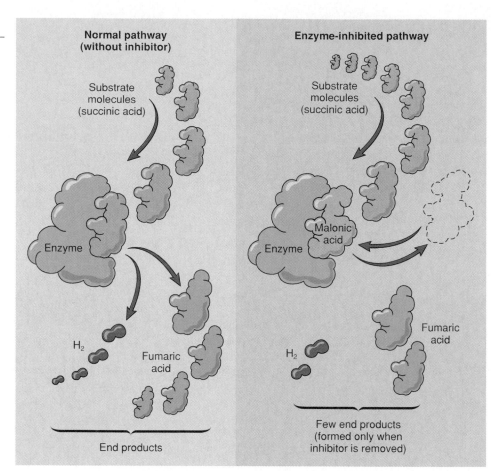

Some inhibitors have a shape that closely resembles the normal substrate of the enzyme. The enzyme is unable to distinguish the inhibitor from the normal substrate and so it combines with either or both. As long as the inhibitor is combined with an enzyme, the enzyme is ineffective in its normal role. Some of these enzyme-inhibitor complexes are permanent. An inhibitor removes a specific enzyme as a functioning part of the cell: the reaction that enzyme catalyzes no longer occurs, and none of the product is formed. This is termed **competitive inhibition** because the inhibitor molecule competes with the normal substrate for the active site of the enzyme (figure 5.9).

We use enzyme inhibition to control disease. The sulfa drugs are used to control a variety of bacteria, such as the bacterium *Streptococcus pyogenes*, the cause of strep throat and scarlet fever. The drug resembles one of the bacterium's necessary substrates and so prevents some of the cell's enzymes from producing an essential cell component. As a result, the bacterial cell dies because its normal metabolism is not maintained. Those that survive become the grandparents of a new population of drug-resistant bacteria. Antibiotics act as agents of natural selection favoring those cells that have the genetic ability to withstand the effects of the drug. Since one essential life characteristic is evolution, the prevention of drug resistance is impossible. The development of resistance can only be slowed, not stopped. Microbes may become resistant to antibiotics in four ways: (1) they can stop producing the molecule that is the target of the drug; (2) they can modify the target; (3) they can become impermeable to the drug; or (4) they can release enzymes that inactivate the antibiotic. Some antibiotics found to be ineffective against drug-resistant bacteria include several of the penicillins, erythromycin, PAS (para-aminosalicylic acid), INH (isoniazid), streptomycin, gentamicin, tetracycline, kanamycin, chloramphenicol, and clindamycin.

SUMMARY

Enzymes are protein catalysts that speed up the rate of chemical reactions without any significant increase in the temperature. They do this by lowering activation energy. Enzymes have a very specific structure that matches the structure of particular substrate molecules. Actually, the substrate molecule comes in contact with only a specific part of the enzyme molecule—the attachment site. The active site of the enzyme is the place where the substrate molecule is changed. The enzyme-substrate complex reacts to form the end product. The protein nature of enzymes makes them sensitive to environmental conditions, such as temperature and pH, that change the structure of proteins. The number and kinds of enzymes are ultimately controlled by the genetic information of the cell. Other kinds of molecules, such as coenzymes, inhibitors, or competing enzymes, can influence specific enzymes. Changing conditions within the cell shift the enzymatic priorities of the cell by influencing the turnover number.

THINKING CRITICALLY

For guidelines to answer this question, visit the Online Learning Center.

The following data were obtained by a number of Nobel-prize-winning scientists from Lower Slobovia. As a member of the group, interpret the data with respect to the following:

1. Enzyme activities
2. Movement of substrates into and out of the cell
3. Competition among different enzymes for the same substrate
4. Cell structure

Data:

a. A lowering of the atmospheric temperature from 22°C to 18°C causes organisms to form a thick, protective coat.
b. Below 18°C, no additional coat material is produced.
c. If the cell is heated to 35°C and then cooled to 18°C, no coat is produced.
d. The coat consists of a complex carbohydrate.
e. The coat will form even if there is a low concentration of simple sugars in the surroundings.
f. If the cell needs energy for growth, no cell coats are produced at any temperature.

CONCEPT MAP TERMINOLOGY

Construct a concept map using these terms.

coenzyme	substrate
enzyme	temperature
enzyme-substrate complex	turnover number
inhibitor	

Visit the Online Learning Center to use an interactive format to construct another concept map.

KEY TERMS

Use interactive flash cards, on the Online Learning Center, to help you learn the meanings of these terms.

activation energy	competitive inhibition
active site	control processes
attachment site	denature
binding site	enzymatic competition
catalyst	enzyme
coenzyme	enzyme-substrate complex

gene-regulator proteins	nutrients
inhibitor	substrate
negative-feedback inhibition	turnover number

LEARNING CONNECTIONS

Visit the Online Learning Center to find additional study resources including those highlighted below.

5.1 How Cells Use Enzymes

Animations and Review
• Thermodynamics
• Enzymes

Experience This!
• Enzymes for your laundry?

1. What is the difference between a catalyst and an enzyme?
2. Describe the sequence of events in an enzyme-controlled reaction.
3. Would you expect a fat and a sugar molecule to be acted on by the same enzyme? Why or why not?
4. Where in a cell would you look for enzymes?

5.2 How Enzymes Speed Chemical Reaction Rates

5. What is turnover number? Why is it important?
6. What is meant by the term *binding site? Active site?*

5.3 How the Environment Affects Enzyme Action

Human Explorations
• Cell chemistry: Thermodynamics

Interactive Concept Maps
• Inhibitors

7. How does changing temperature affect the rate of an enzyme-controlled reaction?
8. What factors in the cell can speed up or slow down enzyme reactions?
9. What is the relationship between vitamins and coenzymes?
10. What effect might a change in pH have on enzyme activity?

5.4 Cellular-Controlling Processes and Enzymes

Interactive Concept Maps
• Text concept map
Review Questions
• Enzymes

11. What is enzyme competition, and why is it important to all cells?
12. Describe the nature and action of an enzyme inhibitor.

Test yourself with multiple choice questions that provide immediate feedback on the Online Learning Center with PowerWeb at

www.mhhe.com/enger11

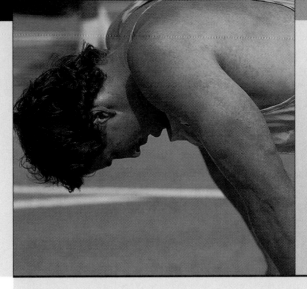

CHAPTER OUTLINE

Biochemical Pathways

GOALS AND OBJECTIVES

Recognize that organisms have several ways to obtain energy.

❒ Describe the differences between autotrophs and heterotrophs.

❒ List the sources of energy used by chemosynthetic and photosynthetic organisms.

Recognize that there are metabolic processes basic to all organisms.

❒ Explain the importance of ATP.

❒ Describe how the proton pump mechanism generates ATP.

Understand the process of aerobic cellular respiration.

❒ Explain the role of oxygen in aerobic respiration.

❒ Describe the reactants and products of glycolysis, the Krebs cycle, and the electron-transport system.

❒ Explain the role of NAD^+ and FAD in cellular respiration.

Understand the process of anaerobic cellular respiration.

❒ Compare the biochemical pathways utilized and the energy yield of aerobic and anaerobic cellular respiration.

❒ Describe two variations of anaerobic respiration.

Recognize the role of glycolysis and the Krebs cycle in fat and protein respiration.

❒ List subunits from fats and proteins that are metabolized by aerobic respiration.

❒ Describe how energy is derived from fats and proteins.

Understand the process of photosynthesis.

❒ Describe the reactants and products of the light-dependent and light-independent reactions of photosynthesis.

❒ Explain the role of pigments in photosynthesis.

❒ Explain the role of glyceraldehyde-3-phosphate in plant metabolism.

6.1 Energy and Organisms

All living organisms require a constant supply of energy to sustain life. They obtain this energy through enzyme-controlled chemical reactions that release the internal potential energy stored in the chemical bonds of molecules (figure 6.1). Burning wood is an example of a chemical reaction that results in the release of energy by breaking chemical bonds. The chemical bonds of large, complex, organic molecules of wood are broken and smaller end products of carbon dioxide (CO_2) and water (H_2O) are produced. There is less potential energy in the chemical bonds of carbon dioxide and water than in the complex organic molecules, and the excess energy is released as light and heat. Living organisms are capable of carrying out a series of oxidation-reduction reactions similar to that seen in burning wood, but in a controlled manner so that the energy is released in the right amount at the right time. By controlling energy-releasing reactions, organisms are able to use the energy to power activities such as reproduction, movement, and growth. • chemical reactions, p. 33

Although there are hundreds of different chemical reactions that take place within organisms, this chapter will focus on the reactions involved in the processes of photosynthesis and respiration. In **photosynthesis,** organisms such as green plants, algae, and certain bacteria trap radiant energy from sunlight and convert it into the energy of chemical bonds in large molecules such as carbohydrates. In **cellular respiration,** organisms control the release of chemical-bond energy from large, organic molecules and use the energy for the many activities necessary to sustain life. All organisms must carry out cellular respiration if they are to survive. Since nearly all organisms use organic molecules as a source of energy, they must either make the organic molecules in a process like photosynthesis or they must obtain energy-containing, organic molecules from their surroundings. Thus, photosynthetic organisms produce food molecules, such as carbohydrates, for themselves as well as for all the other organisms that feed upon them. There are many variations of cellular respiration. Some organisms require the presence of oxygen and carry out a form of respiration called **aerobic cellular respiration.** Other organisms carry out a form of respiration that does not require the presence of oxygen; this form is called **anaerobic cellular respiration.**

Organisms that are able to use basic energy sources such as sunlight to make energy-containing organic molecules from inorganic raw materials are called **autotrophs** (*auto* = self; *troph* = feeding). There are also prokaryotic organisms that use inorganic chemical reactions as a source of energy to make larger, organic molecules. This process is known as **chemosynthesis.** Therefore, there are at least two kinds of autotrophs: Those that use light are called *photosynthetic* autotrophs and those that use inorganic chemical reactions are called *chemosynthetic* autotrophs. All other organisms require organic molecules as food and are called **heterotrophs** (*hetero* = other, *troph* = feeding). Heterotrophs get their energy from the chemical bonds of food molecules such as carbohydrates, fats, and proteins, which they must obtain from their sur-

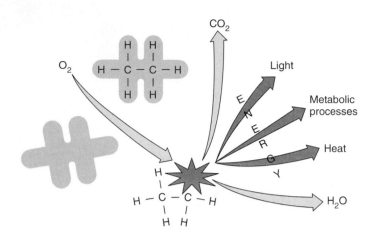

Figure 6.1

Life's Energy: Chemical Bonds
All living things utilize the energy contained in chemical bonds. As organisms break down molecules, the energy released can be used for metabolic processes such as movement, growth, and reproduction. Some organisms, such as fireflies and certain bacteria, are able to use chemical bond energy to release light (bioluminesce). In all cases, there is a certain amount of heat released when chemical bonds are broken.

roundings. Within eukaryotic cells, certain biochemical processes are carried out in specific organelles. Chloroplasts are the sites of photosynthesis, and mitochondria are the sites of most of the reactions of cellular respiration (figure 6.2). Because prokaryotic cells lack mitochondria and chloroplasts, they carry out photosynthesis and cellular respiration within the cytoplasm or on the inner surfaces of the cell membrane or on other special membranes. Table 6.1 provides a summary of the concepts just discussed and how they are related to one another. • organelles, p. 75

6.2 Common Mechanisms Used in Processing Energy and Matter

Organisms use several general mechanisms in the process of obtaining energy. Once you understand the underlying mechanisms you can apply them to specific biochemical processes.

Biochemical Pathways

When wood is burned, all the energy of the chemical bonds in the wood is released and a large amount of heat and light is released. In living things, the same amount of energy is released but it is released in a series of small steps. Each of the steps is controlled by an enzyme. Each step begins with a substrate that is converted to a product, which in turn becomes the substrate for a different enzyme. Such a series of enzyme-controlled reactions is often called a **biochemical pathway** or a **metabolic pathway.** The processes of photosynthesis, respiration, protein synthesis, and many other cellular activities

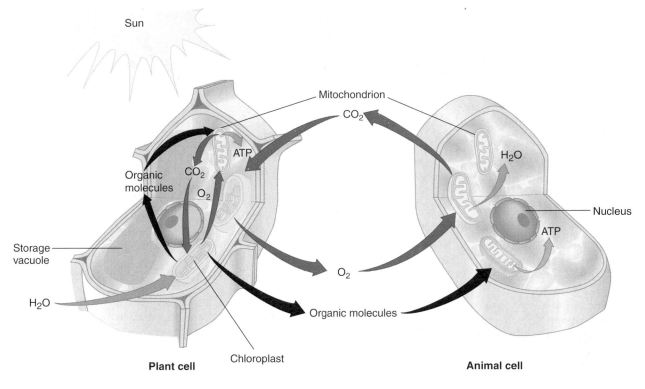

Figure 6.2

Biochemical Pathways that Involve Energy Transformation
Photosynthesis and cellular respiration both involve a series of chemical reactions that control the flow of energy. Organisms that contain photosynthetic machinery are capable of using light, water, and carbon dioxide to produce organic molecules such as sugars, proteins, lipids, and nucleic acids. Oxygen is also released as a result of photosynthesis. In aerobic cellular respiration, organic molecules and oxygen are used to provide the energy to sustain life. Carbon dioxide and water are also released during aerobic respiration.

Table 6.1

SUMMARY OF BIOCHEMICAL PATHWAYS, ENERGY SOURCES, AND KINDS OF ORGANISMS

Biochemical Pathways	Autotroph or Heterotroph	Energy Source	Kinds of Organisms	Notes
Photosynthesis	Autotroph	Light	Prokaryotic—certain bacteria	Prokaryotic photosynthesis is somewhat different from eukaryotic photosynthesis and does not take place in chloroplasts.
			Eukaryotic—plants and algae	Eukaryotic photosynthesis takes place in chloroplasts.
Chemosynthesis	Autotroph	Inorganic chemical reactions	Prokaryotic—certain bacteria and archaea	There are many different types of chemosynthesis.
Cellular respiration	Autotroph and heterotroph	Oxidation of large organic molecules	Prokaryotic—bacteria and archaea	There are many different forms of respiration. Some organisms use aerobic cellular respiration; others use anaerobic cellular respiration.
			Eukaryotic—plants, animals, fungi, algae, protozoa	Most respiration in eukaryotic organisms takes place in mitochondria and is aerobic.

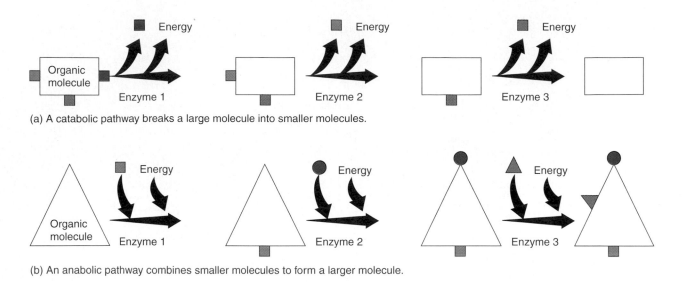

(a) A catabolic pathway breaks a large molecule into smaller molecules.

(b) An anabolic pathway combines smaller molecules to form a larger molecule.

Figure 6.3

Biochemical Pathways

Biochemical pathways are the result of a series of enzyme-controlled reactions. In each step, a substrate is acted upon by an enzyme to produce a product. The product then becomes the substrate for the next enzyme in the chain of reactions. Such pathways can be used to break down molecules, build up molecules, release energy, and perform many other actions.

consist of a series of enzyme-controlled reactions or biochemical pathways. Metabolic pathways that result in the breakdown of compounds are generally referred to as **catabolism.** Metabolic pathways that result in the synthesis of new, larger compounds are known as **anabolism.** • enzymes, p. 92

One of the amazing facts of nature is that most organisms use the same basic biochemical pathways. So, if you study aerobic cellular respiration in an elephant it will be essentially the same as that in a shark, petunia, or earthworm. However, since the kinds of enzymes an organism is able to produce depends on the genes that it has, we should expect some variation in the details of biochemical pathways in different organisms. The fact that so many organisms use essentially the same biochemical processes is a strong argument for the idea of evolution from a common ancestor. Once a successful biochemical strategy evolved, the genes and the pathway were retained by evolutionary descendents with slight modifications of the scheme. Figure 6.3 illustrates the nature of biochemical pathways.

Generating Energy in a Useful Form: ATP

The transfer of chemical energy within living things is handled by a molecule known as *a*denosine *tri*phosphate (ATP). Chemical energy is stored when ATP is made and released when it is broken apart. An ATP molecule is formed by combining adenine (a nitrogenous base), ribose (a sugar), and 3 phosphates (figure 6.4). If only 1 phosphate is present the molecule is known as *a*denosine *m*ono*p*hosphate (AMP). When a second phosphate group is added to the AMP, a molecule of *a*denosine *di*phosphate (ADP) is formed. The ADP, with the addition of more energy, is able to bond to a

Figure 6.4

Adenosine Triphosphate (ATP)

The ATP molecule is an energy carrier. A molecule of ATP consists of several subunits: a molecule of adenine, a molecule of ribose, and 3 phosphate groups. The 2 end phosphate groups are bonded together by high-energy bonds. These bonds are broken easily, so they release a great amount of energy. Because they are high-energy bonds, they are represented by curved, solid lines.

Figure 6.5

ATP Is the Power Supply for Cells
When rechargeable batteries in a flashlight have been drained of their power, they can be recharged by placing them in a specially designed battery charger. This enables the right amount of power from a power plant to be packed into the batteries for reuse. Cells operate in much the same manner. When the cell's "batteries," ATP, are drained while powering a job like muscle contraction, the discharged "batteries," ADP, can be recharged back to full ATP power.

third phosphate group and form ATP. (The addition of phosphate to a molecule is called a *phosphorylation chemical reaction.*) The bonds holding the last 2 phosphates to the molecule are easily broken to release energy for energy-requiring cellular processes. Because the bond between these phosphates is so easy for a cell to use, it is called a **high-energy phosphate bond.** These bonds are often shown as solid, curved lines in diagrams. Both ADP and ATP, because they contain high-energy bonds, are very unstable molecules and readily lose their phosphates. When this occurs, the energy held in the high-energy bonds of the phosphate can be transferred to a lower-energy molecule or released to the environment. (AMP does not have high-energy bonds and is not involved in energy transfer reactions, although it is important in other kinds of reactions.) Within a cell, enzymes speed this release of energy as ATP is broken down to ADP and P (phosphate). When the bond holding the third phosphate of an ATP molecule is broken, energy is released for use in other activities. • **organic chemistry, p. 42**

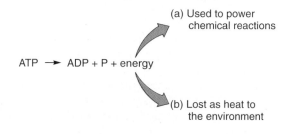

When energy is being harvested from a chemical reaction or another energy source like sunlight, it is stored when a phosphate is attached to an ADP to form ATP.

An analogy that might be helpful is to think of each ATP molecule used in the cell as a rechargeable battery—like those used to power small toys and electronic equipment. Each battery contains just the right amount of energy to power the job. When the power has been drained, it can be recharged numerous times before it must be recycled (figure 6.5).

Electron Transport

Another important concept that can be applied to many different biochemical pathways is the mechanism of *electron transport.* Since the electrons of an atom are on the exterior of the atom, those in the outer energy level may be lost more easily to the surroundings, particularly if they receive additional energy. When electrons receive additional energy, they

Step 1. Electron transport
Electrons give up their energy as they move through a series of electron-transport reactions.

Step 2. Proton gradient established
Energy from the electron-transport reactions is used to pump protons (H+) across a membrane.

Step 3. Proton gradient used to synthesize ATP
When protons move back through special protein in the membrane, enzymes capture their energy and use it to synthesize ATP from ADP and P.

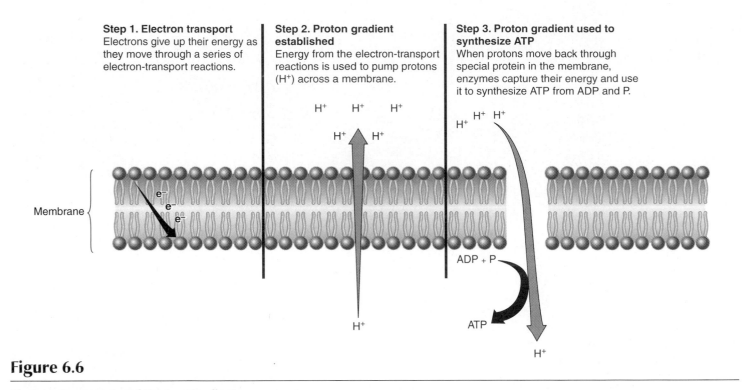

Figure 6.6

Electron Transport and Proton Gradient
The transport of high-energy electrons through a series of electron carriers can allow the energy to be released in discrete, manageable packets. In some cases, the energy given up is used to pump protons (H+) from one side of a membrane to the other and a proton concentration gradient is established. When the protons flow back through the membrane, enzymes in the membrane can capture energy and form ATP.

move to a higher energy level. When they fall back to their original position, they give up that energy. This activity takes place whenever electrons gain or lose energy. In living things, such energy changes are harnessed by special molecules that capture such "excited" electrons and transfer them to other chemical reactions. These electron transfer reactions are commonly called *oxidation-reduction reactions*. Oxidation-reduction (redox) reactions are electron transfer reactions in which the molecules losing electrons become oxidized and those gaining electrons become reduced. This process is not difficult to understand if you think about it in simple terms. The molecule that loses the electron loses energy and the molecule that gains the electron gains energy. • redox, p. 35

There are many different electron acceptors or carriers in cells. However, there are three that are particularly important when discussing photosynthesis and cellular respiration: *ni*coti*n*amide *a*denine *d*inucleotide (NAD+), *n*icotinamide *a*denine *d*inucleotide *p*hosphate (NADP+), and *f*lavin *a*denine *d*inu*c*leotide (FAD). Because they typically work in conjunction with enzymes, they are often called *coenzymes*. The vitamin niacin is needed to make NAD+ and NADP+ and the vitamin riboflavin is needed to make FAD. Since NAD+, NADP+, FAD, and similar molecules accept and release electrons, they often are involved in oxidation-reduction reactions. When NAD+, NADP+, and FAD accept electrons, they become negatively charged. Because of this they readily pick up hydrogen ions (protons), so when they become reduced they are shown as NADH, NADPH, and FADH2. Therefore, it is also possible to think of these molecules as hydrogen carriers. In many bio-

chemical pathways, there is a series of oxidation-reduction reactions (electron-transport reactions) in which each step results in the transfer of a small amount of energy from a higher-energy molecule to a lower-energy molecule (figure 6.6). Thus, electron transport is often tied to the formation of ATP.

Proton Pump

In many of the oxidation-reduction chemical reactions that take place in cells, the electrons that are transferred come from hydrogen atoms. A hydrogen nucleus (proton) is formed whenever electrons are stripped from hydrogen atoms. When these higher-energy electrons are transferred to lower-energy states, it is often the case that protons are pumped across membranes, creating a region with a high concentration of protons on one side of the membrane. Therefore, this process is referred to as a *proton pump*. The "pressure" created by this high concentration of protons is released when protons flow through special pores in the membrane back to the side from which they were pumped. As they pass through the pores, an enzyme, ATPase (a phosphorylase), speeds the formation of an ATP molecule by bonding a phosphate to an ADP molecule. Thus, the establishment of a proton gradient is an important step for the production of much of the ATP produced in cells (see figure 6.6).

These four general concepts—biochemical pathways, ATP production, electron transport, and the proton pump—are all interrelated. In the sections that follow, we will use these concepts to describe particular aspects of photosynthesis and respiration.

6.3 Aerobic Cellular Respiration— An Overview

Aerobic cellular respiration is a specific series of enzyme-controlled chemical reactions in which oxygen is involved in the breakdown of glucose into carbon dioxide and water and the chemical-bond energy from glucose is released to the cell in the form of ATP. While the actual process of aerobic cellular respiration involves many enzyme-controlled steps, the net result is that a reaction between sugar and oxygen results in the formation of carbon dioxide and water with the release of energy. The following equation summarizes this process:

$$\text{glucose} + \text{oxygen} \rightarrow \overset{\text{carbon}}{\text{dioxide}} + \text{water} + \text{energy}$$

$$C_6H_{12}O_6 + 6\ O_2 \rightarrow 6\ CO_2 + 6\ H_2O + \text{energy (ATP + heat)}$$

Covalent bonds are formed by atoms sharing pairs of fast-moving, energetic electrons. Therefore, the covalent bonds in the sugar glucose contain chemical potential energy. Of all the covalent bonds in glucose (O—H, C—H, C—C), those easiest to get at are the C—H and O—H bonds on the outside of the molecule. When these bonds are broken, two things happen:

1. The energy of the electrons can be used to phosphorylate ADP molecules to produce higher-energy ATP molecules and
2. Hydrogen ions (protons) are released.

The ATP is used to power the metabolic activities of the cell. The chemical activities that remove electrons from glucose result in the glucose being *oxidized*.

These high-energy electrons must be controlled. If they were allowed to fly about at random, they would quickly combine with other molecules, causing cell death. Electron transfer molecules like NAD$^+$ and FAD temporarily hold the electrons and transfer them to other electron transfer molecules. ATP is formed when these transfers take place. Once energy has been removed from electrons for ATP production, the electrons must be placed in a safe location. In *aerobic* cellular respiration, these electrons are ultimately attached to oxygen. Oxygen serves as the final resting place of the less-energetic electrons. When the electrons are added to oxygen, it becomes a negatively charged ion, O$^=$. Since the oxygen has gained electrons, it has been *reduced*. So, *in the aerobic cellular respiration of glucose, glucose is oxidized and oxygen is reduced*. If something is oxidized (loses electrons), something else must be reduced (gains electrons). A molecule cannot simply lose its electrons—they have to go someplace! Eventually the positively charged hydrogen ions that were released

from the glucose molecule combine with the negatively charged oxygen ion to form water.

Once all the hydrogens have been stripped off the glucose molecule, the remaining carbon and oxygen atoms are rearranged to form individual molecules of CO_2. The oxidation-reduction reaction is complete. All the hydrogen originally a part of the glucose has been moved to the oxygen to form water. All the remaining carbon and oxygen atoms of the original glucose are now in the form of CO_2. The energy released from this process is used to generate ATP (figure 6.7).

In cells, these oxidation reactions take place in a particular order and in particular places within the cell. In eukaryotic cells, the process of releasing energy from food molecules begins in the cytoplasm and is completed in the

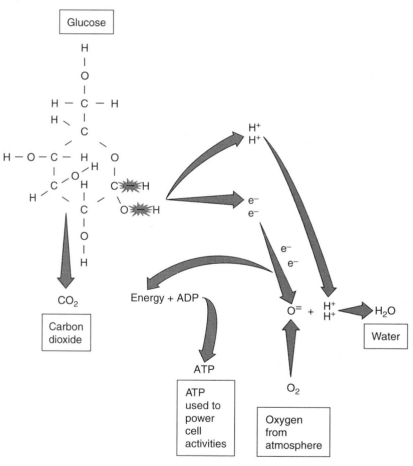

Figure 6.7

Aerobic Cellular Respiration and Oxidation-Reduction Reaction
During aerobic cellular respiration, a series of oxidation–reduction reactions takes place. A sugar molecule loses electrons. This is the "oxidation" portion of the reaction. When the electrons are removed, the sugar is unable to stay together and breaks into smaller units. The reduction part of the reaction occurs when the electrons that were removed are picked up and attached to another molecule. When they are acquired, these electrons can become involved in the formation of new chemical bonds. In the case of aerobic cellular respiration, the electrons are eventually picked up by oxygen and the negatively charged oxygen attracts 2 positively charged hydrogen ions (H$^+$) to form water.

Glucose   +   specific sequence of reactions controlled by enzymes   + O_2 $\longrightarrow$ H_2O + CO_2 + ATP

Figure 6.8

Aerobic Cellular Respiration: Overview

This sequence of reactions in the aerobic oxidation of glucose is an overview of the energy–yielding reactions of a cell. Glycolysis, the Krebs cycle, and the electron-transport system are each a series of enzyme-controlled reactions that extract energy from the chemical bonds in a glucose molecule. During glycolysis, glucose is split into pyruvic acid and ATP and electrons are released. During the Krebs cycle, pyruvic acid is further broken down to carbon dioxide with the release of ATP and the release of electrons. During the electron-transport system, oxygen is used to accept electrons and water is produced. A very large amount of ATP is also produced. Glycolysis takes place in the cytoplasm of the cell. Pyruvic acid enters mitochondria where the Krebs cycle and electron-transport system take place.

mitochondrion. There are three distinct enzymatic pathways involved (figure 6.8):

1. Glycolysis (glyco = sugar; lysis = to split) is a series of enzyme-controlled reactions that takes place in the cytoplasm of cells, which results in the breakdown of glucose with the release of electrons and the formation of ATP. During glycolysis, the 6-carbon sugar, glucose, is split into two smaller, 3-carbon molecules, which undergo further modification to form pyruvic acid or pyruvate.* Enough energy is released to produce 2 ATP molecules. Some of the bonds holding hydrogen atoms to the glucose molecule are broken and the electrons are picked up by electron carrier molecules (NAD+) and transferred to a series of electron transfer reactions known as the electron-transport system (ETS).

2. The **Krebs cycle** is a series of enzyme-controlled reactions that takes place inside the mitochondrion, which completes the breakdown of pyruvic acid with the release of carbon dioxide, electrons, and ATP. During the Krebs cycle, the pyruvic acid molecules produced from glycolysis are further broken down. During these reactions, the remaining hydrogens are removed from the pyruvic acid and their electrons are picked up by the electron carriers NAD+ and FAD. These electrons are sent to the electron-transport system. A small amount of ATP is also formed during the Krebs cycle. The carbon and oxygen atoms that are the remains of the pyruvic acid molecules are released as carbon dioxide (CO_2).

3. The **electron-transport system (ETS)** is a series of enzyme-controlled reactions that converts the kinetic energy of hydrogen electrons to ATP. The electrons are carried to the electron-transport system from glycolysis and the Krebs cycle by NADH and $FADH_2$. The electrons are transferred through a series of oxidation-reduction reactions involving enzymes until eventually

---

* Several different ways of naming organic compounds have been used over the years. For our purposes, pyruvic acid and pyruvate are really the same basic molecule although, technically, pyruvate is what is left when pyruvic acid has lost its hydrogen ion: pyruvic acid → H+ + pyruvate. You also will see terms such as *lactic acid* and *lactate* and *citric acid* and *citrate* and many others used in a similar way.

the electrons are accepted by oxygen atoms to form oxygen ions ($O^=$). During this process, a great deal of ATP is produced. The ATPs are formed as a result of a proton gradient established when the energy of electrons is used to pump protons across a membrane. The subsequent movement of protons back across the membrane results in ATP formation. The negatively charged oxygen atoms attract two positively charged hydrogen ions to form water (H_2O).

We can summarize aerobic respiration as follows: *Glucose* enters glycolysis and is broken down to pyruvic acid, which enters the Krebs cycle where the pyruvic acid molecules are further dismantled. The remains of the pyruvic acid molecules are released as *carbon dioxide*. The electrons and hydrogen ions released from glycolysis and the Krebs cycle are transferred by NADH and $FADH_2$ to the electron-transport system, where the electrons are transferred to *oxygen* available from the atmosphere. When hydrogen ions attach to oxygen ions, *water* is formed. *ATP* is formed during all three stages of aerobic cellular respiration, but the majority comes from the electron transfer system.

6.4 Aerobic Cellular Respiration

The section on aerobic cellular respiration and the section on photosynthesis are divided into two levels: Fundamental Description and Detailed Description. It is a good idea to begin with the simplest description and add layers of understanding as you go to additional levels. Ask your instructor which level is required for your course of study.

Fundamental Description

1. Glycolysis

Glycolysis is a series of enzyme-controlled reactions that takes place in the cytoplasm. During glycolysis, a 6-carbon sugar molecule (glucose) has energy added to it from 2 ATP molecules. Adding this energy makes some of the bonds of the glucose molecule unstable and the glucose molecule is more easily broken down. After passing through several more enzyme-controlled reactions, the 6-carbon glucose is broken down to two 3-carbon molecules known as glyceraldehyde-3-phosphate (also known as phosphoglyceraldehyde*), which undergo additional reactions to form pyruvic acid ($CH_3COCOOH$).

Enough energy is released by this series of reactions to produce 4 ATP molecules. Because 2 ATP molecules were used to start the reaction and 4 were produced, there is a *net gain* of 2 ATPs from the glycolytic pathway (figure 6.9). During the process of glycolysis, some hydrogens and their electrons are removed from the organic molecules being

processed and picked up by the electron transfer molecule NAD^+ to form NADH. Because hydrogen acceptors typically work with specific enzymes, they are often called *coenzymes*. Enough hydrogens are released during glycolysis to form 2 NADHs. The NADH with its extra electrons contains a large amount of potential energy that can be used to make ATP in the electron-transport system. The job of the coenzyme NAD^+ is to safely transport these energy-containing electrons and protons to the electron-transport system. Once they have dropped off their electrons, the oxidized NAD^+s are available to pick up more electrons and repeat the job.

The following is a generalized reaction that summarizes the events of glycolysis:

glucose + 2 ATP + 2 NAD^+ → 4 ATP + 2 NADH + 2 pyruvic acid

2. Krebs Cycle

The series of reactions known as the Krebs cycle takes place within the mitochondria of cells. It gets its name from its discoverer, Hans Krebs, and the fact that the series of reactions begin and end with the same molecule. The Krebs cycle is also known as the citric acid cycle and the tricarboxylic acid cycle (TCA).

$C_6H_{12}O_6$ (glucose) + 2 NAD^+ + 2 ADP + 2 P →
→ 2 $CH_3COCOOH$ (pyruvic acid) + 2 NADH + energy (2 ATP + heat)

Figure 6.9

Glycolysis: Fundamental Description
Glycolysis is the biochemical pathway many organisms use to oxidize glucose. During this sequence of chemical reactions, the 6-carbon molecule of glucose is oxidized. As a result pyruvic acid is produced, electrons are picked up by NAD^+, and ATP is produced.

---

*As with many things in science, the system for naming organic chemical compounds has changed. In the past the term *phosphoglyceraldehyde* commonly was used for this compound and was used in previous editions of this text. However, today the most commonly used term is *glyceraldehyde-3-phosphate*. In order to more accurately reflect current usage, the term glyceraldehyde-3-phosphate is used in this edition.

The 3-carbon pyruvic acid molecules released from glycolysis enter the mitochondria, are acted upon by specific enzymes, and are converted to 2-carbon **acetyl** molecules. At the time the acetyl is produced, 2 hydrogens are attached to NAD^+ to form NADH. The carbon atom that was removed is released as carbon dioxide. The acetyl molecule proceeds through the Krebs cycle. During the Krebs cycle (figure 6.10), the acetyl is completely oxidized. The remaining hydrogens and their electrons are removed. Most of the electrons are picked up by NAD^+ to form NADH, but at one point in the process FAD picks up electrons to form $FADH_2$. Regardless of which electron carrier is being used, the electrons are sent to the electron-transport system. The remaining carbon and oxygen atoms are combined to form CO_2. As in glycolysis, enough energy is released to generate 2 ATP molecules. At the end of the Krebs cycle, the acetyl has been completely broken down (oxidized) to CO_2. The energy in the molecule has been transferred to ATP, NADH, or $FADH_2$. Also, some of the energy has been released as heat. For each of the acetyl molecules that enters the Krebs cycle, 1 ATP, 3 NADHs, and 1 $FADH_2$ are formed. If we count the NADH produced when acetyl is formed, there is a total of 4 NADHs produced for each pyruvic acid that enters a mitochondrion. The following is a generalized equation that summarizes those reactions:

$$\text{pyruvic acid} + ADP + 4\ NAD^+ + FAD \longrightarrow 3\ CO_2 + 4\ NADH + FADH_2 + ATP$$

3. Electron-Transport System

During the three steps of aerobic cellular respiration, cells generate the greatest amount of ATP from the electron-transport system (figure 6.11). During this stepwise sequence of oxidation-reduction reactions, the energy from the NADH and $FADH_2$ molecules generated in glycolysis and the Krebs cycle is used to produce ATP. Iron-containing *cytochrome* (*cyto* = cell; *chrom* = color) enzyme molecules are located on the membranes of the mitochondrion. The energy-rich electrons are passed (*transported*) from one cytochrome to another, and the energy is used to pump protons (hydrogen ions) from one side of the membrane to the other. The result of this is a higher concentration of hydrogen ions on one side of the membrane. As the concentration of hydrogen ions increases on one side, a proton gradient builds up. Because of this concentration gradient, when a membrane channel is opened, the protons flow back to the side from which they were pumped. As they pass through the channels, a phosphorylase enzyme (ATPase) speeds the formation of an ATP molecule by bonding a phosphate to an ADP molecule (phosphorylation). When all of the electrons and hydrogen ions are accounted for, a total of 32 ATPs are formed from the electrons and hydrogens removed from the original glucose molecule.

The following is a generalized reaction that summarizes the events of the electron-transport system:

$$6\ O_2 + 8\ NADH + 4\ FADH_2 + 32\ ADP \longrightarrow 8\ NAD^+ + 4\ FAD + 32\ ATP + 12\ H_2O$$

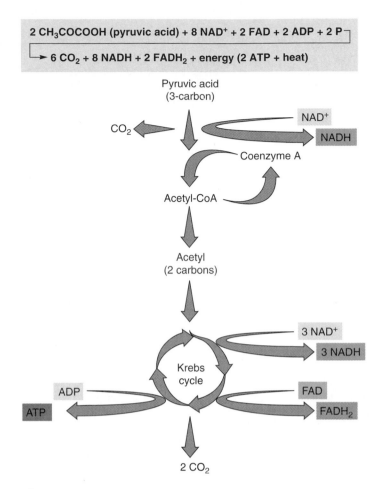

$$2\ CH_3COCOOH\ \text{(pyruvic acid)} + 8\ NAD^+ + 2\ FAD + 2\ ADP + 2\ P \longrightarrow 6\ CO_2 + 8\ NADH + 2\ FADH_2 + \text{energy (2 ATP + heat)}$$

Figure 6.10

Krebs Cycle: Fundamental Description

The Krebs cycle is the biochemical pathway performed in the mitochondria of most cells to complete the oxidation of glucose. During this sequence of chemical reactions, a pyruvic acid molecule produced from glycolysis is stripped of its hydrogens. The hydrogens are picked up by NAD^+ and FAD for transport to the ETS. The remaining atoms are reorganized into molecules of carbon dioxide. Enough energy is released during the Krebs cycle to form 2 ATPs. Because 2 pyruvic acid molecules were produced from glycolysis, the Krebs cycle must be run twice in order to complete their oxidation (once for each pyruvic acid).

Detailed Description

1. Glycolysis

The first stage of the cellular respiration process takes place in the cytoplasm. This first step, known as glycolysis, consists of the enzymatic breakdown of a glucose molecule without the use of molecular oxygen. Because no oxygen is required, glycolysis is called an anaerobic process. The glycolysis pathway can be divided into two general sets of reactions. The first reactions make the glucose molecule unstable and later oxidation-reduction reactions are used to synthesize ATP and capture hydrogens.

The addition of some energy to the glucose molecule must take place in order to start glycolysis because glucose is

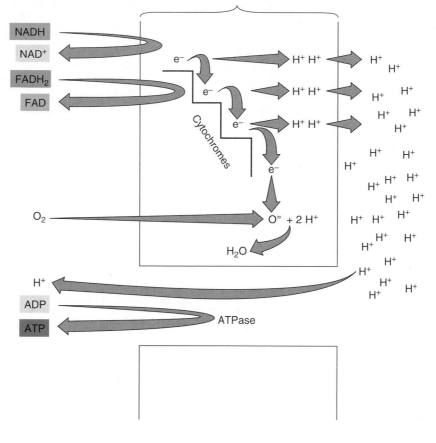

Figure 6.11

The Electron-Transport System: Fundamental Description

The electron-transport system (ETS) is a series of oxidation reduction reactions also known as the cytochrome system. The electrons are passed through a series of enzymes until they are picked up by oxygen. The energy the electrons give up is used to pump protons (H^+) across a membrane in the mitochondrion. When protons flow back through the membrane, enzymes in the membrane cause the formation of ATP. The protons eventually combine with the oxygen that has gained electrons and water is produced.

a very stable molecule and will not automatically break down to release energy. In the process of glycolysis, the initial glucose molecule gains a phosphate to become glucose-6-phosphate, which is converted to fructose-6-phosphate. When a second phosphate is added, fructose-1,6-bisphosphate (P—C_6—P) is formed. This 6-carbon molecule is unstable and breaks apart to form two 3-carbon, glyceraldehyde-3-phosphate molecules.

Each of the two glyceraldehyde-3-phosphate molecules acquires a second phosphate from a phosphate supply normally found in the cytoplasm. Each molecule now has 2 phosphates attached (P—C_3—P). A series of reactions follows in which energy is released by breaking chemical bonds that hold the phosphates to glyceraldehyde-3-phosphate. The energy and the phosphates are used to produce ATP. Since there are 2 glyceraldehyde-3-phosphate molecules, each with 2 phosphates, a total of 4 ATPs are produced. Because 2 ATPs were utilized to start the process, a *net* yield of 2 ATPs results. In addition, 4 hydrogen atoms detach from the carbon skeleton and their electrons are

transferred to NAD^+ to form NADH, which transfers the electrons to the electron-transport system. The 3-carbon pyruvic acid molecules that remain are the raw material for the Krebs cycle. Since glycolysis occurs in the cytoplasm and the Krebs cycle takes place inside mitochondria, the pyruvic acid must enter the mitochondrion before it can be broken down further.

In summary, the process of glycolysis takes place in the cytoplasm of a cell, where glucose ($C_6H_{12}O_6$) enters a series of reactions that:

1. Requires the use of 2 ATPs
2. Ultimately results in the formation of 4 ATPs
3. Results in the formation of 2 NADHs
4. Results in the formation of 2 molecules of pyruvic acid ($CH_3COCOOH$)

Because 2 molecules of ATP are used to start the process and a total of 4 ATPs are generated, each glucose molecule that undergoes glycolysis produces a net yield of 2 ATPs. (Figure 6.12)

Glucose
(C — C — C — C — C — C)

ATP
ADP
Hexokinase

Glucose-6-phosphate
(C — C — C — C — C — C — **P**)

Phosphoglucoisomerase

Fructose-6-phosphate
(C — C — C — C — C — C — **P**)

ATP
ADP
Phosphofructokinase

Fructose-1,6-bisphosphate
(**P** — C — C — C — C — C — C — **P**)

Aldolase

Glyceraldehyde-3-phosphate + Dihydroxyacetone phosphate
(C — C — C — **P**)   (C — C — C — **P**)

Triose phosphate isomerase

P
NAD⁺
NADH
Glyceraldehyde-3-phosphate dehydrogenase

1, 3-Bisphosphoglycerate
(**P** — C — C — C — **P**)

ADP
ATP
Phosphoglycerate kinase

3-Phosphoglycerate
(C — C — C — **P**)

Phosphoglycerate mutase

2-Phosphoglycerate
(C — C — C — **P**)

Enolase

Phosphoenolpyruvate
(C — C — C — **P**)

ADP
ATP
Pyruvate kinase

Pyruvate (pyruvic acid)
(C — C — C)

Figure 6.12

Glycolysis: Detailed Description

Glycolysis is a process that takes place in the cytoplasm of cells. It does not require the presence of oxygen, so this part of the respiration process is anaerobic. During the first few steps, phosphates are added from ATP and ultimately the 6-carbon sugar is split into two 3-carbon compounds. During the final steps in the process, NAD⁺ accepts electrons and hydrogen to form NADH and ATP is formed. Two ATPs form for each of the 3-carbon molecules that is processed in glycolysis. Since there are two 3-carbon compounds, a total of 4 ATPs are formed. However, since 2 ATPs were used to start the process, there is a net gain of 2 ATPs. Pyruvic acid (pyruvate) is left at the end of glycolysis. In aerobic respiration, it will enter the mitochondrion and proceed through the Krebs cycle. The NADHs also enter the mitochondrion and go into the electron-transport system.

2. The Krebs Cycle

After pyruvate (pyruvic acid) enters the mitochondrion, it is first acted on by an enzyme in conjunction with a molecule known as coenzyme A (figure 6.13). This results in three significant products. Hydrogen atoms are removed and NADH is formed, a carbon is removed and carbon dioxide is formed, and a 2-carbon acetyl molecule is formed, which temporarily attaches to coenzyme A to produce acetyl-coenzyme A. (These and subsequent reactions of the Krebs cycle take place in the fluid between the membranes of the mitochondrion.) The acetyl coenzyme A enters the series of reactions known as the Krebs cycle. During the Krebs cycle, the acetyl will be systematically dismantled. Its hydrogen atoms will be removed and the remaining carbons will be released as carbon dioxide. The first step in this process involves the acetyl coenzyme A. The acetyl portion of the complex is transferred to a 4-carbon compound called *oxaloacetate* (*oxaloacetic acid*) and a new 6-carbon citrate molecule (citric acid) is formed. The coenzyme A is released to participate in another reaction with pyruvic acid. This newly formed citrate is broken down in a series of reactions that ultimately produces oxaloacetate, which was utilized in the first step of the cycle. (Hence, the name Krebs cycle, citric acid cycle, or tricarboxylic acid cycle). The series of compounds formed during this cycle are called *keto acids* (not to be confused with ketone bodies). In the process, electrons are removed and, along with protons, become attached to coenzymes. Most become attached to NAD⁺ but some become attached to FAD. As the molecules move through the Krebs cycle, enough energy is released to allow the synthesis of 1 ATP molecule for each acetyl that enters the cycle. The ATP is formed from ADP and a phosphate already present in the mitochondria. For each pyruvate molecule that enters a mitochondrion and is processed through the Krebs cycle, 3 carbons are released as 3 carbon dioxide molecules, 5 pairs of hydrogen atoms are removed and become attached to hydrogen carriers, and 1 ATP molecule is generated. When both pyruvate molecules have been processed through the Krebs cycle: (1) all the original carbons from the glucose have been released into the atmosphere as 6 carbon dioxide molecules; (2) all the hydrogen originally found on the

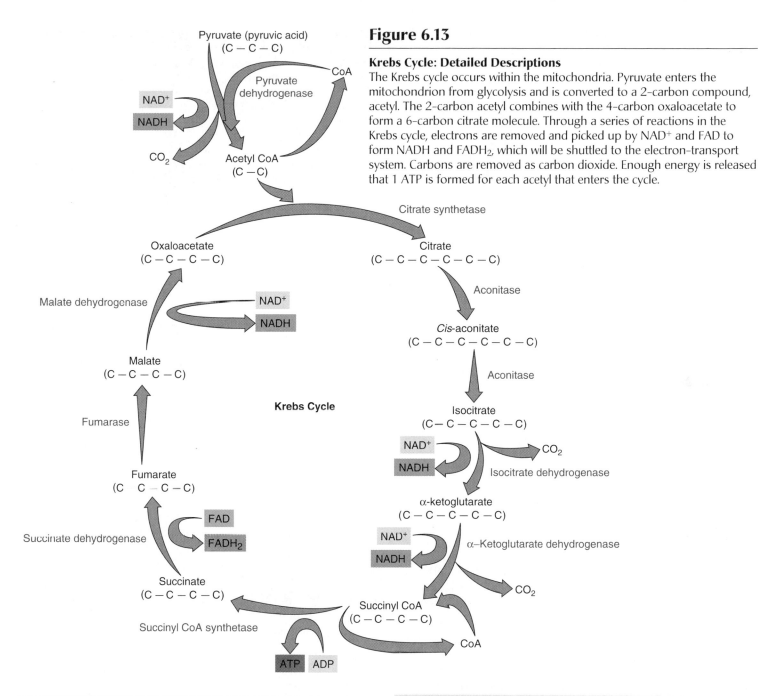

Figure 6.13

Krebs Cycle: Detailed Descriptions

The Krebs cycle occurs within the mitochondria. Pyruvate enters the mitochondrion from glycolysis and is converted to a 2-carbon compound, acetyl. The 2-carbon acetyl combines with the 4-carbon oxaloacetate to form a 6-carbon citrate molecule. Through a series of reactions in the Krebs cycle, electrons are removed and picked up by NAD^+ and FAD to form NADH and $FADH_2$, which will be shuttled to the electron-transport system. Carbons are removed as carbon dioxide. Enough energy is released that 1 ATP is formed for each acetyl that enters the cycle.

glucose has been transferred to either NAD^+ or FAD to form NADH or $FADH_2$; and (3) 2 ATPs have been formed from the addition of phosphates to ADPs (figure 6.13).

In summary, the Krebs cycle takes place within the mitochondria. For each pyruvic acid molecule that enters a mitochondrion and changed to acetyl that is processed through the Krebs cycle:

1. The 3 carbons of the pyruvate are released as carbon dioxide (CO_2).
2. Five pairs of hydrogens become attached to hydrogen carriers (4 NADH and 1 $FADH_2$).
3. One ATP is generated.

3. The Electron-Transport System

The series of reactions in which energy is transferred from the electrons and protons carried by NADH and $FADH_2$ is known as the electron-transport system (ETS) (figure 6.14). This is the final stage of aerobic cellular respiration and is dedicated to generating ATP. The reactions that make up the electron-transport system are a series of oxidation-reduction reactions in which the electrons are passed from one electron-carrier molecule to another until they ultimately are accepted by oxygen atoms. The negatively charged oxygen combines with the hydrogen ions to form water. It is this step that makes the process aerobic. Keep in mind that potential energy increases whenever things experiencing a repelling force are pushed together, such as adding the third phosphate to an

Figure 6.14

The Electron-Transport System: Detailed Description

Most of the ATP produced by aerobic cellular respiration comes from the ETS. NADH and $FADH_2$ deliver electrons to the enzymes responsible for the ETS. There are several protein complexes in the inner membrane of the mitochondrion, each of which is responsible for a portion of the reactions that yield ATP. The energy of electrons is given up in small amounts and used to pump protons into the intermembrane space. When these protons flow back through special pores in the membrane, ATPase produces ATP. The electrons eventually are transferred to oxygen and the negatively charged oxygen ions accept protons to form water.

ADP molecule. Potential energy also increases whenever things that attract each other are pulled apart, as in the separating of the protons from the electrons.

Let's now look in just a bit more detail at what happens to the electrons and protons that are carried to the electron-transport systems by NADH and $FADH_2$ and how these activities are used to produce ATP. The mitochondrion consists of two membranes—an outer enclosing membrane and an inner folded membrane. The reactions of the ETS are associated with this inner membrane. Within the structure of the membrane are several enzyme complexes that perform particular parts of the ETS reactions (see figure 6.14). The production of ATPs involves two separate but connected processes. Electrons carried by NADH enter reactions in complex I, where they lose some energy and are eventually picked up by a coenzyme (coenzyme Q). Electrons from $FADH_2$ enter complex II and also are eventually transferred to coenzyme Q. Coenzyme Q transfers the electrons to complex III. In complex III, the electrons lose additional energy and are transferred to cytochrome c, which transfers electrons to complex IV. In complex IV, the electrons are eventually transferred to oxygen. As the electrons lose energy in complex I, complex III, and complex IV, additional protons are pumped into the intermembrane space. When these protons flow down the concentration gradient through special channels in the membrane, phosphorylase enzymes (ATPase) in the membrane are able to use the energy to generate ATP.

A total of 12 pairs of electrons and hydrogens are transported to the ETS from glycolysis and the Krebs cycle for each glucose that enters the process. In eukaryotic organisms, the pairs of electrons can be accounted for as follows: 4 pairs are carried by NADH and were generated during glycolysis outside the mitochondrion, 8 pairs are carried as NADH and were generated within the mitochondrion; and 2 pairs are carried by $FADH_2$ and were generated within the mitochondrion.

- For each of the 8 NADHs generated within the mitochondrion, enough energy is released to produce 3 ATP molecules. Therefore, 24 ATPs are released from these electrons carried by NADH.
- In eukaryotic cells, the electrons released during glycolysis are carried as NADH and converted to $FADH_2$ in order to shuttle them into the mitochondria. Once they are inside the mitochondria, they follow the same pathway as the other $FADH_2$s.

The electrons carried by $FADH_2$ are lower in energy. When these electrons go through the series of oxidation-reduction reactions, they release enough energy to produce a total of 8 ATPs. Therefore, we have a grand total of 32 ATPs produced from the hydrogen electrons that enter the ETS.

Finally, if we do a complete accounting of all the ATPs produced, we come up with a grand total of 36 ATPs: 32 from the ETS, 2 from glycolysis, and 2 from the Krebs cycle.

In summary, the electron-transport system takes place within the mitochondrion where:

1. Oxygen is used up as the oxygen atoms receive the hydrogens from NADH and $FADH_2$ to form water (H_2O).
2. NAD^+ and FAD are released to be used over again.
3. 32 ATPs are produced.

6.5 Aerobic Cellular Respiration in Prokaryotes

The preceding discussion dealt with the process of aerobic cellular respiration in eukaryotic organisms. However, there are some prokaryotes that utilize aerobic cellular respiration as well. Since prokaryotes do not have mitochondria, it follows that there would be some differences between what they do and what eukaryotes do. The primary difference involves the electrons carried from glycolysis to the electron-transport system. In eukaryotes, the electrons are carried by NADH and transferred to FAD to form $FADH_2$ in order to get the electrons across the outer membrane of the mitochondrion. Since $FADH_2$ results in the production of fewer ATPs than NADH, there is a cost to the eukaryotic cell of getting the electrons into the mitochondrion. This transfer is not necessary in prokaryotes, so they are able to produce a theoretical 38 ATPs for each glucose metabolized rather than the 36 ATPs produced by eukaryotes.

6.6 Anaerobic Cellular Respiration

While aerobic cellular respiration is the fundamental process by which most organisms generate ATP, there are some organisms that do not have the necessary enzymes to carry out the Krebs cycle and ETS. Most of these are prokaryotic organisms but there are certain eukaryotic organisms like yeasts that also can live in the absence of oxygen and do not use the Krebs cycle and ETS. Even within organisms there may be differences in the metabolic activities of cells. Some of their cells may be unable to perform aerobic respiration while others may be able to survive for periods of time without it. However, all of these cells still have a need for a constant supply of ATP. An organism that does not require O_2 as its final electron acceptor is called *anaerobic* (*an* = without; *aerob* = air) and performs anaerobic cellular respiration. Although they do not utilize oxygen, some anaerobic organisms are capable of using other inorganic or organic molecules as their final electron acceptors. The acceptor molecule could be sulfur, nitrogen, or other inorganic atoms or ions. It could also be an organic molecule such as pyruvic acid ($CH_3COCOOH$). Anaerobic pathways that oxidize glucose to generate ATP energy by using an organic molecule as the ultimate hydrogen acceptor are called **fermentation.** Anaerobic cellular respiration results in the release of less ATP and heat energy than released in aerobic cellular respiration. Anaerobic respiration is the incomplete oxidation of glucose:

$$C_6H_{12}O_6 + (H^+ \text{ \& } e^- \text{ acceptor}) \longrightarrow$$
$$\longrightarrow \text{smaller hydrogen-containing molecules + energy (ATP + heat)}$$

Many organisms that perform anaerobic cellular respiration use the glycolysis pathway as a way to obtain energy from sugar molecules. Typically, glucose proceeds through the glycolysis pathway, producing pyruvic acid. The pyruvic acid then undergoes one of several alternative changes depending on the kind of organism and the specific enzymes it

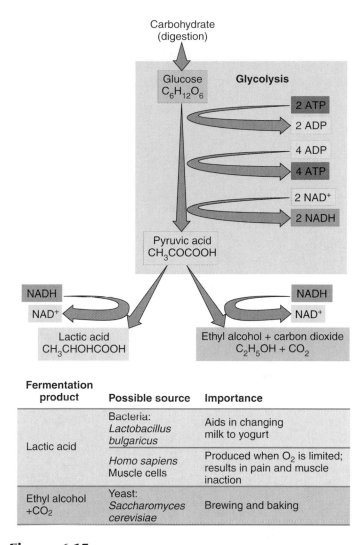

Fermentation product	Possible source	Importance
Lactic acid	Bacteria: *Lactobacillus bulgaricus*	Aids in changing milk to yogurt
	Homo sapiens Muscle cells	Produced when O_2 is limited; results in pain and muscle inaction
Ethyl alcohol +CO_2	Yeast: *Saccharomyces cerevisiae*	Brewing and baking

Figure 6.15

Fermentations
The highlighted portion of this figure is a simplified version of glycolysis. Many organisms can carry out the process of glycolysis and derive energy from it by transferring the electrons removed during glycolysis to pyruvic acid. The ultimate end product is determined by the kinds of enzymes the specific organism can produce. The synthesis of these various molecules is the organism's particular way of oxidizing NADH to NAD^+ and reducing pyruvic acid to a new end product.

possesses. Some organisms are capable of returning the electrons removed from sugar in the earlier stages of glycolysis to the pyruvic acid formed at the end of glycolysis. When this occurs, the pyruvic acid is converted into lactic acid, acetone, or other organic molecules (figure 6.15). Those organisms that do alcoholic fermentation have the enzymes necessary to convert pyruvic acid to ethyl alcohol (ethanol) and carbon dioxide.

Although many different products can be formed from pyruvic acid, we will look at only two anaerobic pathways in detail. **Alcoholic fermentation** is the anaerobic respiration pathway that yeast cells follow when oxygen is lacking in their environment. In this pathway, the pyruvic acid is converted to ethanol (a 2-carbon alcohol, C_2H_5OH) and carbon dioxide. Yeast cells then are able to generate only 4 ATPs from glycolysis. The cost for glycolysis is still 2 ATPs; thus, for each glucose a yeast cell oxidizes, it profits by 2 ATPs.

Although during alcoholic fermentation yeasts get ATP and discard the waste products ethanol and carbon dioxide, these waste products are useful to humans. In making bread, the carbon dioxide is the important end product; it becomes trapped in the bread dough and makes it rise. When this happens, we say the bread is *leavened*. Dough that has not undergone this process is called *unleavened*. The alcohol produced by the yeast evaporates during the baking process. In the brewing industry, ethanol is the desirable product produced by yeast cells. Champagne, other sparkling wines, and beer are products that contain both carbon dioxide and alcohol. The alcohol accumulates, and the carbon dioxide in the bottle makes them sparkling (bubbly) beverages. In the manufacture of many wines, the carbon dioxide is allowed to escape, so these wines are not sparkling but are "still" wines.

In **lactic acid fermentation,** the pyruvic acid (CH_3CO-COOH) that results from glycolysis is converted to lactic acid ($CH_3CHOHCOOH$) by the transfer of electrons that had been removed from the original glucose. In this case, the net profit is again only 2 ATPs per glucose. The buildup of the waste product, lactic acid, eventually interferes with normal metabolic functions and the bacteria die. We use the lactic acid waste product from these types of anaerobic bacteria when we make yogurt, cultured sour cream, cheeses, and other fermented dairy products. The lactic acid makes the milk protein coagulate and become pudding-like or solid. It also gives the products their tart flavor, texture, and aroma.

In the human body, different cells have different metabolic capabilities. Nerve cells must have a constant supply of oxygen to conduct aerobic cellular respiration. Red blood cells lack mitochondria and must rely on the anaerobic process of lactic acid fermentation to provide themselves with energy. Muscle cells can do either. As long as oxygen is available to skeletal muscle cells, they function aerobically. However, when oxygen is unavailable—because of long periods of exercise or heart or lung problems that prevent oxygen from getting to the skeletal muscle cells—the cells make a valiant effort to meet energy demands by functioning anaerobically.

While skeletal muscle cells are functioning anaerobically, they accumulate lactic acid. This lactic acid must ultimately be metabolized, which requires oxygen. Therefore, the accumulation of lactic acid represents an *oxygen debt* that must be paid at some time in the future. It is the lactic acid buildup that makes the muscles tired when we exercise (figure 6.16). When the lactic acid concentration becomes great enough, lactic acid fatigue results. Its symptoms are cramping of the muscles and pain. Because of the pain, we generally stop the activity before the muscle cells die. As a person cools down after a period of exercise, breathing and heart rate stay high until the oxygen debt is repaid and the level of oxygen in muscle cells returns to normal. During this period, the lactic acid that has accumulated is being converted back into pyruvic acid. The pyruvic acid can now continue through the Krebs cycle and the ETS as oxygen becomes available. In addition to what is happening in the muscles, much of the lactic acid is transported by the bloodstream to the liver, where about 20% is metabolized through the Krebs cycle and 80% is resynthesized into glucose.

6.7 Metabolic Processing of Other Molecules

Up to this point we have described only the methods and pathways that allow organisms to release the energy tied up in carbohydrates (sugars). Frequently, cells lack sufficient carbohydrates for their energetic needs but have other materials from which energy can be removed. Fats and proteins, in addition to carbohydrates, make up the diet of many organisms. These three foods provide the building blocks for the cells, and all can provide energy. Carbohydrates can be digested to simple sugars, proteins can be digested to amino acids, and fats can be digested to glycerol and fatty acids. The basic pathways organisms use to extract energy from fat and protein are the same as for carbohydrates: glycolysis, the Krebs cycle, and the electron-transport system. However, there are some additional steps necessary to get fats and proteins ready to enter these pathways and there are several different points in glycolysis and the Krebs cycle where fats and proteins enter to be respired. They are summarized here.

Fat Respiration

A neutral fat (also known as triglyceride) is a large molecule that consists of a molecule of glycerol with 3 fatty acids attached to it. Before fats can be broken down to release energy, they must be converted to smaller units by digestive processes. Several enzymes are involved in these steps. The first step in this process is to break the bonds between the glycerol and fatty acids. Glycerol is a 3-carbon molecule that is converted into glyceraldehyde-3-phosphate. Since glyceraldehyde-3-phosphate is one of the steps in glycolysis, it can enter the glycolysis pathway (figure 6.17). The remaining fatty acids are often long molecules (typically 14 to 20 carbons long) that also must be processed before they can be further metabolized. First they need to enter the mitochondrion, where subsequent reactions take place. Once inside the mitochondrion, each long chain of carbons that makes up the carbon skeleton is hydrolyzed (split by the addition of a water molecule) into 2-carbon fragments. Next, each of the 2-carbon fragments is converted into acetyl. The acetyl molecules are carried into the Krebs cycle by coenzyme A molecules. Once in the Krebs cycle, they proceed through the Krebs cycle just like the acetyls from glucose.

By following the glycerol and each 2-carbon fragment through the cycle, you can see that each molecule of fat has the potential to release several times as much ATP as does a

Glucose

Glycolysis

Pyruvic acid

Oxygen available—aerobic   **Oxygen limited—anaerobic**

Water

Carbon dioxide

Maximum ATP generated

Lactic acid

Minimum ATP generated

Figure 6.16

Oxygen Debt

When oxygen is available to all cells, the pyruvic acid from glycolysis is converted into acetyl-CoA, which is sent to the Krebs cycle, and the hydrogens pass through the electron-transport system. When oxygen is not available in sufficient quantities (because of a lack of environmental oxygen or a temporary inability to circulate enough oxygen to cells needing it), some of the pyruvic acid from glycolysis is converted to lactic acid. The lactic acid builds up in cells when this oxygen debt occurs. It is the presence of this lactic acid that results in muscle fatigue and a burning sensation.

molecule of glucose. Each glucose molecule has 6 pairs of hydrogen, whereas a typical molecule of fat has up to 10 times that number. This is why fat makes such a good long-term energy storage material. It is also why it takes so long for people on a weight-reducing diet to remove fat! It takes time to use all the energy contained in the fatty acids. On a weight basis, there are twice as many calories in a gram of fat as there are in a gram of carbohydrate.

As you have just seen, fats are an excellent source of energy. The storage of fat is an important process for storing energy and other kinds of molecules can be converted to fat. You already know that people can get fat from eating sugar. Again, the same set of chemical reactions is involved in the conversion of carbohydrates to fats. Notice in figure 6.17 that both carbohydrates and fats can enter the Krebs cycle and release energy. Although people require both fats and carbohydrates in their diets, they need not be in precise ratios; the body can make some interconversions. This means that people who eat excessive amounts of carbohydrates will deposit body fat. It also means that people who starve can generate glucose by breaking down fats and using the glycerol to synthesize glucose. • nutrition p. 373

Protein Respiration

Proteins can also be catabolized and interconverted just as fats and carbohydrates are (figure 6.17). The first step in utilizing protein for energy is to digest the protein into individual amino acids. Each amino acid then needs to have the amino group ($—NH_2$) removed. The remaining carbon skeleton, a keto acid, enters the respiratory cycle as acetyl, pyruvic acid, or as one of the other types of molecules found in the Krebs cycle. As the acids progress through the Krebs cycle, the electrons are removed and sent to the ETS where their energy is converted into the chemical-bond energy of ATP. The amino group that was removed from the amino acid is converted into ammonia. Some organisms excrete ammonia directly; others convert ammonia into other nitrogen-containing compounds, such as urea or uric acid. All of these molecules are toxic and must be eliminated. They are transported in the blood to the kidneys, where they are eliminated. In the case of a high-protein diet, increasing fluid intake will allow the kidneys to efficiently remove the urea or uric acid. • kidneys, p. 357

When proteins are eaten, they are digested into their component amino acids. These amino acids are then available to be used to construct other proteins. Proteins cannot be

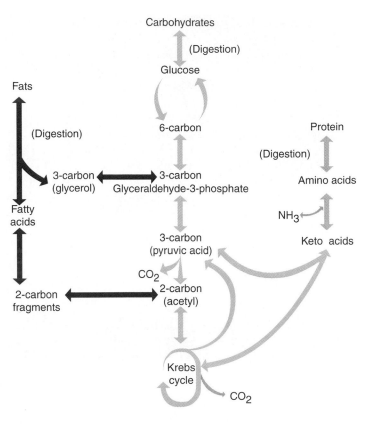

Figure 6.17

The Interconversion of Fats, Carbohydrates, and Proteins
Cells do not necessarily utilize all food as energy. One type of food can be changed into another type to be used as raw materials for construction of needed molecules or for storage. Notice that many of the reaction arrows have two heads (i.e., these reactions can go in either direction). For example, glycerol can be converted into glyceraldehyde-3-phosphate and glyceraldehyde-3-phosphate can become glycerol.

stored. If they or their component amino acids are not needed immediately, they will be converted into fat, carbohydrates, or energy. This presents a problem for individuals who do not have ready access to a continuous source of amino acids in their diet (i.e., individuals on a low-protein diet). If they do not have a source of dietary protein, they must dismantle proteins from important cellular components to supply the amino acids they need. This is the reason why proteins and amino acids are considered an important daily food requirement.

One of the most important concepts you need to recognize from this discussion is that carbohydrates, fats, and proteins can all be used to provide energy. The fate of any type of nutrient in a cell depends on the momentary needs of the cell. An organism whose daily food-energy intake exceeds its daily energy expenditure will convert only the necessary amount of food into energy. The excess food will be interconverted according to the enzymes present and the needs of the organism at that time. In fact, glycolysis and the Krebs cycle allow molecules of the three major food types (carbohydrates, fats, and proteins) to be interchanged.

As long as a person's diet has a certain minimum of each of the three major types of molecules, the cell's meta-

bolic machinery can manipulate molecules to satisfy its needs. If a person is on a starvation diet, the cells will use stored carbohydrates first. Once the carbohydrates are gone (about 2 days), cells will begin to metabolize stored fat. When the fat is gone (a few days to weeks), the proteins will be used. A person in this condition is likely to die.

6.8 Photosynthesis—An Overview

Ultimately, the energy to power all organisms comes from the sun. **Chlorophyll** is a green pigment that absorbs light energy for the process of photosynthesis. Through the process of photosynthesis, plants, algae, and certain bacteria transform light energy to chemical-bond energy in the form of ATP and then use ATP to produce complex organic molecules such as glucose. It is from these organic molecules that organisms obtain energy through the process of cellular respiration. In algae and the leaves of green plants, photosynthesis occurs in cells that contain organelles called **chloroplasts**. Chloroplasts have two distinct regions within them: the grana and stroma. A **granum** consists of stacks of membranous sacks containing chlorophyll, and the **stroma** consists of the intervening spaces between membranes (figure 6.18).

The following equation summarizes the chemical reactions green plants and many other photosynthetic organisms use to make ATP and organic molecules:

light energy + carbon dioxide + water → glucose + oxygen

light energy +   6 CO_2 + 6 H_2O → $C_6H_{12}O_6$ +   6 O_2

There are three distinct events in the photosynthetic pathway:

1. **Light-capturing events.** In eukaryotic cells, the process of photosynthesis takes place within chloroplasts. Each chloroplast is surrounded by membranes and contains the green pigment, chlorophyll, along with other pigments. Chlorophyll and other pigments absorb specific wavelengths of light. When specific amounts of light are absorbed by the photosynthetic pigments, electrons become "excited." With this added energy, these excited electrons are capable of entering into the chemical reactions responsible for the production of ATP. These reactions take place within the grana of the chloroplast.

2. **Light-dependent reactions.** The light-dependent reactions utilize the excited electrons produced by the light-capturing activities. The light-dependent reactions are also known as the *light reactions*. During these reactions, "excited" electrons from the light-capturing reactions are used to do two different things. Some of the "excited" electrons are used to make ATP, while others are used to split water into hydrogen and oxygen. The oxygen from the water is released to the environment as O_2 molecules and the hydrogens are transferred to the electron carrier coenzyme, NADP⁺, to produce NADPH. (NADP⁺ is similar to NAD⁺,

which was discussed in the section on Cellular Respiration.) These reactions also take place in the grana of the chloroplast. However, the NADPH and ATP leave the grana and enter the stroma where the light-independent reactions take place.

3. **Light-independent reactions.** These reactions are also known as the *dark reactions,* since light is not needed for the reactions to take place. During these reactions, ATP and NADPH from the light-dependent reactions are used to attach CO_2 to a 5-carbon molecule, already present in the cell, to manufacture new, larger organic

Figure 6.18

Photosynthesis and the Structure of a Leaf

The leaves of plants are composed of layers of cells that contain chloroplasts, which contain chlorophyll. The chlorophyll molecules are actually located within chloroplasts in stacks of membranous sacks called *grana.* The fluid-filled space in which the grana are located is called the *stroma* of the chloroplast.

Figure 6.19

Photosynthesis Overview

Photosynthesis is a complex biochemical pathway in plants, algae, and certain bacteria. The upper portion of this figure shows the overall process. Sunlight, along with CO_2 and H_2O, is used to make organic molecules such as sugar. The lower portion illustrates the three parts of the process: (1) the light-capturing events, (2) the light-dependent reactions, and (3) the light-independent reactions. Notice that the end products of the light-dependent reactions, NADPH and ATP, are necessary to run the light-independent reactions, while the water and carbon dioxide are supplied from the environment.

molecules. Ultimately, glucose ($C_6H_{12}O_6$) is produced. These light-independent reactions take place in the stroma in either the light or dark as long as ATP and NADPH are available from the light-dependent stage. The ADP and the $NADP^+$ produced during the light-independent reactions are recycled back to the light-dependent reactions to be used over again.

So we can summarize the process of photosynthesis as follows. During the light-capturing events, *light energy* is captured by chlorophyll and other pigments, resulting in "excited" electrons. The energy of these excited electrons is used during the light-dependent reactions to disassociate *water* molecules into hydrogen and oxygen and the *oxygen* is released. Also during the light-dependent reactions, ATP is produced and $NADP^+$ picks up hydrogen released from water to form NADPH. During the light-independent reactions, ATP and NADPH are used to help combine *carbon dioxide* with a 5-carbon molecule so that ultimately organic molecules like *glucose* are produced (figure 6.19).

6.9 Photosynthesis

Fundamental Description

1. Light-Capturing Events

Light energy is used to drive photosynthesis during the light-capturing stage. Visible light is a combination of many different wavelengths of light seen as different colors. Some of

these colors are seen when white light is separated into its colors to form a rainbow. The colors of the electromagnetic spectrum that provide the energy for photosynthesis are correlated with different kinds of light-energy-absorbing pigments. The green chlorophylls are the most familiar and abundant. There are several types of this pigment. The two most common types are chlorophyll *a* and chlorophyll *b*. Both absorb strongly in the red and blue portions of the electromagnetic spectrum, although in slightly different portions of the spectrum (figure 6.20). These pigments reflect green light. That is why we see chlorophyll-containing plants as predominantly green. Other pigments common in plants are called **accessory pigments.** These include the *carotenoids* (yellow, red, and orange). They absorb mostly blue and blue-green light while reflecting the oranges and yellows. While the presence of these pigments is generally masked by the presence of chlorophyll, in the fall, when chlorophyll disintegrates, the reds, oranges, and yellows show through. Accessory pigments are also responsible for the brilliant colors of vegetables such as carrots, tomatoes, eggplant, and peppers. Photosynthetic bacteria and various species of algae have other kinds of accessory pigments not found in plants. Having a combination of different pigments, each of which absorbs a portion of the light hitting it, allows the organism to capture much of the visible light that falls on it.

Any cell with chloroplasts can carry on photosynthesis. However, in most plants, leaves are specialized for photosynthesis and contain cells that have high numbers of chloroplasts (see figure 6.18). Chloroplasts are membranous, saclike organelles containing many thin, flattened sacks.

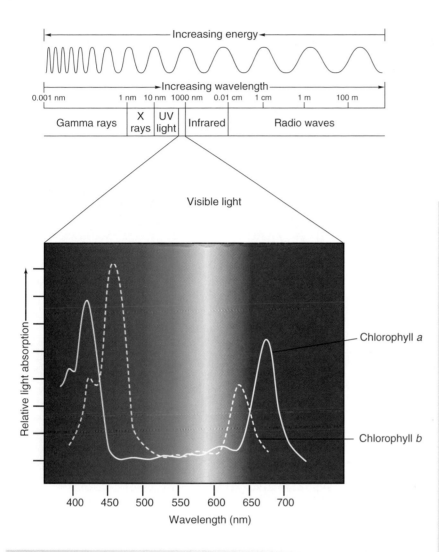

Figure 6.20

The Visible Light Spectrum and Chlorophyll
Light is a form of electromagnetic energy that can be thought of as occurring in waves. The shorter the wavelength the greater the energy it contains. Humans are capable of only seeing waves that are between about 400 and 740 nm (nanometers) long. Chlorophyll *a* (the solid graph line) and chlorophyll *b* (the dotted graph line) absorb different wavelengths of light energy.

trons are passed through a series of electron-transport steps, which result in protons being pumped into the cavity of the thylakoid. When the protons pass back out through the membrane to the outside of the thylakoid, ATP is produced. This is very similar to the reactions that happen in the electron-transport systems of aerobic cellular respiration. In addition, some of the excited electrons are used to split water molecules, resulting in the production of hydrogen ions, electrons, and oxygen gas, O_2. The coenzyme, $NADP^+$ picks up the electrons and becomes reduced to NADPH. The hydrogen ions attach because when $NADP^+$ accepts electrons, it becomes negatively charged ($NADP^-$). The positively charged H^+ are attracted to the negatively charged $NADP^-$. The oxygen remaining from the splitting of water molecules is released into the atmosphere or can be used by the cell in aerobic cellular respiration, which takes place in the mitochondria of plant cells. The ATP and NADPH molecules move from the grana, where the light-dependent reactions take place, to the stroma where the light-independent reactions take place.

A generalized reaction that summarizes the light-dependent reactions follows:

$$\text{excited electrons} + H_2O + ADP + NADP^+ \longrightarrow ATP + NADPH + O_2$$

3. Light-Independent Reactions

The ATP and NADPH provide electrons and hydrogens needed to build large, organic molecules from carbon dioxide. The light-independent reactions are a series of oxidation-reduction reactions that combine hydrogen from water (carried by NADPH) with carbon dioxide from the atmosphere to form simple organic molecules such as sugar. As CO_2 diffuses into the chloroplasts, the enzyme **ribulose bisphosphate carboxylase (RuBisCo)** speeds the combining of the CO_2 with an already-present, 5-carbon carbohydrate, **ribulose.** NADPH then donates its hydrogens and electrons to complete the reduction of the molecule. The resulting 6-carbon molecule is immediately split into two 3-carbon molecules of **glyceraldehyde-3-phosphate.** Some of the glyceraldehyde-3-phosphate molecules are converted through a

These sacks, called **thylakoids,** contain chlorophylls, accessory pigments, electron-transport molecules, and enzymes. They are stacked in groups called *grana* (singular, *granum*). The fluid-filled spaces between the grana are called the stroma of the chloroplast. The structure of the chloroplast is directly related to both the light-capturing and the energy-conversion steps of photosynthesis. In the light-capturing process, the pigments (e.g., chlorophyll), which are embedded in the membranes of the thylakoids, capture light energy and some of the electrons of pigments become "excited." The chlorophylls and other pigments involved in trapping sunlight energy and storing it are arranged into clusters called **photosystems.** By clustering the pigments, photosystems serve as energy-gathering or -concentrating mechanisms that allow light to be collected more efficiently to "excite" the electrons to higher energy levels.

The light-energy-capturing events result in:

photons of light energy → excited electrons from chlorophyll

2. Light-Dependent Reactions

The light-dependent reactions of photosynthesis take place in the thylakoid membranes inside the chloroplast. The "excited" or energized electrons from the light-capturing events are passed to protein molecules in the thylakoid. The elec-

Figure 6.21

Photosynthesis: Fundamental Description
The process of photosynthesis involves light-trapping events by chlorophyll and other pigments. The excited electrons are used in the light-dependent reactions to split water, releasing hydrogens and oxygen. The hydrogens are picked up by $NADP^+$ to form NADPH and the oxygen is released into the atmosphere. Excited electrons are also used to produce ATP. The ATP and NADPH leave the thylakoid and enter the stroma of the chloroplast where they are used in the light-independent reactions to incorporate carbon dioxide into organic molecules. During the light-independent reactions, carbon dioxide is added to a 5-carbon ribulose molecule to form a 6-carbon compound that splits into glyceraldehyde-3-phosphate . Some of the glyceraldehyde-3-phosphate is used to regenerate ribulose and some is used to make other organic molecules. The ADP and $NADP^+$ released from the light-independent reaction stage return to the thylakoid to be used in the synthesis of ATP and NADPH again. Therefore, each stage is dependent on the other.

series of reactions into ribulose. Thus, these reactions constitute a cycle in which carbon dioxide and hydrogens are added and glyceraldehyde-3-phosphate and the original 5-carbon ribulose are produced. The plant can use surplus glyceraldehyde-3-phosphate for the synthesis of glucose. The plant can also use glyceraldehyde-3-phosphate to construct a wide variety of other organic molecules (e.g., proteins, nucleic acids), provided there are a few additional raw materials, such as minerals and nitrogen-containing molecules (figure 6.21).

The activities of the light-independent reactions can be summarized as follows:

$$\text{ATP} + \text{NADPH} + \text{ribulose} + \text{CO}_2 \rightarrow$$
$$\rightarrow \text{ADP} + \text{NADP}^+ + \text{complex organic molecule} + \text{ribulose}$$

Detailed Description

1. Light-Capturing Events
The energy of light comes in discrete packages called *photons*. Photons of light having different wavelengths have different amounts of energy. A photon of red light has a different amount of energy from a photon of blue light. Pigments of different kinds are able to absorb photons of certain wavelengths of light. Chlorophyll absorbs red and blue light best and reflects green light. When a chlorophyll molecule is struck by and absorbs a photon of the correct wavelength, its electrons become excited to a higher energy level. The energy of these excited electrons is used to split water molecules and also to form ATP. The reactions that result in the production of ATP

Figure 6.22

Photosystems II and I and How They Interact: Detailed Description
While light energy strikes and is absorbed by both photosystem II and I, what happens and how they interconnect are not the same. Notice that the electrons released from photosystem II end up in the chlorophyll molecules of photosystem I. The electrons that replace those "excited" out of the reaction center in photosystem II come from water.

and the splitting of water take place in the thylakoids of chloroplasts. There are many different molecules involved and the majority are embedded in the membrane of the thylakoid. The various molecules involved in these reactions are referred to as *photosystems*. A photosystem is composed of (1) an *antenna complex*, (2) a *reaction center*, and (3) other enzymes necessary to store the light energy as ATP and NADPH.

The antenna complex is a network of hundreds of chlorophyll and accessory pigment molecules whose role is to capture photons of light energy and transfer the energy to a specialized portion of the photosystem known as the reaction center. When light shines on the antenna complex and strikes a chlorophyll molecule, an electron becomes excited. The energy of the excited electron is passed from one pigment to another through the antenna complex network. This series of excitations continues until the combined energies from several excitations are transferred to the reaction center, which consists of a complex of chlorophyll *a* and protein molecules. An electron is excited and passed to a primary electron acceptor molecule, oxidizing the chlorophyll and reducing the acceptor. Ultimately, the oxidized chlorophyll then has its electron replaced with another electron from a different electron donor. Exactly where this replacement electron comes from is the basis upon which two different photosystems have been identified—photosystem I and photosystem II.

In summary, the light–capturing reactions take place in the thylakoids of the chloroplast:

1. Chlorophyll and other pigments of the antenna complex capture light energy and produce excited electrons.
2. The energy is transferred to the reaction center.
3. Excited electrons from the reaction center are transferred to a primary electron acceptor molecule.

2. Light-Dependent Reactions

Both photosystems I and II have antenna complexes and reaction centers and provide excited electrons to primary electron acceptors. However, each has slightly different enzymes and other proteins associated with it, so each does a slightly different job. In actuality, photosystem II occurs first and feeds its excited electrons to photosystem I (figure 6.22). One special attribute of photosystem II is that there is an enzyme in the thylakoid membrane responsible for splitting water molecules ($H_2O \rightarrow 2\,H + O$). The oxygen is released into the environment as O_2 and the electrons of the hydrogens are used to replace the electrons that previously had been lost by the chlorophyll. The remaining hydrogen ions (protons) are released to participate in other reactions. Thus, in a sense the light energy trapped by the antenna complex is used to split water into H and O. The excited electrons from photosystem II are sent through a series of electron-transport reactions in which they give up some of their energy. This is similar to the electron-transport system described for aerobic cellular respiration. After passing through the electron-transport system, the electrons are accepted by chlorophyll molecules in photosystem I. While the electron-transport activity is happening, protons are pumped from the stroma into the space inside the thylakoid. Eventually these protons will move back across the membrane. When they do, ATPase is used to produce ATP (ADP is phosphorylated to produce ATP). So a second result of this process is that the energy of sunlight has been used to produce ATP.

The connection between photosystem II and photosystem I involves the transfer of electrons from photosystem II to photosystem I. These electrons are important because photons (from sunlight) are exciting electrons in the reaction center of photosystem I and the electrons from photosytem II replace those lost from photosystem I.

In photosystem I, light is trapped and the energy is absorbed in the same manner as occurred in photosystem II. However, this system does not have the enzyme involved in splitting water into oxygen, protons, and electrons; therefore, no O_2 is released from photosystem I. The high-energy electrons leaving the reaction center of photosystem I make their way through a different series of oxidation-reduction reactions. During these reactions, the electrons are picked up by $NADP^+$, which is reduced to NADPH (see figure 6.22). Thus, the primary result of photosystem I is the production of NADPH.

In summary, the light-dependent reactions of photosynthesis take place in the thylakoids of the chloroplast:

1. Excited electrons from photosystem II are passed through an electron-transport chain and ultimately enter photosystem I.
2. The electron-transport system is used to establish a proton gradient that produces ATP.
3. Excited electrons from photosystem I are transferred to $NADP^+$ to form NADPH.
4. In photosystem II, an enzyme splits water into hydrogen and oxygen. The oxygen is released as O_2.
5. Electrons from the hydrogen of water replace the electrons lost by chlorophyll in photosystem II.

---

3. Light-Independent Reactions

This major series of reactions takes place within the stroma of the chloroplast. The materials needed for the light-independent reactions are ATP, NADPH, CO_2, and a 5-carbon starter molecule called *ribulose*. The first two ingredients (ATP and NADPH) are made available from the light-dependent reactions, photosystems II and I. The carbon dioxide molecules come from the atmosphere and the ribulose starter molecule is already present in the stroma of the chloroplast from previous reactions.

Carbon dioxide is said to undergo *carbon fixation* through the **Calvin cycle** (named after its discoverer, Melvin Calvin). In the Calvin cycle, ATP and NADPH from the

light-dependent reactions are used along with carbon dioxide to synthesize larger, organic molecules. As with most metabolic pathways, the synthesis of organic molecules during the light-independent reactions requires the activity of several enzymes to facilitate the many steps in the process. The fixation of carbon begins with carbon dioxide combining with a 5-carbon molecule *ribulose* to form an unstable 6-carbon molecule. This reaction is carried out by the enzyme ribulose bisphosphate carboxylase (RuBisCo), reportedly the most abundant enzyme on the planet. The newly formed 6-carbon molecule immediately breaks down into two 3-carbon molecules, each of which then undergoes a series of reactions that involve a transfer of energy from ATP and a transfer of hydrogen from NADPH. The result of this series of reactions is two glyceraldehyde-3-phosphate molecules. Because glyceraldehyde-3-phosphate contains 3 carbons and is formed as the first stable compound in this type of photosynthesis, this is sometimes referred to as the C3 photosynthetic pathway. Outlooks 6.1 describes some other forms of photosynthesis that do not use the C3 pathway. Some of the glyceraldehyde-3-phosphate is used to synthesize glucose and other organic molecules and some is used to regenerate the 5-carbon ribulose molecule, so this pathway is a cycle (figure 6.23).

The general chemical equation for the light-independent reactions follows:

$$CO_2 + ATP + NADH + \text{5-carbon starter (ribulose)} \longrightarrow \text{glyceraldehyde-3-phosphate} + NADP^+ + ADP + P$$

In summary, the reactions of the light-independent reactions take place in the stroma of chloroplasts:

1. ATP and NADPH from the light-dependent reactions leave the grana and enter the stroma.
2. The energy of ATP is used in the Calvin cycle to combine carbon dioxide to a 5-carbon starter molecule (ribulose) to form a 6-carbon molecule.
3. The 6-carbon molecule immediately divides into two 3-carbon molecules of glyceraldehyde-3-phosphate.
4. Hydrogens from NADPH are transferred to molecules in the Calvin cycle.
5. The 5-carbon ribulose is regenerated.
6. ADP and $NADP^+$ are returned to the light-dependent reactions.

Glyceraldehyde-3-phosphate: The Product of Photosynthesis

The 3-carbon glyceraldehyde-3-phosphate is the actual product of the process of photosynthesis. However, many textbooks show the generalized equation for photosynthesis as:

$$CO_2 + H_2O + light \rightarrow C_6H_{12}O_6 + O_2$$

making it appear as if a 6-carbon sugar (hexose) is the end product. The reason a hexose ($C_6H_{12}O_6$) is usually listed as the end product is simply because, in the past, the simple sugars were easier to detect than was glyceraldehyde-3-

Figure 6.23

The Calvin Cycle: Detailed Description

During the Calvin cycle, ATP and NADPH from the light-dependent reactions are used to attach CO_2 to a 5-carbon ribulose molecule. The 6-carbon molecule formed immediately breaks down into two 3-carbon molecules. Some of the glyceraldehyde-3-phosphate formed is used to produce glucose and other more complex organic molecules. In order to accumulate enough carbon to make a new glucose molecule, the cycle must turn six times. The remaining glyceraldehyde-3-phosphate is used to regenerate the 5-carbon ribulose to start the process again.

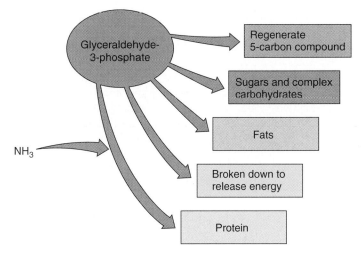

Figure 6.24

Uses of Glyceraldehyde-3-phosphate

The glyceraldehyde-3-phosphate that is produced as the end product of photosynthesis is used for a variety of things. The plant cell can make simple sugars, complex carbohydrates, or even the original 5-carbon starter from it. It can also serve as an ingredient of lipids and amino acids (proteins). In addition, it provides a major source of metabolic energy when it is sent through the respiratory pathway.

phosphate. If a plant goes through photosynthesis and produces 12 glyceraldehyde-3-phosphates, 10 of the 12 are rearranged by a series of complex chemical reactions to regenerate the 5-carbon ribulose needed to operate the light-independent reaction stage. The other two glyceraldehyde-3-phosphates can be considered profit from the process. The excess glyceraldehyde-3-phosphate molecules are frequently changed into a hexose. So, those who first examined photosynthesis chemically saw additional sugars as the product and did not realize that glyceraldehyde-3-phosphate is the initial product.

There are a number of things the cell can do with glyceraldehyde-3-phosphate in addition to manufacturing hexose (figure 6.24). Many other organic molecules can be constructed using glyceraldehyde-3-phosphate as the basic construction unit. Glyceraldehyde-3-phosphate can be converted to glucose molecules, which can be combined to form complex carbohydrates such as starch for energy storage or cellulose for cell wall construction. In addition, other simple sugars can be used as building blocks for ATP, RNA, DNA, or other carbohydrate-containing materials.

The cell can convert the glyceraldehyde-3-phosphate into lipids, such as oils for storage, phospholipids for cell membranes, or steroids for cell membranes. The glyceraldehyde-3-phosphate can serve as the carbon skeleton for the construction of amino acids needed to form proteins. Almost any molecule that a green plant can manufacture begins with this glyceraldehyde-3-phosphate molecule. Finally (and this is easy to overlook), glyceraldehyde-3-phosphate can be broken

OUTLOOKS 6.1

The Evolution of Photosynthesis

When we compare the process of photosynthesis in prokaryotes and eukaryotes, it is amazing that the processes are so similar. The evolution of the process of photosynthesis goes back over 2 billion years, when all life on Earth was prokaryotic and would have occurred in organisms that were aquatic. (There were no terrestrial organisms at the time.) Today we find some bacteria that perform a kind of photosynthesis that does not result in the release of oxygen. In general, these bacteria produce ATP but do not break down water to produce oxygen. Perhaps these are the descendents of the first organisms to carry out a photosynthetic process and oxygen-releasing photosynthesis developed from these earlier forms of photosynthesis.

Evidence from the fossil record shows that beginning at approximately 2 billion years ago, oxygen was present in the atmosphere. Since eukaryotic organisms had not yet developed, the organisms responsible for producing this oxygen would have been prokaryotic. Today we find many kinds of cyanobacteria that perform photosynthesis in essentially the same way as plants, although they use a somewhat different kind of chlorophyll. As a matter of fact, it is assumed that the chloroplasts of eukaryotes are evolved from photosynthetic bacteria. Initially, the first eukaryotes to perform photosynthesis would have been various kinds of simple members of the kingdom Protista (various kinds of algae). When we look at the algae today, we find that certain kinds of algae (red algae, brown algae, green algae) have specific kinds of chlorophylls and other accessory pigments different from the others. Since the group known as the green algae has essentially the same chlorophylls as plants, it is assumed that plants are derived from this aquatic group. • endosymbiosis, p. 447

But the evolution of the process of photosynthesis did not stop once plants came on the scene. Most plants perform photosynthesis in the manner described in this chapter. Light energy is used to generate ATP and NADPH, which are used in the Calvin cycle to incorporate carbon dioxide into glyceraldehyde-3-phosphate. Because the primary product of this form of photosynthesis is a 3-carbon molecule of glyceraldehyde-3-phosphate, it is often called C3 photosynthesis. Among plants, there are two interesting variations of photosynthesis that utilize this same basic process but add interesting twists to the process.

C4 photosynthesis is common in plants like grasses that are typically subjected to high light levels. In these plants, carbon dioxide does not directly enter the Calvin cycle. Instead, the fixation of carbon is carried out in two steps and two different kinds of cells participate. It appears that this adaptation allows plants to more efficiently trap carbon dioxide from the atmosphere under high light conditions. Specialized cells in the leaf capture carbon dioxide and convert a 3-carbon compound to a 4-carbon compound. This 4-carbon compound then releases the carbon dioxide to other cells that use it in the normal Calvin cycle typical of the light-independent reactions. Because a 4-carbon molecule is formed to "store" carbon, this process is known as C4 photosynthesis.

Another variation of photosynthesis is known as CAM (Crassulacean acid metabolism), since this mechanism was first discovered in members of the Crassulaceae family of plants. (A common example, *Crassula*, is known as the jade plant.) CAM photosynthesis is a modification of the basic process of photosynthesis that allows photosynthesis to occur in arid environments while reducing the potential for water loss. In order for plants to take up carbon dioxide, small holes in the leaves (stomates) must be open to allow carbon dioxide to enter. However, relative humidity is low during the day and plants would tend to lose water if their stomates were open. CAM photosynthesis works as follows: During the night, the stomates open and carbon dioxide can enter the leaf. The chloroplasts trap the carbon dioxide by binding it to an organic molecule, similar to what happens in C4 plants. The next morning, when it is light (and drier), the stomates close. During the day, the chloroplasts can capture light and do the light-dependent reactions. They then utilize the carbon stored the previous night to do the light-independent reactions.

---

down during cellular respiration. Cellular respiration releases the chemical-bond energy from glyceraldehyde-3-phosphate and other organic molecules and converts it into the energy of ATP. This conversion of chemical-bond energy enables the plant cell and the cells of all organisms to do things that require energy, such as grow and move materials.

6.10 Other Aspects of Plant Metabolism

Photosynthetic organisms are able to manufacture organic molecules from inorganic molecules. Once they have the basic carbon skeleton, they have the ability to manufacture a variety of other complex molecules for their own needs— fats, proteins, and complex carbohydrates are some of the more common. However, plants produce a wide variety of other molecules for specific purposes. Among the molecules they produce are compounds that are toxic to animals that would use plants as food. Many of these compounds have been discovered to be useful as medicines. Digitalis from the foxglove plant causes the hearts of animals that eat the plant to malfunction. However, it can be used as a medicine in humans who have certain heart ailments. Molecules that paralyze animals have been used in medicine to treat specific ailments and relax muscles so that surgery is easier to perform. Still others have been used as natural insecticides.

Another important group of organic molecules from plants is called *vitamins*. Vitamins are organic molecules that we cannot manufacture but must have in small amounts to maintain good health. The vitamins we get from plants were manufactured by them for their own purposes. By definition, they are not vitamins to the plant because the plant made them for its own use. However, since we cannot make them, we rely on plants to synthesize these important molecules for us and we consume them when we eat foods containing them.

6.11 Interrelationships between Autotrophs and Heterotrophs

Earlier in this chapter, the differences between autotrophs and heterotrophs were described. Autotrophs are able to capture energy to manufacture new organic molecules from inorganic molecules. Heterotrophs must have organic molecules as starting points. However, it is important to recognize that all organisms must do some form of respiration. Plants and other autotrophs obtain energy from food molecules, in the same manner that animals and other heterotrophs do—by processing organic molecules through the respiratory pathways. This means that plants, like animals, require oxygen for the ETS portion of aerobic cellular respiration.

Many people believe that plants only give off oxygen and never require it. This is incorrect! Plants do give off oxygen in the light-dependent reactions of photosynthesis, but in aerobic cellular respiration they use oxygen as does any other organism. During their life spans, green plants give off more oxygen to the atmosphere than they take in for use in respiration. The surplus oxygen given off is the source of oxygen for aerobic cellular respiration in both plants and animals. Animals are not only dependent on plants for oxygen, but are ultimately dependent on plants for the organic molecules necessary to construct their bodies and maintain their metabolism (figure 6.25). An understanding of respiration and photosynthesis has led to many practical applications (How Science Works 6.1).

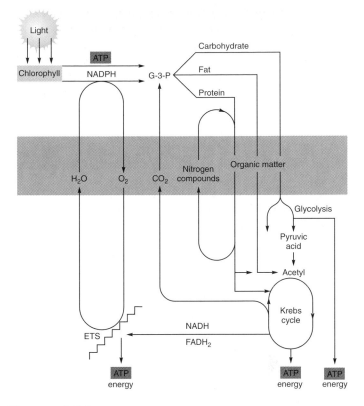

Figure 6.25

The Interdependence of Photosynthesis and Respiration
Plants use the end products of plant and animal respiration–carbon dioxide, water, and nitrogen compounds—to produce various foods. Plants and animals use the end products of plant photosynthesis—food and oxygen—as sources of energy. Therefore, plants are dependent on animals and animals are dependent on plants. The materials that link the two processes are seen in the colored bar.

HOW SCIENCE WORKS 6.1

Applying Knowledge of Biochemical Pathways

As people developed a better understanding of the processes of aerobic cellular respiration, anaerobic cellular respiration, and photosynthesis, several practical applications of this knowledge have developed:

- Although for centuries people have fermented beverages like beer and wine, they were often plagued by sour products that were undrinkable. Once people understood that there were yeasts that produced alcohol under *anaerobic* conditions and bacteria that converted alcohol to acetic acid under *aerobic* conditions, it was a simple task to prevent acetic acid production by preventing oxygen from getting to the fermenting mixture.
- When it was discovered that many kinds of grasses engage in a slightly different form of photosynthesis than most other plants, it was possible to develop herbicides that interfered with the unique photosynthetic pathways of grasses and produce grass killers.
- When it was discovered that the bacterium that causes gas gangrene is anaerobic and is, in fact, poisoned by the presence of oxygen, various oxygen therapies were developed to help cure patients with gangrene. In some cases, persons with gangrene are placed in special chambers with high oxygen levels under pressure. In other cases, only the affected part of the body is enclosed. Under such conditions the gangrene-causing bacteria die or are inhibited.
- Since many disease-causing organisms are prokaryotic and have somewhat different pathways and enzymes than do eukaryotic organisms, it is possible to develop molecules that selectively interfere with the enzymes of prokaryotes without affecting eukaryotes like humans. We call these molecules *antibiotics*.

Thus, animals supply raw materials—CO_2, H_2O, and nitrogen—needed by plants, whereas plants supply raw materials—sugar, oxygen, amino acids, fats, and vitamins—needed by animals. This constant cycling is essential to life on Earth. As long as the sun shines and plants and animals remain in balance, the food cycles of all living organisms will continue to work properly.

SUMMARY

In the process of aerobic cellular respiration, organisms convert foods into energy (ATP) and waste materials (carbon dioxide and water). There are three distinct metabolic pathways involved in aerobic cellular respiration: glycolysis, Krebs cycle, and the electron-transport system. Glycolysis takes place in the cytoplasm of the cell, and the Krebs cycle and electron-transport system take place in mitochondria. Organisms that have oxygen (O_2) can perform aerobic cellular respiration, which employs glycolysis, the Krebs cycle, and the electron-transport system (ETS). Organisms that do not utilize oxygen perform anaerobic cellular respiration (fermentation) and are only able to utilize the glycolysis pathway. Aerobic cellular respiration yields much more ATP than anaerobic cellular respiration. Glycolysis and the Krebs cycle serve as a molecular interconversion system: fats, proteins, and carbohydrates are interconverted according to the needs of the cell.

Sunlight supplies the essential initial energy for making the large organic molecules necessary to maintain the forms of life we know. In the light-capturing events of photosynthesis, plants use chlorophyll to trap the energy of sunlight using photosystems. During the light-dependent reactions, they manufacture a source of chemical energy, ATP, and a source of hydrogen, NADPH. Atmospheric oxygen is released in this stage. In the light-independent reactions of photosynthesis, the ATP energy is used in a series of reactions (the Calvin cycle) to join the hydrogen from the NADPH to a molecule of carbon dioxide and form a simple carbohydrate, glyceraldehyde-3-phosphate. In subsequent reactions, plants use the glyceraldehyde-3-phosphate as a source of energy and raw materials to make complex carbohydrates, fats, and other organic molecules. With the addition of ammonia, plants can form proteins.

THINKING CRITICALLY

For guidelines to answer this question, visit the Online Learning Center.

Both plants and animals carry on metabolism. From a metabolic point of view, which of the two is more complex? Include in your answer the following topics:

1. Cell structure
2. Biochemical pathways
3. Enzymes
4. Organic molecules
5. Autotrophy and heterotrophy

CONCEPT MAP TERMINOLOGY

Construct a concept map using these terms.

aerobic cellular respiration
anabolism
anaerobic cellular respiration
catabolism
fermentation

hydrogen ion and
    electron acceptor
oxidation-reduction
photosynthesis

Visit the Online Learning Center to use an interactive format to construct another concept map.

KEY TERMS

Use interactive flash cards, on the Online Learning Center, to help you learn the meanings of these terms.

accessory pigments
acetyl
adenosine triphosphate (ATP)
aerobic cellular respiration
alcoholic fermentation
anabolism
anaerobic cellular respiration
autotrophs
biochemical pathway
Calvin cycle
catabolism
cellular respiration
chemosynthesis
chlorophyll
chloroplast
electron-transport system (ETS)
FAD (flavin adenine dinucleotide)
fermentation
glyceraldehyde-3-phosphate
glycolysis
grana

heterotrophs
high-energy phosphate bond
Krebs cycle
lactic acid fermentation
light-capturing events
light-dependent reactions
light-independent reactions
metabolic pathway
NAD+ (nicotinamide adenine
    dinucleotide)
NADP+ (nicotinamide adenine
    dinucleotide phosphate)
photosynthesis
photosystem
pyruvate
pyruvic acid
ribulose
ribulose bisphosphate
    carboxylase (RuBisCo)
stroma
thylakoid

LEARNING CONNECTIONS

Visit the Online Learning Center to find additional study resources including those highlighted below.

6.1   Energy and Organisms

1. How do autotrophs and heterotrophs differ?
2. What is chemosynthesis?
3. How are respiration and photosynthesis related to autotrophs and heterotrophs?

6.2   Common Mechanisms Used in Processing Energy and Matter

Animations and Review
• Pathways

4. What is a biochemical pathway? Give two examples.
5. In what way does ATP differ from other organic molecules?
6. What is a proton pump and how is it related to ATP production?
7. List three electron transport molecules.

6.3 Aerobic Cellular Respiration—An Overview

8. Aerobic cellular respiration occurs in three stages. Name these and briefly describe what happens in each stage.
9. Which cellular organelle is involved in the process of respiration?

6.4 Aerobic Cellular Respiration

Labeling Exercise
- Aerobic cellular respiration
- Cristae

Interactive Concept Map
- Cellular respiration

10. For glycolysis, Krebs cycle, and the electron-transport system, list two molecules that enter and two molecules that leave each pathway.
11. How is each of the following involved in aerobic cellular respiration: NAD+, pyruvic acid, oxygen, and ATP?

6.5 Aerobic Cellular Respiration in Prokaryotes

12. How is aerobic cellular respiration different between prokaryotic and eukaryotic organisms?

6.6 Anaerobic Cellular Respiration

Case Study
- The Case of Rachael Martin and lactic acid

13. Pyruvic acid can be converted into a variety of molecules. Name three.
14. Why does aerobic respiration yield more energy than anaerobic respiration?

6.7 Metabolic Processing of Other Molecules

15. Describe how glycolysis and the Krebs cycle can be used to obtain energy from fats and proteins.

6.8 Photosynthesis—An Overview

16. Photosynthesis is a biochemical pathway that involves three kinds of activities. Name these and explain how they are related to each other.
17. Which cellular organelle is involved in the process of photosynthesis?

6.9 Photosynthesis

Experience This!
- Collecting by-products of photosynthesis

18. How do photosystem I and photosystem II differ in the kinds of reactions that take place?
19. What does an antenna complex do? How is it related to the reaction center?

6.10 Other Aspects of Plant Metabolism

20. Is vitamin C a vitamin for an orange tree?

6.11 Interrelationships between Autotrophs and Heterotrophs

Interactive Concept Map
- Photosynthesis and respiration in the cell
- Relationship of Photosynthesis and Respiration

21. Even though animals do not photosynthesize, they rely on the sun for their energy. Why is this so?

Test yourself with multiple choice questions that provide immediate feedback on the Online Learning Center with PowerWeb at

www.mhhe.com/enger11

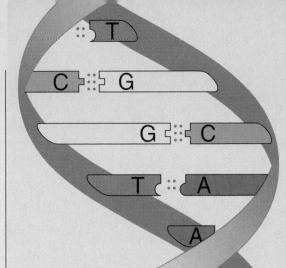

CHAPTER

7

DNA and RNA:
The Molecular Basis of Heredity

GOALS AND OBJECTIVES

Understand the typical flow of genetic information in a cell.

❑ Define *gene, transcription,* and *translation.*

❑ Describe how the processes of transcription and translation relate.

Understand how DNA and RNA control transcription and translation.

❑ State the nucleotides commonly found in DNA and RNA.

❑ Apply the base-pairing rules to predict the nucleotide structure of RNA.

❑ Explain the use of mRNA, rRNA, and tRNA in the process of translation.

❑ Accurately use the codon table to predict the amino acid sequence of a protein.

Understand how mutations affect protein synthesis.

❑ Provide examples of silent mutations.

❑ Provide examples of insertions, deletions, and frameshift mutations.

Learn the importance of controlling gene expression.

❑ State why single cellular and multicellular organisms control gene expression.

❑ Explain how promoters, transcription factors, and splicing affect transcription.

Understand the relationship between DNA replication and cell division.

❑ Describe DNA replication using base-pairing rules and DNA polymerase.

❑ Describe how DNA is organized differently in various types of cells.

❑ Explain how daughter cells inherit a replicated DNA molecule.

Understand the implications that the DNA code is common to all organisms.

❑ Explain how DNA from one organism is used in another organism.

❑ Explain how DNA can be used to uniquely identify individuals.

❑ Identify potential medical treatments based on DNA technology.

Cells: Anatomy and Action

PART TWO

133

7.1 The Central Dogma

Proteins play a critical role in how cells successfully meet the challenges of living. Cells use proteins to maintain their shape and to speed up important chemical reactions such as photosynthesis and respiration. A cell will not live long if it cannot reliably create the proteins that it needs for survival.

This chapter looks at how cells reliably make proteins. To place these ideas in the proper context, remember that some proteins are enzymes that aid cells by catalyzing chemical reactions. These chemical reactions occur after the enzyme binds its substrate at the enzyme's active site. The enzyme's active site matches the substrate molecule in size, shape, and chemical properties. The size, shape, and chemical properties of an enzyme's active site are due to the combination of the enzyme's amino acids, which are the individual subunits of the enzyme. For the cell to reliably make an enzyme, the cell must be able to control the placement of amino acids in a protein during the synthesis of enzymes. • enzymes, p. 92

This control comes from the genetic information stored in the cell's **deoxyribonucleic acid (DNA)** molecule(s). The DNA molecule contains a type of blueprint for making the many different types of proteins that the cell needs. The portion of the DNA strand that codes for a particular protein is called a **gene.** The set of ideas that describes how the cell uses the information stored in DNA is called the **central dogma.** The first step of the central dogma is called **transcription.** Transcription uses DNA as a template to copy genetic information into the form of RNA. In turn, RNA is involved in **translation.** Translation synthesizes the protein using RNA as a template (figure 7.1). Understanding how the steps of the central dogma are carried out is important because we can then understand how mutations arise. Ultimately, mutations can affect the entire organism.

To understand how DNA is able to control the synthesis of enzymes, we need to look at how the structure of proteins and nucleic acids relates to their function—for example, the structure of an enzyme's active site and its function of binding a substrate. Another important structure/function relationship is found in the double-stranded DNA molecule. DNA has four properties that enable it to function as genetic material. It can (1) *store information* that deter-

mines the characteristics of cells and organisms; (2) use this information to *direct the synthesis* of structural and regulatory proteins essential to the operation of the cell or organism; (3) *mutate*, or chemically change, and transmit these changes to future generations; and (4) *replicate* by directing the manufacture of copies of itself.

7.2 Nucleic Acid Structure and Function

Nucleic acid molecules are enormous, complex polymers made up of monomers. Each monomer is a **nucleotide.** Each nucleotide is composed of a sugar molecule containing five carbon atoms, a phosphate group, and a molecule containing nitrogen, which will be referred to as a *nitrogenous base* (figure 7.2). It is possible to classify nucleic acids into two main groups based on the kinds of sugar used in the nucleotides (i.e., DNA and RNA). • macromolecules, p. 46

In cells, DNA is the nucleic acid that functions as the original blueprint for the synthesis of proteins. DNA contains the sugar **deoxyribose,** phosphates, and a unique sequence of the nitrogenous bases **adenine** (A), **guanine** (G), **cytosine** (C), and **thymine** (T). **Ribonucleic acid (RNA)** is a type of nucleic acid that is directly involved in **protein synthesis.** RNA contains the sugar **ribose,** phosphates, and the nitrogenous bases adenine (A), guanine (G), cytosine (C), and **uracil** (U). DNA and RNA share the nitrogenous bases A, G, and C. Thymine is usually only present in DNA and uracil is usually only present in RNA.

DNA and RNA differ significantly in one other respect: DNA is actually a double molecule. It consists of two flexible strands held together by attractive forces, called *hydrogen bonds*, between paired nitrogenous bases. The two strands are twisted about each other in a coil or double helix (figure 7.3). As the nitrogenous bases pair with each other in the DNA helix, they "fit" into each other like two jigsaw puzzle pieces that interlock. This arrangement is stabilized by weak chemical forces called *hydrogen bonds.* The four bases always pair in a definite way: adenine (A) with thymine (T), and guanine (G) with cytosine (C). The number of hydrogen bonds determines the pairing sets. Adenine (A) and thymine (T) both form two hydrogen bonds, while guanine (G) and cytosine (C) form three hydrogen bonds. Notice that the large molecules (A and

Figure 7.1

Central Dogma Concept Map
DNA → (transcription) → RNA → (translation) → different types of proteins. This concept map outlines the steps of the central dogma. Genetic information in the cell is transferred from DNA to RNA by the process of transcription. The genetic information stored in RNA is then used by the process of translation to produce proteins. Transcription and translation are catalyzed by enzymes. The enzymes involved in this process were themselves produced by the processes of transcription and translation.

G) pair with the small ones (T and C), thus keeping the DNA double helix a constant width. The bases that pair are said to be **complementary bases** and this bonding pattern is referred to as *base-pairing rules*.

The base-pairing rules are followed throughout the steps of the central dogma. In transcription, nucleotides from DNA pair with nucleotides with RNA. In this case, guanine (G) still pairs with cytosine (C). However, because RNA does not contain thymine (T), adenine (A) in DNA pairs with uracil (U) in RNA. Later in the translation step of the central dogma, RNA nucleotides pair with RNA nucleotides. Again, guanine (G) pairs with cytosine (C) and adenine (A) pairs with uracil (U). The base-pairing rules for each of these situations are summarized in table 7.1.

Table 7.1

BASE-PAIRING RULES

DNA Paired to DNA		DNA Paired to RNA		RNA Paired to RNA	
G	C	G	C	G	C
C	G	C	G	C	G
A	T	A	U	A	U
T	A	T	A	U	A

Different types of nucleic acids are capable of base-pairing with each other by following the base-pairing rules. These rules are used throughout the process of the central dogma to allow information to go from DNA to RNA and then ultimately to the amino acid sequence of proteins. In general, the rules are that G and C pair, and A and T (or U) pair. Remember that thymine is only found in DNA and uracil is found only in RNA.

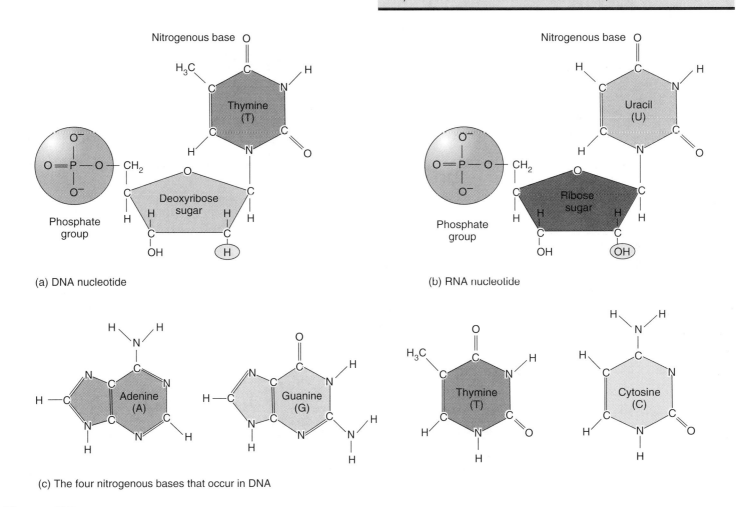

(a) DNA nucleotide

(b) RNA nucleotide

(c) The four nitrogenous bases that occur in DNA

Figure 7.2

Nucleotide Structure

The nucleotide is the basic structural unit of all nucleic acids and consists of a sugar, a nitrogenous base, and a phosphate group. Part (*a*) shows a thymine DNA nucleotide and (*b*) shows a uracil RNA nucleotide. Notice how both DNA and RNA nucleotides consist of the same three parts—a sugar, a nitrogenous base, and a phosphate group. One chemical difference between DNA and RNA is in the sugar. Notice that the DNA nucleotide has a circled H extending from one of its carbon atoms. In RNA, this same carbon atom has an OH group. The lack of this oxygen atom in DNA is why DNA is called deoxyribonucleic acid. (*c*) DNA and RNA also differ in the kinds of bases present in their nucleotides. In DNA, the nitrogenous bases can be adenine, guanine, thymine, or cytosine. In RNA, the nitrogenous bases can be adenine, guanine, uracil, or cytosine.

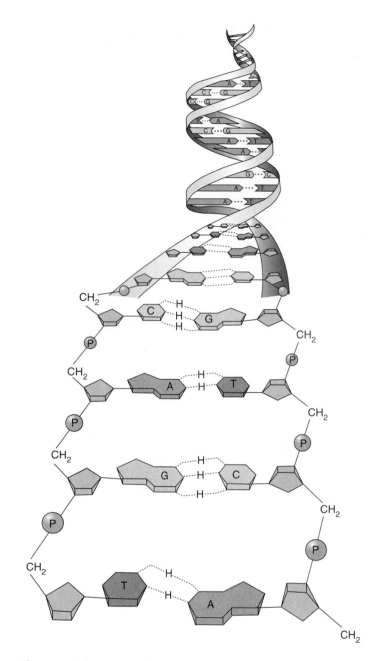

Figure 7.3

Double-Stranded DNA
Polymerized deoxyribonucleic acid (DNA) is a helical molecule. While the units of each strand are held together by covalent bonds, the two parallel strands are interlinked by hydrogen bonds between the paired nitrogenous bases like jigsaw puzzle pieces. A and T pair with two hydrogen bonds. G and C pair with three hydrogen bonds.

The genetic information of DNA is in the form of a chemical code. The **DNA code** is the *order* of the nucleotides. When the coded information is **expressed**, it guides the assembly of particular amino acids into a specific protein. You can "write" a message in the form of a stable DNA molecule by combining the four different DNA nucleotides (A, T, G, C) in particular sequences. In the cell, the four DNA nucleotides are used in a cellular alphabet that only consists of four letters. The letters of this alphabet are arranged in sets of three to construct words that are used to determine which amino acid is needed during translation. Each word, or **codon,** in the mRNA is always three letters (nucleotides) long and later is responsible for the placement of one amino acid in a protein. It is the sequence of nucleotides (A, T, G, C) in DNA that ultimately dictates which amino acids are used to synthesize a protein. The sequence of nucleotides and the base-pairing rules allow nucleic acids to control protein synthesis.

The fact that DNA has a sequence of nucleotides and is double-stranded allows it to fulfill all of its functions as genetic material. The linear sequence of **nitrogenous bases** in DNA allows the *storage of information* and *directs the synthesis* of proteins through base-pairing rules. The implication of this process is that changes in the linear sequence of nitrogenous bases can result in **mutations,** or changes in the amino acid sequence, used to create a protein. These changes can alter how that protein functions (How Science Works 7.1).

Transcription

Transcription is the process of using DNA as a **template** to synthesize RNA. DNA functions in a manner that is similar to a reference library that does not allow its books to circulate. Information from the original copies of the books must be copied for use outside of the library so that the originals are not damaged or destroyed. A temporary copy is made of the necessary information in a DNA molecule so that protein synthesis can occur. This operation is called *transcription (scribe* = to write), which means to transfer data from one form to another. In this case, the data are copied from DNA language to RNA language. The same base-pairing rules that were discussed earlier in this section apply to the process of transcription. Using this process, the genetic information stored as a DNA chemical code is carried in the form of an RNA molecule. Later, this RNA molecule will be used as a template for creating a protein because it is RNA that is used to guide the assembly of amino acids into structural, carrier, and regulatory proteins. Without the process of transcription, genetic information would be useless in directing cell functions.

Although many types of RNA can be synthesized, the three most important are **messenger RNA (mRNA), transfer RNA (tRNA),** and **ribosomal RNA (rRNA).** Each type of RNA is used for a unique purpose within the central dogma. Messenger RNA (mRNA) carries the blueprint for making the necessary enzyme. Transfer RNA (tRNA) and ribosomal

HOW SCIENCE WORKS 7.1

Of Men (and Women!), Microbes, and Molecules

As early as the 1920s, scientists didn't really understand the molecular basis of heredity. They partly understood genetics in terms of the odds that a given trait would be passed on to an individual in the next generation. This "probability" model of genetics left some questions unanswered:

- What is the nature of the genetic information?
- How does the cell use the genetic information?

As is often the case in science, serendipity played a large role in answering these questions. In 1928, a medical doctor, Frederick Griffith, was studying different bacterial strains that caused pneumonia. One of the bacterial strains killed mice very quickly and was extremely virulent. The other strain was not virulent. Griffith observed something unexpected when *dead* bacterial cells of the virulent strain were mixed with *living* cells of the less-virulent strain: The less-virulent strain took on the virulent characteristics of the dead strain. This observation was the first significant step in understanding the molecular basis of genetics because it provided scientists with a situation wherein the scientific method could be applied to ask questions and take measurements about the molecular basis of genetics. Until this point, scientists had lacked a method to provide supporting data.

This spurred the scientific community for the next 14 years to search for the identity of the "genetic molecule." A common hypothesis was that the genetic molecule would be one of the macromolecules—carbohydrates, lipids, proteins, or nucleic acids. During that period, many advances were made in how researchers studied cells. Many of the top minds in the field had formulated the hypothesis that protein was the genetic molecule. They had very good support for this hypothesis, too. Their argument boiled down to two ideas: The first idea is that proteins are found everywhere in the cell. It follows that if proteins were the genetic information, they would be found wherever that information would be used. The other reason is that proteins are structurally and chemically very complex. They are made up of 20 different monomers (amino acids) and come in a wide variety of sizes and shapes. This complexity could be used to account for all the genetic variety we observe in nature. On the other hand, very few, if any, scientists seriously considered the notion that DNA was the heritable material. After all, it was only found in the nucleus and it only consisted of four different monomers (nucleotides). How could this molecule account for the genetic complexity of life?

In 1944, Oswald Avery and his colleagues provided the first evidence that DNA was the genetic molecule. They performed an experiment similar to Griffith's, except they used purified samples of protein, DNA, lipids, and carbohydrates. The scientific community was highly skeptical of these results for two reasons: (1) They hadn't expected this result. They expected the genetic molecule to be protein. More importantly, (2) they didn't know how to explain how DNA could function as the genetic molecule. Because of this mind set, Avery's data was largely disregarded on

the rationale that his samples were impure. This objection was a factor that he controlled for in his scientific design. He reported over 99% purity in the tested DNA samples. It took 8 additional years and a different type of experiment to concretely establish DNA as the genetic molecule.

Alfred Hershey and Martha Chase carried out this definitive experiment in 1952. Their experiment was attractive because it used a relatively simple genetic system—a bacterial phage. A phage is a type of virus that uses a bacterial cell as its host and only consists of DNA and protein. Hershey and Chase hypothesized that the phage genetic information needed to enter the bacterial cell to create new phage. By radioactively labeling the DNA and the protein of the phage, Hershey and Chase were able to track that the DNA went into the bacterial cell, while very little protein did. They reasoned that DNA must be the genetic information.

The scientific community then turned toward the issue of determining how DNA could work as the heritable material. Scientists expected that the genetic molecule would have to do a number of things such as store information, disseminate information, be able to mutate, and be able to replicate itself. Their hypothesis was that the answer was hidden in the structure of the DNA molecule itself.

Investigation of how DNA functioned as the cell's genetic information took a wide variety of different strategies. Some scientists looked at DNA from different organisms. They found that in nearly every organism, the guanine (G) and cytosine (C) nucleotides were present in equal amounts. The same held true for adenine (A) and thymine (T). Later, this provided the basis for establishing the nucleic acid base-pairing rules.

Rosalind Franklin carried out another technique called X-ray crystallography. X-ray crystallography provides clues about molecular structure by recording the reflection pattern of X rays that have been fired at a crystallized chemical sample. Franklin was able to determine the following information from her experiments:

- DNA's helical shape
- DNA's width
- DNA's composition of two parallel strands
- DNA's repeating motifs that occur along the length of the molecule

Finally, two young scientists, James Watson and Francis Crick, put it all together. They simply listened to and read the information that was being discussed in the scientific community. Their key role was the assimilation of all the data. They recognized the importance of the X-ray crystallography data in conjunction with the organic structures of the nucleotides and the data that established the base-pairing rules. Together, they created a model for the structure of DNA that could account for all the things that a genetic molecule must do. They published an article describing this model in 1952. Ten years later, they were awarded the Nobel Prize for their work.

RNA (rRNA) are used later for the actual process of assembling the amino acids (table 7.2).

The actual process of transcription begins in the nucleus by separating the two strands of the double-stranded DNA with an enzyme called **RNA polymerase.** Separating the two strands of DNA exposes their nitrogenous bases so that other enzymes that carry out transcription can read one of the two DNA strands. The strand of DNA that serves as a template for the synthesis of RNA is the **coding strand.** The coding strand is only read in one direction. The strand of DNA that is not read directly by the enzymes is the *non-coding* strand of DNA. The RNA polymerase enzyme catalyzes the dehydration reactions between individual nucleotides so that a large RNA molecule is produced (figure 7.4). • enzymes, p. 92

Table 7.2	
TYPES OF RNA	
Type of RNA	**Description**
mRNA	A linear molecule that carries the gene's information from the cell's nucleus to the cytoplasm
tRNA	An adaptor molecule that aids in the pairing of amino acids to the correct codons of the mRNA
rRNA	Important part of ribosomal molecular structure and the ribosome's ability to carry out translation

(a)   Newly forming RNA

(b)

(c)   RNA

Figure 7.4

Transcription of an RNA Molecule

This summary illustrates the basic events that occur during transcription. (*a*) An enzyme attaches to the DNA at a point that allows it to separate the complementary strands. (*b* As RNA polymerase moves down the DNA strand, new complementary RNA nucleotides are base-paired to one of the exposed DNA strands. The base-paired RNA nucleotides are linked together by RNA polymerase to form a new RNA molecule that is complementary to the nucleotide sequence of the DNA. (*c*) The newly formed (transcribed) RNA is then separated from the DNA molecule and used by the cell.

Translation

Translation is the process of using the information in RNA to direct the ordered assembly of amino acids. The word "translation" refers to the fact that nucleic acid language is being changed to protein language. To translate mRNA language into protein language, a translation dictionary is necessary. Table 7.3 shows an amino acid–mRNA nucleic acid dictionary. A 3-nucleotide combination is correlated to the single amino acid that is required in the process of translation. Each 3-nucleotide combination is called a *codon*. Consider that each codon codes for one, and only one, amino acid. The codon UUU corresponds to only phenylalanine (Phe). However, notice that more than one mRNA codon may code for the same amino acid. Phenylalanine (Phe) can be coded for by the UUU codon and the UUC codon. This is possible because there are only 20 amino acids and 64 different codons (table 7.4). The 64 codons listed in table 7.3 are all the possible ways that the four different nucleotides of RNA can be arranged in combinations of three.

The construction site of the protein molecules (i.e., the translation site) is on the **ribosome** in the cytoplasm of the cell. The ribosome is a cellular organelle that serves as the meeting place for mRNA and the tRNAs that carry amino acid building blocks. There are many ribosomes in a cell. The mRNA and the tRNAs were synthesized by the process of transcription in the cell's nucleus and then moved to the cell's cytoplasm. Ribosomes can be found free in the cytoplasm or attached to the endoplasmic reticulum (ER). Proteins destined to be part of the cell membrane or packaged for export from the cell are synthesized on ribosomes attached to the endoplasmic reticulum. Proteins that are to perform their function in the cytoplasm are synthesized on unattached ribosomes.

While we have initially predicted the amino acid sequence of the protein by using the codon table, the cellular process of translation is a bit more involved. The process of translation can be broken down into three basic processes, (1) initiation; (2) elongation; and (3) termination. The processes of initiation and termination happen only once for

Table 7.3

AMINO ACID–NUCLEIC ACID DICTIONARY

Table 7.4

THE 20 COMMON AMINO ACIDS AND THEIR ABBREVIATIONS

These are the 20 common amino acids used in the protein synthesis operation of a cell. Each has a known chemical structure.

Amino Acid	Three-Letter Abbreviation	Amino Acid	Three-Letter Abbreviation
alanine	Ala	leucine	Leu
arginine	Arg	lysine	Lys
asparagine	Asn	methionine	Met
aspartic acid	Asp	phenylalanine	Phe
cysteine	Cys	proline	Pro
glutamic acid	Glu	serine	Ser
glutamine	Gln	threonine	Thr
glycine	Gly	tryptophan	Trp
histidine	His	tyrosine	Tyr
isoleucine	Ile	valine	Val

each protein molecule that is synthesized. The second step, elongation, happens once for each amino acid that is needed in the protein.

Initiation

Amino acids are not nucleic acids, so their chemical properties and behaviors are not the same. Amino acids will not "play" by the nucleic acid base-pairing rules. As a result, the cell needs a way to match up amino acids with the three-letter codons in the mRNA. The cell's machinery has compensated for this by matching up the different amino acids with an adaptor molecule that can base-pair with the nucleotides of the codons. This adaptor molecule is called *tRNA*.

Protein synthesis does not begin with the very first nucleotide of the mRNA. Rather, protein synthesis starts somewhere in the middle of the mRNA molecule. For this reason, the cell needs a signal that says, "Start protein synthesis here!" That signal is the 3-nucleotide sequence, AUG, aptly named the *start*, or **initiator codon** (figure 7.5). The initiator codon always brings in the methionine (Met) amino acid. After the start codon, the nucleotides are read sequentially in sets of 3 to bring in the correct amino acids.

The correct amino acid is correctly paired to each codon because of the tRNA molecule to which the amino acid is attached. The tRNA has bases that match the codon of the mRNA. The trio of bases in the tRNA that is involved in this base-pairing is called the **anticodon**.

Elongation

Once protein synthesis is initiated, the ribosome, mRNA, and tRNA undergo a repetitive series of events to bring in each subsequent amino acid of the protein (figure 7.6). The

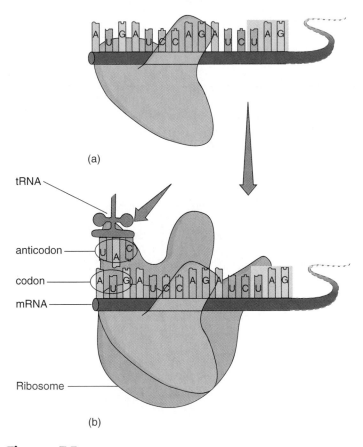

(a)

(b)

Figure 7.5

Initiation

(*a*) An mRNA molecule is positioned in the ribosome so that two codons are in position for transcription. The first of these two codons (AUG) is the initiation codon. (*b*) The tRNA lines up with mRNA over the AUG codon using base-pairing rules between the tRNA and the mRNA.

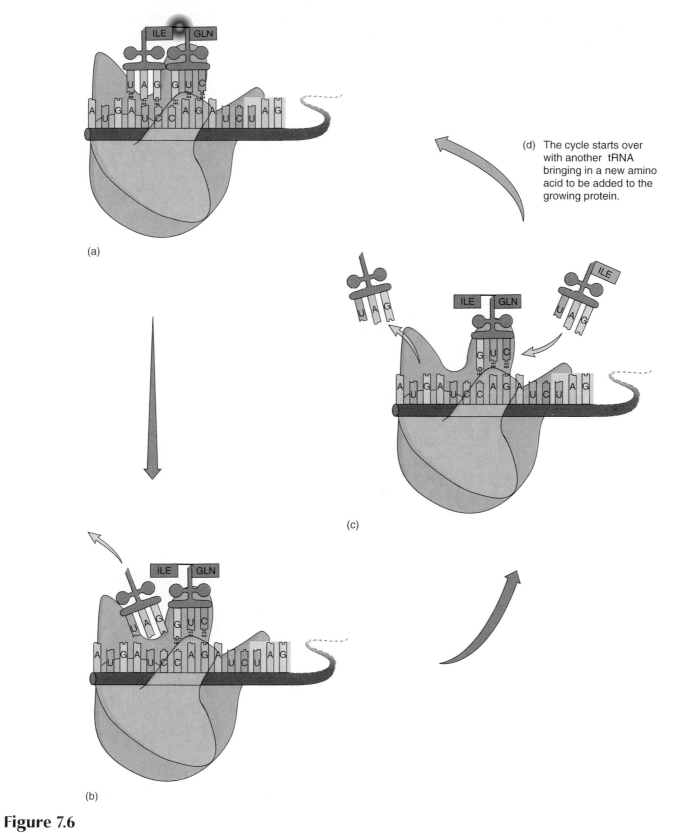

(d) The cycle starts over with another tRNA bringing in a new amino acid to be added to the growing protein.

(a)

(c)

(b)

Figure 7.6

Elongation

(*a*) The two tRNAs align the amino acids (leu and tyr) so that they can be chemically attached to one another by a peptide bond. (*b*) Once the bond is formed, (*b*) the first tRNA detaches from its position on the mRNA. (*c*) The ribosome moves down one codon on the mRNA. Another tRNA now aligns so that the next amino acid (gly) can be added to the growing protein. (*d*) The process continues with a new tRNA, a new amino acid, and formation of a new peptide bond.

order for the entry of amino acids into the growing protein is dictated by base-pairing rules between the codon of the mRNA and the anticodon of the tRNA. The steps of protein elongation are summarized here and are then explained in a bit more detail:

1. The next tRNA, with its attached amino acid, moves into position and pairs with the codon on the mRNA on the ribosome.
2. The amino acid is chemically bonded to the growing protein.
3. The empty tRNA now moves out of position and back into the cell's cytoplasm.
4. The ribosome shifts down the mRNA to use the information in the next codon.

During elongation, the ribosome actually works with 2 side-by-side codons at a time. At the beginning of the elongation cycle, 1 codon is empty of tRNA and the other codon has the forming protein attached to the last tRNA used to bring in an amino acid. With this combination of components in place, step 1 of elongation occurs. The new tRNA with the correct amino acid moves into position. Once the new tRNA is in position, the ribosome actually holds 2 tRNAs in place for a moment. The older tRNA holds the growing protein and the newer tRNA holds the next amino acid to be added to the growing protein. The ribosome moves the growing protein from the older of the 2 tRNAs to the amino acid on the newest tRNA. After this occurs, the older and empty tRNA moves out of position with regard to the ribosome. The ribosome then shifts down the mRNA one codon so that the beginning position of elongation is available again (see figure 7.6). This cycle continues until protein synthesis is terminated.

Termination

The mRNA coding for a protein is not read to its end to finish protein synthesis. Just as there was initiation codon for starting protein synthesis, there are also **termination codons** that allow protein synthesis to end in the middle of the mRNA molecule. By closely looking at table 7.3, you will see that 3 different codons can stop protein synthesis. When any of the three codons (UAA, UAG, or UGA) appear during the elongation process, a release factor rather than a new amino acid is brought into place on the ribosome (figure 7.7). The forming protein is released because no free amino acid is present. Once the protein is released, the ribosome in turn releases the mRNA. This mRNA can be used to make another protein or can be degraded by the cell's enzymes.

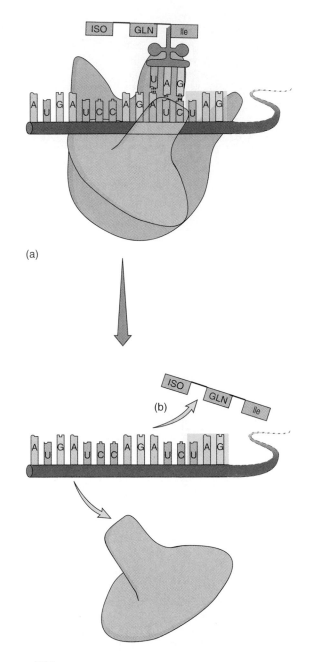

(a)

(b)

Figure 7.7

Termination

(a) A release factor will move into position over a termination codon UAG. (b) The ribosome releases the completed amino acid chain. The ribosome disassembles and the mRNA can be used by another ribosome to synthesize another protein.

mRNA codon   Amino acid

(a)

Glycine
placed in
protein

(b)

Alanine
Translation placed in
protein

(c)

Glycine
placed in
protein

Figure 7.8

DNA Mutations Can Be Silent

A nucleotide substitution changes the protein only if the changed codon results in a different amino acid being substituted into a protein chain. This feature of DNA serves to better ensure that the synthesized protein will be functional. (*a*) In the example provided, the original codon, GGA, calls for Gly. (*b*) An observable mutation changes the nucleotide in the second position of the codon, which now reads GCA. GCA calls for Ala. (*c*) A *silent* mutation is shown where the third position of the codon is changed. The codon GGG calls for the same amino acid as the original version (GGA). Because the proteins produced in the first and third examples will be identical in amino acid sequence, they will function the same also.

7.3 Mutations

Any change in the nucleotide sequence of DNA—the genetic information—is a mutation. Some mutations result in changes in the proteins produced and some do not. As stated earlier, as a genetic material, DNA must be able to change, or mutate. Different types of changes can occur and these changes have a potential impact on protein synthesis.

Normal Sequence	THE ONE BIG FLY HAD ONE RED EYE
Kind of Mutation	*Sequence Change*
Missense	THQ ONE BIG FLY HAD ONE RED EYE
Nonsense	THE ONE BIG
Frameshift	
Due to insertion	THE ONE QBI GFL YHA DON ERE DEY E
Due to deletion	TEO NEB IGF LYH ADO NER EDE YE
Deletion	THE ONE BIG HAD ONE RED EYE
Duplication	THE ONE BIG FLY FLY HAD ONE RED EYE
Insertion	THE ONE BIG WET FLY HAD ONE RED EYE
Expanding mutation:	
Parents	THE ONE BIG FLY HAD ONE RED EYE
Children	THE ONE BIG FLY FLY FLY HAD ONE RED EYE
Grandchildren	THE ONE BIG FLY FLY FLY FLY FLY FLY HAD ONE RED EYE

Figure 7.9

Classes of Mutations

A sentence comprised of three-letter words can provide an analogy to the effect of mutations on a gene's nucleotide sequence. Examine each type of mutation to see how the information in the sentence has been confused by the change.

DNA strikes a good balance between stability and flexibility. Every change in DNA does not cause a change in a protein. For example, the codon GGX codes for the amino acid, Gly, regardless of the nucleotide found in the X position (table 7.3). This is illustrated in figure 7.8 with another example. This means that a large number of changes in DNA sequence may not result in changes in proteins; these are called *silent mutations*. However, not all such changes can be compensated for by the flexibility codon system, and an altered protein may be produced.

Mutagenic agents are substances or conditions that can cause the sequence of DNA to change. Agents known to cause damage to DNA are certain viruses (e.g., papillomavirus), weak or "fragile" spots in the DNA, radiation (X rays or UV light), and chemicals found in foods and other products such as nicotine in tobacco. All have been studied extensively and there is little doubt that they cause mutations.

What types of DNA mutations can be the most devastating to protein synthesis? Consider that any change in the start codon will prevent the protein from being correctly synthesized. Also consider that inserting or deleting a single nucleotide in the protein coding sequence will cause the ribosome to read the wrong set of 3 nucleotides as the codon. This type of mutation is called a *frameshift mutation*. Figure 7.9 provides examples of the types of mutations that are commonly observed in DNA sequences.

To identify the impact that a mutation can have on a protein and, ultimately, the whole organism, consider sickle-cell anemia. In some individuals, a single nucleotide of the gene may be changed. This type of mutation is called a **point mutation.** An example of the effects of altered DNA can be seen in human red blood cells. Red blood cells contain the oxygen-carrier molecule, hemoglobin. Normal hemoglobin molecules are composed of approximately 145 amino acids in four chains—two alpha and two beta. The nucleotide sequence of the gene for the beta chain is known, as is the amino acid sequence for this chain. In normal individuals, the amino acid sequence of the hemoglobin protein begins like this:

Val-His-Leu-Thr-Pro-Glu-Glu-Lys . . .

As a result of a single nucleotide in the DNA sequence, the resulting mutation is a new amino acid sequence in all the red blood cells:

Val-His-Leu-Thr-Pro-Val-Glu-Lys . . .

This single nucleotide change, which causes a single amino acid to change, may seem minor. However, it is the cause of sickle-cell anemia, a disease that affects the red blood cells by changing them from a circular to a sickle shape when oxygen levels are low (figure 7.10). When this sickling occurs, the red blood cells do not flow smoothly through capillaries. Their irregular shapes cause them to clump, clogging the blood vessels. This prevents them from delivering their oxygen load to the oxygen-demanding tissues. A number of physical disabilities may result, including weakness, brain damage, pain and stiffness of the joints, kidney damage, rheumatism, and, in severe cases, death.

Changes in the structure of DNA may have harmful effects on the next generation if they occur in the sex cells. Sex cells like sperm and eggs transmit genetic information from one generation in a species to the next. Mutations that occur to DNA molecules can only be passed on to the next generation when the mutation is present in cells like sperm and eggs that are used to start the next generation.

Some damage to DNA is so extensive that the entire strand of DNA is broken, resulting in the synthesis of abnormal proteins or a total lack of protein synthesis. A number of experiments indicate that many street drugs such as LSD (lysergic acid diethylamide) are mutagenic agents that cause DNA to break.

(a)   (b)

Figure 7.10

Normal and Sickled Red Blood Cells
(*a*) A normal red blood cell is shown in comparison with (*b*) a cell having the sickle shape. This sickling is the result of a single amino acid change in the hemoglobin molecule.

7.4 Controlling Gene Expression

With a solid understanding of transcription and translation, we can begin to look at the implications of that system of information exchange within the context of the cell. At a higher level of organization, you can also look at the central dogma in the context of an entire multicellular organism. One of the first things to consider is that no cell is able to survive with the ability to make just one protein. This means that even the simplest of cells have a large bank of protein blueprints to draw from. This bank of blueprints, the **genome,** is similar to a cook having many recipes to draw from in a single cookbook. To function correctly, cells need to be able to access and use their genetic information in an efficient and successful manner.

Starting Transcription

Just like many recipes can exist on a single page of a cookbook, the blueprints in a genome are not always on separate molecules of DNA. A single molecule of DNA may have many thousands of protein-coding regions. Because there are so many genes in a cell and the codes for many different proteins can be found on a single molecule of DNA, the cell is faced with a number of problems:

1. How does the cell know where to start making the mRNA?
2. How does the cell know which strand of the double-stranded DNA molecule is used for a template of transcription?

Figure 7.11

Gene Structure
There are four general regions of a gene. (*a*) The promoter region attracts the RNA polymerase to the site of the gene and then directs the RNA polymerase to proceed along the DNA strand in the correct direction to find the gene. (*b*) The protein coding region contains the information needed to correctly create a protein—this includes the start codon, the stop codon, and all the codon that are located between them. (*c*) The termination sequence signals the RNA polymerase to end mRNA transcription so that the mRNA can leave the nucleus and enter the cytoplasm for use in the next step of the central dogma, translation. (*d*) The regulatory region can be found all around the promoter region, the protein-coding region, and the termination sequence. The regulatory regions attract and bind proteins that will help or slow down the process of transcription.

3. How does the cell know when to stop making the mRNA?

To understand the answers to these questions, it is necessary to know that there is more information stored in a gene than the portion that codes for a protein. Genes contain four general types of information that is organized into four regions: (1) The **promoter region,** (2) the **protein-coding region,** (3) the **termination sequence,** and (4) the **regulatory regions** (figure 7.11). We have already discussed the protein-coding region, how its information is organized into codons, and that it contains very specific stop and start signals.

The promoter region contains the answer to the first two questions. The promoter region of a gene is a specific sequence of DNA that is located a number of nucleotides upstream of the region of DNA that codes for the protein. The unique sequence of the promoter is recognized by the enzyme responsible for transcription, RNA polymerase. The promoter of a gene acts in a manner similar to runway lights for an airplane. The promoter provides a signal that indicates, "Here's a gene and it is this way!" Once the RNA polymerase has "landed" onto the promoter and is moving in the right direction toward the gene, only one of the strands of DNA can be used as a template for making the mRNA. This is due to the chemical details of how nucleotides are added to a growing strand of RNA and how two strands of nucleic acids orient to each other.

RNA polymerase finishes making an RNA molecule when it encounters a termination sequence—it releases the mRNA that is being synthesized and the DNA template. The termination sequence is a sequence of DNA that is located a number of nucleotides after the protein-coding region. The mRNA that has been created for translation actually spans the area shown in figure 7.11 that starts between the promoter and the protein-coding region and ends at the termi-

nation sequence. This mRNA is then used in translation to direct protein synthesis.

The general scheme presented here is that genetic information flows from DNA to RNA and is used to make protein. In the case of some viruses, genetic information is stored in RNA and is used to make DNA. In this instance, the flow of genetic information is RNA to DNA then to mRNA and protein (Outlooks 7.1).

The Right Amount at the Right Time

This basic understanding of how a gene is organized with a promoter, a protein-coding region, and a termination sequence is vital to understanding how cells are able to control **gene expression.** A gene is expressed when its product, usually a protein, is produced for use by the cell. Control of gene expression involves making sure that a gene is used at the appropriate time and that its protein is being made in adequate amounts.

Regulatory regions are important in determining when and how much a gene expresses itself. The regulatory regions surround the other three regions of the gene. They work by binding specific proteins that are made by the cell. These proteins are called **transcription factors.** These proteins are unique in two ways: (1) They are only present in the cell under certain conditions, such as when a gene's protein must be produced; and (2) they have the effect of making the promoter region more noticeable or more invisible to the RNA polymerase.

Splicing

One of the most significant differences between prokaryotic and eukaryotic transcription is that the protein-coding region of the prokaryotic DNA is continuous, while in eukaryotic cells, it is not (figure 7.12). A eukaryotic gene begins

OUTLOOKS 7.1

HIV Infection (AIDS) and Reverse Transcriptase

AIDS is an acronym for acquired immunodeficiency syndrome and is caused by a retrovirus called *human immunodeficiency virus*. HIV is a spherical virus with an outer membrane, an inside protein coat, and an RNA core. Its genetic material is RNA, not DNA. Genes are carried from one generation to the next as RNA molecules. This is not the case in humans and most other organisms where DNA is the genetic material.

Human sequence:

$$DNA \quad \rightarrow \quad RNA \quad \rightarrow \quad protein$$
$$transcriptase$$

However, once having entered a suitable, susceptible host cell, HIV must convert its RNA to a DNA genome in order to integrate into the host cell's chromosome. Only then can it become an active, disease-causing parasite. This conversion of RNA to DNA is contrary, reverse, or retro to the RNA-forming process controlled by the enzyme transcriptase. Humans do not have the genetic capability to manufacture the enzyme necessary to convert RNA to DNA—reverse transcriptase.

Retrovirus sequence:

$$RNA \quad \rightarrow \quad DNA$$
$$reverse\ transcriptase$$

When HIV carries out gene replication and protein synthesis the process is diagrammed:

$$RNA \quad \rightarrow \quad DNA \quad \rightarrow \quad RNA \quad \rightarrow protein$$
reverse   transcriptase   other enzymes
transcriptase   from host cell   from host cell

This has two important implications. First, the presence of reverse transcriptase in a human can be looked upon as an indication of retroviral infection because it is not manufactured by human cells. However, because HIV is only one of several types of retroviruses, the presence of the enzyme in an individual does not necessarily indicate an HIV infection. It does indicate "some type" of retroviral infection. Second, interference with reverse transcriptase will frustrate the virus's attempt to integrate into the host chromosome.

Treatments for HIV take advantage of a vulnerable point in the retrovirus's life cycle. The drugs fall into three basic categories: (1) blockers, (2) reverse transcriptase inhibitors, and (3) protease inhibitors. Drugs classified as blockers prevent the virus from entering the cell. Reverse transcriptase inhibitors interfere with the crucial step of converting the virus's RNA genetic information from RNA to DNA. Protease inhibitors interfere with the modification of viral proteins after translation. This prevents a complete virus from completely forming and infecting another cell. A treatment regime can include a drug from each of these categories. This combination of drugs is called a *cocktail* and is much more effective against HIV than taking any one of the drugs alone.

However, HIV mutates very quickly because its genetic information is stored as a single-stranded molecule of RNA. This high mutation rate quickly produces strains of the HIV virus that are resistant to these drugs. • **evolution, p. 222**

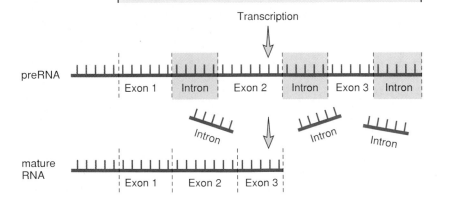

Figure 7.12

Transcription of mRNA in Eukaryotic Cells

This is a summary of the events that occur in the nucleus during the manufacture of mRNA in a eukaryotic cell. Notice that the original nucleotide sequence is first transcribed into an RNA molecule that is later "clipped" and then rebonded to form a shorter version of the original. It is during this time that the introns are removed.

with a promoter region and an initiation code and ends with a termination code and region. However, the intervening gene sequence contains patches of nucleotides that apparently do not code for protein but do serve important roles in maintaining the cell. If they were used in protein synthesis, the resulting proteins would be worthless. To remedy this problem, eukaryotic cells prune these segments from the mRNA after transcription. When such split genes are transcribed, RNA polymerase synthesizes a strand of pre-mRNA that initially includes copies of both exons (meaningful mRNA coding sequences) and introns (intervening DNA sequences). Soon after its manufacture, this pre-mRNA mole-

Figure 7.13

DNA Replication

These illustrations summarize the basic events that occur during DNA replication. (*a*) Enzymes break apart the two strands of DNA. (*b* and *c*) As the DNA strands are separated, new DNA nucleotides are added to the new strands by DNA polymerase. The new DNA strands are synthesized according to base-pairing rules for nucleic acids. (*d*) By working in two directions at once along the DNA strand, the cell is able to more quickly replicate the DNA. Each new daughter cell will receive one of these copies.

cule has the meaningless introns clipped out and the exons spliced together into the final version, or mature mRNA, which is used by the cell. In humans, it has been found that the exons of a single gene may be spliced together in three different ways, resulting in the production of three different, mature messenger RNAs. This means that a single gene can be responsible for the production of three different proteins. Learning this information has led geneticists to revise their estimate of the total number of genes found in the human genome from 100,000 to an estimated 30,000.

7.5 Nucleotide Sequences— The Cell's Legacy

We have looked at how cells use genetic information in DNA to make proteins, but cells use this information in a second, *very different* manner: They pass it on to their offspring. The process of passing DNA on to the next generation of cells helps to ensure that the offspring cells will have the genetic information necessary to carry out the needed chemical reactions of life.

DNA Replication

DNA replication occurs in cells in preparation for cell division. Without replication, daughter cells would not receive the library of information required to sustain life. Because all cells must maintain a complete set of genetic material, there must be a doubling of DNA to produce enough copies to pass on to the offspring. DNA replication is the process of duplicating the genetic material prior to its distribution to daughter cells. When a cell divides into two *daughter cells*, each new cell must receive a complete and exact copy of the parent cell's genetic information or it will not be able to manufacture all the proteins vital to its existence.

DNA replication involves a few general steps:

1. The DNA replication process begins as an enzyme breaks the attachments between the paired bases of the two strands of DNA. In eukaryotic cells, this occurs in hundreds of different positions forming replication bubbles along the length of the DNA. In prokaryotic cells, this occurs in only one spot, the origin of replication (figure 7.13*a*).
2. As the two DNA strands are separated in both directions along the DNA molecule, new DNA

nucleotides bond to the nucleotides on the existing DNA strands using the base-pairing rules and an enzyme called DNA polymerase (figure 7.13b).

3. DNA polymerase sequentially attaches each incoming nucleotide to the previous one by chemical bonds between the phosphate of one nucleotide and the deoxyribose of the next (figure 7.13c).

4. Eventually all the replication bubbles combine and a new strand of DNA is formed on each of the original strands (figure 7.13d). The result is two identical double-stranded DNA molecules.

Note that the overall pattern is that the new complementary strand of DNA forms on each of the old DNA strands, resulting in the formation of two double-stranded, duplex DNA molecules. In this way, the exposed nitrogenous bases of the original DNA serve as a template, or pattern, for the formation of the new DNA. As the new DNA is completed, it twists into its double-helix shape.

The completion of the DNA replication process yields two double helices that are identical in their nucleotide sequences. Half of each is new; half is the original parent DNA molecule. The DNA replication process is highly accurate. It has been estimated that there is only one error made for every 2×10^9 nucleotides. A human cell contains 46 chromosomes consisting of about 3 billion base pairs. This averages to about five errors per cell! Don't forget that this figure is an estimate. Whereas some cells may have five errors per replication, others may have more, and some may have no errors at all. It is also important to note that some errors can be major and deadly, whereas others are insignificant. Because this error rate is so small, DNA replication is considered to be essentially error-free. Following DNA replication, the cell now contains two complete copies of genetic information and is ready to begin the process of distributing one set of genetic information to each of its two daughter cells.

The distribution of DNA involves splitting the cell and distributing a set of genetic information to the two new daughter cells. In this way, each new cell has the necessary information to control its activities. The mother cell ceases to exist when it divides. Its contents are divided between the two smaller daughter cells.

A cell does not really die when it reproduces itself; it merely starts over again. This is called the *life cycle* of a cell. A cell can divide and redistribute its genetic information to the next generation in a number of ways. • mitosis, p. 159, meiosis, p. 170.

DNA and Chromosomes

After DNA is replicated, cells wind the DNA up like thread on a bobbin or spool so that the huge bulk of DNA is more manageable. Eukaryotic and prokaryotic cells do this in different ways.

The genetic material of humans and other eukaryotic organisms consists of strands of coiled, double-stranded DNA,

---

OUTLOOKS 7.2

Telomeres

The ends of a chromosome contain a special sequence of nucleotides called a **telomere.** In humans, these chromosome "caps" contain the nucleotide base-pair sequence:

TTAGGG
AATCCC

This sequence is repeated many times over. Telomeres are very important segments of the chromosome. They are required for chromosome replication; they protect the chromosome from being destroyed by dangerous DNAase enzymes (enzymes that destroy DNA), and they keep chromosomes from bonding to one another end to end. Evidence shows that the loss of telomeres is associated with cell "aging," whereas not removing them has been linked to cancer. Every time a cell reproduces itself, it loses some of its telomeres. However, in cells that have the enzyme telomerase, new telomeres are added to the end of the chromosome each time the cells divide. Therefore, cells that have telomerase do not age as other cells do, and cancer cells are immortal as a result of this enzyme. This enables them to maintain, if not increase, the number of telomeres from one cell generation to the next.

---

which has histone proteins attached along its length. These coiled DNA strands with attached proteins become visible during cell division and are called **nucleoproteins** or **chromatin fibers.** The histone protein and DNA are not arranged randomly, but come together in a highly organized pattern. The double-stranded DNA spirals around repeating clusters of eight histone spheres. Histone clusters with their encircling DNA are called **nucleosomes** (figure 7.14a). When eukaryotic chromatin fibers coil into condensed, highly knotted bodies, they are seen easily through a microscope after staining them with dye. Condensed like this, a chromatin fiber is referred to as a **chromosome** (figure 7.14b; Outlooks 7.2).

The genetic material in bacteria is also double-stranded DNA, but the ends of the molecule are connected to form a loop and they do not form condensed chromosomes (figure 7.15). However, prokaryotic cells have an attached protein called *HU protein*. In certain bacteria, there is an additional loop of DNA called a *plasmid*. Plasmids are considered extra DNA because they appear not to contain genes that are required for the normal function of the cell. However, they can play two important roles in bacteria: (1) Some plasmids have genes that enable the cell to resist certain antibiotics such as the penicillins. The gene may be for the production of the enzyme beta lactamase (formerly known as penicillinase), which is capable of destroying certain forms of penicillin. It is important to note that plasmids are capable of carrying many kinds of genes. We tend to

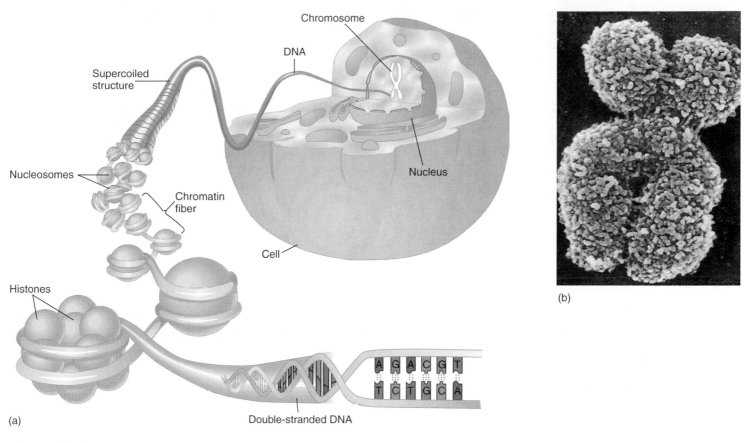

(b)

Figure 7.14

Eukaryotic Genome Packaging

(a) Eukaryotic cells contain DNA in their nuclei that takes the form of a double-stranded helix. The two strands fit together and are bonded by weak hydrogen bonds formed between the complementary, protruding nitrogenous bases according to the base-pairing rule. To form a chromosome, the DNA molecule is wrapped around a group of several histone proteins. Together the histones and the DNA form a structure called the nucleosome. The nucleosomes are packaged to form a chromosome. (b) During certain stages in the reproduction of a eukaryotic cell, the nucleoprotein coils and "supercoils," forming tightly bound masses. When stained, these are easily seen through the microscope. In their supercoiled form, they are called *chromosomes*, meaning "colored bodies."

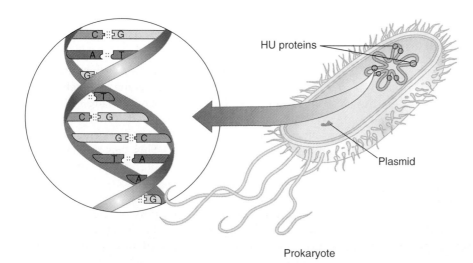

Prokaryote

Figure 7.15

Prokaryotic Genome Packaging

The nucleic acid of prokaryotic cells does not have the histone protein; rather, it has proteins called *HU* proteins that serve a similar function to histones. In addition, the ends of the giant nucleoprotein molecule overlap and bind with one another to form a loop. The additional small loop of DNA is a plasmid, which contains genes that are not essential for the daily life of the cell.

be most concerned about plasmids that carry antibiotic resistance because of the medical implications.

A second important role of plasmids enables the cell to become involved in *genetic recombination*—the transfer of genes from one cell (the donor) to another (the recipient). By transferring genes from one cell to another, cells that receive the genes can become genetically diverse and are more likely to survive threatening environmental hazards.

Each chromatin strand is different because each strand has a different chemical code. Coded DNA serves as a central cell library. Tens of thousands of messages are in this storehouse of information. This information tells the cell such things as (1) how to produce enzymes required for the digestion of nutrients; (2) how to manufacture enzymes that will metabolize the nutrients and eliminate harmful wastes; (3) how to repair and assemble cell parts; (4) how to reproduce healthy offspring; (5) when and how to react to favorable and unfavorable changes in the environment; and (6) how to coordinate and regulate all of life's essential functions. If any of these functions are not performed properly, the cell will die. The importance of maintaining essential DNA in a cell becomes clear when we consider cells that have lost it. For example, human red blood cells lose their nuclei as they become specialized to carry oxygen and carbon dioxide throughout the body. Without DNA, they are unable to manufacture the essential cell components needed to sustain themselves. They continue to exist for about 120 days, functioning only on enzymes manufactured earlier in their lives. When these enzymes are gone, the cells die. Because these specialized cells begin to die the moment they lose their DNA, they are more accurately called *red blood corpuscles* (RBCs): "little dying red bodies."

Chromosomal Aberrations

A **chromosomal aberration** is a major change in DNA that can be observed at the level of the chromosome. There are four types of aberrations: inversions, translocations, duplications, and deletions. An *inversion* occurs when a chromosome is broken and a piece becomes reattached to its original chromosome but in reverse order—it has been cut out and flipped around. A *translocation* occurs when one broken segment of DNA becomes integrated into a different chromosome. *Duplications* occur when a portion of a chromosome is replicated and attached to the original section in sequence. *Deletion* aberrations result when a broken piece becomes lost or is destroyed before it can be reattached.

7.6 Using DNA to Our Advantage

A powerful new science of gene manipulation—**biotechnology**—suggests genetic diseases can be controlled or cured in the future. Since 1953, when the structure of the DNA molecule was first described, there has been a rapid succession of advances in the field of genetics. It is now possible to transfer DNA from one organism to another. This has made possible the manufacture of human genes and gene products by bacteria.

Genetic Engineering and Biotechnology

Biotechnology includes the use of a method of splicing genes from one organism into another, resulting in a new form of DNA called **recombinant DNA.** Organisms with these genetic changes are referred to as **genetically *m*odified (GM),** or **transgenic organisms.** These organisms or their offspring have been engineered so that they contain genes from at least one unrelated organism, which could be a virus, a bacterium, a fungus, a plant, or an animal.

As this highly sophisticated procedure has been refined, it has become possible to quickly and accurately splice genes from a variety of species into host bacteria, making possible the synthesis of large quantities of medically important products. For example, recombinant DNA procedures are responsible for the production of human insulin, used in the control of diabetes; interferon, used as an antiviral agent; human growth hormone, used to stimulate growth in children lacking this hormone; and somatostatin, a brain hormone also implicated in growth. Over 200 such products have been manufactured using these methods.

Revolutionary possibilities open up with the manipulation of DNA. These methods enable cells to produce molecules that they would not normally make. Some research laboratories have even spliced genes into laboratory-cultured human cells. Should such a venture prove to be practical, genetic diseases such as sickle-cell anemia could be controlled. The process of recombinant DNA gene splicing also enables cells to be more efficient at producing molecules that they normally synthesize.

Some of the likely rewards are (1) production of additional, medically useful proteins; (2) mapping of the locations of genes on human chromosomes; (3) a more complete understanding of how genes are regulated; (4) production of crop plants with increased yields; and (5) development of new species of garden plants (figure 7.16).

The discovery of the structure of DNA nearly 50 years ago seemed very far removed from the practical world. The importance of this "pure" or "basic" research is just now being realized. Many companies are involved in recombinant DNA research with the aim of alleviating or curing disease.

The field of **bioengineering** is advancing as quickly as is the electronics industry. The first bioengineering efforts focused on developing genetically modified (GM) crops that had improvements over past varieties, such as increased resistance to infectious plant disease. This was primarily accomplished through selective breeding and irradiation of cells to produce desirable mutations. The second wave of research involved directly manipulating DNA using the more sophisticated techniques such as the polymerase chain reaction (PCR), genetic fingerprinting, and cloning. Genetic engineers identify and isolate sequences of nucleotides from a living or dead cell and install it into another living cell. This

Figure 7.16

Using our understanding of how DNA codes for proteins allows scientists to develop treatments for disease as welll as develop crops with greater yield and nutritional value.

has resulted in improved food handling and processing, such as slower ripening in tomatoes. Currently, crops are being genetically manipulated to manufacture large quantities of specialty chemicals such as antibiotics, steroids, and other biologically useful, organic chemicals.

Although some of these chemicals have been produced in small amounts from genetically engineered microorganisms, crops such as turnips, rice, soybeans, potatoes, cotton, corn, and tobacco can generate tens or hundreds of kilograms of specialty chemicals per year. Many of these GM crops also have increased nutritional value and yet can be cultivated using traditional methods. Such crops have the potential of supplying the essential amino acids, fatty acids, and other nutrients now lacking in the diets of people in underdeveloped or developing nations. Researchers have also shown, for example, that turnips can produce interferon (an antiviral agent), tobacco can create antibodies to fight

human disease, oilseed rape plants can serve as a source of human brain hormones, and potatoes can synthesize human serum albumin that is indistinguishable from the genuine human blood protein.

Gene cloning is accomplished by using enzymes that are naturally involved in the DNA replication process and other enzymes that are naturally produced by bacteria. When genes are spliced from different organisms into host cells, the host cell replicates these new "foreign" genes and synthesizes proteins encoded by them. Gene splicing begins with the laboratory isolation of DNA from an organism that contains the desired gene—for example, from human cells that contain the gene for the manufacture of insulin. If the gene is short enough and its base sequence is known, it can be synthesized in the laboratory from separate nucleotides. If the gene is too long and complex, it is cut from the chromosome with enzymes called *restriction endonucleases,* so named because these enzymes (-ases) only cut DNA (nucle-) at certain base sequences (restricted in their action) and work from the middle or inside of (endo-) the DNA. In contrast, a restriction exonuclease digests DNA from the outside (exo-) or from the end of the DNA molecule. *Endo* nucleases act like molecular scissors that do not cut the DNA straight across, but in a zig-zag pattern that leaves one strand slightly longer than its complement. The short nucleotide sequence that sticks out and remains unpaired is called a *sticky end* because it can be reattached to another complementary strand.

This isolated gene with its "sticky end" is most frequently spliced into microbial DNA. The host DNA is opened up with the proper restriction endonuclease and ligase (i.e., tied together) enzymes that are used to attach the sticky ends into the host DNA. This gene-splicing procedure can be performed with small loops of bacterial DNA that are not part of the main chromosome. These small DNA loops are called *plasmids.*

Once the splicing is completed, the plasmids can be inserted into the bacterial host cell by treating the cell with special chemicals that encourage it to take in these large chunks of DNA. A more efficient alternative is to splice the desired gene into the DNA of a bacterial virus so that it can carry the new gene into the bacterium as it infects the host cell. Once inside the host cell, the genes can be replicated, along with the rest of the DNA, to clone the "foreign" gene, or they can begin to synthesize the encoded protein.

With another genetic engineering accomplishment, *genetic fingerprinting*, it is possible to show the nucleotide sequence differences between individuals since no two people have the same nucleotide sequences. While this sounds like an easy task, the presence of many millions of base pairs in a person's chromosomes makes this a lengthy, and sometimes impractical, process. Therefore, scientists don't really do a complete fingerprint but focus only on certain shorter, repeating patterns in the DNA. By focusing on these shorter, repeating nucleotide sequences, it is possible to determine whether samples from two individuals have these same repeating segments. Scientists use a small number of sequences that are known to vary a great deal among individuals, and compare those to get a certain probability of a match. The more similar the sequences, the more likely the two samples are from the same person. The less similar the sequences, the less likely the two samples are from the same person. In criminal cases, DNA samples from the crime site can be compared to those taken from suspects. If 100% of the short, repeating sequence matches, it is highly probable that the suspect was at the scene of the crime and may be the guilty party. This same procedure can also be used to confirm the identity of a person, as in cases of amnesia, murder, or accidental death (figure 7.17).

In 1989, the American Association for the Advancement of Science named DNA polymerase "Molecule of the Year." The value of this enzyme in the *polymerase chain reaction* (PCR) is so great that it could not be ignored. Just what is the PCR, how does it work, and what can you do with it?

PCR is a laboratory procedure for copying selected segments of DNA. A single cell can provide enough DNA for analysis and identification! Having a large number of copies of a "target sequence" of nucleotides enables biochemists to more easily work with DNA. This is like increasing the one "needle in the haystack" to such large numbers (100 billion in only a matter of hours) that they're not hard to find, recognize, and work with. The types of specimens that can be used include semen, hair, blood, bacteria, protozoa, viruses, mummified tissue, and frozen cells. The process requires the DNA specimen, free DNA nucleotides, synthetic "primer" DNA, DNA polymerase, and simple lab equipment, such as a test tube and a source of heat.

Having decided which target sequence of nucleotides (which "needle") is to be replicated, scientists heat the specimen of DNA to separate the coding and non-coding strands. Molecules of synthetic "primer" DNA are added to the specimen. These primer molecules are specifically designed to attach to the ends of the target sequence. Next, a mixture of triphosphorylated nucleotides is added so that they can become the newly replicated DNA. The presence of the primer, attached to the DNA and added nucleotides, serves as the substrate for the **DNA polymerase.** Once added, the polymerase begins making its way down the length of the DNA from one attached primer end to the other. The enzyme bonds the new DNA nucleotides to the strand, replicating the molecule as it goes. It stops when it reaches the other end, having produced a new copy of the target sequence. Because the DNA polymerase will continue to operate as long as enzymes and substrates are available, the process continues, and in a short time there are billions of small pieces of DNA, all replicas of the target sequence.

So what, you say? Well, consider the following. Using the PCR, scientists have been able to:

1. More accurately diagnose such diseases as sickle-cell anemia, cancer, Lyme disease, AIDS, and Legionnaires' disease;
2. Perform highly accurate tissue typing for matching organ-transplant donors and recipients;
3. Help resolve criminal cases of rape, murder, assault, and robbery by matching suspect DNA to that found at the crime scene;
4. Detect specific bacteria in environmental samples;
5. Monitor the spread of genetically engineered microorganisms in the environment;
6. Check water quality by detecting bacterial contamination from feces;
7. Identify viruses in water samples;
8. Identify disease-causing protozoa in water;
9. Determine specific metabolic pathways and activities occurring in microorganisms;
10. Determine races, distribution patterns, kinships, migration patterns, evolutionary relationships, and rates of evolution of long-extinct species;
11. Accurately settle paternity suits;
12. Confirm identity in amnesia cases;

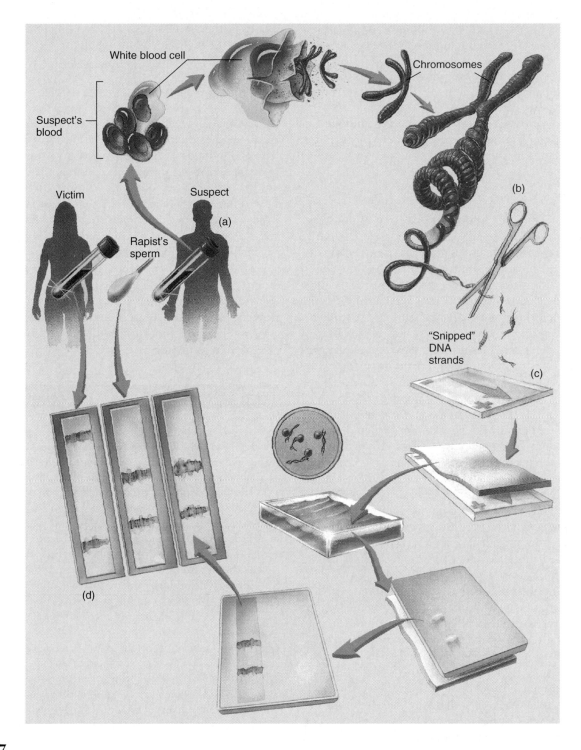

Figure 7.17

DNA Fingerprinting

Because every person's DNA is unique (a), when samples of an individual's DNA are subjected to restriction enzymes, the cuts will occur in different places and DNA chunks of different sizes will result. (b) Restriction enzymes have the ability to cut DNA at places where specific sequences of nucleotides occur. When the chunks are caused to migrate across an electrophoresis gel (c), the smaller fragments migrate farther than the larger fragments, producing a pattern known as a "DNA fingerprint" (d). Because of individual differences in DNA sequences, these sites vary from person to person. As a result, the DNA fingerprint that separates DNA fragments on the basis of size can appear different from one person to another. Several controls are done in this type of experiment. One of the controls is the presence of the victim's DNA. Why is this important for the interpretation of this test?

13. Identify a person as a relative for immigration purposes;
14. Provide the basis for making human antibodies in specific bacteria;
15. Possibly provide the basis for replicating genes that could be transplanted into individuals suffering from genetic diseases; and
16. Identify nucleotide sequences peculiar to the human genome (an application is currently underway as part of the Human Genome Project).

Gene Therapy

The fields of biotechnology and bioengineering allow scientists and medical doctors to work together and potentially cure genetic disorders. Unlike contagious diseases, genetic diseases cannot be caught or transmitted because they are caused by a genetic predisposition for a particular disorder—not a separate, disease-causing organism like a bacterium. Gene therapy involves inserting genes, deleting genes, or manipulating the action of genes in order to cure or lessen the effect of genetic diseases. These activities are very new and experimental. While these lines of investigation create hope for many, there are many problems that must be addressed before gene therapy becomes a reliable treatment for many disorders.

The strategy for treating someone with gene therapy varies depending on the individual's disorder. When designing a gene therapy treatment, scientists have to ask themselves, "Exactly what is the problem?" Is the mutant gene not working at all? Is it working normally, but is there too little activity? Is there too much protein being made? Or possibly, is the problem that the gene is acting in a unique and new manner? If there is no gene activity or too little gene activity, the scientists need to somehow introduce a more active version of the gene. If there is too much activity or if the problem is caused by the gene having a new activity, this excess activity must first be stopped and then the normal activity restored.

To stop a mutant gene from working, scientists must change it. This typically involves inserting a mutation into the protein-coding region of the gene or the promoter region that is necessary to activate the gene. Scientists have used **transposons** and some types of viruses to do this in organisms other than humans for some time now. The difficulty in this technique is to mutate only that one gene without disturbing the other genes of the cells and creating more mutations in other genes. Once the mutant gene is silenced, the scientists begin the work of introducing a "good" copy of the gene. Again there are many difficulties in this process:

- Scientists must find a way of returning the corrected DNA to the cell.
- The corrected DNA must be made a part of the cell's DNA so that it is passed on with each cell division, it doesn't interfere with other genes, and can be transcribed by the cell as needed.
- Finally, cells containing the corrected DNA must be reintroduced to the patient.

Many of these techniques are experimental and the medical community is still evaluating their usefulness in treating many disorders as well as the risks that these techniques pose to the patient.

Recently, the first efforts to determine the **human genome,** the entire human DNA sequence, were completed. Many scientists feel that advances in medical treatments will occur more quickly by having this information available. Three new fields of biology have grown out of these efforts: **genomics, transcriptomics,** and **proteomics.** Genomics is the study of the DNA sequence and looks at the significance of how different genes and DNA sequences are related to each other. When genes are identified from the DNA sequence, transcriptomics looks at when, where, and how much a gene is expressed. Finally, proteomics examines the proteins that are predicted from the DNA sequence. From these types of studies, we are able to identify gene families that can be used to determine how humans have evolved on a molecular level, how genes are used in an organism throughout its body and over its lifespan, and how to identify common themes from one protein to the next.

SUMMARY

The successful operation of a living cell depends on its ability to accurately use the genetic information found in its DNA. The enzymes that can be synthesized using the information in DNA are responsible for the efficient control of a cell's metabolism. However, the production of protein molecules is under the control of the nucleic acids, the primary control molecules of the cell. The structure of the nucleic acids, DNA and RNA, determines the structure of the proteins, whereas the structure of the proteins determines their function in the cell's life cycle. Protein synthesis involves the decoding of the DNA into specific protein molecules and the use of the intermediate molecules, mRNA and tRNA, at the ribosome. Errors in any of the codons of these molecules may produce observable changes in the cell's functioning and can lead to cell death.

DNA replication results in an exact doubling of the genetic material. The process virtually guarantees that identical strands of DNA will be passed on to the next generation of cells.

Methods of manipulating DNA have led to the controlled transfer of genes from one kind of organism to another. This has made it possible for bacteria to produce a number of human gene products.

THINKING CRITICALLY

For guidelines to answer this question, visit the Online Learning Center.

An 18-year-old college student reported that she had been raped by someone she identified as a "large, tanned white man." A student in her biology class fitting that description was said by eyewitnesses to have been, without a doubt, in the area at approximately the time of the crime. The suspect was apprehended and upon investigation was found to look very much like someone who lived in the area and who had a previous record of criminal sexual assaults. Samples of semen from the woman's vagina were taken during a physical exam after the rape. Cells were also taken from the suspect. He was brought to trial but was found to be innocent of the crime based on evidence from the criminal investigations laboratory. His alibi—that he had been working alone on a research project in the biology lab—held up. Without PCR genetic fingerprinting, the suspect would surely have been wrongly convicted, based solely on circumstantial evidence provided by the victim and the "eyewitnesses."

Place yourself in the position of the expert witness from the criminal laboratory who performed the PCR genetic fingerprinting tests on the two specimens. The prosecuting attorney has just asked you to explain to the jury what led you to the conclusion that the suspect could not have been responsible for this crime. Remember, you must explain this to a jury of 12 men and women who, in all likelihood, have little or no background in the biological sciences. Please, tell the whole truth and nothing but the truth.

CONCEPT MAP TERMINOLOGY

Construct a concept map using these terms.

base-pairing
complementary bases
DNA polymerase
DNA repair
mutation
replication
template

Visit the Online Learning Center to use an interactive format to construct another concept map.

KEY TERMS

Use interactive flash cards, on the Online Learning Center, to help you learn the meanings of these terms.

adenine
anticodon
bioengineering
biotechnology
central dogma
chromatin fibers
chromosomal aberrations
chromosome
coding strand
codon
complementary base
cytosine
deoxyribonucleic acid (DNA)
deoxyribose
DNA code
DNA polymerase
DNA replication
expressed
gene
gene expression
genetically *modified* (GM)
genome
genomics
guanine
human genome
initiator codon
messenger RNA (mRNA)
mutagenic agent
mutation
nitrogenous base
nucleic acids
nucleoproteins
nucleosomes
nucleotide
point mutation
promoter region

protein-coding region
protein synthesis
proteomics
recombinant DNA
regulatory region
ribonucleic acid (RNA)
ribose
ribosomal RNA (rRNA)
RNA polymerase
telomere
template
termination codon
termination sequence
thymine
transcription
transcription factors
transcriptomics
transfer RNA (tRNA)
transgenic organisms
translation
transposons
uracil

LEARNING CONNECTIONS

Visit the Online Learning Center to find additional study resources including those highlighted below.

7.1   The Central Dogma

Interactive Concept Maps
• The central dogma

1. What is the product of transcription? Translation?
2. What is a gene?

7.2   Nucleic Acid Structure and Function

Animations and Review
• DNA structure
• Transcription
• Translation

3. What are the differences among a nucleotide, a nitrogenous base, and a codon?
4. What are the differences between DNA and RNA?
5. List the sequence of events that takes place when a DNA message is translated into protein.
6. If a DNA nucleotide sequence is CATAAAGCA, what is the mRNA nucleotide sequence that would base-pair with it?
7. What amino acids would occur in the protein chemically coded by the sequence of nucleotides in question 6?
8. How do tRNA, rRNA, and mRNA differ in function?

7.3   Mutations

Interactive Concept Map
• Effects of mutations

9. Both chromosomal and point mutations occur in DNA. In what ways do they differ? How is this related to recombinant DNA?
10. What is a silent mutation? Provide an example.

7.4   Controlling Gene Expression

Animations and Review
• Gene activity

11. Provide two examples of why it is advantageous for a cell to control gene expression.
12. Provide two examples of how a cell uses transcription to control gene expression.

7.5 Nucleotide Sequences—The Cell's Legacy

Animations and Review
• DNA replication

13. What is polymerase and how does it function?
14. Why is DNA replication necessary?
15. How does DNA replication differ from the manufacture of an RNA molecule?

7.6 Using DNA to Our Advantage

Experience This!
• Engineering agriculture
• New careers

Case Study
• *The New York Times:* Iceland Sells Genes for Profit

Food for Thought
• DNA dragnet

Interactive Concept Map
• Text Concept Map

16. Provide two examples of agricultural benefits developed from DNA technology.
17. Provide two examples of medical benefits developed from DNA technology.

Test yourself with multiple choice questions that provide immediate feedback
on the Online Learning Center with PowerWeb at

www.mhhe.com/enger11

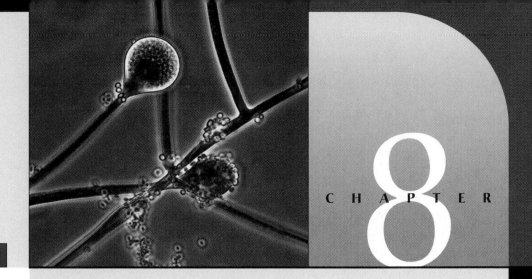

CHAPTER

8

Mitosis
The Cell-Copying Process

GOALS AND OBJECTIVES

Know the purpose of cell division.
❏ List three important purposes of cell division.
❏ Define the term *differentiation*.
❏ Explain the significance of stem cells.

Understand the events of cell division.
❏ Name the stages in the life cycle of a cell.
❏ Name the stage in the cell cycle during which DNA replication occurs.

Understand the stages of mitosis and the events that occur in each stage.
❏ List the stages of mitosis in their proper order.
❏ Describe the events that occur in each stage of mitosis.

Know how plant and animal cells differ in the way they carry out mitosis.
❏ Explain how animals and plants differ in how they carry out mitosis.
❏ Define the terms *cell wall* and *cell plate*.

Understand the nature of cancer as it relates to mitosis.
❏ Define the terms *tumor*, *benign*, *malignant*, and *metastasize*.
❏ Explain the general methods used to control cancer.
❏ List three carcinogenic agents.

Cell Division and Heredity

PART THREE

157

8.1 The Importance of Cell Division

The process of cell division replaces dead cells with new ones, repairs damaged tissues, and allows living organisms to grow. For example, you began as a single cell that resulted from the union of a sperm and an egg. One of the *first* activities of this single cell was to divide. As this process continued, the number of cells in your body increased, so that as an adult your body consists of several trillion cells. The *second* function of cell division is to maintain the body. Certain cells in your body, such as red blood cells and cells of the gut lining and skin, wear out. As they do, they must be replaced with new cells. Altogether, you lose about 50 million cells per second; this means that millions of cells are dividing in your body at any given time. A *third* purpose of cell division is repair. When a bone is broken, the break heals because cells divide, increasing the number of cells available to knit the broken pieces together. If some skin cells are destroyed by a cut or abrasion, cell division produces new cells to repair the damage.

During cell division, two events occur. The replicated genetic information of a cell is equally distributed to two daughter nuclei in a process called **mitosis**. As the nucleus goes through its division, the cytoplasm also divides into two new cells. This division of the cell's cytoplasm is called **cytokinesis**—cell splitting. Each new cell gets one of the two daughter nuclei so that both have a complete set of genetic information.

8.2 The Cell Cycle

All cells go through the same basic life cycle, but they vary in the amount of time they spend in the different stages. A generalized picture of a cell's life cycle may help you understand it better (figure 8.1). Once begun, cell division is a continuous process without a beginning or an end. It is a cycle in which cells continue to grow and divide. There are five stages to the life cycle of a eukaryotic cell: (1) G_1, gap (growth)—phase one; (2) S, synthesis; (3) G_2, gap (growth)—phase two; (4) cell division (mitosis and cytokinesis); and (5) G_0, gap (growth)—mitotic dormancy or differentiated.

During the G_0 phase, cells are not considered to be in the cycle of division but become differentiated or specialized in their function. It is at this time that they "mature" to play the role specified by the genes they express. Whereas some cells entering the G_0 phase remain there more-or-less permanently (e.g., nerve cells), others have the ability to move back into the cell cycle of mitosis—G_1, S, and G_2—with ease (e.g., bone cells).

The first three phases of the cell cycle—G_1, S, and G_2—occur during a period of time known as *interphase*. **Interphase** is the stage between cell divisions. During the G_1 stage, the cell grows in volume as it produces tRNA, mRNA, ribosomes, enzymes, and other cell components. During the S stage, DNA replication occurs in preparation for the distribution of genes to daughter cells. During the G_2 stage that follows, final preparations are made for mitosis with the synthesis of spindle-fiber proteins. • DNA replication, p. 147

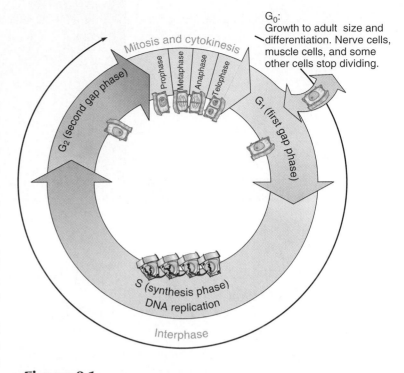

Figure 8.1

The Cell Cycle
During the cell cycle, tRNA, mRNA, ribosomes, and enzymes are produced in the G_1 stage for everyday processes. DNA replication occurs in the S stage. Proteins required for the spindles are synthesized in the G_2 stage. The nucleus is replicated in mitosis and two cells are formed by cytokinesis. Once some organs, such as the brain, have completely developed, certain types of cells, such as nerve cells, enter the G_0 stage. The time periods indicated are relative and vary depending on the type of cell and the age of the organism.

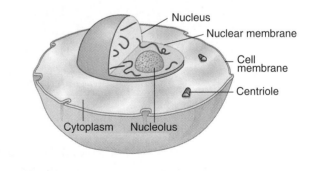

Figure 8.2

Interphase
Growth and the production of necessary organic compounds occur during this phase. If the cell is going to divide, DNA replication also occurs during interphase. The individual chromosomes are not visible, but a distinct nuclear membrane and nucleolus are present. (Some cells have more than one nucleolus.)

During interphase, the cell is not dividing but is engaged in metabolic activities such as muscle-cell contractions, photosynthesis, or glandular-cell secretion. During interphase, the nuclear membrane is intact and the individual chromosomes are not visible (figure 8.2). The individual chromatin strands are too thin and tangled to be seen. Remember that **chromosomes** include various kinds of histone proteins as well as DNA, the cell's genetic information. The double helix of DNA and the nucleosomes are arranged as a chromatid, and there are two attached chromatids for each replicated chromosome. It is these chromatids (chromosomes) that will be distributed during mitosis.

8.3 The Stages of Mitosis

All stages in the life cycle of a cell are continuous; there is no precise point when the G_1 stage ends and the S stage begins, or when the interphase period ends and mitosis begins. Likewise, in the individual stages of mitosis there is a gradual transition from one stage to the next. However, for purposes of study and communication, scientists have divided the process into four stages based on recognizable events. These four phases are prophase, metaphase, anaphase, and telophase.

Prophase

As the G_2 stage of interphase ends, mitosis begins. **Prophase** is the first stage of mitosis. One of the first noticeable changes is that the individual chromosomes become visible (figure 8.3). The thin, tangled chromatin present during interphase gradually coils and thickens, becoming visible as separate chromosomes. The DNA portion of the chromosome has genes that are arranged in a specific order. Each chromosome carries its own set of genes that is different from the sets of genes on other chromosomes.

As prophase proceeds, and as the chromosomes become more visible, we recognize that each chromosome is made of two parallel, threadlike parts lying side by side. Each parallel thread is called a **chromatid** (figure 8.4). These chromatids were formed during the S stage of interphase, when DNA synthesis occurred. The two identical chromatids are attached at a genetic region called the **centromere.** The centromere consists of a portion of the DNA that is not replicated during prophase, but remains base-paired as in the original double-stranded DNA. The centromere is vital to the cell division process. Without the centromere, cells will not complete mitosis and will die.

Several other events occur as the cell proceeds to the late prophase stage (figure 8.5). One of these events is the replication of the **centrioles**. Remember that human and many other eukaryotic cells contain centrioles, microtubule-containing organelles located just outside the nucleus. As they replicate, they move to the poles of the cell. As the centrioles move to the poles, the microtubules are assembled into the *spindle*. The **spindle** is an array of microtubules extending from pole to pole that is used in the movement of chromosomes. • **microtubules, p. 80**

In most eukaryotic cells, as prophase is occurring, the nuclear membrane is gradually disassembled. It is present at the beginning of prophase but disappears by the time this stage is completed. In addition to the disassembled nuclear membrane, the nucleoli within the nucleus disappear. Because of the disassembly of the nuclear membrane, the chromosomes are free to

Figure 8.3

Early Prophase
Chromosomes begin to appear as thin tangled threads and the nucleolus and nuclear membrane are present. The two sets of microtubules known as the centrioles begin to separate and move to opposite poles of the cell. A series of fibers known as the spindle will shortly begin to form.

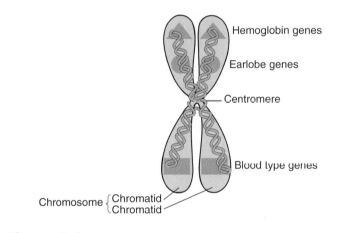

Figure 8.4

Chromosomes
During interphase, when chromosome replication occurs, the original double-stranded DNA unzips to form two identical double strands that remain attached at the centromere. Each of these double strands is a chromatid. The two identical chromatids of the chromosome are sometimes termed a dyad, to reflect that there are two double-stranded DNA molecules, one in each chromatid. The DNA contains the genetic data. (The examples presented here are for illustrative purposes only. Do not assume that the traits listed are actually located in the positions shown on these hypothetical chromosomes.)

move anywhere within the cytoplasm of the cell. As prophase progresses, the chromosomes move through the cell and become attached to the spindle fibers at their centromeres. In prophase, they are distributed randomly throughout the cytoplasm. As the chromosomes move from their random position to the central portion of the cell, the cell enters the next stage of mitosis.

Metaphase

During **metaphase**, the second stage in mitosis, the chromosomes align at the equatorial plane. There is no nucleus present during metaphase, and the spindle, which started to form

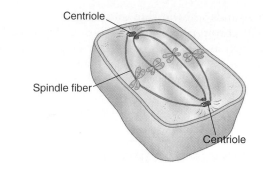

Figure 8.5

Late Prophase
In late prophase, the chromosomes appear as two chromatids (a dyad) connected at a centromere. The nucleolus and the nuclear membrane have disassembled. The centrioles have moved farther apart, and the spindle is produced.

during prophase, is completed. The centrioles are at the poles, and the microtubules extend between them to form the spindle. Then the chromosomes are their most tightly coiled and continue to move until all their centromeres align themselves along the equatorial plane at the equator of the cell (figure 8.6). At this stage in mitosis, each chromosome still consists of 2 chromatids attached at a centromere. In a human cell, there are 46 chromosomes, or 92 chromatids, aligned at the cell's equatorial plane during metaphase.

If we view a cell in the metaphase stage from the side (figure 8.6), it is an equatorial view. In this view, the chromosomes appear as if they were in a line. If we view the cell from the pole, it is a *polar view* (figure 8.7). Chromosomes viewed from this direction look like hot dogs scattered on a plate. In late metaphase, each chromosome splits as the centromeres replicate and the cell enters the next phase, anaphase.

Anaphase

Anaphase is the third stage of mitosis. The nuclear membrane is still absent and the spindle extends from pole to pole. The 2 chromatids within the chromosome separate as they move along the spindle fibers toward opposite ends of the poles (figure 8.8). As this separation of chromatids occurs, the chromatids are called *daughter chromosomes*. The two **daughter chromosomes** of one double chromosome contain identical genetic information.

What is the mechanism that causes the 2 chromatids of a chromosome to move from the metaphase equator? The centromere is the *place* on a chromsome where the 2 chromatids are held together. At the centromere is a complex protein structure called the *kinetochore* (*kin* = chore). One of the proteins of the kinetochore is named a *kinesin*, or *motor protein*. The motor protein is responsible for the movement of the chromatids. Once lined up at the equator, the end of one of a microtubular spindle fiber is fit into the kinetochore and attached to the motor protein. With the aid of ATP, components of the spindle protein (tubulin) are clipped off

Figure 8.6

Metaphase
During metaphase the chromosomes are pulled by the spindle and align at the equatorial plane. Notice that each chromosome still consists of two chromatids.

Figure 8.7

The Equatorial Plane of Metaphase–Polar View
This polar view shows how the chromosomes spread out on the equatorial plane.

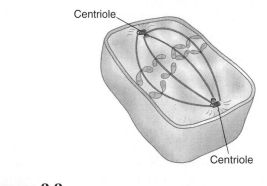

Figure 8.8

Anaphase
The pairs of chromatids separate after the centromeres replicate. The chromatids, now called daughter chromosomes, are separating and moving toward the poles and the cell will begin cytokinesis.

from the attached end while being held by the motor protein. This is like "nibbling" a strand of spaghetti into your mouth. You pull in the strand with your lips while biting the end off with your teeth. Since the spindle is shortened from the equator, the chromatids move toward the poles.

The 4 chromosomes shown in figure 8.8 moving to one pole would have exactly the same genetic information as the 4 moving to the opposite pole. It is the alignment of the chromosomes in metaphase, and their separation in anaphase, that cause this type of distribution. At the end of anaphase, there are two identical groups of chromosomes—one group at each pole. The next stage completes the mitosis process.

Telophase

Telophase is the last stage in mitosis. It is during anaphase that a second important event occurs, cytokinesis. *Cytokinesis* (cytoplasm splitting) divides the cytoplasm of the original cell so that two smaller, separate **daughter cells** result. Daughter cells are two cells formed by cell division that have identical genetic information. Each set of chromosomes unwinds and becomes enclosed by a nuclear membrane and the nucleoli reappear. Now the two cells have identical **daughter nuclei** (figure 8.9). In addition, the microtubules are disassembled, so the spindle disappears. With the formation of the daughter nuclei, mitosis, the first process in cell division, is completed, and the second process, cytokinesis, can occur,

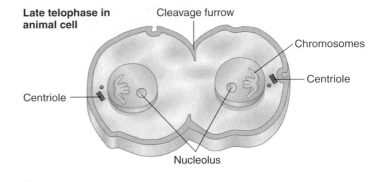

Figure 8.9

Telophase
During telophase the spindle disassembles and the nucleolus and nuclear membrane form. Daughter cells are formed as a result of the division of the cytoplasm.

from which two smaller daughter cells are formed. Each of the newly formed daughter cells then enters the G_1 stage of interphase. These cells can grow, replicate their DNA, and enter another round of mitosis and cytokinesis to continue the cell cycle (table 8.1).

Table 8.1

REVIEW OF THE STAGES OF MITOSIS

Interphase		As the cell moves from G_0 into mitosis, the chromosomes replicate during the S phase of interphase.
Prophase		The replicated chromatin begins to coil into recognizable chromosomes; the nuclear membrane fragments; centrioles move to form the cell's poles; spindle fibers form.
Metaphase		Chromosomes move to the equator of the cell and attach at their centromeres to the spindle fibers.
Anaphase		Centromeres complete DNA replication allowing the chromatids to separate toward the poles.
Telophase		Two daughter cells are formed from the dividing cells; the nuclear membranes and nucleoli re-form; spindle fibers fragment; the chromosomes unwind and change from chromosomes to chromatin.

(a) Animal cell

Interphase   Early prophase   Late prophase   Metaphase

(b) Animal cell

(c) Plant cell

Interphase   Early prophase   Late prophase   Metaphase

(d) Plant cell

Figure 8.10

A Comparison of Plant and Animal Mitosis
(*a*) Drawing of mitosis in an animal cell. (*b*) Photographs of mitosis in a whitefish blastula. (*c*) Drawing of mitosis in a plant cell. (*d*) Photographs of mitosis in an onion root tip.

8.4 Plant and Animal Cell Differences

Cell division is similar in plant and animal cells (figure 8.10). However, there are some minor differences. One difference concerns the centrioles. Centrioles are essential in animal cells, but they are not usually found in plant cells. However, by some process, plant cells do produce a spindle. There is also a difference in the process of cytokinesis (figure 8.11). In animal cells, cytokinesis results from a **cleavage furrow**. This is an indentation of the cell membrane of an animal cell that pinches the cytoplasm into two parts as if a string were tightened about its middle. In an animal cell, cytokinesis begins at the cell

Early telophase, animal cell

Centriole

Cleavage furrow

Early telophase, plant cell

Cell plate

Figure 8.11

Cytokinesis: Plant and Animal
In animal cells there is a pinching in of the cytoplasm that eventually forms two daughter cells. Daughter cells in plants are formed when a cell plate separates the cell into two cells.

Figure 8.10

A Comparison of Plant and Animal Mitosis—*Continued*

membrane and proceeds to the center. In plant cells, a **cell plate** begins at the center and proceeds to the cell membrane, resulting in a cell wall that separates the two daughter cells.

8.5 Differentiation

Because of the two processes in cell division, mitosis and cytokinesis, the daughter cells have the same genetic composition. You received a set of genes from your father in his sperm, and a set of genes from your mother in her egg. By mitosis, this cell formed two daughter cells. This process was repeated, and there were four cells, all of which had the same genes. All the trillions of cells in your body were formed by the process of cell division. This means that, except for mutations, all the cells in your body have the same genes.

While the cells in your body are genetically the same, they do not all have the same function. There are nerve cells, muscle cells, bone cells, skin cells, and many other types. How is it possible that cells with the same genes can be dif-

ferent? Think of the genes in a cell as individual recipes in a cookbook. You could give a copy of the same cookbook to 100 people and, though they all have the same book, each person could prepare a different dish. If you use the recipe to make a chocolate cake, you ignore the directions for making salads, fried chicken, and soups, although these recipes are in the book.

It is the same with cells. Although some genes are used by all cells, some cells activate only certain genes. Muscle cells produce proteins capable of contraction. Most other cells do not use these genes. Pancreas cells use genes that result in the formation of digestive enzymes, but they don't produce contractile proteins. **Differentiation** is the process of forming specialized cells within a multicellular organism.

Stem Cells

There are several types of tissue in the human body (i.e., muscle and nervous tissue). Connective tissue plays many important roles including (1) binding and support, (2) protection,

Figure 8.12

Multinucleated Cells
Many types of fungi, including bread molds, water molds, *Penicillium,* and *Aspergillus* are composed of multinucleated cells. As the organism grows, nuclei undergo mitosis, but cytokinesis does not occur. As a result, each cell contains tens of nuclei.

(3) insulation, and (4) transport. It wraps and cushions organs, insulates, provides a place to store energy (fat or adipose), and serves as a major component of cartilage and bone. It also makes up the cells found in blood, including red and white cells. The sources of red and white blood cells are undifferentiated cells called **stem cells**. Stem cells are found in the circulating blood and the red bone marrow of certain bones including the pelvis. People with certain diseases (i.e., chronic lymphocytic leukemia, CLL) or those receiving chemotherapy or radiation for cancer, lose stem cells and therefore have lower numbers of vital red and white blood cells. However, it may be possible to replace these cells with healthy stem cells from a donor provided there is a match (i.e., donor and recipient have the same histocompatability [HLA] antigens). There are two sources of such stem cells, bone marrow and circulating blood. Stem cells can be collected during an outpatient procedure by using a hypodermic needle and syringe. The sterile needle is inserted into the pelvis and marrow is withdrawn. Stem cells can also be taken in a manner similar to whole blood donation in a procedure called *apheresis*. During apheresis, the donor's blood is passed through a machine that separates the cells from their plasma. The stem cells are separated out from other types of blood cells. Once complete, the remaining cells and plasma are returned to the donor. This may take several hours. To improve the harvest, a stem cell growth factor can be given to the donor prior to apheresis. This stimulates mitosis of the stem cells and increases their numbers. Once harvested, the stem cells are isolated and concentrated for injection into the recipient. If successful, the donated stem cells will repopulate the recipient's marrow, continue to undergo mitosis, and differentiate into healthy mature cells. • HLA antigens, p. 69
Embryonic stem cells are cells that have the ability to differentiate into any type of cell including liver cells, skin cells, and brain cells. This makes them extremely valuable as replacement cells for individuals who have lost cells as a result of illness or injury. To obtain stem cells, researchers use donated sperm and an egg and fertilize them in a Petri dish. The fertilized egg or zygote is allowed to undergo mitosis for several days until it becomes a ball of cells. Researchers then take some of these cells and culture them to produce embryonic stem cells. Another source of stem cells is the umbilical cord of newborns. Since embryonic stem cells can differentiate into any other type of human cell, they could potentially be provided to people to resolve many problems. Such problems might include the repair of spinal injuries to restore function to paralyzed limbs, the replacement of useless scar tissue after a heart attack, or the addition of cells for production of chemicals lacking as a result of a genetic or other disease (ex., Dopamine in Parkinson's patients, insulin in diabetics, specific receptor sites that help people resist HIV or bubonic plague infections). The use of embryonic stem cells might also be used to provide sickle-cell patients with the healthy cells they need to produce hemoglobin with the proper amino acid sequence that prevents them from sickling under low oxygen conditions.

Some cells, such as muscle and nerve cells, naturally lose their ability to divide; they remain in the G_0 phase of interphase. Other cells retain their ability to divide as their form of specialization. Cells that line the digestive tract or form the surface of your skin are examples of dividing cells. In growing organisms such as infants or embryos, most cells are capable of division and divide at a rapid rate. In older organisms, many cells lose their ability to divide as a result of differentiation, and the frequency of cell division decreases. As the organism ages, the lower frequency of cell division may affect many bodily processes, including healing. In some older people, there may be so few cells capable of dividing that a broken bone may never heal. Recall from chapter 7 that the loss of telomeres is associated with cell aging. It is also possible for a cell to undergo mitosis but not cytokinesis. In many types of fungi the cells undergo mitosis but not cytokinesis, which results in multinucleated cells (figure 8.12).

8.6 Abnormal Cell Division

As we have seen, cells become specialized for a particular function. Each cell type has its cell-division process regulated so that it does not interfere with the activities of other cells or the whole organism. Some cells, however, may begin to divide as if they were "newborn" or undifferentiated cells. Sometimes this division occurs in an uncontrolled fashion.

For example, when human white blood cells are grown outside the body under special conditions, they develop a regular cell-division cycle. The cycle is determined by the DNA of the cells. However, white blood cells in the human body may increase their rate of mitosis as a result of other influences. Disease organisms entering the body, tissue damage, and changes in cell DNA all may alter the rate at which white blood cells divide. An increase in white blood cells in response to the invasion of disease organisms is valuable because these white blood cells are capable of destroying the disease-causing organisms. • circulatory system, p. 342

On the other hand, an uncontrolled mitosis in white blood cells causes *leukemia*. In some forms, this condition causes a general weakening of the body because the excess number of white blood cells diverts necessary nutrients from other cells of the body and interferes with their normal activities. It takes a lot of energy to keep these abnormal cells alive.

When such uncontrolled mitotic division occurs, a group of cells forms what is known as a *tumor*. A **tumor** is a mass of undifferentiated cells not normally found in a certain portion of the body. A **benign tumor** is a cell mass that does not fragment and spread beyond its original area of growth. A benign tumor can become harmful by growing large enough to interfere with normal body functions. Some tumors are malignant. **Malignant tumors** are nonencapsulated growths of tumor cells that are harmful; they may spread or invade other parts of the body. Cells of these tumors move from the original site (**metastasize**) and establish new colonies in other regions of the body (figure 8.13). Cells break off from the original tumor and enter the blood stream. When they get stuck to the inside of a capillary, these cells move through the wall of the blood vessel and invade the tissue, where they begin to reproduce by mitosis. This tumor causes new blood vessels to grow into this new site, which will carry nutrients to this growing mass. These vessels can also bring even more spreading cells to the new tumor site. **Cancer** is the term used to refer to any abnormal growth of cells that has a malignant potential. Agents responsible for causing cancer are called **carcinogens** (table 8.2).

Figure 8.13

Skin Cancer
Malignant melanoma is a type of skin cancer. It forms as a result of a mutation in pigmented skin cells. These cells divide repeatedly giving rise to an abnormal mass of pigmented skin cells. Only the dark area in the photograph is the cancer; the surrounding cells have the genetic information to develop into normal, healthy skin. This kind of cancer is particularly dangerous because the cells break off and spread to other parts of the body (metastasize).

Once cancer has been detected, the tumor might be eliminated. If the cancer is confined to a few specific locations, it may be possible to surgically remove it. Many cancers of the skin or breast are dealt with in this manner. However, in some cases surgery is impractical. If the tumor is located where it can't be removed without destroying healthy tissue, surgery may not be used. For example, removing certain brain cancers can severely damage the brain. In such cases, other methods may be used to treat cancer such as chemotherapy and radiation.

Chemotherapy—How It Works

Chemotherapy uses various types of chemicals to destroy mitotically dividing cancer cells. This treatment may be used even when physicians do not know exactly where the cancer cells are located. Many types of leukemia, testicular cancer, and lymphoma are successfully treated with chemotherapy. There are four generally recognized types of chemotherapeutic drugs. *Antimetabolites* appear to the cancer cell as normal nutrients, but in reality they are compounds that will fatally interfere with the cell's metabolic pathways. Methotrexate appears to be the normal substrate for an enzymatic reaction required to produce the nitrogenous bases adenine and guanine. When this medication is given, cancer cells are prevented from synthesizing new DNA. *Topoisomerase inhibitors* are drugs that prevent the DNA of cancer cells from "unzipping" so that DNA replication can occur. Doxorubicin is such a medication. *Alkylating agents* such as cyclophosphamide and chlorambucil form chemical bonds within the DNA of cancer cells resulting in breaks and other damage not easily repaired. The *plant alkaloids* such as vinblastine disrupt the spindle apparatus, thus disrupting the normal separation of chromatids at the time of anaphase. • enzyme specificity, p. 93

Table 8.2

FACTORS ASSOCIATED WITH CANCER

Radiation
X rays and gamma rays
Ultraviolet light (UV-A from tanning lamps; UV-B,
  the cause of sunburn)

Sources of Carcinogens
Nicotine (nitrosamines)
Nickel
Arsenic
Benzene
Dioxin
Asbestos
Uranium
Tar
Cadmium
Chromium
Polyvinyl chloride (PVC)

Diet
Alcohol
Smoked meats and fish
Food containing nitrates (e.g., bacon)

Viruses
Hepatitis B virus (HBV) and liver cancer
Herpes simplex virus (HSV) type II and uterine cancer
Epstein-Barr virus and Burkitt's lymphoma
Human T-cell lymphotropic virus (HTLV-1) and lymphomas and
  leukemias
Papillomavirus

Hormonal Imbalances
Diethylstilbestrol (DES)
Oral contraceptives

Types of Genetic and Familial Cancers
Chronic myelogenous leukemia
Acute leukemias
Retinoblastomas
Certain skin cancers
Breast
Endometrial
Colorectal
Stomach
Prostate
Lung

Most common cancers are not able to be controlled with chemotherapy alone and must be used in combination with radiation. Chemotherapy also has negative effects on normal cells. It lowers the body's immune reaction because it decreases the body's ability to reproduce new white blood cells by mitosis. Chemotherapy interferes with the body's normal defense mechanisms. Therefore cancer patients undergoing chemotherapy must be given antibiotics. The an-

HOW SCIENCE WORKS 8.1

Total Body Radiation to Control Leukemia

Leukemia is a kind of cancer caused by uncontrolled growth of white blood cells. Patients with leukemia have cancer of blood-forming cells located in their bone marrow; however, not all of these cells are cancerous. It is possible to separate the cancerous from the noncancerous bone marrow cells. A radiation therapy method prescribed for some patients involves the removal of some of their bone marrow and isolation of the noncancerous cells for laboratory growth. After these healthy cells have been cultured and increased in number, the patient's whole body is exposed to high doses of radiation sufficient to kill all the cancerous cells remaining in the bone marrow. Because this treatment is potentially deadly, the patient is kept isolated from all harmful substances and infectious microbes. They are fed sterile food, drink sterile water, and breathe sterile air while being closely monitored and treated with antibiotics. The cultured noncancerous cells are injected back into the patient. As if they had a memory of their own, they migrate back to their origins in the bone marrow, establish residence, and begin cell division all over again.

tibiotics help them defend against dangerous bacteria that might invade their bodies. Other side effects include intestinal disorders and loss of hair, which are caused by damage to healthy cells in the intestinal tract and the skin that divide by mitosis.

Radiation—How It Works

Radiation therapy uses powerful X rays or gamma rays. This therapy may be applied from the outside or by implanting radioactive "seeds" into the tumor. Because this treatment damages surrounding healthy cells, it is used very cautiously especially when surgery is impractical (How Science Works 8.1).

What most likely prevents normal cells from becoming tumor cells is the fact that genetic damage or errors are repaired. Damaged cells are put into a *repair cycle* with the assistance of the "guardian of the genome," the tumor-suppressor p53 gene. This gene stops a damaged cell just before the S phase so that it can be repaired. In fact, p53 may be directly involved with the DNA repair process allowing cells to continue dividing on a regulated basis and differentiating according to their genetic plan. Individuals with mutations of the p53 gene are more susceptible to many cancers including retinoblastoma, breast cancer, and leukemia. Over a thousand different mutations have been identified in p53.

It was commonly thought that radiation therapy was effective because cancer cells divide more rapidly than other cells. This would mean that their DNA spent more time coiled up as chromosomes. Chromosomes were thought to be bigger and better targets than chromatin. Radiation would be more likely to hit the target and destroy the chromosome, causing cancer cell death. Current theory proposes that radiation most likely destroys cancer cells by inducing a process called *apoptosis*. **Apoptosis** is also known as "programmed cell death," that is, death that has a genetic basis and is not the result of injury. Apoptosis normally occurs in many cells of the body because they might be harmful or it takes too much energy to maintain them. Think of apoptosis as cell suicide. When they go through apoptosis, cells make the ultimate sacrifice for the good of the rest of the body. During menstruation, cells lining the uterus undergo apoptosis, thus enabling the uterus to be renewed for a possible pregnancy. Cells damaged as a result of viral infection regularly kill themselves, thus helping prevent the spread of the virus to other healthy cells of that tissue.

Radiation stimulates a variety of cellular events that can activate apoptosis in cells with severe genetic damage, or that might undergo uncontrolled mitosis leading to the formation of a tumor. When apoptosis is initiated, the cell's DNA is cut into pieces and the cytoplasm and nucleus shrink. This is followed by its engulfment by phagocytes. In this manner, cells that are potentially dangerous to the entire body (tumor cells) are killed before they cause serious harm.

As a treatment for cancer, radiation is dangerous for the same reasons that it is beneficial. In cases of extreme exposure to radiation, people develop what is called *radiation sickness*. The symptoms of this disease include loss of hair, bloody vomiting and diarrhea, and a reduced white blood cell count. These symptoms occur in parts of the body where mitosis is common. The lining of the intestine is constantly being lost as food travels through and it must be replaced by the process of mitosis. Hair growth is the result of the continuous division of cells at the roots. White blood cells are also continuously reproduced in the bone marrow and lymph nodes. When radiation strikes these rapidly dividing cells and kills them, the lining of the intestine wears away and bleeds, hair falls out, and few new white blood cells are produced to defend the body against infection.

SUMMARY

Cell division is necessary for growth, repair, and reproduction. Cells go through a cell cycle that includes cell division (mitosis and cytokinesis) and interphase. Interphase is the period of growth and preparation for division. Mitosis is divided into four stages: prophase, metaphase, anaphase, and telophase. During mitosis, two daughter nuclei are formed from one parent nucleus. These nuclei have identical sets of chromosomes and genes that are exact copies of those of the parent. Although the process of mitosis has been presented as a series of phases, you should realize that it is a continuous, flowing process from prophase through telophase. Following mitosis, cytokinesis divides the cytoplasm, and the cell returns to interphase.

The regulation of mitosis is important if organisms are to remain healthy. Regular divisions are necessary to replace lost cells and allow for growth. However, uncontrolled cell division may result in cancer and disruption of the total organism's well-being.

THINKING CRITICALLY

For guidelines to answer this question, visit the Online Learning Center.

One "experimental" cancer therapy utilizes laboratory-generated antibodies to an individual's own unique cancer cells. Radioisotopes such as alpha-emitting radium 223 are placed in "cages" and attached to the antibodies. When these immunotherapy medications are given to a patient, the short-lived killer isotopes attach to only the cancer cells. They release small amounts of radiation and for short distances; therefore they cause little harm to healthy cells and tissues before their destructive powers are dissipated. Review the material on cell membranes, antibodies, cancer, and radiation and explain the details of this treatment to a friend. (You might explore the Internet for further information.)

CONCEPT MAP TERMINOLOGY

Construct a concept map using these terms.

apoptosis	interphase
benign	malignant
cell cycle	mitosis
differentiation	tumor

Visit the Online Learning Center to use an interactive format to construct another concept map.

KEY TERMS

Use interactive flash cards, on the Online Learning Center, to help you learn the meanings of these terms.

anaphase	daughter chromosomes
apoptosis	daughter nuclei
benign tumor	differentiation
cancer	interphase
carcinogens	malignant tumors
cell plate	metaphase
centrioles	metastasize
centromere	mitosis
chromatid	prophase
chromosomes	spindle
cleavage furrow	stem cell
cytokinesis	telophase
daughter cells	tumor

LEARNING CONNECTIONS

Visit the Online Learning Center to find additional study resources including those highlighted below.

8.1   The Importance of Cell Division

Animations and Review
- Prokaryotes
- Chromosomes

1. What is the purpose of mitosis?
2. Define the term *cytokinesis*.

8.2   The Cell Cycle

3. What is meant by the cell cycle?
4. What types of activities occur during interphase?

8.3   The Stages of Mitosis

Animations and Review
- Mitosis/cell cycle

Labeling Exercises
- Mitosis overview I
- Mitosis overview II
- Plant cell mitosis

Interactive Concept Maps
- Events during mitosis

5. Name the four stages of mitosis and describe what occurs in each stage.
6. During which stage of a cell's cycle does DNA replication occur?
7. At what phase of mitosis does the DNA become most visible?
8. List five differences between an interphase cell and a cell in mitosis.

8.4   Plant and Animal Cell Differences

Interactive Concept Map
- Mitotic differences between plants and animals

9. What are the differences between plant and animal mitosis?
10. What is the difference between cytokinesis in plants and animals?

8.5   Differentiation

11. How is it possible that cells with the same genes can be different?
12. What does cell specialization mean?
13. What are stem cells and what role do they play?

8.6   Abnormal Cell Division

Experience This!
- Learning about cancer

14. Why can radiation be used to control cancer?
15. List three factors associated with the development of cancer.

Test yourself with multiple choice questions that provide immediate feedback on the Online Learning Center with PowerWeb at

www.mhhe.com/enger11

CHAPTER 9

Meiosis: Sex–Cell Formation

PART THREE   Cell Division and Heredity

Goals and Objectives

Know the steps in meiosis.
❒ Explain what happens when a sex cell is made.
❒ Define the terms *haploid* and *diploid*.
❒ Diagram the stages of meiosis.
❒ Define the terms *synapes*, *crossing-over*, *segregation*, and *independent assortment*.
❒ Explain the genetic advantage to sexual reproduction.
❒ Explain how one person can make many different types of sex cells.

Know how meiosis occurs abnormally.
❒ List two ways in which genetic abnormalities occur.
❒ Define the terms *nondisjunction*, *monosomy*, and *trisomy*.

Understand how gametes are formed and unite at fertilization.
❒ Explain why brothers and sisters of the same birthparents can be so different.

Understand the difference between meiosis and mitosis.
❒ Explain the difference between sexual and asexual reproduction.

9.1 Sexual Reproduction and Meiosis

In addition to mitosis, certain cells are able to divide their nuclei by another method called **meiosis**. Pronounced my-oh-sis, it is from the Greek word *meion* meaning "smaller"—that is, the results of this type of cell division are small cells, no doubt referring to sperm cells. In many sexually reproducing organisms such as humans, meiosis takes place in cells that have become differentiated as reproductive tissue—the **gonads.** The gonads in human females are known as **ovaries** and in males, the **testes.** Ovarian and testicular cells that divide by meiosis become reproductive cells called **gametes**—the **eggs** and **sperm.** These gametes are also referred to as *germ cells.* Organs that produce gametes are also found in algae and plants. Some of these are very simple. In algae such as *Spirogyra,* individual cells become specialized for gamete production. In plants, the structures are very complex. In flowering plants, the **pistil** produces eggs or ova, and the **anther** produces pollen, which contains sperm. Meiosis is a much more complex process than mitosis. Each cell that divides by this process undergoes two divisions: *meiosis I* and *meiosis II,* before becoming four cells (figure 9.1). **Sexual reproduction** is the formation of a new individual by the union of two such sex cells. The uniting of an egg and sperm (gametes) is known as **fertilization.**

In many organisms the **zygote,** which results from the union of an egg and a sperm, divides repeatedly by mitosis to form the complete organism. In chapter 8, we discussed the cell cycle and pointed out that it is a continuous process, without a beginning or an end. The process of mitosis followed by growth is important in the life cycle of any organism. Thus, the *cell cycle* is part of an organism's *life cycle* (figure 9.2).

Notice in figure 9.2 that the zygote and its descendants have two sets of chromosomes. However, the male gamete and the female gamete each contain only one set of chromosomes. These sex cells are said to be **haploid,** and contain a single copy of the genome. The haploid number of chromosomes is noted as *n*. A zygote contains two sets and is said to be **diploid.**

Meiosis
Sex cell formation

Single adult cell	→	Single gamete (sperm or egg)
Contains 2 sets of chromosomes	→	Contains 1 set of chromosomes
2*n*	→	*n*

The diploid number of chromosomes is noted as 2*n* (*n* + *n* = 2*n*). Diploid cells have two sets of chromosomes—one set from each parent. Remember, a chromosome is composed of 2 chromatids, each containing double-stranded DNA. These 2 chromatids are attached to each other at a point called the *centromere.* In a diploid nucleus, the chromosomes occur as **homologous chromosomes**—a pair of chromosomes in a diploid cell that contain similar genes throughout their length. One of the chromosomes of a homologous pair was donated by the father, the other by the mother (figure 9.3). Different species of organisms vary in the number of chromosomes they contain. Table 9.1 lists several different organisms and their haploid and diploid chromosome numbers.

It is necessary for organisms that reproduce sexually to form gametes having only one set of chromosomes. If gametes contained two sets of chromosomes, the zygote resulting from

Figure 9.1

Meiosis
The cell division process of meiosis occurs in organisms that reproduce sexually. Meiosis occurs in two stages. The first stage, meiosis I, results in the formation of two cells. After each of these cells divides during meiosis II, four gametes are produced.

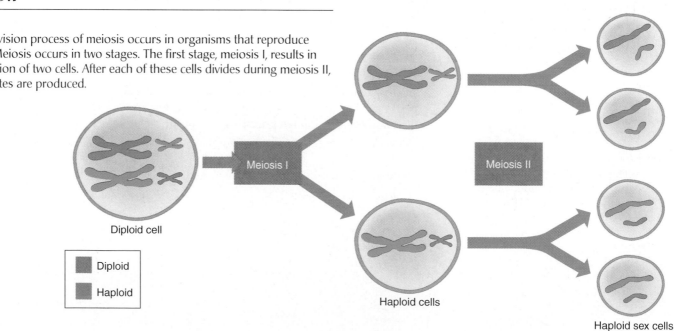

Diploid cell

Meiosis I

Meiosis II

Haploid cells

Haploid sex cells

Diploid

Haploid

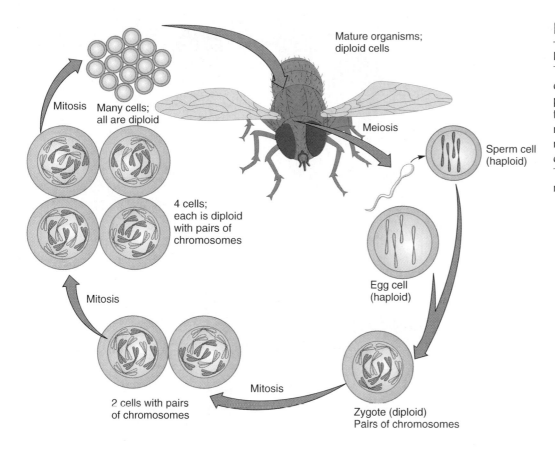

Mature organisms;
diploid cells

Meiosis

Sperm cell
(haploid)

Egg cell
(haploid)

Zygote (diploid)
Pairs of chromosomes

Mitosis

2 cells with pairs
of chromosomes

Mitosis

4 cells;
each is diploid
with pairs of
chromosomes

Mitosis   Many cells;
all are diploid

Mitosis

Figure 9.2

Life Cycle
The cells of this adult fruit fly have 8 chromosomes in their nuclei. In preparation for sexual reproduction, the number of chromosomes must be reduced by half so that fertilization will result in the original number of 8 chromosomes in the new individual. The offspring will grow and produce new cells by mitosis.

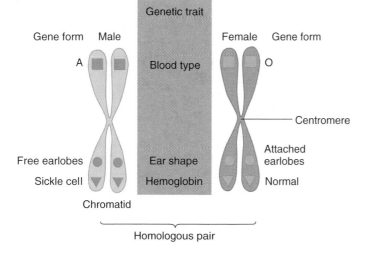

Gene form   Male   Genetic trait   Female   Gene form

A   Blood type   O

                                              Centromere

Free earlobes   Ear shape   Attached earlobes
Sickle cell   Hemoglobin   Normal

Chromatid

Homologous pair

Figure 9.3

A Pair of Homologous Chromosomes
A pair of chromosomes of similar size and shape that have genes for the same traits are said to be homologous. Notice that the genes may not be identical but code for the same type of information. Homologous chromosomes are of the same length, have the same types of genes in the same sequence, and have their centromeres in the same location—one came from the male parent and the other was contributed by the female parent.

Table 9.1

CHROMOSOME NUMBERS

Organism	Haploid Number	Diploid Number
Mosquito	3	6
Fruit fly	4	8
Housefly	6	12
Toad	18	36
Cat	19	38
Human	**23**	**46**
Hedgehog	23	46
Chimpanzee	24	48
Horse	32	64
Dog	39	78
Onion	8	16
Kidney bean	11	22
Rice	12	24
Tomato	12	24
Potato	24	48
Tobacco	24	48
Cotton	26	52

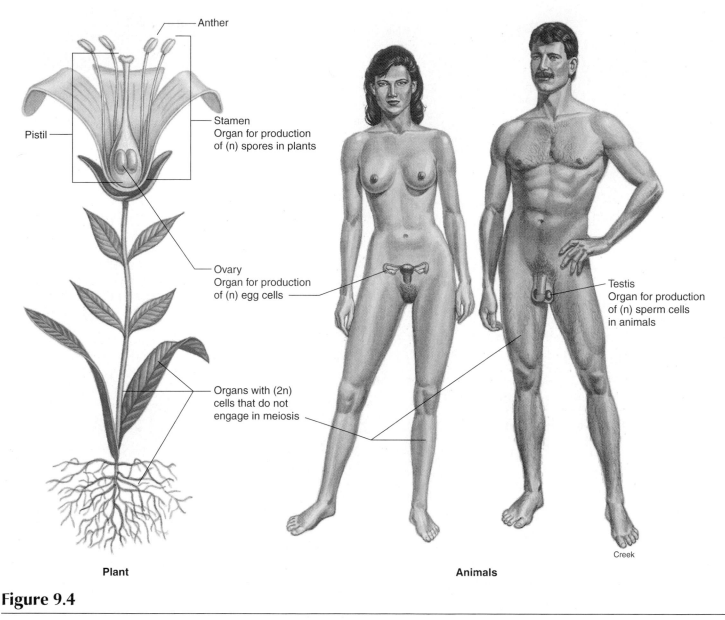

Plant   **Animals**

Figure 9.4

Haploid and Diploid Cells
Both plants and animals produce cells with a haploid number of chromosomes. The male anther in plants and the testes in animals produce haploid male cells, sperm. In both plants and animals, the ovaries produce haploid female cells, eggs.

their union would have four sets of chromosomes. The number of chromosomes would continue to double with each new generation. However, this does not usually happen; the number of chromosomes remains constant generation after generation. Because cell division by mitosis and cytokinesis results in cells that have the same number of chromosomes as the parent cell, two questions arise: how are sperm and egg cells formed, and how do they get only half the chromosomes of the diploid cell? The answers lie in the process of *meiosis*, the specialized pair of cell divisions that reduce the chromosome number from diploid (2n) to haploid (n). One of the major functions of meiosis is to produce cells that have one set of genetic information. Therefore, when fertilization occurs, the zygote will have two sets of chromosomes, as did each parent (figure 9.4).

To illustrate meiosis in this chapter, we have chosen to show a cell that has only 4 chromosomes (figure 9.5). (In reality, humans have 46 chromosomes, or 23 pairs.) The haploid number of chromosomes in this cell is 2, and these haploid cells contain only one complete set of 2 chromosomes. You can see that there are 4 chromosomes in this cell—2 from the mother and 2 from the father. A closer look at figure 9.5 shows you that there are only 2 types of chromosomes, but 2 of each type:

1. Long chromosomes consisting of chromatids attached at centromeres near the center (one green)
2. Short chromosomes consisting of chromatids attached near the center (one blue)

Figure 9.5

Prophase I (Early)
During prophase I, the cell is preparing for division. A unique event that occurs in prophase I is the synapsis of the chromosomes. Notice that the nuclear membrane is no longer apparent and that the paired homologs are free to move about the cell.

Nucleus

Figure 9.6

Prophase I (Late)

We can talk about the number of chromosomes in two ways. We can say that our hypothetical diploid cell has 4 replicated chromosomes, or we can say that it has two pairs of homologous chromosomes.

Haploid cells, on the other hand, do not have homologous chromosomes. They have one of each type of chromosome. The whole point of meiosis is to distribute the chromosomes and the genes they carry so that each daughter cell gets one member of each homologous pair. In this way, each daughter cell gets one complete set of genetic information.

Figure 9.7

Spindle fibers

Metaphase I
Notice that the homologous chromosome pairs are arranged on the equatorial plane in the synapsed condition. This cell shows one way the chromosomes could be lined up; however, a second arrangement is possible. How would they appear in this alternative arrangement?

9.2 The Mechanics of Meiosis I

Meiosis is preceded by an interphase stage during which DNA replication occurs. In a sequence of events called *meiosis I*, members of homologous pairs of chromosomes divide into two complete sets. This is sometimes called a **reduction division,** a type of cell division in which daughter cells get only half the chromosomes from the parent cell ($2n \rightarrow n$). The division begins with replicated chromosomes composed of 2 chromatids. The sequence of events in meiosis I is artificially divided into four phases: prophase I, metaphase I, anaphase I, and telophase I.

Figure 9.8

Anaphase I
During this phase, one member of each homologous pair is segregated from the other member of the pair. Notice that the centromeres of the chromosomes do not replicate.

Prophase I

During prophase I, the cell is preparing itself for division (figure 9.6). The chromatin material coils and thickens into chromosomes, the nucleoli disappear, the nuclear membrane is disassembled, and the spindle begins to form. The spindle is formed in animals when the centrioles move to the poles. There are no centrioles in plant cells, but the spindle does form. However, there is an important difference between the prophase stage of mitosis and prophase I of meiosis. During prophase I, homologous chromosomes recognize one another by their centromeres, move through the cell toward one another, and come to lie next to each other in a process called **synapsis.** While the chromosomes are synapsed, a unique event called *crossing-over* may occur. **Crossing-over** is the exchange of equivalent sections of DNA on homologous chromosomes. We will fit crossing-over into the whole picture of meiosis later. • genetic diversity, p. 222

Metaphase I

The synapsed pair of homologous chromosomes now moves into position on the equatorial plane of the cell. In this stage, the centromere of each chromosome attaches to the spindle. The synapsed homologous chromosomes move to the equator of the cell as single units. How they are arranged on the equator (which one is on the left and which one is on the right) is determined by chance (figure 9.7). In the cell in figure 9.7, 2 green chromosomes from the father and 2 blue chromosomes from the mother are lined up on the left. They could have aligned themselves in several other ways.

Anaphase I

Anaphase I is the stage during which homologous chromosomes separate (figure 9.8). During this stage, the chromosome number is *reduced from diploid to haploid.* The two members of each pair of homologous chromosomes move away from each other toward opposite poles. The centromeres do not replicate during this phase. The direction each takes is determined by how each pair was originally arranged on the spindle. Each chromosome is independently attached to a spindle fiber at its centromere. Unlike the anaphase stage of mitosis, *the centromeres that hold the chromatids together do not divide during anaphase I of meiosis (the chromosomes are still in their replicated form).* Each chromosome still consists of 2 chromatids. Because the homologous chromosomes and the genes they carry are being separated from one another, this process is called **segregation.** The way in which a single pair of homologous chromosomes segregates does not influence how other pairs of homologous chromosomes segregate. That is, each pair segregates independently of other pairs. This is known as **independent assortment** of chromosomes.

Telophase I

Telophase I consists of changes that return the cell to an interphaselike condition (figure 9.9). The chromosomes uncoil and become long, thin threads, the nuclear membrane reforms around them, and nucleoli reappear. During this activity, cytokinesis divides the cytoplasm into two separate cells.

Because of meiosis I, the total number of chromosomes is divided equally, and each daughter cell has one member of each homologous chromosome pair. This means that the genetic data each cell receives is one-half the total, but each cell continues to have a complete set of the genetic information. Each individual chromosome is still composed of 2 chromatids joined at the centromere, and the chromosome number is reduced from diploid ($2n$) to haploid (n). In the cell we have been using as our example, the number of chromosomes is reduced from 4 to 2. The four pairs of chromosomes have been distributed to the two daughter cells. Depending on the type of cell, there may be a time following telophase I when a cell engages in normal metabolic activity that corresponds to an interphase stage. *However, the chromosomes do not replicate before the cell enters meiosis II.* Figure 9.10 shows the events in meiosis I.

Figure 9.9

Telophase I
What activities would you expect during the telophase stage of cell division? What term is used to describe the fact that the cytoplasm is beginning to split the parent cell into two daughter cells?

Nuclear envelope

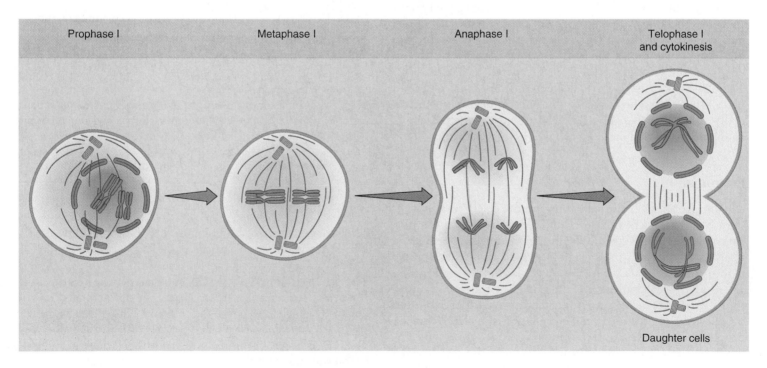

| Prophase I | Metaphase I | Anaphase I | Telophase I and cytokinesis |

Daughter cells

Figure 9.10

Meiosis I
The stages in meiosis I result in reduction division. This reduces the number of chromosomes in the parental cell from the diploid number to the haploid number in each of the two daughter cells.

9.3 The Mechanics of Meiosis II

Meiosis II includes four phases: prophase II, metaphase II, anaphase II, and telophase II. The two daughter cells formed during meiosis I continue through meiosis II so that, usually, four cells result from the two divisions. As you follow meiosis II, recognize that the events in this division sequence are the same as those that occur in mitosis.

Prophase II

Prophase II is similar to prophase in mitosis; the nuclear membrane is disassembled, nucleoli disappear, and the spindle apparatus begins to form. However, it differs from prophase I because these cells are haploid, not diploid (figure 9.11). Also, synapsis, crossing-over, segregation, and independent assortment do not occur during meiosis II.

Metaphase II

The metaphase II stage is typical of any metaphase stage because the chromosomes attach by their centromeres to the spindle at the equatorial plane of the cell. Because pairs of chromosomes are no longer together in the same cell, each chromosome moves as a separate unit (figure 9.12).

Anaphase II

Anaphase II differs from anaphase I because during anaphase II the centromere of each chromosome divides, and the chromatids, now called *daughter chromosomes*, move to the poles as in mitosis (figure 9.13). Remember, there are no paired homologs in this stage; therefore, segregation and independent assortment cannot occur.

Figure 9.12

Metaphase II
During metaphase II, each chromosome lines up on the equatorial plane. Each chromosome is composed of two chromatids (a replicated chromosome) joined at a centromere. How does metaphase II of meiosis compare to metaphase I of meiosis?

Figure 9.13

Anaphase II
Anaphase II is very similar to the anaphase of mitosis. The centromere of each chromosome divides and one chromatid separates from the other. As soon as this happens, we no longer refer to them as chromatids; we now call each strand of nucleoprotein a chromosome.

Figure 9.11

Prophase II
The two daughter cells are preparing for the second division of meiosis. Study this diagram carefully. Can you list the events of this stage?

Telophase II

During telophase II, the cell returns to a nondividing condition. As cytokinesis occurs, new nuclear membranes form, chromosomes uncoil, nucleoli re-form, and the spindles disappear (figure 9.14). This stage is followed by differentiation; the four cells mature into gametes—either sperm or eggs. The events of meiosis II are summarized in figure 9.15.

In many organisms, including humans, egg cells are produced in such a manner that three of the four cells resulting from meiosis in a female disintegrate. However, because the one that survives is randomly chosen, the likelihood of any one particular combination of genes being formed is not affected. The whole point of learning the mechanism of meiosis is to see how variation happens (table 9.2).

Figure 9.14

Telophase II
During the telophase II stage, what events would you expect?

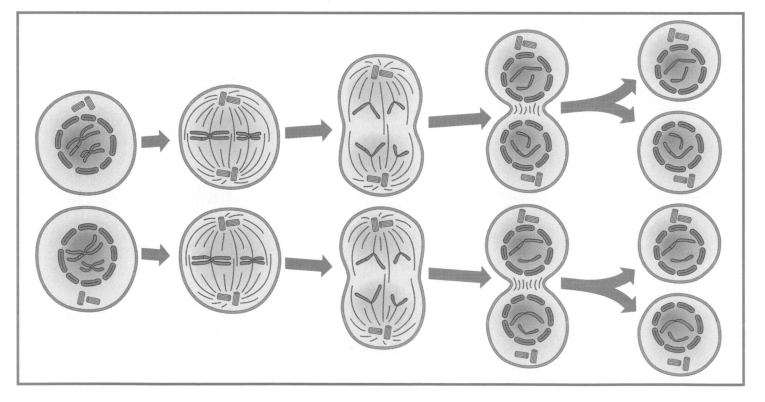

Figure 9.15

Meiosis II
During meiosis II, the centromere of each chromosome replicates and each chromosome divides into separate chromatids. Four haploid cells are produced, each having one chromatid of each kind.

Table 9.2

REVIEW OF THE STAGES OF MEIOSIS

Interphase		As the diploid ($2n$) cell moves from G_0 into meiosis, the chromosomes replicate during the S phase of interphase.
Prophase I		The replicated chromatin begins to coil into recognizable chromosomes and the homologous chromosomes synapse; chromatids may cross over; the nuclear membrane and nucleoli fragment; centrioles move to form the cell's poles; spindle fibers are formed.
Metaphase I		Synapsed homologous chromosomes align as pairs along the equatorial plane and attach to the spindle fibers at their centromeres; each pair positions itself independently of all others.
Anaphase I		Homologous pairs of chromosomes separate from one another as they move toward the poles of the cell.
Telophase I		The two newly forming daughter cells are now haploid (n) since each only contains one of each pair of homologous chromosomes; the nuclear membranes and nucleoli re-form; spindle fibers fragment; the chromosomes unwind and change from chromosomes (composed of 2 chromatids) to chromatin.
Prophase II		Each of the two haploid (n) daughter cells from meiosis I undergoes chromatin coiling to form chromosomes, each of which is composed of 2 chromatids; the nuclear membrane fragments; centrioles move to form the cell's poles; spindle fibers form.
Metaphase II		Chromosomes attach to the spindle fibers at the centromeres and move to the equator of the cell.
Anaphase II		Centromeres complete DNA replication allowing the chromatids to separate toward the poles.
Telophase II		Four haploid (n) cells are formed from the division of the two meiosis I cells; the nuclear membranes and nucleoli re-form; spindle fibers fragment; the chromosomes unwind and change from chromosomes to chromatin; these cells become the sex cells (egg or sperm) of higher organisms.

9.4 Sources of Variation

The formation of a haploid cell by meiosis and the combination of two haploid cells to form a diploid cell by sexual reproduction results in variety in the offspring. Five factors influence genetic variation in offspring: mutations, crossing-over, segregation, independent assortment, and fertilization.

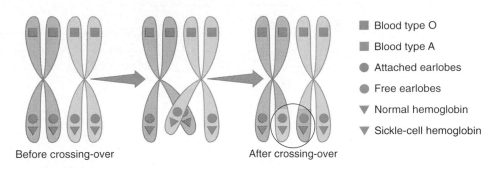

Before crossing-over   After crossing-over

- ■ Blood type O
- ■ Blood type A
- ● Attached earlobes
- ● Free earlobes
- ▼ Normal hemoglobin
- ▼ Sickle-cell hemoglobin

Figure 9.16

Synapsis and Crossing-Over
While pairs of homologous chromosomes are in synapsis, one part of one chromatid can break off and be exchanged for an equivalent part of its homologous chromatid.

Mutation

Several types of mutations were discussed in chapter 7: point mutations and chromosomal mutations. In point mutations, a change in a DNA nucleotide results in the production of a different protein. In chromosomal mutations, genes are rearranged. By causing the production of different proteins, both types of mutations increase variation. The second source of variation is crossing-over. • DNA mutation, p. 141

Crossing-Over

Crossing-over occurs during meiosis I while homologous chromosomes are synapsed. Crossing-over is the exchange of a part of a chromatid from one homologous chromosome with an equivalent part of a chromatid from the other homologous chromosome. This exchange results in a new gene combination. Remember that a chromosome is a double strand of DNA. To break a chromosome, bonds between sugars and phosphates are broken. This is done at the same spot on both chromatids, and the two pieces switch places. After switching places, the two pieces of DNA are bonded together by re-forming the bonds between the sugar and the phosphate molecules.

Examine figure 9.16 carefully to note precisely what occurs during crossing-over. This figure shows a pair of homologous chromosomes close to each other. Notice that each gene occupies a specific place on the chromosome. This is the *locus,* a place on a chromosome where a gene is located. Homologous chromosomes contain an identical order of genes. For the sake of simplicity, only a few loci are labeled on the chromosomes used as examples. Actually, the chromosomes contain hundreds or possibly thousands of genes.

What does crossing-over have to do with the possible kinds of cells that result from meiosis? Consider figure 9.17. Notice that without crossing-over, only two kinds of genetically different gametes result. Two of the four gametes have one type of chromosome, whereas the other two have the other type of chromosome. With crossing-over, four genetically different gametes are formed. With just one crossover, we double the number of kinds of gametes possi-

ble from meiosis. Because crossing-over can occur at almost any point along the length of the chromosome, great variation is possible. In fact, crossing-over can occur at a number of different points on the same chromosome; that is, there can be more than one crossover per chromosome pair (figure 9.18).

Crossing-over helps explain why a child can show a mixture of family characteristics (figure 9.19). If the blue chromosome was the chromosome that a mother received from her mother, the child could receive some genetic information not only from the mother's mother, but also from the mother's father. When crossing-over occurs during the meiotic process, pieces of genetic material are exchanged between the chromosomes. This means that genes that were originally on the same chromosome become separated. They are moved to their synapsed homologue, and therefore into different gametes. The closer two genes are to each other on a chromosome (i.e., the more closely they are *linked*), the more likely they will stay together and not be separated during crossing-over. Thus, there is a high probability that they will be inherited together. The farther apart two genes are, the more likely it is that they will be separated during crossing-over. This fact enables biologists to construct chromosome maps (How Science Works 9.1).

Segregation

After crossing-over has taken place, segregation occurs. This involves the separation and movement of homologous chromosomes to the poles. Let's say a person has a normal form of the gene for insulin production on one chromosome and an abnormal form of this gene on the other. Such a person would produce enough insulin to be healthy and would not be diabetic. When this pair of chromosomes segregates during anaphase I, one daughter cell receives a chromosome with a normal gene for insulin production and the second daughter cell receives a chromosome with an abnormal gene

Figure 9.17

Variations Resulting from Crossing-Over
The cells on the left resulted from meiosis without crossing-over; those on the right had one crossover. The symbols noted on each chromosome stand for the traits noted below the drawings. Compare the results of meiosis in both cases.

Parts "crossed over"

▌ Normal insulin
⦙ Light-skin color

▌ Diabetes
⦙ Dark-skin color

for diabetes. The process of segregation causes genes to be separated from one another so that they have an equal chance of being transmitted to the next generation. If the mate also has one normal gene for insulin production and one abnormal for diabetes, that person also produces two kinds of gametes.

Both of the parents have normal insulin production. If one or both of them contributed a gene for normal insulin production during fertilization, the offspring would produce enough insulin to be healthy. However if, by chance, both parents contributed the gamete with the abnormal gene for diabetes, the child would be a diabetic. Thus, parents may produce offspring with traits different from their own. In this variation, no new genes are created; they are simply redistributed in a fashion that allows for the combination of genes in the offspring to be different from the parents' gene combinations. This will be explored in greater detail in chapter 10.

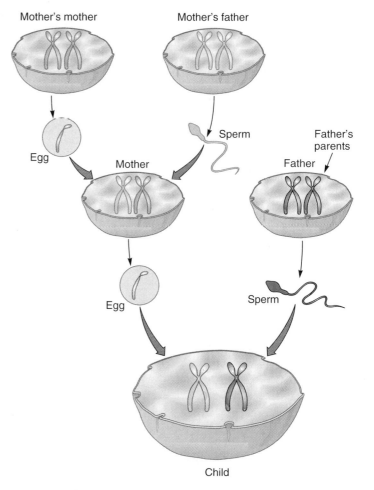

Mother's mother   Mother's father

Egg   Sperm

Mother   Father's parents   Father

Egg   Sperm

Child

Figure 9.19

Mixing of Genetic Information Through Several Generations
The mother of this child has information from both of her parents. The child receives a mixture of this information from the mother. Note that only the maternal line has been traced in this diagram. Can you imagine how many more combinations would result after including the paternal heritage?

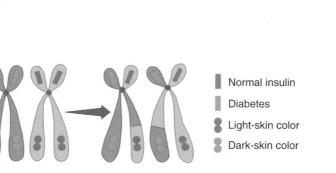

▌ Normal insulin
▌ Diabetes
⦙ Light-skin color
⦙ Dark-skin color

Figure 9.18

Multiple Crossovers
Crossing-over can occur several times between the chromatids of one pair of homologous chromosomes. List the new combinations of genes on each chromatid that have resulted from the crossing-over.

HOW SCIENCE WORKS 9.1

The Human Genome Project

The Human Genome Project (HGP) was first proposed in 1986 by the U.S. Department of Energy (DOE), and was co-sponsored soon after by the National Institutes of Health (NIH). These agencies were the main research agencies within the U.S. government responsible for developing and planning the project. Later, a private U.S. corporation, Celera Genomics, joined the effort as a competitor. It is one of the most ambitious projects ever undertaken in the biological sciences. The goal was nothing less than the complete characterization of the genetic makeup of humans. The project was completed early in 2003 when the complete nucleotide sequence of all 23 pairs of human chromosomes was determined. It is as if we have now identified all the words in the human gene dictionary, but we only have definitions for about half. With this knowledge in hand, scientists will define the rest of the words (learn what the gene does) and produce a map of each of the chromosomes that will show the names and places of all our genes. This international project involving about 100 laboratories worldwide took only 16 years to complete. Work began in many of these labs in 1990. Powerful computers are used to store and share the enormous amount of information derived from the analyses of human DNA. To get an idea of the size of this project, consider this: A human Y chromosome (one of the smallest of the human chromosomes) is estimated to be composed of 59 million nitrogenous bases. The larger X chromosome may be composed of 164 million nitrogenous base pairs! The entire human genome consists of 3.12 gigabases!

Two kinds of work progressed simultaneously. First, *physical maps* were constructed by determining the location of specific "markers" (known sequences of bases) and their closeness to genes (see box figure 9.A, chromosome 21). A kind of chromosome map already exists that shows patterns of colored bands on chromosomes, a result of chromosome-staining

50 million bases

Coxsackie and adenovirus receptor	Myeloproliferative syndrome, transient
Amyloidosis, cerebroarterial, Dutch type	Leukemia, transient, of Down syndrome
Alzheimer disease, APP-related	Enterokinase deficiency
Schizophrenia, chronic	Multiple carboxylase deficiency
Usher syndrome, autosomal recessive	T-cell lymphoma invasion and metastasis
Amytrophic lateral sclerosis	Mycobacterial infection, atypical
Oligomycin sensitivity	Down syndrome (critical region)
Jervell and Lange-Nielsen syndrome	Autoimmune polyglandular disease, type I
Long QT syndrome	Bethlem myopathy
Down syndrome cell adhesion molecule	Epilepsy, progressive myoclonic
Homocystinuria	Holoprosencephaly, alobar
Cataract, congenital, autosomal dominant	Knobloch syndrome
Deafness, autosomal recessive	Hemolytic anemia
Myxovirus (influenza) resistance	Breast cancer
Leukemia, acute myeloid	Platelet disorder, with myeloid malignancy

Figure 9.A

Genes Known to Be on Human Chromosome Number 21
The gene map shows the appropriate positions of several genes known to be on human chromosome number 19. This chromosome has 50 million bases.

procedures. Using these banded chromosomes, the markers were then related to these colored bands on a specific region of a chromosome. Work is continuing on the HGP to identify the location of specific genes. Each year a more complete picture is revealed.

The second kind of work was for labs to determine the exact order of nitrogenous bases of the DNA for each chromosome. Techniques exist for determining base sequences, but it is a time-consuming job to sort out the several million bases that can be found in any one chromosome. Coming from behind with new, speedier techniques, Celera Genomics was able to catch up to NIH labs and completed its sequencing at almost the same time. The benefit of having these two organizations as competitors is that when they finished their research, they could compare and contrast results. Amazingly, the discrepancies between their findings were declared insignificant. It was originally estimated, for example, that there were between 100,000 and

Independent Assortment

So far in discussing variety, we have dealt with only one pair of chromosomes, which allows two varieties of gametes. Now let's consider how variation increases when we add a second pair of chromosomes (figure 9.20).

In figure 9.20, chromosomes carrying ear shape information always separate from each other. The second pair of chromosomes with the information for Tourette syndrome also separates. Because the pole to which a chromosome moves is a chance event, half the time the chromosomes

divide so that free ear lobes and no Tourettes move in one direction, whereas attached ear lobes and Tourettes move in the opposite direction. Half the time, free earlobes and Tourettes go together and attached earlobes and no Tourettes go to the other pole. With three pairs of homologous chromosomes, there are eight possible kinds of cells with respect to chromosome combinations resulting from meiosis. See if you can list them. The number of possible chromosomal combinations of gametes is found by the expression 2^n, where n equals the number of pairs of chromosomes. With three

140,000 genes in the human genome. However, when the results were compared, the evidence from both organizations indicated that there are only about 30,000 protein-coding genes—only about twice as many as in a worm or a fly. However, the genes are more complex, with more alternative splicing, generating a larger number of protein products for each gene. In humans it appears that each gene is capable of coding for between three and four different proteins. Knowing this information provides insights into the evolution of humans and the mutation rates of males versus females. This will make future efforts to work with the genome through bioengineering much easier.

When the physical maps are finally completed for all of the human genes, it will be possible to examine a person's DNA and identify genetic abnormalities. This could be extremely useful in diagnosing diseases and providing genetic counseling to those considering having children. This kind of information would also identify human genes and proteins as drug targets and create possibilities for new gene therapies. Once it is known where an abnormal gene is located and how it differs in base sequence from the normal DNA sequence, steps could be taken to correct the abnormality. Completing the HGP will also result in the discovery of new families of proteins and will help explain basic physiological and cell biological processes common to many organisms, thus increasing the breadth and depth of our understanding of basic biology.

However, there is also a concern that as knowledge of our genetic makeup becomes easier to determine, some people may attempt to use this information for profit or political power. This is a real concern because some health insurance companies refuse to insure people with "preexisting conditions" or those at "genetic risk" for certain abnormalities. They fail to realize that in people between the age of 5 and 50, such "conditions" or mutations are normal. Refusing to provide coverage would save these companies the expense of future medical bills incurred by "less-than-perfect" people. Another fear is that attempts may be made to "breed out" certain genes and people from the human population in order to create a "perfect race."

Here are some other intriguing findings from the human genome and the genome identification projects of other organisms:

- Hundreds of human genes appear likely to have resulted from transfer of genes from bacteria to humans at some point in the vertebrate lineage.
- Human genes are not scattered at random among the human chromosomes. "Forests" or "clusters" of genes are found on certain chromosomes, separated by "deserts" of genes. For example, chromosomes 17 and 19 are forested with thousands of genes while chromosome 18 has many fewer genes.
- Humans are about 99.9% identical at the DNA level! Scientists believe that there is virtually no scientific reason for the concept of "race" since there is much greater variation within a so-called race than there is between the so-called races.
- Rice appears to have about 50,000 genes.
- Chimpanzees have 98–99% of the same genes as humans.
- Roundworms have about 26,000 genes.
- Fruits flies contain an estimated 13,600 genes.
- Yeast cells have about 6,241 genes.
- There are numerous and virtually identical genes found in many organisms that appear to be very distantly related— for example, mice, humans, and yeasts.
- Genes jump around (transposons) within the chromosomes more than scientists ever thought.
- The mutation rate of male humans is about twice that of females.

In the last 25 years, the genome sequences of 599 viruses and viroids, 205 naturally occurring plasmids, 185 organelles, 31 Eubacteria, 7 Archaea, 1 fungus, 2 animals, and 1 plant have been deciphered. This is only the beginning.

With the completing of the HGP, scientists look forward to:

- Determining what makes one person more or less susceptible to disease than another;
- Identifying exactly what makes one person different from another;
- Developing tests that can reveal genetic disease or increased susceptibility to disease;
- Developing drugs to treat such diseases; and
- Developing gene therapies to replace the causative gene from an individual. • **Mendel, p. 194**

pairs of chromosomes, n equals 3, and so $2^n = 2^3 = 2 \times 2 \times 2 = 8$. With 23 pairs of chromosomes, as in the human cell, $2^n = 2^{23} = 8,388,608$. More than 8 million kinds of sperm cells or egg cells are possible from a single human parent organism. This number is actually smaller than the maximum variety that could be produced because it only takes into consideration the variety generated as a result of independent assortment. This huge variation is possible because each pair of homologous chromosomes assorts independently of the other pairs of homologous chromosomes (independent assortment). In addition to this variation, crossing-over creates new gene combinations, and mutation can cause the formation of new genes, thereby increasing this number greatly.

Fertilization

Because of the large number of possible gametes resulting from independent assortment, segregation, mutation, and crossing-over, an incredibly large number of types of offspring can result (How Science Works 9.2). Because human males

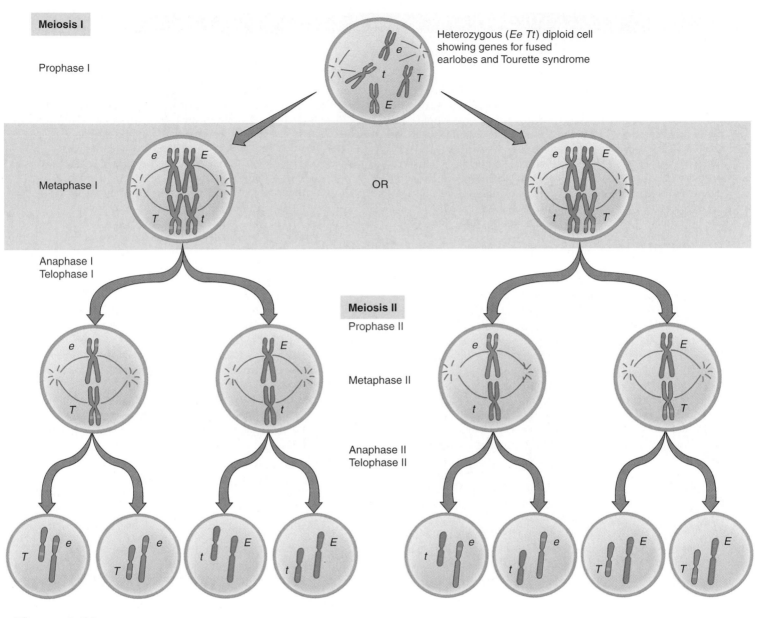

Meiosis I

Prophase I

Heterozygous (*Ee Tt*) diploid cell showing genes for fused earlobes and Tourette syndrome

Metaphase I

OR

Anaphase I
Telophase I

Meiosis II

Prophase II

Metaphase II

Anaphase II
Telophase II

Figure 9.20

The Independent Orientations of Homologous Chromosome Pairs

The orientation of one pair of chromosomes on the equatorial plane does not affect the orientation of a second pair of chromosomes. Note in the figure the different ways that these two chromosomes can line up. How one homologous pair lines up does not influence how another positions itself. Just because the longer pair in the left portion of the figure is lined up with the dark chromosome on the left and the light on the right does not mean that the shorter pair has to line up with its dark on the left and light on the right. In this figure, *E* = free ear lobes, *e* = attached ear lobes; *T* = displays Tourette syndrome, and *t* = no Tourette syndrome.

can produce millions of genetically different sperm and females can produce millions of genetically different eggs, the number of kinds of offspring possible is infinite for all practical purposes. With the possible exception of identical twins, every human that has ever been born is genetically unique (refer to chapter 21).

9.5 Nondisjunction and Chromosomal Abnormalities

In the normal process of meiosis, diploid cells have their number of chromosomes reduced to haploid. This involves segregating homologous chromosomes into separate cells during the first meiotic division. Occasionally, a pair of homologous chromosomes does not segregate properly during gametogenesis

HOW SCIENCE WORKS 9.2

Cloning—What This Term Really Means

The term **clone** refers to exact copies of biological entities such as genes, organisms, or cells. The term refers to the outcome, not the way that the results are achieved. Many whole organisms "clone" themselves (i.e., bacteria undergo asexual binary fission and many plants reproduce "clonally" through asexual means, sometimes referred to as *vegetative regeneration*). One process of making exact copies of the cells of a cat, sheep, or other organism in the laboratory is technically known as "*somatic cell nuclear transfer.*" The purpose of somatic cell nuclear transfer is not the "cloning" of a human or other adult but the establishment of a clonal cell line. In this technique, the nucleus of a somatic (body) cell of an adult donor is transferred into an oocyte, or "egg cell," whose own nucleus has been removed. Should such hybrid cells be cultured *in vitro* (*vitro* = glass; ex., in test tubes) for a few weeks, the result is the production of a line of cloned cells (i.e., human embryonic stem cells). These can be used for a variety of medically valuable therapies or research (i.e., investigating certain genetic or other diseases).

Using the other technique, *reproductive cloning,* the hybrid cell can be implanted into the uterus of a host or surrogate mother where it may develop into a genetic copy or clone of the whole individual who donated the nucleus.

(a)

(b)

Figure 9.B

"CC" the cloned kitty with surrogate mother and her genetic mother, "Rainbow"

Out of 87 implanted cloned embryos, CC (*Copy Cat!*) is the only one to survive—comparable to the success rate in sheep, mice, cows, goats, and pigs. (*a*) Notice that she is completely unlike her tabby surrogate mother. (*b*) "Rainbow" is her genetic donor, and both are female calico domestic shorthair cats.

(Source: *Nature*, AOP, Published online: 14 February 2002; DOI: 10.1038/nature723)

and both chromosomes of a pair end up in the same gamete. This kind of division is known as **nondisjunction** (figure 9.21). As you can see in this figure, two cells are missing a chromosome and the genes that were carried on it. This usually results in the death of the cells. The other cells have a double dose of 1 chromosome. Apparently, the genes of an organism are bal-

anced against one another. A double dose of some genes and a single dose of others results in abnormalities that may lead to the death of the cell. Some of these abnormal cells, however, do live and develop into sperm or eggs. If one of these abnormal sperm or eggs unites with a normal gamete, the offspring will have an abnormal number of chromosomes. There will be

Gametogenesis

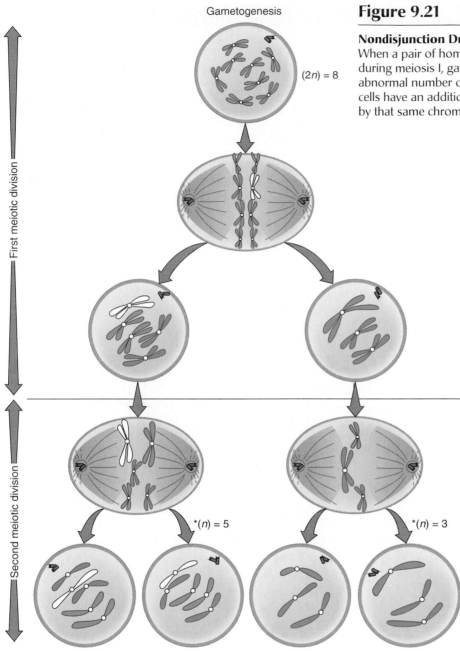

First meiotic division

Second meiotic division

$(2n) = 8$

*$(n) = 5$ *$(n) = 3$

* Should have been (n) = 4.

Figure 9.21

Nondisjunction During Gametogenesis

When a pair of homologous chromosomes fails to separate properly during meiosis I, gametogenesis results in gametes that have an abnormal number of chromosomes. Notice that two of the highlighted cells have an additional chromosome, whereas the other two are deficient by that same chromosome.

picture of an individual's chromosomal makeup is referred to as that person's *karyotype*.

One example of the effects of nondisjunction is the condition known as **Down syndrome**. If a gamete with 2 number 21 chromosomes has been fertilized by another containing the typical one copy of chromosome number 21, the resulting zygote would have 47 chromosomes (e.g., 24 from the female plus 23 from the male parent) (figure 9.23). The child who developed from this fertilization would have 47 chromosomes in every cell of his or her body as a result of mitosis, and thus would have the symptoms characteristic of Down syndrome. These may include thickened eyelids, some mental impairment, and faulty speech (figure 9.24). Premature aging is probably the most significant impact of this genetic disease. On the other hand, a child born with only 1 chromosome 21 rarely survives.

It was thought that the mother's age at childbirth played an important part in the occurrence of trisomies such as Down syndrome. In women, gametogenesis begins early in life, but cells destined to become eggs are put on hold during meiosis I (see chapter 21). Beginning at puberty and ending at menopause, one of these cells completes meiosis I monthly. This means that cells released for fertilization later in life are older than those released earlier in life. Therefore, it was believed that the chances of abnormalities such as nondisjunction increase as the age of the mother increases. However, the evidence no longer supports this age-egg link. Currently, the increase in frequency of trisomies with age has been correlated with a decrease in the activity of a woman's immune system. As she ages, her immune system is less likely to recognize the difference between an abnormal and a normal embryo. This means that she is more likely to carry an abnormal fetus to full term.

Figure 9.25 illustrates the frequency of occurrence of Down syndrome at different ages in women. Notice that the frequency increases very rapidly after age 37. For this reason,

three of one of the kinds of chromosomes instead of the normal two, a condition referred to as **trisomy**. Should the other cell survive and become involved in fertilization, it will only have one of the pair of homologous chromosomes, a condition referred to as **monosomy**. All the cells that develop by mitosis from such zygotes will be either trisomic or monosomic.

It is possible to examine cells and count chromosomes. Among the easiest cells to view are white blood cells. They are dropped onto a microscope slide so that the cells are broken open and the chromosomes are separated. Photographs are taken of chromosomes from cells in the metaphase stage of mitosis. The chromosomes in the pictures can then be cut and arranged for comparison to known samples (figure 9.22). This

Figure 9.22

Human Male and Female Chromosomes

The randomly arranged chromosomes shown in the circle simulate metaphase cells spattered onto a microscope slide (*a*). Those in parts (*b*) and (*c*) have been arranged into homologous pairs. Part (*b*) shows a male karyotype with an X and Y chromosome and (*c*) shows a female karyotype with two X chromosomes.

(a)

(b)

(c)

Figure 9.23

Chromosomes from an Individual Displaying Down Syndrome

Notice that each pair of chromosomes has been numbered and that the person from whom these chromosomes were taken has an extra chromosome number 21. The person with this trisomic condition could display a variety of physical characteristics, including slightly slanted eyes, flattened facial features, a large tongue, and a tendency toward short stature and fingers. Most individuals also display mental retardation.

Figure 9.24

Down Syndrome

Every cell in a Downic child's body has one extra chromosome. With special care, planning, and training, people with this syndrome can lead happy, productive lives.

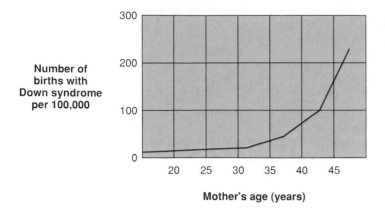

Figure 9.25

Down Syndrome as a Function of a Mother's Age
Notice that as the age of the female increases, the frequency of Downic children increases only slightly until the age of approximately 37. From that point on, the rate increases drastically. This increase may be because older women experience fewer miscarriages of abnormal embryos.

many physicians encourage couples to have their children in their early to mid-twenties and not in their late thirties or early forties. Physicians normally encourage older women who are pregnant to have the cells of their fetus checked to see if they have the normal chromosome number. It is important to know that the male parent can also contribute the extra chromosome 21. However, it appears that this occurs less than 30% of the time.

Sometimes a portion of chromosome 14 may be cut out and joined to chromosome 21. The transfer of a piece of one nonhomologous chromosome to another is called a chromosomal **translocation**. A person with this 14/21 translocation is monosomic and has only 45 chromosomes; one 14 and one 21 are missing and replaced by the translocated 14/21. Statistically, about 15% of the children of carrier mothers inherit the 14/21 chromosome and have Down syndrome. Fewer of the children born to fathers with the 14/21 translocation inherit the abnormal chromosome and are Downic.

Whenever an individual is born with a chromosomal abnormality such as a monosomic or a trisomic condition, it is recommended that both parents have a karyotype evaluation in an attempt to identify the possible source of the problem. This is not to fix blame but to provide information on the likelihood that a next pregnancy would also result in a child with a chromosomal abnormality. Other examples of trisomy are described in chapter 21, Human Reproduction, Sex, and Sexuality.

9.6 Chromosomes and Sex Determination

You already know that there are several different kinds of chromosomes, that each chromosome carries genes unique to it, and that these genes are found at specific places on the chromosome. Furthermore, diploid organisms have homolo-

OUTLOOKS 9.1

The Birds and the Bees . . . and the Alligators

The determination of the sex of an individual depends on the kind of organism you are! For example, in humans, the physical features that result in maleness are triggered by a gene on the Y chromosome. Lack of a Y chromosome results in an individual that is female. In other organisms, sex may be determined by other combinations of chromosomes or environmental factors.

Organism	Sex Determination
Birds	Chromosomally determined: XY individuals are female.
Bees	Males (the drones) are haploid and females (workers or queens) are diploid.
Certain species of alligators, turtles, and lizards	Egg incubation temperatures cause hormonal changes in the developing embryo; higher incubation temperatures cause the developing brain to shift sex in favor of the individual becoming a female. (Placing a drop of the hormone estrogen on the developing egg also causes the embryo to become female!)
Boat shell snails	Males can become females but will remain male if they mate and remain in one spot.
Shrimp, orchids, and some tropical fish	Males convert to females; on occasion females convert to males, probably to maximize breeding.
African reed frog	Females convert to males, probably to maximize breeding.

gous pairs of chromosomes. Genes determine sexual characteristics in the same manner as other types of characteristics. In many organisms, sex-determining genes are located on specific chromosomes known as **sex chromosomes**. All other chromosomes not involved in determining the sex of an individual are known as **autosomes**. In humans and all other mammals, and in some other organisms (e.g., fruit flies), the sex of an individual is determined by the presence of a certain chromosome combination. The genes that determine maleness are located on a small chromosome known as the Y *chromosome*. This Y chromosome behaves as if it and another larger chromosome, known as the X *chromosome*, were homologs, and they pair up during meiosis even though they are not identical. Males have one X and one Y chromosome. Females have two X chromosomes. Some animals have their sex determined in a completely different way. In bees, for example, the females are diploid and the males are haploid. Other plants and animals have still other chromosomal mechanisms for determining their sex (Outlooks 9.1).

Table 9.3

A COMPARISON OF MITOSIS AND MEIOSIS

Mitosis	Meiosis
1. One division completes the process.	1. Two divisions are required to complete the process.
2. Chromosomes do not synapse.	2. Homologous chromosomes synapse in prophase I.
3. Homologous chromosomes do not cross over.	3. Homologous chromosomes do cross over.
4. Centromeres divide in anaphase.	4. Centromeres divide in anaphase II, but not in anaphase I.
5. Daughter cells have the same number of chromosomes as the parent cell ($2n \rightarrow 2n$ or $n \rightarrow n$).	5. Daughter cells have half the number of chromosomes as the parent cell ($2n \rightarrow n$).
6. Daughter cells have the same genetic information as the parent cell.	6. Daughter cells are genetically different from the parent cell.
7. Results in growth, replacement of worn-out cells, and repair of damage.	7. Results in sex cells.

9.7 A Comparison of Mitosis and Meiosis

Some of the similarities and differences between mitosis and meiosis were pointed out earlier in this chapter. Study table 9.3 to familiarize yourself with the differences between these two processes.

SUMMARY

Meiosis is a specialized process of cell division resulting in the production of four cells, each of which has the haploid number of chromosomes. The total process involves two sequential divisions during which one diploid cell reduces to four haploid cells. Because the chromosomes act as carriers for genetic information, genes separate into different sets during meiosis. Crossing-over and segregation allow hidden characteristics to be displayed, whereas independent assortment allows characteristics donated by the mother and the father to be mixed in new combinations.

Together, crossing-over, segregation, and independent assortment ensure that all sex cells are unique. Therefore when any two cells unite to form a zygote, the zygote will also be one of a kind. The sex of many kinds of organisms is determined by specific chromosome combinations. In humans, females have two X chromosomes; males have an X and a Y chromosome.

THINKING CRITICALLY

For guidelines to answer this question, visit the Online Learning Center.

Assume that corn plants have a diploid number of only 2. In the following figure, the male plant's chromosomes are diagrammed on the left, and those of the female are diagrammed on the right.

Diagram sex-cell formation in the male and female plant. How many variations in sex cells can occur and what are they?

What variations can occur in the production of chlorophyll and starch in the descendants of these parent plants?

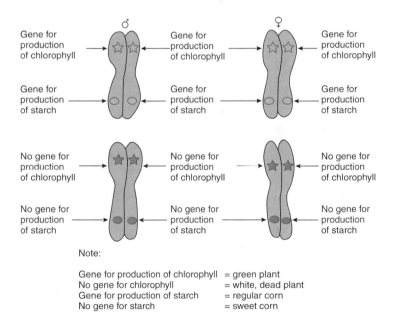

Note:

Gene for production of chlorophyll	= green plant
No gene for chlorophyll	= white, dead plant
Gene for production of starch	= regular corn
No gene for starch	= sweet corn

CONCEPT MAP TERMINOLOGY

Construct a concept map using these terms.

age	reduction division
Down syndrome	segregation
meiosis I	synapsis
nondisjunction	trisomy

Visit the Online Learning Center to use an interactive format to construct another concept map.

KEY TERMS

Use interactive flash cards, on the Online Learning Center, to help you learn the meanings of these terms.

anther	monosomy
autosomes	nondisjunction
clone	ovaries
crossing-over	pistil
diploid	reduction division (also meiosis)
Down syndrome	segregation
egg	sex chromosomes
fertilization	sexual reproduction
gamete	sperm
gonad	synapsis
haploid	testes
homologous chromosomes	translocation
independent assortment	trisomy
meiosis	zygote

LEARNING CONNECTIONS

Visit the Online Learning Center to find additional study resources including those highlighted below.

9.1   Sexual Reproduction and Meiosis

Animations and Review
• Evolution of sex

1. How do haploid cells differ from diploid cells?
2. Why is meiosis necessary in organisms that reproduce sexually?
3. Define the terms *zygote, fertilization,* and *homologous chromosomes.*
4. Diagram fertilization as it would occur between a sperm and an egg with the haploid number of 3.

9.2   The Mechanics of Meiosis I

Labeling Exercises
• Meiosis I

5. Diagram the metaphase I stage of a cell with the diploid number of 8.
6. What is unique about prophase I?

9.3   The Mechanics of Meiosis II

Animations and Review
• Meiosis

Interactive Concept Map
• Meiosis I and meiosis II

Experience This!
• Models of meiosis

7. During which phase do daughter chromosomes form?
8. Why does synapse not occur during meiosis II?

9.4   Sources of Variation

Animations and Review
• Recombination

9. How much variation as a result of independent assortment can occur in cells with the following number of diploid numbers: 2, 4, 6, 8, and 22?
10. What are the major sources of variation in the process of meiosis?

9.5   Nondisjunction and Chromosomal Abnormalities

Animations and Review
• Introduction
• Abnormal chromosomes

Interactive Concept Map
Human Explorations
• Exploring meiosis: Down syndrome

11. Define the term *nondisjunction.*
12. What is the difference between monosomy and trisomy?

9.6   Chromosomes and Sex Determination

Animations and Review
• Sex chromosomes

13. What are the differences between autosomes and sex chromosomes?
14. What sex chromosomes do most human females carry? Human males?

9.7   A Comparison of Mitosis and Meiosis

Animations and Review
• Review of cell division

Interactive Concept Maps
• Mitosis versus meiosis

15. Can a haploid cell undergo meiosis?
16. List three differences between mitosis and meiosis.

Test yourself with multiple choice questions that provide immediate feedback
on the Online Learning Center with PowerWeb at

www.mhhe.com/enger11

Mendelian Genetics

GOALS AND OBJECTIVES

Understand the concepts of genotype and phenotype.

❒ Explain how a person can have the allele for a particular trait but not show it.

Understand the basics of Mendelian genetics.

❒ Determine if the children of a father and a mother with a certain gene combination will automatically show that trait.

❒ Explain how people inherit varying degrees of traits such as skin color.

Know single-gene and double-factor genetic methods of inheritance.

❒ Explain the likelihood that a particular trait will be passed on to the next generation.

❒ Determine the chances that children will carry two particular genes.

Understand how a person's sex can influence the expression of his or her genes.

❒ Explain why men and women inherit some traits differently.

Appreciate how genes and their alleles interact.

❒ Define the concepts of dominant alleles and recessive alleles, codominant alleles, and X-linkage to explain inheritance patterns.

PART THREE   Cell Division and Heredity

10.1 Genetics, Meiosis, and Cells

Why do you have a particular blood type or hair color? Why do some people have the same skin color as their parents and others have a skin color different from that of their parents? Why do flowers show such a wide variety of colors? Why is it that generation after generation of plants, animals, and microbes look so much like members of their own kind? These questions and many others can be better answered if you have an understanding of genetics.

A **gene** is a portion of DNA that determines a characteristic. Through meiosis and reproduction, genes can be transmitted from one generation to another. The study of genes, how genes produce characteristics, and how the characteristics are inherited is the field of biology called **genetics**. The first person to systematically study inheritance and formulate laws about how characteristics are passed from one generation to the next was an Augustinian monk named Gregor Mendel (1822–1884). Mendel's work was not generally accepted until 1900, when three men, working independently, rediscovered some of the ideas that Mendel had formulated more than 30 years earlier. Because of his early work, the study of the pattern of inheritance that follows the laws formulated by Gregor Mendel is often called **Mendelian genetics**.

To understand this chapter, you need to know some basic terminology. One term that you have already encountered is *gene*. Mendel thought of a gene as a *particle* that could be passed from the parents to the **offspring** (*children, descendants,* or *progeny*). Today we know that genes are actually composed of specific sequences of DNA nucleotides. His particle concept was not entirely inaccurate, because a particular gene is located at a specific place on a chromosome called its **locus** (*locus* = location; plural, loci).

Another important idea to remember is that all sexually reproducing organisms have a diploid (2n) stage. Because gametes are haploid (n) and most organisms are diploid, the conversion of diploid to haploid cells during meiosis is an important process. • haploid/diploid, p. 170

$$2(n) \text{ cell} \rightarrow \text{meiosis} \rightarrow (n) \text{ gametes}$$

The diploid cells have two sets of chromosomes—one set inherited from each parent.

$$n + n \text{ gametes} \rightarrow \text{fertilization} \rightarrow 2n \text{ cell}$$

Therefore, they have 2 chromosomes of each kind and have 2 genes for each characteristic. When sex cells are produced by meiosis, reduction division occurs, and the diploid number is reduced to haploid. Therefore, the sex cells produced by meiosis have only 1 chromosome of each of the homologous pairs that were in the diploid cell that began meiosis. Diploid organisms usually result from the fertilization of a haploid egg by a haploid sperm. Thus, they inherit 1 gene of each type from each parent. For example, each of us has 2 genes for earlobe shape: one came with our father's sperm, the other with our mother's egg (figure 10.1).

(a)   (b)

Figure 10.1

Genes Control Structural Features
Whether your earlobe is free (a) or attached (b) depends on the genes you have inherited. As genes express themselves, their actions affect the development of various tissues and organs. Some people's earlobes do not separate from the sides of their heads in the same manner as do those of others. How genes control this complex growth pattern and why certain genes function differently than others is yet to be clarified.

10.2 Single-Gene Inheritance Patterns

In diploid organisms there may be two different forms of the gene. In fact, there may be *several* alternative forms or *alleles* of each gene within a population. **Alleles** are two or more different forms of a gene located at the same locus. In people, for example, there are 2 alleles for earlobe shape. One allele produces an earlobe that is fleshy and hangs free, whereas the other allele produces a lobe that is attached to the side of the face and does not hang free. The type of earlobe that is present is determined by the type of allele (gene) received from each parent and the way in which these alleles interact with one another. Alleles are located on the same pair of homologous chromosomes—one allele on each chromosome. These alleles are also at the same specific location, or locus (figure 10.2).

The **genome** is a set of all the genes necessary to specify an organism's complete list of characteristics. The term genome is used in two ways. It may refer to the diploid (2n) or haploid (n) number of chromosomes in a cell. Be sure to clarify how this term is used by your instructor. The **genotype** of an organism is a listing of the genes present in that organism. It consists of the cell's DNA code; therefore, you cannot see the genotype of an organism. It is not yet possible to know the complete genotype of most organisms, but it is

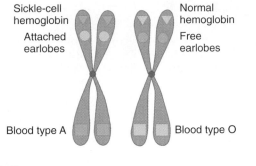

Sickle-cell hemoglobin

Attached earlobes

Blood type A

Normal hemoglobin

Free earlobes

Blood type O

Figure 10.2

A Pair of Homologous Chromosomes

Homologous chromosomes contain genes for the same characteristics at the same place. Note that the attached-earlobe allele is located at the ear-shape locus on one chromosome, and the free-earlobe allele is located at the ear-shape locus on the other member of the homologous pair of chromosomes. The other two genes are for hemoglobin structure (alleles for normal and sickled cells) and blood type (alleles for blood types A and O). The examples presented here are for illustrative purposes only. We do not really know if these particular genes are on these chromosomes. It is hoped that the Human Genome Project, described in chapter 9, will resolve this problem.

often possible to figure out the genes present that determine a particular characteristic. For example, there are three possible genotypic combinations of the 2 alleles for earlobe shape. Genotypes are typically represented by upper- and lowercase letters. In the case of the earlobe trait, the allele for free earlobes is designated "*E*," whereas that for attached earlobes is "*e*." A person's genotype could be (1) two alleles for attached earlobes, (2) one allele for attached earlobes and 1 allele for free earlobes, or (3) two alleles for free earlobes.

How would individuals with each of these three genotypes look? The way each combination of alleles *expresses* (shows) itself is known as the **phenotype** of the organism. The phrase **gene expression** refers to the degree to which a gene goes through transcription and translation to show itself as an observable feature of the individual.

A person with 2 alleles for attached earlobes will have earlobes that do not hang free. A person with 1 allele for attached earlobes and 1 allele for free earlobes will have a phenotype that exhibits free earlobes. An individual with 2 alleles for free earlobes will also have free earlobes. Notice that there are three genotypes, but only two phenotypes. The individuals with the free-earlobe phenotype can have different genotypes.

Alleles	Genotypes	Phenotypes
E = free earlobes	*EE*	Free earlobes
e = attached earlobes	*Ee*	Free earlobes
	ee	Attached earlobes

The expression of some genes is directly influenced by the presence of other alleles. For any particular pair of alleles in an individual, the 2 alleles from the two parents are either

identical or not identical. Persons are **homozygous** for a trait when they have the combination of two identical alleles for that particular characteristic; for example, *EE* and *ee*. A person with 2 alleles for freckles is said to be homozygous for that trait. A person with 2 alleles for no freckles is also homozygous. If an organism is homozygous, the characteristic expresses itself in a specific manner. A person homozygous for free earlobes has free earlobes, and a person homozygous for attached earlobes has attached earlobes.

Individuals are designated as **heterozygous** when they have two different allelic forms of a particular gene, for example, *Ee*. The heterozygous individual received one form of the gene from one parent and a different allele from the other parent. For instance, a person with 1 allele for freckles and 1 allele for no freckles is heterozygous. If an organism is heterozygous, these two different alleles interact to determine a characteristic.

Dominant and Recessive Alleles

Often, one allele in the pair expresses itself more than the other. A **dominant allele** masks the effect of other alleles for the trait. For example, if a person has 1 allele for free earlobes and 1 allele for attached earlobes, that person has a phenotype of free earlobes. We say the allele for free earlobes is dominant. Dominant alleles are typically represented with a capital letter. A **recessive allele** is one that, when present with another allele, has its actions overshadowed by the other; it is masked by the effect of the other allele. Recessive alleles are typically represented with a lowercase letter. Having attached earlobes is the result of having a combination of two recessive characteristics. A person with 1 allele for free earlobes and 1 allele for attached earlobes has a phenotype of free earlobes. The expression of recessive alleles is only noted when the organism is homozygous for the recessive alleles. If you have attached earlobes, you have 2 alleles for that trait. A **carrier** is any person who is heterozygous for a trait. In this situation, the recessive allele is hidden—that is, it does not express itself enough to be a phenotype.

Don't think that recessive alleles are necessarily bad. *The term "recessive" has nothing to do with the significance or value of the allele—it simply describes how it can be expressed. Recessive alleles are not less likely to be inherited but must be present in a homozygous condition to express themselves. Also, recessive alleles are not necessarily less frequent in the population* (see table 11.1). Sometimes the physical environment determines whether or not dominant or recessive genes function. For example, in humans, genes for freckles do not show themselves fully unless a person's skin is exposed to sunlight (figure 10.3).

Codominance

In cases of dominance and recessiveness, one allele of the pair clearly overpowers the other. Although this is common, it is not always the case. In some combinations of alleles,

Figure 10.3

The Environment and Gene Expression

The expression of many genes is influenced by the environment. The allele for dark hair in the cat is sensitive to temperature and expresses itself only in the parts of the body that stay cool. The allele for freckles expresses itself more fully when a person is exposed to sunlight.

there is a **codominance.** This is a situation in which both alleles in a heterozygous condition express themselves.

A classic example of codominance in plants involves the color of the petals of snapdragons. There are 2 alleles for the color of these flowers. Because neither allele is recessive, we cannot use the traditional capital and small letters as symbols for these alleles. Instead, the allele for white petals is given the symbol F^W, and the one for red petals is given the symbol F^R (figure 10.4). There are three possible combinations of these 2 alleles:

Genotype	Phenotype
$F^W F^W$	White flower
$F^R F^R$	Red flower
$F^R F^W$	Pink flower

Notice that there are only two different alleles, red and white, but there are three phenotypes, red, white, and pink. Both the red-flower allele and the white-flower allele partially express themselves when both are present, and this results in pink.

A human example involves the genetic abnormality, *sickle-cell disease* (see figure 7.10). Having the two recessive alleles for sickle-cell hemoglobin (Hb^S and Hb^S) can result in abnormally shaped red blood cells. This occurs because the hemoglobin molecules are synthesized with the wrong amino acid sequence. These abnormal hemoglobin molecules tend to attach to one another in long, rodlike chains when oxygen is in short supply—that is, with exercise, pneumonia, emphysema. These rodlike chains distort the shape of the red blood cells into a sickle shape. When these abnormal red blood cells change shape, they clog small blood vessels. The sickled red cells are also destroyed more rapidly than normal cells. This results in a shortage of red blood cells, a condition known as anemia, and an oxygen deficiency in the tissues

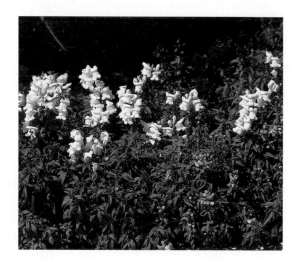

Figure 10.4

A Case of Codominance

The colors of these snapdragons are determined by two alleles for petal color, F^W and F^R. There are three different phenotypes because of the way in which the alleles interact with one another. In the heterozygous condition, neither of the alleles dominates the other.

that have become clogged. People with sickle-cell anemia may experience pain, swelling, and damage to organs such as the heart, lungs, brain, and kidneys. • **sickle cell,** and **DNA, p. 144**

Sickle-cell anemia can be lethal in the homozygous recessive condition. In the homozygous dominant condition ($Hb^A Hb^A$), the person has normal red blood cells. In the heterozygous condition ($Hb^A Hb^S$), patients produce both kinds of red blood cells. When the amount of oxygen in the blood falls below a certain level, those able to sickle will distort. However, when this occurs, most people heterozygous for the trait do not show severe symptoms. Therefore, these

X chromosome — labels:
- Ichthyosis (dry scaly skin)
- Duchenne muscular dystrophy
- Retinosis pigmentosa (deposit of pigment in retina of eye leading to blindness)
- Night blindness
- Centromere
- Ocular albinism (no eye pigment)
- Absence of sweat glands
- X-linked cleft palate
- Testicular feminization (cells do not respond to testosterone—develops female characteristics but has testes)
- Split hand/foot deformity
- Fragile X (leads to mental retardation)
- Hemophilia (blood will not clot)
- Color deficiency (blindness)

Y chromosome — labels:
- Stature- and height-promoting genes
- SRY—testes-determining factor
- Skeletal abnormalities
- Promotes spermatogenesis

Figure 10.5

Sex Chromosomes

Why is the Y chromosome so small? Is there an advantage to a species in having one sex chromosome deficient in genes? One hypothesis answers yes! Consider the idea that, with genes for supposedly "female" characteristics eliminated from the Y chromosome, crossing-over and recombining with "female" genes on the X chromosome during meiosis could help keep sex traits separated. Males would be males and females would stay females. The chances of "male-determining" and "female-determining" genes getting mixed onto the same chromosome would be next to impossible because they would not even exist on the Y chromosome.

alleles are related to one another in a codominant fashion. However, under the right circumstances, being heterozygous can be beneficial. A person with a single sickle-cell allele is more resistant to malaria than a person without this allele.

Genotype	Phenotype
$Hb^A\ Hb^A$	Normal hemoglobin and nonresistance to malaria
$Hb^A\ Hb^S$	Normal hemoglobin and resistance to malaria
$Hb^S\ Hb^S$	Resistance to malaria but death from sickle-cell anemia

Originally, sickle-cell anemia was found at a high frequency in parts of the world where malaria was common, such as tropical regions of Africa and South America. Today, however, this genetic disease can be found anywhere in the world. In the United States, it is most common among black populations whose ancestors came from equatorial Africa.

X-Linked Genes

Pairs of alleles located on nonhomologous chromosomes separate independently of one another during meiosis when the chromosomes separate into sex cells. Because each chromosome has many genes on it, these genes tend to be inherited as a group. Genes located on the same chromosome that tend to be inherited together are called a **linkage group.** The process of crossing-over, which occurs during prophase I of meiosis I, may split up these linkage groups. Crossing-over happens between homologous chromosomes donated by the mother and the father and results in a mixing of genes. The closer two genes are to each other on a chromosome, the more probable it is that they will be inherited together.
• determination of sex, p. 186

People and many other organisms have two types of chromosomes. **Autosomes** (22 pairs) are not involved in sex determination and have the same kinds of genes on both members of the homologous pair of chromosomes. **Sex chromosomes** are a pair of chromosomes that control the sex of an organism. In humans, and some other animals, there are two types of sex chromosomes—the X chromosome and the Y chromosome. The Y chromosome is much shorter than the X chromosome and has fewer genes for traits than found on the X chromosome (figure 10.5). One genetic trait that is located on the Y chromosome contains the testis-determining gene—*SRY* (Sex-determining Region Y). Females are normally produced when two X chromosomes are present. Males are usually produced when one X chromosome and one Y chromosome are present.

Genes found together on the X chromosome are said to be **X-linked.** Because the Y chromosome is shorter than the

X chromosome, it does not have many of the alleles that are found on the comparable portion of the X chromosome. Therefore, in a man, the presence of a single allele on his only X chromosome will be expressed, regardless of whether it is dominant or recessive. A Y-linked trait in humans is the *SRY* gene. This gene controls the differentiation of the embryonic gonad to a male testis. By contrast, more than 100 genes are on the X chromosome. Some of these X-linked genes can result in abnormal traits such as *color deficiency, hemophilia, brown teeth,* and at least two forms of *muscular dystrophy* (Becker's and Duchenne's).

Table 10.1

DOMINANT AND RECESSIVE TRAITS IN PEA PLANTS

Characteristic	Dominant Allele	Recessive Allele
Plant height	Tall	Dwarf
Pod shape	Full	Constricted
Pod color	Green	Yellow
Seed surface	Round	Wrinkled
Seed color	Yellow	Green
Flower color	Purple	White

10.3 Mendel's Laws of Heredity

Heredity problems are concerned with determining which alleles are passed from the parents to the offspring and how likely it is that various types of offspring will be produced. The first person to develop a method of predicting the outcome of inheritance patterns was Mendel, who performed experiments concerning the inheritance of certain characteristics in garden pea (*Pisum sativum*) plants. From his work, Mendel concluded which traits were dominant and which were recessive. Some of his results are shown in table 10.1.

What made Mendel's work unique was that he studied only one trait at a time. Previous investigators had tried to follow numerous traits at the same time. When this was attempted, the total set of characteristics was so cumbersome to work with that no clear idea could be formed of how the offspring inherited traits. Mendel used traits with clear-cut alternatives, such as purple or white flower color, yellow or green seed pods, and tall or dwarf pea plants. He was very lucky to have chosen pea plants in his study because they naturally self-pollinate. When self-pollination occurs in pea plants over many generations, it is possible to develop a population of plants that is homozygous for a number of characteristics. Such a population is known as a *pure line.*

Mendel took a pure line of pea plants having purple flower color, removed the male parts (anthers), and discarded them so that the plants could not self-pollinate. He then took anthers from a pure-breeding white-flowered plant and pollinated the antherless purple flower. When the pollinated flowers produced seeds, Mendel collected, labeled, and planted them. When these seeds germinated and grew, they eventually produced flowers.

You might be surprised to learn that all the plants resulting from this cross had purple flowers. One of the prevailing hypotheses of Mendel's day would have predicted that the purple and white colors would have blended, resulting in flowers that were lighter than the parental purple flowers. Another hypothesis would have predicted that the offspring would have had a mixture of white and purple flowers. The unexpected result—all the offspring produced flowers like those of one parent and no flowers like those of the other—caused Mendel to examine other traits as well and formed the basis for much of the rest of his work. He repeated his experiments using pure strains for other traits.

Pure-breeding tall plants were crossed with pure-breeding dwarf plants. Pure-breeding plants with yellow pods were crossed with pure-breeding plants with green pods. The results were all the same: The offspring showed the characteristics of one parent and not the other.

Next, Mendel crossed the offspring of the white-purple cross (all of which had purple flowers) with each other to see what the third generation would be like. Had the characteristic of the original white-flowered parent been lost completely? This second-generation cross was made by pollinating these purple flowers that had one white parent among themselves. The seeds produced from this cross were collected and grown. When these plants flowered, three-fourths of them produced purple flowers and one-fourth produced white flowers.

After analyzing his data, Mendel formulated several genetic laws to describe how characteristics are passed from one generation to the next and how they are expressed in an individual.

Mendel's **law of dominance**   When an organism has two different alleles for a given trait, the allele that is expressed, overshadowing the expression of the other allele, is said to be *dominant*. The gene whose expression is overshadowed is said to be *recessive*.

Mendel's **law of segregation**   When gametes are formed by a diploid organism, the alleles that control a trait separate from one another into different gametes, retaining their individuality.

Mendel's **law of independent assortment**   Members of one gene pair separate from each other independently of the members of other gene pairs.

At the time of Mendel's research, biologists knew nothing of chromosomes or DNA or of the processes of mitosis and meiosis. Mendel assumed that each gene was separate from other genes. It was fortunate for him that most of the characteristics he picked to study were found on separate chromosomes. If two or more of these genes had been located on the same chromosome (*linked genes*), he probably would not have been able to formulate his laws. The discovery of chromosomes and DNA have led to modifications in Mendel's laws, but it was Mendel's work that formed the foundation for the science of genetics.

HOW SCIENCE WORKS 10.1

Cystic Fibrosis—What's the Probability?

One in every 20 persons has a defective allele that causes cystic fibrosis. *Cystic fibrosis (CF)* is among the most common lethal genetic disorders that affect Caucasians in the United States. CF affects nearly 30,000 people in North America. About 5% of American whites are asymptomatic carriers of the recessive allele. About 1,000 new cases are identified in the United States each year. The gene for CF occurs on chromosome 7 and is responsible for the manufacture of cystic fibrosis transmembrane regulator (CFTR) protein. The most common of dozens of possible mutations in this gene is a deletion mutation responsible for the loss of a single amino acid in CFTR. As a result, CFTR is unable to control the movement of chloride ions through chloride-permeable cell membrane channels. This results in the malfunction of sweat glands in the skin and secretion of excess chloride ions, which causes a salty taste of the skin of children with CF. It is also responsible for other symptoms including:

1. Mucus filling the bronchioles, resulting in blocked breathing and frequent upper respiratory infections;
2. Bile duct clogging, which interferes with digestion and liver function;
3. Mucus clogging the pancreas ducts, preventing the flow of digestive enzymes into the intestinal tract;
4. Bowel obstructions caused by thickened stools; and
5. Sterility in males due to absence of vas deferens development and, on occasion, female sterility due to dense mucus that blocks sperm from reaching eggs.

Until recently, the test used to detect this gene in people with a family history of cystic fibrosis was only 50% accurate and could not detect the gene in persons with no family history of the disease. A new test is 75% accurate in detecting people with the gene regardless of their family history. This new test enables physicians to screen prospective parents to better determine the probability that they will have an offspring with cystic fibrosis. Confirmation of CF requires additional testing.

To reduce the risk of serious pulmonary complications, CF patients require daily airway clearance therapy. This can be accomplished by a therapy called "cupping" or with the use of mechanical devices. Cupping is done by "cupping" the hands and tamping the back of the patient to loosen the thick secretions formed in the lungs. Mechanical methods include The Vest™ airway clearance system, an easy-to-use medical device for children and adults. Designed for in-home use, The Vest system is quiet, comfortable, and portable. Its features promote user independence and simplify treatment routines. The Vest system uses a technology called high-frequency chest wall oscillation (HFCWO). The system has an inflatable vest connected by tubes to a generator. During therapy, the vest inflates and deflates rapidly, applying gentle pressure to the chest wall. This works to loosen and thin mucus and to move it toward the larger airways, where it can be cleared by coughing or suctioning (see figure).

Figure 1

10.4 Probability Versus Possibility

In order to solve heredity problems, you must have an understanding of probability. **Probability** is the chance that an event will happen, and is often expressed as a percentage or a fraction. *Probability* is not the same as *possibility*. It is possible to toss a coin and have it come up heads. But the probability of getting a head is more precise than just saying it is possible to get a head. The probability of getting a head is 1 out of 2 (½ or 0.5 or 50%) because there are two sides to the coin, only one of which is a head. Probability can be expressed as a fraction:

$$\text{probability} = \frac{\text{the number of events that can produce a given outcome}}{\text{the total number of possible outcomes}}$$

What is the probability of cutting a deck of cards and getting the ace of hearts? The number of times that the ace of hearts can occur is 1. The total number of possible outcomes (number of cards in the deck) is 52. Therefore, the probability of cutting an ace of hearts is ½₂.

What is the probability of cutting an ace? The total number of aces in the deck is 4, and the total number of cards is 52. Therefore, the probability of cutting an ace is ⁴⁄₅₂ or ¹⁄₁₃.

It is also possible to determine the probability of two independent events occurring together. *The probability of two or more events occurring simultaneously is the product of their individual probabilities.* (How Science Works 10.1). If you throw a pair of dice, it is possible that both will be 4s. What is the probability that both will be 4s? The probability of one die being a 4 is ⅙. The probability of the other die being a 4 is also ⅙. Therefore, the probability of throwing two 4s is:

$$1/6 \times 1/6 = 1/36$$

10.5 Steps in Solving Heredity Problems: Single-Factor Crosses

The first type of problem we will consider is the easiest type, a single-factor cross. A **single-factor cross** (sometimes called a monohybrid cross: *mono* = one; *hybrid* = combination) is a genetic cross or mating in which a single characteristic is followed from one generation to the next. For example, in humans, the allele for *Tourette syndrome (TS)* is inherited as an autosomal dominant allele.

For centuries, people displaying this genetic disorder were thought to be possessed by the devil since they displayed such unusual behaviors. These motor and verbal behaviors or *tics* are involuntary and range from mild (e.g., leg tapping, eye blinking, face twitching) to the more violent forms such as the shouting of profanities, head jerking, spitting, compulsive repetition of words, or even barking like a dog. The symptoms result from an excess production of the brain messenger, dopamine.

If both parents are heterozygous (have 1 allele for Tourette and 1 allele for no Tourette syndrome), what is the probability that they can have a child without Tourette syndrome? With Tourette syndrome?

Steps in Solving Heredity Problems—Single-Factor Crosses

Five basic steps are involved in solving a heredity problem.

Step 1: Assign a Symbol for Each Allele.

Usually a capital letter is used for a dominant allele and a small letter for a recessive allele. Use the symbol *T* for Tourette and *t* for no Tourette.

Allele	Genotype	Phenotype
T = Tourette	*TT*	Tourette syndrome
t = normal	*Tt*	Tourette syndrome
	tt	Normal

Step 2: Determine the Genotype of Each Parent and Indicate a Mating.

Because both parents are heterozygous, the male genotype is *Tt*. The female genotype is also *Tt*. The × between them is used to indicate a mating.

$$Tt \times Tt$$

Step 3: Determine All the Possible Kinds of Gametes Each Parent Can Produce.

Remember that gametes are haploid; therefore, they can have only 1 allele instead of the 2 present in the diploid cell. Because the male has both the Tourette syndrome allele and the normal allele, half his gametes will contain the Tourette syndrome allele and the other half will contain the normal allele. Because the female has the same genotype, her gametes will be the same as his.

For genetic problems, a *Punnett square* is used. A **Punnett square** is a box figure that allows you to determine the

probability of genotypes and phenotypes of the progeny of a particular cross. Remember, because of the process of meiosis, each gamete receives only 1 allele for each characteristic listed. Therefore, the male will produce sperm with either a *T* or a *t*; the female will produce ova with either a *T* or a *t*. The possible gametes produced by the male parent are listed on the left side of the square and the female gametes are listed on the top. In our example, the Punnett square would show a single dominant allele and a single recessive allele from the male on the left side. The alleles from the female would appear on the top.

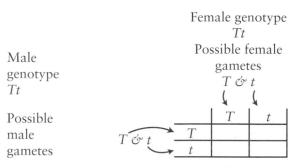

Step 4: Determine All the Gene Combinations That Can Result When These Gametes Unite.

To determine the possible combinations of alleles that could occur as a result of this mating, simply fill in each of the empty squares with the alleles that can be donated from each parent. Determine all the gene combinations that can result when these gametes unite.

Step 5: Determine the Phenotype of Each Possible Gene Combination.

In this problem, three of the offspring, *TT*, *Tt*, and *Tt*, have Tourette syndrome. One progeny, *tt*, is normal. Therefore, the answer to the problem is that the probability of having offspring with Tourette syndrome is ¾; for no Tourette syndrome, it is ¼.

Take the time to learn these five steps. All single-factor problems can be solved using this method; the only variation in the problems will be the types of alleles and the number of possible types of gametes the parents can produce. Now let's consider a problem in which one parent is heterozygous and the other is homozygous for a trait.

Problem: Dominant/Recessive PKU

Some people are unable to convert the amino acid phenylalanine into the amino acid tyrosine. The buildup of phenylalanine in the body prevents the normal development of the nervous system. Such individuals suffer from phenylketonuria (PKU) and may become mentally retarded (figure 10.6). The normal condition is to convert phenylalanine to tyrosine. It is

Proteins from food

Phenylalanine → Phenylpyruvic acid

1. Toxic to nerve cells

Gene → Enzyme phenylalanine hydroxylase

Absence of enzyme encourages phenylpyruvic acid production and prevents tyrosine production

Other sources of tyrosine → Tyrosine → Thyroxine (results in normal growth)

2. Less thyroxine produced (may result in abnormal growth)

Less melanin produced   3. Light-skin pigment

Melanin (skin pigment)

Figure 10.6

Phenylketonuria

PKU is an autosomal recessive disorder located on chromosome 12. This diagram shows how the normal pathways work (these are shown in gray). If the enzyme phenylalanine hydroxylase is not produced because of a mutated gene, the amino acid phenylalanine cannot be broken down, and is converted into phenylpyruvic acid, which accumulates in body fluids. In blue, there are three major results: (1) mental retardation because phenylpyruvic acid kills nerve cells, (2) abnormal body growth because less of the growth hormone thyroxine is produced, and (3) pale skin pigmentation because less melanin is produced (abnormalities are shown in color). It should also be noted that if a woman who has PKU becomes pregnant, her baby is likely to be born retarded. Although the embryo may not have the genetic disorder, the phenylpyruvic acid produced by the pregnant mother will damage the developing brain cells. This is called *maternal PKU*.

---

dominant over the condition for PKU. If one parent is heterozygous and the other parent is homozygous for PKU, what is the probability that they will have a child who is normal? A child with PKU?

Step 1:
Use the symbol N for normal and *n* for PKU.

Allele	Genotype	Phenotype
N = normal	NN	Normal metabolism of phenylalanine
n = PKU	Nn	Normal metabolism of phenylalanine
	nn	PKU disorder

Step 2:

$$Nn \times nn$$

Step 3:

	n
N	
n	

Step 4:

	n
N	Nn
n	*nn*

Step 5:
In this problem, ½ of the progeny will be normal and ½ will have PKU.

Problem: Codominance

If a pink snapdragon is crossed with a white snapdragon, what phenotypes can result, and what is the probability of each phenotype?

Step 1:

$$F^W = \text{white flowers} \qquad F^R = \text{red flowers}$$

Genotype	Phenotype
$F^W F^W$	White flower
$F^W F^R$	Pink flower
$F^R F^R$	Red flower

Step 2:

$$F^R F^W \times F^W F^W$$

Step 3:

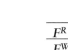

	F^W
F^R	
F^W	

Step 4:

	F^W
F^R	$F^W F^R$ Pink flower
F^W	$F^W F^W$ White flower

Step 5:
This cross results in two different phenotypes—pink and white. No red flowers can result because this would require that both parents be able to contribute at least one red allele. The white flowers are homozygous for white, and the pink flowers are heterozygous.

Problem: X-Linked

In humans, the gene for normal color vision is dominant and the gene for color deficiency is recessive. Both genes are X-linked. People who are color blind are not really blind, but should more appropriately be described as having "color defective vision." A male who has normal vision mates with a female who is heterozygous for normal color vision. What type of children can they have in terms of these traits, and what is the probability for each type?

Step 1:
This condition is linked to the X chromosome, so it has become traditional to symbolize the allele as a superscript on the letter X. Because the Y chromosome does not contain a homologous allele, only the letter Y is used.

$$X^N = \text{normal color vision}$$
$$X^n = \text{color-deficient}$$
$$Y = \text{male (no gene present)}$$

Genotype	Phenotype
$X^N Y$	Male, normal color vision
$X^n Y$	Male, color-deficient
$X^N X^N$	Female, normal color vision
$X^N X^n$	Female, normal color vision
$X^n X^n$	Female, color-deficient

Step 2:

Male's genotype = $X^N Y$ (normal color vision)
Female's genotype = $X^N X^n$ (normal color vision)

$$X^N Y \times X^N X^n$$

Step 3:
The genotype of the gametes are listed in the Punnett square:

	X^N	X^n
X^N		
Y		

Step 4:
The genotypes of the probable offspring are listed in the body of the Punnett square:

	X^N	X^n
X^N	$X^N X^N$	$X^N X^n$
Y	$X^N Y$	$X^n Y$

Step 5:
The phenotypes of the offspring are determined:

Normal female	Carrier female
Normal male	Color-deficient male

10.6 The Double-Factor Cross

A **double-factor cross** is a genetic study in which two pairs of alleles are followed from the parental generation to the offspring. Sometimes this type of cross is referred to as a dihybrid (di = two; hybrid = combination) cross. This problem is solved in basically the same way as a single-factor cross. The main difference is that in a double-factor cross you are working with two different characteristics from each parent.

It is necessary to use Mendel's law of independent assortment when considering double-factor problems. Recall that according to this law, members of one allelic pair sepa-rate from each other independently of the members of other pairs of alleles. This happens during meiosis when the chromosomes segregate. (Mendel's law of independent assortment applies only if the two pairs of alleles are located on separate chromosomes. We will assume this is so in double-factor crosses.)

In humans, the allele for free earlobes is dominant over the allele for attached earlobes. The allele for dark hair dominates the allele for light hair. If both parents are heterozygous for earlobe shape and hair color, what types of offspring can they produce, and what is the probability for each type?

Step 1:
Use the symbol E for free earlobes and e for attached earlobes. Use the symbol D for dark hair and d for light hair.

E = free earlobes D = dark hair
e = attached earlobes d = light hair

Genotype	Phenotype
EE	Free earlobes
Ee	Free earlobes
ee	Attached earlobes
DD	Dark hair
Dd	Dark hair
dd	Light hair

Step 2:
Determine the genotype for each parent and show a mating. The male genotype is $EeDd$, the female genotype is $EeDd$, and the $\times$ between them indicates a mating.

$$EeDd \times EeDd$$

Step 3:
Determine all the possible gametes each parent can produce and write the symbols for the alleles in a Punnett square. Because there are two pairs of alleles in a double-factor cross, each gamete must contain 1 allele from each pair—one from the earlobe pair (either E or e) and one from the hair color pair (either D or d). In this example, each parent can produce four different kinds of gametes. The four squares on the left indicate the gametes produced by the male; the four on the top indicate the gametes produced by the female.

To determine the possible gene combinations in the gametes, select 1 allele from one of the pairs of alleles and match it with 1 allele from the other pair of alleles. Then match the second allele from the first pair of alleles with each of the alleles from the second pair. This may be done as follows:

	ED	Ed	eD	ed
ED				
Ed				
eD				
ed				

Step 4:

Determine all the gene combinations that can result when these gametes unite. Fill in the Punnett square.

	ED	Ed	eD	ed
ED	EEDD	EEDd	EeDD	EeDd
Ed	EEDd	EEdd	EeDd	Eedd
eD	EeDD	EeDd	eeDD	eeDd
ed	EeDd	Eedd	eeDd	eedd

Step 5:

Determine the phenotype of each possible gene combination. In this double-factor problem there are 16 possible ways in which gametes can combine to produce offspring. There are four possible phenotypes in this cross. They are represented in the following chart.

Genotype	Phenotype	Symbol
EEDD or EEDd or EeDD or EeDd	Free earlobes/dark hair	*
EEdd or Eedd	Free earlobes/light hair	∧
eeDD or eeDd	Attached earlobes/dark hair	"
eedd	Attached earlobes/light hair	+

	ED	Ed	eD	ed
ED	EEDD *	EEDd *	EeDD *	EeDd *
Ed	EEDd *	Eedd ∧	EeDd *	Eedd ∧
eD	EeDD *	EeDd *	eeDD "	eeDd "
ed	EeDd *	Eedd ∧	eeDd "	eedd +

The probability of having a given phenotype is:

⁹⁄₁₆ free earlobes, dark hair
³⁄₁₆ free earlobes, light hair
³⁄₁₆ attached earlobes, dark hair
¹⁄₁₆ attached earlobes, light hair

For our next problem, let's say a man with attached earlobes is heterozygous for hair color and his wife is homozygous for free earlobes and light hair. What can they expect their offspring to be like?

This problem has the same characteristics as the previous problem. Following the same steps, the symbols would be the same, but the parental genotypes would be as follows:

$$eeDd \times EEdd$$

The next step is to determine the possible gametes that each parent could produce and place them in a Punnett square. The male parent can produce two different kinds of gametes, eD and ed. The female parent can produce only one kind of gamete, Ed.

	Ed
eD	
ed	

If you combine the gametes, only two kinds of offspring can be produced:

	Ed
eD	EeDd
ed	Eedd

They should expect either a child with free earlobes and dark hair or a child with free earlobes and light hair.

10.7 Alternative Inheritance Situations

So far we have considered a few straightforward cases in which a characteristic is determined by simple dominance and recessiveness between two alleles. Other situations, however, may not fit these patterns. Some genetic characteristics are determined by more than 2 alleles; moreover, some traits are influenced by gene interactions and some traits are inherited differently, depending on the sex of the offspring.

Multiple Alleles

So far we have discussed only traits that are determined by 2 alleles; for example, A, a. However, there can be more than 2 different alleles for a single trait. All the various forms of the same gene (alleles) that control a particular trait are referred to as **multiple alleles**. However, one person can have only a maximum of 2 of the alleles for the characteristic. A good example of a characteristic that is determined by multiple alleles is the ABO blood type. There are 3 alleles for blood type:

*Allele**

I^A = blood has type A antigens on red blood cell surface
I^B = blood has type B antigens on red blood cell surface
i = blood type O has neither type A nor type B antigens on surface of red blood cell

In the ABO system, A and B show *codominance* when they are together in the same individual, but both are dominant over the O allele. These 3 alleles can be combined as pairs in six different ways, resulting in four different phenotypes:

Genotype	Phenotype
$I^A I^A$	Blood type A
$I^A i$	Blood type A
$I^B I^B$	Blood type B
$I^B i$	Blood type B
$I^A I^B$	Blood type AB
ii	Blood type O

Multiple-allele problems are worked as single-factor problems.

*The symbols, I and i stand for the technical term for the antigenic carbohydrates attached to red blood cells, the immunogens. These alleles are located on human chromosome 9. The ABO system is not the only one used to type blood. Others include the Rh, MNS, and Xg systems.

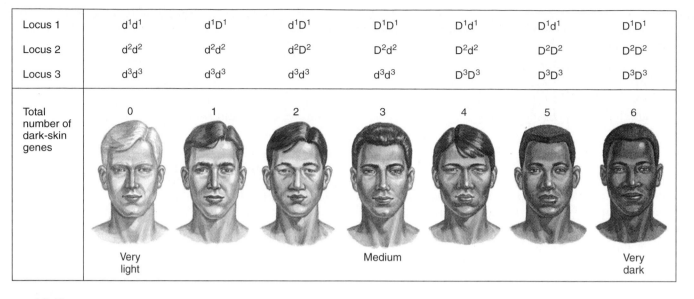

Locus 1	d^1d^1	d^1D^1	d^1D^1	D^1D^1	D^1d^1	D^1d^1	D^1D^1
Locus 2	d^2d^2	d^2d^2	d^2D^2	D^2d^2	D^2d^2	D^2D^2	D^2D^2
Locus 3	d^3d^3	d^3d^3	d^3d^3	d^3d^3	D^3D^3	D^3D^3	D^3D^3
Total number of dark-skin genes	0	1	2	3	4	5	6

Very light   Medium   Very dark

Figure 10.7

Polygenic Inheritance
Skin color in humans is an example of polygenic inheritance. The darkness of the skin is determined by the number of dark-skin genes a person inherits from his or her parents.

Polygenic Inheritance and Genetic Heterogeneity

Thus far we have considered phenotypic characteristics that are determined by alleles at a specific, single place on homologous chromosomes. However, some characteristics are determined by the interaction of genes at several different loci (on different chromosomes or at different places on a single chromosome). This is called **polygenic inheritance**. The fact that a phenotypic characteristic can be determined by many different alleles for a particular characteristic is referred to as **genetic heterogeneity**. A number of different pairs of alleles may combine their efforts to determine a characteristic. Skin color in humans is a good example of this inheritance pattern. According to some experts, genes for skin color are located at a minimum of three loci. At each of these loci, the allele for dark skin is dominant over the allele for light skin. Therefore a wide variety of skin colors is possible depending on how many dark-skin alleles are present (figure 10.7).

Polygenic inheritance is very common in determining characteristics that are quantitative in nature. In the skin-color example, and in many others as well, the characteristics cannot be categorized in terms of *either/or,* but the variation in phenotypes can be classified as *how much* or *what amount* (Outlooks 10.1). For instance, people show great variations in height. There are not just tall and short people—there is a wide range. Some people are as short as 1 meter, and others are taller than 2 meters. This quantitative trait is probably determined by a number of different genes. Intelligence also varies significantly, from those who are severely retarded to those who are geniuses. Many of these traits may be influenced by outside environmental factors such as diet, disease, accidents, and social factors. These are just a few examples of polygenic inheritance patterns.

Pleiotropy

Even though a single gene produces only one type of mRNA during transcription, it often has a variety of effects on the phenotype of the person. This is called *pleiotropy*. **Pleiotropy** (*pleio* = changeable) is a term used to describe the multiple effects that a gene may have on the phenotype. A good example of pleiotropy has already been discussed—that is, PKU. In PKU a single gene affects many different chemical reactions that depend on the way a cell metabolizes the amino acid phenylalanine commonly found in many foods (refer to figure 10.6). Another example is *Marfan syndrome* (figure 10.8), a disease suspected to have occurred in former U.S. president, Abraham Lincoln. Currently there is no DNA test to definitely diagnose people who have Marfan. Therefore, others speculated to have had Marfan include Flo Hyman (Olympic volleyball player), Jonathan Larson (playwright and author of *Rent*), John Tavener (classical musician), Paganini (classical musician), and Vincent Schiavelli (TV actor of *X-files*). This syndrome is a disorder of the body's connective tissue but can also have effects in many other organs. Symptoms generally include:

Skeletal

- Long arms and legs; disproportionate in length to the body
- Abnormally long fingers
- Skinniness
- Scoliosis of the spine
- Abnormally shaped chest: pectus excavatum (funnel chest)—chest caves in; or pectus carinatum (pigeon chest)—chest protrudes outward

continued on page 202

OUTLOOKS 10.1

The Inheritance of Eye Color

It is commonly thought that eye color is inherited in a simple dominant/recessive manner. Brown eyes are considered dominant over blue eyes. The real pattern of inheritance, however, is considerably more complicated than this. Eye color is determined by the amount of a brown pigment, known as melanin, present in the iris of the eye. If there is a large quantity of melanin present on the anterior surface of the iris, the eyes are dark. Black eyes have a greater quantity of melanin than brown eyes.

If a large amount of melanin is not present on the anterior surface of the iris, the eyes will appear blue, not because of a blue pigment but because blue light is returned from the iris (see illustration). The iris appears blue for the same reason that deep bodies of water tend to appear blue. There is no blue pigment in the water, but blue wavelengths of light are returned to the eye from the water. People appear to have blue eyes because the blue wavelengths of light are reflected from the iris.

Just as black and brown eyes are determined by the amount of pigment present, colors such as green, gray, and hazel are produced by the various amounts of melanin in the iris. If a very small amount of brown melanin is present in the iris, the eye tends to appear green, whereas relatively large amounts of melanin produce hazel eyes.

Several different genes are probably involved in determining the quantity and placement of the melanin and, therefore, in determining eye color. These genes interact in such a way that a wide range of eye color is possible. Eye color is probably determined by polygenic inheritance, just as skin color and height are. Some newborn babies have blue eyes that later become brown. This is because they have not yet begun to produce melanin in their irises at the time of birth.

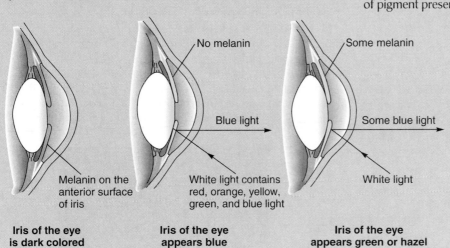

Iris of the eye is dark colored — Melanin on the anterior surface of iris

Iris of the eye appears blue — No melanin; Blue light; White light contains red, orange, yellow, green, and blue light

Iris of the eye appears green or hazel — Some melanin; Some blue light; White light

(a) (b) (c)

Figure 10.8

Marfan Syndrome
It is estimated that about 40,000 (the incidence is 1 out of 10,000) people in the United States have this autosomal dominant abnormality. Notice the lanky appearance to the body and face of this person with Marfan syndrome (a). Photos (b) and (c) illustrate their unusually long fingers.

Eye problems
- Myopia (near-sightedness)
- Dislocated lenses

Heart and aortic problems
- Weak or defective heart valves—mitral valve prolapse
- Endocarditis
- Aneurysms
- Aortic dissections (ripping of the vessel wall)

Lung and breathing problems
- Collapsed lungs
- Sleep apnea

10.8 Environmental Influences on Gene Expression

Maybe you assumed that the dominant allele would always be expressed in a heterozygous individual. It is not so simple! Here, as in other areas of biology, there are exceptions. For example, the allele for six fingers (*polydactylism*) is dominant over the allele for five fingers in humans. Some people who have received the allele for six fingers have a fairly complete sixth finger; in others, it may appear as a little stub. In another case, a dominant allele causes the formation of a little finger that cannot be bent like a normal little finger. However, not all people who are believed to have inherited that allele will have a stiff little finger. In some cases, this dominant characteristic is not expressed or perhaps only shows on one hand. Thus, there may be variation in the degree to which an allele expresses itself *in an individual*. Geneticists refer to this as *variable expressivity*. A good example of this occurs in the genetic abnormality *neurofibromatosis type 1 (NF1)* (figure 10.9). In some cases it may not be expressed *in the population* at all. This is referred to as a *lack of penetrance*. Other genes may be interacting with these dominant alleles, causing the variation in expression.
- Central Dogma, p. 134

Both internal and external environmental factors can influence the expression of genes. For example, at conception, a male receives genes that will eventually determine the pitch of his voice. However, these genes are expressed differently after puberty. At puberty, male sex hormones are released. This internal environmental change results in the deeper male voice. A male who does not produce these hormones retains a higher-pitched voice in later life. Another characteristic whose expression is influenced by internal gene-regulating mechanisms is that of male-pattern baldness (figure 10.10).

A comparable situation in females occurs when an abnormally functioning adrenal gland causes the release of large amounts of male hormones. This results in a female with a deeper voice. Also recall the genetic disease PKU. If children with *phenylketonuria (PKU)* are allowed to eat foods containing the amino acid phenylalanine, they will be-

Figure 10.9

Neurofibromatosis 1
This abnormality is seen in many forms including benign fibromatous skin tumors, "café au lait" spots, nodules in the iris, and possible malignant tumors. It is extremely variable in its expressivity (i.e., the traits may be almost unnoticeable or extensive). An autosomal dominant trait, it is the result of a mutation and the production of a protein (neurofibromin) that normally would suppress the activity of a gene that causes tumor formation.

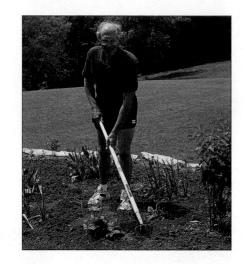

Figure 10.10

Baldness and the Expression of Genes
It is a common misconception that males have genes for baldness and females do not. Male-pattern baldness is a sex-influenced trait in which both males and females may possess alleles coding for baldness. These genes are turned on by high levels of the hormone testosterone. This is another example of an internal gene-regulating factor.

come mentally retarded. However, if the amino acid phenylalanine is excluded from the diet, and certain other dietary adjustments are made, the person will develop normally. NutraSweet is a phenylalanine-based sweetener, so people with this genetic disorder must use caution when buying products that contain it.

Diet is an external environmental factor that can influence the phenotype of an individual. *Diabetes mellitus*, a metabolic disorder in which glucose is not properly metabolized and is passed out of the body in the urine, has a genetic basis. Some people who have a family history of diabetes are thought to have inherited the trait for this disease. Evidence indicates that they can delay the onset of the disease by reducing the amount of sugar in their diet. This change in the external environment influences gene expression in much the same way that sunlight affects the expression of freckles in humans (see figure 10.3).

SUMMARY

Genes are units of heredity composed of specific lengths of DNA that determine the characteristics an organism displays. Specific genes are at specific loci on specific chromosomes. The phenotype displayed by an organism is the result of the effect of the environment on the ability of the genes to express themselves.

Diploid organisms have 2 genes for each characteristic. The alternative forms of genes for a characteristic are called alleles. There may be many different alleles for a particular characteristic. Organisms with two identical alleles are homozygous for a characteristic; those with different alleles are heterozygous. Some alleles are dominant over other alleles that are said to be recessive.

Sometimes 2 alleles express themselves, and often a gene has more than one recognizable effect on the phenotype of the organism. Some characteristics may be determined by several different pairs of alleles. In humans and some other animals, males have an X chromosome with a normal number of genes and a Y chromosome with fewer genes. Although they are not identical, they behave as a pair of homologous chromosomes. Because the Y chromosome is shorter than the X chromosome and has fewer genes, many of the recessive characteristics present on the X chromosome appear more frequently in males than in females, who have two X chromosomes.

THINKING CRITICALLY

For guidelines to answer this question, visit the Online Learning Center.

The breeding of dogs, horses, cats, and many other domesticated animals is done with purposes in mind (i.e., producing offspring that have specific body types, colors, behaviors, or athletic abilities). For example, horses can be bred to produce offspring that have a particular way of moving on foot, such as trotting or pacing. If a person buys a horse that has inherited the genes that make it a trotter, it will be a frustrating experience to both trainer and horse trying to teach it to pace. Trotters are genetically programmed to trot and pacers are programmed to pace. Every living thing is programmed to be good at something. Some people are going to have the muscle development to be great baseball players while others couldn't hit the ball. Some have great mathematical skills while others have a tough time adding 2 + 2. How do you think you have been genetically programmed? What are your strengths? As parent or child, what frustrations have you experienced in teaching and learning? What are the difficulties in determining which of your traits are genetic and which are not?

CONCEPT MAP TERMINOLOGY

Construct a concept map using these terms.

law of independent assortment	recessive allele
locus	single-factor inheritance
offspring	X-linked trait
probability	

Visit the Online Learning Center to use an interactive format to construct another concept map.

KEY TERMS

Use interactive flash cards, on the Online Learning Center, to help you learn the meanings of these terms.

alleles	law of segregation
autosomes	linkage group
carrier	locus (loci)
codominance	Mendelian genetics
dominant allele	multiple alleles
double-factor cross	offspring
gene	phenotype
gene expression	pleiotropy
genetic heterogeneity	polygenic inheritance
genetics	probability
genome	Punnett square
genotype	recessive allele
heterozygous	sex chromosomes
homozygous	single-factor cross
law of dominance	X-linked gene
law of independent assortment	

LEARNING CONNECTIONS

Visit the Online Learning Center to find additional study resources including those highlighted below.

10.1   Genetics, Meiosis, and Cells

Animations and Review
- Introduction

1. How many kinds of gametes are possible with each of the following genotypes?
   a. *Aa*
   b. *AaBB*
   c. *AaBb*
   d. *AaBbCc*
2. What is meant by the symbols *n* and *2n*?

10.2   Single-Gene Inheritance Patterns

3. Distinguish between phenotype and genotype.
4. What is the difference between dominant and codominant?

10.3 Mendel's Laws of Heredity

Interactive Concept Map
• Mendel's laws

5. In your own words describe Mendel's law of segregation.
6. What does it mean when geneticists use the term *independent assortment*?

10.4 Probability versus Possibility

7. What is the difference between probability and possibility?
8. In what mathematical forms might probability be expressed?

10.5 Steps in Solving Heredity Problems: Single-Factor Crosses

Animations and Review
• Monohybrid cross

Interactive Concept Map
• Steps to solving genetics problems

9. What is a Punnett square?
10. What types of symbols are typically used to express genotypes?

10.6 The Double-Factor Cross

Animations and Review
• Dihybrid cross

11. What is the probability of each of the following sets of parents producing the given genotypes in their offspring?

Parents	Offspring
a. *AA × aa*	*Aa*
b. *Aa × Aa*	*Aa*
c. *Aa × Aa*	*aa*
d. *AaBb × AaBB*	*AABB*
e. *AaBb × AaBB*	*AaBb*
f. *AaBb × AaBb*	*AABB*

12. If an offspring has the genotype *Aa*, what possible combinations of parental genotypes exist?

10.7 Alternative Inheritance Situations

Animations and Review
• Beyond Mendel
Experience This!
• Chart your own pedigree
Case Study
• Should you need a license to be a parent?

13. In humans, the allele for albinism is recessive to the allele for normal skin pigmentation.
   a. What is the probability that a child of a mother and a father who are heterozygous will be albino?
   b. If a child is normal, what is the probability that he or she is a carrier of the albino allele?
14. In certain pea plants, the allele *T* for tallness is dominant over *t* for shortness.
   a. If a homozygous tall and homozygous short plant are crossed, what will be the phenotype and genotype of the offspring?
   b. If both individuals are heterozygous, what will be the phenotypic and genotypic ratios of the offspring?

15. What is the probability of a child having type AB blood if one of the parents is heterozygous for A blood and the other is heterozygous for B? What other genotypes are possible in this child?

10.8 Environmental Influences on Gene Expression

16. What is meant by *expressivity*?
17. Define the phrase *lack of penetrance*.

Test yourself with multiple choice questions that provide immediate feedback
on the Online Learning Center with PowerWeb at

www.mhhe.com/enger11

CHAPTER

11

Diversity Within Species

GOALS AND OBJECTIVES

Recognize how the concepts of *species, gene pool,* **and** *population* **are related.**

❒ Explain the difference between the biological species concept and the
morphological species concept.

❒ State why all organisms of a species are not the same.

❒ Distinguish between gene pool and genetic diversity.

❒ List three methods used to distinguish species from one another.

**Know the factors that can cause differences in genetic diversity of different
populations of the same species.**

❒ Explain how each of the following affects the genetic diversity within
populations: mutation, sexual reproduction, population size, and migration.

❒ Describe three processes that could result in different populations of the same
species having different gene combinations.

**Learn the processes used to produce specific varieties of domesticated plants
and animals.**

❒ Relate cloning and hybridization to asexual and sexual reproduction.

❒ Explain how hybrid plants and animals are produced.

❒ Describe how genetic engineering differs from the development of intraspecific
hybrids and clones.

❒ Describe the value and potential danger of the practice of monoculture.

Recognize that population genetics principles apply to human populations.

❒ Describe why certain diseases are more common in some groups of people than in
others.

❒ Describe how a lack of understanding about population genetics contributed to
the eugenics movements.

PART FOUR · Evolution and Ecology

205

11.1 Populations and Species

To understand the principles of genetics in chapter 10, we concerned ourselves with small numbers of organisms having specific genotypes. When these organisms reproduced, we could predict the probability of an allele being passed to the next generation. Plants, animals, and other kinds of organisms, however, don't exist as isolated individuals but as members of populations. Since populations typically consist of large numbers of individuals each with its own set of alleles, populations contain many more possible alleles than a few individuals involved in a breeding experiment. Before we go any further, we need to develop a clear understanding of two terms that are used throughout this chapter, *population* and *species*.

The concepts of population and species are interwoven: A **population** is considered to be all the organisms of the same species found within a specific geographic region. A population is primarily concerned with numbers of organisms in a particular place at a particular time. A standard definition for species is that a **species** is a population of all the organisms potentially capable of breeding naturally among themselves and having offspring that also interbreed. *An individual organism is not a species but is a member of a species.* This definition of a species is often called the **biological species concept** and involves an understanding that organisms of different species do not interchange genetic information. Most populations consist of a portion of the members of a species, as when we discuss the wolves of Yellowstone National Park or the dandelion population in a city park. At other times it is possible to consider all the members of a species as being one large population, as when we talk about the human population of the world or the current numbers of the endangered whooping crane.

11.2 The Species Problem

A clear understanding of the concept of a species is important as we begin to consider how genetic material is passed around within populations as sexual reproduction takes place. If you examine the chromosomes of reproducing organisms, you find that they are identical in number and size and usually carry very similar groups of genes. In the final analysis, the biological species concept assumes that the genetic similarity of organisms is the best way to identify a species regardless of where or when they exist.

Often, organisms that are known to belong to distinct species differ in one or more ways that allow us to recognize them as separate species. Therefore, it is common to differentiate species on the basis of key structural characteristics. This method of using structural characteristics to identify species is called the **morphological species concept.** Structural differences are useful but not foolproof ways to distinguish species. However, we must rely on such indirect ways to identify species because we cannot possibly test every individual by breeding it with another to see if they will have fer-

tile offspring. Furthermore, many kinds of organisms reproduce primarily by asexual means. Because organisms that reproduce exclusively by asexual methods do not exchange genes with any other individuals, they do not fit our *biological species* definition very well.

Conversely, the structure or behavior of an organism may mislead people into assuming that two organisms are different species when actually they represent the extremes of variation within a species. Many plants have color variations or differences in leaf shape that cause them to look quite different although they are members of the same species. The eastern gray squirrel has black members within the species that many people assume to be a different species because they are so different in color. A good example of the genetic diversity within a species is demonstrated by the various breeds of dogs. A Great Dane does not look very much like a Pekinese. However, mating can occur between these two very different-appearing organisms (figure 11.1). They are of the same species.

Several other techniques are also used to identify species. Among animals, *differences in behavior* are often useful in identifying species. Some species of birds and insects are very similar structurally but can be easily identified by differences in the nature of their songs. Because it is often difficult to distinguish among bacteria, fungi, and other microorganisms on the basis of structure, *differences in metabolism* that result in the presence or absence of specific chemicals within the organism are often used to help distinguish among species.

Because we use many standards—biological, morphological, behavioral, and metabolic differences—to identify different species and none of these standards is perfect, we frequently have situations wherein individuals of two recognized species interbreed to a certain degree. Dogs, coyotes, and wolves have long been considered separate species. Differences in behavior and social systems tend to prevent mating among these three species. Wolves typically compete with coyotes and kill them when they are encountered. However, natural dog-coyote, wolf-coyote, and wolf-dog hybrids occur and the young are fertile (How Science Works 11.1). In fact, people have purposely encouraged mating between dogs and wolves for a variety of reasons. It is commonly thought that dogs are descendants of wolves that have been domesticated, so it should not be surprising that mating between wolves and dogs is easy to accomplish. The question then becomes, because matings do occur and the offspring are fertile, "Should dogs and wolves be considered members of the same species?" There is no simple answer to the question.

The species concept is an attempt to define groups of organisms that are *reproductively isolated* and, therefore, constitute a distinct unit of evolution. We must accept that some species will be completely isolated from other closely related species and will fit the definition well; some will have occasional exchange of genetic material between species and will not fit the definition as well; and some groups will interbreed so much that they must be considered distinct populations of

(a)

(b)

(c)

Figure 11.1

Genetic Diversity in Dogs
Although these four breeds of dogs look quite different, they all have the same number of chromosomes and are capable of interbreeding. Therefore, they are members of the same species. The considerable difference in phenotypes is evidence of the genetic variation among breeds—(*a*) golden retriever, (*b*) dalmatian, (*c*) dingo, (*d*) Pekinese.

(d)

HOW SCIENCE WORKS 11.1

Legal Implications of Defining a Species

The red wolf (*Canis rufus*) is listed as an endangered species, so the U.S. Fish and Wildlife Service has instituted a captive breeding program to preserve the animal and reintroduce it to a suitable habitat in the southeastern United States, where it was common into the 1800s. Biologists have long known that red wolves will hybridize with both the coyote, *Canis latrans,* and the gray wolf, *Canis lupus,* and many suspect that the red wolf is really a hybrid between the gray wolf and the coyote. Gray wolf–coyote hybrids are common in nature where one or the other species is rare. Some have argued that the red wolf does not meet the definition of a species and should not be protected under the Endangered Species Act.

Museums have helped shed light on this situation by providing skulls of all three kinds of animals preserved in the early 1900s. It is known that during the early 1900s as the number of red wolves in the southeastern United States declined, they readily interbred with coyotes, which were very common. The gray wolf had been exterminated by the early 1900s. Some scientists believe that the skulls of the few remaining "red wolves"

might not represent the true red wolf but a "red wolf" with many coyote characteristics. Studies of the structure of the skulls of red wolves, coyotes, and gray wolves show that the red wolves were recognizably different and intermediate in structure between coyotes and gray wolves. This supports the hypothesis that the red wolf is a distinct species.

DNA studies were performed using material from preserved red wolf pelts. The red wolf DNA was compared to coyote and gray wolf DNA. These studies show that red wolves contain DNA sequences typical of both gray wolves and coyotes but do not appear to have distinct base sequences found only in the red wolf. This evidence supports the opinion that the red wolf is not a species but may be a population that resulted from hybridization between gray wolves and coyotes.

There is still no consensus on the status of the red wolf. Independent researchers disagree with one another and with Fish and Wildlife Service scientists, who have been responsible for developing and administering a captive breeding program and planning reintroductions of the red wolf.

---

the same species. Throughout the next several chapters we will use the term *species*, as a population of reproductively isolated individuals, complete with the flaws and shortcomings of this definition, because it is a useful way to identify groups of organisms that have great genetic similarity and maintain a certain degree of genetic separateness from all similar organisms.

There is one other thing you need to be careful about when using the word *species*. It is both a singular and plural word so you can talk about a single species or you can talk about several species. The only way you can tell how the word is being used is by assessing the context of the sentence.

11.3 The Gene Pool Concept

We have just related the species concept to genetic similarity; however, you know that not all individuals of a species are genetically identical. Any one organism has a specific genotype consisting of all the genetic information that organism has in its DNA. It can have a maximum of 2 different alleles for a characteristic because it has inherited an allele from each parent. In a population, however, there may be many more than 2 alleles for a specific characteristic. In humans, there are 3 alleles for blood type (A, B, and O) within the population, but an individual can have only up to 2 of the alleles. Because, theoretically, all organisms of a species are able to exchange genetic material, we can think of all the genetic information of all the individuals of the same species as a giant **gene pool. • genetics, p. 190**

Because each individual organism is like a container of a particular set of these alleles, the gene pool contains much more genetic variation than any one of the individuals. The gene pool is like a refrigerator full of cartons of different kinds of milk—chocolate, regular, skim, buttermilk, low-fat, and so on. If you were blindfolded and reached in with both hands and grabbed two cartons, you might end up with two chocolate, a skim and a regular, or one of the many other possible combinations. The cartons of milk represent different alleles, and the refrigerator (gene pool) contains more genetic diversity than could be determined by randomly selecting two cartons of milk at a time.

Individuals of a species usually are not found evenly distributed within a geographic region but occur in clusters as a result of factors such as geographic barriers that restrict movement or the local availability of resources. Local populations with distinct genetic combinations may differ quite a bit from one place to another. There may be differences in the kinds of alleles and the numbers of each kind of allele in different populations of the same species. Note in figure 11.2 that within

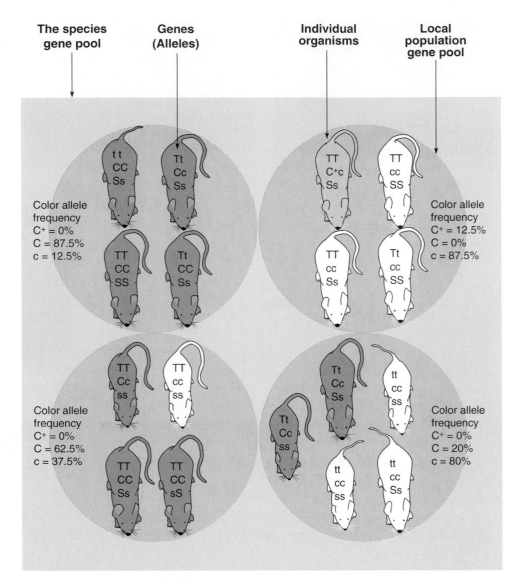

Figure 11.2

Genes, Populations, and Gene Pools

Each individual shown here has a specific combination of alleles that constitutes its genotype. The frequency of a specific allele varies from one local population to another. Each local population has a gene pool that is somewhat different from the others. The gene pool of the species consists of all the individuals of all the separate populations and includes all the individuals present. Assume that T = long tail; t = short tail; C^+ = gray color; C = brown color; c = white color; S = large size; and s = small size. Notice how differences in the frequency of particular alleles in local populations affects the appearance of the individuals. Local populations that show differences from other populations are often referred to as *subspecies*.

the gene pool of the species there are 3 possible alleles for color (C^+, C, and c). However, the frequencies of these alleles are different in the four local populations and this difference in allele frequency is reflected in the colors seen in the individuals of the population.

Because organisms tend to interbreed with other organisms located close by, local collections of alleles tend to

Northern water snake

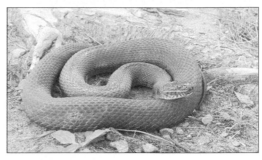

Lake Erie water snake

Figure 11.3

Subspecies of the Water Snake *Nerodia sepidon*
The water snake *Nerodia sepidon* is found throughout the eastern part of the United States and extends into Canada. The northern water snake subspecies is widespread in the northcentral United States and adjacent Canada and is generally brown with lighter diamond–shaped patches. The Lake Erie water snake subspecies is limited to the islands in the western section of Lake Erie. It is generally a solid color without the lighter patches. The difference in color patterns is related to different allele frequencies for color pattern. A third subspecies, the midland water snake, is found south of the habitat of the northern water snake.

remain the same unless, in some way, alleles are added to or subtracted from this local population. Water snakes are found throughout the eastern portion of the United States (figure 11.3). The Lake Erie water snake, which is confined to the islands in western Lake Erie, is one of the several distinct populations within this species. The northern water snakes of the mainland have light and dark bands. The island populations do not have this banded coloration. Most island individuals have alleles for solid coloration; very few individuals have alleles for banded coloration. The island snakes are geographically isolated from the main gene pool and mate only with one another. Thus, the different color patterns shown by island snakes and mainland snakes result from a high incidence of solid-color alleles in the island populations and a high incidence of banded-color alleles in the mainland populations.

Within a population, genetic material is repackaged into new individuals from one generation to the next. Often there is very little adding or subtracting of genetic material from a local population of organisms, and a widely distributed species will consist of a number of more or less separate groups that are known as **subspecies, races, breeds, strains,** or **varieties.** All of these terms are used to describe different forms of organisms that are all members of the same species. However, certain terms are used more frequently than others, depending on one's field of interest. For example, dog breeders use the term *breed,*

OUTLOOKS 11.1

Biology, Race, and Racism

The concept of racial difference among groups of people must be approached carefully. Two distortions can occur when people use the term *race.* First, the designation of race focuses on differences, most of which are superficial. Skin color, facial features, and the texture of the hair are examples. Although these examples are easy to see, they are arbitrary, and emphasis on them tends to obscure the fact that humans are all fundamentally the same, with minor variations in the frequency of certain alleles.

A second problem with the concept of race is that it is very difficult to separate genetic from cultural differences among people. People tend to equate cultural characteristics with genetic differences. Culture is learned and, therefore, is an acquired characteristic not based on the genes a person inherits. Cultures do differ, but these differences cannot be used as a basis for claiming genetic distinctions.

horticulturalists use the term *variety,* microbiologists use the term *strain,* and anthropologists use the term *race* (Outlooks 11.1). The most general and widely accepted term is *subspecies.*

11.4 Describing Genetic Diversity

Throughout the next three chapters you will need to watch several terms carefully. **Genetic diversity** is a term used to describe genetic differences among members of a population. High genetic diversity indicates many different kinds of alleles for each characteristic, and low genetic diversity indicates that nearly all the individuals in the population have the same alleles. In general, the term **gene frequency** is used in a general sense to convey the idea that there are genetic differences between populations. For example, one could say that there are differences in gene frequency between the northern water snake and the Lake Erie water snake. The term **allele frequency** is more properly used when specifically discussing how common a particular form of a gene (allele) is compared to other forms. For example, we could say that the frequency of the blond hair allele is high in northern Europe and low in Africa.

Figure 11.4

Gene Frequency Differences Among Humans
Different physical characteristics displayed by people from different parts of the world are an indication that gene frequencies differ as well.

Allele frequency is commonly stated in terms of a percentage or decimal fraction (e.g., 10% or 0.1; 50% or 0.5). It is a mathematical statement of how frequently a particular allele is found in a population. It is possible for two populations of the same species to have all the same alleles but with very different frequencies.

As an example, all humans are of the same species and, therefore, constitute one large gene pool found throughout the Earth. There are, however, many distinct local populations scattered across the surface of the Earth. These more localized populations show many distinguishing characteristics that have been perpetuated from generation to generation. In Africa, alleles for dark skin, tightly curled hair, and a flat nose have very high frequencies. In Europe, the frequencies of alleles for light skin, straight hair, and a narrow nose are the highest. People in Asia tend to have moderately colored skin, straight hair, and broad noses (figure 11.4). All three of these populations have alleles for dark skin and light skin, straight hair and curly hair, narrow noses and broad noses. The three differ, however, in the frequencies of these alleles. Once a particular mixture of alleles is present in a population, that mixture tends to maintain itself unless something is operating to change the frequencies. In other words, allele frequencies are not

going to change without reason. With the development of transportation, more people have moved from one geographic area to another, and human allele frequencies have begun to change. Ultimately, as barriers to interracial marriage (both geographic and sociological) are leveled, the human gene pool will show fewer and fewer geographically distinct populations.

For some reason, people tend to think that the frequency of alleles has something to do with dominance or recessiveness. This is not true. Often in a population, recessive alleles are more frequent than their dominant counterparts. Straight hair, blue eyes, and light skin are all recessive characteristics, yet they are quite common in the populations of certain European countries. See table 11.1 for other examples.

What really determines the frequency of an allele in a population is the value that the allele has to the organisms possessing it. The dark-skin alleles are valuable to people living under the bright sun in tropical Africa. These alleles are less valuable to those living in the less intense sunlight of the cooler European countries. This idea of the value of alleles and how this affects allele frequency will be dealt with more fully when the process of natural selection is discussed in chapter 12.

Table 11.1

RECESSIVE TRAITS WITH A HIGH FREQUENCY OF EXPRESSION

Many recessive characteristics are extremely common in some human populations. The corresponding dominant characteristic is also shown here.

Recessive	Dominant
Light skin color	Dark skin color
Straight hair	Curly hair
Five fingers	Six fingers
Type O blood	Type A or B blood
Normal hip joints	Dislocated hip birth defect
Blue eyes	Brown eyes
Normal eyelids	Drooping eyelids
No tumor of the retina	Tumor of the retina
Normal fingers	Short fingers
Normal thumb	Extra joint in the thumb
Normal fingers	Webbed fingers
Ability to smell	Inability to smell
Normal tooth number	Extra teeth
Presence of molars	Absence of molars

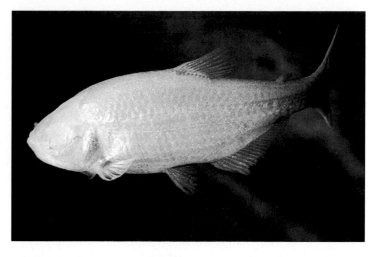

Figure 11.5

Specialized Local Populations
Populations of fish and other animals that live in caves where there is no light often lack pigment and functional eyes. If, at some time in the past, the genetic material that directed the production of these traits was lost or mutated, it would not have negatively affected the organism. In fact, it may be an advantage to not spend the energy to produce such unneeded traits. Hence, in many cave populations there is a low frequency of genes for the production of eyes and pigment.

11.5 Why Genetically Distinct Populations Exist

Many species have wide geographic distribution with reasonably distinct subspecies. There are several reasons why these subspecies developed.

Adaptation to Local Environmental Conditions

Because individual organisms within a population are not genetically identical, some individuals may possess genetic combinations that are particularly valuable for survival in the local environment. As a result, some individuals find the environment less hostile than do others. The individuals with unfavorable genetic combinations leave the population more often, either by death or migration, and remove their genes from the population. Therefore, local populations that occupy sites that differ greatly from conditions at other locations would be expected to consist of individuals having gene combinations suited to local conditions. For example, White Sands National Monument in New Mexico has extensive dunes of white gypsum sand. Several animals that live there, like lizards and mice, have very light coloring that allows them to blend in with their surroundings. Other populations of the same species that do not live in such a white environment do not have the light coloring.

Many kinds of animals that live in caves lack eyes and pigment. A blind fish living in a lake is at a severe disadvantage. A blind fish living in a cave where there is no light, however, is not at the same disadvantage. Thus, these two environments might allow or encourage characteristics to be present in the two populations at different frequencies. Since it takes energy to produce eyes and pigment, individual cave fishes that do not produce these characteristics may have an advantage over those individuals that continue to spend the energy to produce these features. In addition, it is possible that eyes could serve as a site for infections and those that lack eyes would again have an advantage over those that have them (figure 11.5).

Founder Effect

A second mechanism that tends to create genetically distinct populations with unique allele frequencies involves the founding of a new population. The collection of alleles from a small founding population is likely to be different from that present in the larger parent population from which it came. After all, a few individuals leaving a population would be unlikely to carry copies of all the alleles found within the original population. They may even carry an unrepresentative mixture of alleles. This situation in which a genetically distinct local population is established by a few colonizing individuals is know as the **founder effect**. For example, it is possible that the Lake Erie water snake discussed earlier was founded by a small number of individuals from the mainland that had a high frequency of alleles for solid coloration rather than the more typical banded pattern. (It is even possible that the island populations could have been founded by one fertilized female.) Once a small founding population establishes itself, it tends to maintain its collection of alleles because the organisms mate only among themselves. This results in a reshuffling of alleles from generation to generation but discourages the introduction of new genetic information into the population.

Genetic Bottleneck

A third cause of local genetically distinct populations relates to the past history of the population. Some local populations, and occasionally entire species, have reduced genetic diversity because their populations were severely reduced in the past. When the size of a population is greatly reduced it is likely that some alleles will be lost from the population. Such a population reduction that results in reduced genetic diversity is called a **genetic bottleneck.** Any subsequent increase in the size of the population by reproduction among the remaining members of the population will not replace the genetic diversity lost. There are thousands of species that are currently undergoing genetic bottlenecks. Although some endangered species were always rare, most have experienced recent reductions in their populations and a reduction in their genetic diversity, which is a consequence of severely reduced population size.

Barriers to Movement

A fourth factor that tends to encourage the maintenance of genetically distinct populations is the presence of barriers to free movement. Animals and plants that live in lakes tend to be divided into small, separate populations by barriers of land. Whenever such barriers exist, there will very likely be differences in the allele frequencies from lake to lake because each lake was colonized separately and their environments are not identical. Other species of organisms like migratory birds (robins, mallard ducks) experience few barriers; therefore, subspecies are quite rare.

11.6 How Genetic Diversity Comes About

A large gene pool with great genetic diversity is more likely to contain some genetic combinations that will allow the organisms to adapt to a new environment. A number of mechanisms introduce this necessary variation into a population.

Mutations

Mutations introduce new genetic information into a population by modifying alleles that are already present. Sometimes a mutation introduces a new allele into the gene pool of a species. At other times, a mutation may introduce an allele that was absent in a local population, although it is present in other populations of the species. All the different alleles for a particular trait originated as a result of mutations some time in the past and have been maintained within the gene pool of the species as they were passed from generation to generation during reproduction. Many mutations are harmful, but very rarely one will occur that is valuable to the organism. If a mutation produces a harmful allele, it will remain uncommon in the population. For example, at some time in the past, mutations occurred in the DNA of certain insect species that made some individuals tolerant to the insecticide DDT, even though the chemical had not yet been invented. These alleles remained very rare in these insect populations until DDT was used. Then, these alleles became very valuable to the insects that carried them. Because insects that lacked the alleles for tolerance died when they came in contact with DDT, more of the DDT-tolerant individuals were left to reproduce the species and, therefore, the DDT-tolerant alleles became much more common in these populations. • **mutations, p. 141**

Sexual Reproduction

Although the process of *sexual reproduction* does not create new alleles, it tends to generate new *genetic combinations* when the genetic information from two individuals mixes during fertilization, generating a unique individual. This doesn't directly change the frequency of alleles within the gene pool, but the new member may have a unique combination of characteristics so superior to those of other members of the population that the new member will be much more successful in producing offspring. In a corn population, there may be alleles for resistance to corn blight (a fungal disease) and resistance to attack by insects. Corn plants that possess both of these characteristics are going to be more successful than corn plants that have only one of these qualities. They will probably produce more offspring (corn seeds) than the others because they will survive fungal and insect attacks. Thus, there will be a change in the gene frequency for these characteristics in future generations (figure 11.6).

Migration

The *migration* of individuals from one genetically distinct population to another is also an important way for alleles to be added to or subtracted from a local population. Whenever an organism leaves one population and enters another, it subtracts its genetic information from the population it left and adds it to the population it joins. If it contains rare alleles, it may significantly affect the allele frequency of both populations. The extent of migration need not be great. As long as alleles are entering or leaving a population, the gene pool will change.

Many captive populations of animals in zoos are in danger of dying out because of severe inbreeding (breeding with near relatives), resulting in reduced genetic diversity. Often when genetic diversity is reduced, deleterious recessive

Sexual reproduction between varieties

Resistant to insect attack

Resistant to fungus attack

Sexual reproduction within variety

Sexual reproduction within variety

Resistant to insect attack

Resistant to fungus attack

Resistant to insect attack and fungus attack

Figure 11.6

New Genetic Combinations

Sexual reproduction can bring about new genetic combinations that are extremely valuable. These valuable new combinations tend to be perpetuated. Plants that are resistant only to insect attack or only to fungus attack produce fewer seeds than plants that are resistant to both insects and fungus. Plants resistant to both produce more seeds, passing that genetic combination on to the next generation.

alleles in closely related mates are likely to be passed to off-spring in a homozygous state resulting in offspring that have reduced chances of survival. Most zoo managers have recognized the importance of increasing genetic diversity in their small populations of captive animals and have instituted programs of loaning breeding animals to distant zoos in an effort to increase genetic diversity. In effect, they are attempting to simulate natural migration that frequently introduces new alleles from distant populations.

The Importance of Population Size

The *size of the population* has a lot to do with how effective any of these mechanisms are at generating diversity within a gene pool. The smaller the population, the less genetic diversity it can contain. Therefore, migrations, mutations, and accidental death can have great effects on the genetic makeup of a small population. For example, if a town has a population of 20 people and only 2 have brown eyes and the rest

Cuttings   A clone

Figure 11.7

Clones
All the plants in the right-hand photograph were produced asexually from cuttings and are identical genetically. The left-hand photograph shows how cuttings are made. The original plant is cut into pieces. Then the cut ends are treated with a growth stimulant and placed in moist sand or other material. Eventually, the pieces will root and become independent plants.

have blue eyes, what happens to those 2 brown-eyed people is more critical than if the town has 20,000 people and 2,000 have brown eyes. Although the ratio of brown eyes to blue eyes is the same in both cases, even a small change in a population of 20 could significantly change the frequency of the brown-eye allele. Often in small populations, random events can significantly alter the gene pool when rare alleles are lost from the population. This process is called *genetic drift* because the changes are not caused by selection. This idea will be discussed in greater detail in chapter 12. • **genetic drift, p. 231**

11.7 Genetic Diversity in Domesticated Plants and Animals

Humans often work with small, select populations of plants and animals in order to artificially construct specific genetic combinations that are useful or desirable. This is particularly true of plants and animals used for food. If we can produce domesticated animals and plants with genetic characteristics for rapid growth, high reproductive capacity, resistance to disease, and other desirable characteristics, we will be better able to supply ourselves with energy in the form of food. Several different processes are used to develop such specialized populations of plants and animals. Most of them have a side effect of reducing genetic diversity.

Cloning

Cloning is the process of reproducing organisms asexually so that large numbers of genetically identical individuals are produced. The individuals produced are called *clones*. • **mitosis, p. 158**

Plants are particularly easy to work with in this manner because we can often increase the numbers of specific organisms by asexual (without sex) reproduction. Potatoes, apple trees, strawberries, and many other plants can be reproduced by simply cutting the original plant into a number of parts and allowing these parts to sprout roots, stems, and leaves. If a single potato has certain desirable characteristics, it may be reproduced asexually. All of the individual plants reproduced asexually. Figure 11.7 shows how a clone is developed.

Cloning of most kinds of domesticated animals is much more difficult than cloning plants. However, in recent years many kinds of animals have been cloned. The process involves the substitution of a nucleus from a mature animal for the nucleus of an egg. This cell is then stimulated to develop as an embryo. There is a high rate of failure but the goals of animal cloning are the same as for cloning in plants: the production of genetically identical individuals. The first such animal produced, a sheep named Dolly, died in 2003. Subsequently, monkeys, mice, and other mammals have been cloned, but the success rate is still very low.

Selective Breeding

Humans can also bring together specific genetic combinations in either plants or animals by selective breeding. Because sexual reproduction tends to generate new genetic combinations rather than preserve desirable combinations, the mating of individual organisms must be controlled to obtain the desirable combination of characteristics. Selective breeding involves the careful selection of individuals with specific desirable characteristics and their controlled mating with the goal of producing a population that has a high proportion of individuals with the desired characteristics.

Through selective breeding, some varieties of chickens have been developed that grow rapidly and are good for meat. Others have been developed to produce large numbers of eggs. Often the development of new varieties of domesticated animals and plants involves the crossing of individuals from different populations. For this technique to be effective, the desirable characteristics in each of the two varieties should have homozygous genotypes. In small, controlled populations it is relatively easy to produce individuals that are homozygous for one specific trait. To make two characteristics homozygous in the same individual is more difficult. Therefore, such varieties are usually developed by crossing two different populations to collect several desirable characteristics in one organism. The organisms that are produced by the controlled breeding of separate varieties of the same species are often referred to as **intraspecific hybrids.**

Occasionally **interspecific hybrids**—hybrids between two different species—are produced as a way of introducing new desirable characteristics into a domesticated organism. Since plants can be reproduced by cloning, it is possible to produce an interspecific hybrid and then reproduce it by cloning. The tangelo is an interspecific hybrid between a tangerine and a grapefruit. An interspecific hybrid between cattle and the American bison was used to introduce certain desirable characteristics into cattle.

Genetic Engineering

In recent years, advances in our understanding of DNA have allowed specific pieces of genetic material to be inserted into cells. This has greatly expanded the ability of people to modify the characteristics of domesticated plants and animals. The primary goal of genetic engineering is to manipulate particular pieces of DNA and transfer them into specific host organisms so that they have particular, valuable characteristics. These topics are dealt with in greater detail in chapter 7.
• genetic engineering, p. 150.

The Impact of Monoculture

Although some of these techniques have been used to introduce new genetic information into domesticated organisms, one of the driving forces behind domestication is to pro-

Figure 11.8

Monoculture
This wheat field is an example of monoculture, a kind of agriculture in which large areas are exclusively planted with a single crop with a very specific genetic makeup. Monoculture makes it possible to use large farm machinery, but it also creates conditions that can encourage the spread of disease because the plants have reduced genetic diversity.

duce organisms that have uniform characteristics. In order to achieve such uniformity it is necessary to reduce genetic diversity. The plants have been extremely specialized through selective breeding to produce the qualities that growers want. Most agriculture in the world is based on extensive plantings of the same varieties of a species over large expanses of land (figure 11.8). This agricultural practice is called **monoculture.** It is certainly easier to manage fields in which there is only one kind of plant growing. This is particularly true today when herbicides, insecticides, and fertilizers are tailored to meet the needs of specific crop species. However, with monoculture comes a significant risk. Because these organisms are so similar, if a new disease comes along, most of them will be affected in the same way and the whole population may be killed or severely damaged.

Our primary food plants are derived from wild ancestors that had genetic combinations that allowed them to compete successfully with other organisms in their environment. When humans reduce genetic diversity by developing special populations with certain desirable characteristics, other valuable genetic information is lost from the gene pool. When we select specific, good characteristics, we often get harmful ones along with them. Therefore, these "special" plants and animals require constant attention. Insecticides, herbicides, cultivation, and irrigation are all used to aid the plants and animals we need. In effect, these plants are able to live only under conditions that people carefully maintain. Furthermore, we plant vast expanses of the same plant, creating tremendous potential for extensive crop loss from diseases.

Because our domesticated organisms are so genetically similar there is a great danger that an environmental change or new disease could cause great damage to our ability to produce food. In order to protect against such disasters, gene banks have been established. Gene banks consist of populations of primitive ancestors of modern domesticated plants and animals. By preserving these organisms, their genetic diversity is available for introduction into our domesticated plants and animals if the need arises.

4 in 1,000 carry the allele

Total U.S. population   **Ashkenazi Jews (world)**   **New York City Jews**

28 in 1,000 carry the allele

33 in 1,000 carry the allele

Frequency of Tay-Sachs allele in three populations

Figure 11.9

The Frequency of Tay-Sachs Allele
The frequency of an allele can vary from one population to another. Genetic counselors use this information to advise people of their chances of having specific alleles and of passing them on to their children.

11.8 Human Population Genetics

At the beginning of this chapter, we pointed out that the human gene pool consists of a number of subgroups. The particular characteristics that set one group apart from another originated many thousands of years ago before travel was as common as it is today, and we still associate certain physical features with certain geographic areas. Although there is much more movement of people and a mixing of racial and ethnic types today, people still tend to have children with others who are of the same social, racial, and economic background and who live in the same locality.

This non-random mate selection can sometimes bring together two individuals who have alleles that are relatively rare. Information about allele frequencies within specific human subpopulations can be very important to people who wish to know the probability of having children with particular harmful genetic abnormalities. This is particularly important if both individuals are descended from a common ancestral tribal, ethnic, or religious group. For example, Tay-Sachs disease causes degeneration of the nervous system and early death of children. Because it is caused by a recessive allele, both parents must pass the allele to their child in order for the child to have the disease. By knowing the frequency of the allele in the background of both parents, we can determine the probability of their having a child with this disease.

Ashkenazi Jews have a higher frequency of this recessive allele than do people of any other group of racial or social origin. The Jewish population of New York City has a slightly higher frequency of this allele than the worldwide population of Ashkenazi Jews (figure 11.9). Therefore, peo-

(a)   (b)

Figure 11.10

Normal and Sickle-Shaped Cells
Sickle-cell anemia is caused by a recessive allele that changes one amino acid in the structure of the oxygen-carrying hemoglobin molecule within red blood cells. (a) Normal cells are disk shaped. (b) In sickled cells, the abnormal hemoglobin molecules tend to stick to one another and distort the shape of the cell when the cells are deprived of oxygen.

ple of this particular background should be aware of the probability that they may have children who will develop Tay-Sachs disease.

Likewise, sickle-cell anemia is more common in people of specific African ancestry than in any other human subgroup (figure 11.10). Because many black slaves came from regions where sickle-cell anemia is common, African Americans should be aware that they might be carrying the allele for this type of defective hemoglobin. If they carry the allele, they should consider their chances of having children with this disease. These and other cases make it very important that trained **genetic counselors** have information about the frequencies of alleles in specific human ethnic groups so that they can help couples with genetic questions.

11.9 Ethics and Human Population Genetics

Misunderstanding the principles of heredity has resulted in bad public policy. Often when there is misunderstanding there is mistrust. Even today, many prejudices against certain genetic conditions persist.

Modern genetics had its start in 1900 with the rediscovery of the fundamental laws of inheritance proposed by Mendel. For the next 40 or 50 years, this rather simple understanding of genetics resulted in unreasonable expectations on the part of both scientists and laypeople. People generally assumed that much of what a person was in terms of structure, intelligence, and behavior was inherited. This led to the passage of **eugenics laws.** Their basic purpose was to eliminate "bad genes" from the human gene pool and encourage "good genes." These laws often prevented the marriage or permitted the sterilization of people who were "known" to have "bad genes" (figure 11.11). Often these laws were promoted as a way to save money because sterilization would prevent the birth of future "defectives" and, therefore, would reduce the need for expensive mental institutions or prisons. These laws were also used by people to legitimize racism and promote prejudice.

The writers of eugenics laws (How Science Works 11.2) overestimated the importance of genetics and underestimated the significance of such environmental factors as disease and poor nutrition. They also overlooked the fact that many genetic abnormalities are caused by recessive alleles. In most cases, the negative effects of recessive alleles can be recognized only in homozygous individuals. Removing only the homozygous individuals from the gene pool would have little influence on the frequency of the "bad genes" in the population. Many recessive alleles would be masked by dominant alleles in heterozygous individuals and would continue to show up in future generations. In addition, we now know that most characteristics are not inherited in a simple dominant/recessive fashion and that often many alleles at different loci cooperate in the production of a phenotypic characteristic. Furthermore, mutations occur constantly, adding new alleles to the mix. Usually these new alleles are recessive and deleterious. So, essentially all individuals are carrying "bad genes."

Today, genetic diseases and the degree to which behavioral characteristics and intelligence are inherited are still important social and political issues. The emphasis, however, is on determining the specific method of inheri-

---

720.301 Sterilization of mental defectives; statement of policy

Sec. 1. It is hereby declared to be the policy of the state to prevent the procreation and increase in number of feeble-minded and insane persons, idiots, imbeciles, moral degenerates and sexual perverts, likely to become a menace to society or wards of the state. The provisions of this act are to be liberally construed to accomplish this purpose. As amended 1962, No. 160, § 1, Eff. March 28,1963.

Figure 11.11

A Eugenics Law
This particular state law was enacted in 1929 and is typical of many such laws passed during the 1920s and 1930s. A basic assumption of this law is that the conditions listed are inheritable; therefore, the sterilization of affected persons would decrease the frequency of these conditions. Prior to 1962, the law also included epileptics. The law was repealed in 1974.

---

tance or the specific biochemical pathways that result in what we currently label as insanity, lack of intelligence, or antisocial behavior. Although progress is slow, several genetic abnormalities have been "cured," or at least made tolerable, by medicines or control of the diet. For example, phenylketonuria (PKU) is a genetic disease caused by an abnormal biochemical pathway. If children with this condition are allowed to eat foods containing the amino acid phenylalanine, they will become mentally retarded. However, if the amino acid phenylalanine is excluded from the diet, and certain other dietary adjustments are made, the person will develop normally. NutraSweet is a phenylalanine-based sweetener, so people with this genetic disorder must use caution when buying products that contain it. This abnormality can be diagnosed very easily by testing the urine of newborn infants. • **PKU, p. 196**

Effective genetic counseling has become the preferred method of dealing with genetic abnormalities. A person known to be a carrier of a "bad gene" can be told the likelihood of passing that characteristic on to the next generation before deciding whether or not to have children. In addition, *amniocentesis* (a medical procedure that samples amniotic fluid) and other tests make it possible to diagnose some genetic abnormalities early in pregnancy. If an abnormality is diagnosed, an abortion can be performed. Because abortion is unacceptable to some people, the counseling process must include a discussion of the facts about an abortion and the alternatives. Although at one time counselors often pushed people toward specific decisions, today it is inappropriate for counselors to be advocates; their role is to provide information that better allows individuals to make the best decisions possible for them.

HOW SCIENCE WORKS 11.2

Bad Science: A Brief History of the Eugenics Movement

- **1885** Francis Galton, cousin to Charles Darwin, proposes that human society could be improved "through better breeding." The term "eugenics" is coined; that is, "the systematic elimination of undesirables to improve humanity." This would be accomplished by breeding those with "desirable" traits and preventing reproduction of those with "undesirable" traits. John Humphrey Noyes, an American sexual libertarian, molds the eugenics concept to justify polygamy. "While the good man will be limited by his conscience to what the law allows, the bad man, free from moral check, will distribute his seed beyond the legal limit."
- **1907** The state of Indiana is the first to pass an involuntary sterilization law.
- **1919** Charles B. Davenport, founder of Cold Springs Harbor Laboratory and of the Eugenics Record Office, "proved" that "pauperism" was inherited. Also "proved that being a naval officer is an inherited trait." He noted that the lack of women in the navy also "proved" that the gene was unique to males.
- **1920** Davenport founds the American Eugenics Society. He sponsored "Fitter Families Contests" held at many state fairs around the country. The society persuaded 20 state governments to authorize the sterilization of men and women in prisons and mental hospitals. The society also put pressure on the federal government to restrict the immigration of "undesirable" races into the United States.
- **1927** Oliver Wendel Holmes argued for the involuntary sterilization of Carrie S. Buck. The 18-year-old Carrie was a resident of the Virginia State Colony for Epileptics and Feeble-Minded and the first person to be selected for sterilization under the law. Holmes won his case and Carrie was sterilized even though it was later revealed that neither she nor her illegitimate daughter, Vivian, were feebleminded.
- **1931** Involuntary sterilization measures were passed by 30 states.
- **1933–1941** Nazi death camps with the mass murder of Jews, Gypsies, Poles, and Russians were established and run resulting in the extermination of millions of people. "Adolf Hitler . . . guided by the nation's anthropologists, eugenists and social philosophers, has been able to construct a comprehensive racial policy of population development and improvement. . . . It sets a pattern. . . . These ideas have met stout opposition in the Rousseauian social philosophy . . . which bases . . . its whole social and political theory upon the patent fallacy of human equality. . . . Racial consanguinity occurs only through endogamous mating or interbreeding within racial stock . . . conditions under which racial groups of distinctly superior hereditary qualities . . . have emerged." (*The New York Times*, August 29, 1935)
- **1972–1973** Up to 4,000 sterilizations still performed in the state of Virginia alone, and the federal government estimated that 25,000 adults were sterilized nationwide.
- **1973** Since March 1973 the American Eugenics Society has called itself The Society for the Study of Social Biology.
- **1987** Eugenic sterilization of institutionalized retarded persons was still permissible in 19 states, but the laws were rarely carried out. Some states enact laws that forbid sterilization of people in state institutions.
- **Present** Some groups and individuals still hold to the concepts of eugenics claiming recent evidence "proves" that traits such as alcoholism, homosexuality, and schizophrenia are genetic and therefore should be eliminated from the population to "improve humanity." However, the movement lacks the organization and legal basis it held in the past. Modern genetic advances such as genetic engineering techniques and the mapping of the human genome provide the possibility of identifying individuals with specific genetic defects. Questions about who should have access to such information and how it could be used causes renewed interest in the eugenics debate.

SUMMARY

All organisms with similar genetic information and the potential to reproduce are members of the same species. A species usually consists of several local groups of individuals known as populations. Groups of interbreeding organisms are members of a gene pool. Although individuals are limited in the number of alleles they can contain, within the population there may be many different kinds of alleles for a trait. Subpopulations may have different gene frequencies from one another.

Genetically distinct populations exist because local conditions may demand certain characteristics, founding populations may have had unrepresentative gene frequencies, and barriers may prevent free flow of genetic information from one locality to another. Distinguishable subpopulations are often known as subspecies, varieties, strains, breeds, or races.

Genetic diversity is generated by mutations, which can introduce new alleles; sexual reproduction, which can generate new genetic combinations; and migration, which can subtract genetic information from or add genetic information to a local population. The size of the population is also important, because small populations typically have reduced genetic diversity.

Knowledge of population genetics is useful for plant and animal breeders and for people who specialize in genetic counseling. The genetic diversity of domesticated plants and animals has been reduced as a result of striving to produce high frequencies of valuable alleles. Clones and intraspecific hybrids are examples. Understanding gene frequencies and how they differ in various populations sheds light on why certain alleles are common in some human populations. Such understanding is also valuable in counseling members of populations with high frequencies of alleles that are relatively rare in the general population.

THINKING CRITICALLY

For guidelines to answer this question, visit the Online Learning Center.

Albinism is a condition caused by a recessive allele that prevents the development of pigment in the skin and other parts of the body. Albinos need to protect their skin and eyes from sunlight. The allele has a frequency of about 0.00005. What is the likelihood that both members of a couple would carry the allele? Why might two cousins or two members of a small tribe be more likely to have the allele than two nonrelatives from a larger population? If an island population has its first albino baby in history, why might it have suddenly appeared? Would it be possible to eliminate this allele from the human population? Would it be desirable to do so?

CONCEPT MAP TERMINOLOGY

Construct a concept map using these terms.

allele frequency	hybrid
breed	monoculture
clone	population
genus	species

Visit the Online Learning Center to use an interactive format to construct another concept map.

KEY TERMS

Use interactive flash cards, on the Online Learning Center, to help you learn the meanings of these terms.

allele frequency	interspecific hybrid
biological species concept	intraspecific hybrid
cloning	monoculture
eugenics laws	morphological species concept
founder effect	population
gene frequency	species
gene pool	subspecies (races, breeds,
genetic bottleneck	strains, or varieties)
genetic counselor	
genetic diversity	

LEARNING CONNECTIONS

Visit the Online Learning Center to find additional study resources including those highlighted below.

11.1 Populations and Species

1. How do the concepts of species and genetically distinct populations differ?
2. Describe the biological species concept.

11.2 The Species Problem

3. List four techniques used to distinguish one species from another.
4. Explain why the phrase "reproductively isolated" and the word "species" are related.

11.3 The Gene Pool Concept

Interactive Concept Map
• Gene pools

5. Give an example of a gene pool containing a number of separate populations.
6. Why can there be greater genetic diversity within a gene pool than in an individual organism?

11.4 Describing Genetic Diversity

7. What is meant by the terms *gene frequency* and *allele frequency*?
8. Give an example of a human characteristic that has a high frequency in Europe and a low frequency in Africa.

11.5 Why Genetically Distinct Populations Exist

9. List four processes that can lead to local genetically distinct populations.
10. In what way are the founder effect and a genetic bottleneck similar in their effect on the genetic diversity of a local population?

11.6 How Genetic Diversity Comes About

Interactive Concept Map
• Creating diversity

11. Why does the size of a population affect the gene pool?
12. List three factors that change allele frequencies in a population.

11.7 Genetic Diversity in Domesticated Plants and Animals

Food for Thought
• Cloning

13. How do the genetic combinations in clones and sexually reproducing populations differ?
14. How is a clone developed? What are its benefits and drawbacks?
15. How is an intraspecific hybrid formed? What are its benefits and drawbacks?
16. Why is genetic diversity in domesticated plants and animals reduced?

11.8 Human Population Genetics

17. Give an example of a human population with a high frequency of a deleterious allele.

11.9 Ethics and Human Population Genetics

Experience This!
• Eugenics where you live

18. What forces maintain racial differences in the human gene pool?
19. What were eugenics laws? List two facts about human genetics that the advocates of eugenics failed to consider.

Test yourself with multiple choice questions that provide immediate feedback
on the Online Learning Center with PowerWeb at

www.mhhe.com/enger11

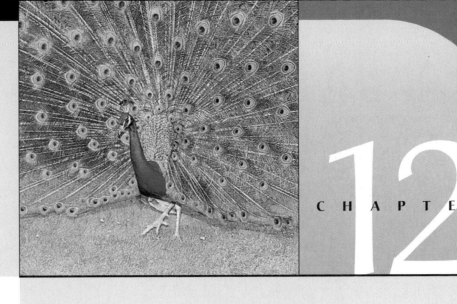

Natural Selection and Evolution

GOALS AND OBJECTIVES

Recognize that evolutionary change is the result of natural selection.

❏ Describe how the concepts of evolution and natural selection are related.

Understand how natural selection works.

❏ Explain why genetic diversity is essential for natural selection to occur.

❏ Describe how individuals produced by sexual reproduction can have fitness different from that of their parents.

❏ Explain how mutation and migration affect the genetic diversity of a population.

❏ Explain why excess reproduction is important to the concept of natural selection.

❏ List and describe three circumstances that can prevent a specific allele from being expressed in the phenotype of an organism.

❏ Describe common misunderstandings about the nature of natural selection.

❏ Explain how survival, reproductive success, and mate selection can alter gene frequency from one generation to the next.

Understand that evolution is the process of changing gene frequencies.

❏ Describe the conditions that can lead to genetic drift.

❏ Explain how natural selection can change the nature of a species.

❏ Provide examples that indicate that evolution is occurring.

Recognize the conditions under which the Hardy-Weinberg concept applies.

❏ List the five conditions necessary to prevent gene frequency changes according to the Hardy-Weinberg concept.

❏ Explain why Hardy-Weinberg conditions rarely occur.

12.1 The Concept of Evolution

In many cultural contexts, the word *evolution* means progressive change. We talk about the evolution of economies, fashion, or musical tastes. From a biological perspective, the word has a more specific meaning. **Evolution** is a change in the frequency of genetically determined characteristics within a population over time. Evolution can be looked at from several points of view. On the one hand, we can look at minor differences in *allele frequency* between populations of the same species that have developed recently, as when we examine genetic differences between subspecies. On the other hand, we can look at major differences that have occurred over long time periods that have resulted in so much genetic change that new *kinds of species* are produced. Regardless of which perspective we use, the mechanisms that bring about changes in allele frequency are the same. The focus of this chapter is on the mechanisms that result in evolutionary change. Chapter 13 focuses on processes that lead to the development of new species. Evolution, from both perspectives, involves changes in characteristics and the genetic information that produces these characteristics over time. There are many different forces that affect the characteristics present in a population or species. A primary process that alters the frequency of specific genetically determined characteristics within a population is the process of natural selection.

12.2 The Role of Natural Selection in Evolution

Evolution results when there are changes in allele frequency in a population. Individual organisms cannot evolve—only populations can. Although evolution is a population process, the mechanisms that bring it about operate at the level of the individual.

There are three factors that interact to determine how a species changes over time: *environmental factors* that affect organisms, *sexual reproduction* among the individuals in the gene pool, and the amount of *genetic diversity* within the gene pool.

The characteristics displayed by a species typically match the demands of the surroundings that the species typically encounters. In general, the reproductive success of any individual within a population is determined by how well the characteristics of the individual match the demands of the environment in which it lives. Biologists refer to the reproductive success of an individual as its **fitness**. Those individuals whose characteristics best fit their environment will be more likely to successfully reproduce. They have greater fitness.

The processes that operate to determine which individuals have the greater reproductive success are collectively known as **natural selection.** Thus, the **theory of natural selection** is the idea that some individuals whose genetic combinations favor life in their surroundings will be more likely to survive, reproduce, and pass their genes on to the next generation than those individuals who have unfavorable genetic combinations. This process results in populations that are genetically adapted to their surroundings. Natural selection is a primary process that brings about evolution by "selecting" which individuals will survive and reproduce and pass their genes to the next generation. The processes of natural selection do not affect genes directly but do so indirectly by selecting individuals for success based on the phenotype displayed. Recall that the characteristics displayed by an organism (phenotype) are related to the genes possessed by the organisms (genotype). It is important to recognize that while natural selection works at the level of the individual, the evolutionary changes that natural selection brings about occur at the population level. By affecting the reproductive success of individuals, natural selection affects allele frequencies within the population. That change in allele frequency is evolution.

It is also important to recognize that when we talk about the characteristics of an organism, we are not just talking about structural characteristics. Behavioral, biochemical, and metabolic characteristics are also important. We often use behavior, DNA differences, and other chemical differences to help us assess evolutionary relationships among existing organisms. However, when looking at historically extinct species, we are usually confined to using structural characteristics to guide our thinking. • speciation, p. 241

Recall that a theory is a well-established generalization supported by many different kinds of evidence. The theory of natural selection was first proposed by Charles Darwin and Alfred Wallace. This theory was clearly set forth in 1859 by Darwin in his book *On the Origin of Species by Means of Natural Selection, or the Preservation of Favored Races in the Struggle for Life* (How Science Works 12.1). Since the time it was first proposed, the theory of natural selection has been subjected to countless tests and remains the core concept for explaining how evolution occurs.

12.3 What Influences Natural Selection?

Now that we have a basic understanding of how natural selection works, we can look in more detail at factors that influence it. *Genetic diversity* within a species, *genetic recombination* as a result of sexual reproduction, the degree of *genetic expression,* and the ability of most species to *reproduce excess offspring* all exert an influence on the process of natural selection.

Mechanisms that Affect Genetic Diversity

In order for natural selection to occur, there must be genetic differences among the many individuals of an interbreeding population of organisms. Consider the implications of a population wherein all organisms are genetically identical. In this case, it does not matter which individuals reproduce because

HOW SCIENCE WORKS 12.1

The Voyage of HMS *Beagle*, 1831–1836

Young Charles Darwin examining specimens on the Galápagos Islands

Probably the most significant event in Charles Darwin's life was his opportunity to sail on the British survey ship HMS *Beagle*. Surveys were common at this time; they helped refine maps and chart hazards to shipping. Darwin was 22 years old and probably would not have had the opportunity had his uncle not persuaded Darwin's father to allow him to take the voyage. Darwin was to be a gentleman naturalist and companion to the ship's captain Robert Fitzroy. When the official naturalist left the ship and returned to England, Darwin became the official naturalist for the voyage. The appointment was not a paid position.

The voyage of the *Beagle* lasted nearly 5 years. During the trip, the ship visited South America, the Galápagos Islands, Australia, and many Pacific Islands. The *Beagle's* entire route is shown on the accompanying map. Darwin suffered greatly from seasickness and, perhaps because of it, he made extensive journeys by mule and on foot some distance inland from wherever the *Beagle* happened to be at anchor. These inland trips provided Darwin the opportunity to make many of his observations. His experience was unique for a man so young and very difficult to duplicate because of the slow methods of travel used at that time.

Although many people had seen the places that Darwin visited, never before had a student of nature collected volumes of information on them. Also, most other people who had visited these faraway places were military men or adventurers who did not recognize the significance of what they saw. Darwin's notebooks included information on plants, animals, rocks, geography, climate, and the native peoples he encountered. The natural history notes he took during the voyage served as a vast storehouse of information that he used in his writings for the rest of his life.

Because Darwin was wealthy, he did not need to work to earn a living and could devote a good deal of his time to the further study of natural history and the analysis of his notes. He was a semi-invalid during much of his later life. Many people think his ill health was caused by a tropical disease he contracted during the voyage of the *Beagle*. As a result of his experiences, he wrote several volumes detailing the events of the voyage, which were first published in 1839 in conjunction with other information related to the voyage of the *Beagle*. His volumes were revised several times and eventually were entitled *The Voyage of the Beagle*. He also wrote books on barnacles, the formation of coral reefs, how volcanos might have been involved in reef formation, and, finally, the *Origin of Species*. This last book, written 23 years after his return from the voyage, changed biological thinking for all time.

The Voyage of HMS *Beagle*, 1831–1836

0 1000 2000 3000
Equatorial scale of miles

the same genes will be passed on to the next generation and natural selection cannot occur. However, when genetic differences exist among individuals in a population and these differences affect fitness, natural selection can take place. Therefore, it is important to identify processes that generate genetic diversity within a population. Genetic diversity within a population is generated in three ways: by *mutation, migration of organisms,* and *sexual reproduction.* • **genetic diversity, p. 210**

Mutations and Migration

Spontaneous mutations are changes in DNA that cannot be tied to a particular causative agent: Mutations may alter existing genes, resulting in the introduction of entirely new genetic information into a gene pool. It is suspected that cosmic radiation or naturally occurring mutagenic chemicals might be the cause of many of these mutations. It is known that subjecting organisms to high levels of radiation or to certain chemicals increases the rate at which mutations occur. It is for this reason that people who work with these types of materials take special safety precautions.

Naturally occurring mutation rates are low. The odds of a gene mutating are on the order of 1 in 100,000. Most of these mutations are harmful. Rarely does a mutation occur that is actually helpful. However, in populations of millions of individuals, each of whom has thousands of genes, over thousands of generations it is quite possible that a new beneficial piece of genetic information could come about as a result of mutation. When we look at the various alleles that exist in humans or in any other organism, we should remember that every allele originated as a modification of a previously existing piece of DNA. For example, the allele for blue eyes may be a mutated brown-eye allele, or blond hair may have originated as a mutated brown-hair allele. When we look at a species such as corn (*Zea mays*), we can see that there are many different alleles for seed color. Each probably originated as a mutation (figure 12.1). Thus, mutations have been very important for introducing new genetic material into species over time.

In order for mutations to be important in the evolution of organisms, they must be in cells that give rise to gametes (eggs or sperm). Mutations in other cells such as those in the skin or liver will only affect those specific cells and will not be passed on to the next generation.

Migration is another way in which new genetic material can enter a population. When individuals migrate into a population from some other population, they may bring alleles that were rare or absent. Similarly, when individuals leave a population, they can remove certain alleles from the population.

Sexual Reproduction and Genetic Recombination

Sexual reproduction is important in generating new genetic combinations in individuals. While sexual reproduction does *not* generate new genetic information, it does allow for the mixing of genes into combinations that did not occur previously. Each individual entering a population by sexual reproduction or migration carries a unique combina-

Figure 12.1

Genetic Diversity in Corn (*Zea mays*)
There are many characteristics of corn that vary considerably. The ears of corn shown here illustrate the genetic diversity in color of the seeds. Although this is only one small part of the genetic makeup of the plant, the diversity is quite large.

tion of genes—half donated by the mother and half donated by the father. During meiosis, unique combinations of alleles are generated in the gametes through crossing-over between homologous chromosomes and independent assortment of nonhomologous chromosomes. This results in millions of possible genetic combinations in the gametes of any individual. When fertilization occurs, one of the millions of possible sperm unites with one of the millions of possible eggs, resulting in a genetically unique individual. The genetic mixing that occurs during sexual reproduction is known as **genetic recombination.** The new individual has a genetic assemblage that is different from that of any other organism that ever existed. When genetic recombination occurs, it is possible that a new combination of alleles would give its bearer a selective advantage, leading to greater reproductive success.

Organisms that primarily use asexual reproduction as opposed to sexual reproduction do not benefit from genetic recombination. In most cases, however, when their life history is studied closely, it is apparent that they also have the ability to reproduce sexually at certain times. Organisms that reproduce exclusively by asexual methods are not able to generate new genetic combinations but still acquire new genetic information through mutations.

It is even possible to have new genetic information enter a species by the process of interspecific hybridization. As we have learned more about the nature of species, it has become clear that genetic exchange has taken place frequently between what were traditionally considered to be distinct species. In some cases, it appears that whole genomes have been added when the cells of two different kinds of cells combined into one cell. • **endosymbiosis p. 440**

The Role of Gene Expression

Even when genes are present, they do not always express themselves in the same way. *In order for genes to be selected for or against, they must be expressed in the phenotype of the individuals possessing them.* Genetic recombination is important because the expression of many alleles is influenced by the presence of other specific alleles for other characteristics.

There are many cases of genetic characteristics being expressed to different degrees in different individuals. Often the reason for this difference in expression is unknown. **Penetrance** is a term used to describe how often an allele expresses itself when present. Some alleles have 100% penetrance, others may only express themselves 80% of the time. For example, there is a dominant allele that causes people to have a stiff little finger. The expression of this trait results in the tendons being attached to the bones of the finger in such a way that the finger does not flex properly. This dominant allele does not express itself in every person who contains it; occasionally parents without the characteristic have children that show the characteristic. **Expressivity** is a term used to describe situations in which the allele is penetrant but is not expressed equally in all individuals who have it. An example of expressivity involves a dominant allele for six fingers. Some people with this allele have an extra finger on each hand; some have an extra finger on only one hand. Furthermore, some sixth fingers are well-formed with normal bones, whereas others are fleshy structures that lack bones.

There are many reasons why a specific allele may not be expressed. Some genetic characteristics can be expressed only during specific periods in the life of an organism. If an organism dies before the characteristic was expressed, it never has the opportunity to contribute to the overall fitness of the organism. Say, for example, a tree has genes for producing very attractive fruit. The attractive fruit is important as a dispersal mechanism because animals select the fruit for food and distribute the seeds as they travel. However, if the tree dies before it can reproduce, the characteristic may never be expressed. By contrast, genes such as those that contribute to heart disease or cancer usually have their effect late in a person's life. Since they were not expressed during the person's reproductive years, they were not selected against because the person reproduced before the effects of the gene were apparent.

In addition, many genes require an environmental trigger to initiate their expression. If the trigger is not encountered, the gene never expresses itself. It is becoming clear that many kinds of human cancers are caused by the presence of genes that require an environmental trigger. Therefore, we seek to identify the triggers and prevent these negative genes from being turned on and causing disease.

When both dominant and recessive alleles are present for a characteristic, the recessive alleles must be present in a homozygous condition before they have an opportunity to express themselves. For example, the allele for albinism is recessive. There are people who carry this recessive allele but never express it because it is masked by the dominant allele for normal pigmentation (figure 12.2).

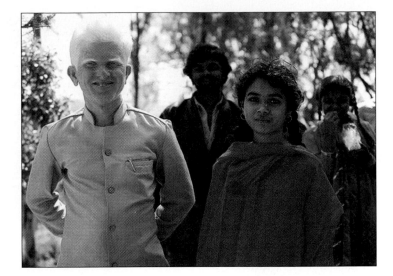

Figure 12.2

Gene Expression

Genes must be expressed to allow the environment to select for or against them. The recessive allele *c* for albinism shows itself only in individuals who are homozygous for the recessive characteristic. The man in this photo is an albino who has the genotype *cc*. The characteristic is absent in those who are homozygous dominant and is hidden in those who are heterozygous. The dark-skinned individuals could be either *Cc* or *CC*. However, because the albino individual cannot produce pigment, characteristics for dark skin and dark hair cannot be expressed.

Some genes may have their expression hidden because the action of a completely unrelated gene is required before they can express themselves. The albino individual shown in figure 12.2 has alleles for dark skin and hair that will never have a chance to express themselves because of the presence of two alleles for albinism. The alleles for dark skin and hair can express themselves only if the person has the ability to produce pigment and albinos lack that ability.

Just because an individual organism has a "good" gene does not guarantee that it will be passed on. The organism may also have "bad" genes in combination with the good, and the "good" characteristics may be overshadowed by the "bad" characteristics. All individuals produced by sexual reproduction probably have certain genetic characteristics that are extremely valuable for survival and others that are less valuable or harmful. However, natural selection operates on the total phenotype of the organism. Therefore, it is the combination of characteristics that is evaluated—not each characteristic individually. For example, fruit flies may show resistance to insecticides or lack of it, may have well-formed or shriveled wings, and may exhibit normal vision or blindness. An individual with insecticide resistance, shriveled wings, and normal vision has two good characteristics and one negative one, but it would not be as successful as an individual with insecticide resistance, normal wings, and normal vision.

Figure 12.3

Reproductive Potential
The ability of a population to reproduce greatly exceeds the number necessary to replace those who die. Here are some examples of the prodigious reproductive abilities of some species.

The Importance of Excess Reproduction

Whenever a successful organism is examined, it can be shown that it reproduces at a rate in excess of that necessary to merely replace the parents when they die (figure 12.3). For example, geese have a life span of about 10 years and, on the average, a single pair can raise a brood of about eight young each year. If these two parent birds and all their offspring were to survive and reproduce at this same rate for a 10-year period, there would be a total of 19,531,250 birds in the family.

However, the size of goose populations and most other populations remains relatively constant over time. Minor changes in number may occur, but if the environment remains constant, a population does not experience dramatic increases in population size. A high deathrate tends to offset the high reproductive rate and population size remains stable. But don't think of this as a "static population." Although the total number of organisms in the species may remain constant, the individuals that make up the population change. It is this extravagant reproduction that provides the large surplus of genetically unique individuals that allows natural selection to take place.

If there are many genetically unique individuals within a population, it is highly probable that some individuals will survive to reproduce even if the environment changes somewhat, although the gene frequency of the population may be changed to some degree. For this to occur, members of the population must be eliminated in a non-random manner. Even if they are not eliminated, some may have greater reproductive success than others. Those individuals with the greatest reproductive success will have more of their genetic information present in the next generation than those that die or do not reproduce very successfully. Those that are the most successful at reproducing are those that are, for the most part, better suited to the environment.

12.4 Common Misunderstandings About Natural Selection

There are several common misinterpretations associated with the process of natural selection. The first involves the phrase "survival of the fittest." Individual survival is certainly important because those that do not survive will not reproduce. However, the more important factor is the number of descendants an organism leaves. An organism that has survived for hundreds of years but has not reproduced has not contributed any of its genes to the next generation and so has been selected against. Therefore, the key to being the fittest is not survival alone, but survival and reproduction of the more fit organisms.

Second, the phrase "struggle for life" in the title of Darwin's book does not necessarily refer to open conflict and fighting. It is usually much more subtle than that. When a resource such as nesting material, water, sunlight, or food is in short supply, some individuals survive and reproduce more effectively than others. For example, many kinds of birds require holes in trees as nesting places (figure 12.4). If these are in short supply, some birds will be fortunate and find a top-quality nesting site, others will occupy less suitable holes, and some many not find any. There may or may not be fighting for possession of a site. If a site is already occupied, a bird may not necessarily try to dislodge its occupant but may just continue to search for suitable but less valuable sites. Those that successfully occupy good nesting sites will be much more successful in raising young than will those that must occupy poor sites or those that do not find any.

Similarly, on a forest floor where there is little sunlight, some small plants may grow fast and obtain light while shading out plants that grow more slowly. The struggle for life in this instance involves a subtle difference in the rate at which the plants grow. But the plants are indeed engaged in a struggle, and a superior growth rate is the weapon for survival.

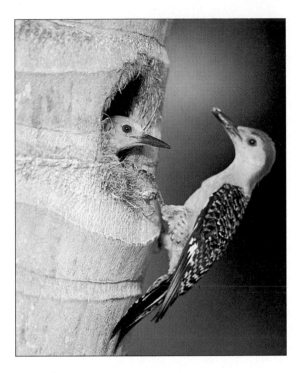

Figure 12.4

Tree Holes as Nesting Sites
Many kinds of birds, like this red-bellied woodpecker, nest in holes in trees. If old and dead trees are not available, they may not be able to breed. Many people build birdhouses that provide artificial "tree holes" to encourage such birds to nest near their homes.

A third common misunderstanding involves the significance of phenotypic characteristics that are gained during the life of an organism, but are not genetically determined. Although such **acquired characteristics** may be important to an individual's success, since they are not genetically determined, they cannot be passed on to future generations through sexual reproduction. Therefore, acquired characteristics are not important to the processes of natural selection. Consider an excellent tennis player's skill. Although this person may have inherited physical characteristics of good eyesight, tallness, and muscular coordination that are beneficial to a tennis player, the ability to play a good game of tennis is acquired through practice, not through genes. An excellent tennis player's offspring will not automatically be excellent tennis players. They may inherit some of the genetically determined physical characteristics necessary to become excellent tennis players, but the skills are still acquired through practice (figure 12.5).

We often desire a specific set of characteristics in our domesticated animals. For example, the breed of dog known as boxers is "supposed" to have short tails. However, the alleles for short tails are rare in this breed. Consequently, the tails of these dogs are amputated—a procedure called docking. Similarly, the tails of lambs are also usually amputated. These acquired characteristics are not passed on to the next generation. Removing the tails of these animals does not remove the genetic information for tail production from their genomes and each generation of puppies and lambs is born with tails.

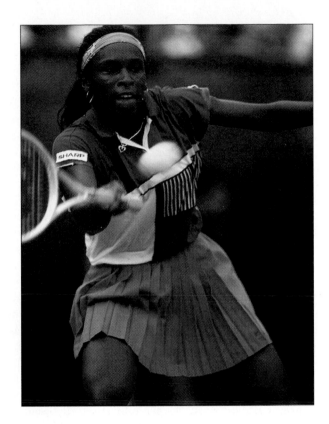

Figure 12.5

Acquired Characteristics
The ability to play an outstanding game of tennis is learned through long hours of practice. The tennis skills this person acquired by practice cannot be passed on to her offspring.

Another common misconception involves understanding the relationship between the mechanism of natural selection (death, reproductive success, mate choice) and the outcomes of the selection process. Although the effects of natural selection are shown at the population level, the actual selecting events take place one at a time at the level of the individual organism.

12.5 Processes that Drive Selection

Several mechanisms allow for selection of certain individuals for successful reproduction. The specific environmental factors that favor certain characteristics are called **selecting agents.** If predators must pursue swift prey organisms, then the faster predators will be selected for, and the selecting agent is the swiftness of available prey. If predators must find prey that are slow but hard to see, then the selecting agent is the camouflage coloration of the prey, and keen eyesight is selected for. If plants are eaten by insects, then the production of toxic materials in the leaves is selected for. All selecting agents influence the likelihood that certain characteristics will be passed on to subsequent generations.

Differential Survival

As stated previously, the phrase "survival of the fittest" is often associated with the theory of natural selection. Although this is recognized as an oversimplification of the concept, survival is an important factor in influencing the flow of genes to subsequent generations. If a population consists of a large number of genetically and phenotypically different individuals it is likely that some of them will possess characteristics that make their survival difficult. Therefore, they are likely to die early in life and not have an opportunity to pass their genes on to the next generation.

Charles Darwin described several species of finches on the Galápagos Islands and scientists have often used these birds in scientific studies of evolution. On one of the islands, scientists studied one of the species of seed-eating ground finches, *Geospiza fortis.* They measured the size of the animals and the size of their bills and related these characteristics to their survival. They found the following: During a drought, the birds ate the smaller, softer seeds more readily than the larger, harder seeds. As the birds consumed the more easily eaten seeds, only the larger, harder seeds remained. During the drought mortality was extremely high. When scientists looked at ground finch mortality, they found that the larger birds with stronger, deeper bills survived better than smaller birds with weaker, narrower bills. They also showed that the offspring of the survivors tended to show larger body and bill size as well. The lack of small, easily eaten seeds resulted in selection for larger birds with stronger bills that could crack open larger, tougher seeds. Table 12.1 shows data on two of the parameters measured in this study.

Table 12.1

CHANGES IN BODY STRUCTURE OF *Geospiza fortis*

	Before Drought	After Drought
Average body weight	16.06 g	17.13 g
Average bill depth	9.21mm	9.70mm

As another example of how differential survival can lead to changed gene frequencies, consider what has happened to many insect populations as we have subjected them to a variety of insecticides. Because there is genetic diversity within all species of insects, an insecticide that is used for the first time on a particular species kills all those exposed individuals that are genetically susceptible. However, individuals with slightly different genetic compositions and those not exposed may not be killed by the insecticide.

Suppose that in a population of a particular species of insect, 5% of the individuals have genes that make them resistant to a specific insecticide. The first application of the insecticide could, therefore, kill a majority of those exposed. However, tolerant individuals and those that escaped exposure would then constitute the remaining breeding population. This would mean that many more insects in the second generation would be tolerant. The second use of the insecticide on this population would not be as effective as the first. With continued use of the same insecticide, each generation would become more tolerant, because the individuals that are not tolerant are being eliminated and those that can tolerate the toxin pass their genes for tolerance on to their offspring.

Many species of insects produce a new generation each month. In organisms with a short generation time, 99% of the population could become resistant to the insecticide in just 5 years. As a result, the insecticide would no longer be useful in controlling the species. As a new factor (the insecticide) was introduced into the environment of the insect, natural selection resulted in a population that was tolerant of the insecticide.

The same kind of selection process has occurred with herbicides. Within the past 50 years, many different kinds of herbicides have been developed to control weeds in agricultural fields. After several years of use, a familiar pattern develops as more and more species of weeds show resistance to the herbicide. Figure 12.6 shows several different kinds of herbicides and the number of weed species that have become resistant over time. In the case of each weed species there has been selection for those individuals that have the genetic information that allows them to tolerate the presence of the herbicide.

Differential Reproductive Rates

Survival alone does not always ensure reproductive success. For a variety of reasons, some organisms may be better able to utilize available resources to produce offspring. If one individual leaves 100 offspring and another leaves only 2, the first organism has passed more copies of its genetic information on to the next generation than has the second. If we assume that all 102 individual offspring have similar survival rates, the first organism has been selected for and its genes have become more common in the subsequent population.

Scientists have conducted studies of the frequencies of genes for the height of clover plants (figure 12.7). Two identical fields of clover were planted and cows were allowed to graze in one of them. Cows acted as a selecting agent by eating the taller plants first. These tall plants rarely got a chance to reproduce. Only the shorter plants flowered and produced seeds. After some time, seeds were collected from both the grazed and ungrazed fields and grown in a greenhouse under identical conditions. The average height of the plants from the ungrazed field was compared to that of the plants from the grazed field. The seeds from the ungrazed field produced some tall, some short, but mostly medium-sized plants. However, the seeds from the grazed field produced many more shorter plants than medium or tall ones. The cows had selectively eaten the tall plants. Because the flowers are at the tip of the plant, tall plants were less likely to successfully reproduce, even though they might have been able to survive grazing by cows.

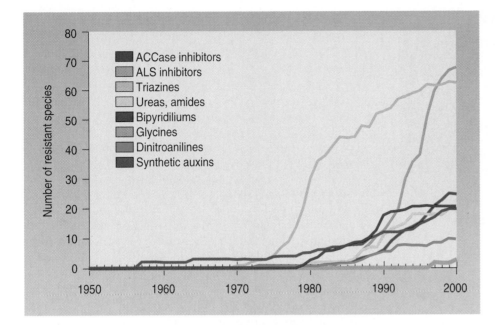

Figure 12.6

Evolutionary Change

Populations of weed plants that have been subjected repeatedly to herbicides often develop resistant populations. Those individual weed plants that were able to resist the effects of the herbicide lived to reproduce and pass their genes for resistance on to their offspring; thus, resistant populations of weeds develop.

Source: Data from Ian Heap, "The International Survey of Herbicide Resistant Weeds." Available at www.weedscience.com.

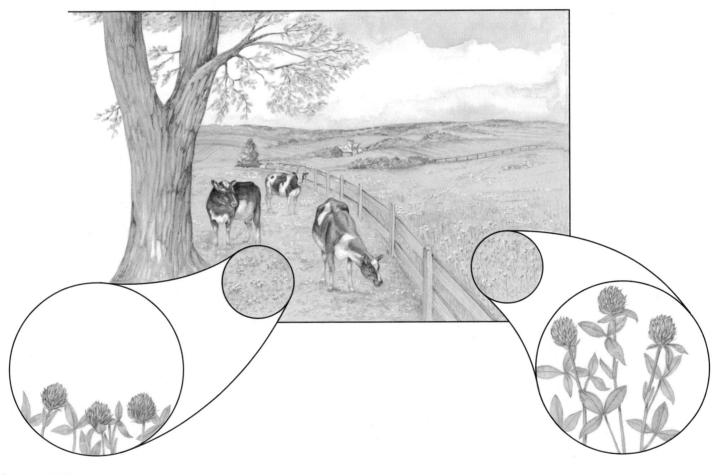

Figure 12.7

Selection for Shortness in Clover

The clover field to the left of the fence is undergoing natural selection: The grazing cattle are eating the tall plants and causing them to reproduce less than do the short plants. The other field is not subjected to this selection pressure, so its clover population has more genes for tallness.

Figure 12.8

Mate Selection

In many animal species the males display very conspicuous characteristics that are attractive to females. Because the females choose the males they will mate with, those males with the most attractive characteristics will have more offspring and, in future generations, there will be a tendency to enhance the characteristic. With peacocks, those individuals with large colorful displays are more likely to mate.

Differential Mate Choice—Sexual Selection

Within animal populations, some individuals may be chosen as mates more frequently than others. This is called **sexual selection.** Obviously, those that are frequently chosen have an opportunity to pass on more copies of their genetic information than those that are rarely chosen. Characteristics of the more frequently chosen individuals may involve general characteristics, such as body size or aggressiveness, or specific conspicuous characteristics attractive to the opposite sex.

For example, male red-winged blackbirds establish territories in cattail marshes where females build their nests. A male will chase out all other males but not females. Some males have large territories, some have small territories, and some are unable to establish territories. Although it is possible for any male to mate, it has been demonstrated that those that have no territory are least likely to mate. Those that defend large territories may have two or more females nesting in their territories and are very likely to mate with those females. It is unclear exactly why females choose one male's territory over another, but the fact is that some males are chosen as mates and others are not.

In other cases, it appears that the females select males that display conspicuous characteristics. Male peacocks have very conspicuous tail feathers. Those with spectacular tails are more likely to mate and have offspring (figure 12.8). Darwin was puzzled by such cases as the peacock in which the large and conspicuous tail should have been a disadvantage to the bird. Long tails require energy to produce, make it more difficult to fly, and make it likely that predators will capture the individual. The current theory that seeks to explain this paradox involves female choice. If the

females have an innate (genetic) tendency to choose the most elaborately decorated males, genes that favor such plumage will be regularly passed on to the next generation.

12.6 Patterns of Selection

Understanding the nature of the environment is key to determining how natural selection will affect how a species will change. In general, three different forms of selection have been identified: stabilizing selection, directional selection, and disruptive selection (figure 12.9).

Stabilizing selection occurs when individuals at the extremes of the range of a characteristic are consistently selected against. This kind of selection is very common. If the environment is stable, then it is logical that the majority of individuals would show characteristics that are consistent with the demands of the environment. For example, for many kinds of animals there may be a range of color possibilities. Suppose a population of mice has mostly brown individuals and a few white and black individuals. If the white and black individuals are more conspicuous and are consistently more likely to be discovered and killed by predators, then the elimination of the extreme forms will result in a continued dominance of the intermediate brown form. Many kinds of marine animals have remained unchanged for thousands of years. The marine environment is relatively constant and probably favors stabilizing selection.

Directional selection occurs when individuals at one extreme of the range of a characteristic are consistently selected for. This kind of selection often occurs when there is a consistent change in the environment in which the organism exists. For example, when a particular insecticide is introduced to control a certain species of pest insect, there is consistent selection for individuals that had alleles for resistance to the insecticide. Because of this, there is a shift in the original allele frequency from one in which the alleles for resistance to the insecticide were rare to one in which the majority of the population has the alleles for resistance. Similarly, changes in climate, such as long periods of drought, can consistently select for individuals that have characteristics that allow them to survive in the dryer environment and a change in allele frequency can result.

Disruptive selection occurs when both extremes of a range for a characteristic are selected for and the intermediate condition is selected against. This kind of selection is likely to happen when there are sharp differences in the nature of the environment where the organisms live. For example, there are many kinds of insects that feed on the leaves of trees. Many of these insects have colors that match the leaves they feed upon. Suppose there is a range of green color from light green to dark green and medium green is the most common. If a particular species of insect had some individuals that feed on plants with dark-green leaves while other individuals feed on plants with light-green leaves, it would be possible that medium green insects would be selected against and the two extremes would be selected for, depending on the kind of plant they were feeding upon.

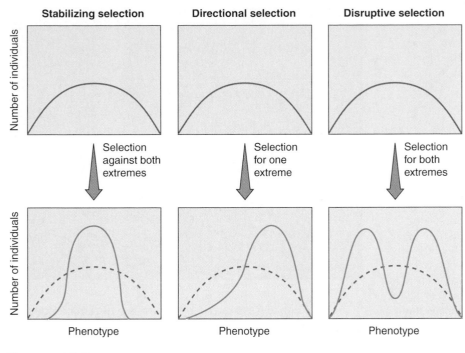

Figure 12.9

Patterns of Selection

In stabilizing selection, both extremes for a characteristic are selected against. Thus, gene frequencies do not change, and the original range of phenotypes is maintained. In directional selection, one extreme is selected for and the other extreme is selected against. This results in a shift in gene frequency and range of phenotype in a direction toward one extreme and away from the other. In disruptive selection, both extremes are selected for and the intermediate condition is selected against. This leads to the development of two distinct phenotypes with different gene frequencies.

12.7 Evolution Without Selection—Genetic Drift

Gene-frequency differences that result from chance are more likely to occur in small populations than in large populations. Since they result from random events, such changes are not the result of natural selection or sexual selection. For example, a population of 10 organisms, of which 20% have curly hair and 80% have straight hair, is significantly changed by the death of 1 curly-haired individual. Such situations in which the frequency of an allele changes significantly but the change is not the result of selection are called **genetic drift.** Often the characteristics affected by genetic drift do not appear to have any particular adaptive value to the individuals in the population. However, in extremely small populations vital genes may be lost.

Occasionally, we see a population that has unusual colors, shapes, or behaviors compared to others of the same species. Such unusual occurrences are typically associated with populations that are small, or passed through a genetic bottleneck in the past. In large populations any unusual shifts in gene frequency in one part of the population usually would be counteracted by reciprocal changes in other parts of the population. However, in small populations the random distribution of genes to gametes may not reflect the percentages present in

the population. For example, consider a situation in which there are 100 plants in a population and 10 have dominant alleles for patches of red color while others do not. If in those 10 plants the random formation of gametes resulted in no red alleles present in the gametes that were fertilized, then the allele could be eliminated. Similarly, if those plants with the red allele all happened to be in a hollow that was subjected to low temperatures, they might be killed by a late frost and would not pass their alleles on to the next generation. Therefore the allele would be lost but the loss would not be the result of natural selection.

Consider the example of cougars in North America. Cougars require a wilderness setting for success. As Europeans settled the land over the past 200 years, the cougars were divided into small populations in those places where relatively undisturbed habitat still existed. The Florida panther is an isolated population of cougars found in the Everglades. The next nearest population of cougars is in Texas. Because the Florida panther is on the endangered species list, efforts have been made to ensure its continued existence in the Everglades. However, the population is small and studies show that it has little genetic diversity. A long period of isolation and a small population created conditions that led to this reduced genetic diversity. The accidental death of a few key individuals could have resulted in the loss of certain genes from the population. The general health of individuals in the population is poor and reproductive success is low.

In 1995, wildlife biologists began a program of introducing individuals from the Texas population into the Florida population. The purpose of the program is to reintroduce genetic diversity that has been lost during the long period of isolation. The program appears to be working, since there has been an increase in genetic diversity within the population (figure 12.10).

Many zoos around the world cooperate in captive breeding programs designed to maintain genetic diversity in the gene pool of endangered species. Some of these species no longer exist in the wild but there are hopes that they may sometime be reintroduced to the wild. For example, at one time the entire California condor population consisted of a few individuals in zoos. Although many individuals of the species still reside in zoos, attempts are being made to reintroduce them to the wild in California and Arizona. Maintaining genetic diversity in the population can be very difficult when the species consists of few individuals. Often the DNA of the animals is characterized and records are kept to assure that the animals that breed are not close relatives. To accomplish their goals, zoos often exchange or loan animals for breeding purposes.

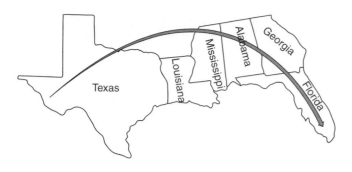

Figure 12.10

The Florida Panther
The Florida panther (cougar) is confined to the Everglades at the southern tip of Florida. The population is small, isolated from other populations of cougars, and prior to 1995 had lost about 50% of its genetic diversity. In 1995, eight female cougars from Texas were introduced into the Everglades to increase the genetic diversity of the Florida panther population. The plan seems to be working. By 2002, the population had increased to about 80 adult cougars and the population showed about 20% genes from the Texas subpopulation. Certain obvious characteristics like kinked tails and cowlicks that were common in the pre-1995 population are now less common.

12.8 Gene-Frequency Studies and Hardy-Weinberg Equilibrium

Throughout this chapter we have made frequent references to changing gene frequencies and allele frequencies. (*Mutations introduce new alleles into the gene pool of a species, causing gene frequencies to change. Successful organisms pass on more of their genetic information to the next generation, causing gene frequencies to change.*) In the early 1900s an English mathematician, G. H. Hardy, and a German physician, Wilhelm Weinberg, recognized that it was possible to apply a simple mathematical relationship to the study of gene frequencies. Their basic idea was that if certain conditions existed, gene frequencies would remain constant, and that the distribution of genotypes could be described by the relationship $A^2 + 2Aa + a^2 = 1$, where A^2 represents the frequency of the homozygous dominant genotype, $2Aa$ represents the frequency of the heterozygous genotype, and a^2 represents the frequency of the homozygous recessive genotype. Constant gene frequencies over several generations would imply that evolution is *not* taking place. Changing gene frequencies would indicate that evolution is taking place.

The conditions necessary for gene frequencies to remain constant are:

1. Mating must be completely random.
2. Mutations must not occur.
3. Migration of individual organisms into and out of the population must not occur.
4. The population must be very large.
5. All genes must have an equal chance of being passed on to the next generation. (Natural selection is not occurring.)

The concept that gene frequencies will remain constant if these five conditions are met has become known as the **Hardy-Weinberg concept.** The Hardy-Weinberg concept is important because it allows a simple comparison of allele frequency to indicate if genetic changes are occurring within the population. Two different populations of the same species can be compared to see if they have the same allele frequencies, or populations can be examined at different times to see if allele frequencies are changing.

Determining Genotype Frequencies

It is possible to apply the Punnett square method from chapter 10 to an entire gene pool to illustrate how the Hardy-Weinberg concept works. Consider a gene pool composed of only two alleles, A and a. Of the alleles in the population 60% (0.6) are A and 40% (0.4) are a. In this hypothetical gene pool, we do not know which individuals are male or female and we do not know their genotypes. With these allele frequencies, how many of the individuals would be homozygous dominant (AA), homozygous recessive (aa), and heterozygous (Aa)? To find the answer, we treat these alleles and their frequencies as if they were individual alleles being distributed into sperm and eggs. The sperm produced by the males of the population will be 60% (0.6) A and 40% (0.4) a. The females will produce eggs with the same relative frequencies. We can now set up a Punnett square as follows:

		Possible female gametes	
		$A = 0.6$	$a = 0.4$
Possible male gametes	$A = 0.6$	Genotype of offspring $AA = 0.6 \times 0.6 = 0.36 = 36\%$	Genotype of offspring $Aa = 0.6 \times 0.4 = 0.24 = 24\%$
	$a = 0.4$	Genotype of offspring $Aa = 0.4 \times 0.6 = 0.24 = 24\%$	Genotype of offspring $aa = 0.4 \times 0.4 = 0.16 = 16\%$

The Punnett square gives the frequency of occurrence of the three possible genotypes in this population: $AA = 36\%$, $Aa = 48\%$, and $aa = 16\%$.

OUTLOOKS 12.1

The Development of New Viral Diseases

Mutation is necessary if evolution is to take place. As parasites and their hosts interact, they are constantly reacting to each other in an evolutionary fashion. Hosts develop new mechanisms to combat parasites and parasites develop new mechanisms to overcome the defenses of hosts. One of the mechanisms employed by viruses involves high rates of mutation. This ability to mutate has resulted in many new, serious human diseases. In addition, many new diseases arise when viruses that cause disease in another animal are able to establish themselves in humans. Many kinds of influenza originated in pigs, ducks, or chickens and were passed to humans through close contact with infected animals or by eating infected animals. In many parts of the world, these domesticated animals live in close contact with humans (often in the same building), making conditions favorable for the transmission of animal viruses to humans.

Each year mutations result in new varieties of influenza and colds that pass through the human population. Occasionally the new varieties are particularly deadly. In 1918, a new variety of influenza virus originated in pigs in the United States and spread throughout the world. During the 1918–1919 influenza pandemic that followed, 20 to 40 million people died. In 1997 in Hong Kong, a new kind of influenza was identified that killed 6 of 18 people infected. When public health officials discovered the virus had come from chickens, they ordered the slaughter of all live chickens in Hong Kong. This stopped the spread of the disease.

In early 2003, an outbreak of a new viral disease known as *severe acute respiratory syndrome (SARS)* originated in China.

SARS is a variation of a coronavirus, a class of virus commonly associated with the common cold, but it causes severe symptoms and if untreated may result in death. In June 2003, the SARS virus was isolated from an animal known as the masked palm civet (*Paguma larvata*). This animal is used for food in China and is a possible source of the virus that caused SARS in humans. However, other animals have also tested positive for the virus. The disease rapidly spread to several countries as people traveled by airplane from China to other parts of the world. Recognition of the seriousness of the disease and isolation of infected persons prevented further spread and this new disease was brought under control. However, if the virus still exists in some unknown wild animal host it could reappear in the future.

In addition to cold and influenza, other kinds of diseases often make the leap from nonhuman to human hosts. AIDS is a new disease caused by the human immunodeficiency viruses (HIV), which have been traced to a similar virus in monkeys. In June 2003, public health officials announced that monkeypox virus was causing disease in people in Wisconsin, Illinois, and Indiana. The monkeypox virus was traced to a shipment of prairie dogs that were sold as pets. It is likely that the prairie dogs became infected with the virus through contact with an infected Gambian giant rat, which was part of a shipment of animals from Ghana in April, 2003. Thus, it appears that the disease entered North America from Africa by way of the Gambian giant rat and was transferred to prairie dogs and subsequently from prairie dogs to people.

If we use the relationship $A^2 + 2Aa + a^2 = 1$, you can see that if $A^2 = 0.36$ then A would be the square root of 0.36, which is equal to 0.6—our original frequency for the A allele. Similarly, $a^2 = 0.16$ and a would be the square root of 0.16, which is equal to 0.4. In addition, $2Aa$ would equal $2 \times 0.6 \times 0.4 = 0.48$. If this population were to reproduce randomly, it would maintain an allele frequency of 60% A and 40% a alleles.

It is important to understand that Hardy-Weinberg equilibrium conditions rarely exist; therefore, we usually see changes in gene frequency over time or genetic differences in separate populations of the same species. If gene frequencies are changing, evolution is taking place. Let's now examine why this is the case.

Why Hardy-Weinberg Conditions Rarely Exist

First of all, *random mating does not occur* for a variety of reasons. Many species are divided into segments that are isolated from one another to some degree so that no mating with other segments occurs during the lifetime of the individuals. In human populations, these isolations may be geographic, political, or social. In addition, some individuals

may be chosen as mates more frequently than others because of the characteristics they display. Therefore, the Hardy-Weinberg conditions often are not met because non-random mating is a factor that leads to changing gene frequencies.

Second, *spontaneous mutations occur*. Totally new kinds of alleles are introduced into a population, or one allele is converted into another currently existing allele. Whenever an allele is changed, one allele is subtracted from the population and a different one is added, thus changing the allele frequency in the gene pool. Mutations in disease-causing organims may have significant impacts (Outlooks 12.1).

Third, *immigration or emigration* of individual organisms is common. When organisms move from one population to another they carry their genes with them. Their genes are subtracted from the population they left and added to the population they enter, thus changing the gene pool of both populations. It is important to understand that migration is common for plants as well as animals. In many parts of the world, severe weather disturbances have lifted animals and plants (or their seeds) and moved them over great distances, isolating them from their original gene pool. In other instances, organisms have been distributed by floating on debris on the surface of the ocean. As an example of how important these mechanisms are, consider the tiny island of Surtsey (3 km²), which emerged from the sea as a volcano near Iceland in 1963. It continued to erupt until 1967. The new island was declared a nature preserve and has been surveyed regularly to record the kinds of organisms present. The nearest possible source of new organisms was about 20 kilometers away. The first living thing observed on the island was a fly seen less than a year after the initial eruption. By 1965, the first flowering plant was found and by 1996, 50 different species of flowering plants had been recorded on the island. In addition, several kinds of sea birds nest on the island.

The fourth problem is that *populations are not infinitely large,* as assumed by the Hardy-Weinberg concept. If numbers are small, random events to a few individual organisms might alter gene frequencies from what was expected. Take coin flipping as an analogy. Coins have two surfaces, so if you flip a coin once, there is a 50:50 chance that the coin will turn up heads. If you flip two coins, you may come up with two heads, two tails, or one head and one tail. Only one of these possibilities gives us the theoretical 50:50 ratio. To come closer to the statistical probability of flipping 50% heads and 50% tails, you would need to flip many coins at the same time. The more coins you flip, the more likely it is that you will end up with 50% of all coins showing heads and the other 50% showing tails. The number of coins flipped is important. The same is true of gene frequencies. The smaller the population, the more likely it is that random events will alter the gene pool—that is, the more likely it is that genetic drift will occur.

Finally, *genes are not all equally likely to be passed to the next generation.* It is important to understand that genes differ in their value to the species. Some genes result in characteristics that are important to survival and reproductive success. Other genes may reduce the likelihood of survival and reproduction. Many animals have cryptic color patterns that make them difficult to see. The genes that determine the cryptic color pattern would be selected for (favored) because animals that are difficult to see are not going to be killed and eaten as often as those that are easy to see. Albinism is the inability to produce pigment so that the individual's color is white. White animals are conspicuous and so we might expect them to be discovered more easily by predators (figure 12.11). Because not all genes have equal value, natural selection will operate and some genes will be more likely to be passed on to the next generation than will others.

Using the Hardy-Weinberg Concept to Show Allele-Frequency Change

Now we can return to our original example of alleles *A* and *a* to show how natural selection based on differences in survival can result in allele-frequency changes in only one generation. Again, assume that the parent generation has the following genotype frequencies: *AA* = 36%, *Aa* =

Figure 12.11

Albino Animal in the Wild
Predators are more likely to spot an albino than a member of the species with normal coloration. Albinism prevents the prey from blending in with its surroundings.

48%, and *aa* = 16%, with a total population of 100,000 individuals. Suppose that 50% of all the individuals having at least one *A* allele do not reproduce because they are more susceptible to disease. The parent population of 100,000 would have 36,000 individuals with the *AA* genotype, 48,000 with the *Aa* genotype, and 16,000 with the *aa* genotype. Because only 50% of those with an *A* allele reproduce, only 18,000 *AA* individuals and 24,000 *Aa* individuals will reproduce. All 16,000 of the *aa* individuals will reproduce, however. Thus, there is a total reproducing population of only 58,000 individuals out of the entire original population of 100,000. What percentage of *A* and *a* will go into the gametes produced by these 58,000 individuals?

The percentage of *A*-containing gametes produced by the reproducing population will be 31% from the *AA* parents and 20.7% from the *Aa* parents (table 12.2). The frequency of the *A* gene in the gametes is 51.7% (31% + 20.7%). The percentage of *a*-containing gametes is 48.3% (20.7% from the *Aa* parents plus 27.6% from the *aa* parents). The original parental allele frequencies were *A* = 60% and *a* = 40%. These have changed to *A* = 51.7% and *a* = 48.3%. More individuals in the population will have the *aa* genotype, and fewer will have the *AA* and *Aa* genotypes.

If this process continued for several generations, the allele frequency would continue to shift until the *A* allele became rare in the population (figure 12.12). This is natural selection in action. Differential reproduction rates have changed the frequency of the *A* and *a* alleles in this population.

Table 12.2

DIFFERENTIAL REPRODUCTION

The percentage of each genotype in the offspring differs from the percentage of each genotype in the original population as a result of differential reproduction.

Original Frequency of Genotypes	Total Number of Individuals Within a Population of 100,000 with Each Genotype	Number of Each Genotype Not Reproducing Subtracted from the Total	Total of Each Genotype in the Reproducing Population of 58,000 Following Selection	New Percentage of Each Genotype in the Reproducing Population
$AA = 36\%$	36,000	36,000 $-18,000$ ———— 18,000	18,000	$\dfrac{18,000}{58,000} = 31.0\%$
$Aa = 48\%$	48,000	48,000 $-24,000$ ———— 24,000	24,000	$\dfrac{24,000}{58,000} = 41.4\%$
$aa = 16\%$	16,000	16,000 $-\quad 0$ ———— 16,000	16,000	$\dfrac{16,000}{58,000} = 27.6\%$
100%	100,000		58,000	100.0%

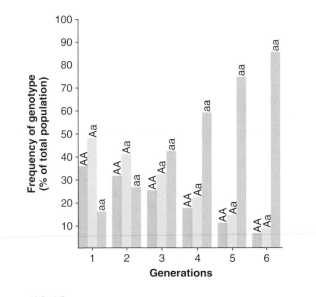

Figure 12.12

Changing Allele Frequency

If 50% of all individuals with the genotypes *AA* and *Aa* do not reproduce in each generation, the frequency of the *a* allele will increase while the frequency of the *A* allele will decrease. Consequently, the *aa* genotype will increase in frequency while that of the *AA* and *Aa* genotypes will decrease.

12.9 A Summary of the Causes of Evolutionary Change

At the beginning of this chapter, evolution was described as the change in allele frequency over time. We can now see that several different mechanisms operate to bring about this change. Mutations can either change one allele into another or introduce an entirely new piece of genetic information into the population. Immigration can introduce new genetic information if the entering organisms have genetic information that was not in the population previously. Emigration and death remove genes from the gene pool. Natural selection systematically filters some genes from the population, allowing other genes to remain and become more common. The primary mechanisms involved in natural selection are differences in deathrates, reproductive rates, and the rate at which individuals are selected as mates (figure 12.13). In addition, gene frequencies are more easily changed in small populations because events such as death, immigration, emigration, and mutation can have a greater impact on a small population than on a large population.

Now that you have an understanding of the mechanisms of natural selection and how natural selection brings about evolution, examine some common myths and misunderstandings about evolution in Outlooks 12.2.

OUTLOOKS 12.2

Common Misconceptions About the Theory of Evolution

1. *Evolution happened only in the past and is not occurring today.* In fact, we see lots of evidence that changes in the frequency of alleles are occurring in the populations of current species (antibiotic resistance, pesticide resistance).
2. *Evolution has a specific goal.* Natural selection selects those organisms that best fit the current environment. As the environment changes, so do the characteristics that have value. Random events such as changes in sea level, major changes in climate such as ice ages, or collisions with asteroids have had major influences on the subsequent natural selection and evolution. Evolution results in organisms that "fit" the current environment.
3. *Changes in the environment cause mutations that are needed to survive under the new environmental conditions.* Mutations are random events and are not necessarily adaptive. However, when the environment changes, mutations that were originally detrimental may have greater value. The genetic information did not change but the environmental conditions did. In some cases, the mutation rate may increase or there may be more frequent genetic exchanges between individuals when the environment changes, but the mutations are still random. They are not directed to a particular goal.
4. *Individual organisms evolve.* Individuals are stuck with the genes they inherited from their parents. Although individuals may adapt by changing their behavior or physiology they cannot evolve; only populations can change gene frequencies.
5. *Today's species can frequently be shown to be derived from other present-day species (apes gave rise to humans).* There are few examples in which it can be demonstrated that one current species gave rise to another. Apes did not become humans, but apes and humans had a common ancestor several million years ago.
6. *Genes that are valuable to an organism's survival become dominant.* An allele that is valuable may be either dominant or recessive. However, if it has a high value for survival it will become *common* (more frequent). Commonness has nothing to do with dominance and recessiveness.

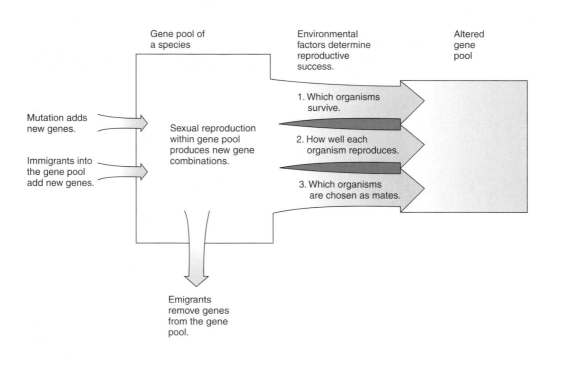

Figure 12.13

Processes that Influence Evolution
Several different processes cause gene frequencies to change. New genetic information enters populations through immigration and mutation. Genetic information leaves populations through death and emigration. Natural selection operates within populations through death, mate selection, and rates of reproduction. Genetic drift can also result in evolutionary change but is not shown in this diagram.

SUMMARY

All populations of sexually reproducing organisms naturally exhibit genetic diversity among individuals as a result of mutations and the genetic recombination resulting from meiosis and fertilization. The genetic differences are reflected in phenotypic differences among individuals. These genetic differences are important in changing environments because natural selection must have genetic diversity to select from. Natural selection by the environment results in better-suited individual organisms having greater numbers of offspring than those that are less well off genetically. Not all genes are equally expressed. Some express themselves only during specific periods in the life of an organism and some may be recessive alleles that show themselves only when in the homozygous state. Characteristics that are acquired during the life of the individual and are not determined by genes cannot be raw material for natural selection. Sexual selection occurs when specific individuals are chosen as mates to the exclusion of others. In addition to natural selection and sexual selection, genetic drift in small populations can lead to evolutionary change.

Selecting agents act to change the gene frequencies of the population if the conditions of the Hardy-Weinberg concept are violated. The conditions required for Hardy-Weinberg equilibrium are random mating, no mutations, no migration, large population size, and no selective advantage for any genes. These conditions are met only rarely, however, so that typically, after generations of time, the genes of the more favored individuals will make up a greater proportion of the gene pool. The process of natural selection allows the maintenance of a species in its environment, even as the environment changes.

THINKING CRITICALLY

For guidelines to answer this question, visit the Online Learning Center.

Penicillin was first introduced as an antibiotic in the early 1940s. Since that time, it has been found to be effective against the bacteria that cause gonorrhea, a sexually transmitted disease. The drug acts on dividing bacterial cells by preventing the formation of a new protective cell wall. Without the wall, the bacteria can be killed by normal body defenses. Recently, a new strain of this disease-causing bacterium has been found. This particular bacterium produces an enzyme that metabolizes penicillin. How can gonorrhea be controlled now that this organism is resistant to penicillin? How did a resistant strain develop? Include the following in your consideration: DNA, enzymes, selecting agents, and gene-frequency changes.

CONCEPT MAP TERMINOLOGY

Construct a concept map using these terms.

differential reproductive success	natural selection
evolution	selecting agent
gene frequency	"survival of the fittest"

Visit the Online Learning Center to use an interactive format to construct another concept map.

KEY TERMS

Use interactive flash cards, on the Online Learning Center, to help you learn the meanings of these terms.

acquired characteristics	Hardy-Weinberg concept
directional selection	natural selection
disrupting selection	penetrance
evolution	selecting agent
expressivity	sexual selection
fitness	spontaneous mutation
genetic drift	stabilizing selection
genetic recombination	theory of natural selection

LEARNING CONNECTIONS

Visit the Online Learning Center to find additional study resources including those highlighted below.

12.1   The Concept of Evolution

1. Describe the biological meaning of the word *evolution*.

12.2   The Role of Natural Selection in Evolution

Animations and Review
- Natural selection
- Adaptation

2. Define *natural selection*.
3. What is fitness and how is it related to reproduction ?

12.3 What Influences Natural Selection?

Interactive Concept Map
• Natural selection

Experience This!
• Genetic variation in the human population

4. What factors can contribute to diversity in the gene pool?
5. Why is over-reproduction necessary for evolution?
6. Why is sexual reproduction important to the process of natural selection?
7. How might a harmful allele remain in a gene pool for generations without being eliminated by natural selection?

12.4 Common Misunderstandings About Natural Selection

8. Why are acquired characteristics of little interest to evolutionary biologists?
9. In what way are the phrases "survival of the fittest" and "struggle for existence" correct? In what ways are they misleading?

12.5 Processes that Drive Natural Selection

Animations and Review
• Other processes

10. A gene pool has equal numbers of alleles *B* and *b*. Half of the *B* alleles mutate to *b* alleles in the original generation. What will the allele frequencies be in the next generation?
11. List three factors that can lead to changed gene frequencies from one generation to the next.
12. Give two examples of selecting agents and explain how they operate.

12.6 Patterns of Selection

13. Distinguish between stabilizing, directional, and disruptive selection

12.7 Evolution Without Selection—Genetic Drift

14. Why is genetic drift more likely in small populations ?
15. Give an example of genetic drift.

12.8 Gene-Frequency Studies and Hardy-Weinberg Equilibrium

Interactive Concept Map
• Hardy-Weinberg assumptions

16. The Hardy-Weinberg concept is only theoretical. What factors do not allow it to operate in a natural gene pool?
17. The smaller the population, the more likely it is that random changes will influence gene frequencies. Why is this true?

12.9 A Summary of the Causes of Evolutionary Change

Interactive Concept Map
• Influences on evolutionary change

Food for Thought
• Creationism

18. Why is each of the following important for an understanding of evolution: mutation, migration, sexual reproduction, selective agents, and population size?

**Test yourself with multiple choice questions that provide immediate feedback
on the Online Learning Center with PowerWeb at**

www.mhhe.com/enger11

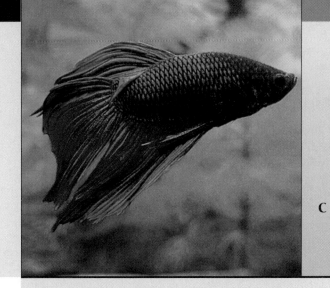

CHAPTER OUTLINE

Speciation and Evolutionary Change

GOALS AND OBJECTIVES

Understand the process of speciation.

❏ Describe the steps necessary for speciation to occur.

❏ Describe how polyploidy can result in the production of new species.

❏ Explain why reproductive isolation is important to the process of speciation.

❏ Describe how a study of chromosomes could determine if a species is a polyploid.

Understand the relationship between natural selection and evolution.

❏ Describe the contributions of the following individuals to evolutionary thought: Lamarck, Buffon, Wallace, and Darwin.

❏ List five assumptions by Darwin that were important to his developing the theory of natural selection.

Appreciate that evolution is a well-supported theory at the center of all biological thinking.

❏ Describe three kinds of evidence that support the concept of evolution.

❏ Describe two examples of new discoveries that have refined our understanding of evolution without refuting it.

❏ Explain three basic evolutionary patterns.

❏ Give examples that explain why the rate of evolution can be different for different species.

Appreciate that evidence indicates that humans have an evolutionary history.

❏ Describe current scientific theories for the evolution of modern humans.

❏ State evidence that suggests humans originated in Africa.

❏ Explain three differences between Australopiths and members of the genus *Homo*.

Evolution and Ecology

PART FOUR

13.1 Species: A Working Definition

Before we consider how new species are produced, let's recall from chapter 11 how one species is distinguished from another. A **species** is commonly defined as a population of organisms whose members have the potential to interbreed naturally to produce fertile offspring but *do not* interbreed with other groups. This is a working definition; it applies in most cases but must be interpreted to encompass some exceptions. There are two key ideas within this definition. First, a species is a population of organisms. An individual—you, for example—is not a species. You can only be a member of a group that is recognized as a species. The human species, *Homo sapiens,* consists of over 6 billion individuals, whereas the endangered California condor species, *Gymnogyps californianus,* consists of about 200 individuals.

Second, the definition involves the ability of individuals within the group to produce fertile offspring. Obviously, we cannot check every individual to see if it is capable of mating with any other individual that is similar to it, so we must make some judgment calls. Do most individuals within the group potentially have the capability of interbreeding to produce fertile offspring? In the case of humans we know that some individuals are sterile and cannot reproduce, but we don't exclude them from the human species because of this. If they were not sterile, they would have the potential to interbreed. We recognize that, although humans normally choose mating partners from their subpopulations, humans from all parts of the world are potentially capable of interbreeding. We know this to be true because of the large number of instances of reproduction involving people of different ethnic and racial backgrounds. The same is true for many other species that have local subpopulations but have a wide geographic distribution.

Another way to look at this question is to think about gene flow. **Gene flow** is the movement of genes from one generation to the next or from one region to another. Two or more populations that demonstrate gene flow between them constitute a single species. Conversely, two or more populations that do not demonstrate gene flow between them, when given the opportunity, are generally considered to be different species. Some examples will clarify this working definition.

The mating of a male donkey and a female horse produces young that grow to be adult mules, incapable of reproduction (figure 13.1). Because mules are nearly always sterile, there can be no gene flow between horses and donkeys and they are considered to be separate species. Similarly, lions and tigers can be mated in zoos to produce offspring. However, this does not happen in nature and so gene flow does not occur naturally; thus they are considered to be two separate species.

Still another way to try to determine if two organisms belong to different species is to determine their *genetic similarity*. The recent advances in molecular genetics allows scientists to examine the sequence of bases in genes present in individuals from a variety of different populations. Those that have a great deal of genetic similarity are assumed to have resulted from populations that have exchanged genes through sexual reproduction in the recent past. If there are significant differences in the genes present in individuals from two populations, they have not exchanged genes recently and are more likely to be members of separate species. Interpretation of the results obtained by examining genetic differences still requires the judgment of experts. It will not unequivocally settle every dispute related to the identification of species, but it is another tool that helps clarify troublesome situations.

Figure 13.1

Hybrid Sterility
Even though they do not do so in nature, (a) horses and (b) donkeys can be mated. The offspring is called a (c) mule and is sterile. Because the mule is sterile, the horse and the donkey are considered to be of different species.

(a)

(b)

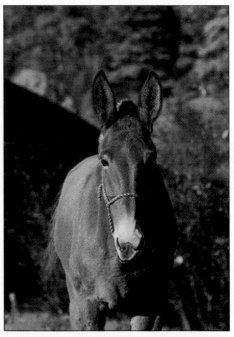

(c)

13.2 How New Species Originate

Speciation is the process of generating new species. When we look at the evolutionary history of living things we see that new species arise continuously. When looking at the past evolution of organisms, we often use **fossils,** which are physical evidences of former life.

The fossil record contains examples of the origin of huge numbers of new species. The fossil record also indicates that most species have gone extinct. There are several mechanisms thought to be involved in generating new species. We will look at two mechanisms that are probably responsible for the vast majority of speciation events: geographic isolation and polyploidy.

Speciation by Geographic Isolation

Geographic isolation occurs when a portion of a species becomes totally isolated from the rest of the gene pool by some geographic change, such as the formation of a mountain range, river valley, desert, or ocean. In order for geographic isolation to lead to speciation, the following steps are necessary: (1) geographic isolation of a subpopulation of a species; (2) genetic divergence of the isolated subpopulation from the rest of the species; and (3) reproductive isolation of the new species from the its parent species.

Geographic isolation can occur for a variety of reasons. The geographic area over which a species can be found is known as its **range.** The range of the human species is the entire world, whereas that of a bird known as a snail kite is a small region of southern Florida. As a species expands its range or environmental conditions change in some parts of a species' current range, portions of the population can become separated from the rest. Thus, many species consist of partially isolated populations that display characteristics significantly different from other local populations. Many of the differences observed may be directly related to adaptations to local environmental conditions. This means that new colonies or isolated populations may have infrequent gene exchange with their geographically distant relatives. As you will recall from chapter 11, these genetically distinct populations are known as *subspecies.*

In this method of speciation it is typical for some **geographic barrier** to totally isolate a subpopulation from the rest of the species. The uplifting of mountains, the rerouting of rivers, and the formation of deserts all may separate one portion of a gene pool from another. For example, two kinds of squirrels are found on opposite sides of the Grand Canyon. Some people consider them to be separate species; others consider them to be different isolated subpopulations of the same species (figure 13.2). Even small changes may cause geographic isolation in species that have little ability to move. A fallen tree, a plowed field, or even a new freeway may effectively isolate populations within such species. Snails in two valleys separated by a high ridge have been found to be closely related but different species. The snails cannot get from one valley to the next because of the height and climatic differences presented by the ridge (figure 13.3).

If two populations of a species are geographically isolated, they are not able to exchange genetic information.

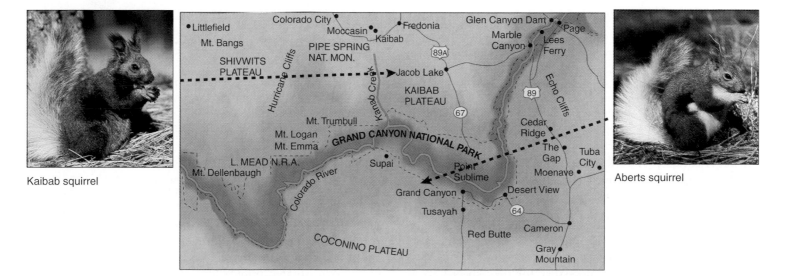

Kaibab squirrel

Aberts squirrel

Figure 13.2

Isolation by Geographic Barriers
These two squirrels are found on opposite sides of the Grand Canyon. Some people consider them to be different species; others consider them to be distinct populations of the same species.

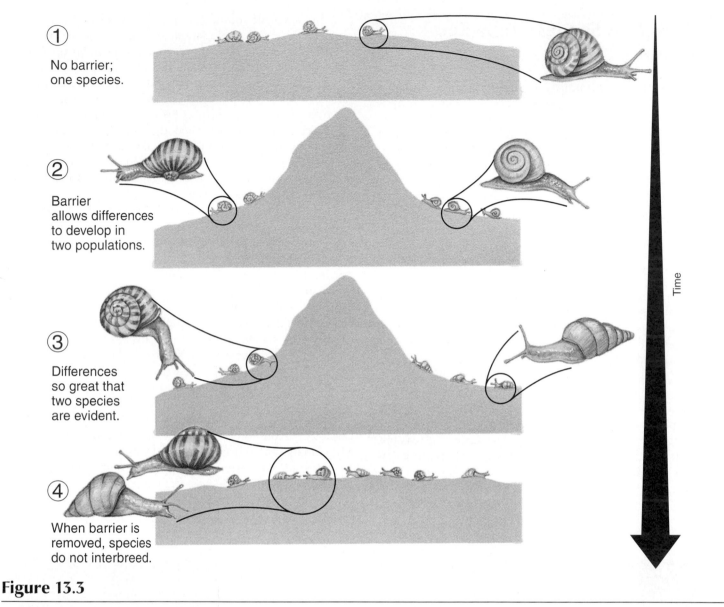

① No barrier; one species.

② Barrier allows differences to develop in two populations.

③ Differences so great that two species are evident.

④ When barrier is removed, species do not interbreed.

Time

Figure 13.3

The Effect of Geographic Isolation

If a single species of snail was to be divided into two different populations by the development of a ridge between them, the two populations could be subjected to different environmental conditions. This could result in a slow accumulation of changes that could ultimately result in two populations that would not be able to interbreed even if the ridge between them were to erode. They would be different species.

However, the separation of a species into two or more isolated subpopulations is not enough to generate new species. Even after many generations of geographic isolation, these separate groups may still be able to exchange genes (mate and produce fertile offspring) if they overcome the geographic barrier, because they have not accumulated enough genetic differences to prevent reproductive success.

Genetic divergence is necessary for new species to develop. Differences in environments and natural selection play very important roles in the process of forming new species. Following separation from the main portion of the gene pool by geographic isolation, the organisms within the small, local population are likely to experience different environ-

mental conditions. If, for example, a mountain range has separated a species into two populations, one population may receive more rain or more sunlight than the other (figure 13.4). These environmental differences act as natural selecting agents on the two gene pools and, acting over a long period of time, account for different genetic combinations in the two places. Furthermore, different mutations may occur in the two isolated populations, and each may generate different random combinations of genes as a result of sexual reproduction. This would be particularly true if one of the populations was very small. As a result, the two populations may show differences in color, height, enzyme production, time of seed germination, or many other characteristics.

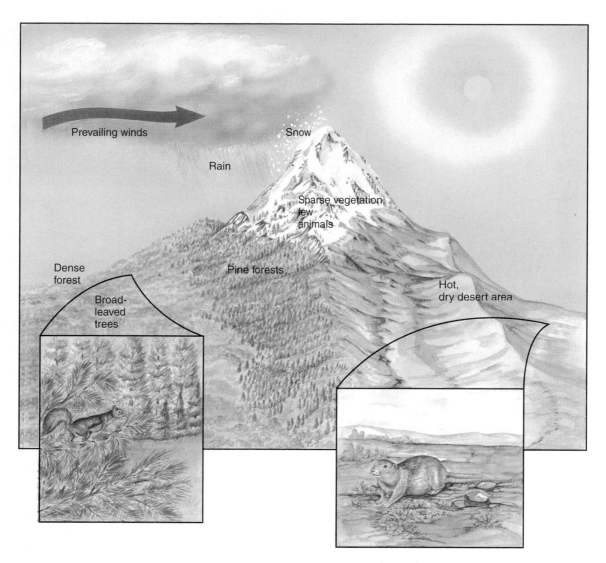

Prevailing winds

Snow

Rain

Sparse vegetation, few animals

Dense forest

Pine forests

Broad-leaved trees

Hot, dry desert area

Figure 13.4

Environmental Differences Caused by Mountain Ranges

Most mountain ranges affect the local environment. Because of the prevailing winds, most rain falls on the windward side of the mountain. This supports abundant vegetation. The other side of the mountain receives much less rain and is drier. Often a desert may exist. Both plants and animals must be adapted to the kind of climate typical for their regions. Cactus and ground squirrels would be typical of the desert and pine trees and tree squirrels would be typical of the windward side of the mountain.

Over a long period of time, the genetic differences that accumulate may result in regional populations called **subspecies** that are significantly modified structurally, physiologically, or behaviorally.

Reproductive isolation or *genetic isolation* has occurred if the genetic differences between the two populations has become so great that reproduction cannot occur between members of the two populations if they are brought together. At this point speciation has occurred. In other words, the process of speciation can begin with the geographic isolation of a portion of the species, but new species are generated only if isolated populations become separate from one another *genetically* and gene flow is not reestablished if the geographic barrier is removed.

Speciation Without Geographic Barriers

It is also possible to envision ways in which speciation could occur without geographic barriers being necessary. Any process that could result in the reproductive isolation of a portion of a species could lead to the possibility of speciation. For example, within populations, some individuals may breed or flower at a somewhat different time of the year. If the difference in reproductive time is genetically based, different breeding populations could be established, which could eventually lead to speciation. Among animals, variations in the genetically determined behaviors related to courtship and mating could effectively separate one species into two or more separate breeding populations. In plants,

Figure 13.5

Polyploidy

Many species of plants have been created by increasing the chromosome number. Many large-flowered varieties have been produced artificially by means of this technique. (*a*) A normal diploid *Hibiscus moscheutos*. (*b*) A polyploid variety of this hibiscus. Note the differences in flower size and petal shape.

(a)   (b)

genetically determined incompatibility of the pollen of one population of flowering plants with the flowers of other populations of the same species could lead to separate species. While there are many examples of these kinds of speciation mechanisms, geographic isolation and polyploidy are considered the primary mechanisms for speciation.

Polyploidy: Instant Speciation

Another important mechanism known to generate new species is polyploidy. **Polyploidy** is a condition of having multiple sets of chromosomes rather than the normal haploid or diploid number. The increase in the number of chromosomes can result from abnormal mitosis or meiosis in which the chromosomes do not separate properly. For example, if a cell had the normal diploid chromosome number of six ($2n = 6$), and the cell went through mitosis but did not divide into two cells, it would then contain 12 chromosomes. It is also possible that a new polyploid species could result from crosses between two species followed by a doubling of the chromosome number. Because the number of chromosomes of the polyploid is different from that of the parent, successful reproduction between the polyploid and the parent species would be difficult. This is because meiosis would result in gametes that had different chromosome numbers from the original, parent organism. In one step, the polyploid could be isolated reproductively from its original species. A single polyploid plant does not constitute a new species. However, because most plants can reproduce asexually, they can create an entire population of organisms that have the same polyploid chromosome number. The members of this population would all have the same chromosome number and would probably be able to undergo normal meiosis and would be capable of sexual reproduction among themselves. In effect, a new species can be created within a couple of generations. Some groups of plants, such as the grasses, may have 50% of their species produced as a result of polyploidy. Many economically important species are polyploids. Cotton, potatoes, sugarcane, wheat, and many garden flowers are examples (figure 13.5). Although it is rare in animals, polyploidy is found in a few groups that typically use asexual reproduction in addition to sexual reproduction. Certain lizards have only female individuals and lay eggs that develop into additional females. Differ-

ent species of these lizards appear to have developed by polyploidy. • species, p. 206

13.3 Maintaining Genetic Isolation Between Species

In order for a new species to continue to exist, it must reproduce but continue to remain genetically distinct from other similar species. The speciation process typically involves the development of **reproductive isolating mechanisms** or **genetic isolating mechanisms.** These mechanisms prevent matings between species and therefore help maintain distinct species. A great many different mechanisms for maintaining reproductive (genetic) isolation are recognized.

Habitat preference or **ecological isolating mechanisms** occur when two species do not have the opportunity to interbreed because they typically live in different ecological settings. In central Mexico, two species of robin-sized birds called *towhees* live in the same general region. However, the collared towhee lives on the mountainsides in the pine forest, while the spotted towhee is found at lower elevations in oak forests. Geography presents no barriers to these birds. They are perfectly capable of flying to each other's habitats, but they do not. Therefore, they are reproductively isolated because of the habitats they prefer. Similarly, areas with wet soil have different species of plants than nearby areas with drier soils.

Seasonal isolating mechanisms (differences in the time of the year at which reproduction takes place) are effective genetic isolating mechanisms. Some plants flower only in the spring of the year, whereas other species that are closely related flower in midsummer or fall; therefore, the two species are not very likely to pollinate one another. Among insects there is a similar spacing of the reproductive periods of closely related species so that they do not overlap.

Behavioral isolating mechanisms occur when inborn behavior patterns prevent breeding between species. The mating calls of frogs and crickets are highly specific. The sound pattern produced by the males is species-specific and invites only females of the same species to engage in mating. The females have a built-in response to the particular species-specific call and only mate with those that produce the correct call.

Figure 13.6

Courtship Behavior (Behavioral Isolating Mechanism)
The dancing of a male prairie chicken attracts female prairie chickens, but not females of other species. This behavior tends to keep prairie chickens reproductively isolated from other species.

The courtship behavior of birds involves both sound and visual signals that are species-specific. For example, groups of male prairie chickens gather on meadows shortly before dawn in the early summer and begin their dances. The air sacs on either side of the neck are inflated so that the bright-colored skin is exposed. Their feet move up and down very rapidly and their wings are spread out and quiver slightly (figure 13.6). This combination of sight and sound attracts females. When the females arrive, the males compete for the opportunity to mate with them. Other related species of birds conduct their own similar, but distinct, courtship displays. The differences among the dances are great enough so that a female can recognize the dance of a male of her own species.

Behavioral isolating mechanisms such as these occur among other types of animals as well. The strutting of a peacock, the fin display of Siamese fighting fish, and the flashing light patterns of "lightning bugs" of different species are all examples of behaviors that help individuals identify members of their own species and prevent different species from interbreeding (figure 13.7).

Mechanical or **morphological isolating mechanisms** involve differences in the structure of organisms. The specific shapes of the structures involved in reproduction may prevent different species from interbreeding. Among insects, the structure of the penis and the reciprocal structures of the female fit like a lock and key and therefore breeding between different species is very difficult. Similarly the shapes of flowers may permit only certain animals to carry pollen from one flower to the next.

Biochemical isolating mechanisms occur when molecular incompatibility prevents successful mating. There are a vast number of biochemical activities that take place around the union of egg and sperm. Molecules on the outside of the egg or sperm may trigger events that prevent their union if

(a)

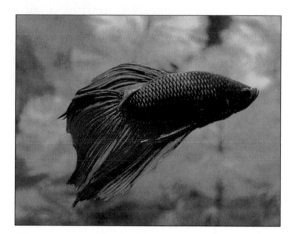

(b)

Figure 13.7

Animal Communication by Displays
Most animals have specific behaviors that they use to communicate with others of the same species. (a) The croaking of a male frog is specific to its species and is different from that of males of other species. (b) The visual displays of Siamese fighting fish are also used to communicate with others of the same species.

they are not from the same species. Among plants it is common to have biochemical interactions between the pollen and receiving flower that prevent the germination of the pollen grain and, therefore, prevent sexual reproduction between two closely related species.

13.4 The Development of Evolutionary Thought

Today, most scientists consider speciation an important first step in the process of evolution. However, this was not always the case. For centuries people believed that the various species of plants and animals were fixed and unchanging—that is, they were thought to have remained unchanged from the time of their creation. Although today we know this is not true, we can understand why people may have thought it was true. Since they knew nothing about DNA, meiosis, or population genetics, they did not have the tools to examine the genetic nature of species. Furthermore, the process of evolution is so slow that the results of evolution were usually not evident during a human lifetime. It is even difficult for modern scientists to recognize this slow change in many kinds of organisms.

In the mid-1700s, Georges-Louis Buffon, a French naturalist, expressed some curiosity about the possibilities of change (evolution) in animals, but he did not suggest any mechanism that would result in evolution. In 1809, Jean-Baptiste de Lamarck, a student of Buffon's, suggested a process by which evolution might occur. He proposed that acquired characteristics could be transmitted to offspring. For example, he postulated that giraffes originally had short necks. Because giraffes constantly stretched their necks to obtain food, their necks got slightly longer. This slightly longer neck acquired through stretching could be passed to the offspring, who were themselves stretching their necks, and over time, the necks of giraffes would get longer and longer. Although we now know Lamarck's theory was wrong (because acquired characteristics are not inherited), it stimulated further thought as to how evolution could occur. All during this period, from the mid-1700s to the mid-1800s, lively arguments continued about the possibility of evolutionary change. Some, like Lamarck and others, thought that change did take place; many others said that it was not possible. It was the thinking of two English scientists that finally provided a mechanism to explain how evolution could occur.

In 1858, Charles Darwin and Alfred Wallace suggested the theory of natural selection as a mechanism for evolution. They based their theory on the following assumptions about the nature of living things:

1. All organisms produce more offspring than can survive.
2. No two organisms are exactly alike.
3. Among organisms, there is a constant struggle for survival.
4. Individuals that possess favorable characteristics for their environment have a higher rate of survival and produce more offspring.
5. Favorable characteristics become more common in the species, and unfavorable characteristics are lost.

Using these assumptions, the Darwin-Wallace theory of evolution by natural selection offers a different explanation for the development of long necks in giraffes (figure 13.8):

1. In each generation, more giraffes would be born than the food supply could support.
2. In each generation, some giraffes would inherit longer necks, and some would inherit shorter necks.
3. All giraffes would compete for the same food sources.

4. Giraffes with longer necks would obtain more food, have a higher survival rate, and produce more offspring.
5. As a result, succeeding generations would show an increase in the number of individuals with longer necks.

This logic seems simple and obvious today, but remember that at the time Darwin and Wallace proposed their theory, the processes of meiosis and fertilization were poorly understood, and the concept of the gene was only beginning to be discussed. Nearly 50 years after Darwin and Wallace suggested their theory, the rediscovery of the work of Gregor Mendel (chapter 10) provided an explanation for how characteristics could be transmitted from one generation to the next. Not only did Mendel's idea of the gene provide a means of passing traits from one generation to the next, it also provided the first step in understanding mutations, gene flow, and the significance of reproductive isolation. All of these ideas are interwoven into the modern concept of evolution. If we look at the same five ideas from the thinking of Darwin and Wallace and update them with modern information, they might look something like this:

1. An organism's capacity to over-reproduce results in surplus organisms.
2. Because of mutation, new genetically determined characteristics enter the gene pool. Because of sexual reproduction, involving meiosis and fertilization, new genetic combinations are present in every generation. These processes are so powerful that each individual in a sexually reproducing population is genetically unique. The genetic information present is expressed in the phenotype of the organism.
3. Resources such as food, soil nutrients, water, mates, and nest materials are in short supply, so some individuals will do without. Other environmental factors, such as disease organisms, predators, or helpful partnerships with other species also affect survival. All these factors that affect survival are called selecting agents.
4. Selecting agents favor individuals with the best combination of genes. They will be more likely to survive and reproduce, passing more of their genes on to the next generation. An organism is selected against if it has fewer offspring than other individuals that have a more favorable combination of genes. It does not need to die to be selected against.
5. Therefore, genes or gene combinations that produce characteristics favorable to survival will become more common in the population and, on the average, members of the species will be better adapted to their environment. • **natural selection, p. 222**

(a) Lamarck theory

Time

(b) Darwin-Wallace theory

Time

Figure 13.8

The Contrasting Ideas of Lamarck and Darwin

(a) Lamarck thought that acquired characteristics could be passed on to the next generation. Therefore, he postulated that as giraffes stretched their necks to get food, their necks got slightly longer. This characteristic was passed on to the next generation, which would have longer necks. (b) The Darwin-Wallace theory states that there is variation within the population and that those with longer necks would be more likely to survive and reproduce and pass their genes for long necks on to the next generation.

13.5 Evolutionary Patterns Above the Species Level

The development of a new species is the smallest irreversible unit of evolution. Because the exact conditions present when a species came into being will never exist again it is unlikely that they will evolve back into an earlier stage in their development. Furthermore, because species are reproductively isolated from one another, they usually do not combine with other species to make something new; they can only diverge (separate) further. When we look at past life and compare it to our current diversity of life, we recognize many different patterns.

Divergent evolution is a basic evolutionary pattern in which individual speciation events cause successive branches in the evolution of a group of organisms. This basic pattern is well illustrated by the evolution of the horse shown in figure 13.9. Each of the many branches of the evolutionary history of the horse began with a speciation event that separated one species into two species as each separately adapted to local conditions. Changes in the environment from moist forests to drier grasslands would have set the stage for change. The modern horse, with its large size, single toe on each foot, and teeth designed for grinding grasses, is thought to be the result of accumulated changes beginning from a small, dog-sized animal with four toes on its front feet, three toes on its hind feet, and teeth designed for chewing leaves and small twigs. Even though we know much about the evolution of the horse, there are still many gaps that need to be filled before we have a complete evolutionary history.

Extinction is loss of a species. It is a common pattern in the evolution of organisms. Notice in figure 13.9 that most of the species that developed during the evolution of the horse are extinct. This is typical. Most of the species of organisms that have ever existed are extinct. Estimates of extinction are around 99%; that is, 99% or more of all the species of organisms that ever existed are extinct. Given this high rate of extinction, we can picture current species of organisms as the product of much evolutionary experimentation,

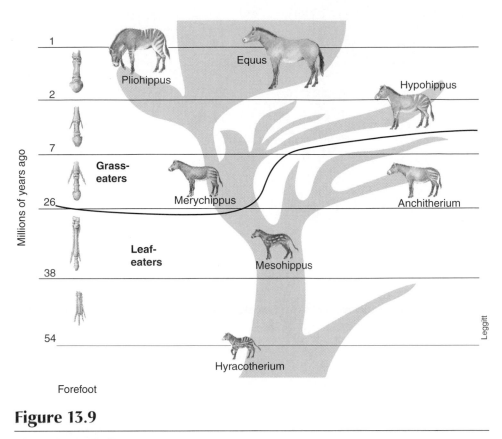

Figure 13.9

Divergent Evolution

In the evolution of the horse, many speciation events followed one after another. What began as a small, leaf-eating, four-toed animal of the forest evolved into a large, grass-eating, single-toed animal of the plains. There are many related species alive today (horse, donkey, zebra) but early ancestral types are extinct.

most of which resulted in failure. This is not the complete picture though. From chapter 12 we recognize that organisms are continually being subjected to selection pressures that lead to a high degree of adaptation to a particular set of environmental conditions. Organisms become more and more specialized. However, the environment does not remain constant and often changes in such a way that the species that were originally present are unable to adapt to the new set of conditions. The early ancestors of the modern horse were well adapted to a moist tropical environment, but when the climate became drier, most were no longer able to survive. Only some kinds had the genes necessary to lead to the development of modern horses.

Furthermore, it is important to recognize that many extinct species were very successful organisms for millions of years. They were not failures for their time but simply did not survive to the present. It is also important to realize that many currently existing organisms will eventually become extinct.

Tracing the evolutionary history of an organism back to its origins is a very difficult task because most of its ancestors no longer exist. We may be able to look at fossils of ex-

tinct organisms but must keep in mind that the fossil record is incomplete and provides only limited information about the biology of the organism represented in that record. We may know a lot about the structure of the bones and teeth or the stems and leaves of an extinct ancestor but know almost nothing about its behavior, physiology, and natural history. Biologists must use a great deal of indirect evidence to piece together the series of evolutionary steps that led to a current species. Figure 13.10 is typical of evolutionary diagrams that help us understand how time and structural changes are related in the evolution of birds, mammals, and reptiles.

The basic pattern is one of divergence with a great deal of extinction. It is also possible to see a pattern in which the same basic body plan is used but modifications of the basic plan occurred as new kinds of organisms developed. For example, reptiles, birds, and mammals typically have two sets of limbs. However, the limbs have been modified in a variety of ways in the different groups. Birds have wings, ichthyosaurs have flippers, and even some snakes show remnants of hind limbs. While divergence and extinction are dominant themes in the evolution of life, it is possible to superimpose several other patterns on this basic design.

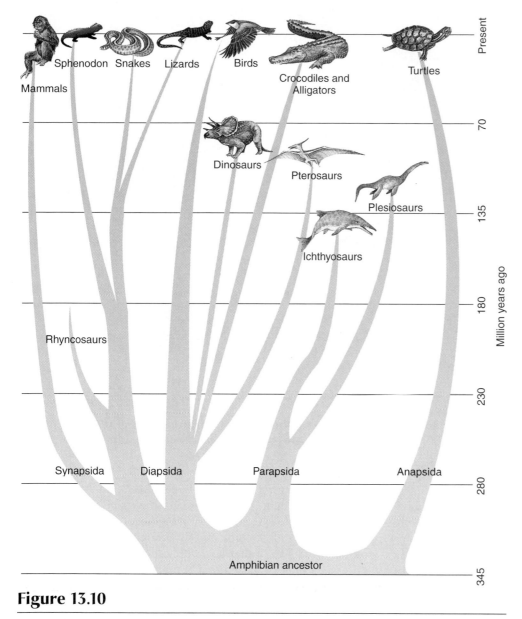

Mammals
Sphenodon   Snakes   Lizards   Birds
Crocodiles and
Alligators
Turtles

Present

70

Dinosaurs
Pterosaurs
Plesiosaurs

135

Ichthyosaurs

180

Million years ago

Rhyncosaurs

230

Synapsida   Diapsida   Parapsida   Anapsida

280

Amphibian ancestor

345

Figure 13.10

An Evolutionary Diagram

This diagram shows how present-day reptiles, birds, and mammals are thought to have evolved from primitive reptilian ancestors. Notice that an extremely long period of time is involved (over 300 million years) and that many of the species illustrated are extinct.

Adaptive radiation is a special evolutionary pattern, characterized by a rapid increase in the number of kinds of closely related species. Adaptive radiation results in an evolutionary explosion of new species from a common ancestor. There are basically two situations that are thought to favor adaptive radiation. One is a condition in which an organism invades a previously unexploited environment. For example, at one time there were no animals on the landmasses of the earth. The amphibians were the first vertebrate animals able to spend part of their lives on land. Fossil evidence shows that a variety of different kinds of amphibians evolved rapidly and exploited several different kinds of lifestyles.

Another good example of adaptive radiation is found among the finches of the Galápagos Islands, located 1,000 kilometers west of Ecuador in the Pacific Ocean. These birds were first studied by Charles Darwin. Because these islands are volcanic and arose from the floor of the ocean, it is assumed that they have always been isolated from South America and originally lacked finches and other land-based birds. It is thought that one kind of finch arrived from South America to colonize the islands and that adaptive radiation from the common ancestor resulted in the many different kinds of finches found on the islands today (figure 13.11). Although the islands are close to one another, they are quite diverse. Some are dry and treeless, some have moist forests, and others have intermediate conditions. Conditions were ideal for several speciation events. Because the islands were separated from one another, the element of geographic isolation was present. Because environmental conditions on the islands were quite different, particular characteristics in the resident birds would have been favored. Furthermore the absence of other kinds of birds meant that there were many lifestyles that had not been exploited.

In the absence of competition, some of these finches took roles normally filled by other kinds of birds elsewhere in the world. Although finches are normally seed-eating birds, some of the Galápagos finches became warblerlike, insect-eaters, others became leaf-eaters, and one uses a cactus spine as a tool to probe for insects.

A second set of conditions that can favor adaptive radiation is one in which a type of organism evolves a new set of characteristics that enable it to displace organisms that previously filled roles in the environment. For example, although amphibians were the first vertebrates to occupy land, they lived only near freshwater where they would not dry out and could lay eggs, which developed in the water. They were replaced by reptiles with such characteristics as dry skin, which prevented the loss of water, and an egg that could develop on land. The adaptive radiation of reptiles was extensive. They invaded most terrestrial settings and even evolved

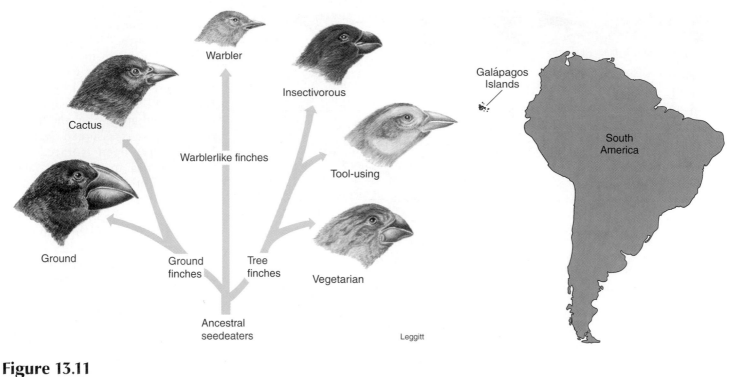

Figure 13.11

Adaptive Radiation

When Darwin discovered the finches of the Galápagos Islands, he thought they might all have derived from one ancestor that arrived on these relatively isolated islands. If they were the only birds to inhabit the islands, they could have evolved very rapidly into the many different types shown here. The drawings show the specializations of beaks for different kinds of food.

forms that flew and lived in the sea. With the extinction of dinosaurs, birds and mammals went through a similar radiation. Perhaps the development of homeothermism (the ability to maintain a constant body temperature) had something to do with the success of birds and mammals. Figure 13.12 shows the sequence of radiations that occurred within the vertebrate group. The number of species of amphibians and reptiles has declined, whereas the number of species of birds and mammals has increased.

Convergent evolution is an interesting evolutionary pattern that involves the development of similar characteristics in organisms of widely different backgrounds. This particular pattern often leads people to misinterpret the evolutionary history of organisms. For example, many kinds of plants that live in desert situations have sharp, pointed structures like spines or thorns and lack leaves during much of the year. Superficially, the sharp, pointed structures may resemble one another to a remarkable degree, but may have a completely different evolutionary history. Some of these structures are modified twigs, others are modified leaves, and still others are modifications of the surface of the stem. The presence of sharp, pointed structures and the absence of leaves are adaptations to a desert type of environment: The thorns and spines discourage herbivores and the absence of leaves reduces water loss.

Another example involves animals that survive by catching insects while flying. Bats, swallows, and dragonflies all obtain food in this manner. They all have wings, good

eyesight or hearing to locate flying insects, and great agility and speed in flight, but they are evolved from quite different ancestors (figure 13.13). At first glance, they may appear very similar and perhaps closely related, but detailed study of their wings and other structures shows that they are quite different kinds of animals. They have simply converged in structure, type of food eaten, and method of obtaining food. Likewise, whales, sharks, and tuna appear to be similar. They have a streamlined shape that aids in rapid movement through the water, a dorsal fin that helps prevent rolling, fins or flippers for steering, and a large tail that provides the power for swimming. They are quite different kinds of animals that happen to live in the open ocean where they pursue other animals as prey. The structural similarities they have are adaptations to being fast-swimming predators.

To help distinguish structures that are the result of convergent evolution and those that have a common ancestry, we often use the terms *homologous* and *analogous structures*. **Homologous structures** are those that have evolved from the same ancestral structure. Thus, the wing of a bat, the front leg of a horse, and the arm of a human show the same basic pattern of bones, but they are extreme modifications of the same basic evolutionary structure. On the other hand, the wing of a butterfly and the wing of a bird have the same function but different evolutionary backgrounds. Such structures are called **analogous structures** and are the result of convergent evolution.

Historical record of the vertebrates

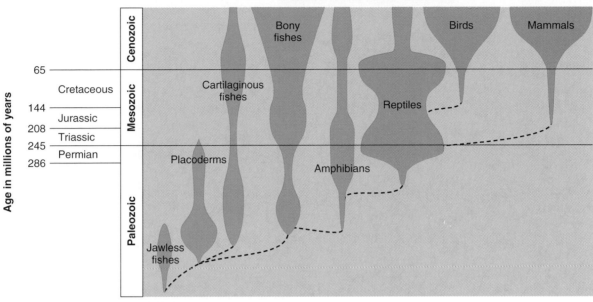

Figure 13.12

Adaptive Radiation in Terrestrial Vertebrates
The amphibians were the first vertebrates to live on land. They were replaced by the reptiles, which were better adapted to land. The reptiles, in turn, were replaced by the adaptive radiation of birds and mammals. (Note: The width of the colored bars indicates the number of species present.)

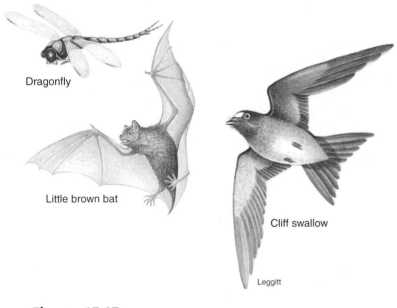

Figure 13.13

Convergent Evolution
All of these animals have evolved wings as a method of movement and capture insects for food as they fly. However, flight originated independently in each of these organisms. Therefore, this is an example of convergent evolution.

13.6 Rates of Evolution

Although it is commonly thought that evolutionary change takes long periods of time, you should understand that rates of evolution can vary greatly. Remember that natural selection is driven by the environment. If the environment is changing rapidly, one would expect rapid changes in the organisms present. Periods of rapid environmental change also result in extensive episodes of extinction. During some periods in the history of the Earth when little environmental change was taking place, the rate of evolutionary change was probably slow. Nevertheless, when we talk about evolutionary time, we are generally thinking in thousands or millions of years. Although both of these time periods are long compared to the human life span, the difference between thousands of years and millions of years in the evolutionary time scale is still significant.

When we examine the fossil record, we can often see gradual changes in physical features of organisms over time. For example, the extinct humanoid fossil *Homo erectus* shows a gradual increase in the size of the cranium, a reduction in the size of the jaw, and the development of a chin over about a million years of time. The accumulation of these changes could result in such extensive change from the original species that we would consider the current organism

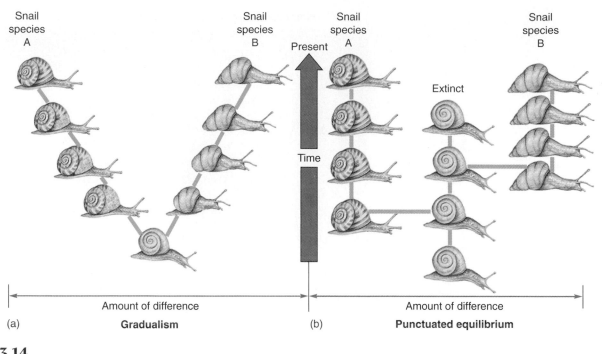

Figure 13.14

Gradualism Versus Punctuated Equilibrium

Gradualism (a) is the evolution of new species from the accumulation of a series of small changes over a long period of time. Punctuated equilibrium (b) is the evolution of new species from a large number of changes in a short period of time. Note that in both instances the ancestral snail has evolved into two species (A and B). However, it is possible that they were produced by different processes.

to be a different species from its ancestor. (Many believe that *Homo erectus* became modern humans, *Homo sapiens.*) This is such a common feature of the evolutionary record that biologists refer to this kind of evolutionary change as **gradualism** (figure 13.14a). Charles Darwin's view of evolution was based on gradual changes in the features of specific species he observed in his studies of geology and natural history. However, as early as the 1940s, some biologists began to challenge gradualism as the only model for evolutionary change. They pointed out that the fossils of some species were virtually unchanged over millions of years. If gradualism were the only explanation for how species evolved, then gradual changes in the fossil record of a species would always be found. Furthermore, some organisms appear suddenly in the fossil record and show rapid change from the time they first appeared. We have many modern examples of rapid evolutionary change. The development of pesticide resistance in insects and antibiotic resistance in various bacteria occurred within our lifetime.

In 1972, two biologists, Niles Eldredge of the American Museum of Natural History and Stephen Jay Gould of Harvard University, proposed the idea of **punctuated equilibrium.** This hypothesis suggests that evolution occurs in spurts of rapid change followed by long periods with little evolutionary change (figure 13.14b). It is important to recognize that the punctuated equilibrium concept suggests a different way of achieving evolutionary change. Rather than

one species slowly accumulating changes to become a different descendant species, rapid evolution of several closely related species from isolated populations would produce a number of species that would compete with one another as the environment changed. Many of these species would become extinct and the fossil record would show change.

Another way to look at these two mechanisms is to assume that both occur. It is clear from the fossil record that there were periods in the past when there was very rapid evolutionary change compared to other times. We also recognize that some environments, like the ocean, are relatively stable while others, like the terrestrial environment, have changed significantly. Many marine organisms have remained unchanged for hundreds of millions of years, while there has been great change in the kinds of terrestrial organisms in the last few million years. So it is possible that both mechanisms (gradualism and punctuated equilibrium) have operated. The important contribution of punctuated equilibrium is that it forces people to recognize that there can be alternative ways of interpreting the fossil record and that the pace of evolution can be quite variable.

At the present time, the scientific community has not resolved these two alternative mechanisms for how evolutionary change occurs. However, both approaches recognize the importance of genetic diversity as the raw material for evolution and the mechanism of natural selection as the process of determining which gene combinations fit the environment. The gradualists point to the fossil record as proof that evolution is a slow, steady process. Those who support punctuated equilibrium point to the gaps in the fossil record as evidence that rapid change occurs. As with most controversies of this nature, more information is required to resolve the question. It will take decades to collect all the information and, even then, the differences of opinion may not be reconciled.

13.7 The Tentative Nature of the Evolutionary History of Organisms

It is important to understand that thinking about the concept of evolution can take us in several different directions. First, it is clear that genetic changes do occur. Mutations introduce new genetic information into a species. This has been demonstrated repeatedly with chemicals and radiation. Our recognition of this danger is evident by the ways we protect ourselves against excessive exposure to mutagenic agents. We also recognize that species can change. We purposely manipulate the genetic constitution of our domesticated plants and animals and change their characteristics to suit our needs. We also recognize that different populations of the same species show genetic differences. Examination of fossils shows that species of organisms that once existed are no longer in existence. We even have historical examples of plants and animals that are now extinct. We can also demonstrate that new species come into existence. This is easiest to do in plants with polyploidy. It is clear from this evidence that species are not fixed, unchanging entities.

However, when we try to piece together the evolutionary history of organisms over long periods of time, we must use much indirect evidence, and it becomes difficult to state definitively the specific sequence of steps that the evolution of a species followed. Although it is clear that evolution occurs, it is not possible to state unconditionally that evolution of a particular group of organisms has followed a specific path. There will always be new information that will require changes in thinking, and equally reputable scientists will disagree on the evolutionary processes or the sequence of events that led to a specific group of organisms. • classification, p. 456

For example, the fossil record provides a great deal of information about the kinds of organisms that have existed in the past. However, the fossil record is not a complete record and new fossils are being discovered every year. There are several reasons why the fossil record is incomplete. First of all the likelihood that an organism will become a fossil is low. Most organisms die and decompose leaving no trace of their existence. (Today, road-killed opossums are not likely to become fossils because they will be eaten by scavengers, repeatedly run over, or decompose by the roadside.) In order to form a fossil the dead organism must be covered over by sediments, or dehydrated or preserved in some other way.

The following factors increase the likelihood that an organism will be found in the fossil record: the presence of hard body parts that resist decomposition; marine organisms that can be covered by sediment on the bottom; recent fossils are less likely to have been destroyed by geological forces; fossils of large organisms are easier to find; and organisms that were extremely common are more likely to show up in the fossil record. For example, trilobites are very common in the fossil record. They were relatively large marine organisms with hard body parts that were extremely abundant. However, fossils of soft-bodied, extremely ancient, worm-like organisms are rare. Finally, the discovery of fossils is often accidental. It is impossible to search through all the layers of sedimentary rock on the entire surface of the Earth. Therefore, there will continue to be additions of new fossils that will extend our information about ancient life into the foreseeable future. But there can be no question that evolution occurred in the past and continues to occur today (How Science Works 13.1: Accumulating Evidence of Evolution).

13.8 Human Evolution

There is intense curiosity about how our species (*Homo sapiens*) came to be and the evolution of the human species remains an interesting and controversial topic. We recognize that humans show genetic diversity, experience mutations, and are subject to the same evolutionary forces as other organisms. We also recognize that some individuals have genes that make them subject to early death or make them unable to reproduce. On the other hand, because all of our close evolutionary relatives are extinct, it is difficult for us to visualize our evolutionary development and we tend to think we are unique and not subject to the laws of nature.

We use several kinds of evidence to try to sort out our evolutionary history. Fossils of various kinds of human and prehuman ancestors have been found, but these are often fragmentary and hard to date. Stone tools of various kinds have also been found that are associated with human and prehuman sites. Finally, other aspects of the culture of our human ancestors have been found in burial sites, cave paintings, and the creation of ceremonial objects. Various methods have been used to age these findings. Some can be dated quite accurately, whereas others are more difficult to pinpoint.

When fossils are examined, anthropologists can identify differences in the structures of bones that are consistent with changes in species. Based on the amount of change they see and the ages of the fossils, these scientists make judgments about the species to which the fossil belongs. As new discoveries are made, opinions of experts will change and our evolutionary history may become more clear as old ideas are replaced. It is also clear from the fossil record that humans are relatively recent additions to the forms of life. Assembling all of these bits of information into a clear picture

HOW SCIENCE WORKS 13.1

Accumulating Evidence of Evolution

The theory of evolution has become the major unifying theory of the biological sciences. Medicine recognizes the dangers of mutations, the similarity in function of the same organ in related species, and the way in which the environment can interfere with the preprogramed process of embryological development. Agricultural science recognizes the importance of selecting specific genetically determined characteristics for passage into new varieties of crop plants and animals. The concepts of mutation, selection, and evolution are so fundamental to understanding what happens in biology that we often forget to take note of the many kinds of observations that support the theory of evolution. The following list describes some of the more important pieces of evidence that support the idea that evolution has been and continues to be a major force in shaping the nature of living things.

1. Species and populations are not genetically fixed. Change occurs in individuals and populations.
   a. Mutations cause slight changes in the genetic makeup of an individual organism.
   b. Different populations of the same species show adaptations suitable for their local conditions.
   c. Changes in the characteristics displayed by species can be linked to environmental changes.
   d. Selective breeding of domesticated plants and animals indicates that the shape, color, behavior, metabolism, and many other characteristics of organisms can be selected for.
   e. Extinction of poorly adapted species is common.

2. All evidence suggests that once embarked on a particular evolutionary road, the system is not abandoned, only modified. New organisms are formed by the modification of ancestral species, not by major changes. The following list supports the concept that evolution proceeds by modification of previously existing structures and processes rather than by catastrophic change.
   a. All species use the same DNA code.
   b. All species use the same left-handed, amino acid building blocks.
   c. It is difficult to eliminate a structure when it is part of a developmental process controlled by genes. Vestigial structures like the human appendix or tailbone are evidence of genetic material from previous stages in evolution.
   d. Embryological development of related animals is similar regardless of the peculiarities of adult anatomy. All vertebrates' embryos have an early stage that contains gill slits.
   e. Species of organisms that are known to be closely related show greater similarity in their DNA than those that are distantly related.

3. Several aspects of the fossil record support the concept of evolution.
   a. The nature of the Earth has changed significantly over time.
   b. The fossil record shows vast changes in the kinds of organisms present on Earth. New species appear and most go extinct. This is evidence that living things change in response to changes in their environment.
   c. The fossils found in old rocks do not reappear in younger rocks. Once an organism goes extinct it does not reappear, but new organisms arise that are modifications of previous organisms.

4. New techniques and discoveries invariably support the theory of evolution.
   a. The recognition that the Earth was formed billions of years ago supports the slow development of new kinds of organisms.
   b. The recognition that the continents of the Earth have separated helps explain why organisms on Australia are so different from elsewhere.
   c. The discovery of DNA and how it works helps explain mutation and allows us to demonstrate the genetic similarity of closely related species.

---

is not possible at this point, but a number of points are well accepted.

1. There is a great deal of fossil evidence that several species of hominids of the genera *Australopithecus* and *Paranthropus* were among the earliest hominid fossils. These organisms are often referred to collectively as australopiths.
2. Based on fossil evidence, it appears that the climate of Africa was becoming drier during the time that hominid evolution was occurring.
3. The earliest *Australopithecus* fossils are from about 4.2 million years ago. Earlier fossils such as *Ardipithecus* may be ancestral to *Australopithecus*. *Australopithecus* and *Paranthropus* were herbivores and walked upright.

Their fossils and the fossils of earlier organisms like *Ardipithecus* are found only in Africa.
4. The australopiths were sexually dimorphic with the males much larger than the females and had relatively small brains (cranial capacity 530 cubic centimeters or less).
5. Several species of the genus *Homo* became prominent in Africa and appear to have made a change from a primarily herbivorous diet to a carnivorous or omnivorous diet.
6. All members of the genus *Homo* have relatively large brains for their body size (cranial capacity 650 cubic centimeters or more) and are associated with various degrees of stone tool construction and use. It is possible that some of the australopiths may have constructed stone tools.

7. Fossils of several later species of the genus *Homo* are found in Africa, Europe, and Asia, but not in Australia or the Americas. Only *Homo sapiens* is found in Australia and the Americas.

8. Since the fossils of *Homo* species found in Asia and Europe are generally younger than the early *Homo* species found in Africa, it is assumed that they moved to Europe and Asia from Africa. Genetic studies also support this idea.

9. Size differences between the sexes are less prominent in members of the genus *Homo* so perhaps there were fewer differences in activities.

When we try to put all of these bits of information together we can construct the following scenario for the evolution of our species. Monkeys, apes, and other primates are adapted to living in forested areas where their grasping hands, opposable thumbs and big toes, and wide range of movement of the shoulders allow them to move freely in the trees. As the climate became drier the forests were replaced by grasslands and, as is always the case, some organisms became extinct and others adapted to the change.

The First Hominids—The Australopiths

Various species of *Australopithecus* and *Paranthropus* were present in Africa from about 4.4 million years ago until about 1 million years ago. It is important to recognize that there are few fossils of these early humanlike organisms and that often they are fragments of the whole organism. This has led to much speculation and argument among experts about the specific position each fossil has in the evolutionary history of humans. However, from examining the fossil bones of the leg, pelvis, and foot, it is apparent that the australopiths were relatively short (males, 1.5 meters or less; females, about 1.1 meters) and stocky and walked upright like humans.

An upright posture had several advantages in a world that was becoming drier. It allowed for more rapid movement over long distances, the ability to see longer distances, and reduced the amount of heat gained from the sun. In addition, upright posture freed the arms for other uses such as carrying and manipulating objects, and using tools. The various species of *Australopithecus* and *Paranthropus* shared these characteristics and, based on the structure of their skulls, jaws, and teeth, appear to have been herbivores with relatively small brains.

Later Hominids—The Genus *Homo*

About 2.5 million years ago the first members of the genus *Homo* appeared on the scene. There is considerable disagreement about how many species there were but *Homo habilis* is one of the earliest. *Homo habilis* had a larger brain (650 cubic centimeters) and smaller teeth than australopiths and made much more use of stone tools. Some people believe that it was a direct descendant of *Australopithecus africanus*.

Many experts believe that *Homo habilis* was a scavenger that made use of group activities, tools, and higher intelligence to hijack the kills made by other carnivores. The higher-quality diet would have supported the metabolic needs of the larger brain.

About 1.8 million years ago, *Homo ergaster* appeared on the scene. It was much larger (up to 1.6 meters) than *H. habilis* (about 1.3 meters) and also had a much larger brain (cranial capacity of 850 cubic centimeters). A little later a similar species (*Homo erectus*) appears in the fossil record. Some people consider *H. ergaster* and *H. erectus* to be variations of the same species. The larger brain of *H. ergaster* and *H. erectus* appears to be associated with extensive use of stone tools. Hand axes were manufactured and used to cut the flesh of prey and crush the bones for marrow. These organisms appear to have been predators, whereas *H. habilis* was a scavenger. The use of meat as food allows animals to move about more freely, because appropriate food is available almost everywhere. By contrast, herbivores are often confined to places that have foods appropriate to their use; fruits for fruit eaters, grass for grazers, forests for browsers, and so forth. In fact, fossils of *H. erectus* have been found in the Middle East and Asia as well as Africa. Most experts think that *H. erectus* originated in Africa and migrated through the Middle East to Asia.

About 800,000 years ago another hominid, classified as *Homo heidelbergensis*, appears in the fossil record. Since fossils of this species are found in Africa, Europe, and Asia, it appears that they constitute a second wave of migration of early *Homo* from Africa to other parts of the world. Both *H. erectus* and *H. heidelbergensis* disappear from the fossil record as two new species (*Homo neanderthalensis* and *Homo sapiens*) become common.

The Neandertals were primarily found in Europe and adjoining parts of Asia and are not found in Africa. Therefore many scientists feel they are descendants of *Homo heidelbergensis*, which was common in Europe.

The Origin of *Homo sapiens*

Homo sapiens is found throughout the world and is now the only hominid species remaining of a long line of ancestors. There are two different theories that seek to explain the origin of *Homo sapiens*. One theory, known as the **out-of-Africa hypothesis**, states that modern humans (*Homo sapiens*) originated in Africa as had several other hominid species and migrated from Africa to Asia and Europe and displaced species such as *H. erectus* and *H. heidelbergensis* that had migrated into these areas previously. The other theory, known as the **multiregional hypothesis**, states that *H. erectus* evolved into *H. sapiens*. During a period of about 1.7 million years, fossils of *Homo erectus* showed a progressive increase in the size of the cranial capacity and reduction in the size of the jaw, so that it becomes difficult to distinguish *H. erectus* from *H. heidelbergensis* and *H. heidelbergensis* from *H. sapiens*. Proponents of this hypothesis believe that *H. heidelbergensis* is not

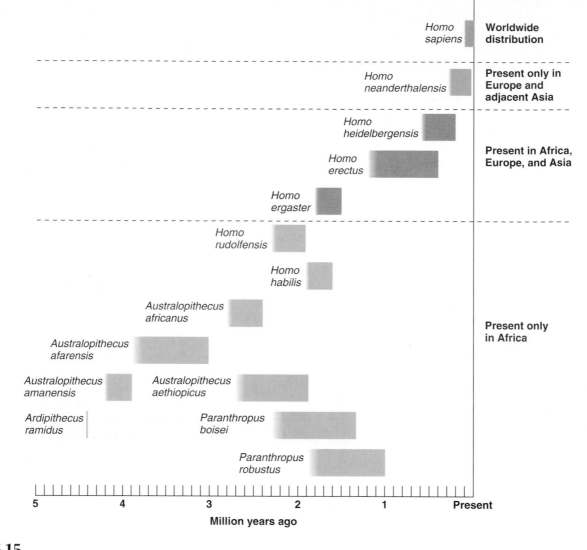

Figure 13.15

Human Evolution

This diagram shows the various organisms thought to be relatives of humans. The bars represent approximate times the species are thought to have existed. Notice that: (1) All species are extinct today except for modern humans, (2) Several different species of organisms coexisted for extensive periods, (3) All the older species are only found in Africa, (4) More recent species of *Homo* are found in Europe and Asia as well as Africa.

a distinct species but an intermediate between the earlier *H. erectus* and *H. sapiens*. According to this theory, various subgroups of *H. erectus* existed throughout Africa, Asia, and Europe and that interbreeding among the various groups gave rise to the various races of humans we see today.

Another continuing puzzle is the relationship of humans that clearly belong to the species *Homo sapiens* with a contemporary group known as Neandertals. Some people consider Neandertals to be a subgroup of *Homo sapiens* specially adapted to life in the harsh conditions found in postglacial Europe. Others consider them to be a separate species of *Homo neanderthalensis*. The Neandertals were muscular, had a larger brain capacity than modern humans, and had many elements of culture, including burials. The cause of their disappearance from the fossil record at about 25,000 years ago remains a mystery. Perhaps climate change to a

warmer climate was responsible. Perhaps contact with *Homo sapiens* resulted in their elimination either through hostile interactions or, if they were able to interbreed with *H. sapiens*, they could have been absorbed into the larger *H. sapiens* population.

Large numbers of fossils of prehistoric humans have been found in all parts of the world. Many of these show evidence of a collective group memory we call *culture*. Cave paintings, carvings in wood and bone, tools of various kinds, and burials are examples. These are also evidence of a capacity to think and invent, and "free time" to devote to things other than gathering food and other necessities of life. We may never know how we came to be, but we will always be curious and will continue to search and speculate about our beginnings. Figure 13.15 summarizes the current knowledge of the historical record of humans and their relatives.

SUMMARY

Populations are usually genetically diverse. Mutations, meiosis, and sexual reproduction tend to introduce genetic diversity into a population. Organisms with wide geographic distribution often show different gene frequencies in different parts of their range. A species is a group of organisms that can interbreed to produce fertile offspring. The process of speciation usually involves the geographic separation of the species into two or more isolated populations. While they are separated, natural selection operates and each population adapts to its environment. If this generates enough change, the two populations may become so different that they cannot interbreed. Similar organisms that have recently evolved into separate species normally have genetic (reproductive) isolating mechanisms to prevent interbreeding. Some of these are habitat preference, seasonal isolating mechanisms, and behavioral isolating mechanisms. Plants have a special way of generating new species by increasing their chromosome numbers as a result of abnormal mitosis or meiosis. Organisms that have multiple sets of chromosomes are called polyploids.

At one time, people thought that all organisms were fixed and unchangeable. Lamarck suggested that change did occur and thought that acquired characteristics could be passed from generation to generation. Darwin and Wallace proposed the theory of natural selection as the mechanism that drives evolution. Evolution is basically a divergent process upon which other patterns can be superimposed. Adaptive radiation is a very rapid divergent evolution; convergent evolution involves the development of superficial similarities among widely different organisms. The rate at which evolution has occurred probably varies. The fossil record shows periods of rapid change interspersed with periods of little change. This has caused some to look for mechanisms that could cause the sudden appearance of large numbers of new species in the fossil record, which challenge the traditional idea of slow, steady change accumulating enough differences to cause a new species to be formed.

The early evolution of humans has been difficult to piece together because of the fragmentary evidence. Beginning about 4.4 million years ago the earliest forms of *Australopithecus* and *Paranthropus* showed upright posture and other humanlike characteristics. The structure of the jaw and teeth indicates that the various kinds of australopiths were herbivores. *Homo habilis* had a larger brain and appears to have been a scavenger. Several other species of the genus *Homo* arose in Africa. These forms appear to have been carnivores. Some of these migrated to Europe and Asia. The origin of *Homo sapiens* is in dispute. It may have arisen in Africa and migrated throughout the world or evolved from earlier ancestors found throughout Africa, Asia, and Europe.

THINKING CRITICALLY

For guidelines to answer this question, visit the Online Learning Center.

Explain how all the following are related to the process of speciation: mutation, natural selection, meiosis, geographic isolation, changes in the Earth, and gene pool.

CONCEPT MAP TERMINOLOGY

Construct a concept map using these terms.

adaptive radiation	gene flow
behavioral isolating mechanisms	genetic isolating mechanisms
	geographic isolation
convergent evolution	seasonal isolating mechanisms
divergent evolution	speciation
ecological isolating mechanisms	species

Visit the Online Learning Center to use an interactive format to construct another concept map.

KEY TERMS

Use interactive flash cards, on the Online Learning Center, to help you learn the meanings of these terms.

adaptive radiation	gradualism
analogous structures	habitat preference
behavioral isolating mechanisms	homologous structures
biochemical isolating mechanism	mechanical (morphological) isolating mechanism
convergent evolution	multiregional hypothesis
divergent evolution	out-of-Africa hypothesis
ecological isolating mechanism	polyploidy
	punctuated equilibrium
extinction	range
fossil	reproductive isolating mechanism
gene flow	
genetic isolating mechanism	seasonal isolating mechanism
geographic barrier	speciation
geographic isolation	species
	subspecies

LEARNING CONNECTIONS

Visit the Online Learning Center to find additional study resources including those highlighted below.

13.1 Species: A Working Definition

Animations and Review
• Introduction to speciation

1. How is the concept of gene flow related to the species concept?

13.2 How New Species Originate

Experience This!
• Observing isolation mechanisms firsthand

2. How does speciation differ from the formation of subspecies or races?
3. Why aren't mules considered a species?
4. Can you always tell by looking at two organisms whether or not they belong to the same species?
5. Why is geographic isolation important in the process of speciation?
6. How does a polyploid organism differ from a haploid or diploid organism?
7. List the series of events necessary for speciation to occur.

13.3 Maintaining Genetic Isolation Between Species

Interactive Concept Map
• Reproductive isolation

8. Describe three kinds of genetic isolating mechanisms that prevent interbreeding between different species.
9. Give an example of seasonal isolating mechanisms, ecological isolating mechanisms, and behavioral isolating mechanisms.

13.4 The Development of Evolutionary Thought

Animations and Review
• Before Darwin
• Evidence for Evolution

10. Why has Lamarck's theory been rejected?
11. List five assumptions about the nature of living things that support the concept of evolution by natural selection.

13.5 Evolutionary Patterns Above the Species Level

12. Describe convergent evolution and adaptive radiation.
13. What are the two dominant evolutionary patterns?
14. How are analogous and homologous structures different?

13.6 Rates of Evolution

15. What is the difference between gradualism and punctuated equilibrium?

13.7 The Tentative Nature of the Evolutionary History of Organisms

16. Why is it difficult to determine the evolutionary history of a species ?

13.8 Human Evolution

Interactive Concept Map
• Human adaptations

Animations and Review
• Primates
• Hominids

17. List three differences between australopiths and members of the genus *Homo*.
18. Compare the out-of-Africa hypothesis with the multiregional hypothesis of the origin of *Homo sapiens*.

Test yourself with multiple choice questions that provide immediate feedback on the Online Learning Center with PowerWeb at

www.mhhe.com/enger11

Ecosystem Organization and Energy Flow

GOALS AND OBJECTIVES

Understand the nature of an ecosystem.

❑ Identify biotic and abiotic environmental factors.
❑ Explain how energy is related to ecosystems.

Recognize the types of relationships that organisms have to each other.

❑ Explain why plants are called producers.
❑ Identify the trophic levels occupied by herbivores and carnivores and why they are called consumers.
❑ Describe the role of decomposers.

Understand energy flow through an ecosystem.

❑ Explain why predators are more rare than herbivores.
❑ Describe why useful energy is lost as energy flows from one trophic level to the next.
❑ Describe the benefits and shortcomings of using each of the following: pyramid of numbers, pyramid of biomass, and pyramid of energy.

Recognize the abiotic and biotic features that are typical of the major biomes of the world.

❑ List typical abiotic and biotic characteristics of different biomes.
❑ Explain why some plants and animals are found only in certain parts of the world.
❑ Describe the effect of temperature and rainfall to the kind of biome that develops.

Understand the concept of succession.

❑ Explain the concept of a climax community.
❑ Describe the effect humans have had on natural climax ecosystems.
❑ Explain why a vacant lot becomes a tangle of plants.

Evolution and Ecology

PART FOUR

259

14.1 Ecology and Environment

Today we hear people from all walks of life using the terms *ecology* and *environment*. Students, homeowners, politicians, planners, and union leaders speak of "environmental issues" and "ecological concerns." Often these terms are interpreted in different ways, so we need to establish some basic definitions.

Ecology is the branch of biology that studies the relationships between organisms and their environments. This is a very simple definition for a very complex branch of science. Most ecologists define the word **environment** very broadly as anything that affects an organism during its lifetime. These environmental influences can be divided into two categories. Other living things that affect an organism are called **biotic factors,** and nonliving influences are called **abiotic factors** (figure 14.1). If we consider a fish in a stream, we can identify many environmental factors that are important to its life. The temperature of the water is extremely important as an abiotic factor, but it may be influenced by the presence of trees (biotic factor) along the stream bank that shade the stream and prevent the Sun from heating it. Obviously, the kind and number of food organisms in the stream are important biotic factors as well. The type of material that makes up the stream bottom and the amount of oxygen dissolved in the water are other important abiotic factors, both of which are related to how rapidly the water is flowing.

As you can see, characterizing the environment of an organism is a complex and challenging process; everything seems to be influenced or modified by other factors. A plant is influenced by many different factors during its lifetime: the types and amounts of minerals in the soil; the amount of sunlight hitting the plant; the animals that eat the plant; and the wind, water, and temperature. Each item on this list can be further subdivided into other areas of study. For instance, water is important in the life of plants, so rainfall is studied in plant ecology. But even the study of rainfall is not simple. The rain could come during one part of the year, or it could be evenly distributed throughout the year. The rainfall could be hard and driving, or it could come as gentle, misty showers of long duration. The water could soak into the soil for later use, or it could run off into streams and be carried away.

Temperature is also very important to the life of a plant. For example, two areas of the world can have the same average daily temperature of 10°C* but not have the same plants because of different temperature extremes. In one area, the temperature may be 13°C during the day and 7°C at night, for a 10°C average. In another area, the temperature may be 20°C in the daytime and only 0°C at night, for a 10°C average. Plants react to extremes in temperature as well as to the daily average. Furthermore, different parts of a plant may respond differently to temperature. Tomato plants will grow at temperatures below 13°C but will not begin to develop fruit below 13°C.

The animals in an area are influenced as much by abiotic factors as are the plants. If nonliving factors do not favor the growth of plants, there will be little food and few

---

*See the metric conversion chart in appendix 1 for conversion to Fahrenheit.

Figure 14.1

Biotic and Abiotic Environmental Factors
(*a*) The woodpecker feeding its young in the hole in this tree is influenced by several biotic factors. The tree itself is a biotic factor as is the disease that weakened it, causing conditions that allowed the woodpecker to make a hole in the rotting wood. (*b*) The irregular shape of the trees is the result of wind and snow, both abiotic factors. Snow driven by the prevailing winds tends to "sandblast" one side of the tree and prevent limb growth.

(a)

(b)

hiding places for animal life. Two types of areas that support only small numbers of living animals are deserts and polar regions. Near the polar regions of the Earth, the low temperature and short growing season inhibits growth; therefore, there are relatively few species of animals with relatively small numbers of individuals. Deserts receive little rainfall and therefore have poor plant growth and low concentrations of animals. On the other hand, tropical rainforests have high rates of plant growth and large numbers of animals of many kinds.

As you can see, living things are themselves part of the environment of other living things. If there are too many animals in an area, they can demand such large amounts of food that they destroy the plant life, and the animals themselves will die. So far we have discussed how organisms interact with their environments in rather general terms. Ecologists have developed several concepts that help us understand how biotic and abiotic factors interrelate in a complex system.

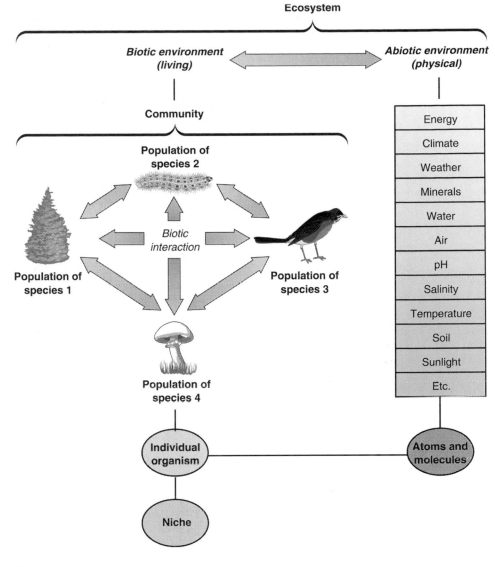

14.2 The Organization of Ecological Systems

Ecologists can study ecological relationships at several different levels of organization. The smallest, independent living unit is the individual *organism*. Groups of organisms of the same species are called **populations.** Interacting populations of different species are called **communities.** And an **ecosystem** consists of all the interacting organisms in an area and their interactions with their abiotic surroundings. Figure 14.2 shows how these different levels of organization are related to one another.

Figure 14.2

Levels of Organization in Ecology
Ecology is the science that deals with the interactions between organisms and their environment. This study can take place at several different levels, from the broad ecosystem level through community interactions to population studies and the study of individual organisms. Ecology also involves study of the physical environment, which makes up the nonliving parts of an ecosystem.

All living things require continuous supplies of energy to maintain life. Therefore, many people like to organize living systems by the energy relationships that exist among the different kinds of organisms present. An ecosystem contains several different kinds of organisms. Those that trap sunlight for photosynthesis, resulting in the production of organic material from inorganic material, are called **producers.** Green plants and other photosynthetic organisms such as algae and cyanobacteria are, in effect, converting sunlight energy into the energy contained within the chemical bonds of organic compounds. There is a flow of energy from the Sun into the living matter of plants. • **photosynthesis, p. 122**

The energy that plants trap can be transferred through a number of other organisms in the ecosystem. Because all of these organisms must obtain energy in the form of organic matter, they are called **consumers.** Consumers cannot capture energy from the Sun as plants do. All animals are consumers. They either eat plants directly or eat other sources of organic matter derived from plants. Each time the energy enters a different organism, it is said to enter a different **trophic level,** which is a step, or stage, in the flow of energy through

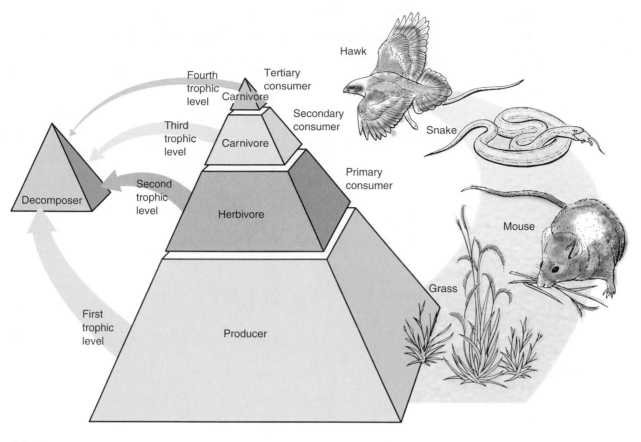

Figure 14.3

The Organization of an Ecosystem
Organisms within ecosystems can be divided into several different trophic levels on the basis of how they obtain energy. Several different sets of terminology are used to identify these different roles. This illustration shows how the different sets of terminology are related to one another.

an ecosystem (figure 14.3). The plants (producers) receive their energy directly from the Sun and are said to occupy the *first trophic level*.

Various kinds of consumers can be divided into several categories, depending on how they fit into the flow of energy through an ecosystem. Animals that feed directly on plants are called **herbivores,** or **primary consumers,** and occupy the *second trophic level*. Animals that eat other animals are called **carnivores,** or **secondary consumers,** and can be subdivided into different trophic levels depending on what animals they eat. Animals that feed on herbivores occupy the *third trophic level* and are known as **primary carnivores.** Animals that feed on the primary carnivores are known as **secondary carnivores** and occupy the *fourth trophic level*. For example, a human may eat a fish that ate a frog that ate a spider that ate an insect that consumed plants for food. This sequence of organisms feeding on one another is known as a **food chain.** Figure 14.4 shows the six different trophic levels in this food chain.

Obviously, there can be higher categories, and some organisms don't fit neatly into this theoretical scheme. Some

animals are carnivores at some times and herbivores at others; they are called **omnivores.** They are classified into different trophic levels depending on what they happen to be eating at the moment. If an organism dies, the energy contained within the organic compounds of its body is finally released to the environment as heat by organisms that decompose the dead body into carbon dioxide, water, ammonia, and other simple inorganic molecules. Organisms of decay, called **decomposers,** are things such as bacteria, fungi, and other organisms that use dead organisms as sources of energy (Outlooks 14.1).

This group of organisms efficiently converts nonliving organic matter into simple inorganic molecules that can be used by producers in the process of trapping energy. Decomposers are thus very important components of ecosystems that cause materials to be recycled. As long as the Sun supplies the energy, elements are cycled through ecosystems repeatedly. Table 14.1 summarizes the various categories of organisms within an ecosystem. Now that we have a better idea of how ecosystems are organized, we can look more closely at energy flow through ecosystems.

Figure 14.4

Trophic Levels in a Food Chain
As one organism feeds on another organism, there is a flow of energy from one trophic level to the next. This illustration shows six trophic levels.

14.3 Characterizing Ecosystems: Energy, Numbers, Biomass

The ancient Egyptians constructed elaborate tombs we call *pyramids*. The broad base of the pyramid is necessary to support the upper levels of the structure, which narrows to a point at the top. This same kind of relationship exists when we look at how the various trophic levels of ecosystems are related to one another. Biologists have adopted this

Table 14.1		
ROLES IN AN ECOSYSTEM		
Classification	**Description**	**Examples**
Producers	Organisms that convert simple inorganic compounds into complex organic compounds by photosynthesis	Trees, flowers, grasses, ferns, mosses, algae, cyanobacteria
Consumers	Organisms that rely on other organisms as food Animals that eat plants or other animals	
Herbivore	Eats plants directly	Deer, goose, cricket, vegetarian human, many snails
Carnivore	Eats meat	Wolf, pike, dragonfly
Omnivore	Eats plants and meat	Rat, most humans
Scavenger	Eats food left by others	Coyote, skunk, vulture, crayfish
Parasite	Lives in or on another organism, using it for food	Tick, tapeworm, many insects
Decomposers	Organisms that return organic compounds to inorganic compounds; important components in recycling	Bacteria, fungi

model as a way to think about how ecosystems are organized (see figure 14.3).

The Pyramid of Energy

A constant source of energy is needed by any living thing. There are two fundamental physical laws of energy that are important when looking at ecological systems from an energy point of view. First of all, the first law of thermodynamics states that energy is neither created nor destroyed. That means that we should be able to describe the amounts of energy in each trophic level and follow energy as it flows through successive trophic levels. The second law of thermodynamics states that when energy is converted from one form to another some energy escapes as heat. This means that as energy passes from one trophic level to the next there will be a reduction in the amount of energy in living things and an increase in the amount of heat in their surroundings. • energy, p. 24

At the base of the energy pyramid is the producer trophic level, which contains the largest amount of energy of any of the trophic levels within an ecosystem. In an ecosystem, the total energy can be measured in several ways. The total producer trophic level can be harvested and burned. The number of calories of heat energy produced by burning is equivalent to the energy content of the organic material of

OUTLOOKS 14.1

Detritus Food Chains

Although most ecosystems receive energy directly from the Sun through the process of photosynthesis, some ecosystems obtain most of their energy from a constant supply of dead organic matter. For example, forest floors and small streams receive a rain of leaves and other bits of material that small animals use as a food source. The small pieces of organic matter, such as broken leaves, feces, and body parts, are known as *detritus*. The insects, slugs, snails, earthworms, and other small animals that use detritus as food are often called *detritivores*. In the process of consuming leaves, detritivores break the leaves and other organic material into smaller particles that may be used by other organisms for food. The smaller size also allows bacteria and fungi to more effectively colonize the dead organic matter, further decomposing the organic material and making it available to still other organisms as a food source. The bacteria and fungi are in turn eaten by other detritus feeders. Some biologists believe that we greatly underestimate the energy flow through detritus food chains.

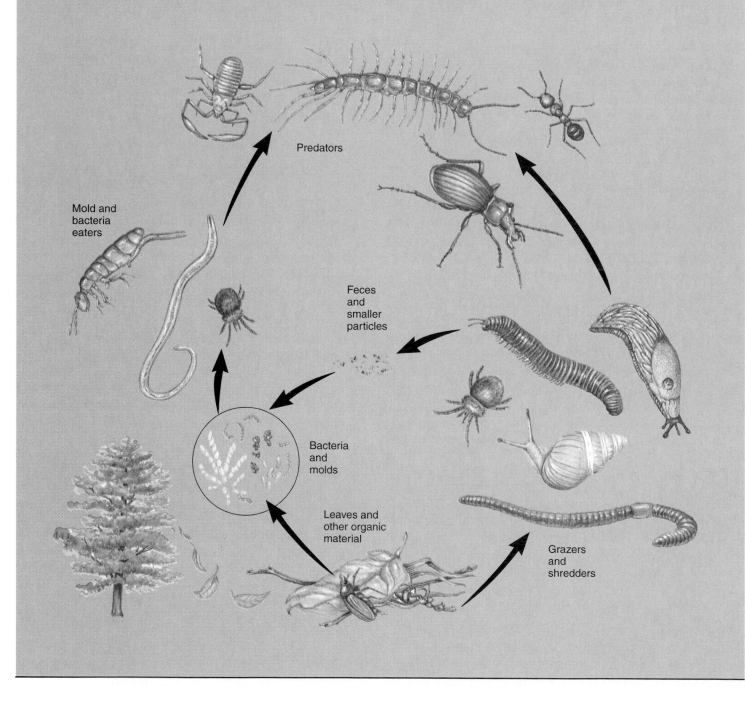

the plants. Another way of determining the energy present is to measure the rate of photosynthesis and respiration and calculate the amount of energy being trapped in the living material of the plants.

Because only the plants, algae, and cyanobacteria in the producer trophic level are capable of capturing energy from the Sun, all other organisms are directly or indirectly dependent on the producer trophic level. The second trophic level consists of herbivores that eat the producers. This trophic level has significantly less energy in it for several reasons. *In general, there is about a 90% loss of energy as we proceed from one trophic level to the next higher level.* Actual measurements will vary from one ecosystem to another. Some may lose as much as 99%, while other more efficient systems may lose only 70%, but 90% is a good rule of thumb. This loss in energy content at the second and subsequent trophic levels is primarily due to the second law of thermodynamics. Think of any energy-converting machine; it probably releases a great deal of heat energy. For example, an automobile engine must have a cooling system to get rid of the heat energy produced. An incandescent lightbulb also produces large amounts of heat. Although living systems are somewhat different, they must follow the same energy rules.

In addition to the loss of energy as a result of the second law of thermodynamics, there is an additional loss involved in the capture and processing of food material by herbivores. Although herbivores don't need to chase their food, they do need to travel to where food is available, then gather, chew, digest, and metabolize it. All these processes require energy.

Just as the herbivore trophic level experiences a 90% loss in energy content, the higher trophic levels of primary carnivores, secondary carnivores, and tertiary carnivores also experience a reduction in the energy available to them. Figure 14.5 shows the flow of energy through an ecosystem. At each trophic level, the energy content decreases by about 90%.

The Pyramid of Numbers

Because it may be difficult to measure the amount of energy in any one trophic level of an ecosystem, people often use other methods to quantify the different trophic levels. One method is to simply count the number of organisms at each trophic level. This generally gives the same pyramid relationship, called a *pyramid of numbers* (figure 14.6). Obviously this is not a very good method to use if the organisms at the different trophic levels are of greatly differing sizes. For example, if you count all the small insects feeding on the leaves of one large tree, you would actually get an inverted pyramid.

The Pyramid of Biomass

Because of the size-difference problem, many people like to use biomass as a way of measuring ecosystems. **Biomass** is usually determined by collecting all the organisms at one trophic level and measuring their dry weight. This eliminates the size-difference problem because all the organisms at each trophic level are weighed. This *pyramid of biomass* also shows the typical 90% loss at each trophic level. Although a biomass pyramid is better than a pyramid of numbers in measuring some ecosystems, it has some shortcomings.

Some organisms tend to accumulate biomass over long periods of time, whereas others do not. Many trees live for hundreds of years; their primary consumers, insects, generally live only one year. Likewise, a whale is a long-lived animal, whereas its food organisms are relatively short-lived. Figure 14.7 shows two biomass pyramids.

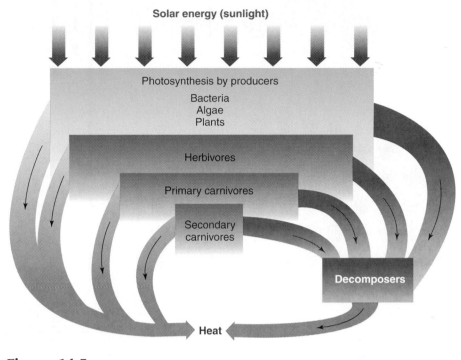

Figure 14.5

Energy Flow Through an Ecosystem
As energy flows from one trophic level to the next, approximately 90% of it is lost. This means that the amount of energy at the producer level must be 10 times larger than the amount of energy at the herbivore level.

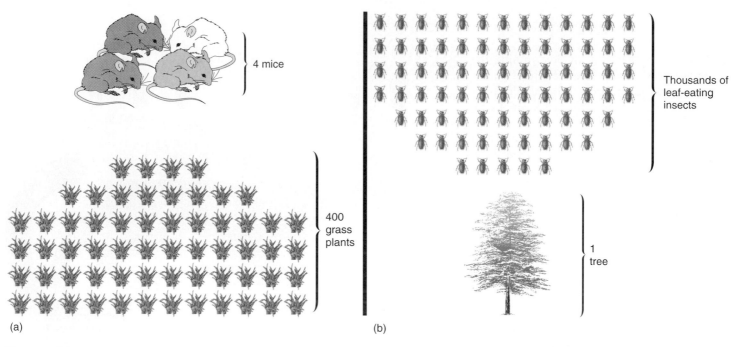

(a)

4 mice

400 grass plants

(b)

Thousands of leaf-eating insects

1 tree

Figure 14.6

A Pyramid of Numbers

One of the easiest ways to quantify the various trophic levels in an ecosystem is to count the number of individuals in a small portion of the ecosystem. As long as all the organisms are of similar size and live about the same length of time, this method gives a good picture of how different trophic levels are related. (a) The relationship between grass and mice is a good example. However, if the organisms at one trophic level are much larger or live much longer than those at other levels, our picture of the relationship may be distorted. (b) This is what happens when we look at the relationship between forest trees and the insects that feed on them. A pyramid of numbers becomes inverted in this instance.

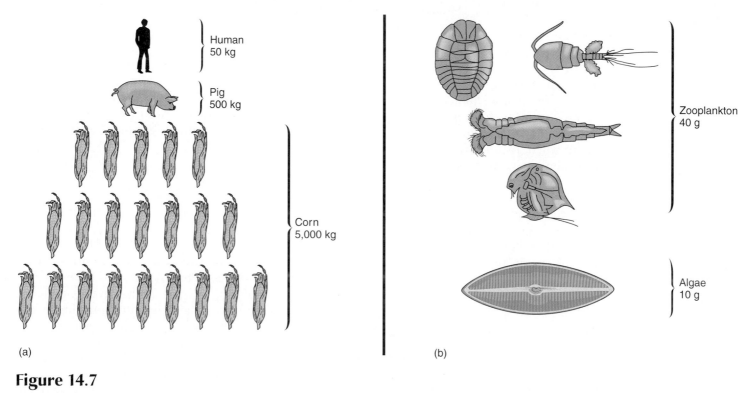

(a)

Human 50 kg

Pig 500 kg

Corn 5,000 kg

(b)

Zooplankton 40 g

Algae 10 g

Figure 14.7

A Pyramid of Biomass

Biomass is determined by collecting and weighing all the organisms in a small portion of an ecosystem. (a) This method of quantifying trophic levels eliminates the problem of different-sized organisms at different trophic levels. However, it does not always give a clear picture of the relationship between trophic levels if the organisms have widely different lengths of life. (b) For example, in aquatic ecosystems, many of the small producers may divide several times per day. The tiny animals (zooplankton) that feed on them live much longer and tend to accumulate biomass over time. The single-celled algae produce much more living material, but are eaten as fast as they are produced and so do not accumulate a large biomass.

14.4 Community Interactions

In the previous section we looked at ecological relationships from the point of view of ecosystems and the way energy flows through them. But we can also study relationships at the community level and focus on the kinds of interactions that take place among organisms (Outlooks 14.2).

As you know from the discussion in the previous section, one of the ways that organisms interact is by feeding on one another. A community includes many different food chains and many organisms may be involved in several of the food chains at the same time, so the food chains become interwoven into a **food web** (figure 14.8). In a community, the

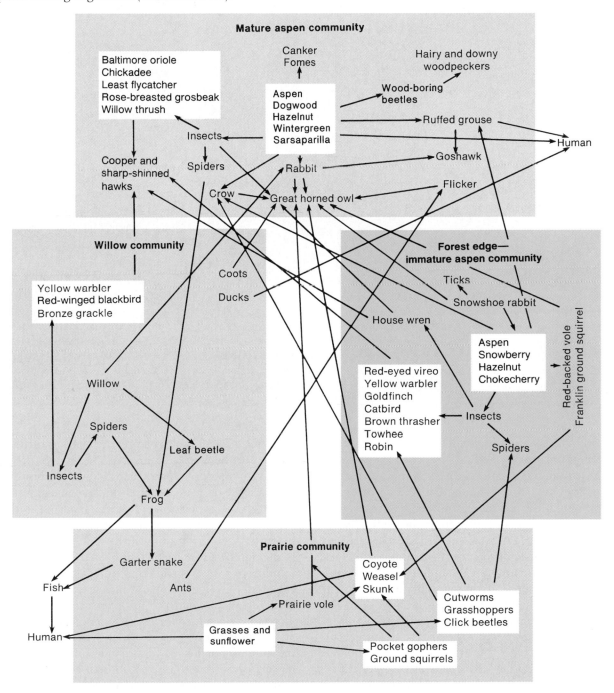

Figure 14.8

A Food Web

When many different food chains are interlocked with one another, a food web results. The arrows indicate the direction of energy flow. Notice that some organisms are a part of several food chains—the great horned owl in particular. Because of the interlocking nature of the food web, changing conditions may shift the way in which energy flows through this system.

From Ralph D. Bird, "Biotic Communities of the Aspen Parkland of Central Canada" in *Ecology*, 11(2): 410. Copyright © Ecological Society of America. Reprinted by permission.

OUTLOOKS 14.2

Zebra Mussels: Invaders from Europe

In the mid–1980s a clamlike organism called the zebra mussel, *Dressenia polymorpha,* was introduced into the waters of the Great Lakes. It probably arrived in the ballast water of a ship from Europe. Ballast water is pumped into empty ships to make them more stable when crossing the ocean. Immature stages of the zebra mussel were probably emptied into Lake St. Clair, near Detroit, Michigan, when the ship discharged its ballast water to take on cargo. This organism has since spread to many areas of the Great Lakes and smaller inland lakes. It has also been discovered in other parts of the United States including the mouth of the Mississippi River. Zebra mussels attach to any hard surface and reproduce rapidly. Densities of more than 20,000 individuals per square meter have been documented in Lake Erie.

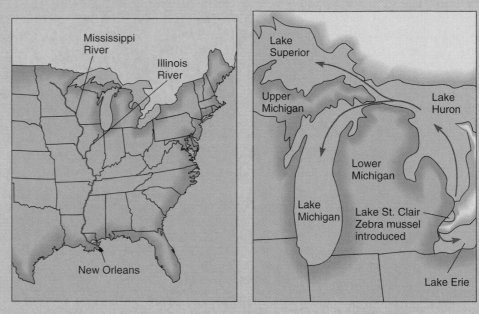

These invaders are of concern for several reasons. First, they coat the intake pipes of municipal water plants and other facilities requiring expensive measures to clean the pipes. Second, they coat any suitable surface, preventing native organisms from using the space. Third, they introduce a new organism into the food chain. Zebra mussels filter small aquatic organisms from the water very efficiently and may remove food organisms required by native species. Their filtering activity has significantly increased the clarity of the water in several areas in the Great Lakes. This can affect the kinds of fish present, because greater clarity allows predator fish to find prey more easily. In addition, the clarity of the water has allowed for greater production of aquatic vegetation and algae since light better penetrates the water to a greater depth. There is concern that zebra mussels have significantly changed the ecological organization of the Great Lakes.

---

interacting food chains usually result in a relatively stable combination of populations.

Although communities are relatively stable we need to recognize that they are also dynamic collections of organisms: As one population increases, another decreases. This might occur over several years, or even in the period of one year. This happens because factors that affect ecosystems are not constant. There may be differences in rainfall throughout the year or changes in the amount of sunlight and in the average temperature. We should expect populations to fluctuate as abiotic factors change. A change in the size of one population will trigger changes in other populations as well. Figure 14.9 shows what happens to the size of a population of deer as the seasons change. The area can support 100 deer from January through February, when plant food for deer is least available. As spring arrives, plant growth increases. It is no accident that deer breed in the fall and give birth in the spring. During the spring producers are increasing, and the area has more available food to support a large deer population. It is also no accident that wolves and other carnivores that feed on deer give birth in the spring. The increased available energy in the form of plants (producers) means

more food for deer (herbivores), which, in turn, means more energy for the wolves (carnivores) at the next trophic level.

If numbers of a particular kind of organism in a community increase or decrease significantly, some adjustment usually occurs in the populations of other organisms within the community. For example, the populations of many kinds of small mammals fluctuate from year to year. This results in changes in the numbers of their predators or the predators must switch to other prey species and impact other parts of the community. As another example, humans have used insecticides to control the populations of many kinds of insects. Reduced insect populations may result in lower numbers of insect-eating birds and affect the predators that use these birds as food. Furthermore the indiscriminate use of insecticides often increases the populations of herbivorous, pest insects because insecticides kill many beneficial predator insects that normally feed on the pest, rather than just the one or two target pest species.

Because communities are complex and interrelated, it is helpful if we set artificial boundaries that allow us to focus our study on a definite collection of organisms. An example of a community with easily determined natural boundaries is a small pond (figure 14.10). The water's edge naturally

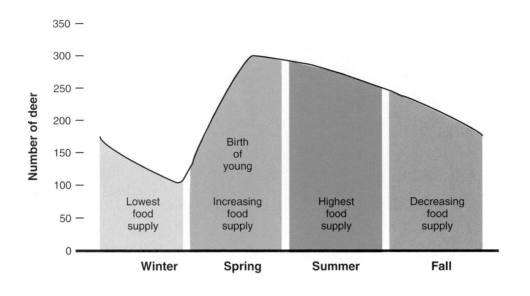

Figure 14.9

Annual Changes in Population Size

The number of organisms living in an area varies during the year. The availability of food is the primary factor determining the size of the population of deer in this illustration, but water availability, availability of soil nutrients, and other factors could also be important.

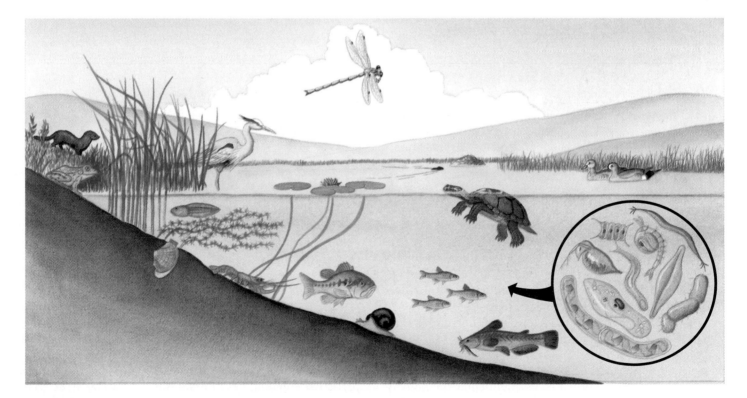

Figure 14.10

A Pond Community

Although a pond would seem to be an easy community to characterize, it interacts extensively with the surrounding land-based communities. Some of the organisms associated with a pond community are always present in the water (fish, pondweeds, clams); others occasionally venture from the water to the surrounding land (frogs, dragonflies, turtles, muskrats); still others are occasional or rare visitors (minks, heron, ducks).

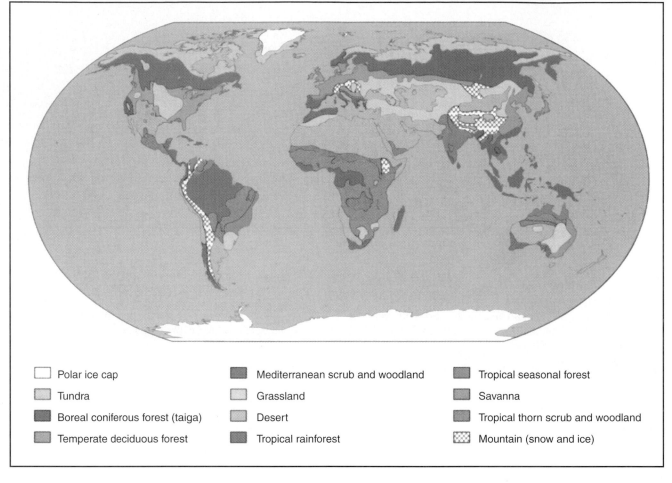

Figure 14.11

Biomes of the World
Major climatic differences determine the kind of vegetation that can live in a region of the world. Associated with specialized groups of plants are particular kinds of animals. These regional ecosystems are called biomes.

☐ Polar ice cap	◼ Mediterranean scrub and woodland	◼ Tropical seasonal forest
☐ Tundra	☐ Grassland	◼ Savanna
◼ Boreal coniferous forest (taiga)	☐ Desert	◼ Tropical thorn scrub and woodland
◼ Temperate deciduous forest	◼ Tropical rainforest	▩ Mountain (snow and ice)

defines the limits of this community. You would expect to find certain animals and plants living in the pond, such as fish, frogs, snails, insects, algae, pondweeds, bacteria, and fungi. But you might ask at this point, What about the plants and animals that live right at the water's edge? That leads us to think about the animals that spend only part of their lives in the water. That awkward-looking, long-legged bird wading in the shallows and darting its long beak down to spear a fish has its nest atop some tall trees away from the water. Should it be considered part of the pond community? Should we also include the deer that comes to drink at dusk and then wanders away? Small parasites could enter the body of the deer as it drinks. The immature parasite will develop into an adult within the deer's body. That same parasite must spend part of its life cycle in the body of a certain snail. Are these parasites part of the pond community? Several animals are members of more than one community. What originally seemed to be a clear example of a community has become

less clear-cut. Although the general outlines of a community can be arbitrarily set for the purposes of a study, we must realize that the boundaries of a community, or any ecosystem for that matter, must be considered somewhat artificial.

14.5 Types of Communities

Ponds and other small communities are parts of large regional terrestrial communities known as *biomes*. **Biomes** are particular communities of organisms that are adapted to particular climate conditions. The primary climatic factors that determine the kinds of organisms that can live in an area are the amount and pattern of *precipitation* and the *temperature* ranges typical for the region. The map in figure 14.11 shows the distribution of the major biomes of the world. Each biome can be characterized by specific climate conditions,

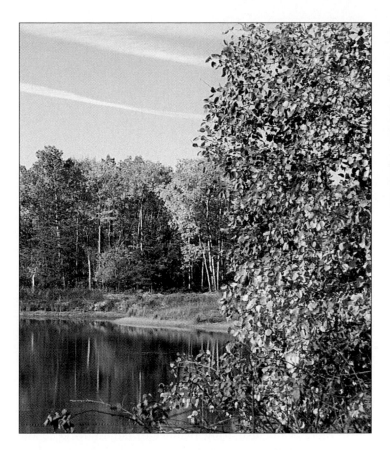

Figure 14.12

Temperate Deciduous Forest Biome
This kind of biome is found in parts of the world that have significant rainfall (75–130 centimeters) and cold weather for a significant part of the year when the trees are without leaves.

particular kinds of organisms, and characteristic activities of the organisms of the region.

Temperate Deciduous Forest

The *temperate deciduous forest* covers a large area from the Mississippi River to the Atlantic Coast, and from Florida to southern Canada. This type of biome is also found in parts of Europe and Asia. Temperate deciduous forests exist in parts of the world that have moderate rainfall (75–130 centimeters per year) spread over the entire year and a relatively long summer growing season (130–260 days without frost). This biome, like other land-based biomes, is named for a major feature of the ecosystem, which in this case happens to be the dominant vegetation. The predominant plants are large trees that lose their leaves more or less completely during the fall of the year and are therefore called deciduous (figure 14.12). The trees typical of this biome are adapted to conditions with significant precipitation and short mild winters. Since the trees are the major producers and new leaves are produced each spring, a major category of primary con-

Figure 14.13

Grassland (Prairie) Biome
This typical short-grass prairie of western North America is associated with an annual rainfall of 30 to 85 centimeters. This community contains a unique grouping of plant and animal species.

sumers in this biome consists of leaf-eating insects. These insects then become food for a variety of birds that typically raise their young in the forest during the summer and migrate to more moderate climates in the fall. Many other animals like squirrels, some birds, and deer use the fruits of the trees as food. Carnivores such as foxes, hawks, and owls eat many of the small mammals and birds typical of the region. Another feature typical of the temperate deciduous forest is an abundance of spring woodland wildflowers that emerge early in the spring before the trees have leafed out. Of course, because the region is so large and has somewhat different climatic conditions in various areas, we can find some differences in the particular species of trees (and other organisms) in this biome. For instance, in Maryland the tulip tree is one of the state's common large trees, while in Michigan it is so unusual that people plant it in lawns and parks as a decorative tree. Aspen, birch, cottonwood, oak, hickory, beech, and maple are typical trees found in this geographic region. Typical animals of this biome are many kinds of leaf-eating insects, wood-boring beetles, migratory birds, skunks, porcupines, deer, frogs, opossums, owls, and mosquitoes. In much of this region, the natural vegetation has been removed to allow for agriculture, so the original character of the biome is gone except where farming is not practical or the original forest has been preserved.

Grassland

The biome located to the west of the temperate deciduous forest in North America is the *grassland* or *prairie* biome (figure 14.13). This kind of biome is also common in parts of Eurasia, Africa, Australia, and South America. The rainfall (30–85 centimeters per year) in grasslands is not adequate to support the growth of trees and the dominant vegetation consists of various species of grasses. It is typical to have long periods during the year when there is no rainfall. Trees are common in this biome only along streams where they can obtain sufficient water. Interspersed among the grasses are many kinds of prairie wildflowers. The dominant animals are those that use grasses as food; large grazing mammals (bison and pronghorn antelope); small insects (grasshoppers and ants); and rodents (mice and prairie dogs). A variety of carnivores (meadowlarks, coyotes, and snakes) feed on the herbivores. Most of the species of birds are seasonal visitors to the prairie. At one time fire was a common feature of the prairie during the dry part of the year.

Today most of the original grasslands, like the temperate deciduous forest, have been converted to agricultural uses. Breaking the sod (the thick layer of grass roots) so that wheat, corn, and other grains can be grown exposes the soil to the wind, which may cause excessive drying and result in soil erosion that depletes the fertility of the soil. Grasslands that are too dry to allow for farming typically have been used as grazing land for cattle and sheep. The grazing of these domesticated animals has modified the natural vegetation as has farming in the moister grassland regions.

Savanna

A biome that is similar to a prairie is a *savanna* (figure 14.14). Savannas are tropical biomes of central Africa, Northern Australia, and parts of South America that have distinct wet and dry seasons. Although these regions may receive 100 centimeters of rainfall per year there is an extended dry season of three months or more. Because of the extended period of dryness the dominant vegetation consists of grasses. In addition, a few thorny, widely spaced drought-resistant trees dot the landscape. Many kinds of grazing mammals are found in this biome—various species of antelope, wildebeest, and zebras in Africa; various kinds of kangaroos in Australia; and a large rodent, the capybara, in South America. Another animal typical of the savanna is the termite, a colonial insect that typically builds mounds above ground.

During the wet part of the season the trees produce leaves, the grass grows rapidly, and most of the animals raise their young. In the African savanna, seasonal migrations of the grazing animals is typical. Many of these tropical grasslands have been converted to grazing for cattle and other domesticated animals.

Desert

Very dry areas are known as *deserts* and are found throughout the world wherever rainfall is low and irregular. Typically the rainfall is less than 25 centimeters per year. Some deserts are extremely hot; others can be quite cold during much of the year. The distinguishing characteristic of desert

Figure 14.14

Savanna Biome
A savanna is likely to develop in areas that have a rainy season and a dry season. During the dry season, fires are frequent. The fires kill tree seedlings and prevent the establishment of forests.

biomes is low rainfall, not high temperature. Furthermore, deserts show large daily fluctuations in air temperature. When the Sun goes down at night, the land cools off very rapidly because there is no insulating blanket of clouds to keep the heat from radiating into space.

A desert biome is characterized by scattered, thorny plants that lack leaves or have reduced leaves (figure 14.15). Because leaves tend to lose water rapidly, the lack of leaves is an adaptation to dry conditions. Under these conditions the stems are green and carry on photosynthesis. Many of the plants, like cacti, are capable of storing water in their fleshy stems. Others store water in their roots. Although this is a very harsh environment, many kinds of flowering plants, insects, reptiles, and mammals can live in this biome. The animals usually avoid the hottest part of the day by staying in burrows or other shaded, cool areas. Staying underground or in the shade also allows the animal to conserve water.

There are also many annual plants but the seeds only germinate and grow following the infrequent rainstorms. When it does rain the desert blooms.

Boreal Coniferous Forest

Through parts of southern Canada, extending southward along the Appalachian and Rocky Mountains of the United States, and in much of northern Europe and Asia we find communities that are dominated by evergreen trees. This is the *taiga, boreal coniferous forest, or northern coniferous forest biome* (figure 14.16). The evergreen trees are especially adapted to withstand long, cold winters with abundant snowfall. Typically the growing season is less than 120 days and rainfall ranges between 40 and 100 centimeters per year. However, because of the low average temperature, evaporation is

Figure 14.15

Desert Biome

The desert gets less than 25 centimeters of precipitation per year, but it contains many kinds of living things. Cacti, sagebrush, lichens, snakes, small mammals, birds, and insects inhabit the desert. All deserts are dry, and the plants and animals show adaptations that allow them to survive under these extreme conditions. In hot deserts where daytime temperatures are high, most animals are active only at night when the air temperature drops significantly.

low and the climate is humid. Most of the trees in the wetter, colder areas are spruces and firs, but some drier, warmer areas have pines. The wetter areas generally have dense stands of small trees intermingled with many other kinds of vegetation and many small lakes and bogs. In the mountains of the western United States, pine trees are often widely scattered and very large, with few branches near the ground. The area has a park-like appearance because there is very little vegetation on the forest floor. Characteristic animals in this biome include mice, snowshoe hare, lynx, bears, wolves, squirrels, moose, midges, and flies. These animals can be divided into four general categories: those that become dormant in winter (insects and bears); those that are specially adapted to withstand the severe winters (snowshoe hare, lynx); those that live in protected areas (mice under the snow); and those that migrate south in the fall (most birds).

Temperate Rainforest

The coastal areas of northern California, Oregon, Washington, British Columbia, and southern Alaska contain an unusual set of environmental conditions that support a *temperate rainforest*. The prevailing winds from the west bring moisture-laden air to the coast. As the air meets the coastal mountains and is forced to rise, it cools and the moisture falls as rain or snow. Most of these areas receive 200 centimeters or more precipitation per year. This abun-

Figure 14.16

Boreal Coniferous Forest Biome

Conifers are the dominant vegetation in most of Canada, in a major part of Russia, and at high elevations in sections of western North America. The boreal coniferous forest biome is characterized by cold winters with abundant snowfall.

dance of water, along with fertile soil and mild temperatures, results in a lush growth of plants.

Sitka spruce, Douglas fir, and western hemlock are typical evergreen coniferous trees in the temperate rainforest. Undisturbed (old growth) forests of this region have trees as old as 800 years that are nearly 100 meters tall. Deciduous trees of various kinds (red alder, big leaf maple, black cottonwood) also exist in open areas where they can get enough light. All trees are covered with mosses, ferns, and other plants that grow on the surface of the trees. The dominant color is green because most surfaces have something photosynthetic growing on them.

When a tree dies and falls to the ground it rots in place and often serves as a site for the establishment of new trees. This is such a common feature of the forest that the fallen, rotting trees are called nurse trees. The fallen tree also serves as a food source for a variety of insects, which are food for a variety of larger animals.

Because of the rich resource of trees, 90% of the original temperate rainforest has already been logged. Many of the remaining areas have been protected because they are home to the endangered northern spotted owl and marbled murrelet (a seabird).

Tundra

North of the coniferous forest biome is an area known as the *tundra* (figure 14.17). It is characterized by extremely long, severe winters and short, cool summers. The growing season is less than 100 days and even during the short summer the nighttime temperatures approach 0°C. Rainfall is low

Figure 14.17

Tundra Biome
The tundra biome is located in northern parts of North America and Eurasia. It is characterized by short, cool summers and long, extremely cold winters. There is a layer of soil below the surface that remains permanently frozen; consequently, no large trees exist in this biome. Relatively few kinds of plants and animals can survive this harsh environment.

(10–25 centimeters per year). The deeper layers of the soil remain permanently frozen, forming a layer called the *permafrost*. Because the deeper layers of the soil are frozen, when the surface thaws the water forms puddles on the surface. Under these conditions of low temperature and short growing season, very few kinds of animals and plants can survive. No trees can live in this region. Typical plants and animals of the area are grasses, sedges, dwarf willow, and some other shrubs, reindeer moss (actually a lichen), caribou, wolves, musk oxen, fox, snowy owls, mice, and many kinds of insects. Many kinds of birds are summer residents only. The tundra community is relatively simple, so any changes may have drastic and long-lasting effects. The tundra is easy to injure and slow to heal; therefore we must treat it gently. The construction of the Alaskan pipeline has left scars that could still be there 100 years from now.

Tropical Rainforest

The *tropical rainforest* is at the other end of the climate spectrum from the tundra. Tropical rainforests are found primarily near the equator in Central and South America, Africa, parts of southern Asia, and some Pacific Islands (figure 14.18). The temperature is high (averaging about 27°C), rain falls nearly every day (typically 200–1,000 centimeters per year), and there

Figure 14.18

Tropical Rainforest Biome
The tropical rainforest is a moist, warm region of the world located near the equator. The growth of vegetation is extremely rapid. There are more kinds of plants and animals in this biome than in any other.

are thousands of species of plants in a small area. Balsa (a very light wood), teak (used in furniture), and ferns the size of trees are examples of plants from the tropical rainforest. Typically, every plant has other plants growing on it. Tree trunks are likely to be covered with orchids, many kinds of vines, and mosses. Tree frogs, bats, lizards, birds, monkeys, and an almost infinite variety of insects inhabit the rainforest. These forests are very dense, and little sunlight reaches the forest floor. When the forest is opened up (by a hurricane or the death of a large tree) and sunlight reaches the forest floor, the opened area is rapidly overgrown with vegetation.

Because plants grow so quickly in these forests, people assume the soils are fertile, and many attempts have been made to bring this land under cultivation. In reality, the soils are poor in nutrients. The nutrients are in the organisms, and as soon as an organism dies and decomposes its nutrients are reabsorbed by other organisms. Typical North American agricultural methods, which require the clearing of large areas, cannot be used with the soil and rainfall conditions of the tropical rainforest. The constant rain falling on these fields quickly removes the soil's nutrients so that heavy applications

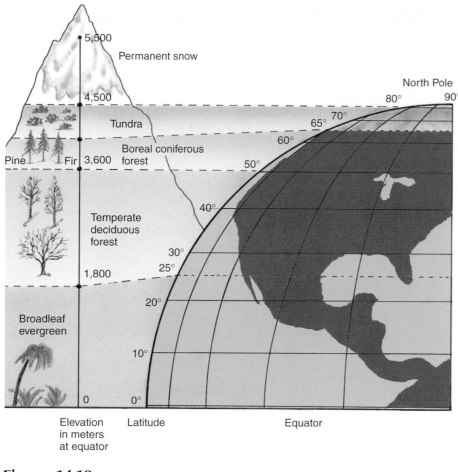

Figure 14.19

Relationship Between Elevation, Latitude, and Vegetation
As one travels up a mountain, the climate changes. The higher the elevation, the cooler the climate. Even in the tropics tall mountains can have snow on the top. Thus, it is possible to experience the same change in vegetation by traveling up a mountain as one would experience traveling from the equator to the North Pole.

of fertilizer are required. Often these soils become hardened when exposed in this way. Although most of these forests are not suitable for agriculture, large expanses of tropical rainforest are being cleared yearly because of the pressure for more farmland in the highly populated tropical countries and the desire for high-quality lumber from many of the forest trees.

The Relationship Between Elevation and Climate

The distribution of terrestrial ecosystems is primarily related to temperature and precipitation. Air temperatures are warmest near the equator and become cooler as the poles are approached. Similarly, air temperature decreases as elevation increases. This means that even at the equator it is possible to have cold temperatures on the peaks of tall mountains. Therefore, as one proceeds from sea level to the tops of mountains, it is possible to pass through a series of biomes that are similar to what one would encounter traveling from the equator to the North Pole (figure 14.19).

14.6 Succession

Each of the communities we have just discussed is relatively stable over long periods of time. A relatively stable, long-lasting community is called a **climax community** (How Science Works 14.1). The word *climax* implies the final step in a series of events. That is just what the word means in this context because communities can go through a series of predictable, temporary stages that eventually result in a long-lasting stable community. The process of changing from one type of community to another is called **succession,** and each intermediate stage leading to the climax community is known as a **successional stage** or **successional community.**

Two different kinds of succession are recognized: **primary succession,** in which a community of plants and animals develops where none existed previously, and **secondary succession,** in which a community of organisms is disturbed by a natural or human-related event (e.g., hurricane, volcano, fire, forest harvest) and returned to a previous stage in the succession. Primary succession is much more difficult to observe than secondary succession because there are relatively few places on earth that lack communities of organisms. The tops of mountains, newly formed volcanic rock, and rock newly exposed by erosion or glaciers can be said to lack life. However, bacteria, algae, fungi, and lichens quickly begin to grow on the bare rock surface, and the process of succession has begun. The first organisms to colonize an area are often referred to as **pioneer organisms,** and the community is called a **pioneer community.**

Lichens are frequently important in pioneer communities. They are unusual organisms that consist of a combination of algae cells and fungi cells—a combination that is very hardy and is able to grow on the surface of bare rock (figure 14.20). Because algae cells are present, the lichen is capable of photosynthesis and can form new organic matter. Furthermore, many tiny consumer organisms can make use of the lichens as a source of food and a sheltered place to live. The action of the lichens also tends to break down the rock surface upon which they grow. This fragmentation of rock by lichens is aided by the physical weathering processes of freezing and thawing, dissolution by water, and wind erosion. Lichens also trap dust particles, small rock particles, and the dead remains of lichens and other organisms that live in and on them. These processes of breaking down rock and trapping particles result in the formation of a thin layer of soil.

HOW SCIENCE WORKS 14.1

The Changing Nature of the Climax Concept

When European explorers traveled across the North American continent they saw huge expanses of land covered by the same kinds of organisms. Deciduous forests in the East, coniferous forests in the North, grasslands in central North America, and deserts in the Southwest. These collections came to be considered the steady-state or normal situation for those parts of the world. When ecologists began to explore the way in which ecosystems developed over time they began to think of these ecosystems as the end point or climax of a long journey beginning with the formation of soil and its colonization by a variety of plants and other organisms.

As settlers removed the original forests or grasslands and converted the land to farming, the original "climax" community was replaced with an agricultural ecosystem. Eventually, as poor farming practices depleted the soil, the farms were abandoned and the land was allowed to return to its "original" condition. This secondary succession often resulted in forests or grasslands that resembled those that had been destroyed. However, in most cases these successional ecosystems contained fewer species and in some cases were entirely different kinds of communities from the originals.

As ecologists recognized that there was not a fixed, predetermined community for each part of the world, they began to modify the way they looked at the concept of climax communities. The concept today is a more plastic one. The term *climax* is still used to talk about a stable stage following a period of change, but ecologists no longer believe that land will eventually return to a "preordained" climax condition. They have also recognized in recent years that the type of climax community that develops depends on many factors other than simply climate. One of these is the availability of seeds to colonize new areas. Two areas with very similar climate and soil characteristics may contain different species because of the seeds available when the lands were released from agriculture. Furthermore, we need to recognize that the only thing that differentiates a "climax" community from a successional one is the time scale over which change occurs. "Climax" communities do not change as rapidly as successional ones. However all communities are eventually replaced, as were the swamps that produced coal deposits, the preglacial forests of Europe and North America, and the pine forests of the northeastern United States.

So what should we do with this concept? Although the climax concept embraces a false notion that there is a specific end point to succession, it is still important to recognize that there is a predictable pattern of change during succession and that later stages in succession are more stable and longer lasting than early stages. Whether we call it a climax community is not really important.

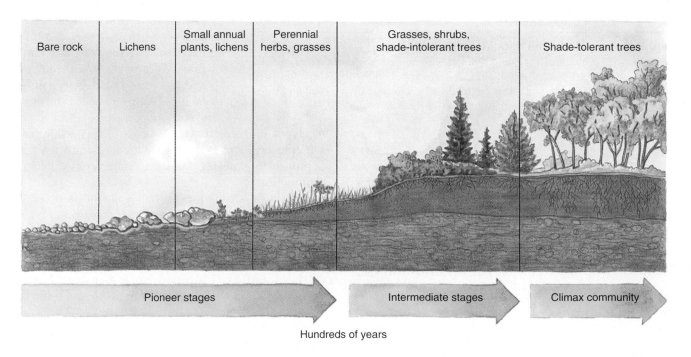

Figure 14.20

Primary Succession
The formation of soil is a major step in primary succession. Until soil is formed, the area is unable to support large amounts of vegetation. The vegetation modifies the harsh environment and increases the amount of organic matter that can build up in the area. The presence of plants eliminates the earlier pioneer stages of succession. If given enough time, a climax community may develop.

Figure 14.21

Succession from a Pond to a Wet Meadow
A shallow pond will slowly fill with organic matter from producers in the pond. Eventually, a floating mat will form over the pond and grasses will become established. In many areas this will be succeeded by a climax forest.

As the soil layer becomes thicker, small plants such as mosses may become established, increasing the rate at which energy is trapped and adding more organic matter to the soil. Eventually, the soil may be able to support larger plants that are even more efficient at trapping sunlight, and the soil-building process continues at a more rapid pace. Associated with the producers in each successional stage are a variety of small animals, fungi, and bacteria. Each change in the community makes it more difficult for the previous group of organisms to maintain itself. Tall plants shade the smaller ones they replaced; consequently, the smaller organisms become less common, and some may disappear entirely. Only shade-tolerant species will be able to grow and compete successfully in the shade of the taller plants. As this takes place we can recognize that one stage has succeeded the other.

Depending on the physical environment and the availability of new colonizing species, succession from this point can lead to different kinds of climax communities. If the area is dry, it might stop at a grassland stage. If it is cold and wet, a coniferous forest might be the climax community. If it is warm and wet, it may be a tropical rainforest. The rate at which this successional process takes place is variable. In some warm, moist, fertile areas the entire process might take place in less than 100 years. In harsh environments, like mountaintops or very dry areas, it may take thousands of years.

Primary succession can also be observed in the progression from an aquatic community to a terrestrial community. Lakes, ponds, and slow-moving parts of rivers accumulate or-

ganic matter. Where the water is shallow, this organic matter supports the development of rooted plants. In deeper water, we find only floating plants like water lilies that send their roots down to the mucky bottom. In shallower water, upright rooted plants like cattails and rushes develop. As the plants contribute more organic matter to the bottom, the water level becomes more shallow. Eventually, a mat of mosses, grasses, and even small trees may develop on the surface along the edge of the water. If this continues for perhaps 100 to 200 years, an entire pond or lake will become filled in. More organic matter accumulates because of the large number of producers and because the depression that was originally filled with water becomes drier. This will usually result in a wet grassland, which in many areas will be replaced by the climax forest community typical of the area (figure 14.21).

Secondary succession occurs when a climax community or one of the successional stages leading to it is changed to an earlier stage. For example, when land is converted to agriculture the original climax vegetation is removed. When agricultural land is abandoned it returns to something like the original climax community. One obvious difference between primary succession and secondary succession is that in the latter there is no need to develop a soil layer. Another difference is that there is likely to be a reservoir of seeds from plants that were part of the original climax community. The seeds may have existed for years in a dormant state or they may be transported to the disturbed site from undisturbed sites that still hold the species typical of the climax

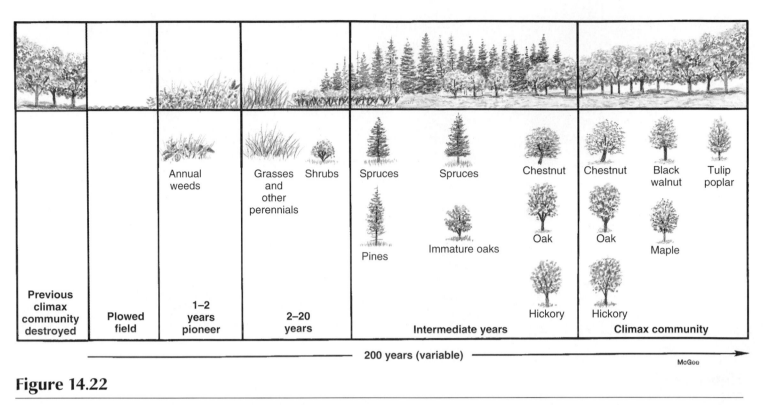

Figure 14.22

Secondary Succession on Land

A plowed field in the southeastern United States shows a parade of changes over time involving plant and animal associations. The general pattern is for annual weeds to be replaced by grasses and other perennial herbs, which are replaced by shrubs, which are replaced by trees. As the plant species change, so do the animal species.

community for the region. If we begin with bare soil the first year, it is likely to be invaded by a pioneer community of weed species that are annual plants. Within a year or two, perennial plants like grasses become established. Because most of the weed species need bare soil for seed germination, they are replaced by the perennial grasses and other plants that live in association with grasses. The more permanent grassland community is able to support more insects, small mammals, and birds than the weed community could. If rainfall is low, succession is likely to stop at this grassland stage. If rainfall is adequate, several species of shrubs and fast-growing trees that require lots of sunlight (e.g., birch, aspen, juniper, hawthorn, sumac, pine, spruce, and dogwood) will become common. As the trees become larger, the grasses fail to get sufficient sunlight and die out. Eventually, shade-tolerant species of trees (e.g., beech, maple, hickory, oak, hemlock, and cedar) will replace the shade-intolerant species, and a climax community results (figure 14.22).

14.7 Human Use of Ecosystems

The extent to which humans use an ecosystem is often tied to its productivity. **Productivity** is the rate at which an ecosystem can accumulate new organic matter. Because plants are the producers, it is their activities that are most

important. Ecosystems in which conditions are most favorable for plant growth are the most productive. Warm, moist, sunny areas with high levels of nutrients in the soil are ideal. Some areas have low productivity because one of the essential factors is missing. Deserts have low productivity because water is scarce, arctic areas because temperature is low, and the open ocean because nutrients are in short supply. Some communities, such as coral reefs and tropical rainforests, have high productivity. Marshes and estuaries are especially productive because the waters running into them are rich in the nutrients that aquatic photosynthesizers need. Furthermore, these aquatic systems are usually shallow so that light can penetrate through most of the water column.

Converting Ecosystems to Human Use

Humans have been able to make use of naturally productive ecosystems by harvesting food from them. However, in most cases, we have altered certain ecosystems substantially to increase productivity for our own purposes. In so doing, we have destroyed the original ecosystem and replaced it with an agricultural ecosystem. For example, nearly all of the Great Plains region of North America has been converted to agriculture. The original ecosystem included the Native Americans who used buffalo as a source of food. There was much grass, many buffalo, and few humans. Therefore, in the Native Americans' pyramid of energy, the base was more than ample. However, with the exploitation and settling of America, the population in North America increased at a rapid rate. The top of the pyramid became larger. The food chain (prairie grass—buffalo—human) could no longer supply the food needs of the growing population. As the top of the pyramid grew, it became necessary for the producer base to grow larger. Because wheat and corn yield more biomass for humans than the original prairie grasses could, the settlers' domestic grain and cattle replaced the prairie grass and buffalo. This was fine for the settlers, but devastating for the buffalo and Native Americans.

In similar fashion the deciduous forests of the East were cut down and burned to provide land for crops. The crops were able to provide more food than did harvesting game and plants from the forest.

Forestry practices often seek to simplify the forest by planting single-species forests of the same age. This certainly makes management and harvest practices easier and more efficient, but these kinds of communities do not contain the variety of plants, animals, fungi, and other organisms typically found in natural ecosystems.

Many biomes, particularly the drier grasslands, cannot support the raising of crops. However, they can still be used as grazing land to raise livestock. Like the raising of crops, grazing often significantly alters the original grassland ecosystem. Some attempts have been made to harvest native species of animals from grasslands, but the species primarily involved are domesticated cattle, sheep, and goats. The substitution of the domesticated animals displaces the animals that are native to the area and also alters the plant community, particularly if too many animals are allowed to graze.

Even aquatic ecosystems have been significantly altered by human activity. Overfishing of many areas of the ocean has resulted in the loss of some important commercial species. For example, the codfishing industry along the east coast of North America has been destroyed by overfishing. Pacific salmon species are also heavily fished and disagreements among the countries that exploit these species may cause the decline of this fishery as well.

Since most agricultural use of ecosystems involves replacing the natural climax community with an artificial early successional stage, it requires considerable effort on our part to prevent succession to the original climax community. This is certainly true if remnants of the original natural community are still locally available to colonize agricultural land. Small woodlots in agricultural areas of the eastern United States serve this purpose. Much of the work and expense of farming (tilling, herbicides) is necessary to prevent succession to the natural climax community. It takes a lot of energy to fight nature.

Human-constructed lakes or farm ponds often have weed problems because they are shallow and provide ideal conditions for the normal successional processes that lead to their being filled in. Often we do not recognize what a powerful force succession is.

The Energy Pyramid and Human Nutrition

Anywhere in the world where the human population increases, natural ecosystems are replaced with agricultural ecosystems. In many parts of the world, the human demand for food is so large that it can be met only if humans occupy the herbivore trophic level rather than the carnivore trophic level. Humans are omnivores that can eat both plants and animals as food, so they have a choice. However, as the size of the human population increases, it cannot afford the 90% loss that occurs when plants are fed to animals that are in turn eaten by humans. In much of the less-developed world, the primary food is grain; therefore, the people are already at the herbivore level. It is only in the developed countries that people can afford to eat large quantities of meat. This is true from both an energy point of view and a monetary point of view. Figure 14.23 shows a pyramid of biomass having a producer base of 100 kilograms of grain. The second trophic level only has 10 kilograms of cattle because of the 90% loss typical when energy is transferred from one trophic level to the next (90% of the corn raised in the United States is used as cattle feed). The consumers at the third trophic level, humans in this case, experience a similar 90% loss. Therefore, only 1 kilogram of humans can be sustained by the two-step energy transfer. There has been a 99% loss in energy: 100 kilograms of grain are necessary to sustain 1 kilogram of humans.

Because much of the world's population is already feeding at the second trophic level, we cannot expect food production to increase to the extent that we could feed 10 times more people than exist today.

It is unlikely that most people will be able to fulfill all their nutritional needs by just eating grains. In addition to calories, people need a certain amount of protein in their diets and one of the best sources of protein is meat. Although protein is available from plants, the concentration is greater from animal sources. People in major parts of Africa and Asia have diets that are deficient in both calories and protein. These people have very little food, and what food they do have is mainly from plant sources. These are also the parts of the world where human population growth is most rapid. In other words, these people are poorly nourished and, as the population increases, they will probably experience greater calorie and protein deficiency. This example reveals that even when people live as consumers at the second trophic level, they may still not get enough food, and if they do, it may not have the protein necessary for good health. It is important to point out that there is currently enough food in the world to feed everyone, but it is not distributed equitably for a variety of reasons. The primary reasons for starvation are political and economic. Wars and civil unrest disrupt the normal food-raising process. People leave their homes and migrate to areas unfamiliar to them. Poor people and poor countries cannot afford to buy food from the countries that have a surplus.

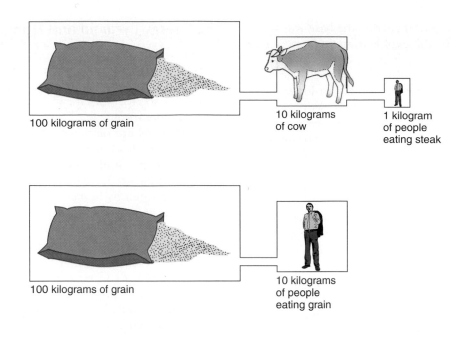

100 kilograms of grain

10 kilograms of cow

1 kilogram of people eating steak

100 kilograms of grain

10 kilograms of people eating grain

Figure 14.23

Human Biomass Pyramids

Because approximately 90% of the energy is lost as energy passes from one trophic level to the next, more people can be supported if they eat producers directly than if they feed on herbivores. Much of the less-developed world is in this position today. Rice, corn, wheat, and other producers provide the majority of food for the world's people.

SUMMARY

Ecology is the study of how organisms interact with their environment. The environment consists of biotic and abiotic components that are interrelated in an ecosystem. All ecosystems must have a constant input of energy from the Sun. Producer organisms are capable of trapping the Sun's energy and converting it into biomass. Herbivores feed on producers and are in turn eaten by carnivores, which may be eaten by other carnivores. Each level in the food chain is known as a trophic level. Other kinds of organisms involved in food chains are omnivores, which eat both plant and animal food, and decomposers, which break down dead organic matter and waste products.

All ecosystems have a large producer base with successively smaller amounts of energy at the herbivore, primary carnivore, and secondary carnivore trophic levels. This is because each time energy passes from one trophic level to the next, about 90% of the energy is lost from the ecosystem. A community consists of the interacting populations of organisms in an area. The organisms are interrelated in many ways in food chains that interlock to create food webs. Because of this interlocking, changes in one part of the community can have effects elsewhere.

Major land-based regional ecosystems are known as biomes. The temperate deciduous forest, boreal coniferous forest, tropical rainforest, grassland, desert, savanna, temperate rainforest, and tundra are examples of biomes. Ecosystems go through a series of predictable changes that lead to a relatively stable collection of plants and animals. This stable unit is called a climax community, and the process of change is called succession.

Humans use ecosystems to provide themselves with necessary food and raw materials. As the human population increases, most people will be living as herbivores at the second trophic level because they cannot afford to lose 90% of the energy by first feeding it to a herbivore, which they then eat. Humans have converted most productive ecosystems to agricultural production and continue to seek more agricultural land as population increases.

THINKING CRITICALLY

For guidelines to answer this question, visit the Online Learning Center.

Farmers are managers of ecosystems. Consider a cornfield in Iowa. Describe five ways in which the cornfield ecosystem differs from the original prairie it replaced. What trophic level does the farmer fill?

CONCEPT MAP TERMINOLOGY

Construct a concept map using these terms.

biome
carnivore
climax community
consumer
decomposer
food chain
food web

herbivore
pioneer organism
primary succession
producer
secondary succession
trophic level

Visit the Online Learning Center to use an interactive format to construct another concept map.

KEY TERMS

Use interactive flash cards, on the Online Learning Center, to help you learn the meanings of these terms.

abiotic factors
biomass
biomes
biotic factors
carnivores
climax community
community
consumers
decomposers
ecology
ecosystem
environment
food chain
food web
herbivores

omnivores
pioneer community
pioneer organisms
population
primary carnivores
primary consumers
primary succession
producers
productivity
secondary carnivores
secondary consumers
secondary succession
succession
successional community (stage)
trophic level

LEARNING CONNECTIONS

Visit the Online Learning Center to find additional study resources including those highlighted below.

14.1 Ecology and Environment

Animations and Review
• Introduction
Interactive Concept Map
• Ecology

1. List three abiotic factors in the environment of a fish.
2. List three biotic factors in the environment of a song bird.
3. What is the difference between the terms *ecosystem* and *environment?*

14.2 The Organization of Ecological Systems

Interactive Concept Map
• Food chains and trophic levels

4. Describe two ways that decomposers differ from herbivores.
5. Name an organism that would occupy each of the following trophic levels: the producer trophic level, the second trophic level, and the third trophic level.
6. What is the difference between an ecosystem and a community?

14.3 Characterizing Ecosystems: Energy, Numbers, Biomass

Interactive Concept Map
• Ecological pyramids
Animations and Review
• Introduction
• Energy flow

7. Describe the flow of energy through an ecosystem.
8. Why is an herbivore biomass usually larger than a carnivore biomass?
9. Can energy be recycled through an ecosystem? Explain why or why not.

14.4 Community Interactions

10. Why is a food web better than a food chain as a way to describe a community?
11. As the number of insects drops in the autumn, what would you expect to happen to bird populations?

14.5 Types of Communities

Animations and Review
• Introduction
• Climate
• Land biomes
• Aquatic systems

Interactive Concept Map
• Temperature and moisture

12. List a predominant abiotic factor in each of the following biomes: temperate deciduous forest, boreal coniferous forest, grassland, desert, tundra, temperate rainforest, tropical rainforest, and savanna.

13. List a dominant producer organism typical of each of the following biomes: temperate deciduous forest, boreal coniferous forest, grassland, desert, tundra, temperate rainforest, and savanna.

14.6 Succession

Animations and Review
• Succession
• Biodiversity

Experience This!
• Trophic levels in the market

14. How does primary succession differ from secondary succession?

15. How does a climax community differ from a successional community?

16. Describe the steps in secondary succession of abandoned farmland to a climax community.

17. Describe the steps in succession of a pond to a climax community.

14.7 Human Use of Ecosystems

18. Why must people who live in countries that are not able to produce surplus food eat grains?

19. Define the term *productivity*.

20. What impact does human use of ecosystems have on succession?

Test yourself with multiple choice questions that provide immediate feedback
on the Online Learning Center with PowerWeb at

www.mhhe.com/enger11

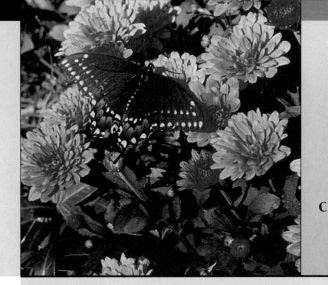

Community Interactions

GOALS AND OBJECTIVES

Understand that organisms interact in a variety of ways within a community.

❒ Describe differences among predation, mutualism, competition, parasitism, and commensalism.

❒ Explain the difference between niche and habitat.

❒ Describe an organism's niche, habitat, and community.

Understand that materials are cycled through organisms and the abiotic parts of ecosystems.

❒ Describe the flow of atoms through the carbon, nitrogen, phophorus, and hydrologic cycles.

❒ Explain why animals must eat.

❒ Describe the importance of bacteria in nutrient cycles.

❒ Explain why carbon and nitrogen must be recycled in ecosystems.

Appreciate that humans alter and interfere with natural ecological processes.

❒ Describe the impact of introduced species, predator control, and habitat destruction on natural communities.

❒ Explain why persistent organic chemicals reach higher levels in carnivores than in herbivores.

❒ Relate extinctions to human activities.

Evolution and Ecology

PART FOUR

15.1 Community, Habitat, and Niche

People approach the study of organism interactions in two major ways. Many people look at interrelationships from the broad ecosystem point of view; others focus on individual organisms and the specific things that affect them in their daily lives. The first approach involves the study of all the organisms that interact with one another—the community—and usually looks at general relationships among them. Chapter 14 described categories of organisms—producers, consumers, and decomposers—that perform different functions in a community.

Another way of looking at interrelationships is to study in detail the ecological relationships of certain species of organisms. Each organism has particular requirements for life and lives where the environment provides what it needs. The environmental requirements of a whale include large expanses of ocean, but with seasonally important feeding areas and protected locations used for giving birth. The kind of place, or part of an ecosystem, occupied by an organism is known as its **habitat.** Habitats are usually described in terms of conspicuous or particularly significant features in the area where the organism lives. For example, the habitat of a prairie dog is usually described as a grassland and the habitat of a tuna is described as the open ocean. The habitat of the fiddler crab is sandy ocean shores and the habitat of various kinds of cacti is the desert. The key thing to keep in mind when you think of habitat is the *place* in which a particular kind of organism lives. In our descriptions of the habitats of organisms, we sometimes use the terminology of the major biomes of the world, such as desert, grassland, or savanna, but it is also possible to describe the habitat of the bacterium *Escherichia coli* as the gut of humans and other mammals, or the habitat of a fungus as a rotting log. Organisms that have very specific places in which they live simply have more restricted habitats.

Each species has particular requirements for life and places specific demands on the habitat in which it lives. The specific functional role of an organism is its **niche.** Its niche is the way it goes about living its life. Just as the word *place* is the key to understanding the concept of habitat, the word *function* is the key to understanding the concept of a niche. To understand the niche of an organism involves a detailed understanding of the impacts an organism has on its biotic and abiotic surroundings as well as all the factors that affect the organism. For example, the niche of an earthworm includes abiotic items such as soil particle size; soil texture; and the moisture, pH, and temperature of the soil. The earthworm's niche also includes biotic impacts such as serving as food for birds, moles, and shrews; as bait for anglers; or as a consumer of dead plant organic matter (figure 15.1). In addition, an earthworm serves as a host for a variety of parasites, transports minerals and nutrients from deeper soil layers to the surface, incorporates organic matter into the soil, and creates burrows that allow air and water to penetrate the soil more easily. And this is only a limited sample of all the aspects of its niche.

Some organisms have rather broad niches; others, with very specialized requirements and limited roles to play, have niches that are quite narrow. The opossum (figure 15.2a) is an animal with a very broad niche. It eats a wide variety of plant and animal foods, can adjust to a wide variety of climates, is used as food by many kinds of carnivores (including humans), and produces large numbers of offspring. By contrast, the koala of Australia (figure 15.2b) has a very narrow niche. It can live only in areas of Australia with specific species of *Eucalyptus* trees because it eats the leaves of only a few kinds of these trees. Furthermore, it cannot tolerate low temperatures and does not produce large numbers of offspring. As you might guess, the opossum is expanding its range, and the koala is endangered in much of its range.

The complete description of an organism's niche involves a very detailed inventory of influences, activities, and impacts. It involves what the organism does and what is done to the organism. Some of the impacts are abiotic, others are biotic. Because the niche of an organism is a complex set of items, it is often easy to overlook important roles played by some organisms.

For example, when Europeans introduced cattle into Australia—a continent where there had previously been no large, hoofed mammals—they did not think about the impact of cow manure or the significance of a group of beetles called *dung beetles.* These beetles rapidly colonize fresh dung and cause it to be broken down. No such beetles existed in Australia; therefore, in areas where cattle were raised, a significant amount of land became covered with accumulated cow dung. This reduced the area where grass could grow and reduced productivity. The problem was eventually solved by the importation of several species of dung beetles from Africa, where large, hoofed mammals are common. The dung beetles made use of what the cattle did not digest, returning it to a form that plants could more easily recycle into plant biomass.

15.2 Kinds of Organism Interactions

One of the important components of an organism's niche is the other living things with which it interacts. When organisms encounter one another in their habitats, they can influence one another in numerous ways. Some interactions are harmful to one or both of the organisms. Others are beneficial. Ecologists have classified kinds of interactions between organisms into several broad categories, which we will discuss here.

Predation

Predation occurs when one animal captures, kills, and eats another animal. The organism that is killed is called the **prey,** and the one that does the killing is called the **predator.** The predator obviously benefits from the relationship; the prey organism is harmed. Most predators are relatively large

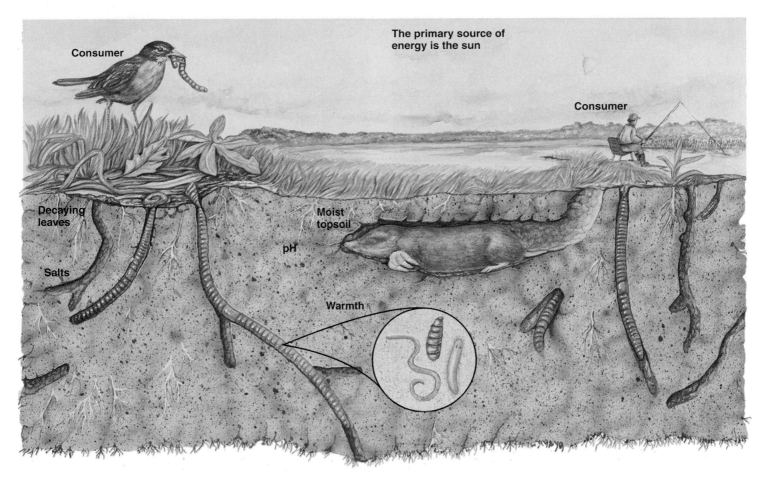

Figure 15.1

The Niche of an Earthworm

The niche of an earthworm involves a great many factors. It includes the fact that the earthworm is a consumer of dead organic matter, a source of food for other animals, a host to parasites, and bait for an angler. Furthermore, that the earthworm loosens the soil by its burrowing and "plows" the soil when it deposits materials on the surface are other factors. Additionally, the pH, texture, and moisture content of the soil have an impact on the earthworm. Keep in mind that this is but a small part of what the niche of the earthworm includes.

(a)

(b)

Figure 15.2

Broad and Narrow Niches

(a) The opossum has a very broad niche. It eats a variety of foods, is able to live in a variety of habitats, and has a large reproductive capacity. It is generally extending its range in the United States.
(b) The koala has a narrow niche. It feeds on the leaves of only a few species of *Eucalyptus* trees, is restricted to relatively warm, forested areas, and is generally endangered in much of its habitat.

(a)   (b)

Figure 15.3

The Predator-Prey Relationship
(*a*) Many predators capture prey by making use of speed. The cheetah can reach estimated speeds of 100 kilometers per hour (about 60 mph) during sprints to capture its prey. (*b*) Other predators, like this veiled chameleon, blend in with their surroundings, lie in wait, and ambush their prey. Because strength is needed to kill the prey, the predator is generally larger than the prey. Obviously, predators benefit from the food they obtain to the detriment of the prey organism.

compared to their prey and have specific adaptations that aid them in catching prey. Many spiders build webs that serve as nets to catch flying insects. The prey are quickly paralyzed by the spider's bite and wrapped in a tangle of silk threads. Other rapidly moving spiders, like wolf spiders and jumping spiders, have large eyes that help them find prey without using webs. Dragonflies patrol areas where they can capture flying insects. Hawks and owls have excellent eyesight that allows them to find their prey. Many predators, like leopards, lions, and cheetahs, use speed to run down their prey; others such as frogs, toads, and many kinds of lizards blend in with their surroundings and strike quickly when a prey organism happens by (figure 15.3).

Many kinds of predators are useful to us because they control the populations of organisms that do us harm. For example, snakes eat many kinds of rodents that eat stored grain and other agricultural products. Many birds and bats eat insects that are agricultural pests. It is even possible to think of a predator as having a beneficial effect on the prey species. Certainly the *individual* organism that is killed is harmed, but the *population* can benefit. Predators can prevent large populations of prey organisms from destroying their habitat by hindering overpopulation of prey species or they can reduce the likelihood of epidemic disease by eating sick or diseased individuals. Furthermore, predators act as selecting agents. The individuals who fall to them as prey are likely to be less well adapted than the ones that escape predation. Predators usually kill slow, unwary, sick, or injured individuals. Thus the genes that may have contributed to slowness, inattention, illness, or the likelihood of being injured are removed from the gene pool and a better-adapted population remains. Because predators eliminate poorly adapted *individuals*, the *species* benefits. What is bad for the individual can be good for the species.

Parasitism

Another kind of interaction in which one organism is harmed and the other aided is the relationship of parasitism. In fact, there are more species of parasites in the world than there are nonparasites, making this a very common kind of relationship. **Parasitism** involves one organism living in or on another living organism from which it derives nourishment. The **parasite** derives the benefit and harms the **host,** the organism it lives in or on (figure 15.4). Many kinds of fungi live on trees and other kinds of plants, including those that are commercially valuable. Dutch elm disease is caused by a fungus that infects the living, sap-carrying parts of the tree. Mistletoe is a common plant that is a parasite on other plants. The mistletoe plant invades the tissues of the tree it is living on and derives nourishment from the tree.

Many kinds of worms, protozoa, bacteria, and viruses are important parasites. Parasites that live on the outside of their hosts are called **external parasites.** For example, fleas live on the outside of the bodies of mammals like rats, dogs, cats, and humans, where they suck blood and do harm to their hosts. At the same time, the host could also have a tapeworm in its intestine. Because the tapeworm lives inside the host, it is called an **internal parasite.** Another kind of parasite that may be found in the blood of rats is the bacterium *Yersinia pestis.* It does little harm to the rat but causes a disease known as *plague* or *black death* if it is transmitted to humans. Because fleas can suck the blood of rats and also live on and bite humans they can serve as carriers of bacteria between rats and humans. An organism that can carry a disease from one individual to another is called a **vector.** During the mid-1300s, when living conditions were poor and rats and fleas were commonly associated with humans, epidemics of plague killed millions of people. In some countries

(a)

(b)

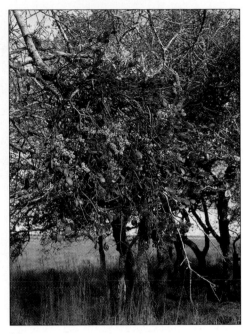
(c)

Figure 15.4

The Parasite–Host Relationship
Parasites benefit from the relationship because they obtain nourishment from the host. Tapeworms (a) are internal parasites in the guts of their host where they absorb food from the host's gut. The lamprey (b) is an external parasite that sucks body fluids from its host. Mistletoe (c) is a photosynthesizing plant that also absorbs nutrients from the tissues of its host tree. The host in any of these three situations may not be killed directly by the relationship, but it is often weakened, thus becoming more vulnerable to predators or diseases. There are more species of parasites in the world than species of organisms that are not parasites.

in western Europe, 50% of the population was killed by this disease. Plague is still a problem today when living conditions are poor and sanitation is lacking. Cases of plague are even found in developed countries like the United States on occasion.

Lyme disease is also a vector-borne disease caused by the bacterium, *Borrelia burgdorferi*, that is spread by certain species of ticks (figure 15.5). Over 80% of the cases are centered in the Northeast (New York, Pennsylvania, Maryland, Delaware, Connecticut, Rhode Island, and New Jersey).

Both predation and parasitism are relationships in which one member of the pair is helped and the other is harmed. But there are many kinds of interactions in which one is harmed and the other aided that don't fit neatly into the categories of interactions dreamed up by scientists. For example, when a cow eats grass, it is certainly harming the grass while deriving benefit from it. We could call cows *grass predators*, but we usually refer to them as *herbivores*. Likewise, such animals as mosquitoes, biting flies, vampire bats, and ticks take blood meals but don't usually live permanently on the host or kill it. Are they temporary parasites or specialized predators? Finally, birds like cowbirds and some species of European cuckoos lay their eggs in the nests of other species of birds, who raise these foster young rather than their own. The adult cowbird and cuckoo often remove eggs from the host nest. In addition, the cowbird or cuckoo offspring eject the eggs or the young of the host-bird species, so that usually only the cowbird or cuckoo chick is raised by the foster parents. This kind of relationship has been called *nest parasitism*, because the host parent birds are not killed and aid the cowbird or cuckoo by raising their young.

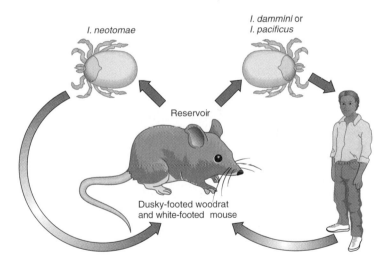

Figure 15.5

Lyme Disease—Hosts, Parasites, and Vectors
Lyme disease is a bacterial disease originally identified in a small number of individuals in the Old Lyme, Connecticut, area. Today it is found throughout the United States and Canada. Once the parasite, *Borrelia burgdorferi*, has been transferred into a suitable susceptible host (e.g., humans, mice, horses, cattle, domestic cats, and dogs), it causes symptoms that have been categorized into three stages. The first-stage symptoms may appear 3 to 32 days after an individual is bitten by an infected tick (various members of the genus *Ixodes*) and include a spreading red rash, headache, nausea, fever, aching joints and muscles, and fatigue. Stage two may not appear for weeks or months after the infection and may affect the heart and nervous system. The third stage may occur months or years later and typically appears as severe arthritis attacks. The main reservoir of the disease is the white-footed mouse and dusky-footed woodrat.

(a)

(b)

Figure 15.6

Commensalism

In the relationship called commensalism, one organism benefits and the other is not affected. (*a*) The remora fish shown here hitchhike a ride on the shark. They eat scraps of food left over by the messy eating habits of the shark. The shark does not seem to be hindered in any way. (*b*) The epiphytic plants growing on this tree do not harm the tree but are aided by using the tree surface as a place to grow.

Commensalism

Both predation and parasitism involve one organism benefiting while the other is harmed. Another common relationship is one in which one organism benefits and the other is not affected. This is known as **commensalism**. For example, sharks often have another fish, the remora, attached to them. The remora has a sucker on the top side of its head that allows it to attach to the shark and get a free ride (figure 15.6*a*). Although the remora benefits from the free ride and by eating leftovers from the shark's meals, the shark does not appear to be troubled by this uninvited guest, nor does it benefit from the presence of the remora.

Another example of commensalism is the relationship between trees and epiphytic plants. **Epiphytes** are plants that live on the surface of other plants but do not derive nourishment from them (figure 15.6*b*). Many kinds of plants (e.g., orchids, ferns, and mosses) use the surfaces of trees as places to live. These kinds of organisms are particularly common in tropical rainforests. Many epiphytes derive benefit from the relationship because they are able to be located in the tops of the trees, where they receive more sunlight and moisture. The trees derive no benefit from the relationship, nor are they harmed; they simply serve as support surfaces for epiphytes.

Mutualism

So far in our examples, only one species has benefited from the association of two species. There are also many situations in which two species live in close association with one another, and both benefit. This is called **mutualism.** One interesting example of mutualism involves digestion in rabbits. Rabbits eat plant material that is high in cellulose even though they do not produce the enzymes capable of breaking down cellulose molecules into simple sugars. They manage to get energy out of these cellulose molecules with the help of special bacteria living in their digestive tracts. The bacteria produce cellulose-digesting enzymes, called *cellulases*, that break down cellulose into smaller carbohydrate molecules that the rabbit's digestive enzymes can break down into still smaller glucose molecules. The bacteria benefit because the gut of the rabbit provides them with a moist, warm, nourishing environment in which to live. The rabbit benefits because the bacteria provide it with a source of food. Termites, cattle, buffalo, and antelope also have collections of bacteria and protozoa living in their digestive tracts that help them digest cellulose.

Another kind of mutualistic relationship exists between flowering plants and bees. Undoubtedly you have observed bees and other insects visiting flowers to obtain nectar from the blossoms (figure 15.7). Usually the flowers are

Figure 15.7

Mutualism
Mutualism is an interaction between two organisms in which both benefit. The plant benefits because cross-fertilization (exchange of gametes from a different plant) is more probable; the butterfly benefits by acquiring nectar for food.

Figure 15.8

Competition
Whenever a needed resource is in limited supply, organisms compete for it. This competition may be between members of the same species (*intraspecific*), illustrated by the vultures shown in the photograph, or may involve different species (*interspecific*).

constructed in such a manner that the bees pick up pollen (sperm-containing packages) on their hairy bodies, which they transfer to the female part of the next flower they visit. Because bees normally visit many individual flowers of the same species for several minutes and ignore other species of flowers, they can serve as pollen carriers between two flowers of the same species. Plants pollinated in this manner produce less pollen than do plants that rely on the wind to transfer pollen. This saves the plant energy because it doesn't need to produce huge quantities of pollen. It does, however, need to transfer some of its energy savings into the production of showy flowers and nectar to attract the bees. The bees benefit from both the nectar and pollen; they use both for food.

Lichens and corals exhibit a more intimate kind of mutualism. In both cases the organisms consist of the cells of two different organisms intermingled with one another. Lichens consist of fungal cells and algal cells in a partnership; corals consist of the cells of the coral organism intermingled with algal cells. In both cases, the algae carry on photosynthesis and provide nutrients and the fungus or coral provides a moist, fixed structure for the algae to live in.

One other term that relates to parasitism, commensalism, and mutualism is *symbiosis*. **Symbiosis** literally means "living together." Unfortunately, this word is used in several ways, none of which is very precise. It is often used as a synonym for mutualism, but it is also often used to refer to commensal relationships and parasitism. The emphasis, however, is on interactions that involve a close physical relationship between the two kinds of organisms.

Competition

So far in our discussion of organism interactions we have left out the most common one. It is reasonable to envision every organism on the face of the Earth being involved in competitive interactions. **Competition** is a kind of interaction between organisms in which both organisms are harmed to some extent. Competition occurs whenever two organisms need a vital resource that is in short supply (figure 15.8). The vital resource could be food, shelter, nesting sites, water, mates, or space. Competition can involve a snarling tug-of-war between two dogs over a scrap of food, or it can be a silent struggle between plants for access to available light. If you have ever started tomato seeds (or other garden plants) in a garden and failed to eliminate the weeds, you have witnessed competition. If the weeds are not removed, they compete with the garden plants for available sunlight, water, and nutrients, resulting in poor growth of both the garden plants and the weeds.

The more similar the requirements of two species of organisms, the more intense the competition. According to the **competitive exclusion principle,** no two species of organisms can occupy the same niche at the same time. If two species of organisms do occupy the same niche, the competition will be so intense that one or more of the following will occur: one will become extinct, one will be forced to migrate to a different area, or the two species may evolve into slightly different niches so that they do not compete.

It is important to recognize that although competition results in harm to both organisms there can still be winners and losers. The two organisms may not be harmed to the same extent with the result that one will have greater access to the limited resource. Furthermore, even the loser can continue to survive if it migrates to an area where competition is less intense or if its species evolves to exploit a different niche. Thus competition provides a major mechanism for natural selection. With the development of slight differences between niches the intensity of competition is reduced. For example, many birds catch flying insects as food. However, they do not compete directly with each other because some feed at night, some feed high in the air, some feed only near the ground, and still others perch on branches and wait for insects to fly past. The insect-eating niche can be further subdivided by specialization on particular sizes or kinds of insects.

Many of the relationships just described involve the transfer of nutrients from one organism to another (predation, parasitism, mutualism). Another important way scientists look at ecosystems is to look at how materials are cycled from organism to organism.

15.3 The Cycling of Materials in Ecosystems

Although some new atoms are being added to the Earth from cosmic dust and meteorites, this amount is not significant in relation to the entire biomass of the Earth. Therefore, the Earth can be considered to be a closed ecosystem as far as matter is concerned. Only sunlight energy comes to the Earth in a continuous stream, and even this is ultimately returned to space as heat energy. However, it is this flow of energy that drives all biological processes. Living systems have evolved ways of using this energy to continue life through growth and reproduction and the continual reuse of existing atoms. In this recycling process, inorganic molecules are combined to form the organic compounds of living things. If there were no way of recycling this organic matter back into its inorganic forms, organic material would build up as the bodies of dead organisms. This is thought to have occurred millions of years ago when the present deposits of coal, oil, and natural gas were formed. Under most conditions decomposers are available to break down organic material to inorganic material that can then be reused by other organisms to rebuild organic material. One way to get an appreciation of how various kinds of organisms interact to cycle materials is to look at a specific kind of atom and follow its progress through an ecosystem.

The Carbon Cycle

Living systems contain many kinds of atoms, but some are more common than others. Carbon, nitrogen, oxygen, hydrogen, and phosphorus are found in all living things and must be recycled when an organism dies. Let's look at some examples of this recycling process. Carbon and oxygen atoms combine to form the molecule carbon dioxide (CO_2), which is a gas found in small quantities in the atmosphere. During photosynthesis, carbon dioxide (CO_2) combines with water (H_2O) to form complex organic molecules like sugar ($C_6H_{12}O_6$). At the same time, oxygen molecules (O_2) are released into the atmosphere (Outlooks 15.1 page 294). • **photosynthesis, p. 120; cellular respiration, p. 109**

The organic matter in the bodies of plants may be used by herbivores as food. When an herbivore eats a plant, it breaks down the complex organic molecules into more simple molecules, like simple sugars, amino acids, glycerol, and fatty acids. These can be used as building blocks in the construction of its own body. Thus, the atoms in the body of the herbivore can be traced back to the plants that were eaten. Similarly, when herbivores are eaten by carnivores, these same atoms are transferred to them. Finally, the waste products of plants and animals and the remains of dead organisms are used by decomposer organisms as sources of carbon and other atoms they need for survival. In addition, all the organisms in this cycle—plants, herbivores, carnivores, and decomposers—obtain energy (ATP [adenosine triphosphate]) from the process of respiration, in which oxygen (O_2) is used to break down organic compounds into carbon dioxide (CO_2) and water (H_2O). Thus, the carbon atoms that started out as components of carbon dioxide (CO_2) molecules have passed through the bodies of living organisms as parts of organic molecules and returned to the atmosphere as carbon dioxide, ready to be cycled again. Similarly, the oxygen atoms (O) released as oxygen molecules (O_2) during photosynthesis have been used during the process of respiration (figure 15.9). • **food chains, p. 262**

The Hydrologic Cycle

Water molecules are the most common molecules in living things and are essential for life. Water molecules are used as raw materials in the process of photosynthesis. The hydrogen atoms (H) from water (H_2O) molecules are added to carbon atoms to make carbohydrates and other organic molecules. Furthermore, the oxygen atoms in water molecules are released during photosynthesis as oxygen molecules (O_2). In addition, all the metabolic reactions that occur in organisms take place in a watery environment. We can trace the movement and reuse of water molecules by picturing a hydrologic cycle (figure 15.10).

Most of the forces that cause water to be cycled do not involve organisms, but are the result of normal physical processes. Because of the kinetic energy possessed by water molecules, at normal Earth temperatures liquid water evaporates into the atmosphere as water vapor. This can occur wherever water is present; it evaporates from lakes, rivers, soil, or the surfaces of organisms. Because the oceans contain most of the world's water, an extremely large amount of water enters the atmosphere from the oceans. In addition, **transpiration** in plants involves the transport of water from

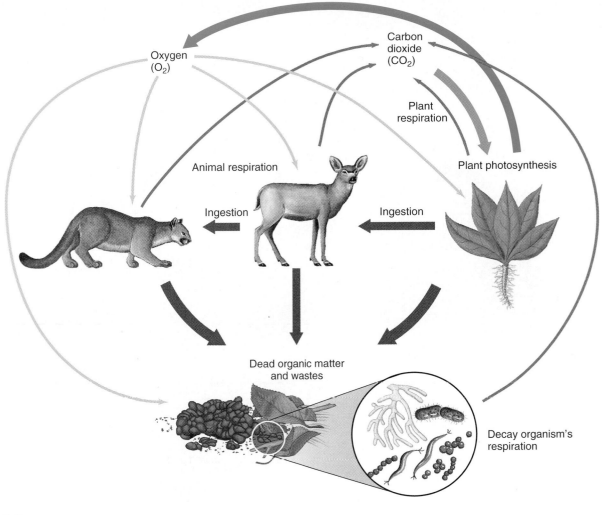

Figure 15.9

The Carbon Cycle
Carbon atoms are cycled through ecosystems. Carbon dioxide (green arrows) produced by respiration is the source of carbon that plants incorporate into organic molecules when they carry on photosynthesis. These carbon–containing organic molecules (black arrows) are passed to animals when they eat plants and other animals. Organic molecules in waste or dead organisms are consumed by decay organisms in the soil when they break down organic molecules into inorganic molecules. All organisms (plants, animals, and decomposers) return carbon atoms to the atmosphere as carbon dioxide when they carry on cellular respiration. Oxygen (blue arrows) is being cycled at the same time that carbon is. The oxygen is released to the atmosphere and into the water during photosynthesis and taken up during cellular respiration.

the soil to leaves, where it evaporates. The movement of water carries nutrients to the leaves and the evaporation of the water assists in the movement of water upward in the stem.

Once the water molecules are in the atmosphere, they are moved by prevailing wind patterns. If warm, moist air encounters cooler temperatures, which often happens over landmasses, the water vapor condenses into droplets and falls as rain or snow. When the precipitation falls on land, some of it runs off the surface, some of it evaporates, and some penetrates into the soil. The water in the soil may be taken up by plants and transpired into the atmosphere, or it may become groundwater. Much of the groundwater also eventually makes its way into lakes and streams and ultimately arrives at the ocean from which it originated.

The Nitrogen Cycle

Another important element for living things is nitrogen (N). Nitrogen is essential in the formation of amino acids, which are needed to form proteins, and in the formation of nitrogenous bases, which are a part of ATP and the nucleic acids DNA and RNA. Nitrogen (N) is found as molecules of nitrogen gas (N_2) in the atmosphere. Although nitrogen gas (N_2) makes up approximately 80% of the Earth's atmosphere, only a few kinds of bacteria are able to convert it into nitrogen compounds that other organisms can use. Therefore, in most terrestrial ecosystems, the amount of nitrogen available limits the amount of plant biomass that can be produced. (Most aquatic ecosystems are limited by the amount of phosphorus

Figure 15.10

The Hydrologic Cycle
The cycling of water through the environment follows a simple pattern. Moisture in the atmosphere condenses into droplets that fall to the Earth as rain or snow. Organisms use some of the water, but much of it flows over the Earth as surface water or through the soil as groundwater. It eventually returns to the oceans, where it evaporates back into the atmosphere to begin the cycle again.

rather than the amount of nitrogen.) Plants utilize several different nitrogen-containing compounds to obtain the nitrogen atoms they need to make amino acids and other compounds (figure 15.11).

Symbiotic nitrogen-fixing bacteria live in the roots of certain kinds of plants, where they convert nitrogen gas molecules into compounds that the plants can use to make amino acids and nucleic acids. The most common plants that enter into this mutualistic relationship with bacteria are legumes such as beans, clover, peas, alfalfa, and locust trees. Some other organisms, such as alder trees and even a kind of aquatic fern can also participate in this relationship. There are also **free-living nitrogen-fixing bacteria** in the soil that provide nitrogen compounds that can be taken up through

the roots, but the bacteria do not live in a close physical union with plants.

Another way plants get usable nitrogen compounds involves a different series of bacteria. Decomposer bacteria convert organic nitrogen-containing compounds into ammonia (NH_3). **Nitrifying bacteria** can convert ammonia (NH_3) into nitrite-containing (NO_2^-) compounds, which in turn can be converted into nitrate-containing (NO_3^-) compounds. Many kinds of plants can use either ammonia (NH_3) or nitrate (NO_3^-) from the soil as building blocks for amino acids and nucleic acids.

All animals obtain their nitrogen from the food they eat. The ingested proteins are broken down into their component amino acids during digestion. These amino acids can

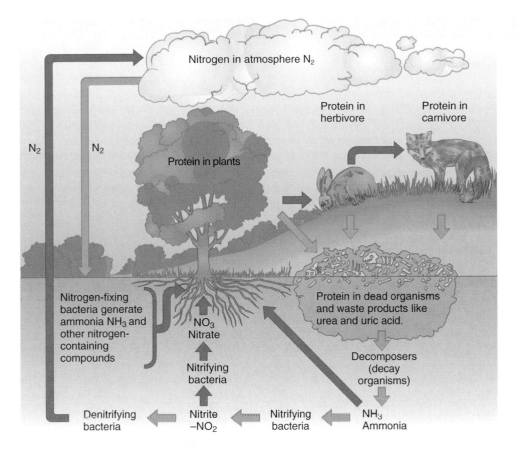

Figure 15.11

The Nitrogen Cycle
Nitrogen atoms are cycled through ecosystems. Atmospheric nitrogen is converted by nitrogen-fixing bacteria to nitrogen-containing compounds that plants can use to make proteins and other compounds. Proteins are passed to other organisms when one organism is eaten by another. Dead organisms and their waste products are acted upon by decay organisms to form ammonia, which may be reused by plants and converted to other nitrogen compounds by nitrifying bacteria. Denitrifying bacteria return nitrogen as a gas to the atmosphere.

then be reassembled into new proteins characteristic of the animal. All dead organic matter and waste products of plants and animals are acted upon by decomposer organisms, and the nitrogen is released as ammonia (NH_3), which can be taken up by plants or acted upon by nitrifying bacteria to make nitrate (NO_3^-).

Finally, the cycling of nitrogen gas (N_2) back to the atmosphere occurs with the action of bacteria called **denitrifying bacteria** that are capable of converting nitrite (NO_2^-) to nitrogen gas (N_2). Thus, in the nitrogen cycle, nitrogen from the atmosphere is passed through a series of organisms, many of which are bacteria, and ultimately returns to the atmosphere to be cycled again. However, there is also a secondary cycle in which nitrogen compounds are recycled without returning to the atmosphere.

Because nitrogen is in short supply in most ecosystems, farmers usually find it necessary to supplement the natural nitrogen sources in the soil to obtain maximum plant growth. This can be done in a number of ways. Alternating nitrogen-producing crops with nitrogen-demanding crops helps maintain high levels of usable nitrogen in the soil. One year a crop such as beans or clover that has symbiotic nitrogen-fixing bacteria in its roots can be planted. The following year the farmer can plant a nitrogen-demanding crop such as corn. The use of manure is another way of improving nitrogen levels. The waste products of animals are broken down by de-composer bacteria and nitrifying bacteria, resulting in enhanced levels of ammonia and nitrate. Finally, the farmer can use industrially produced fertilizers containing ammonia or nitrate. These compounds can be used directly by plants or converted into other useful forms by nitrifying bacteria.

The Phosphorus Cycle

Phosphorus is another kind of atom common in the structure of living things. It is present in many important biological molecules such as DNA and in the membrane structure of cells. In addition, the bones and teeth of animals contain significant quantities of phosphorus. The ultimate source of phosphorus atoms is rock. In nature, new phosphorus compounds are released by the erosion of rock and dissolved in water. Plants use the dissolved phosphorus compounds to construct the molecules they need. Animals obtain the phosphorus they need when they consume plants or other animals. When an organism dies or excretes waste products, decomposer organisms recycle the phosphorus compounds back into the soil. Phosphorus compounds that are dissolved in water are ultimately precipitated as deposits. Geologic processes elevate these deposits and expose them to erosion, thus making these deposits available to organisms. Waste products of animals often have significant amounts of phosphorus. In places where large numbers of seabirds or bats congregate for hundreds of

OUTLOOKS 15.1

Carbon Dioxide and Global Warming

Humans have significantly altered the carbon cycle. As we burn fossil fuels, the amount of carbon dioxide in the atmosphere continually increases. Carbon dioxide allows light to enter the atmosphere but does not allow heat to exit. Because this is similar to what happens in a greenhouse, carbon dioxide and other gases that have similar effects are called greenhouse gases. Therefore, increased amounts of carbon dioxide in the atmosphere could lead to a warming of the Earth. Many are concerned that increased carbon dioxide levels will lead to a warming of the planet and, thus, cause major changes in weather and climate, leading to the flooding of coastal cities and major changes in agricultural production. The Intergovernmental Panel on Climate Change (IPCC), established by the United Nations, concluded that there has been an increase in the average temperature of the Earth and humans are the cause because of the burning of fossil fuels and the destruction of forests. There is no doubt that the amount of carbon dioxide in the atmosphere has been increasing. Despite this fact and the conclusions of the IPCC, there is still controversy about this topic and some doubt that warming is occurring. At a meeting in Kyoto, Japan, in 1998 many countries agreed to reduce the amount of carbon dioxide and other greenhouse gases they release into the atmosphere. However, emissions of carbon dioxide are directly related to economic activity and the energy usage that fuels it. Therefore, it remains to be seen if countries will meet their goals or will succumb to economic pressures to allow continued use of large amounts of fossil fuels. However, some countries have sought to control the change in the amount of carbon dioxide in the atmosphere by planting millions of trees or preventing the destruction of forests. The thought is that the trees carry on photosynthesis, grow, and store carbon in their bodies, leading to reduced carbon dioxide levels. At the same time people in other parts of the world continue to destroy forests at a rapid rate. Tree planting does not offset deforestation. In addition, the trees that have been planted will ultimately die and decompose, releasing carbon dioxide back into the atmosphere, so it is not clear that this is an effective means of reducing atmospheric carbon dioxide over the long term.

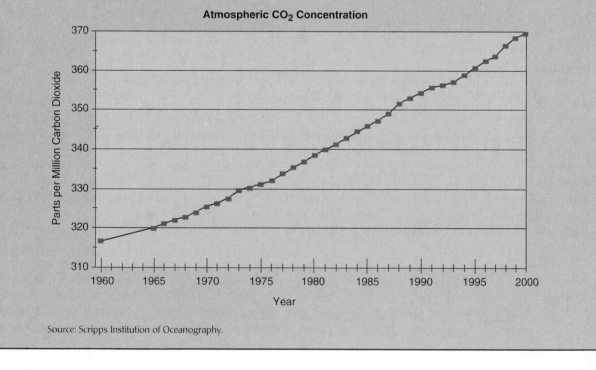

Source: Scripps Institution of Oceanography.

years, the thickness of their droppings (called *guano*) can be a significant source of phosphorus for fertilizer (figure 15.12). In many soils, phosphorus is in short supply and must be provided to crop plants to get maximum yields. Phosphorus is also in short supply in aquatic ecosystems.

Fertilizers usually contain nitrogen, phosphorus, and potassium compounds. The numbers on a fertilizer bag indicate the percentage of each in the fertilizer. For example, a 6-24-24 fertilizer has 6% nitrogen, 24% phosphorus, and 24% potassium compounds. In addition to carbon, nitrogen, and phosphorus, potassium and other elements are cycled within ecosystems. In an agriculture ecosystem, these elements are removed when the crop is harvested. Therefore farmers must not only return the nitrogen, phosphorus, and potassium, but they must also analyze for other less prominent elements and add them to their fertilizer mixture as well. Aquatic ecosystems are also sensitive to nutrient levels. High levels of nitrates or phosphorus

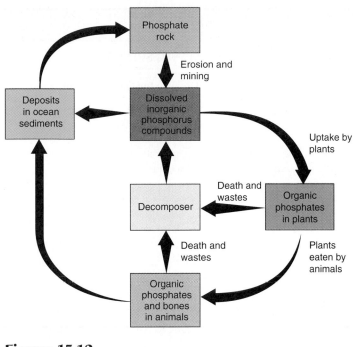

Figure 15.12

Phosphorus Cycle
The source of phosphorus is rock which, when dissolved, provides a source of phosphorus used by plants and animals.

compounds often result in rapid growth of aquatic producers. In aquaculture, such as that used to raise catfish, fertilizer is added to the body of water to stimulate the production of algae, which is the base of many aquatic food chains.

15.4 The Impact of Human Actions on Communities

As you can see from this discussion and from the discussion of food webs in chapter 14, all organisms are associated in a complex network of relationships. A community consists of all these sets of interrelations. Therefore, before one decides to change a community, it is wise to analyze how the organisms are interrelated. This is not always an easy task because there is much we still do not know about how organisms interact and how they utilize molecules from their environment. Several lessons can be learned from studying the effects of human activity on communities.

Introduced Species

One of the most far-reaching effects humans have had on natural ecosystems involves the introduction of foreign species. Most of these introductions have been conscious decisions. Nearly all of our domesticated plants and animals are introductions from elsewhere. Cattle, horses, pigs, goats, and many introduced grasses have significantly altered the original ecosystems present in the Americas. Nearly all of our agriculturally important plants and animals are not native to North America. (Corn, beans, sunflowers, squash,

and the turkey are exceptions.) Cattle have replaced the original grazers on grasslands. Pigs have become a major problem in Hawaii and many other places in the world where they destroy the natural ecosystem by digging up roots and preventing the reproduction of native plants. The introduction of grasses as food for cattle has resulted in the decline of many native species of grasses and other plants that were originally part of grassland ecosystems. In Australia the introduction of domesticated plants and animals, and wild animals such as rabbits and foxes, has severely reduced the populations of many native marsupial mammals.

Accidental introductions have also significantly altered ecosystems. Chestnut blight essentially eliminated the American chestnut from the forests of eastern North America. Similarly, a fungal disease (Dutch elm disease) has severely reduced the number of elms in forests.

Predator Control

During the formative years of wildlife management, it was thought that populations of game species could be increased if the populations of their predators were reduced. Consequently, many states passed laws that encouraged the killing of foxes, eagles, hawks, owls, coyotes, cougars, and other predators that use game animals as a source of food. Often bounties were paid to people who killed these predators. In South Dakota, it was decided to increase the pheasant population by reducing the numbers of foxes and coyotes. However, when the supposed predator populations were significantly reduced, there was no increase in the pheasant population. There was rapid increase in the rabbit and mouse populations, however, and they became serious pests. Evidently the foxes and coyotes were major factors in keeping rabbit and mouse populations under control but had only a minor impact on pheasants.

The absence of predators can lead to many kinds of problems with prey species. In many metropolitan areas deer have become pests. This is due to several reasons, including the fact that there are no predators, and hunting (predation by humans) is either not allowed or is impractical because of the highly urbanized nature of the area. Some municipalities have instituted programs of chemical birth control for their deer populations. In parts of Florida, increased numbers of alligators present a danger, particularly to pets and children. Hunting is now allowed in an effort to control the numbers of alligators because humans are the only effective predators of large alligators. Only a few years ago the alligator was on the endangered species list and all hunting was suspended. Similarly, in Yellowstone National Park, elk, bison, and moose populations have become very large because hunting is not allowed and predators are in low numbers. In 1995 wolves were reintroduced to the park in the hope that they would help bring the elk and moose populations under control. This was a controversial decision because ranchers in the vicinity do not want a return of large predators that might prey on their livestock. They are also opposed to having bison, many of which carry a disease that can affect cattle, stray onto their land. The wolf populations have increased significantly in Yellowstone and are having an effect on the populations of bison, elk, and moose. Regardless of the politics involved in the decision, Yellowstone is in a more natural condition today with wolves present than it was prior to 1995.

By contrast, the state of Alaska instituted a project to kill wolves because they believe the wolves are reducing caribou populations below optimal levels. Caribou hunting is an important source of food for Alaskan natives, and hunters who visit the state provide a significant source of income. Many groups oppose the killing of wolves in Alaska. They consider the policy misguided and believe it will not have a positive effect on the caribou population. They also object to the killing of wolves on ethical grounds.

Habitat Destruction

Some communities are fragile and easily destroyed by human activity, whereas others seem able to resist human interference. Communities with a wide variety of organisms that show a high level of interaction are more resistant than those with few organisms and little interaction. In general, the more complex an ecosystem is, the more likely it is to recover after being disturbed. The tundra biome is an example of a community with relatively few organisms and interactions. It is not very resistant to change, and because of its slow rate of repair, damage caused by human activity may persist for hundreds of years.

Some species are more resistant to human activity than others. Rabbits, starlings, skunks, and many kinds of insects and plants are able to maintain high populations despite human activity. Indeed, some may even be encouraged by human activity. By contrast, whales, condors, eagles, and many plant and insect species are not able to resist human interference very well. For most of these endangered species it is not humans acting directly with the organisms that cause their endangerment. Very few organisms have been driven to extinction by hunting or direct exploitation. Usually the cause of extinction or endangerment is an indirect effect of habitat destruction as humans exploit natural ecosystems. As humans convert land to farming, grazing, commercial forestry, development, and special wildlife management areas, the natural ecosystems are disrupted, and plants and animals with narrow niches tend to be eliminated because they lose critical resources in their environment. Table 15.1 lists several endangered species and the probable causes of their difficulties.

Pesticide Use

Humans have developed a variety of chemicals to control specific pest organisms. One of the first that was used widely was the insecticide DDT. DDT is an abbreviation for the chemical name dichlorodiphenyltrichloroethane. DDT is one of a group of organic compounds called *chlorinated hydro-*

Table 15.1

ENDANGERED AND THREATENED SPECIES

Species	Reason for Endangerment
Hawaiian crow *Corvis hawaiinsis*	Predation by cat and mongoose; disease; habitat destruction
Sonora chub *Gila ditaenia*	Competition with introduced species
Black-footed ferret *Mustela nigripes*	Poisoning of prairie dogs (their primary food)
Snail kite *Rostrhamus sociabilis*	Specialized eating habits (only eat apple snails); draining of marshes
Grizzly bear *Ursus arctos*	Loss of wilderness areas
California condor *Gymnogyps californianus*	Slow breeding; lead poisoning
Ringed sawback turtle *Graptemys oculifera*	Modification of habitat by construction of reservoir that reduced their primary food source
Scrub mint *Dicerandra frutescens*	Conversion of habitat to citrus groves and housing

carbons. Because DDT is a poison that was used to kill a variety of insects, it was called an **insecticide.** Another term that is sometimes used is **pesticide,** which implies that the poison is effective against pests. Although it is no longer used in the United States (its use was banned in the early 1970s), DDT is still manufactured and used in many parts of the world, including Mexico.

DDT was a valuable insecticide for the U.S. Armed Forces during World War II. It was sprayed on clothing and dusted on the bodies of soldiers, refugees, and prisoners to kill body lice and other insects. Lice, besides being a nuisance, carry the bacteria that can cause a disease known as *typhus fever.* When bitten by a louse, a person can develop typhus fever. Because body lice could be transferred from one person to another by contact or by wearing infested clothing, DDT was important in maintaining the health of millions of people. Because DDT was so useful in controlling these insects, people envisioned the end of pesky mosquitoes and flies, as well as the elimination of many disease-carrying insects.

Although DDT was originally very effective, many species of insects developed a resistance to it. The genetic diversity present in all species is related to their ability to respond to many environmental factors, including manufactured ones such as DDT. When DDT or any pesticide is applied to a population of insects, susceptible individuals die, and those with some degree of resistance have a greater chance of living. Now the reproducing population consists of many individuals that have resistant genes, which are passed on to the offspring. When this happens repeatedly over a long time, a resistant population develops, and the insecticide is no longer useful.

DDT and other pesticides act as selecting agents, killing the normal insects but allowing the resistant individuals to live. This happened in the orange groves of California, where many populations of pests became DDT-resistant. Similarly, throughout the world, DDT was used (and in many areas is still used) to control malaria-carrying mosquitoes. Many of these populations have become resistant to DDT and other kinds of insecticides. The people who anticipated the elimination of insect pests did not reckon with the genetic diversity of the gene pools of these insects.

Another problem associated with pesticide use is the effects of pesticides on valuable nontarget organisms. Because many of the insects we consider pests are herbivores, you can expect that carnivores in the community use the pest species as prey, and parasites use the pest as a host. These predators and parasites have important roles in controlling the numbers of a pest species.

Generally, predators and parasites reproduce more slowly than their prey or host species. Because of this, the use of a nonspecific pesticide may indirectly make controlling a pest more difficult. If such a pesticide is applied to an area, the pest is killed but so are its predators and parasites. Because the herbivore pest reproduces faster than its predators and parasites, the pest population rebounds quickly, unchecked by natural predation and parasitism. This may necessitate more frequent and more concentrated applications of pesticides. This has actually happened in many cases of pesticide use; the pesticides made the problem worse, and the chemicals became increasingly costly to apply.

Today, a more enlightened approach to pest control involves *integrated pest management*, which uses a variety of approaches to reduce pest populations. Integrated pest management may involve the use of pesticides as part of a pest control program, but it will also include strategies such as encouraging the natural enemies of pests, changing farming practices to discourage pests, changing the mix of crops grown, and accepting low levels of crop damage as an alternative to costly pesticide applications.

Biomagnification

Another problem associated with the use of persistent chemicals involves their effect on the food chain. DDT was a very effective insecticide because it is extremely toxic to insects but not very toxic to birds and mammals. It is also a very stable compound, which means that once it was applied, it remained effective for a long time. It sounds like an ideal insecticide. What went wrong? Why was its use banned?

When DDT was sprayed over an area, it fell on the insects and on the plants that the insects used for food. Even-

tually the DDT entered the insect either directly through the body wall or through its food. When DDT enters the body of an insect, it interferes with the insect's normal metabolism. If small quantities are taken in, the insect can digest and break down the DDT just like any other large organic molecule. Because DDT is soluble in fat or oil, the DDT or its breakdown products are stored in the fat deposits of the insect.

Some insects can break down and store all the DDT they encounter and, therefore, they survive. If an area has been lightly sprayed with DDT, some insects die, some are able to tolerate the DDT, and others break down and store nonlethal quantities of DDT. As much as one part DDT per 1 million parts of insect tissue can be stored in this manner. This is not much DDT! It is equivalent to one drop of DDT in 100 railroad tank cars. However, when an aquatic area is sprayed with a small concentration of DDT, many kinds of organisms in the area can accumulate tiny quantities in their bodies. Even algae and protozoa found in aquatic ecosystems accumulate persistent pesticides. They may accumulate concentrations in their cells that are 250 times more concentrated than the amount sprayed on the ecosystem. The algae and protozoa are eaten by insects, which in turn are eaten by frogs, fish, or other carnivores.

The concentration in frogs and fish may be 2,000 times what was sprayed. The birds that feed on the frogs and fish may accumulate concentrations that are as much as 80,000 times the original amount. Because DDT is relatively stable and is stored in the fat deposits of the organisms that take it in, what was originally a dilute concentration becomes more concentrated as it moves up the food chain.

Before DDT was banned, many animals at higher trophic levels died as a result of lethal concentrations of pesticide accumulated from the food they ate. Each step in the food chain accumulated some DDT and, therefore, higher trophic levels had higher concentrations. This process is called **biomagnification** (figure 15.13). Even if they were not killed directly by DDT, many birds at higher trophic levels, such as eagles, pelicans, and osprey, suffered reduced populations because the DDT interfered with the female birds' ability to produce eggshells. Thin eggshells are easily broken, and thus no live young hatched. Both the bald eagle and the brown pelican were placed on the endangered species list because their populations dropped dramatically as a result of DDT poisoning. The ban on DDT use in the United States and Canada has resulted in an increase in the populations of both kinds of birds; the status of the bald eagle has been upgraded from endangered to threatened and that of the brown pelican has been upgraded to threatened in all states except Texas, Louisiana, Mississippi, and California.

Another widely used group of synthetic compounds of environmental concern are polychlorinated biphenyls (PCBs). PCBs are highly stable compounds that resist changes from heat, acids, bases, and oxidation. These characteristics made PCBs desirable for industrial use, but also

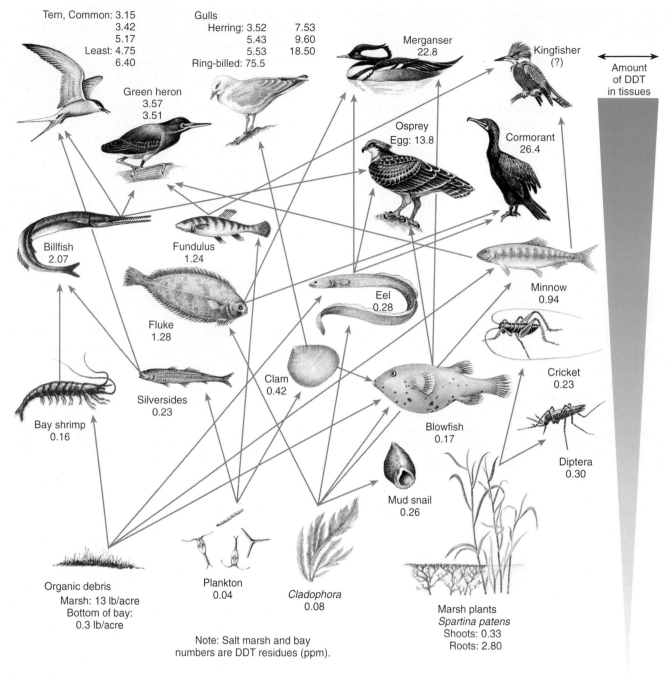

Figure 15.13

The Biomagnification of DDT

All the numbers shown are in parts per million (ppm). A concentration of one part per million means that in a million equal parts of the organism, one of the parts would be DDT. Notice how the amount of DDT in the bodies of the organisms increases as we go from producers to herbivores to carnivores. Because DDT is persistent, it builds up in the top trophic levels of the food chain.

made them persistent pollutants when released into the environment. About half the PCBs were used in transformers and electrical capacitors. Other uses included inks, plastics, tapes, paints, glues, waxes, and polishes. PCBs are harmful to fish and other aquatic forms of life because they interfere with reproduction. In humans, PCBs produce liver ailments and skin lesions. In high concentrations, they can damage the nervous system and are suspected carcinogens. In 1970, PCB production was limited to cases where satisfactory substitutes were not available. Today, substitutes have been found for nearly all the former uses of PCBs (How Science Works 15.1).

HOW SCIENCE WORKS 15.1

Herring Gulls as Indicators of Contamination in the Great Lakes

Herring gulls nest on islands and other protected sites throughout the Great Lakes region. Because they feed primarily on fish, they are near the top of the aquatic food chains and tend to accumulate toxic materials from the food they eat. Eggs taken from nests can be analyzed for a variety of contaminants.

Since the early 1970s, the Canadian Wildlife Service has operated a herring gull monitoring program to assess trends in the levels of various contaminants in the eggs of herring gulls. In general, the levels of contaminants have declined as both the Canadian and U.S. governments have taken action to stop new contaminants from entering the Great Lakes. The figure shows

the trends for PCBs. PCBs are a group of organic compounds, some of which are much more toxic than others. They were used as fire retardants, lubricants, insulation fluids in electrical transformers, and in some printing inks. Both Canada and the United States have eliminated most uses of PCBs.

The data collected show a downward trend in the amount of PCBs present in the eggs of herring gulls. Long-term studies like this one are very important in showing slow, steady responses to changes in the environment. Without such long-term studies we would be less sure of the impact of environmental clean-up activities.

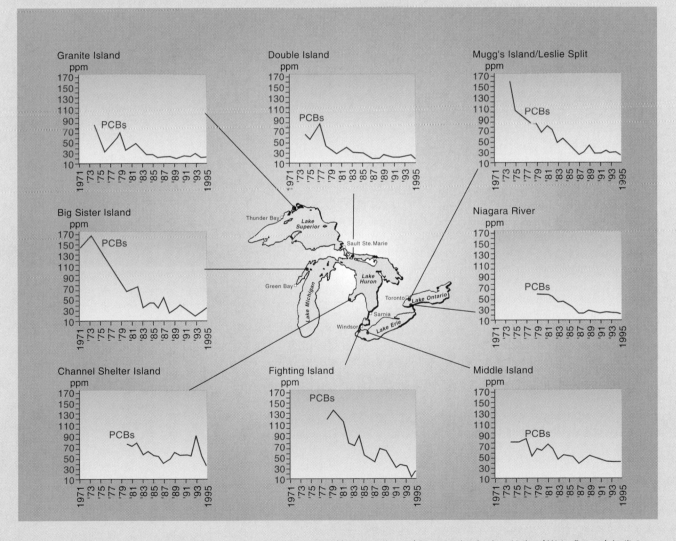

Source: The Canada Centre for Inland Waters, World Wide Web Servers operated by Computing and Programming Services, National Water Research Institute, Environment Canada. http://www.cciw.ca/green-lane/wildlife/gl-factsheet/herring

SUMMARY

Each organism in a community occupies a specific space known as its habitat and has a specific functional role to play, known as its niche. An organism's habitat is usually described in terms of some conspicuous element of its surroundings. The niche is difficult to describe because it involves so many interactions with the physical environment and other living things.

Interactions between organisms fit into several categories. Predation is one organism benefiting (predator) at the expense of the organism killed and eaten (prey). Parasitism is one organism benefiting (parasite) by living in or on another organism (host) and deriving nourishment from it. Organisms that carry parasites from one host to another are called vectors. Commensal relationships exist when one organism is helped but the other is not affected. Mutualistic relationships benefit both organisms. Symbiosis is any interaction in which two organisms live together in a close physical relationship. Competition causes harm to both of the organisms involved, although one may be harmed more than the other and may become extinct, evolve into a different niche, or be forced to migrate.

Many atoms are cycled through ecosystems. The carbon atoms of living things are trapped by photosynthesis, passed from organism to organism as food, and released to the atmosphere by respiration. Water is necessary as a raw material for photosynthesis and as the medium in which all metabolic reactions take place. Water is cycled by the physical processes of evaporation and condensation. Nitrogen originates in the atmosphere, is trapped by nitrogen-fixing bacteria, passes through a series of organisms, and is ultimately released to the atmosphere by denitrifying bacteria. Phosphorus compounds are found in rocky deposits. Erosion of rock and dissolving of phosphorus compounds make phosphorus available to plants. Animals obtain phosphorus in the food they eat. Phosphorus in waste products may be recycled or be deposited in sediments, which may be subjected to erosion at some future date.

Organisms within a community are interrelated in sensitive ways; thus, changing one part of a community can lead to unexpected consequences. Introduction of foreign species, predator-control practices, habitat destruction, pesticide use, and biomagnification of persistent toxic chemicals all have caused unanticipated changes in communities.

THINKING CRITICALLY

For guidelines to answer this question, visit the Online Learning Center.

This is a thought puzzle. Place the following items on a sheet of paper so that they show levels of interaction. Which is the most important item? Which items are dependent on others? Here are the pieces:

- People are starving.
- Commercial fertilizer production requires temperatures of 900°C.
- Geneticists have developed plants that grow very rapidly and require high amounts of nitrogen to germinate during the normal growing season.
- Fossil fuels are stored organic matter.
- The rate of the nitrogen cycle depends on the activity of bacteria.
- The Sun is expected to last for several million years.
- Crop rotation is becoming a thing of the past.
- The clearing of forests for agriculture changes the weather in the area.

CONCEPT MAP TERMINOLOGY

Construct a concept map using these terms.

competition	parasite
epiphytes	predator
host	prey
mutualism	producers
niche	symbiotic nitrogen-fixing
nitrogen cycle	bacteria

Visit the Online Learning Center to use an interactive format to construct another concept map.

KEY TERMS

Use interactive flash cards, on the Online Learning Center, to help you learn the meanings of these terms.

biomagnification	niche
commensalism	nitrifying bacteria
competition	parasite
competitive exclusion principle	parasitism
denitrifying bacteria	pesticide
epiphyte	predation
external parasite	predator
free-living nitrogen-fixing	prey
bacteria	symbiosis
habitat	symbiotic nitrogen-fixing
host	bacteria
insecticide	transpiration
internal parasite	vector
mutualism	

LEARNING CONNECTIONS

Visit the Online Learning Center to find additional study resources including those highlighted below.

15.1   Community, Habitat, and Niche

1. List 10 items that are a part of your niche.
2. What is the difference between habitat and niche?

15.2   Kinds of Organism Interactions

Animations and Review
• Species interactions

Interactive Concept Map
• Interactions

Experience This!
• Observing organism interactions

3. What do parasites, commensal organisms, and mutualistic organisms have in common? How are they different?
4. Describe two situations in which competition may involve combat and two that do not involve combat.

15.3   The Cycling of Materials in Ecosystems

Animations and Review
• Nutrient cycles

Interactive Concept Maps
• Carbon cycle
• Nitrogen cycle
• Hydrologic cycle

Case Study
• Averting disaster in Biosphere 2

5. Trace the flow of carbon atoms through a community that contains plants, herbivores, decomposers, and parasites.
6. Describe four different roles that bacteria play in the nitrogen cycle.
7. Describe the flow of water through the hydrologic cycle.
8. List three ways the carbon and nitrogen cycles are similar and three ways they differ.
9. Describe the major processes that make phosphorus available to plants.

15.4   The Impact of Human Actions on Communities

Animations and Review
• Pesticides
• Biomagnification
• Alternatives

Case Studies
• Columbia River tragedy
• Killer seaweed begins U.S. invasion

Food for Thought
• The wolf in Yellowstone National Park

10. Why do DDT and PCBs increase in concentration in the bodies of organisms at higher trophic levels?
11. How have past practices of predator control and habitat destruction negatively altered biological communities?
12. List three introduced species that have become pests and explain why they became pests.

Test yourself with multiple choice questions that provide immediate feedback on the Online Learning Center with PowerWeb at

www.mhhe.com/enger11

Population Ecology

GOALS AND OBJECTIVES

Recognize that populations vary in gene frequency, age distribution, sex ratio, size, and density.

❏ State how age distribution, sex ratio, and density can affect the rate of population growth.

Understand why the size of a population tends to increase.

❏ Describe and draw the stages of a typical population growth curve.

❏ Identify key components that cause population growth.

❏ Identify the factors that ultimately limit population size.

❏ State the importance of the birthrates and deathrates to population growth.

Recognize that human populations obey the same rules of growth as populations of other types of organisms.

❏ State why the human population must have an upper limit.

❏ List methods that would effectively control human population size.

Evolution and Ecology

PART FOUR

16.1 Population Characteristics

A **population** is a group of organisms of the same species located in the same place at the same time. Examples are the number of dandelions in your front yard, the rat population in the sewers of your city, or the number of people in your biology class. On a larger scale, all the people of the world constitute the human population. The terms *species* and *population* are interrelated because a species is a population—the largest possible population of a particular kind of organism. The term *population,* however, is often used to refer to portions of a species by specifying a space and time. For example, the size of the human population in a city changes from hour to hour during the day and varies according to where you set the boundaries of the city. Because each local population is a small portion of a species, we should expect various populations of the same species to show many kinds of differences.

Gene frequency is a measure of how often a specific gene shows up in the gametes of a population and is one of the ways in which populations differ. Chapter 11 on population genetics introduced you to this concept. Two populations of the same species often have quite different gene frequencies. For example, many populations of mosquitoes have high frequencies of insecticide-resistant genes, whereas others do not. The frequency of the genes for tallness in humans is greater in certain African tribes than in any other human population. Figure 16.1 shows that the frequency of the allele for type B blood differs significantly from one human population to another.

Because members of a population are of the same species, sexual reproduction can occur, and genes can flow from one generation to the next. Genes can also flow from one place to another as organisms migrate or are carried from one geographic location to another. **Gene flow** is used to refer to both the movement of genes within a species because of migration and the movement from one generation to the next as a result of gene replication and sexual reproduction. Typically both happen together as individuals migrate to new regions and subsequently reproduce, passing their genes to the next generation in the new area.

Age distribution is the number of organisms of each age in the population. In addition, organisms are often grouped into the following categories:

1. Prereproductive juveniles—insect larvae, plant seedlings, or babies
2. Reproductive adults—mature insects, plants producing seeds, or humans in early adulthood
3. Postreproductive adults no longer capable of reproduction—annual plants that have shed their seeds, salmon that have spawned, and many elderly humans

A population is not necessarily divided into equal thirds (figure 16.2). In some situations, a population may be made up of a majority of one age group. If the majority of the population is prereproductive, then a rapid increase in the size of the population should be anticipated in the future as the prereproductive individuals reach sexual maturity. If a majority of the population is reproductive, the population should be growing rapidly. If the majority of the population is postreproductive, a population decline should be anticipated. Many organisms that live only a short time and have high reproductive rates can have age distributions that change significantly in a matter of weeks or months. For example, many birds have a flurry of reproductive activity during the summer months. Therefore, if you sample the population of a specific species of bird at different times during the summer

Figure 16.1

Distribution of the Allele for Type B Blood
The allele for type B blood is not evenly distributed in the world. This map shows that the type B allele is most common in parts of Asia and has been dispersed to the Middle East and parts of Europe and Africa. There has been very little flow of the allele to the Americas.

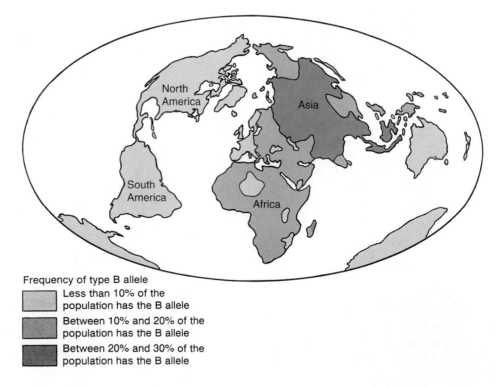

Frequency of type B allele

☐ Less than 10% of the population has the B allele

☐ Between 10% and 20% of the population has the B allele

■ Between 20% and 30% of the population has the B allele

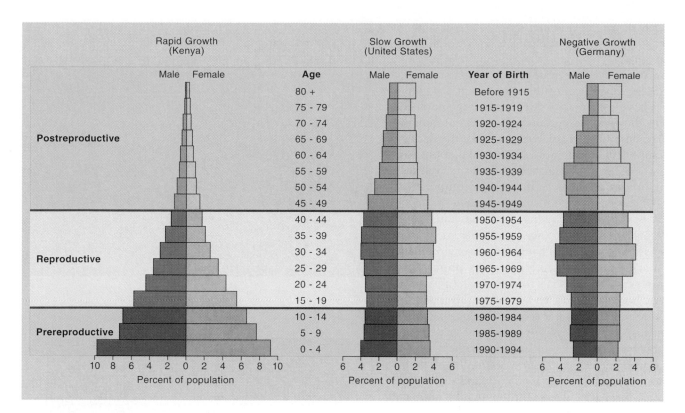

Figure 16.2

Age Distribution in Human Populations

The relative number of individuals in each of the three categories (prereproductive, reproductive, and postreproductive) can give a good clue to the future of the population. Kenya has a large number of young individuals who will become reproducing adults. Therefore this population would be projected to grow rapidly. The United States has a declining proportion of prereproductive individuals but a relatively large reproductive population. Therefore it will continue to grow for a time but will probably stabilize in the future. Germany's population has a large proportion of postreproductive individuals and a small proportion of prereproductive individuals. Its population is declining.

Source: Data from Population Reference Bureau

you would have widely different proportions of reproductive and prereproductive individuals.

The **sex ratio** is the number of males in a population compared to the number of females. In bird and mammal species where strong pair-bonding occurs, the sex ratio may be nearly one to one (1:1). Among mammals and birds that do not have strong pair-bonding, sex ratios may show a larger number of females than males. This is particularly true among game species, where more males than females are shot, resulting in a higher proportion of surviving females. Because one male can fertilize several females, the population can remain large even though the females outnumber the males. However, if the population of these managed game species becomes large enough to cause a problem, it becomes necessary to harvest some of the females as well because their number determines how much reproduction can take place. In addition to these examples, many species of animals like bison, horses, and elk have mating systems in which one male maintains a harem of females. The sex ratio in these small groups is quite different from a 1:1 ratio (figure 16.3). There are very few situations in which the number of males exceeds the number of females. In some human and other populations, there may be

Figure 16.3

Sex Ratio in Elk

Some male animals defend a harem of females; therefore the sex ratio in these groups is several females per male.

sex ratios in which the males dominate if female mortality is unusually high or if some special mechanism separates most of one sex from the other.

Population density is the number of organisms of a species per unit area. Regardless of the sex ratio of a population, most species have the capacity to generate large numbers of offspring. Depending on the reproductive success of individuals in the population and the resources available, the density of populations can vary considerably. Some populations are extremely concentrated in a limited space; others are well dispersed. As the population density increases, competition among members of the population for the necessities of life increases. This increases the likelihood that some individuals will explore new habitats and migrate to new areas. Increases in the intensity of competition that cause changes in the environment and lead to dispersal are often referred to as **population pressure.** The dispersal of individuals to new areas can relieve the pressure on the home area and lead to the establishment of new populations. Among animals, it is often the juveniles who participate in this dispersal process. For example, female bears generally mate every other year and abandon their nearly grown young the summer before the next set of cubs is to be born. The abandoned young bears tend to wander and disperse to new areas. Similarly, young turtles, snakes, rabbits, and many other common animals disperse during certain times of the year. That is one of the reasons you see so many road-killed animals in the spring and fall.

If dispersal cannot relieve population pressure, there is usually an increase in the rate at which individuals die because of predation, parasitism, starvation, and accidents. In plant populations, dispersal is not very useful for relieving population density; instead, the death of weaker individuals usually results in reduced population density. In the lodgepole pine, seedlings become established in areas following fire and dense thickets of young trees are established. As the stand ages, many small trees die and the remaining trees grow larger as the population density drops (figure 16.4).

16.2 Reproductive Capacity

Sex ratios and age distributions within a population have a direct bearing on the rate of reproduction. Each species has an inherent **reproductive capacity** or **biotic potential,** which is the theoretical maximum rate of reproduction. Generally, this biotic potential is many times larger than the number of offspring needed simply to maintain the population. For example, a female carp may produce 1 million to 3 million eggs in her lifetime. This is her reproductive capacity. However, only two or three of these offspring ever develop into sexually mature adults. Therefore, her reproductive rate is two or three offspring per lifetime and is much smaller than her reproductive potential.

(a)

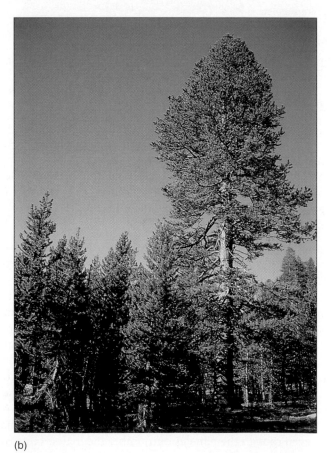
(b)

Figure 16.4

Changes in Population Density
This population of lodgepole pine seedlings consists of a large number of individuals very close to one another (a). As the trees grow, many of the weaker trees will die, the distance between individuals will increase, and the population density will be reduced (b).

A high reproductive capacity is valuable to a species because it provides many slightly different individuals for the environment to select among. With most plants and animals, many of the potential gametes are never fertilized. An oyster may produce a million eggs a year, but not all of them are fertilized, and most that are fertilized die. An apple tree with thousands of flowers may produce only a few apples because the pollen that contains the sperm cells was not transferred to the female part of each flower in the process of pollination. Even after the new individuals are formed, mortality is usually high among the young. Most seeds that fall to the earth do not grow, and most young animals die. But, usually, enough survive to ensure continuance of the species. Organisms that reproduce in this way spend large amounts of energy on the production of gametes and young, without caring for the young. Thus, the probability that any individual will reach reproductive age is small.

A second way of approaching reproduction is to produce relatively fewer individuals but provide care and protection that ensure a higher probability that the young will become reproductive adults. Humans generally produce a single offspring per pregnancy, but nearly all of them live. In effect, energy has been channeled into the care and protection of the young produced rather than into the production of incredibly large numbers of potential young. Even though fewer young are produced by animals like birds and mammals, their reproductive capacity still greatly exceeds the number required to replace the parents when they die.

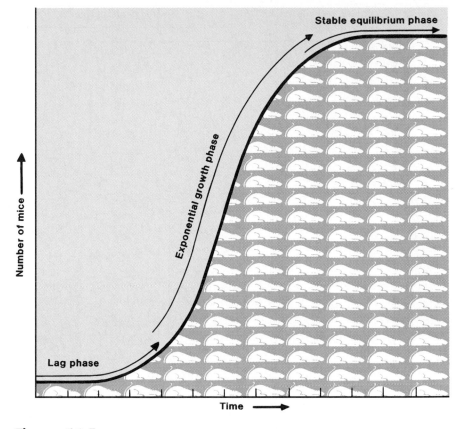

Figure 16.5

A Typical Population Growth Curve
In this mouse population, the period of time in which there is little growth is known as the lag phase. This is followed by a rapid increase in population as the offspring of the originating population begin to reproduce themselves; this is known as the exponential growth phase. Eventually the population reaches a stable equilibrium phase, during which the birthrate equals the deathrate.

16.3 The Population Growth Curve

Because most species of organisms have a high reproductive capacity, there is a tendency for populations to grow if environmental conditions permit. The change in the size of a population depends on the rate at which new organisms enter the population compared to the rate at which they leave. The number of new individuals added to the population by reproduction per thousand individuals is called **natality.** The number of individuals leaving the population by death per thousand individuals is called **mortality.** If a species enters a previously uninhabited area, its population will go through a typical pattern of growth. Figure 16.5 shows a graph of change in population size over time known as a **population growth curve.**

When a small number of organisms (two mice) first invade an area, there is a period of time before reproduction takes place during which the population remains small and relatively constant. This part of the population growth curve is known as the **lag phase.** During the lag phase both natality and mortality are low. The lag phase occurs because reproduction is not an instantaneous event. Even after animals enter an area they must mate and produce young. This may take days or years depending on the animal. Similarly, new plant introductions must grow to maturity, produce flowers, and set seed. Some annual plants may do this in less than a year, whereas some large trees may take several years of growth before they produce flowers.

In organisms that take a long time to mature and produce young, such as elephants, deer, and many kinds of

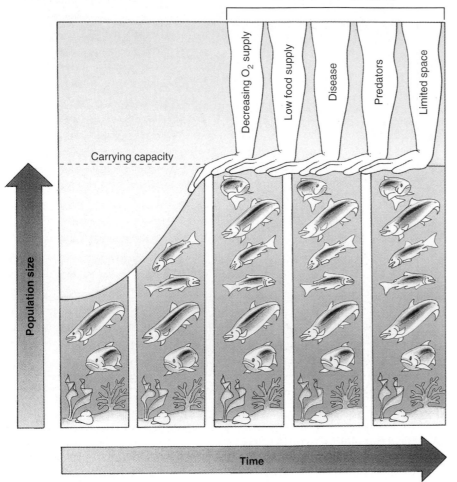

Environmental resistance

Decreasing O₂ supply

Low food supply

Disease

Predators

Limited space

Carrying capacity

Population size

Time

Figure 16.6

Carrying Capacity
A number of factors in the environment, such as food, oxygen supply, diseases, predators, and space determine the maximum number of organisms that can be sustained in a given area—the carrying capacity of that area. The environmental factors that limit populations are collectively known as environmental resistance.

plants, the lag phase may be measured in years. With the mice in our example, it will be measured in weeks. Once the first litter of young is produced the population is likely to increase rapidly. The first litter of young will be able to reproduce in a matter of weeks. If the usual litter size for a pair of mice is 4, the 4 would produce 8, which in turn would produce 16, and so forth. Furthermore, the original parents will probably produce an additional litter or two during this time period. Now we have several pairs of mice reproducing more than just once. With several pairs of mice reproducing, natality increases and mortality remains low; therefore the population begins to grow at an ever-increasing (accelerating) rate. This portion of the population growth curve is known as the **exponential growth phase.**

The number of mice (or any other organism) cannot continue to increase at a faster and faster rate because, even-

tually, something in the environment will become limiting and cause an increase in the number of deaths. For animals, food, water, or nesting sites may be in short supply, or predators or disease may kill many individuals. Plants may lack water, soil nutrients, or sunlight. Eventually, the number of individuals entering the population will come to equal the number of individuals leaving it by death or migration, and the population size becomes stable. Often there is both a decrease in natality and an increase in mortality at this point. This portion of the population growth curve is known as the **stable equilibrium phase.** It is important to recognize that this is still a population with births, deaths, migration, and a changing mix of individuals; however the size of the population is stable.

16.4 Population-Size Limitations

Populations cannot continue to increase indefinitely; eventually, some factor or set of factors acts to limit the size of a population, leading to the development of a stable equilibrium phase or even to a reduction in population size. The identifiable factors that prevent unlimited population growth are known as **limiting factors.** All the different limiting factors that act on a population are collectively known as **environmental resistance,** and the maximum sustainable population for an area is known as the **carrying capacity.** In general, organisms that are small and have short life spans tend to have fluctuating populations and do not reach a carrying capacity, whereas large organisms that live a long time tend to reach a population size that can be sustained over an extended period (figure 16.6). A forest ecosystem contains populations of many insect species that fluctuate widely and rarely reach a carrying capacity, but the number of a specific species of tree or large animals such as owls or deer is relatively constant. Each is at the carrying capacity of the ecosystem for its species.

Carrying capacity is not an inflexible number, however. Often such environmental differences as successional changes, climate variations, disease epidemics, forest fires, or floods can change the carrying capacity of an area for specific species. In aquatic ecosystems one of the major factors that determine the carrying capacity is the amount of nutrients in the water. In areas where nutrients are abundant, the numbers of various kinds of organisms are high. Often nutrient levels fluctuate with changes in current or runoff from the land, and plant and animal populations fluctuate as well.

1 fish—2 kg

2 fish—1.3 kg
0.7 kg
_____
2.0 kg

Figure 16.7

The Effect of Biomass on Carrying Capacity

Each aquarium can support a biomass of 2 kilograms of fish. The size of the population is influenced by the body size of the fish in the population.

8 fish each 250 g = 2 kg

6 fish—2 (500 g)
4 (250 g)
_____
2,000 g or 2.0 kg

In addition, a change that negatively affects the carrying capacity for one species may increase the carrying capacity for another. For example, the cutting down of a mature forest followed by the growth of young trees increases the carrying capacity for deer and rabbits, which use the new growth for food, but decreases the carrying capacity for squirrels, which need mature, fruit-producing trees as a source of food and old, hollow trees for shelter.

Wildlife management practices often encourage modifications to the environment that will increase the carrying capacity for the designated game species. The goal of wildlife managers is to have the highest sustainable population available for harvest by hunters. Typical habitat modifications include creating water holes, cutting forests to provide young growth, and encouraging the building of artificial nesting sites.

In some cases the size of the organisms in a population also affects the carrying capacity. For example, an aquarium of a certain size can support only a limited number of fish, but the size of the fish makes a difference. If all the fish are tiny, a large number can be supported, and the carrying capacity is high; however, the same aquarium may be able to support only one large fish. In other words, the biomass of the population makes a difference (figure 16.7). Similarly, when an area is planted with small trees, the population size is high. But as the trees get larger, competition for nutrients and sunlight becomes more intense, and the number of trees declines while the biomass increases.

16.5 Categories of Limiting Factors

Limiting factors can be placed in four broad categories:

1. Availability of raw materials
2. Availability of energy

3. Production and disposal of waste products
4. Interaction with other organisms

Availability of raw materials is an extremely important limiting factor. For example, plants require magnesium for the manufacture of chlorophyll, nitrogen for protein production, and water for the transport of materials and as a raw material for photosynthesis. If these substances are not present in the soil, the growth and reproduction of plants is inhibited. However, if fertilizer supplies these nutrients, or if irrigation is used to supply water, the effects of these limiting factors can be removed, and some other factor becomes limiting. For animals, the amount of water, minerals, materials for nesting, suitable burrow sites, or food may be limiting factors. Food for animals really fits into both this category and the next because it supplies both raw materials and energy.

Availability of energy is a major type of limiting factor because all living things need a source of energy. The amount of light available is often a limiting factor for plants, which require light as an energy source for photosynthesis. Because all animals use other living things as sources of energy and raw materials, a major limiting factor for any animal is its food source.

Accumulation of waste products can significantly limit the size of populations in certain instances. It does not usually limit plant populations because they produce relatively few wastes. However, the buildup of high levels of self-generated waste products is a problem for bacterial populations and populations of tiny aquatic organisms. As wastes build up, they become more and more toxic, and eventually reproduction stops, or the population may even die out. For example, when a few bacteria are introduced into a solution containing a source of food, they go through the kind of population growth curve typical of all organisms. As expected, the number of bacteria begins to increase following a

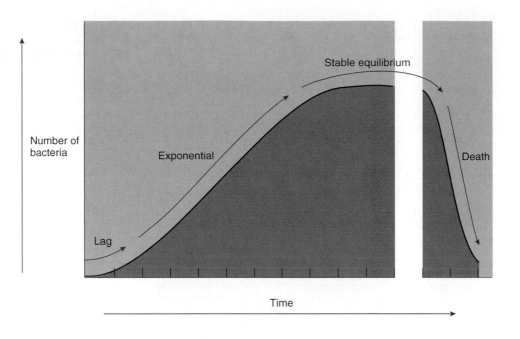

Number of bacteria

Lag

Exponential

Stable equilibrium

Death

Time

Figure 16.8

Bacterial Population Growth Curve
The rate of increase in the size of the population of these bacteria is typical of population growth in a favorable environment. When the environmental conditions change as a result of an increase in the amount of waste products, the population first levels off, then begins to decrease. This period of decreasing population size is known as the death phase.

lag phase, increases rapidly during the exponential growth phase, and eventually reaches stability in the stable equilibrium phase. But as waste products accumulate, the bacteria literally drown in their own wastes. When space for disposal is limited, and no other organisms are present that can convert the harmful wastes to less harmful products, a population decline known as the **death phase** follows (figure 16.8).

Wine makers deal with this situation. When yeasts ferment the sugar in grape juice, they produce ethyl alcohol. When the alcohol concentration reaches a certain level, the yeast population stops growing and eventually declines. Therefore wine can naturally reach an alcohol concentration of only 12% to 15%. To make any drink stronger than that (of a higher alcohol content), water must be removed (to distill) or alcohol must be added (to fortify).

In small aquatic pools like aquariums, it is often difficult to keep populations of organisms healthy because of the buildup of ammonia in the water from the waste products of the animals. This is the primary reason that activated charcoal filters are commonly used in aquariums. The charcoal removes many kinds of toxic compounds and prevents the buildup of waste products.

Organism interactions are also important in limiting population size. As we learned in chapter 15, organisms influence each other in many ways. Some organisms are harmed and others benefit. The population size of any organism is negatively affected by parasitism, predation, or competition. Parasitism and predation usually involve interactions between two different species, although cannibalism of others

of the same species does occur in some animals. Competition among members of the same species is often extremely intense. This is true for all kinds of organisms, not just animals.

On the other hand, many kinds of organisms perform services for others that have beneficial effects on the population. For example, decomposer organisms destroy toxic waste products, thus benefiting populations of animals. They also recycle materials needed for the growth and development of all organisms. Mutualistic relationships benefit both populations involved. The absence of such beneficial organisms would be a limiting factor.

Often, the population sizes of two kinds of organisms are interdependent because each is a primary limiting factor of the other. This is most often seen in parasite-host relationships and predator-prey relationships. A good example is the relationship of the lynx (a predator) and the varying hare (the prey) as it was studied in Canada. The varying hare has a high reproductive capacity. In peak reproductive years, a female varying hare can produce 16 to 18 young. As with many animals, a primary cause of death is predation. The varying hare population is a good food source for a variety of predators including the lynx. When the population of varying hares increases it provides an abundant source of food for the lynx and the size of the lynx population rises, and when the population of hares decreases so does that of the lynx. This pattern repeats itself in approximately a 10-year cycle (figure 16.9). Biologists have studied this pattern for many years seeking to determine what triggers the fluctuations in the populations of these two animals.

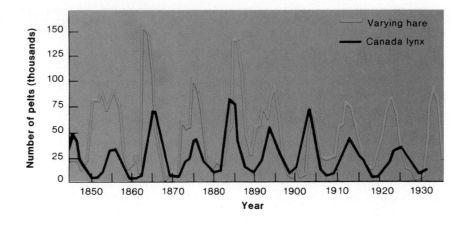

Figure 16.9

Organism Interaction

The interaction between predator and prey species is complex and often difficult to interpret. These data were collected from the records of the number of pelts purchased by the Hudson Bay Company. It shows that the two populations fluctuate, with changes in the lynx population usually following changes in varying hare populations.

Source: Data from D. A. MacLulich, *Fluctuations in the Numbers of the Varying Hare (Lepus americanus)*, University of Toronto Press, 1937, reprinted 1974. Toronto, Canada.

Recent studies indicate that one of the causes of the decline in varying hare populations is a reduction in their reproductive rate. The causes of this reduction may be related to a variety of factors including: reduced quality of food and higher levels of stress resulting from greater difficulty in finding food and avoiding predators. With reduced reproduction and continued high predation the varying hare population drops. With reduced numbers of hares to serve as food, the lynx population drops. Eventually the reproductive rate of hares increases and the population rebounds followed by a rebound in the lynx population as well. It appears that both food availability and predation are important limiting factors that determine the size of the varying hare population and the number of varying hares is a primary limiting factor for the lynx.

Extrinsic and Intrinsic Limiting Factors

Some factors that control populations come from outside the population and are known as **extrinsic factors.** Predators, loss of a food source, lack of sunlight, or accidents of nature are all extrinsic factors. However, the populations of many kinds of organisms appear to be regulated by factors from within the populations themselves. Such limiting factors are called **intrinsic factors.** For example, a study of rats under crowded living conditions showed that as conditions became more crowded, abnormal social behavior became common. There was a decrease in litter size, fewer litters per year were produced, mothers were more likely to ignore their young, and many young were killed by adults. Thus changes in the behavior of the members of the rat population itself resulted in lower birthrates and higher deathrates, leading to a reduction in the population growth rate. As another example, trees that are stressed by physical injury or disease often produce extremely large numbers of seeds (offspring) the follow-

ing year. The trees themselves alter their reproductive rate. The opposite situation is found among populations of white-tailed deer. It is well known that reproductive success is reduced when the deer experience a series of severe winters. When times are bad, the female deer are more likely to have single offspring rather than twins.

Density-Dependent and Density-Independent Limiting Factors

Many populations are controlled by limiting factors that become more effective as the density of the population increases. Such factors are referred to as **density-dependent factors.** Many of the factors we have already discussed are density-dependent. For example, the larger a population becomes, the more likely it is that predators will have a chance to catch some of the individuals. A prolonged period of increasing population allows the size of the predator population to increase as well. Large populations with high population density are more likely to be affected by epidemics of parasites than are small populations of widely dispersed individuals because dense populations allow for the easy spread of parasites from one individual to another. The rat example discussed previously is another good example of a density-dependent factor operating because the amount of abnormal behavior increased as the density of the population increased. In general, whenever there is competition among members of a population, its intensity increases as the population density increases. Large organisms that tend to live a long time and have relatively few young are most likely to be controlled by density-dependent factors.

Density-independent factors are population-controlling influences that are not related to the density of the population. They are usually accidental or occasional extrinsic factors in

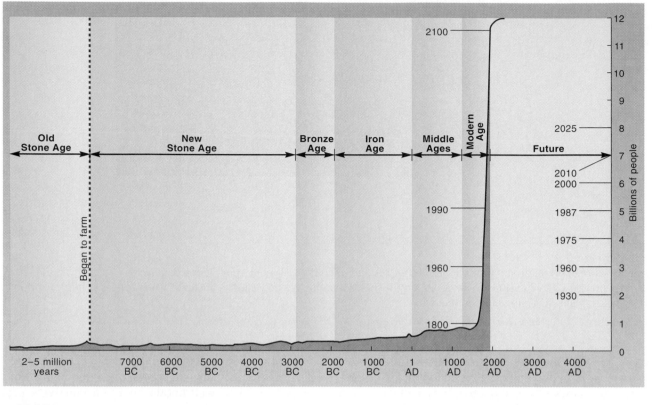

Figure 16.10

Human Population Growth

The number of humans doubled from A.D. 1800 to 1930 (from 1 billion to 2 billion), doubled again by 1975 (4 billion), and is projected to double again (8 billion) by about the year 2025. How long can the human population continue to double before the Earth's ultimate carrying capacity is reached?

Source: Data from Jean Van Der Tak, et al., "Our Population Predicament: A New Look," in *Population Reference Bureau*, Washington, D.C.

nature that happen regardless of the density of a population. A sudden rainstorm may drown many small plant seedlings and soil organisms. Many plants and animals are killed by frosts that come late in spring or early in the fall. A small pond may dry up, resulting in the death of many organisms. The organisms most likely to be controlled by density-independent factors are small, short-lived organisms that can reproduce very rapidly.

So far we have looked at populations primarily from a nonhuman point of view. Now it is time to focus on the human species and the current problem of world population growth.

16.6 Limiting Factors to Human Population Growth

Today we hear differing opinions about the state of the world's human population. On one hand we hear that the population is growing rapidly. By contrast we hear that some countries are afraid that their populations are shrinking. Other countries are concerned about the aging of their populations because birthrates and deathrates are low. In maga-

zines and on television we see that there are starving people in the world. At the same time we hear discussions about the problem of food surpluses and obesity in many countries. Some have even said that the most important problem in the world today is the rate at which the human population is growing; others maintain that the growing population will provide markets for goods and be an economic boon. How do we reconcile this mass of conflicting information?

It is important to realize that human populations follow the same patterns of growth and are acted upon by the same kinds of limiting factors as are populations of other organisms. When we look at the growth of the human population over the past several thousand years, we see a pattern that resembles the lag and exponential growth phases of a population growth curve. It is estimated that the human population remained low and constant for thousands of years but has increased rapidly in the past few hundred years (figure 16.10). For example, it has been estimated that when Columbus discovered America, the Native American population was about 1 million and was at or near its carrying capacity. Today, the population of North America is over 300 million people. Does this mean that humans are different from other animal species? Can the human population continue to grow forever?

HOW SCIENCE WORKS 16.1

Thomas Malthus and His Essay on Population

In 1798 Thomas Robert Malthus, an Englishman, published an essay on human population. It presented an idea that was contrary to popular opinion. His basic thesis was that human population increased in a geometric or exponential manner (2, 4, 8, 16, 32, 64, etc.), whereas the ability to produce food increased only in an arithmetic manner (1, 2, 3, 4, 5, 6, etc.). The ultimate outcome of these different rates would be that population would outgrow the ability of the land to produce food. He concluded that wars, famines, plagues, and natural disasters would be the means of controlling the size of the human population. His predictions were hotly debated by the intellectual community of his day. His assumptions and conclusions were attacked as erroneous and against the best interest of society. At the time he wrote the essay, the popular opinion was that human knowledge and "moral constraint" would be able to create a world that would supply all human needs in abundance. One of Malthus's basic postulates was that "commerce between the sexes" (sexual intercourse) would continue unchanged; other philosophers of the day believed that sexual behavior would take less procreative forms and human population would be limited. Only within the

past 50 years, however, have really effective conception-control mechanisms become widely accepted and used, and they are used primarily in developed countries.

Malthus did not foresee the use of contraception, major changes in agricultural production techniques, or the exporting of excess people to colonies in the Americas. These factors, as well as high deathrates, prevented the most devastating of his predictions from coming true. However, in many parts of the world today, people are experiencing the forms of population control (famine, epidemic disease, wars, and natural disasters) predicted by Malthus in 1798. Many people believe that his original predictions were valid—only his time scale was not correct—and that we are seeing his predictions come true today.

Another important impact of Malthus's essay was the effect it had on the young Charles Darwin. When Darwin read it, he saw that what was true for the human population could be applied to the whole of the plant and animal kingdoms. As over-reproduction takes place, there would be increased competition for food, resulting in the death of the less fit organisms. This was an important part of the theory he called *natural selection*.

The human species is no different from other animals. It has an upper limit set by the carrying capacity of the environment but the human population has been able to increase astronomically because technological changes and displacement of other species has allowed us to shift the carrying capacity upward. Much of the exponential growth phase of the human population can be attributed to the removal of diseases, improvement in agricultural methods, and replacement of natural ecosystems with artificial agricultural ecosystems. But even these conditions have their limits. There will be some limiting factors that eventually cause a leveling off of our population growth curve. We cannot increase beyond our ability to get raw materials and energy, nor can we ignore the waste products we produce or the other organisms with which we interact.

Available Raw Materials

To many of us, raw materials consist simply of the amount of food available, but we should not forget that in a technological society, iron ore, lumber, irrigation water, and silicon chips are also raw materials. However, most people of the world have much more basic needs. For the past several decades, large portions of the world's population have not had enough food. Although it is biologically accurate to say that the world can currently produce enough food to feed all the people of the world, there are many reasons why people can't get food or won't eat it. Many cultures have food taboos or traditions that prevent the use of some available food sources. For example, pork is forbidden in some cultures. Certain groups of people find it almost impossible to digest milk. Some African cultures

use a mixture of cow's milk and cow's blood as food, which people of other cultures might be unable to eat.

In addition, there are complex political, economic, and social issues related to the production and distribution of food. In some cultures, farming is a low-status job, which means that people would rather buy their food from someone else than grow it themselves. This can result in under-utilization of agricultural resources. Food is sometimes used as a political weapon when governments want to control certain groups of people. But probably most important is the fact that transportation of food from centers of excess to centers of need is often very difficult and expensive. Those societies with excess food have been unwilling to bridge these political and economic gaps.

However, a more fundamental question is whether the world can continue to produce enough food. In 2003, the world population was growing at a rate of 1.3% per year. This amounts to about 150 new people added to the world population every minute. This would result in an increase from our current population of about 6.3 billion to over 9 billion by 2050. With a continuing increase in the number of mouths to feed, it is unlikely that food production will be able to keep pace with the growth in human population (How Science Works 16.1). A primary indicator of the status of the world food situation is the amount of grain produced for each person in the world (per capita grain production). World per capita grain production peaked in 1984. The less-developed nations of the world have a disproportionately large increase in population and a decline in grain production because they are less able to afford costly fertilizer, machinery, and the energy necessary to run the machines and irrigate the land to produce their own grain.

Availability of Energy

The availability of energy is the second broad limiting factor that affects human populations as well as other kinds of organisms. All species on Earth ultimately depend on sunlight for energy—including the human species. Whether one produces electrical power from a hydroelectric dam, burns fossil fuels, or uses a solar cell, the energy is derived from the Sun. Energy is needed for transportation, building and maintaining homes, and food production. It is very difficult to develop unbiased, reasonably accurate estimates of global energy "reserves" in the form of petroleum, natural gas, and coal. Therefore, it is difficult to predict how long these "reserves" might last. We do know, however, that the quantities are limited and that the rate of use has been increasing.

If the less-developed countries were to attain a standard of living equal to that of the developed nations, the global energy "reserves" would disappear overnight. Because the United States constitutes approximately 4.6% of the world's population and consumes approximately 25% of the world's energy resources, raising the standard of living of the entire world population to that of the United States would result in a tremendous increase in the rate of consumption of energy and reduce theoretical reserves drastically. Humans should realize that there is a limit to our energy resources; we are living on solar energy that was stored over millions of years, and we are using it at a rate that could deplete it in hundreds of years. Will energy availability be the limiting factor that determines the ultimate carrying capacity for humans, or will problems of waste disposal predominate?

Production of Wastes

One of the most talked-about aspects of human activity is the problem of waste disposal. Not only do we have normal biological wastes, which can be dealt with by decomposer organisms, but we generate a variety of technological wastes and by-products that cannot be efficiently degraded by decomposers. Most of what we call pollution results from the waste products of technology. The biological wastes usually can be dealt with fairly efficiently by building wastewater treatment plants and other sewage facilities. Certainly these facilities take energy to run, but they rely on decomposers to degrade unwanted organic matter to carbon dioxide and water. Earlier in this chapter we discussed the problem that bacteria and yeasts face when their metabolic waste products accumulate. In this situation, the organisms so "befoul their nest" that their wastes poison them. Are humans in a similar situation on a much larger scale? Are we dumping so much technological waste, much of which is toxic, into the environment that we are being poisoned? Some people believe that disregard for the quality of our environment will be a major factor in decreasing our population growth rate. In any case, it makes good sense to do everything possible to stop pollution and work toward cleaning our nest.

Interactions with Other Organisms

The fourth category of limiting factors that determine carrying capacity is interaction among organisms. Humans interact with other organisms in as many ways as other animals do. We have parasites and occasionally predators. We are predators in relation to a variety of animals, both domesticated and wild. We have mutualistic relationships with many of our domesticated plants and animals because they could not survive without our agricultural practices and we would not survive without the food they provide. Competition is also very important. Insects and rodents compete for the food we raise, and we compete directly with many kinds of animals for the use of ecosystems.

As humans convert more and more land to agriculture and other purposes, many other organisms are displaced. Many of these displaced organisms are not able to compete successfully and must leave the area, have their populations reduced, or become extinct. The American bison (buffalo), African and Asian elephants, panda, and grizzly bear are a few species that are considerably reduced in number because they were not able to compete successfully with the human species. The passenger pigeon, Carolina parakeet, and great auk are a few that have become extinct. Our parks and natural areas have become tiny refuges for plants and animals that once occupied vast expanses of the world. If these refuges are lost, many organisms will become extinct. What today might seem to be an insignificant organism that we can easily do without may tomorrow be seen as a link to our very survival. We humans have been extremely successful in our efforts to convert ecosystems to our own uses at the expense of other species.

Competition with one another (intraspecific competition), however, is a different matter. Because competition is negative to both organisms, competition between humans harms humans. We are not displacing another species, we are displacing some of our own kind. Certainly, when resources are in short supply, there is competition. Unfortunately, it is usually the young that are least able to compete, and high infant mortality is the result.• organism interactions, p. 284

Control of Human Population Is a Social Problem

Humans are different from most other organisms in a fundamental way: We are able to predict the outcome of a specific course of action. Current technology and medical knowledge are available to control human population and improve the health and well-being of the people of the world. Why then does the human population continue to grow, resulting in human suffering and stressing the environment in which we live? Because we are social animals that have freedom of choice, we frequently do not do what is considered "best" from an unemotional, unselfish, biological point of view. People make decisions based on historical, social, cultural, ethical, and personal considerations. What is best for the population as a whole may be bad for you as an individual (Outlooks 16.1).

OUTLOOKS 16.1

Government Policy and Population Control

The actions of government can have a significant impact on the population growth patterns of nations. Some countries have policies that encourage couples to have children. The U.S. tax code indirectly encourages births by providing tax advantages to the parents of children. Some countries in Europe are concerned about the lack of working-age people in the future and are considering ways to encourage births.

China and India are the two most populous countries in the world. Both have over 1 billion people. However, China has taken steps to control its population and now has a total fertility rate of 1.8 children per woman while India has a total fertility rate of 3.2. The total fertility rate is the average number of children born to a woman during her lifetime. The differences in total fertility rates between these two countries are the result of different policy decisions over the last 50 years. The history of China's population policy is an interesting study of how government policy affects reproductive activity among its citizens. When the People's Republic of China was established in 1949, the official policy of the government was to encourage births because more Chinese would be able to produce more goods and services, and production was the key to economic prosperity. The population grew from 540 million to 614 million between 1949 and 1955 while economic progress was slow. Consequently, the government changed its policy and began to promote population control.

The first family-planning program began in 1955, as a means of improving maternal and child health. Birthrates fell (see graph). But other social changes resulted in widespread famine and increased deathrates and low birthrates in the late 1950s and early 1960s.

The present family-planning policy began in 1971 with the launching of the *wan xi shao* campaign. Translated, this phrase means "later" (marriages), "longer" (intervals between births), and "fewer" (children). As part of this program the legal ages for marriage were raised. For women and men in rural areas, the ages were raised to 23 and 25, respectively; for women and men in urban areas the ages were raised to 25 and 28, respectively. These policies resulted in a reduction of birthrates by nearly 50% between 1970 and 1979.

An even more restrictive one child campaign was begun in 1978–1979. The program offered incentives for couples to restrict their family size to one child. Couples enrolled in the program would receive free medical care, cash bonuses for their work, special housing treatment, and extra old-age benefits.

(continued on page 316)

Source: Data from H. Yuan Tien, "China's Demographic Dilemmas," in *Population Bulletin,* 1992, Population Reference Bureau, Inc., Washington, D.C., and Natural Family Planning Commission of China; and more recent data taken from Population Reference Bureau, Inc.

OUTLOOKS 16.1 (continued)

Those who broke their pledge were penalized by the loss of these benefits as well as other economic penalties. By the mid-1980s less than 20% of the eligible couples were signing up for the program. Rural couples, particularly, desired more than one child. In fact, in a country where over 60% of the population is rural, the rural total fertility rate was 2.5 children per woman. (The total fertility rate is the number of children born per woman per lifetime.) In 1988 a second child was sanctioned for rural couples if their first child was a girl, which legalized what had been happening anyway.

The programs appear to have had an effect because the current total fertility rate has fallen to 1.7 children per woman. Replacement fertility, the total fertility rate at which the population would eventually stabilize, is 2.1 children per woman per lifetime. Furthermore, over 80% of couples use contraception. Abortion is also an important aspect of this program, with a ratio of more than 600 abortions per 1,000 live births.

By contrast, during the same 50 years India has had little success in controlling its population. In 2000 a new plan was unveiled that includes the goal of bringing the total fertility rate from 3.2 children per woman to 2 (replacement rate) by 2010. In the past, the emphasis of government programs was on meeting goals of sterilization and contraceptive use but this has not been successful. Today less than 50% of couples use contraceptives. This new plan will emphasize improvements in the quality of life of the people. The major thrusts will be to reduce infant and maternal death, immunize children against preventable disease, and encourage girls to attend school. It is hoped that improved health will remove the perceived need for large numbers of births. Currently, less than 50% of the women in India can read and write. The emphasis on improving the educational status of women is related to the experiences of other developing countries. In many other countries it has been shown that an increase in the education level of women has been linked to lower fertility rates.

It seems overly optimistic to think that the total fertility rate can be reduced from 3.2 to 2 in just 10 years, but programs that emphasize improvements in maternal and child health and increasing the educational level of women have been very effective in reducing total fertility in other countries.

The biggest problems associated with control of the human population are not biological problems but, rather, require the efforts of philosophers, theologians, politicians, and sociologists. As population increases, so will political, social, and biological problems; individual freedom will diminish, intense competition for resources will intensify, and famine and starvation will become even more common. The knowledge and technology necessary to control the human population are available, but the will is not. What will eventually limit the size of our population? Will it be lack of resources, lack of energy, accumulated waste products, competition among ourselves, or rational planning of family size?

Recent studies of the changes in the population growth rates of different countries indicates that a major factor determining the size of families is the educational status of women. Regardless of other cultural differences, as girls and women become educated they have fewer children. Several reasons have been suggested for this trend. Higher levels of education allow women to get jobs with higher pay, which makes them less dependent on males for their support. Being able to read may lead to better comprehension of how methods of birth control work. Regardless of the specific reasons, improving educational levels of women has now become a major technique employed by rapidly growing countries that hope to eventually control their populations.

SUMMARY

A population is a group of organisms of the same species in a particular place at a particular time. Populations differ from one another in gene frequency, age distribution, sex ratio, and population density. Organisms typically have a reproductive capacity that exceeds what is necessary to replace the parent organisms when they die. This inherent capacity to overreproduce causes a rapid increase in population size when a new area is colonized. A typical population growth curve consists of a lag phase in which population rises very slowly, followed by an exponential growth phase in which the population increases at an accelerating rate, followed by a leveling off of the population in a stable equilibrium phase as the carrying capacity of the environment is reached. In some populations, a fourth phase may occur, known as the death phase. This is typical of bacterial and yeast populations.

The carrying capacity is the maximum sustainable number of organisms an area can support. It is set by a variety of limiting factors. Availability of energy, availability of raw materials, accumulation of wastes, and interactions with other organisms are all categories of limiting factors. Because organisms are interrelated, population changes in one species sometimes affect the size of other populations. This is particularly true when one organism uses another as a source of food. Some limiting factors operate from outside the population and are known as extrinsic factors; others are properties of the species itself and are called intrinsic factors. Some limiting factors become more intense as the density of the population increases; these are known as density-dependent factors. Other limiting factors that are more accidental and not related to population density are called density-independent factors.

Humans as a species have the same limits and influences that other organisms do. Our current problems of food production, energy needs, pollution, and habitat destruction are outcomes of uncontrolled population growth. However, humans can reason and predict, thus offering the possibility of population control through conscious population limitation.

THINKING CRITICALLY

For guidelines to answer this question, visit the Online Learning Center.

If you return to figure 16.10, you will note that it has very little in common with the population growth curve shown in figure 16.5. What factors have allowed the human population to grow so rapidly? What natural limiting factors will eventually bring this population under control?

What is the ultimate carrying capacity of the world? What alternatives to the natural processes of population limitation could bring human population under control?

Consider the following in your answer: reproduction, death, diseases, food supply, energy, farming practices, food distribution, cultural biases, and anything else you consider relevant.

CONCEPT MAP TERMINOLOGY

Construct a concept map using these terms.

biotic potential	mortality
carrying capacity	natality
exponential growth phase	population density
extrinsic limiting factors	sex ratio
intrinsic limiting factors	stable equilibrium phase
lag phase	

Visit the Online Learning Center to use an interactive format to construct another concept map.

KEY TERMS

Use interactive flash cards, on the Online Learning Center, to help you learn the meanings of these terms.

age distribution	lag phase
biotic potential	limiting factors
carrying capacity	mortality
death phase	natality
density-dependent factors	population
density-independent factors	population density
environmental resistance	population growth curve
exponential growth phase	population pressure
extrinsic factors	reproductive capacity
gene flow	sex ratio
gene frequency	stable equilibrium phase
intrinsic factors	

LEARNING CONNECTIONS

Visit the Online Learning Center to find additional study resources including those highlighted below.

16.1 Population Characteristics

Animations and Review
- Introduction
- Characteristics

Interactive Concept Map
- Population characteristics

1. Describe four ways in which two populations of the same species can differ.
2. How is population pressure related to population density?

16.2 Reproductive Capacity

3. Why do populations grow?
4. In what way do the activities of species that produce few young differ from those that produce huge numbers of offspring?
5. All organisms over-reproduce. What advantages does this give to the species? What disadvantages?

16.3 The Population Growth Curve

Animations and Review
- Growth

Interactive Concept Map
- Growth curve

6. Draw a population growth curve. Label the lag, exponential growth, and stable equilibrium phases.
7. What causes a lag phase in a population growth curve? An exponential growth phase? A stable equilibrium phase?

16.4 Population-Size Limitations

8. How is the carrying capacity of an environment related to the stable equilibrium phase of a population growth curve?

9. Give an example of how the carrying capacity of an environment can be related to the size of individual organisms.

16.5 Categories of Limiting Factors

Animations and Review
• Size regulation

10. List four kinds of limiting factors that help set the carrying capacity for a species.

11. Differentiate between density-dependent and density-independent limiting factors. Give an example of each.

12. Differentiate between intrinsic and extrinsic limiting factors. Give an example of each.

16.6 Limiting Factors to Human Population Growth

Animations and Review
• Food needs
• Food production
• Food quality
• Future prospects

Experience This!
• Plot the growth of a population

13. As the human population continues to grow, what should we expect to happen to other species?

14. How does the population growth curve of humans compare with that of other kinds of animals?

Test yourself with multiple choice questions that provide immediate feedback on the Online Learning Center with PowerWeb at

www.mhhe.com/enger11

Evolutionary and Ecological Aspects of Behavior

GOALS AND OBJECTIVES

Understand that instinctive and learned behaviors have evolutionary and ecological significance.

❑ Explain how instinctive and learned behaviors differ.

❑ Describe the characteristics of animals in which instinctive behaviors predominate.

❑ Describe the characteristics of animals in which learned behaviors have a dominant role.

❑ Give examples of instinctive behaviors and state how you know they are instinctive.

❑ Give examples of behaviors that have both instinctive and learned components.

Recognize that there are different kinds of learning.

❑ Explain the nature and significance of habituation.

❑ Describe the different kinds of association learning.

❑ State the nature and significance of imprinting, exploratory learning, and insight.

❑ Provide examples of human behaviors that illustrate habituation, association, imprinting, and insight.

Recognize the relationship between behavior patterns and the ecological niche of an organism.

❑ Describe how territoriality and dominance hierarchies allocate resources.

❑ Explain the adaptive value of specific behaviors such as communication, food storage, navigation, a time sense, care of the young, and hibernation in particular ecological settings.

❑ Explain why the behavioral evolution of social animals is different from that of nonsocial animals.

17.1 The Adaptive Nature of Behavior

Behavior is how an animal acts, what it does, and how it does it. When we observe behavior, we must keep in mind that behavior is adaptive just like any other characteristic displayed by an animal. Behaviors are important for survival and appropriate for the environment in which an animal lives. As with the highly adaptive structural characteristics we see in animals, behaviors are the result of a long evolutionary process.

Both plants and animals must solve the same kinds of problems. For example, they must obtain nutrients, avoid being eaten, and reproduce. While behaviors are important for animals, plants, for the most part, must rely on structures, physiological changes, or chance to accomplish the same ends. Many kinds of animals use various behaviors (searching, flying, walking, chewing) to find and process the food they need, while plants can only obtain nutrients that happen to be located near their roots. A rabbit can run away from a predator, a plant cannot. But plants may have thorns or toxic compounds in their leaves that discourage animals from eating them. Some animals engage in elaborate courtship behaviors to identify the species and sex of a potential mate. Most plants rely on a much more random method for transferring male gametes to the female plant, although they may use the behavior of insects and other animals to transfer pollen.

Before we go much further, we need to discuss how animals generate specific behaviors and the two major kinds of behaviors: instinctive and learned. Both instinct and learning are involved in the behavior patterns of most organisms. We recognize **instinctive behavior** as behavior that is inborn, automatic, and inflexible; whereas **learned behavior** requires experience, produces new behaviors, and can be changed. Most animals (jellyfish, clams, insects, worms, frogs, turtles, etc.) have a high proportion of instinctive behavior and very little learning; some, like many birds and mammals, are able to demonstrate a great deal of learned behavior in addition to many instinctive behaviors.

Since the majority of the behaviors of most animals are instinctive, inherited behaviors, they evolved just like the structures of animals. In this respect, behavioral characteristics are no different from structural characteristics. However, the evolution of behavior is much more difficult to study, because behavior is temporary and it is difficult to find fossils that allow us to follow the development of behavior the way fossils allow us to follow changes in structures. Fossils of footprints, and nests, and the presence of specific structures like grinding teeth, do tell us something about the behavior of extinct animals, but these represent only a few fragments of the total behavior that must have been a part of the life of extinct animals. When we compare the behavior of closely related animals living today, it is often possible to see inherited behaviors that are slightly different from one another, just as the wings of different species of birds are slightly different but based on the same pattern. • evolution, p. 222.

Figure 17.1

Animal Behavior
A baby herring gull causes its parent to regurgitate food onto the ground by pecking at the red spot on its parent's bill. This behavior is instinctive. The parent then picks up the food and feeds it to the baby.

17.2 Interpreting Behavior

When you watch a bird or squirrel or any other animal, its activities appear to have a purpose. Birds search for food, take flight as you approach, and build nests in which to raise young. Usually the nests are inconspicuous or placed in difficult-to-reach spots. Likewise, squirrels collect and store nuts and acorns, "scold" you when you get too close, and learn to visit sites where food is available. All of these activities are adaptive and contribute to the individual's success. Birds that do not take flight at the approach of another animal will be eaten by predators. Squirrels that do not find deposits of food will be less likely to survive, and birds that build obvious nests on the ground will be more likely to lose their young to predators. However, we need to take care not to attach too much meaning to what animals do. The behaviors of nonhuman animals may not be the result of the same thought processes and motivations that humans experience.

It is not always easy to identify the significance of a behavior without careful study of the behavior pattern and the impact it has on other organisms (How Science Works 17.1). For example, a baby herring gull pecks at a red spot on its parent's bill. What possible value can this behavior have for either the chick or the parent? If we watch, we see that when the chick pecks at the spot, the parent regurgitates food onto the ground, and the chick feeds (figure 17.1). This looks like a simple behavior, but there is more to it than meets the eye. Why did the chick peck to begin with? How did it "know" to peck at that particular spot? Why did the pecking cause the parent to regurgitate food? These questions are not easy to answer. Many people assume that the actions have the

HOW SCIENCE WORKS 17.1

Observation and Ethology

Ethology is the study of the behavior of animals in their natural environment. Such studies do not lend themselves easily to experimentation because it is difficult to manipulate conditions to fit the criteria of an experimental design. Therefore, most studies of the behavior of wild animals involve careful, extended observation to determine the significance of specific animal behaviors. Often the meaning of behaviors can be understood only if the observer is able to identify the individuals involved. For most of us, all robins, raccoons, and opossums look alike, making it difficult for us to understand behaviors we see between individuals.

Ethologists have used various techniques to allow easy identification of individuals. Some animals have individual variations that allow identification. Wild horses differ in size and color patterns, whales have differently shaped tails, and gorillas have size, color, and facial differences. It is the observational skill of the scientist that allows recognition of unique features. One

scientist observing chipmunks was able to identify many individuals by specific characteristics of the tail, such as the shape and size of the tail, how the tail was carried, and minor differences in color.

In studies of bird behavior, scientists have placed colored bands on birds' legs that can be viewed through powerful binoculars. This allows investigators to determine which birds in a group are dominant, how they move through their environment, or how long they live. The attachment of small radio transmitters to animals as small as pigeons and as large as bears and sea turtles has allowed investigators to locate and follow individuals over large areas.

Although various techniques can be used to identify individual animals, they are only tools that help investigators answer questions and decipher meaning from the behaviors they observe.

same motivation and direction as similar human behaviors, but this is not necessarily a correct assumption. For example, when a human child points to a piece of candy and makes appropriate noises, it is indicating to its parent that it wants some candy. Is that what the herring gull chick is doing? We don't know. Although both kinds of young may get food, we don't know what the baby gull is thinking because we can't ask it.

17.3 The Problem of Anthropomorphism

Poets, composers, and writers have often described birdsong as the act of a joyful songster. But is that bird singing on a warm, sunny spring day making that beautiful sound because it is so happy? Students of animal behavior do not accept this idea and have demonstrated that a bird sings primarily to attract mates and tell other birds to keep out of its territory. The barbed stinger of a honeybee remains in your skin after you are stung, and the bee tears the stinger out of its body when it flies away. The damage to its body is so great that it dies. Has the bee performed a noble deed of heroism and self-sacrifice? Was it defending its hive from you? We need to know a great deal more about the behavior of bees to understand the value of such behavior to the success of the bee species. The fact that bees are social animals like us makes it particularly tempting to think that they are doing things for the same reasons we are.

The idea that we can ascribe human feelings, meanings, and emotions to the behavior of animals is called **anthropomorphism.** But we cannot crawl inside the brain of another animal and see what it is thinking so we need to avoid these anthropomorphic assumptions when attempting to understand the behavior of nonhuman animals. The fable of the

Figure 17.2

An Example of Anthropomorphism
In many ways we give human meaning to the actions of animals. The ant is portrayed as an industrious individual that prepares for the future, and the grasshopper as a lazy fellow that sits in the sun and sings all day. This view of animal behavior is anthropomorphic.

grasshopper and the ant is another example of inappropriately crediting animals with human qualities. The ant is pictured as an animal that, despite temptations, works hard from morning until night, storing away food for the winter (figure 17.2). The grasshopper, on the other hand, is represented as a lazy good-for-nothing that fools away the summer singing, when it really ought to be saving up for the tough times ahead. If one is looking for analogies to human behavior, these are pretty good illustrations. But they really are not accurate statements about the lives of the animals from an ecological point of view. Ant colonies may live through the winter on stored food and all the grasshoppers may die. However, the grasshopper has left eggs in sheltered spots and these will hatch the next spring. Both the ant and the grasshopper are very successful organisms, but each has a different way of satisfying its needs and ensuring that some of its offspring will be able to provide another generation of organisms. One method of survival is not necessarily better than another, as long as both animals are successful. This is what the study of behavior is all about—looking at the activities of an organism during its entire life and determining the value of the behavior in the ecological niche of the organism. The scientific study of the nature of behavior and its ecological and evolutionary significance in its natural setting is known as **ethology.**

17.4 Instinct and Learning

Instinctive and learned behaviors have already been defined, but a more complete understanding of these two kinds of behaviors is necessary to better understand how the behavior of an animal fits with its environment and how differences in behavior relate to the evolutionary history of an animal.

Instinctive Behavior

Instinctive behaviors are automatic, preprogrammed, and genetically determined. Such behaviors are found in a wide range of organisms from simple one-celled protozoans to complex vertebrates. These behaviors are performed correctly the first time without previous experience when the proper stimulus is given. A **stimulus** is some change in the internal or external environment of the organism that causes it to react. The reaction of the organism to the stimulus is called a **response.** • nervous system, p. 388

An organism can respond only to stimuli it can recognize. For example, it is difficult for us as humans to appreciate what the world seems like to a bloodhound. The bloodhound is able to identify individuals by smell, whereas we have great difficulty detecting, let alone distinguishing, many odors. Some animals, like dogs, deer, and mice are color-blind and are able to see only shades of gray. Others, such as honeybees, see ultraviolet light, which is invisible to us. Some birds and other animals are able to detect the magnetic field of the Earth. And some, such as rattlesnakes, are able to detect infrared radiation (heat) from distant objects.

In our example of the herring-gull chick, the red spot on the bill of the adult bird serves as a stimulus to the chick. The chick responds to this spot in a genetically programmed way. The behavior is innate—it is done correctly the first time without prior experience. The pecking behavior of the chick is in turn the stimulus for the adult bird to regurgitate food. It is obvious that these behaviors have adaptive value for the survival of the young because they leave little to chance. The young will get food from the adult automatically. Instinctive behavior has great value to a species because it allows correct, appropriate, and necessary behavior to occur without prior experience.

Instinctive behavior can be very effective for the survival of individuals of a species if it is involved in fundamental, essential activities that rarely require modification. The major

drawback of instinctive behavior is that it cannot be modified when a new situation presents itself. Instinctive behavior makes up most of the behavior in animals that have short life cycles, simple nervous systems, and little contact with parents. Over long periods of evolutionary time, genetically determined behaviors have been selected for and have been useful to most of the individuals of the species. However, some instances of inappropriate behavior may be generated by unusual stimuli or unusual circumstances in which the stimulus is given. For example, many kinds of insects fly toward a source of light. Over the millions of years of insect evolution, this has been a valuable and useful behavior. It allows them to easily find their way to open space. However, the human species invented artificial lights and transparent windows that generate totally inappropriate behavior. You have all seen insects drawn to lights at night or observed insects inside houses constantly flying against window panes through which sunlight is entering. This mindless, mechanical behavior seems incredibly stupid to us but, although *individual* animals die, the general behavior pattern of flying toward light is still valuable because the *majority* of individuals of the species do not encounter artificial lights or get trapped inside houses, and complete their life cycles normally.

Geese and other birds sit on nests. This behavior serves to keep the eggs warm and protect them from other dangers. During this incubation period, eggs may roll out of the nest as the parents get on and off the nest. When this happens, certain species of geese display a behavior that involves rolling eggs back into the nest. If the developing young within the egg are exposed to extremes of heat or cold they will be killed. Thus, the egg-rolling behavior has a significant adaptive value. If, however, the egg is taken from the goose when it is in the middle of egg-rolling behavior, the goose will continue its egg rolling until it gets back to the nest, even though there is no egg to roll (figure 17.3). This is typical of the inflexible nature of instinctive behaviors. It was also discovered that many other somewhat egg-shaped structures would serve as a stimulus to generate the same behavior. For example, beer cans and baseballs were good triggers for egg-rolling behavior. So not only was the bird unable to stop the egg-rolling behavior in midstride, but several nonegg objects generated inappropriate behavior because they had approximately the correct shape.

Some activities are so complex that it seems impossible for an organism to be born with such abilities. For example, you have seen spiderwebs in fields, parks, or vacant lots. You may have even watched a spider spin its web. This is not just a careless jumble of silk threads. A web is so precisely made that you can recognize what species of spider made it. But web spinning is not a learned ability. A spider has no opportunity to learn how to spin a web because it never observes others doing it. Furthermore, spiders do not practice several times before they get a proper, workable web. It is as if a "program" for making a particular web is in the spider's "computer" (figure 17.4). Many species of spiders appear to be unable to repair defective webs. When a web is damaged, they typically start from the beginning and build an entirely

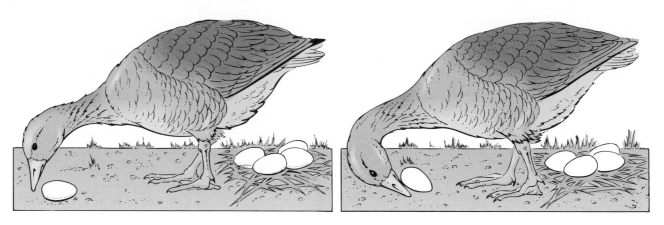

Figure 17.3

Egg-Rolling Behavior in Geese
Geese will use a specific set of head movements to roll any reasonably round object back to the nest. There are several components to this instinctive behavior, including recognition of the object and head-tucking movements. If the egg is removed during the head-tucking movements, the behavior continues as if the egg were still there.

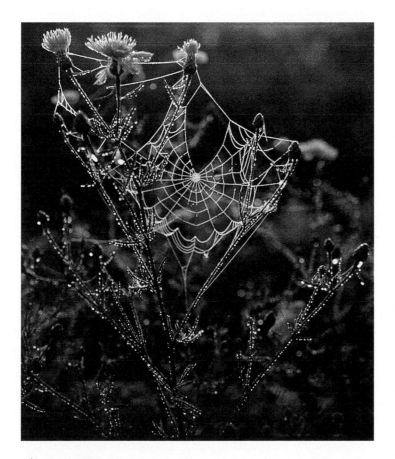

Figure 17.4

Inflexible Instinctive Behavior
The kind of web constructed by a spider is determined by instinctive behavior patterns. The spider cannot change the fundamental form of the web. If the web is damaged, the spider generally rebuilds it from the beginning rather than repair it.

new web. This inability to adapt as circumstances change is a prominent characteristic of instinctive behavior.

Could these behavior patterns be the result of natural selection? It is well established that many kinds of behaviors are controlled by genes. The "computer" in our example is really the DNA of the organism, and the "program" consists of a specific package of genes. Through the millions of years that spiders have been in existence, natural selection has modified the web-making program to refine the process. Certain genes of the program have undergone mutation, resulting in changes in behavior. Imagine various ancestral spiders, each with a slightly different program. The inherited program that gives individuals the best chance of living long enough to produce a new generation is the program selected for and most likely to be passed on to the next generation.

Learned Behavior

The alternative to preprogrammed, instinctive behavior is learned behavior. **Learning** is a change in behavior as a result of experience. (Your behavior will be different in some way as a result of reading this chapter.)

Learning becomes more significant in long-lived animals that care for their young. Animals that live many years are more likely to benefit from an ability to recognize previously encountered situations and modify their behavior accordingly. Furthermore, because the young spend time with their parents they can imitate their parents and develop behaviors that are appropriate to local conditions. These behaviors take time to develop but have the advantage of adaptability. In order for learning to become a dominant feature of an animal's life, the animal must also have a memory which requires a relatively large brain in which to store the new information it is learning. This is probably why learning is a major part of life for only a few kinds of animals like the vertebrates. In humans, it is clear that nearly all behavior is learned. Even such important behaviors as walking, communicating, feeding oneself, and sexual intercourse must be learned.

17.5 Kinds of Learning

Scientists who study learning recognize that there are different kinds that can be subdivided into several categories: *habituation, association, imprinting,* and *insight.*

Habituation

Habituation is a change in behavior in which an animal ignores an insignificant stimulus after repeated exposure to it. There are many examples of this kind of learning. Typically, wild animals flee from humans. Under many conditions this is a valuable behavior. However, in situations where wild animals frequently encounter humans and never experience negative outcomes, they may "learn" to ignore humans. Many wild animals such as the deer, elk, and bears in parks have been habituated to the presence of humans and behave in a way that would be totally inappropriate in areas near the park where hunting is allowed. Similarly, loud noises will startle humans and other animals. However, constant exposure to such sounds results in the individuals ignoring the sound. As a matter of fact, the sound may become so much a part of the environment that the cessation of the sound will evoke a response. This kind of learning is valuable because the animal does not waste time and energy responding to a stimulus that will not have a beneficial or negative impact on the animal. Animals that are continually responding to inconsequential stimuli have less time to feed and may miss other more important stimuli.

Association

Association occurs when an animal makes a connection between a stimulus and an outcome. Associating a particular outcome with a particular stimulus is important to survival because it allows an animal to avoid danger or take advantage of a beneficial event. If this kind of learning allows the animal to get more food, avoid predators, or protect its young more effectively, it will be advantageous to the species. The association of certain shapes, colors, odors, or sounds with danger is especially valuable. There are three common kinds of association learning: *classical conditioning, operant (instrumental) conditioning,* and *observational learning (imitation).*

Classical Conditioning

Classical conditioning occurs when an involuntary, natural, reflexive response to a natural stimulus is transferred from the natural stimulus to a new stimulus. The response produced by the new stimulus is called a **conditioned response.** During the period when learning is taking place, the new stimulus is given *before* or *at the same time as* the normal stimulus.

A Russian physiologist, Ivan Pavlov (1849–1936), was investigating the physiology of digestion when he discovered that dogs can transfer a normal response to a new stimulus. He was studying the production of saliva by dogs and he knew that a natural stimulus, such as the presence or smell of food, would cause the dogs to start salivating. Then he rang a bell just prior to the presentation of the food. After a training period, the dogs would begin to salivate when the bell was rung even though no food was presented. The natural response (salivating) was transferred from the natural stimulus (smell or taste of food) to the new stimulus (the sound of a bell). Animals can also be conditioned unintentionally. Many pets anticipate their mealtimes because their owners go through a certain set of behaviors, such as going to a cupboard or opening a can of pet food prior to putting food in the dish. It is doubtful that this kind of learning is a common occurrence in wild animals, because it is hard to imagine such tightly controlled sets of stimuli in nature.

Operant Conditioning

Operant (instrumental) conditioning also involves the association of a particular outcome with a specific stimulus, but differs from classical conditioning in several ways. First, during operant conditioning the animal learns to repeat acts that bring good results and avoid those that bring bad results. Second, a reward or punishment is received *after* the animal has engaged in a particular behavior. Third, the response is typically a more complicated behavior than a simple reflex. A reward that encourages a behavior is known as positive reinforcement and a punishment that discourages a behavior is known as negative reinforcement. The training of many kinds of animals involves this kind of conditioning. If a dog being led on a chain is given the command "heel" and is then vigorously jerked into the correct position by its master it eventually associates the word "heel" with assuming the correct position at the knee. This is negative reinforcement because the animal avoids the unpleasantness of being jerked about if it assumes the correct position. Similarly, petting or giving food to a dog when it has done something correctly will positively reinforce the desired behavior. For example, pushing the dog into the sitting position on the command "sit" and rewarding the dog when it performs the behavior on command is positive reinforcement.

Wild animals have many opportunities to learn through positive or negative reinforcement. As animals encounter the same stimulus repeatedly there is an opportunity to associate the stimulus with a particular outcome. For example, many kinds of birds eat berries and other small fruits. If a distinctly colored berry has a good flavor, birds will return repeatedly and feed from that source. Pigeons in cities have learned to associate food with people in parks. They can even identify specific individuals who regularly feed them. Their behavior is reinforced by being fed. Many birds in urban areas have associated automobiles with food, and are seen picking smashed insects from the grills and bumpers of cars. When a car drives into the area, birds immediately examine it for food.

In some of our national parks, bears have associated backpacks with food. In some cases, attempts have been

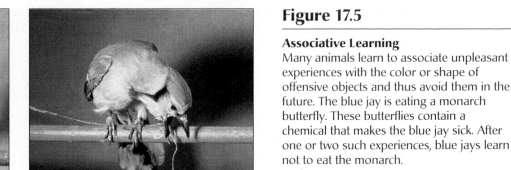

Figure 17.5

Associative Learning
Many animals learn to associate unpleasant experiences with the color or shape of offensive objects and thus avoid them in the future. The blue jay is eating a monarch butterfly. These butterflies contain a chemical that makes the blue jay sick. After one or two such experiences, blue jays learn not to eat the monarch.

made to use negative reinforcement to condition these bears to avoid humans. Bears that are repeat offenders are often killed. Conversely, in areas where bears are hunted they have generally been conditioned to avoid contact with humans. If certain kinds of fruits or insects have unpleasant tastes, animals will learn to associate the bad tastes with the colors and shapes of the offending objects and avoid them in the future (figure 17.5). Each species of animal has a distinctive smell. If a deer or rabbit has several bad experiences with a predator that has a particular smell, it can avoid places where the smell of the predator is present.

Animals also engage in trial-and-error learning, which involves elements of conditioning. When confronted with a particular problem, they will try one option after another until they achieve a positive result. Once they have solved the problem, they can use the same solution repeatedly. For example, if a squirrel has a den in a hollow tree on one side of a stream and is attracted to a source of food on the other side, it may explore several routes to get across the stream. It may jump from a tree on one side of the stream to another on the opposite side. It may run across a log that spans the stream. It may wade a shallow portion of the stream. Once it has found a good pathway, it is likely to use the same pathway repeatedly. Many hummingbirds visit many different flowers during the course of a day. When they have found a series of nectar-rich flowers, they will follow a particular route and visit the same flowers several times a day.

Observational Learning

In animals that participate in social groups, imitation is possible. **Observational learning (imitation)** is a form of associative learning that consists of a complex set of associations formed while watching another animal being rewarded or punished after performing a particular behavior. In this case, the animal is not receiving the reward or punishment itself but is observing the "fruits" of the behavior of other animals. Subsequently, the observer may show the same behavior. It is likely that conditioning is involved in imitation, since when an animal imitates a beneficial behavior it is rewarded. Observing a negative outcome to another animal is also beneficial because it allows the observer to avoid negative consequences. Many kinds of young birds and mammals follow their parents and sample the same kinds of foods their parents eat. If the foods taste good, they are positively reinforced. They may also observe warning and avoidance behaviors associated with particular predators and mimic these behaviors when the predator is present. For example, crows will "mob" predators such as hawks and owls. As young birds observe older crows cawing loudly and chasing an owl, the young crows learn to perform the same behavior. They associate a certain kind of behavior ("mobbing") with a certain kind of stimulus (owls or hawks).

Exploratory Learning

Animals are constantly moving about and sampling their environment. In animals that rely primarily on instinctive behaviors, this movement increases the likelihood that they will find food or other valuable resources but does not result in learning because they do not increase their efficiency at finding resources. In animals that have a significant amount of memory, it is possible for them to store information about their surroundings as they wander about. In some cases, the new information may have immediate value. For example, many kinds of birds and mammals have a memory of where certain physical features are located in their environment. They may find a food source during their wanderings and can return to it repeatedly in the future. When you put up a bird feeder, it does not take very long before many birds are visiting the feeder on a regular basis. Birds certainly find suitable nesting sites by exploring and then store the location in their memory. Even invertebrate animals with limited ability to store information engage in exploratory learning. For example, in the spring of the year a queen bumblebee will fly about examining holes in the ground. Eventually she will find a hole in which she will lay eggs and begin to raise her first brood of young. Once she has selected a site she must learn to recognize that particular spot so she can return to it each time she leaves to find food, or her young will die.

In other cases, the information learned may not be used immediately but could be of use in the future. If an animal has an inventory of its environment, it can call on the inventory to solve problems later in life. Many kinds of animals hide food items when food is plentiful and are able to find these hidden sources later when food is needed. Even if they don't remember exactly where the food is hidden, if they always hide food in a particular kind of place they are likely to be able to find it at a later date. (For example, if you need to drive a car that you have never seen before, you would know that you need to use the key and you would search in a particular place in the car for the place to insert the key.) Having a general knowledge of its environment is very useful to an animal.

Many kinds of small mammals such as mice and ground squirrels avoid predators by scurrying under logs or other objects or into holes in the ground. Experiments with mice and owl predators show that mice that have developed a familiarity with their surroundings are more likely to escape predators than are those that are unfamiliar with their surroundings.

Imprinting

Imprinting is a special kind of irreversible learning in which a very young animal is genetically primed to learn a specific behavior in a very short period during a specific time in its life. The time during which the learning is possible is known as the **critical period**. This type of learning was originally recognized by Konrad Lorenz (1903–1989) in his experiments with geese and ducks. He determined that, shortly after hatching, a duckling would follow an object if the object was fairly large, moved, and made noise. In one of his books, Lorenz described himself as squatting on the lawn one day, waddling and quacking, followed by newly hatched ducklings (figure 17.6). He was being a "mother duck." He was surprised to see a group of tourists on the other side of the fence watching him in amazement. They couldn't see the ducklings hidden by the tall grass. All they could see was this strange performance by a big man with a beard!

Ducklings will follow only the object on which they were originally imprinted. Under normal conditions, the first large, noisy, moving object newly hatched ducklings see is their mother. Imprinting ensures that the immature birds will follow her and may learn appropriate feeding, defensive tactics, and other behaviors by example. Because they are always near their mother, she can also protect them from enemies or bad weather. If animals imprint on the wrong objects, they are not likely to survive. Since these experiments by Lorenz in the early 1930s, we have discovered that many young animals can be imprinted on several types of stimuli and that there are responses other than following.

The way song sparrows learn their song appears to be a kind of imprinting. It has been discovered that the young birds must hear the correct song during a specific part of their youth or they will never be able to perform the song correctly as adults. This is true even if later in life they are surrounded by other adult song sparrows that are singing the correct song. Furthermore, the period of time when they learn the song is prior to the time they begin singing. Recognizing and performing the correct song is important because it has particular meaning to other song sparrows. For males, it conveys the information that a male song sparrow has a space reserved for himself. For females, the male's song is an

Figure 17.6

Imprinting
Imprinting is a special kind of irreversible learning that occurs during a very specific part of the life of an animal. These geese have been imprinted on Konrad Lorenz and exhibit the "following response" typical of this type of behavior.

announcement of the location of a male of the correct species that could be a possible mate.

Mother sheep and many other kinds of mammals imprint on the odor of their offspring. They are able to identify their offspring among a group of lambs and will allow only their own lambs to suck milk. Shepherds have known for centuries that they can sometimes get a mother that has lost her lambs to accept an orphan lamb if they place the skin of the mother's dead lamb over the orphan.

Many fish appear to imprint on odors in the water. Salmon are famous for their ability to return to the freshwater streams where they were hatched. They will jump waterfalls and use specially constructed fish ladders to get around dams. Fish that are raised in artificial hatcheries can be imprinted on minute amounts of special chemicals and be induced to return to any stream that contains the chemical.

Insight

Insight is a special kind of learning in which past experiences are reorganized to solve new problems. When you are faced with a new problem, whether it is a crossword puzzle, a math problem, or any one of a hundred other everyday problems, you sort through your past experiences and locate those that apply. You may not even realize that you are doing it, but you put these past experiences together in a new way that may give the solution to your problem. Because this process is internal and can be demonstrated only through some response, it is very difficult to understand exactly what goes on during insight learning. Behavioral scien-

Table 17.1

THE SIGNIFICANCE OF VARIOUS KINDS OF LEARNING

Kind of Learning	Defining Characteristic	Ecological Significance	Example
Habituation	An animal ignores a stimulus to which it is continually subjected.	An animal does not waste time or energy by responding to unimportant stimuli.	Wild animals raised in the presence of humans "lose their fear" of humans.
Association	An animal learns that a particular outcome is connected with a particular stimulus.	An animal can avoid danger or anticipate beneficial events by connecting a particular outcome with a specific stimulus when that stimulus is frequently tied to a particular outcome.	Many mammalian predators associate prey with high-pitched, squealing sounds.
Classical conditioning	A new stimulus is presented with a natural stimulus. The animal transfers its response from the natural stimulus to the new stimulus.	Classical conditioning probably does not happen in natural settings.	Pets can anticipate when they will be fed because food-preparing behavior of their owner occurs before food is presented.
Operant conditioning	A new stimulus elicits a response. The animal is rewarded or punished following its response. Eventually the animal associates the reward or punishment with the stimulus and responds appropriately to the stimulus.	Animals can avoid harmful situations that are associated with particular stimuli. They can learn to repeat behaviors that bring benefits.	Animals that associate a shape, color, smell, or sound with a predator can take refuge. Animals that associate a shape, color, smell, or sound with the presence of food will find food more readily.
Observational learning (imitation)	An animal imitates the behaviors of others.	The animal can acquire the "knowledge" of others by watching. This can be more rapid than learning by trial and error.	Young animals run when their mothers do or feed on the food their parents do.
Exploratory learning	An animal moves through and observes elements of its environment.	Information stored in memory may be valuable later.	Awareness of hiding places could allow an animal to escape a predator.
Imprinting	An animal learns specific, predetermined activity at a specific time in life.	The animal gains a completely developed behavior that has immediate value to survival.	Many kinds of newborn animals will follow their mothers.
Insight	An animal understands the connection between things it had no way of experiencing previously.	Information stored in an animal's memory can be used to solve new problems.	The manufacture and use of tools by humans and some other primates suggests insight is involved.

tists have explored this area for many years, but the study of insight learning is still in its infancy.

Insight learning is particularly difficult to study because it is impossible to know for sure whether a novel solution to a problem is the result of "thinking it through" or an accidental occurrence. For example, a small group of Japanese macaques (monkeys) was studied on an island. They were fed by simply dumping food, such as sweet potatoes or wheat, onto the beach. Eventually, one of the macaques discovered that she could get the sand off the sweet potato by washing it in a nearby stream. She also discovered that she could sort the wheat from the sand by putting the mixture into water because the wheat would float. Are these examples of insight learning? We will probably never know, but it is tempting to think so. In addition, in the colony of macaques the other individuals soon began to display the same behavior, probably because they were imitating the female that first made the discovery.

Table 17.1 summarizes the significance of each of the kinds of learning.

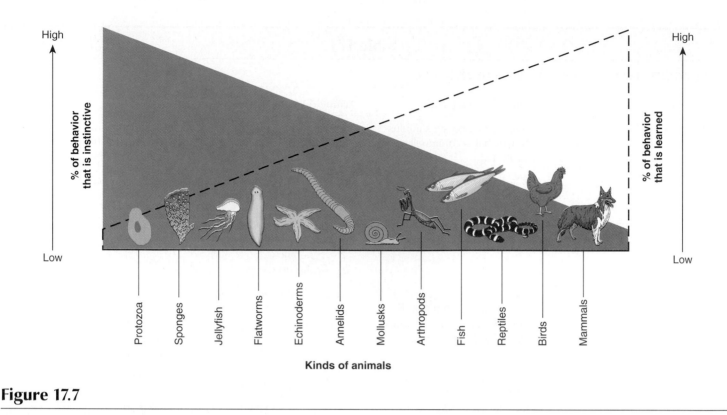

Figure 17.7

The Distribution of Learned and Instinctive Behaviors
Different groups of animals show different proportions of instinctive and learned behaviors in their behavior patterns.

17.6 Instinct and Learning in the Same Animal

It is important to recognize that all animals have both learned and instinctive behaviors and that one behavior pattern may include both instinctive and learned elements. For example, biologists have raised young song sparrows in the absence of any adult birds so there was no song for the young birds to imitate. These isolated birds would sing a series of notes similar to the normal song of the species, but not exactly correct. Birds from the same nest that were raised with their parents developed a song nearly identical to that of their parents. If bird songs were totally instinctive, there would be no difference between these two groups. It appears that the basic melody of the song was inherited by the birds and that the refinements of the song were the result of experience. Therefore, the characteristic song of that species was partly learned behavior (a change in behavior as a result of experience) and partly unlearned (instinctive). This is probably true of the behavior of many organisms; they show complex behaviors that are a combination of instinct and learning. It is important to note that many kinds of birds learn most of their songs with very few innate components. Mocking-birds are very good at imitating the songs of a wide variety of bird species found in their local region.

This mixture of learned and instinctive behavior is not the same for all species. Many invertebrate animals rely on instinct for the majority of their behavior patterns, whereas many of the vertebrates (particularly birds and mammals) make use of a great deal of learning (figure 17.7).

When a behavior pattern has both instinctive and learned components, it is typical that the learned components of an animal's behavior have particular value for the animal's survival. For example, most of the behavior of a honeybee is instinctive; however, when it finds a new food source, it is able to remember the route between the hive and the food source. The style of nest built by a bird is instinctive, but the skill with which it builds may improve with experience. The food-searching behavior of birds is probably instinctive, but the ability to modify the behavior to exploit unusual food sources such as bird feeders is learned. On the other hand, honeybees cannot be taught to make products other than honey and beeswax, a robin will not build a nest in a birdhouse, and most insect-eating birds will not learn to visit bird feeders. Table 17.2 compares instinctive behaviors and learned behaviors.

Table 17.2

A COMPARISON OF INSTINCT AND LEARNING

Instinct	Learning
Animal is born with the behavior.	Animal is not born with the behavior.
Instinctive behavior is genetically determined.	Learned behavior is not genetically determined but the way in which learning occurs is at least partly hereditary.
No experience is required; behavior is done correctly the first time.	Performance improves with experience; behavior requires practice.
Behavior cannot be changed.	Behavior can be changed.
Memory is not important.	Memory is important.
Instinctive behavior can evolve as gene frequencies change.	Changes in learned behaviors are not the result of genetic changes.
Instinct is typical of simple animals that have short lives and little contact with their parents.	Learning is typical of more complex animals that have long lives and extensive contact with parents.
Instinctive behaviors can only be passed from parent to offspring by genetic means.	Learning allows acquired behaviors to be passed from parent to offspring by cultural means.

17.7 What About Human Behavior?

We tend to think of ourselves as being different from other animals, and we are. However, it is important to recognize that we are different only in the degree to which we employ these different kinds of behavior. Humans have few behaviors that can be considered instincts. We certainly have reflexes that cause us to respond appropriately without thinking. Touching a hot object and rapidly pulling your hand away is a good example. Newborns grasp objects and hang on tightly with both their hands and feet. This kind of grasping behavior in our primitive ancestors would have allowed the child to hang onto its mother's hair as the mother and child traveled from place to place. But do we have more complicated instinctive behaviors? Although nearly all behavior other than reflexes is learned, newborn infants display several behaviors that could be considered instinctive. If you stroke the side of an infant's face, the child will turn its head toward the side touched and begin sucking movements. This is not a simple reflex behavior but rather requires the coordination of several sets of muscles and certainly involves the brain. It is hard to see how this could be a learned behavior because the child does the behavior without prior experience. Therefore it is probably instinctive. This behavior may be associated with nursing, because carrying the baby on its back would place the cheek of the child against the breast of the mother. Other mammals, even those whose eyes do not open for several days following birth, are able to find nipples and begin nursing shortly after birth.

Habituation is a common experience. We readily ignore sounds that are continuous such as the sound of air conditioning equipment or the background music typical of shopping malls. Teachers recognize that it is important to change activities regularly if they are going to keep the attention of their students.

Associative learning is extremely common in humans. We associate smells with certain kinds of food, sirens with emergency vehicles, and words with their meanings. Much of the learning that we do is by association. We also often use positive and negative reinforcement as ways to change behavior. We seek to reward appropriate behavior and punish inappropriate behavior. Much of the positive and negative reinforcement can be accomplished without having the actual experience because we can visualize possible consequences of our behavior. Adults routinely describe consequences for children so that children will not experience particularly harmful effects. "If you don't study for your biology exam you will probably fail it."

Imprinting in humans is more difficult to demonstrate, but there are some instances in which imprinting may be taking place. Bonding between mothers and infants is thought to be an important step in the development of the mother-child relationship. Most mothers form very strong emotional attachments to their children and, likewise, the children are attached to their mothers, sometimes literally, as they seek to maintain physical contact with their mothers. However, it is very difficult to show what is actually happening at this early time in the life of a child.

Another interesting possibility is the language development of children. All children learn whatever languages are spoken where they grow up. If multiple languages are spoken, they will learn them all and they learn them easily. However, adults have more difficulty learning new languages, and they often find it impossible to "unlearn" previous languages, so they speak new languages with an accent. This appears to meet the definition of imprinting. Learning takes place at a specific time in life (critical period), the kind of learning is preprogrammed, and what is learned cannot be unlearned. Recent research using tomographic images of the brain shows that those who learned a second language as adults use two different parts of the brain for language—one part for the native language or languages they learned as children and a different part for their second language.

Insight is what our species prides itself on. We are thinking animals. **Thinking** is a mental process that involves memory and an ability to reorganize information. A related aspect of our thinking nature is our concept of self. We can project ourselves into theoretical situations and measure our success by thinking without needing to experience the situation. For example, we can look at the width of a small stream and estimate our ability to jump across it without needing to do the task. We are able to mentally measure our personal abilities against potential tasks and decide if we should attempt them. In the process of thinking, we come up with new solutions to problems. We invent new objects, new languages, new culture, and new challenges to solve. However, how much of what we think is really completely new, and how much is imitation? As mentioned earlier, association is a major core of our behavior, but we also are able to use past experiences, stored in our large brains, to provide clues to solving new problems.

17.8 Selected Topics in Behavioral Ecology

Of the examples used so far in this chapter, some involved laboratory studies, some were field studies, and some included aspects of both. Often these studies overlap with the field of psychology. This is particularly true for many of the laboratory studies. You can see that the science of animal behavior is a broad one that draws on information from several fields of study and can be used to explore many kinds of questions. The topics that follow avoid the field of psychology and concentrate on the significance of behavior from ecological and evolutionary points of view.

Now that we have some understanding of how organisms generate behavior, we can look at a variety of behaviors in several kinds of animals and see how they are significant for animals in their ecological niches.

Reproductive Behavior

Reproductive behavior of many kinds of animals has been studied a great deal. Although each species of animal has its own specific behaviors, there are certain components of reproductive behavior that are common to nearly all kinds of animals. In order for an animal to reproduce, several events must occur: a suitable mate must be located, mating and fertilization must take place, and the young must be provided for.

Finding Each Other

In order to reproduce, an animal must find individuals of the same species that are of the opposite sex. Several techniques are used for this purpose. Different species of animals employ different methods, but most involve the production of signals that can be interpreted by others of the same species. Depending on the species, any of the senses (sound, sight, touch, smell, or taste) may be used to identify the species, sex, and sexual receptiveness of another animal. For instance, different species of frogs produce distinct calls. The call is a code system that delivers a very private message because it is meant for only one species. It is, however, meant for any member of that species near enough to hear. The call produced by male frogs, which both male and female frogs can receive by hearing, results in frogs of both sexes congregating in a limited area. Once they gather in a small pond, it is much easier to have the further communication necessary for mating to take place.

Many other animals including most birds, insects like crickets, reptiles like alligators, and some mammals produce sounds that are important for bringing individuals together for mating.

Chemicals can also serve to attract animals. **Pheromones** are chemicals produced by animals and released into the environment that trigger behavioral or developmental changes in other animals of the same species. They have the same effect as sounds made by frogs or birds; they are just a different code system. The classic example of a pheromone is the chemical that female moths release into the air. The large, fuzzy antennae of the male moths can receive the chemical in unbelievably tiny amounts. The male then changes his direction of flight and flies upwind to the source of the pheromone, which is the female (figure 17.8). Some of these sex-attractant pheromones have been synthesized in the laboratory. One of these, called Disparlure, is widely used to attract and trap male gypsy moths. Gypsy moth larvae cause considerable damage to trees when they feed on the leaves. In order to develop a control strategy it is necessary to have an estimate of the size of the gypsy moth population. The sex attractant is used to lure male moths into traps. The number captured is used to estimate the size of the population and this information is used to determine when control measures should be initiated in order to prevent large population outbreaks.

Most mammals rely on odors. Females typically produce distinct odors when they are in breeding condition. When males happen on such an odor trail, they follow it to the female. Many reptiles also produce distinctive odors.

Visual cues are also important for many species. Brightly colored birds, insects, fish, and many other animals often use specific patches of color for species identification. Conspicuous movements may also be used to attract the attention of a member of the opposite sex.

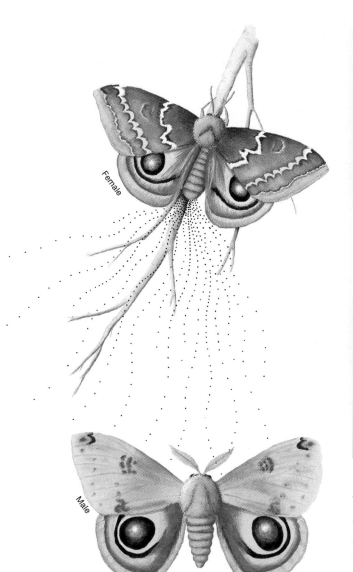

Figure 17.8

Chemical Communication
The female moth signals her readiness to mate and attracts males by releasing a pheromone that attracts males from long distances downwind.

Figure 17.9

Firefly Communication
The pattern of light flashes, the location of light flashes, and the duration of light flashes all help fireflies identify members of the opposite sex who are of the appropriate species.

Courtesy of James E. Lloyd.

The firefly is probably the most familiar organism that uses light signals to bring males and females together. Several different species may live in the same area, but each species flashes its own code. The code is based on the length of the flashes, their frequency, and their overall pattern (figure 17.9). There is also a difference between the signals given by males and those given by females. For the most part, males are attracted to and mate with females of their own species. Once male and female animals have attracted one another's attention, the second stage in successful reproduction takes place. However, in one species of firefly, the female has the remarkable ability to signal the correct answering code to species other than her own. After she has mated with a male of her species, she will continue to signal to passing males of other species. She is not hungry for sex, she is just hungry. The luckless male who responds to her "come-on" is going to be her dinner.

Assuring Fertilization

The second important activity in reproduction is fertilizing eggs. Many marine organisms simply release their gametes into the sea simultaneously and allow fertilization and further development to take place without any input from the parents. Sponges, jellyfishes, and many other marine animals fit in this category. Other aquatic animals congregate so that the chances of fertilization are enhanced by the male and female being near one another as the gametes are shed. This is typical of many fish and some amphibians, such as frogs. Internal fertilization, in which the sperm are introduced into the reproductive tract of the female, occurs in most terrestrial animals. Some spiders and other terrestrial animals produce packages of sperm that the female picks up with her reproductive structures. Many of these mating behaviors require elaborate, species-specific communication prior to the mating act. Several examples were given in the previous paragraphs.

Raising the Young

A third element in successful reproduction is providing the young with the resources they need to live to adulthood. Many invertebrate animals spend little energy on the care of the young, leaving them to develop on their own. Usually the young become free-living larvae that eat and grow rapidly. In some species, females make preparations for the young by laying their eggs in suitable sites. Many insects lay their eggs on the particular species of plant that the larva will use as food as it develops. Parasitic species seek out the required host in which to lay their eggs. The eggs of others may be placed in spots that provide safety until the young hatch from the egg. Turtles, many fish, and some insects fit this category. In most of these cases, however, the female lays large numbers of eggs, and most of the young die before reaching adulthood. This is an enormously expensive process: the female invests considerable energy in the production of the eggs but has a low success rate. Nevertheless, most species of animals utilize this strategy, so it has proven to be very successful.

An alternative strategy to this "wasteful" loss of potential young is to produce fewer young but invest large amounts of energy in their care. This is not necessarily a more efficient system, since it still requires large investments by the parents in the production and care of the young. However, this strategy is typical of birds and mammals. Parents build nests, share in the feeding and protection of the young, and often assist the young in learning appropriate behaviors. The activity of caring for the young involves many complex behavior patterns. It appears that most animals that feed and raise young are able to recognize their own young from those of other nearby families and may even kill the young of another family unit. Elaborate greeting ceremonies are usually performed when animals return to the nest or the den. Perhaps this has something to do with being able to identify individual young. Often this behavior is shared among adults as well. This is true for many colonial nesting birds, such as gulls and penguins, and for many carnivorous mammals, such as wolves, dogs, and hyenas.

Care of the young also occurs in several species of fish and reptiles such as alligators. Certain kinds of insects, such as bees, ants, and termites, have elaborate social organizations in which one or a few females produce large numbers of young that are cared for by sterile offspring of the fertile females. Some of the female's offspring will be fertile, reproducing individuals.

Territorial Behavior

One kind of behavior pattern that is often tied to successful reproduction is territorial behavior. A **territory** is the space used for food, mating, or other purposes, that an animal defends against others of the same species. The behaviors involved in securing and defending the territory are called **territorial behaviors.** A territory has great importance because it reserves exclusive rights to the use of a certain space for an individual. • competition, p. 289

When territories are first being established, there is often much conflict between individuals. This eventually gives way to the use of a series of signals that define the territory and communicate to others that the territory is occupied. The male redwing blackbird has red shoulder patches, but the female does not. The male will perch on a high spot, flash his red shoulder patches, and sing to other males that happen to venture into his territory. Most other males get the message and leave his territory; those that do not leave, he attacks. He will also attack a stuffed, dead male redwing blackbird in his territory, or even a small piece of red cloth. Clearly, the spot of red is the characteristic that stimulates the male to defend his territory. Once the initial period of conflict is over, the birds tend to respect one another's boundaries. All that is required is to frequently announce that the territory is still occupied. This is accomplished by singing from some conspicuous position in the territory. After the territorial boundaries are established, little time is required to prevent other males from venturing close. Thus the animal may spend a great deal of time and energy securing the territory initially, but doesn't need to expend much to maintain it.

With many kinds of animals, the possession of a territory is often a requirement for reproductive success. Not all male redwing blackbirds are successful in obtaining territories. During the initial period, when fighting is common, some birds regularly win and maintain their territories. Some lose and must choose a less favorable territory or go without. Because females choose which male's territory they will build their nest in, males that do not have territories are much less likely to fertilize females. In a way, then, territorial behavior has the effect of allocating breeding space. This kind of behavior is widespread in the animal kingdom and can be seen in such diverse groups as insects, spiders, fish, reptiles, birds, and mammals.

Many seabird colonies have extremely small nest territories. Each territory is just beyond the reach of the bills of the neighbors (figure 17.10). Trespassers are severely punished. Within a gull colony, each nest is in a territory of about 1 square meter. When one gull walks or lands on the territory of another, the defender walks toward the other in the upright threat posture. The head is pointed down with the neck stretched outward and upward. The folded wings are raised slightly. The upright threat posture is one of a number of movements that signal what an animal is likely to do in the near future. The bird is communicating an intention to do something, to fight in this case, but it may not follow through. If the invader shows no sign of retreating, then one or both gulls may start pulling up the grass very vigorously with their beaks. This seems to make no sense. The gulls were ready to fight one moment; the next moment they apparently have forgotten about the conflict and are pulling grass. But the struggle has not been forgotten: pulling grass

Figure 17.10

Territorial Behavior
Colonial nesting seabirds typically have very small nest territories. Each territory is just out of pecking range of the neighbors.

Figure 17.11

A Dominance Hierarchy
Many animals maintain order within their groups by establishing a dominance hierarchy. For example, whenever you see a group of cows or sheep walking in single file, it is likely that the dominant animal is at the head of the line and the lowest-ranking individual is at the end.

is an example of redirected aggression. In **redirected aggression,** the animal attacks something other than the natural opponent. Even though the aggression is not directed at the opponent, it is still a threatening activity that may deter an interloper from continuing to trespass. If the intruding gull doesn't leave at this point, there will be an actual battle. (A person who pounds the desk during an argument is showing similar behavior. Look for examples of this behavior in your neighborhood cats and dogs—maybe even in yourself!)

Many carnivorous mammals like foxes, weasels, cougars, coyotes, and wolves use urine or other scents to mark the boundaries of their territories. One of the primary values of the territory for these animals is the food contained within the large space they defend. These territories may include several square kilometers of land. Many other kinds of animals are territorial but use other signaling methods to maintain ownership of their territories. For example, territorial fish use color patterns and threat postures to defend their territories. Crickets use sound and threat postures. Male bullfrogs engage in shoving matches to displace males who invade their small territories along the shoreline.

Dominance Hierarchy

In social animals, a kind of organization known as a *dominance hierarchy* is often observed. A **dominance hierarchy** is a relatively stable, mutually understood order of priority within the group. One individual in the group dominates all others. A second-ranking individual dominates all but the highest-ranking individual, and so forth, until the lowest-ranking individual must give way to all others within the group. Generally, the initial establishment of a dominance

hierarchy involves fighting. The best fighters take the top places. However, once the hierarchy is established, it results in a more stable social unit with little conflict, since the positions of the individuals within the hierarchy are reinforced by various kinds of threats without the need for fighting. This kind of behavior is seen in barnyard chickens and is known as a *pecking order*. Figure 17.11 shows a dominance hierarchy; the lead animal has the highest ranking and the last animal has the lowest ranking.

A dominance hierarchy gives certain individuals preferential treatment when resources are scarce. The dominant individual will have first choice of food, mates, shelter, water, and other resources because of its position. Animals low in the hierarchy may be malnourished or fail to mate in times of scarcity. In many social animals, like wolves, usually only the dominant male and female reproduce. If the dominant male and female achieved their status because they have superior characteristics, it is likely that their favorable genes will be passed on to the next generation. Poorly adapted animals with low rank may never reproduce. Such a hierarchy frequently results in low-ranking individuals emigrating from the area. Such migrating individuals are often subject to heavy predation.

Avoiding Periods of Scarcity

Resource allocation becomes most critical during periods of scarcity. In some areas, the dry part of the year is the most stressful. In temperate areas, winter reduces many sources of food and forces organisms to adjust. Animals have several ways of coping with seasonal stress. Some animals simply avoid the stress. In areas where drought occurs, many animals become inactive until water becomes available. Frogs, toads, and many insects remain inactive (estivate) underground during long periods of drought and emerge to mate when it rains. Hibernation in mammals is a response to cold, seasonal temperatures in which the body temperature drops and there is a physiological slowing of all body processes. The slowing of body processes allows an animal to survive on food it has stored within its body prior to the onset of severe environmental conditions and low food availability. Hibernation is typical of bats, marmots, and some squirrels. Animals like insects, reptiles, and amphibians are not able to regulate their temperatures like birds and mammals. Because their body temperatures drop when the environmental temperature drops, their activities are slowed during the winter and they require less food.

Other animals have built-in behavior patterns that cause them to store food during seasons of plenty for periods of scarcity. These behaviors are instinctive and are seen in a variety of animals. Squirrels bury nuts, acorns, and other seeds. (They also plant trees because they never find all the seeds they bury.) Chickadees stash seeds in cracks and crevices when the food is plentiful and spend many hours during the winter exploring similar places for food. Some of the food they find is food they stored. Honeybees store honey, which allows them to live through the winter when nectar is not available. This requires a rather complicated set of behaviors that coordinates the activities of thousands of bees in the hive.

Navigation and Migration

Because animals move from place to place to meet their needs it is useful to be able to return to a nest, water hole, den, or favorite feeding spot. This requires some sort of memory of their surroundings (a mental map) and a way of determining direction. Often it is valuable to have information about distance as well. Direction can be determined by such things as magnetic fields, identifying landmarks, scent trails, or reference to the Sun or stars. If the Sun or stars are used for navigation, some sort of time sense is also needed because these bodies move in the sky.

In honeybees, navigation also involves communication among the various individuals that are foraging for nectar. The bees are able to communicate information about the direction and distance of the nectar source from the hive. If the source of nectar is some distance from the hive, the scout bee performs a "wagging dance" in the hive. The bee walks in a straight line for a short distance, wagging its rear end from side to side. It then circles around back to its starting position and walks the same path as before (figure 17.12). This dance is repeated many times. The direction of the straight-path portion of the dance indicates the direction of the nectar relative to the position of the Sun. For instance, if the bee walks straight upward on a vertical surface in the hive, that

Leggitt

Figure 17.12

Honeybee Communication and Navigation
The direction of the straight, tail-wagging part of the dance of the honeybee indicates the direction to a source of food. The angle that this straight run makes with the vertical is the same angle the bee must fly in relation to the Sun to find the food source. The length of the straight run and the duration of each dance cycle indicate the flying time necessary to reach the food source.

tells the other bees to fly directly toward the Sun. If the path is 30 degrees to the left of vertical, the source of the nectar is 30 degrees to the left of the Sun's position.

The duration of the entire dance and the number of waggles in the straight-path portion of the dance are positively correlated with the length of time the bee must fly to get to the nectar source. So the dance is able to communicate the duration of flight as well as the direction. Because the recruited bees have picked up the scent of the nectar source from the dancer, they also have information about the kind of flower to visit when they arrive at the correct spot. Because the Sun is not stationary in the sky, the bee must constantly adjust its angle to the Sun. It appears that they do this with some kind of internal clock. Bees that are prevented from going to the source of nectar or from seeing the Sun will still fly in the proper direction sometime later, even though the position of the Sun is different.

Animals often use the ability to sense changes in time to prepare for seasonal changes. In areas away from the equator, the length of the day changes as the seasons change. The length of the day is called the **photoperiod**. Many birds prepare for migration and have their migration direction determined by the changing photoperiod. For example, in the fall of the year many birds instinctively change their behavior, store up fat, and begin to migrate from northern areas to areas closer to the equator. This seasonal migration allows them to avoid the harsh winter conditions signaled by the shortening of days.

The return migration in the spring is triggered by the lengthening photoperiod. This migration certainly requires a lot of energy, but it allows many birds to exploit temporary food resources in the north during the summer months.

Like honeybees, some daytime-migrating birds use the Sun to guide them. We need two instruments to navigate by the Sun—an accurate clock and a sextant for measuring the angle between the Sun and the horizon. Can a bird perform such measurements without instruments when we, with our much bigger brains, need these instruments to help us? It is unquestionably true! For nighttime migration, some birds use the stars to help them find their way. In one interesting experiment, warblers, which migrate at night, were placed in a planetarium. The pattern of stars as they appear at any season could be projected onto a large domed ceiling. During autumn, when these birds would normally migrate southward, the stars of the autumn sky were shown on the ceiling. The birds responded with much fluttering activity at the south side of the cage, as if they were trying to migrate southward. Then the experimenters tried projecting the stars of the spring sky, even though it was autumn. Now the birds tended to try to fly northward, although there was less unity in their efforts to head north; the birds seemed somewhat confused. Nevertheless, the experiment showed that the birds recognized star patterns and were influenced by them.

There is evidence that some birds navigate by compass direction—that is, they fly as if they had a compass in their heads. They seem to be able to sense magnetic north. Their ability to sense magnetic fields was proven at the U.S. Navy's test facility in Wisconsin. The weak magnetism radiated from this test site has changed the flight pattern of migrating birds, but it is yet to be proved that birds use the magnetism of the Earth to guide their migration. Homing pigeons are famous for their ability to find their way home. They make use of a wide variety of clues, but it has been shown that one of the clues they use involves magnetism. Birds with tiny magnets glued to the sides of their heads were very poor navigators; others with nonmagnetic objects attached to the sides of their heads did not lose their ability to navigate.

Biological Clocks

As mentioned earlier, bees, birds, and probably most other animals have internal clocks. In the case of bees, the clock allows them to predict the position of the Sun. In the case of birds and mammals, the changing length of day allows them to time their migration, mating, food-storing behavior, or time for entering hibernation. So some clocks are annual clocks, whereas others are daily clocks. For instance, you have a daily clock. Travelers who fly partway around the world by nonstop jet plane need some time to recover from "jet lag." Their digestion, sleep, or both, may be upset. Their discomfort is not caused by altitude, water, or food, but by having rapidly crossed several time zones. There is a great difference in the time as measured by the Sun or local clocks and that measured by the body; the body's clock adjusts more slowly.

There are many examples of animal behaviors that are timed, some of which show a great deal of precision. In the animal world, mating is the most obviously timed event. In the Pacific Ocean, off some of the tropical islands, lives a marine worm known as the *palolo worm*. Its habit of making a well-timed brief appearance in enormous swarms is a striking example of a biological clock phenomenon. At mating time, these worms swarm into the shallows of the islands and discharge sperm and eggs. There are so many worms that the sea looks like noodle soup. The people of the islands find this an excellent time to change their diets. They dip up the worms much as North Americans dip up smelt or other small fish that are making a spawning run. The worms appear around the third quarter of the Moon in October or November, the time varying somewhat according to local environmental conditions. It appears that they have an annual cycle for mating but that a monthly, or lunar, cycle is superimposed on the annual cycle. Because these animals are marine worms it is unclear whether they are responding to the Moon or the tidal effects of the Moon or something else entirely.

Social Behavior

Many species of animals are characterized by interacting groups called **societies**, in which there is division of labor. Societies differ from simple collections of organisms by the greater specialization and division of labor in the roles displayed by the individuals in the group. The individuals performing one function cooperate with others having different special abilities. As a result of specialization and cooperation, the society has characteristics not found in any one member of the group: the whole is more than the sum of its parts. But if cooperation and division of labor are to occur, there must be communication among individuals and coordination of effort.

Honeybees, for example, have an elaborate communication system and are specialized for specific functions. A few individuals known as *queens* and *drones* specialize in reproduction, whereas large numbers of sterile *worker* honeybees are involved in collecting food, defending the hive, and caring for the larvae. These roles are rigidly determined by inherited behavior patterns. Each worker honeybee has a specific task, and all tasks must be fulfilled for the group to survive and prosper. As they age, the worker honeybees move through a series of tasks over a period of weeks. When they first emerge from their wax cells, they clean the cells. Several days later, their job is to feed the larvae. Next they build cells. Later they become guards that challenge all insects that land near the entrance to the hive. Finally they become foragers who find and bring back nectar and pollen to feed the other bees in the hive and to be stored for the winter. Foraging is usually the last job before the worker honeybee dies. Although this progression of tasks is the usual order, workers can shift from their main task to others if there is a need. Both the tasks performed and the progression of tasks are instinctively (genetically) determined (figure 17.13).

Figure 17.13

Honeybee Society
Within the hive the queen lays eggs that the sterile workers care for. The workers also clean and repair the hive and forage for food.

Figure 17.14

African Wild Dog Society
African wild dogs hunt in groups and share food that they bring back to the den. Only the dominant male and female mate and raise offspring.

A hive of bees may contain thousands of individuals, but under normal conditions only the queen bee and the male drones are capable of reproduction. None of the thousands of workers who are also females will reproduce. This does not seem to make sense because they appear to be giving up their chance to reproduce and pass their genes on to the next generation. Is this some kind of self-sacrifice (altruistic behavior) on the part of the workers, or is there another explanation? In general, the workers in the hive are the daughters or sisters of the queen and therefore share a large number of her genes. This means that they are really helping a portion of their genes get to the next generation by assisting in the raising of their own sisters, some of whom will become new queens. This argument has been used to partially explain behaviors in societies that might be bad for the individual but advantageous for the society as a whole.

Animal societies exhibit many levels of complexity and types of social organization differ from species to species. Some societies show little specialization of individuals other than that determined by sexual differences or differences in physical size and endurance. The African wild dog illustrates such a flexible social organization. These animals are nomadic and hunt in packs. Although an individual wild dog can kill prey about its own size, groups are able to kill fairly large animals if they cooperate in the chase and the kill, which often involves a chase of several kilometers. Young pups are unable to follow the pack. When adults return from a successful hunt, they regurgitate food if the proper begging signal is presented to them (figure 17.14). Therefore, the young and the adults that remained behind to guard the young are fed by the hunters. The young are the responsibility of the entire pack, which cooperates in their feeding and protection. During the time that the young are at the den site, the pack must give up its nomadic way of life. Therefore, the young are born during the time of year when prey are most abundant. Only one or two of the females in the pack have young each year. If every female had young, the pack couldn't feed them all. At about two months of age, the young begin to travel with the pack, and the pack can return to its nomadic way of life.

In many ways honeybee and African wild dog societies are similar. Not all females reproduce, the raising of young is a shared responsibility, and there is some specialization of roles. The analysis and comparison of animal societies has led to the thought that there may be fundamental processes that shape all societies. The systematic study of all forms of social behavior, both human and nonhuman, is called **sociobiology.**

How did various types of societies develop? What selective advantage does a member of a social group have? In what ways are social groups better adapted to their environment than nonsocial organisms? How does social organization affect the way populations grow and change? These are difficult questions because, although evolution occurs at the population level, it is individual organisms that are selected. Thus, we need new ways of looking at evolutionary processes when describing the evolution of social structures.

The ultimate step is to analyze human societies according to sociobiological principles. Such an analysis is difficult and controversial, however, because humans have a much greater ability to modify behavior than do other animals. However, when we look at human social behavior we see some clear parallels between human and nonhuman social behaviors. This implies that there are certain fundamental similarities among social organisms regardless of their species. Do we see territorial behavior in humans? "No trespassing" signs and fences between neighboring houses seem to be clear indications of territorial behavior in our social species. Do groups of humans have dominance hierarchies? Most business, government, and social organizations have a clear dominance hierarchy in which those at the top get more resources (money, prestige) than those lower in the organization. Do human societies show division of labor? Our societies clearly benefit from the specialized skills of certain individuals. Do humans treat their own children differently from other children? Studies of child abuse indicate that abuse is more common between stepparents and their nongenetic stepchildren than between parents and their biological children. Although these few examples do not prove that human societies follow certain rules typical of other animal societies, it bears further investigation. Sociobiology will continue to explore the basis of social organization and behavior and will continue to be an interesting and controversial area of study.

SUMMARY

Behavior is how an organism acts, what it does, and how it does it. The kinds of responses that organisms make to environmental changes (stimuli) may be simple reflexes, very complex instinctive behavior patterns, or learned responses.

From an evolutionary viewpoint, behaviors represent adaptations to the environment. They increase in complexity and variety the more highly specialized and developed the organism is. All organisms have inborn or instinctive behavior, but higher animals also have one or more ways of learning. These include habituation, association, exploration, imprinting, and insight. Communication for purposes of courtship and mating is accomplished by sounds, visual displays, touch, and chemicals like pheromones. Many animals have special behavior patterns that are useful in the care and raising of the young.

Territorial behavior is used to obtain exclusive use of an area and its resources. Both dominance hierarchies and territorial behavior are involved in the allocation of scarce resources. To escape from seasonal stress, some animals estivate or hibernate, others store food, and others migrate. Migration to avoid seasonal extremes involves a timing sense and some way of determining direction. Animals navigate by means of sound, celestial light cues, and magnetic fields.

Societies consist of groups of animals in which individuals specialize and cooperate. Sociobiology attempts to analyze all social behavior in terms of evolutionary principles, ecological principles, and population dynamics.

THINKING CRITICALLY

For guidelines to answer this question, visit the Online Learning Center.

If you were going to teach an animal to communicate a message new to that animal, what message would you select? How would you teach the animal to communicate the message at the appropriate time?

CONCEPT MAP TERMINOLOGY

Construct a concept map using these terms.

association
imprinting
insight
instinctive behavior
learning

response
stimulus
territorial behavior
thinking

Visit the Online Learning Center to use an interactive format to construct another concept map.

KEY TERMS

Use interactive flash cards, on the Online Learning Center, to help you learn the meanings of these terms.

anthropomorphism
association
behavior
classical conditioning
conditioned response
critical period
dominance hierarchy
ethology
habituation
imprinting
insight
instinctive behavior
learned behavior
learning

observational learning
  (imitation)
operant (instrumental)
  conditioning
pheromone
photoperiod
redirected aggression
response
society
sociobiology
stimulus
territorial behavior
territory
thinking

LEARNING CONNECTIONS

Visit the Online Learning Center to find additional study resources including those highlighted below.

17.1   The Adaptive Nature of Behavior

Experience This!
• Observing types of behavior

1. List two ways in which learned and instinctive behaviors differ.
2. Why is it more difficult to study the evolution of behavior than the evolution of structural characteristics?

17.2   Interpreting Behavior

Case Study
• Cicada Woodstock

3. List two animal behaviors and state the significance of the behavior to the reproductive success of the animal.

17.3   The Problem of Anthropomorphism

4. Why do students of animal behavior reject the idea that a singing bird is happy?
5. Define the terms *ethology* and *anthropomorphism*.

17.4   Instinct and Learning

Animations and Review
• Nature vs. nurture
• Innate behavior
• Learning

6. Briefly describe an example of unlearned behavior in a particular animal. Explain why you know it is unlearned.
7. Briefly describe an example of learned behavior in a particular animal. Explain why you know it is learned.

17.5   Kinds of Learning

Interactive Concept Map
• Learned behavior

8. Give an example of a conditioned response. Can you describe one that is not mentioned in this chapter?

9. How are classical conditioning and operant conditioning different?
10. What is imprinting, and what value does it have to an organism?
11. Give an example of habituation in a wild animal.
12. In what kinds of animals is observational learning common?
13. Why is insight difficult to demonstrate in animals?

17.6   Instinct and Learning in the Same Animal

14. Which one of the following animals—goose, shark, or grasshopper—is likely to have the highest proportion of learning in its behavior? List two reasons for your answer.

17.7   What About Human Behavior?

Interactive Concept Map
• Human behavior

Food for Thought
• Behavior disordered students

15. Give an example of instinctive behavior in humans.
16. Give an example of habituation, association, and imprinting in humans.

17.8   Selected Topics in Behavioral Ecology

Animations and Review
• Aggression
• Navigation
• Altruism and sociality

17. Describe why communication is important to successful reproduction.
18. Describe two alternative strategies for assuring reproductive success.
19. How do territorial behavior and dominance hierarchies provide certain individuals with an advantage?
20. What distinguishes societies from simple aggregations of individuals?
21. How do animals use chemicals, light, and sound to communicate?
22. What is sociobiology?

Test yourself with multiple choice questions that provide immediate feedback on the Online Learning Center with PowerWeb at

www.mhhe.com/enger11

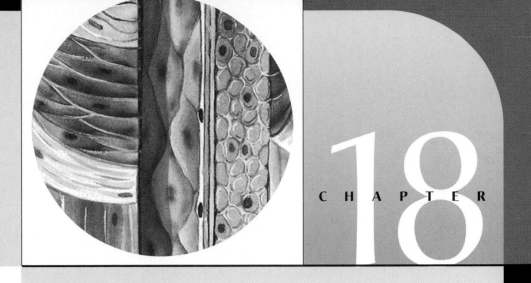

Materials Exchange in the Body

GOALS AND OBJECTIVES

Understand how the concept of surface area-to-volume ratio is related to human physiology.

❏ Explain why cells are small.

❏ Describe structural characteristics that increase the surface area of the gut.

❏ Explain why the lungs, gut, and kidneys have large numbers of capillaries and a large surface area.

Recognize that several mechanisms are involved in maintaining homeostasis.

❏ Describe how the circulatory system and respiratory system interact to maintain pH and oxygen levels.

❏ Explain how the circulatory system and sweat glands interact to maintain a constant body temperature.

❏ Describe how the processes of breathing, circulation, and exercise are interrelated.

❏ Explain the function of breathing to oxygen and carbon dioxide exchange.

❏ Describe the role of the kidneys in regulating blood pH.

❏ Describe how the kidneys regulate the salt and water content of the body.

Understand how the circulatory, respiratory, digestive, and excretory systems function.

❏ Explain how the kidneys are able to rid the body of waste products.

❏ Describe the functions of the liver that are related to digestion.

❏ Explain why a strongly pumping heart and open arteries and veins are essential to good health.

❏ Explain the role of various organs of the digestive system in the enzymatic, mechanical, and chemical digestion of foods.

Physiological Processes

PART FIVE

339

18.1 Exchanging Materials: Basic Principles

Living things are complex machines with many parts that must work together in a coordinated fashion. All systems are integrated and affect one another in many ways. Many of the activities of an organism involve exchanging materials between the organism and its environment. This exchange requires regulation of different systems and specialization of some body surfaces to allow for more efficient exchange of materials.

Homeostasis

Homeostasis is the maintenance of a constant internal environment as a result of monitoring and modifying the functioning of various systems.

When you run up a hill, your leg and arm muscles move in a coordinated way to provide power. They burn fuel (glucose) for energy and produce carbon dioxide and lactic acid as waste products, which tend to lower the pH of the blood. Your heart beats faster to provide oxygen and nutrients to the muscles, you breathe faster to supply the muscles with oxygen and to get rid of carbon dioxide, and the blood vessels in the muscles dilate to allow more blood to flow to them. As you run you generate excess heat. As a result, more blood flows to the skin to get rid of the heat and sweat glands begin to secrete, thus cooling the skin. All of these automatic internal adjustments help the body maintain a constant level of oxygen, carbon dioxide, and glucose in the blood; constant pH; and constant body temperature. To explore the various mechanisms that help organisms maintain homeostasis, we will begin at the cellular level.

Limitations to Cell Size

Cells are highly organized units that require a constant flow of energy in order to maintain themselves. The energy they require is provided in the form of nutrient molecules that enter the cell. Oxygen is required for the efficient release of energy from the large organic molecules that serve as fuel. Inevitably, as oxidation takes place, waste products form that are useless or toxic. These must be removed from the cell. All these exchanges of food, oxygen, and waste products must take place through the cell surface.

As a cell grows its volume increases, and the amount of metabolic activity required to maintain it rises. The quantity of materials that must be exchanged between the cell and its surroundings increases. Thus, growth cannot continue indefinitely. The ultimate size of a cell is limited by one or more of the following interrelated factors:

1. *The strength of the membrane:* As the cell increases in size, the membrane will eventually not be strong enough to withstand the forces caused by the mass of material inside it. If you had a balloon and kept adding water to it, eventually the balloon would burst. Similarly, dams have failed when too much water was accumulated behind them.

2. *The cell surface area:* If materials are to enter a cell they must pass through a surface. The cell membrane is a selectively permeable barrier to the passage of material in and out of the cell. The amount of surface will determine how much material can pass. If you were to pour water through two coffee filters of different size, the one with the largest surface area would allow the water to pass through more rapidly.

3. *The surface area-to-volume ratio:* The metabolic needs of a cell are determined by its volume. However, the ability to exchange materials between the cell and its surroundings is determined by its surface area. As a cell increases in size, its volume increases faster than its surface area. This relationship between surface area and volume is often expressed as the **surface area-to-volume ratio.**

Assume that we have a cube 1 centimeter on a side, as shown in figure 18.1. This cube will have a volume of 1 cubic centimeter (1 cm^3). Each side of the cube will have a surface area of 1 square centimeter (1 cm^2) and, because there are six surfaces on a cube, it will have a total surface area of 6 square centimeters (6 cm^2). It has a surface area-to-volume ratio of 6:1 (6 cm^2 of surface to 1 cm^3 of volume). If we increase the size of the cube so that each side has an area of 4 square centimeters, the total surface area of the cube will be 24 square centimeters (6 surfaces $\times$ 4 cm^2 per square = 24 cm^2). However, the volume now becomes 8 cubic centimeters, because each side of this new, larger cube is 2 centimeters ($2 \text{ cm} \times 2 \text{ cm} \times 2 \text{ cm} = 8 \text{ cm}^3$). Therefore, the surface area-to-volume ratio is 24:8, which reduces to 3:1. So you can see that as an object increases in size its volume increases faster than its surface area.

The ability to transport materials into or out of a cell is determined by its surface area, whereas its metabolic demands are determined by its volume. So the larger a cell becomes, the more difficult it is to satisfy its needs. Some cells overcome this handicap by having highly folded cell membranes that substantially increase their surface areas. This is particularly true of cells that line the intestine or are involved in the transport of large numbers of nutrient molecules. These cells have many tiny, folded extensions of the cell membrane called **microvilli** (figure 18.2).

In a similar way, the structure of an automobile radiator increases the efficiency of heat exchange between the engine and the air. The radiator has many fins attached to tubes through which a coolant fluid is pumped. Because of the large surface area provided by the fins, heat from the engine can be efficiently radiated away.

In addition to the limitation that surface area presents to the transport of materials, large cells also have a problem with diffusion. The molecular process of diffusion is quite rapid over short distances but becomes very slow over longer distances. Diffusion is generally insufficient to handle the

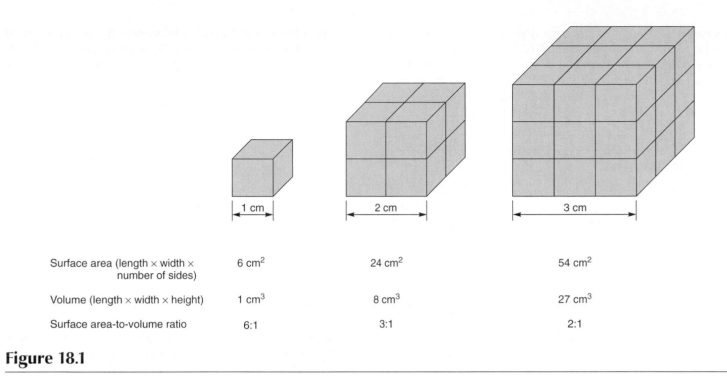

	1 cm	2 cm	3 cm
Surface area (length × width × number of sides)	6 cm²	24 cm²	54 cm²
Volume (length × width × height)	1 cm³	8 cm³	27 cm³
Surface area-to-volume ratio	6:1	3:1	2:1

Figure 18.1

The Surface Area-to-Volume Ratio
As the size of an object increases, its volume increases faster than its surface area. Therefore, the surface area-to-volume ratio decreases.

(a)

(b)

Figure 18.2

Intestinal Cell Surface Folding
Intestinal cells that are in contact with the food in the gut have highly folded surfaces. The tiny projections of these cells are called microvilli. These can be clearly seen in the photomicrograph in (a). The drawing in (b) shows that only one surface has these projections.

needs of cells if it must take place over a distance of more than 1 millimeter. The center of the cell would die before it received the molecules it needed if the distance were greater. Because of this and the problems presented by the surface area-to-volume ratio, it is understandable that the basic unit of life, the cell, must remain small. • **diffusion, p. 69**

All single-celled organisms are limited to a small body size because they handle the exchange of molecules through their cell membranes. Large, multicellular organisms consist of a multitude of cells, many of which are located far from the surface of the organism. Each cell within a multicellular organism must solve the same materials-exchange problems as single-celled organisms. Large organisms have several interrelated systems that are involved in the exchange and transport of materials so that each cell can meet its metabolic needs. Diffusion, facilitated diffusion, and active transport are all involved in moving molecules across cell membranes. These topics are presented in chapter 4.

18.2 Circulation

Large, multicellular organisms like humans consist of trillions of cells. Because many of these cells are buried within the organism far from the body surface, there must be some sort of distribution system to assist them in solving their materials-exchange problems. The primary mechanism used is the circulatory system.

The circulatory system consists of several fundamental parts. **Blood** is the fluid medium that assists in the transport of materials and heat. The **heart** is a pump that forces the fluid blood from one part of the body to another. The heart pumps blood into **arteries,** which distribute blood to organs. It flows into successively smaller arteries until it reaches tiny vessels called **capillaries,** where materials are exchanged between the blood and tissues through the walls of the capillaries. The blood flows from the capillaries into **veins** that combine into larger veins that ultimately return the blood to the heart from the tissues.

The Nature of Blood

Blood is a fluid that consists of several kinds of cells suspended in a watery matrix called **plasma.** This fluid plasma also contains many kinds of dissolved molecules. The primary function of the blood is to transport molecules, cells, and heat from one part of the body to another. The major kinds of molecules that are distributed by the blood are respiratory gases (oxygen and carbon dioxide), nutrients of various kinds, waste products, disease-fighting antibodies, and chemical messengers (hormones). Blood has special characteristics that allow it to distribute respiratory gases very efficiently. Although little oxygen is carried as free, dissolved oxygen in the plasma, *red blood cells (RBCs)* contain **hemoglobin,** an iron-containing molecule, to which oxygen molecules readily bind. This allows for much more oxygen to be carried than could be possible if it were simply dissolved in the blood. Because hemoglobin is inside red blood cells, it is possible to assess certain kinds of health problems by counting the number of red blood cells. If the number is low, the person will not be able to carry oxygen efficiently and will tire easily. This condition, in which a person has reduced oxygen-carrying capacity, is called **anemia.** Anemia can also result when a person does not get enough iron. Because iron is a central atom in hemoglobin molecules, people with an iron deficiency are not able to manufacture sufficient hemoglobin. They can be anemic even though their number of red blood cells may be normal.

Red blood cells are also important in the transport of carbon dioxide. Carbon dioxide is produced as a result of normal aerobic cellular respiration of food materials in the cells of the body. If it is not eliminated, it causes the blood to become more acidic (lowers its pH), eventually resulting in death. Carbon dioxide can be carried in the blood in three forms: about 7% is dissolved in the plasma, about 23% is carried attached to hemoglobin molecules, and 70% is carried as bicarbonate ions. An enzyme in red blood cells known as **carbonic anhydrase** assists in converting carbon dioxide into bicarbonate ions (HCO_3^-), which can be carried as dissolved ions in the plasma of the blood. The following reversible chemical equation shows the changes that occur. • pH, p. 37

$$CO_2 + H_2O \longleftrightarrow H_2CO_3 \longleftrightarrow H^+ + HCO_3^-$$

When the blood reaches the lungs, dissolved carbon dioxide is lost from the plasma, and carbon dioxide is released from the hemoglobin molecules as well. In addition, the bicarbonate ions reenter the red blood cells and can be converted back into molecular carbon dioxide by the same enzyme-assisted process that converts carbon dioxide to bicarbonate ions. The importance of this mechanism will be discussed later when the exchange of gases at the lung surface is described (Gas Exchange section).

Heat is also transported by the blood. Heat is generated by metabolic activities and must be lost from the body. To handle excess body heat, blood is shunted to the surface of the body, where heat can be radiated away. In addition, humans and some other animals have the ability to sweat. The evaporation of sweat from the body surface also gets rid of excess heat. If the body is losing heat too rapidly, blood flow is shunted away from the skin, and metabolic heat is conserved. Vigorous exercise produces an excess of heat so that, even in cold weather, blood is shunted to the skin and the skin feels hot.

The plasma also carries nutrient molecules from the gut to other locations where they are modified, metabolized, or incorporated into cell structures. Amino acids and simple sugars are carried as dissolved molecules in the blood. Lipids, which are not water soluble, are combined with proteins and carried as suspended particles, called lipoproteins. Most lipids do not enter the bloodstream directly from the small intestine but are carried to the bloodstream by the lymphatic system. Other organs, like the liver, manufacture or modify molecules for use elsewhere; therefore they must constantly receive raw materials and distribute their products to the cells that need them through the transportation function of the blood.

In addition, many different kinds of hormones are produced by the brain, reproductive organs, digestive organs, and glands of the body. These are secreted into the bloodstream and transported throughout the body. Tissues with appropriate receptors bind to these molecules and respond to these chemical messengers.

Table 18.1

THE COMPOSITION OF BLOOD

Component	Quantity Present	Function
Plasma	55%	Maintain fluid nature of blood
Water	91.5%	
Protein	7.0%	
Other materials	1.5%	
Cellular material	45%	
Red blood cells (erythrocytes)	4.3–5.8 million/mm^3	Carry oxygen and carbon dioxide
White blood cells (leukocytes)	5–9 thousand/mm^3	Immunity
Lymphocytes	25%–30% of white cells present	
Monocytes	3%–7% of white cells present	
Neutrophils	57%–67% of white cells present	
Eosinophils	1%–3% of white cells present	
Basophils	Less than 1% of white cells present	
Platelets	250–400 thousand/mm^3	Clotting

Neutrophils   Eosinophils   Basophils

Lymphocytes   Monocytes   Platelets   Erythrocytes

The Immune System

Table 18.1 lists the variety of cells found in blood. Whereas the red, hemoglobin-containing erythrocytes serve in the transport of oxygen and carbon dioxide, the *white blood cells (WBCs)* carried in the blood are involved in defending against harmful agents. These cells help the body resist many diseases. They constitute the core of the **immune system.** The various WBCs participate in providing immunity in several ways. First, immune system cells are able to recognize cells and molecules that are foreign to the body. If a molecule is recognized as foreign, certain WBCs produce *antibodies (immunoglobulins)* that attach to the foreign materials. The foreign molecules that stimulate the production of antibodies are called *antigens (immunogens).* When harmful microorganisms (e.g., bacteria, viruses, fungi), cancer cells, or toxic molecules enter the body, other WBCs (1) recognize, (2) boost their abilities to respond to, (3) move toward, and (4) destroy the problem causers. *Neutrophils, eosinophils, basophils,* and *monocytes* are specific kinds of WBCs capable of engulfing foreign material, a process called phagocytosis. Thus, they are often called *phagocytes.* Although most can move from the bloodstream into the surrounding tissue, monocytes undergo such a striking increase in size that they are given a different name—*macrophages.* Macrophages can be found throughout the body and are the most active of the phagocytes.

The other major type of white cells, *lymphocytes,* work with phagocytes to provide protection. The two major types are *T-lymphocytes (T-cells)* and *B-lymphocytes (B-cells).* T-cells are involved in a cell-mediated immune response in which cells directly attack potentially dangerous objects. This highly complex response involves the release of chemical messengers that coordinate the response, an increase in the population of T-cells and B-cells, and stimulation of B-cell and macrophage activities. Some T-cells are capable of killing dangerous cells by destroying their cell membranes. T-cells are primarily involved in fighting infections of viruses, fungi, protozoa, worms, and cancer cells.

B-cells are involved in antibody-mediated immunity in which B-cells produce antibody molecules that are released into the bloodstream and are distributed to all parts of the body. These antibodies attach to the foreign molecules, causing them to clump together. This clumping may destroy their harmful properties, make them more susceptible to chemical attack, or make them more recognizable for phagocytes. Many kinds of diseases caused by viruses and bacteria are controlled by antibodies produced by B-cells.

Another kind of cellular particle in the blood is the platelet. These are fragments of specific kinds of white blood cells and are important in blood clotting. They collect at the site of a wound where they break down, releasing molecules. This begins a series of reactions that results in the formation of fibers that trap blood cells and form a plug in the opening of the wound.

The Heart

Blood can perform its transportation function only if it moves. The organ responsible for providing the energy to pump the blood is the heart. In order for a fluid to flow through a tube, there must be a pressure difference between the two ends of the tube. Water flows through pipes because it is under pressure. Because the pressure is higher behind a faucet than at the spout, water flows from the spout when the faucet is opened. The circulatory system can be analyzed from the same point of view. The heart is a muscular pump that provides the pressure necessary to propel the blood throughout the body. It must continue its cycle of contraction and relaxation, or blood stops flowing and body cells are unable to obtain nutrients or get rid of wastes. Some cells, such as brain cells, are extremely sensitive to having

their flow of blood interrupted because they require a constant supply of glucose and oxygen. Others, such as muscle cells or skin cells, are much better able to withstand temporary interruptions of blood flow.

The hearts of humans, other mammals, and birds consist of four chambers and four sets of valves that work together to ensure that blood flows in one direction only. Two of these chambers, the right and left **atria** (singular, *atrium*), are relatively thin-walled structures that collect blood from the major veins and empty it into the larger, more muscular ventricles (figure 18.3). Most of the flow of blood from the atria to the ventricles is caused by the lowered pressure produced within the ventricles as they relax. The contraction of the thin-walled atria assists in emptying them more completely.

The right and left **ventricles** are chambers that have powerful muscular walls whose contraction forces blood to flow through the arteries to all parts of the body. The valves between the atria and ventricles, known as **atrioventricular valves,** are important one-way valves that allow the blood to flow from the atria to the ventricles but prevent flow in the opposite direction. Similarly, there are valves in the aorta and pulmonary artery, known as **semilunar valves.** The **aorta** is the large artery that carries blood from the left ventricle to the body, and the **pulmonary artery** carries blood from the right ventricle to the lungs. The semilunar valves prevent blood from flowing back into the ventricles. If the atrioventricular or semilunar valves are damaged or function improperly, the efficiency of the heart as a pump is diminished, and the person may develop an enlarged heart or other symptoms. Malfunctioning heart valves are often diagnosed because they cause abnormal sounds as the blood passes through them. These sounds are referred to as heart murmurs. Similarly, if the muscular walls of the ventricles are weakened because of infection, damage from a heart attack, or lack of exercise, the pumping efficiency of the heart is reduced and the person develops symptoms that may include chest pain, shortness of breath, or fatigue. The pain is caused by an increase in the amount of lactic acid in the heart muscle because the muscle is not getting sufficient blood to satisfy its needs. It is important to understand that the muscle of the heart receives blood from coronary arteries that are branches of the aorta. It is not nourished by the blood that flows through its chambers. If heart muscle does not get sufficient oxygen for a period of time, the portion of the heart muscle not receiving blood dies. Shortness of breath and fatigue result because the heart is not able to pump blood efficiently to the lungs, muscles, and other parts of the body.

The right and left sides of the heart have slightly different jobs because they pump blood to different parts of the body. The right side of the heart receives blood from the general body and pumps it through the pulmonary arteries to the lungs, where exchange of oxygen and carbon dioxide takes place and the blood returns from the lungs to the left atrium. This is called **pulmonary circulation.** The larger, more powerful left side of the heart receives blood from the lungs, delivers it through the aorta to all parts of the body,

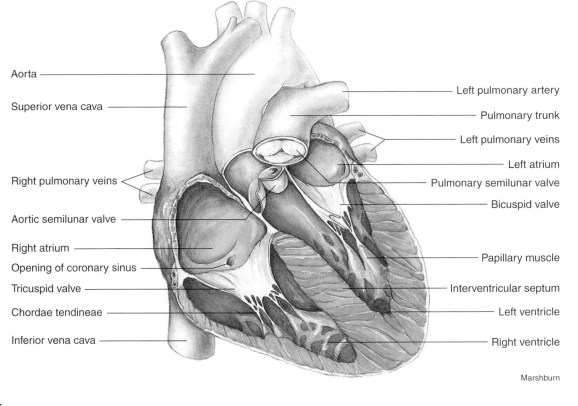

Aorta

Superior vena cava

Right pulmonary veins

Aortic semilunar valve

Right atrium

Opening of coronary sinus

Tricuspid valve

Chordae tendineae

Inferior vena cava

Left pulmonary artery

Pulmonary trunk

Left pulmonary veins

Left atrium

Pulmonary semilunar valve

Bicuspid valve

Papillary muscle

Interventricular septum

Left ventricle

Right ventricle

Marshburn

Figure 18.3

The Anatomy of the Heart
The heart consists of two thin-walled chambers called atria that contract to force blood into the two ventricles. When the ventricles contract, the atrioventricular valves (bicuspid and tricuspid) close, and blood is forced into the aorta and pulmonary artery. Semilunar valves in the aorta and pulmonary artery prevent the blood from flowing back into the ventricles when they relax.

and returns it to the right atrium by way of veins. This is known as **systemic circulation.** Both circulatory pathways are shown in figure 18.4. The systemic circulation is responsible for gas, nutrient, and waste exchange in all parts of the body except the lungs.

Arteries and Veins

Arteries and veins are the tubes that transport blood from one place to another within the body. Figure 18.5 compares the structure and function of arteries and veins. Arteries carry blood away from the heart because it is under considerable pressure from the contraction of the ventricles. The contraction of the walls of the ventricles increases the pressure in the arteries. A typical pressure recorded in a large artery while the heart is contracting is about 120 millimeters of mercury. This is known as the **systolic blood pressure.** The pressure recorded while the heart is relaxing is about 80 millimeters of mercury. This is known as the **diastolic blood pressure.** A blood pressure reading includes both numbers and is recorded as 120/80. (Originally, blood pressure was measured by how high the pressure of the blood would cause a column of mercury [Hg] to rise in a tube. Although the de-

vices used today have dials or digital readouts and contain no mercury, they are still calibrated in mmHg.)

The walls of arteries are relatively thick and muscular, yet elastic. Healthy arteries have the ability to expand as blood is pumped into them and return to normal as the pressure drops. This ability to expand absorbs some of the pressure and reduces the peak pressure within the arteries, thus reducing the likelihood that they will burst. If arteries become hardened and less resilient, the peak blood pressure rises and they are more likely to rupture. The elastic nature of the arteries is also responsible for assisting the flow of blood. When they return to normal from their stretched condition they give a little push to the blood that is flowing through them.

Blood is distributed from the large aorta through smaller and smaller blood vessels to millions of tiny capillaries. Some of the smaller arteries, called **arterioles,** may contract or relax to regulate the flow of blood to specific parts of the body. Major parts of the body that receive differing amounts of blood, depending on need, are the digestive system, muscles, and skin. When light-skinned people *blush*, it is because many arterioles in the skin have expanded, allowing a large volume of blood to flow to the capillaries of the

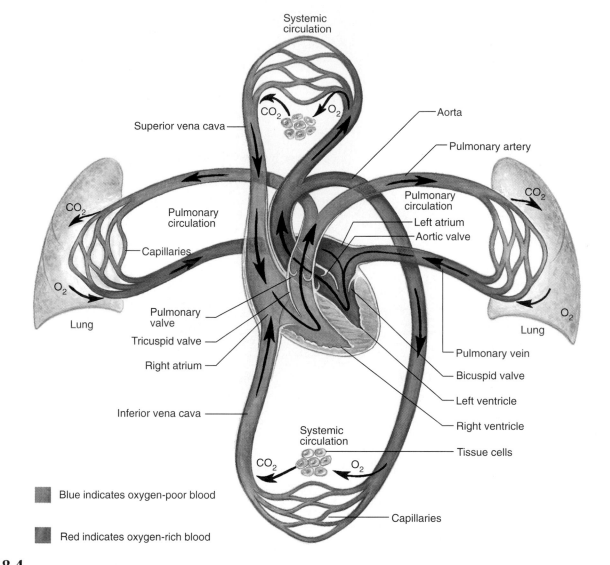

Figure 18.4

Pulmonary and Systemic Circulation
The right ventricle pumps blood that is poor in oxygen to the two lungs by way of the pulmonary arteries, where it receives oxygen and turns bright red. The blood is then returned to the left atrium by way of four pulmonary veins. This part of the circulatory system is known as pulmonary circulation. The left ventricle pumps oxygen-rich blood by way of the aorta to all parts of the body except the lungs. This blood returns to the right atrium, depleted of its oxygen, by way of the superior vena cava from the head region and the inferior vena cava from the rest of the body. This portion of the circulatory system is known as systemic circulation.

skin. Because the blood is red, their skin reddens. Similarly, when people exercise, there is an increased blood flow to muscles to accommodate their increased metabolic needs for oxygen and glucose and to get rid of wastes. Exercise also results in an increased flow of blood to the skin, which allows for heat loss. At the same time, the amount of blood flowing to the digestive system is reduced. Athletes do not eat a full meal before exercising because the additional flow of blood to the digestive system reduces the amount of blood available to go to muscles and lungs needed for vigorous exercise. Muscular cramps may result if insufficient blood is getting to the muscles.

Veins collect blood from the capillaries and return it to the heart. The pressure in these blood vessels is very low. Some of the largest veins may have a blood pressure of 0.0 mmHg for brief periods. The walls of veins are not as muscular as those of arteries. Because of the low pressure, veins must have valves that prevent the blood from flowing backward, away from the heart. Veins are often found at the surface of the body and are seen as blue lines. *Varicose veins* result when veins contain faulty valves that do not allow efficient return of blood to the heart. Therefore, blood pools in these veins, and they become swollen, bluish networks.

Artery

Vein

Epithelial lining

Connective tissue

Elastic tissue

Muscle layers

Valve

Blood flow

Figure 18.5

The Structure of Arteries and Veins
The walls of arteries are much thicker than the walls of veins. (The pressure in arteries is much higher than the pressure in veins.) The pressure generated by the ventricles of the heart forces blood through the arteries. Veins often have very low pressure. The valves in the veins prevent the blood from flowing backward, away from the heart.

Because pressure in veins is so low, muscular movements of the body are important in helping return blood to the heart. When muscles of the body contract, they compress nearby veins, and this pressure pushes blood along in the veins. Because the valves allow blood to flow only toward the heart, this activity acts as an additional pump to help return blood to the heart. People who sit or stand for long periods without using their leg muscles tend to have a considerable amount of blood pool in the veins of their legs and lower body. Thus, less blood may be available to go to the brain and the person may faint.

Although the arteries are responsible for distributing blood to various parts of the body and arterioles regulate where blood goes, it is the function of capillaries to assist in the exchange of materials between the blood and cells.

Capillaries

Capillaries are tiny, thin-walled tubes that receive blood from arterioles. They are so small that red blood cells must go through them in single file. They are so numerous that

each cell in the body has a capillary located near it. It is estimated that there are about 1,000 square meters of surface area represented by the capillary surface in a typical human. Each capillary wall consists of a single layer of cells and therefore presents only a thin barrier to the diffusion of materials between the blood and cells. It is also possible for liquid to flow through tiny spaces between the individual cells of most capillaries (figure 18.6). The flow of blood through these smallest blood vessels is relatively slow. This allows time for the diffusion of such materials as oxygen, glucose, and water from the blood to surrounding cells, and for the movement of such materials as carbon dioxide, lactic acid, and ammonia from the cells into the blood.

In addition to molecular exchange, considerable amounts of water and dissolved materials leak through the small holes in the capillaries. This liquid is known as **lymph.** Lymph is produced when the blood pressure forces water and some small dissolved molecules through the walls of the capillaries. Lymph bathes the cells but must eventually be returned to the circulatory system by lymph vessels or swelling will occur. Return is accomplished by the **lymphatic system,** a collection of thin-walled tubes that branch throughout the

Figure 18.6

Capillaries

Capillaries are tiny blood vessels. Exchange of cells and molecules can occur between blood and tissues through their thin walls. Molecules diffuse in and out of the blood, and cells such as monocytes can move from the blood through the thin walls into the surrounding tissue. There is also a flow of liquid through holes in the capillary walls. This liquid, called lymph, bathes the cells and eventually enters small lymph vessels that return lymph to the circulatory system near the heart.

body. These tubes collect lymph that is filtered from the circulatory system and ultimately empty it into major blood vessels near the heart. As the lymph makes its way back to the circulatory system, it is filtered by lymph nodes that contain large numbers of white blood cells that remove microorganisms and foreign particles. There are many lymph nodes located throughout the body. The tonsils and adenoids are large masses of lymph node tissue. The spleen also contains large numbers of white blood cells and serves to filter the blood. The thymus gland is located over the breastbone and is large and active in children. Its primary function is to produce T-lymphocytes that are distributed throughout the body and establish themselves in lymph nodes. The thymus shrinks in size in adulthood, but the descendants of the T-lymphocytes it produced earlier in life are still active throughout the lymphatic system. Figure 18.7 shows the structure of the lymphatic system.

Some of this leakage through the capillary walls is normal, but the flow is subject to changes in pressure inside the capillaries and in the tissues, and changes in the permeability of the capillary wall. If pressure inside the capillary increases, more fluid may leak from the capillaries into the tissues and cause swelling. This swelling is called *edema*, and it is common in circulatory disorders. Another cause of edema is an increase in the permeability of the capillaries. This is commonly associated with injury to a part of the body: A sprained ankle and smashed thumb are examples you have probably experienced.

18.3 Gas Exchange

The exchange of gases between the exterior of an organism and the cells that need to obtain oxygen and get rid of carbon dioxide is a good example of how the circulatory system is involved in distributing materials throughout the body. In addition, the concept of homeostasis and the importance of a large surface area are easy to visualize. The significance of

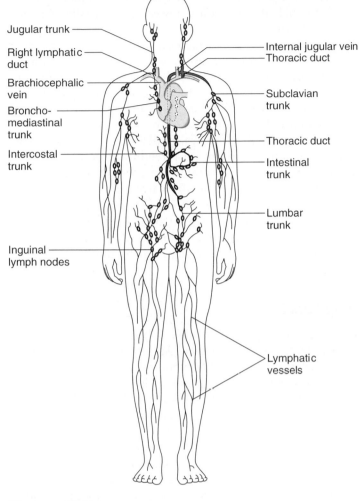

Figure 18.7

The Lymphatic System
The lymphatic system consists of many ducts that transport lymph fluid back toward the heart. Along the way the lymph is filtered in the lymph nodes, and bacteria and other foreign materials are removed before the lymph is returned to the circulatory system near the heart.

each of these factors is easily seen in the structure and function of the different parts of the respiratory system.

Respiratory Anatomy

The **lungs** are organs of the body that allow gas exchange to take place between the air and blood. Associated with the lungs is a set of tubes that conducts air from outside the body to the lungs. The single, large-diameter **trachea** is supported by rings of cartilage that prevent its collapse. It branches into two major **bronchi** (singular, *bronchus*) that deliver air to smaller and smaller branches. Bronchi are also supported by cartilage. The smallest tubes, known as **bronchioles,** contain smooth muscle and are therefore capable of constricting. Finally, the bronchioles deliver air to clus-

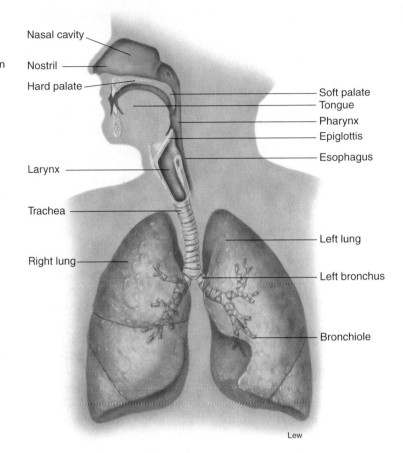

Figure 18.8

Respiratory Anatomy
Although the alveoli of the lungs are the places where gas exchange takes place, there are many other important parts of the respiratory system. The nasal cavity cleans, warms, and humidifies the air entering the lungs. The trachea is also important in cleaning the air going to the lungs.

ters of tiny sacs, known as **alveoli** (singular, *alveolus*), where the exchange of gases takes place between the air and blood.

The nose, mouth, and throat are also important parts of the air-transport pathway because they modify the humidity and temperature of the air and clean the air as it passes. The lining of the trachea contains cells with cilia that beat in a direction to move mucus and foreign materials from the lungs. The foreign matter may then be expelled by swallowing, coughing, or other means. Figure 18.8 illustrates the various parts of the respiratory system.

Breathing System Regulation

Breathing is the process of moving air in and out of the lungs. It is accomplished by the movement of a muscular organ known as the **diaphragm,** which separates the chest cavity and the lungs from the abdominal cavity. In addition, muscles located between the ribs (*intercostal* muscles) are attached to the ribs in such a way that their contraction causes

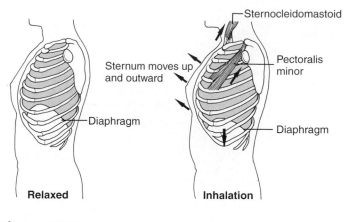

Figure 18.9

Breathing Movements
During inhalation, the diaphragm and external intercostal muscles between the ribs contract, causing the volume of the chest cavity to increase. During a normal exhalation, these muscles relax, and the chest volume returns to normal.

the chest wall to move outward and upward, which increases the size of the chest cavity. During inhalation, the diaphragm moves downward and the external intercostal muscles of the chest wall contract, causing the volume of the chest cavity to increase. This results in a lower pressure in the chest cavity compared to the outside air pressure. Consequently, air flows from the outside high-pressure area through the trachea, bronchi, and bronchioles to the alveoli. During normal relaxed breathing, exhalation is accomplished by the chest wall and diaphragm simply returning to their normal, relaxed positions. Muscular contraction is not involved (figure 18.9).

However, when the body's demand for oxygen increases during exercise, the only way that the breathing system can respond is by exchanging the gases in the lungs more rapidly. This can be accomplished both by increasing the breathing rate and by increasing the volume of air exchanged with each breath. Increase in volume exchanged per breath is accomplished in two ways. First, the muscles of inhalation can contract more forcefully, resulting in a greater change in the volume of the chest cavity. In addition, the lungs can be emptied more completely by contracting the muscles of the abdomen, which forces the abdominal contents upward against the diaphragm and compresses the lungs. A set of internal intercostal muscles also helps compress the chest. You are familiar with both mechanisms: When you exercise you breathe more deeply and more rapidly.

Several homeostatic mechanisms can cause changes in the rate and depth of breathing, but the primary mechanism involves the amount of carbon dioxide present in the blood. Carbon dioxide is a waste product of aerobic cellular respiration and becomes toxic in high quantities because it combines with water to form carbonic acid:

$$CO_2 + H_2O \longrightarrow H_2CO_3$$

As mentioned previously, if carbon dioxide cannot be eliminated, the pH of the blood is lowered. Eventually, this may result in death.

Exercising causes an increase in the amount of carbon dioxide in the blood because muscles are oxidizing glucose more rapidly. This lowers the pH of the blood. Certain brain cells and specialized cells in the aortic arch and carotid arteries are sensitive to changes in blood pH. When they sense a lower blood pH, nerve impulses are sent more frequently to the diaphragm and intercostal muscles. These muscles contract more rapidly and more forcefully, resulting in more rapid, deeper breathing. Because more air is being exchanged per minute, carbon dioxide is lost from the lungs more rapidly. When exercise stops, blood pH rises, and breathing eventually returns to normal (figure 18.10). Bear in mind, however, that moving air in and out of the lungs is of no value unless oxygen is diffusing into the blood and carbon dioxide is diffusing out.

Lung Function

The lungs are organs that allow blood and air to come in close contact with each other. Air flows in and out of the lungs during breathing. The blood flows through capillaries in the lungs and is in close contact with the air in the cavities of the lungs. For oxygen to enter the body or carbon dioxide to exit the body the molecules must pass through a surface. The efficiency of exchange is limited by the surface area available. This problem is solved in the lungs by the large number of tiny sacs, the alveoli. Each alveolus is about 0.25 to 0.5 millimeters across. However, alveoli are so numerous that the total surface area of all these sacs is about 70 square meters—comparable to the floor space of many standard-sized classrooms! Associated with these alveoli are large numbers of capillaries (figure 18.11). The walls of both the capillaries and alveoli are very thin, and the close association of alveoli and capillaries in the lungs allows the easy diffusion of oxygen and carbon dioxide across these membranes.

Another factor that increases the efficiency of gas exchange is that both the blood and air are moving. Because blood is flowing through capillaries in the lungs, the capillaries continually receive new blood that is poor in oxygen and high in carbon dioxide. As blood passes by the alveoli, it is briefly exposed to the gases in the alveoli, where it gains oxygen and loses carbon dioxide. Thus, blood that leaves the lungs is high in oxygen and low in carbon dioxide. Although the movement of air in the lungs is not in one direction, as is the case with blood, the cycle of inhalation and exhalation allows air that is high in carbon dioxide and low in oxygen to exit the body and brings in new air that is rich in oxygen and low in carbon dioxide. This oxygen-rich blood is then sent to the left side of the heart and pumped throughout the body.

Any factor that interferes with the flow of blood or air or alters the effectiveness of gas exchange in the lungs reduces the efficiency of the organism. A poorly pumping

Figure 18.10

The Control of Breathing Rate

The rate of breathing is controlled by specific cells in the brain that sense the pH of the blood. When the amount of CO_2 increases, the pH drops (becomes more acid) and the brain sends more frequent messages to the diaphragm and intercostal muscles, causing the breathing rate to increase. More rapid breathing increases the rate at which CO_2 is lost from the blood; thus the blood pH rises (becomes less acid) and the breathing rate decreases.

heart sends less blood to the lungs, and the person experiences shortness of breath as a symptom. Similarly, diseases like *asthma*, which causes constriction of the bronchioles, reduce the flow of air into the lungs and inhibit gas exchange.

Any process that reduces the number of alveoli will also reduce the efficiency of gas exchange in the lungs. *Emphysema* is a progressive disease in which some of the alveoli are lost. As the disease progresses, those afflicted have less and less respiratory surface area and experience greater and

greater difficulty getting adequate oxygen, even though they may be breathing more rapidly. Often emphysema is accompanied by an increase in the amount of connective tissue and the lungs do not stretch as easily, further reducing the ability to exchange gases.

The breathing mechanism is designed to get oxygen into the bloodstream so that it can be distributed to the cells that are carrying on the oxidation of food molecules, such as glucose and fat. Obtaining food molecules involves a variety of organs and activities associated with the digestive system.

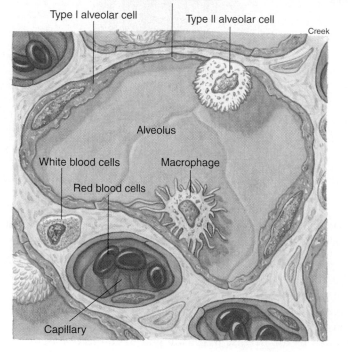

Figure 18.11

The Association of Capillaries with Alveoli
The exchange of gases takes place between the air-filled alveolus and the blood-filled capillary. The capillaries form a network around the saclike alveoli. The thin walls of the alveolus and capillary are in direct contact with one another; their combined thickness is usually less than 1 micrometer (a thousandth of a millimeter).

18.4 Obtaining Nutrients

All cells must have a continuous supply of nutrients that provides the energy they require and the building blocks needed to construct the macromolecules typical of living things. The specific functions of various kinds of nutrients are discussed in chapter 19. This section will deal with the processing and distribution of different kinds of nutrients.

The digestive system consists of a muscular tube with several specialized segments. In addition, there are glands that secrete digestive juices into the tube. Four different kinds of activities are involved in getting nutrients to the cells that need them: mechanical processing, chemical processing, nutrient uptake, and chemical alteration.

Mechanical and Chemical Processing

The digestive system is designed as a disassembly system. Its purpose is to take large chunks of food and break them down to smaller molecules that can be taken up by the circulatory system and distributed to cells. The first step in this process is mechanical processing.

It is important to grind large particles into small pieces by chewing in order to increase their surface areas and allow for more efficient chemical reactions. It is also important to add water to the food, which further disperses the particles and provides the watery environment needed for these chemical reactions. Materials must also be mixed so that all the molecules that need to interact with one another have a good chance of doing so. The oral cavity and the stomach are the major body regions involved in reducing the size of food particles. The teeth are involved in cutting and grinding food to increase its surface area. This is another example of the surface area-to-volume concept presented at the beginning of this chapter. The watery mixture that is added to the food in the oral cavity is known as *saliva*, and the three pairs of glands that produce saliva are known as **salivary glands.** Saliva contains the enzyme **salivary amylase,** which initiates the chemical breakdown of starch. Saliva also lubricates the oral cavity and helps bind food before swallowing.

In addition to having taste buds that help identify foods, the tongue performs the important service of helping position the food between the teeth and pushing it to the back of the throat for swallowing. The oral cavity is very much like a food processor in which mixing and grinding take place. Figure 18.12 describes and summarizes the functions of these and other structures of the digestive system.

Once the food has been chewed, it is swallowed and passes down the esophagus to the stomach. The process of swallowing involves a complex series of events. First, a ball of food, known as a *bolus*, is formed by the tongue and moved to the back of the mouth cavity. Here it stimulates the walls of the throat, also known as the **pharynx.** Nerve endings in the lining of the pharynx are stimulated, causing a reflex contraction of the walls of the esophagus, which transports the bolus to the stomach. Because both food and air pass through the pharynx, it is important to prevent food from getting into the lungs. During swallowing the larynx is pulled upward. This causes a flap of tissue called the *epiglottis* to cover the opening to the trachea and prevent food from entering the trachea. In the stomach, additional liquid, called **gastric juice,** is added to the food. Gastric juice contains enzymes and hydrochloric acid. The major enzyme of the stomach is **pepsin,** which initiates the chemical breakdown of protein. The pH of gastric juice is very low, generally around pH 2. Consequently, very few kinds of bacteria or protozoa emerge from the stomach alive. Those that do have special protective features that allow them to survive as they pass through the stomach. The entire mixture is churned by the contractions of the three layers of muscle in the stomach wall. The combined activities of enzymatic breakdown, chemical breakdown by hydrochloric acid, and mechanical processing by muscular movement result in a thoroughly mixed liquid called *chyme*. Chyme eventually leaves the stomach through a valve known as the **pyloric sphincter** and enters the small intestine (How Science Works 18.1).

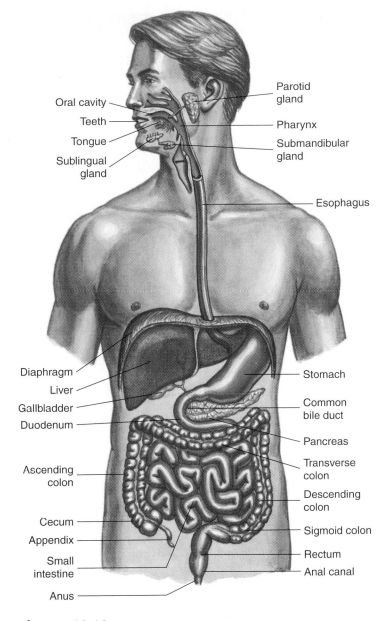

Figure 18.12

The Digestive System
The teeth, tongue, and enzymes from the salivary glands modify the food before it is swallowed. The stomach adds acid and enzymes and further changes the texture of the food. The food is eventually emptied into the duodenum, where the liver and pancreas add their secretions. The small intestine also adds enzymes and is involved in absorbing nutrients. The large intestine is primarily involved in removing water.

The first part of the **small intestine** is known as the **duodenum.** In addition to producing enzymes, the duodenum secretes several kinds of hormones that regulate the release of chyme from the stomach and the release of secretions from the pancreas and liver. The **pancreas** produces a number of different digestive enzymes and also secretes large amounts of bicarbonate ions, which neutralize the acids that enter from the stomach so that the pH of the

HOW SCIENCE WORKS 18.1

An Accident and an Opportunity

On the morning of June 6, 1822, on Mackinac Island in northern Lake Huron, a 19-year-old French-Canadian fur trapper named Alexis St. Martin was shot in the stomach by the accidental discharge from a shotgun. The army surgeon at Fort Mackinac, Michigan, Dr. William Beaumont, was called to attend the wounded man. Part of the stomach and body wall had been shot away and parts of St. Martin's clothing were imbedded in the wound. Although Dr. Beaumont did not expect St. Martin to live, he quickly cleaned the wound, pushed portions of the lung and stomach that were protruding back into the cavity, and dressed the wound. Finding St. Martin alive the next day, Beaumont was surprised and encouraged to do what he could to extend his life. In fact, Beaumont cared for St. Martin for 2 years. When the wound was completely healed the stomach had fused to the body wall and a hole allowed Beaumont to look into the stomach. Fortunately for St. Martin, a flap of tissue from the lining of the stomach closed off the opening so that what he ate did not leak out.

Beaumont found that he could look through the opening and observe the activities in the stomach and recognized that this presented an opportunity to study the function of the stomach in a way that had not been done before. He gathered gastric juice, had its components identified, introduced food into the hole with a string attached so that he could retrieve the food particles that were partially digested for examination, and observed the effect of emotion on digestion. He discovered many things that were new to science and contrary to the teachings of the time. He recounted many of his observations and experiments in his journal. "I consider myself but a humble inquirer after truth—a simple experimenter. And if I have been led to conclusions opposite to the opinions of many who have been considered luminaries of physiology, and, in some instances, from all the professors of this science, I hope the claim of sincerity will be conceded to me, when I say that such difference of opinion has been forced upon me by the convictions of experiment, and the fair deductions of reasoning."

The following are some of his important discoveries.

1. He measured the temperature of the stomach and found that it does not heat up when food is introduced as was thought at the time. "But from the result of a great number of experiments and examinations, made with a view to asserting the truth of this opinion, in the empty and full state of the organ, . . . I am convinced that there is no alteration of temperature. . . ."

2. He found that pure gastric juice contains large amounts of hydrochloric acid. This was contrary to the prevailing opinion that gastric juice was simply water. "I think I am warranted, from the result of all the experiments, in saying, that the gastric juice, so far from being 'inert as water,' as some authors assert, is the most general solvent in nature, of alimentary matter—even the hardest bone cannot withstand its action."

3. He observed that gastric juice is not stored in the stomach but is secreted when food is eaten. "The gastric juice does not accumulate in the cavity of the stomach, until alimentary matter is received, and excite its vessels to discharge their contents, for the immediate purpose of digestion."

4. He realized that digestion begins immediately when food enters the stomach. The prevailing opinion of the day was that nothing happened for an hour or more. "At 2 o'clock P.M.—twenty minutes after having eaten an ordinary dinner of boiled, salted beef, bread, potatoes, and turnips, and drank [sic] a gill of water, I took from his stomach, through the artificial opening, a gill of the contents. . . . Digestion had evidently commenced, and was perceptibly progressing, at the time."

5. He discovered that food in the stomach satisfies hunger even though it is not eaten. "To ascertain whether the sense of hunger would be allayed without food being passed through the oesophagus [sic], he fasted from breakfast time, til 4 o'clock, P.M., and became quite hungry. I then put in at the aperture, three and a half drachms of lean, boiled beef. The sense of hunger immediately subsided, and stopped the borborygmus, or croaking noise, caused by the motion of the air in the stomach and intestines, peculiar to him since the wound, and almost always observed when the stomach is empty."

St. Martin did not take kindly to these probings and twice ran away from Beaumont's care back to Canada where he married, had two children, and resumed his former life as a voyageur and fur trapper. He did not die until the age of 83, having lived over 60 years with a hole in his stomach.

---

duodenum is about pH 8. The **liver** is a large organ in the upper abdomen that performs several functions. One of its functions is the secretion of **bile**. When bile leaves the liver, it is stored in the **gallbladder** prior to being released into the duodenum. When bile is released from the gallbladder, it assists mechanical mixing by breaking large fat globules into smaller particles. This process is called *emulsification*.

Emulsification is important because fats are not soluble in water, yet the reactions of digestion must take place in a water solution. Bile causes large globules of fat to be broken into much smaller units (increasing the surface area-to-volume ratio) much as soap breaks up fat particles into smaller globules that are suspended in water and washed away. The activity of bile is important for the further digestion of fats in the intestine.

Along the length of the intestine, additional watery juices are added until the mixture reaches the **large intestine**. The large intestine is primarily involved in reabsorbing the water that has been added to the food tube along with saliva, gastric juice, bile, pancreatic secretions, and intestinal

Table 18.2

DIGESTIVE ENZYMES AND THEIR FUNCTIONS

Enzyme	Site of Production	Molecules Altered	Molecules Produced
Salivary amylase	Salivary glands	Starch	Smaller polysaccharides (many sugar molecules attached together)
Pepsin	Stomach lining	Proteins	Peptides (several amino acids)
Gastric lipase	Stomach lining	Fats	Fatty acids and glycerol
Chymotrypsin	Pancreas	Polypeptides (long chains of amino acids)	Peptides
Trypsin	Pancreas	Polypeptides	Peptides
Carboxypeptidase	Pancreas	Peptides	Smaller peptides and amino acids
Pancreatic amylase	Pancreas	Polysaccharides	Disaccharides
Pancreatic lipase	Pancreas	Fats	Fatty acids and glycerol
Nuclease	Pancreas	Nucleic acids	Nucleotides
Aminopeptidase	Intestinal lining	Peptides	Smaller peptides and amino acids
Dipeptidase	Intestinal lining	Dipeptides	Amino acids
Lactase	Intestinal lining	Lactose	Glucose and galactose
Maltase	Intestinal lining	Maltose	Glucose
Sucrase	Intestinal lining	Sucrose	Glucose and fructose
Nuclease	Intestinal lining	Nucleic acids	Nucleotides

juices. The large intestine is also home to a variety of different kinds of bacteria. Most live on the undigested food that makes it through the small intestine. Some provide additional benefit by producing vitamins that can be absorbed from the large intestine. A few kinds may cause disease.

Several different kinds of enzymes have been mentioned in this discussion. Each is produced by a specific organ and has a specific function. Chapter 5 introduced the topic of enzymes and how they work. Some enzymes, such as those involved in glycolysis, the Krebs cycle, and protein synthesis are produced and used inside cells; others, such as the digestive enzymes, are produced by cells and secreted into the digestive tract. Digestive enzymes are simply a special class of enzymes and have the same characteristics as the enzymes you studied previously. They are protein molecules that speed up specific chemical reactions and are sensitive to changes in temperature or pH. The various digestive enzymes, the sites of their production, and their functions are listed in table 18.2.

Nutrient Uptake

The process of digestion results in a variety of simple organic molecules that are available for absorption from the tube of the gut into the circulatory system. As we move simple sugars, amino acids, glycerol, and fatty acids into the circulatory system, we encounter another situation where surface area is important. The amount of material that can be taken up is limited by the surface area available. This problem is solved by increasing the surface area of the intestinal tract in several ways. First, the small intestine is a very long tube; the longer the tube, the greater the internal surface area. In a typical adult human it is about 3 meters long. In addition to length, the lining of the intestine consists of millions of fingerlike projections called **villi,** which increase the surface area. When we examine the cells that make up the villi, we find that they also have folds in their surface membranes. All of these characteristics increase the surface area available for the transport of materials from the gut into the circulatory system (figure 18.13). Scientists estimate that the cumulative effect of all of these features produces a total intestinal surface area of about 250 square meters. That is equivalent to about the area of a basketball court.

The surface area by itself would be of little value if it were not for the intimate contact of the circulatory system with this lining. Each villus contains several capillaries and a branch of the lymphatic system called a **lacteal.** The close association between the intestinal surface and the circulatory and lymphatic systems allows for the efficient uptake of nutrients from the cavity of the gut into the circulatory system.

Several different kinds of processes are involved in the transport of materials from the intestine to the circulatory system. Some molecules, such as water and many ions, simply diffuse through the wall of the intestine into the circulatory system. Other materials, such as amino acids and simple sugars, are assisted across the membrane by carrier molecules. Fatty acids and glycerol are absorbed into the intestinal lining cells where they are resynthesized into fats and enter lacteals in the villi. Because the lacteals are part of the lymphatic system, which eventually empties its contents into the circulatory system, fats are also transported by the blood. They just reach the blood by a different route.

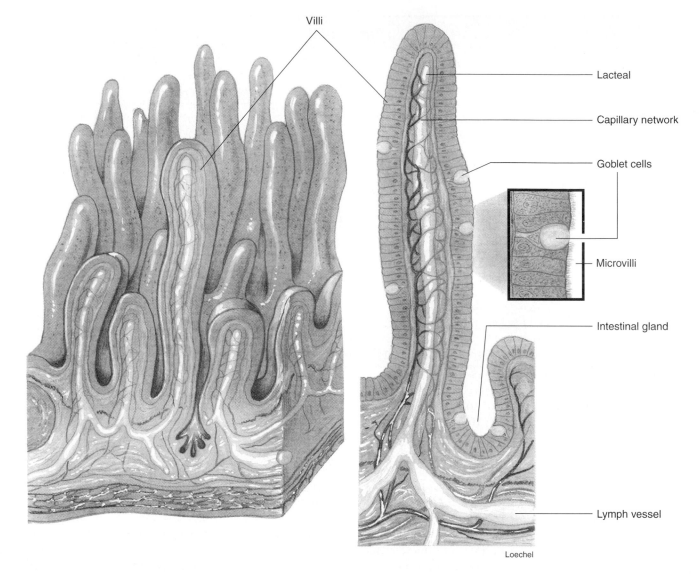

Villi

Lacteal

Capillary network

Goblet cells

Microvilli

Intestinal gland

Lymph vessel

Loechel

Figure 18.13

The Exchange Surface of the Intestine

The surface area of the intestinal lining is increased by the many fingerlike projections known as villi. Within each villus are capillaries and lacteals. Most kinds of materials enter the capillaries, but most fat-soluble substances enter the lacteals, giving them a milky appearance. Lacteals are part of the lymphatic system. Because the lymphatic system empties into the circulatory system, fat-soluble materials also eventually enter the circulatory system. The close relationship between the vessels and the epithelial lining of the villus allows for efficient exchange of materials from the intestinal cavity to the circulatory system.

Chemical Alteration: The Role of the Liver

When the blood leaves the intestine, it flows directly to the liver through the **hepatic portal vein.** Portal veins are blood vessels that collect blood from capillaries in one part of the body and deliver it to a second set of capillaries in another part of the body without passing through the heart. Thus, the hepatic portal vein collects nutrient-rich blood from the intestine and delivers it directly to the liver. As the blood flows through the liver, enzymes in the liver cells modify many of the molecules and particles that enter them. One of the functions of the liver is to filter any foreign organisms

from the blood that might have entered through the intestinal cells. It also detoxifies many dangerous molecules that might have entered with the food.

Many foods contain toxic substances that could be harmful if not destroyed by the liver. Ethyl alcohol is one obvious example. Many plants contain various kinds of toxic molecules that are present in small quantities and could accumulate to dangerous levels if the liver did not perform its role of detoxification.

In addition, the liver is responsible for modifying nutrient molecules. The liver collects glucose molecules and synthesizes glycogen, which can be stored in the liver for later use. When glucose is in short supply, the liver can convert some of its stored glycogen back into glucose. Although amino acids are not stored, the liver can change the relative numbers of different amino acids circulating in the blood. It can remove the amino group from one kind of amino acid and attach it to a different carbon skeleton, generating a different amino acid. The liver is also able to take the amino group off amino acids so that what remains of the amino acid can be used in aerobic cellular respiration. The toxic amino groups are then converted to urea by the liver. Urea is secreted back into the bloodstream and is carried to the kidneys for disposal in the urine.

18.5 Waste Disposal

Because cells are modifying molecules during metabolic processes, harmful waste products are constantly being formed. Urea is a common waste; many other toxic materials must be eliminated as well. Among these are large numbers of hydrogen ions produced by metabolism. This excess of hydrogen ions must be removed from the bloodstream. Other molecules, such as water and salts, may be consumed in excessive amounts and must be removed. The primary organs involved in regulating the level of toxic or unnecessary molecules are the **kidneys** (figure 18.14).

Kidney Structure

The kidneys consist of about 2.4 million tiny units called **nephrons.** At one end of a nephron is a cup-shaped structure called **Bowman's capsule,** which surrounds a knot of capillaries known as a **glomerulus** (figure 18.15). In addition to Bowman's capsule, a nephron consists of three distinctly different regions: the **proximal convoluted tubule,** the **loop of Henle,** and the **distal convoluted tubule.** The distal convoluted tubule of a nephron is connected to a collecting duct that transports fluid to the ureters, and ultimately to the urinary bladder, where it is stored until it can be eliminated.

Kidney Function

As in the other systems discussed in this chapter, the excretory system involves a close connection between the circulatory system and a surface. In this case, the large surface is provided by the walls of the millions of nephrons, which are surrounded by capillaries. Three major activities occur at these surfaces: filtration, reabsorption, and secretion. The glomerulus presents a large surface for the filtering of material from the blood to Bowman's capsule. Blood that enters the glomerulus is under pressure from the muscular contraction of the heart. The capillaries of the glomerulus are quite porous and provide a large surface area for the movement of water and small dissolved molecules from the blood into Bowman's capsule. Normally, only the smaller molecules, such as glucose, amino acids, and ions, are able to pass

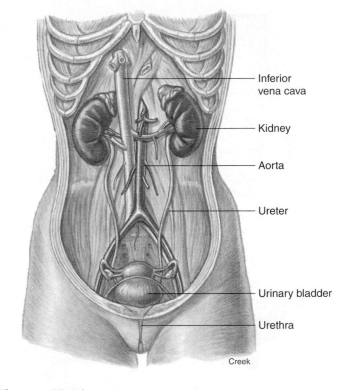

Figure 18.14

The Urinary System
The primary organs involved in removing materials from the blood are the kidneys. The urine produced by the kidneys is transported by the ureters to the urinary bladder. From the bladder, the urine is emptied to the outside of the body by way of the urethra.

through the glomerulus into the Bowman's capsule at the end of the nephron. The various kinds of blood cells and larger molecules like proteins do not pass out of the blood into the nephron. This physical filtration process allows many kinds of molecules to leave the blood and enter the nephron. The volume of material filtered in this way through the approximately 2.4 million nephrons of our kidneys is about 7.5 liters per hour. Because your entire blood supply is about 5 to 6 liters, there must be some method of recovering much of this fluid.

Surrounding the various portions of the nephron are capillaries that passively accept or release molecules on the basis of diffusion gradients. The walls of the nephron are made of cells that actively assist in the transport of materials. Some molecules are reabsorbed from the nephron and picked up by the capillaries that surround them, whereas other molecules are actively secreted into the nephron from the capillaries. Each portion of the nephron has cells with specific secretory abilities.

The proximal convoluted tubule is primarily responsible for reabsorbing valuable materials from the fluid moving through it. Molecules like glucose, amino acids, and sodium ions are actively transported across the membrane of the proximal convoluted tubule and returned to the blood. In

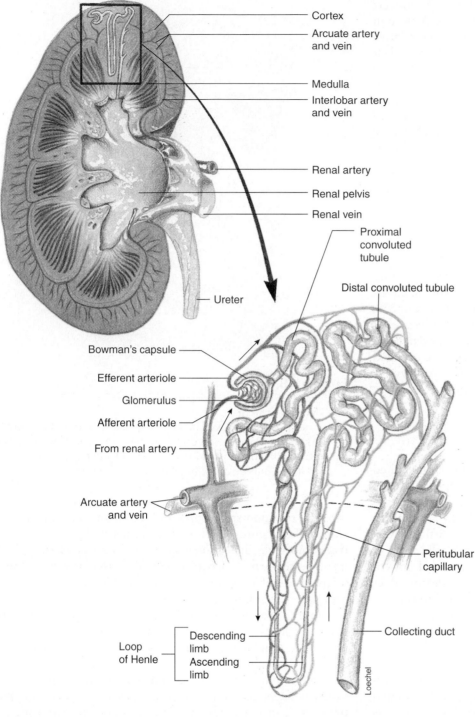

Cortex

Arcuate artery
and vein

Medulla

Interlobar artery
and vein

Renal artery

Renal pelvis

Renal vein

Proximal
convoluted
tubule

Distal convoluted tubule

Ureter

Bowman's capsule

Efferent arteriole

Glomerulus

Afferent arteriole

From renal artery

Arcuate artery
and vein

Peritubular
capillary

Loop
of Henle

Descending
limb
Ascending
limb

Collecting duct

Loechel

Figure 18.15

The Structure of the Nephron

The nephron and the closely associated blood vessels create a system that allows for the passage of materials from the circulatory system to the nephron by way of the glomerulus and Bowman's capsule. Materials are added to and removed from the fluid in the nephron via the tubular portions of the nephron.

addition, water moves across the membrane because it follows the absorbed molecules and diffuses to the area where water molecules are less common. By the time the fluid has reached the end of the proximal convoluted tubule, about 65% of the fluid has been reabsorbed into the capillaries surrounding this region.

The next portion of the tubule, the loop of Henle, is primarily involved in removing additional water from the nephron. Although the details of the mechanism are complicated, the principles are rather simple. The cells of the ascending loop of Henle actively transport sodium ions from the nephron into the space between nephrons where sodium ions accumulate in the fluid that surrounds the loop of Henle. The collecting ducts pass through this region as they carry urine to the ureters. Because the area these collecting ducts pass through is high in sodium ions, water within the collecting ducts diffuses from the ducts and is picked up by surrounding capillaries. However, the ability of water to pass through the wall of the collecting duct is regulated by hormones. Thus, it is possible to control water loss from the body by regulating the amount of water lost from the collecting ducts. For example, if you drank a liter of water or some other liquid, the excess water would not be allowed to leave the collecting duct and would exit the body as part of the urine. However, if you were dehydrated, most of the water passing through the collecting ducts would be reabsorbed, and very little urine would be produced. The primary hormone involved in regulating water loss is the antidiuretic hormone (ADH). When the body has excess water, cells in the hypothalamus of the brain respond and send a signal to the pituitary and only a small amount of ADH is released and water is lost. When you are dehydrated these same brain cells cause

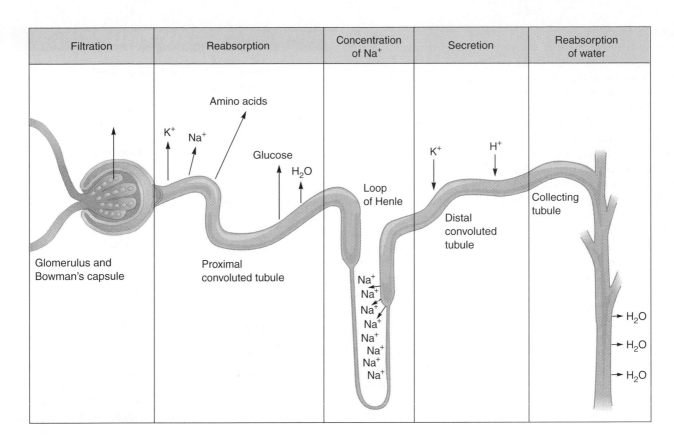

Filtration	Reabsorption	Concentration of Na$^+$	Secretion	Reabsorption of water

Figure 18.16

Specific Functions of the Nephron

Each portion of the nephron has specific functions. The glomerulus and Bowman's capsule accomplish the filtration of fluid from the bloodstream into the nephron. The proximal convoluted tubule reabsorbs a majority of the material filtered. The loop of Henle concentrates Na$^+$ so that water will move from the collecting tubule. The distal convoluted tubule regulates pH and ion concentration by differential secretion of K$^+$ and H$^+$.

more ADH to be released and water leaves the collecting duct and is returned to the blood.

The distal convoluted tubule is primarily involved in fine-tuning the amounts of various kinds of molecules that are lost in the urine. Hydrogen ions (H$^+$), sodium ions (Na$^+$), chloride ions (Cl$^-$), potassium ions (K$^+$), and ammonium ions (NH$_4^+$) are regulated in this way.

Some molecules that pass through the nephron are relatively unaffected by the various activities going on in the kidney. One of these is urea, which is filtered through the glomerulus into Bowman's capsule. As it passes through the nephron, much of it stays in the tubule and is eliminated in the urine. Many other kinds of molecules, such as minor metabolic waste products and some drugs, are also treated in this manner. Figure 18.16 summarizes the major functions of the various portions of the kidney tubule system.

SUMMARY

The body's various systems must be integrated in such a way that the internal environment stays relatively constant. This concept is called homeostasis. This chapter surveys four systems of the body—the circulatory, respiratory, digestive, and excretory systems—and describes how they are integrated. All of these systems are involved in the exchange of materials across cell membranes. Because of problems of exchange, cells must be small. Exchange is limited by the amount of surface area present, so all of these systems have special features that provide large surface areas to allow for necessary exchanges.

The circulatory system consists of a pump, the heart, and blood vessels that distribute the blood to all parts of the body. The blood is a carrier fluid that transports molecules and heat. The exchange of materials between the blood and body cells takes place through the walls of the capillaries. Because the flow of blood can be regulated by the contraction of arterioles, blood can be sent to different parts of the body at different times. Hemoglobin in red blood cells is very important in the transport of oxygen. Carbonic anhydrase is an enzyme in red blood cells that converts carbon dioxide into bicarbonate ions that can be easily carried by the blood.

The respiratory system consists of the lungs and associated tubes that allow air to enter and leave the lungs. The diaphragm and muscles of the chest wall are important in the process of breathing. In the lungs, tiny sacs called alveoli provide a large surface area in association with capillaries, which allows for rapid exchange of oxygen and carbon dioxide.

The digestive system is involved in disassembling food molecules. This involves several processes: grinding by the teeth and stomach, emulsification of fats by bile from the liver, addition of water to dissolve molecules, and enzymatic action to break complex molecules into simpler molecules for absorption. The intestine provides a large surface area for the absorption of nutrients because it is long and its wall contains many tiny projections that increase surface area. Once absorbed, the materials are carried to the liver, where molecules can be modified.

The excretory system is a filtering system of the body. The kidneys consist of nephrons into which the circulatory system filters fluid. Most of this fluid is useful and is reclaimed by the cells that make up the walls of these tubules. Materials that are present in excess or those that are harmful are allowed to escape. Some molecules may also be secreted into the tubules before being eliminated from the body.

THINKING CRITICALLY

For guidelines to answer this question, visit the Online Learning Center.

It is possible to keep a human being alive even if the heart, lungs, kidneys, and digestive tract are not functioning by using heart-lung machines in conjunction with kidney dialysis and intravenous feeding. This implies that the basic physical principles involved in the functioning of these systems is well understood because the natural functions can be duplicated with mechanical devices. However, these machines are expensive and require considerable maintenance. Should society be spending money to develop smaller, more efficient mechanisms that could be used to replace diseased or damaged hearts, lungs, and kidneys? Debate this question.

CONCEPT MAP TERMINOLOGY

Construct a concept map using these terms.

alveoli	pepsin
bile	salivary amylase
capillaries	small intestine
emphysema	surface area-to-volume ratio
microvilli	villi
nephron	

Visit the Online Learning Center to use an interactive format to construct another concept map.

KEY TERMS

Use interactive flash cards, on the Online Learning Center, to help you learn the meanings of these terms.

alveoli	lacteals
anemia	large intestine
aorta	liver
arteries	loop of Henle
arterioles	lung
atria	lymph
atrioventricular valves	lymphatic system
bile	microvilli
blood	nephrons
Bowman's capsule	pancreas
breathing	pepsin
bronchi	pharynx
bronchioles	plasma
capillaries	proximal convoluted tubule
carbonic anhydrase	pulmonary artery
diaphragm	pulmonary circulation
diastolic blood pressure	pyloric sphincter
distal convoluted tubule	salivary amylase
duodenum	salivary glands
gallbladder	semilunar valves
gastric juice	small intestine
glomerulus	surface area-to-volume ratio
heart	systemic circulation
hemoglobin	systolic blood pressure
hepatic portal vein	trachea
homeostasis	veins
immune system	ventricles
kidneys	villi

LEARNING CONNECTIONS

Visit the Online Learning Center to find additional study resources including those highlighted below.

18.1 Exchanging Materials: Basic Principles

1. List three reasons why cells must be small.
2. Give an example of homeostasis.

18.2 Circulation

Animations and Review
- Breathing

Labeling Exercises
- Veins
- Artery
- Capillary bed
- External anatomy of the heart
- Internal anatomy of the heart
- Blood circuits
- Plaque

Interactive Concept Map
- Heart

Case Studies
- Heart transplant
- Transplant

3. What are the functions of the heart, arteries, veins, arterioles, blood, and capillaries?
4. How do red blood cells assist in the transportation of oxygen and carbon dioxide?
5. How does the function of T-lymphocytes differ from that of B-lymphocytes?
6. What is the function of a phagocyte?

18.3 Gas Exchange

Labeling Exercises
- Respiratory tract I
- Respiratory tract II

Case Study
- Breathing liquids: Reality or science fiction?

Food for Thought
- Smoking ban

7. Describe the mechanics of breathing.
8. How are blood pH and breathing interrelated?
9. Why does a poorly functioning heart result in a person experiencing shortness of breath? (Hint: Think about carbon dioxide.)

18.4 Obtaining Nutrients

Animations and Review
- Introduction
- Human digestion
- Mouth and esophagus
- Stomach
- Small intestine

- Accessory organs
- Large intestine

Labeling Exercises
- Swallowing
- Digestive system I
- Digestive system II
- Digestive system III
- Small intestine villi

Interactive Concept Map
- Digestion

Experience This !
- Solubility and digestion

10. Describe three ways in which the digestive system increases its ability to absorb nutrients.
11. List three functions of the liver.
12. Name five digestive enzymes and describe their functions.
13. What is the role of bile in digestion?
14. How is fat absorption different from absorption of carbohydrate and protein?

18.5 Waste Disposal

Animations and Review
- Human excretion
- Kidney anatomy

Labeling Exercises
- Nephron anatomy I
- Nephron anatomy II

15. What are the functions of the glomerulus, proximal convoluted tubule, loop of Henle, and distal convoluted tubule?
16. Describe how the kidney regulates the amount of water in the body.

Test yourself with multiple choice questions that provide immediate feedback on the Online Learning Center with PowerWeb at

www.mhhe.com/enger11

Nutrition
Food and Diet

GOALS AND OBJECTIVES

Know the functions of the six types of nutrients.

❐ Point out the value of recommended dietary allowances.

❐ Explain why deficiencies of certain nutrients result in ill health.

Understand the value of a balanced diet consisting of each of the food groups.

❐ Explain why grains should make up the bulk of your diet.

❐ Describe why some protein sources do not contain all of the essential amino acids.

❐ Describe why some nutrients should be limited in order to maintain good health.

Know that a calorie is a measure of energy.

❐ Recognize that exercise is important in expending the energy gained by eating.

❐ Give examples of foods that have more calories than others.

❐ Explain how metabolic rate is related to diet and weight control.

Understand that there is a great deal of individual variation in the basal metabolic rate and the voluntary activity of people.

❐ Explain how people's food needs change at different points in their lives.

❐ Recognize that overeating and under-exercising has resulted in a U.S. population in which approximately 60% is overweight or obese.

Understand that eating has a strong psychological motivation.

❐ Identify the signs and symptoms of the common eating disorders that affect health.

Physiological Processes

PART FIVE

19.1 Living Things as Chemical Factories: Matter and Energy Manipulators

Organisms maintain themselves by constantly processing molecules to provide building blocks for new living material and energy to sustain themselves. Autotrophs can manufacture organic molecules from inorganic molecules, but heterotrophs must consume organic molecules to get what they need. All molecules required to support living things are called **nutrients**. Some nutrients are inorganic molecules such as calcium, iron, or potassium; others are organic molecules such as carbohydrates, proteins, fats, and vitamins. All heterotrophs obtain the nutrients they need from food and each kind of heterotroph has particular nutritional requirements. This chapter deals with the nutritional requirements of humans. • evolution of heterotrophs, p. 444

The word *nutrition* is used in two related contexts. First, nutrition is a branch of science that seeks to understand food, its nutrients, how the nutrients are used by the body, and how inappropriate combinations or quantities of nutrients lead to ill health. The word **nutrition** is also used in a slightly different context to refer to all the processes by which we take in food and utilize it, including ingestion, digestion, absorption, and assimilation. **Ingestion** involves the process of taking food into the body through eating. **Digestion** involves the breakdown of complex food molecules to simpler molecules. **Absorption** involves the movement of simple molecules from the digestive system to the circulatory system for dispersal throughout the body. **Assimilation** involves the modification and incorporation of absorbed molecules into the structure of the organism.

Many of the nutrients that enter living cells undergo chemical changes before they are incorporated into the body. These interconversion processes are ultimately under the control of the genetic material, DNA. It is DNA that codes the information necessary to manufacture the enzymes required to extract energy from chemical bonds and to convert raw materials (nutrients) into the structure (anatomy) of the organism.

The food and drink consumed from day to day constitute a person's **diet**. It must contain the minimal nutrients necessary to manufacture and maintain the body's structure (bones, skin, tendon, muscle, etc.) and regulatory molecules (enzymes and hormones), and to supply the energy (ATP [adenosine triphosphate]) needed to run the body's machinery. If the diet is deficient in nutrients, or if a person's body cannot process nutrients efficiently, a dietary deficiency and ill health may result. A good understanding of nutrition can promote good health and involves an understanding of the energy and nutrient content in various foods.

19.2 Kilocalories, Basal Metabolism, and Weight Control

The unit used to measure the amount of energy in foods is the **kilocalorie (kcal)**. The amount of energy needed to raise the temperature of 1 *kilogram* of water 1°C is 1 kilocalorie. Remember that the prefix *kilo-* means "1,000 times" the value listed. Therefore, a kilocalorie is 1,000 times more heat energy than a **calorie**, which is the amount of heat energy needed to raise the temperature of 1 *gram* of water 1°C. However, the amount of energy contained in food is usually called a Calorie with a capital C. This is unfortunate because it is easy to confuse a Calorie, which is really a kilocalorie, with a calorie. Most books on nutrition and dieting use the term Calorie to refer to *food calories*. The energy requirements in kilocalories for a variety of activities are listed in table 19.1 (How Science Works 19.1). • temperature, heat, and calories, p. 27

Significant energy expenditure is required for muscular activity. However, even at rest, energy is required to maintain breathing, heart rate, and other normal body functions. The rate at which the body uses energy when at rest is known as the **basal metabolic rate (BMR)**. The basal metabolism of most people requires more energy than their volun-

---

Table 19.1

TYPICAL ENERGY REQUIREMENTS FOR COMMON ACTIVITIES

Light Activities, 120–150 kcal/h	Light-to-Moderate Activities, 150–300 kcal/h	Moderate Activities, 300–400 kcal/h	Heavy Activities, 400–600 kcal/h
Dressing	Sweeping floors	Pulling weeds	Chopping wood
Typing	Painting	Walking behind lawn mower	Shoveling snow
Studying	Store clerking	Walking 3.5–4 mph on level	Walking 5 mph
Standing	Bowling	surface	Walking up hills
Slow walking	Walking 2–3 mph	Calisthenics	Cross-country skiing
Sitting activities	Canoeing 2.5–3 mph	Canoeing 4 mph	Swimming
	Bicycling on level surface	Doubles tennis	Jogging 5 mph
	at 5.5 mph	Volleyball	Bicycling 11–12 mph
		Golf (no cart)	or in hilly terrain

HOW SCIENCE WORKS 19.1

Measuring the Caloric Value of Foods

A *bomb calorimeter* is an instrument that is used to determine the heat of combustion or caloric value of foods. To operate the instrument, a small sample is formed into a pellet and sealed inside the bomb, a strong container. The bomb is filled with 30 atmospheres of oxygen and is then placed in a surrounding jacket that is filled with water. The sample is electrically ignited and as it aerobically reacts with the oxygen, the bomb releases heat to the surrounding water jacket. As the heat is released, the temperature of the water rises. Recall that a calorie is the amount of energy (or heat) needed to increase the temperature of 1 gram of water 1°C. Therefore, if 1 gram of food results in a water temperature increase of 13°C for each gram of water, that food has the equivalent of 13 calories.

tary muscular activity. Much of this energy is used to keep the body temperature constant. A true measurement of basal metabolic rate requires a measurement of oxygen used over a specific period under controlled conditions. There are several factors that affect an individual's basal metabolic rate. Children have higher basal metabolic rates and the rate declines throughout life. Elderly people have the lowest basal metabolic rate. In general, males have higher metabolic rates than women. Height and weight are also important. The larger a person the higher their metabolic rate. With all of these factors taken into account, most young adults would fall into the range of 1,200 to 2,200 kilocalories for a basal metabolic rate. Some other factors are: climate (cold climate = higher BMR), altitude (higher altitude = higher BMR), physical condition (regular exercise raises BMR for some time following exercise), hormones (thyroid-stimulating hormones, growth-stimulating hormones, and androgens raise BMR), previous diet (malnourished or starving persons typically have lower BMR), percent of weight that is fat (fat tissue has a lower metabolic rate than lean tissue), and time of the year (people have higher BMR during the colder part of the year).

Because few of us rest 24 hours a day, we normally require more than the energy needed for basal metabolism. One of these requirements is the amount of energy needed to process the food we eat. This is called **specific dynamic action (SDA)** and is equal to approximately 10% of your total daily kilocalorie intake.

In addition to basal metabolism and specific dynamic action, the activity level of a person determines the number of kilocalories needed. A good general indicator of the number of kilocalories needed above basal metabolism is the type

Table 19.2

ADDITIONAL KILOCALORIES AS DETERMINED BY OCCUPATION

Occupation	Kilocalories Needed per Day Above Basal Metabolism*
Sedentary (student)	500–700
Light work (business person)	750–1,200
Moderate work (laborer)	1,250–1,500
Heavy work (professional athlete)	1,550–5,000 and up

*These are general figures and will vary from person to person depending on the specific activities performed in the job.

of occupation a person has (table 19.2). Since most adults are relatively sedentary, they would receive adequate amounts of energy if women consumed 2,200 kilocalories and men consumed 2,900 kilocalories per day. Since approximately 60% of Americans are overweight or obese, the U.S. Department of Agriculture has developed a program aimed at educating people about the health hazards of obesity. One of the problems associated with obesity is identification—developing a good definition that can be easily understood. Table 19.3 shows guidelines for determining whether you are overweight or not. It is based on a specific method for determining *body mass index*—appropriate body weight compared to height. **Body mass index (BMI)** is calculated by determining a person's weight (without clothing) in kilograms and barefoot height in meters. The body mass index is

Table19.3

ARE YOU OVERWEIGHT?

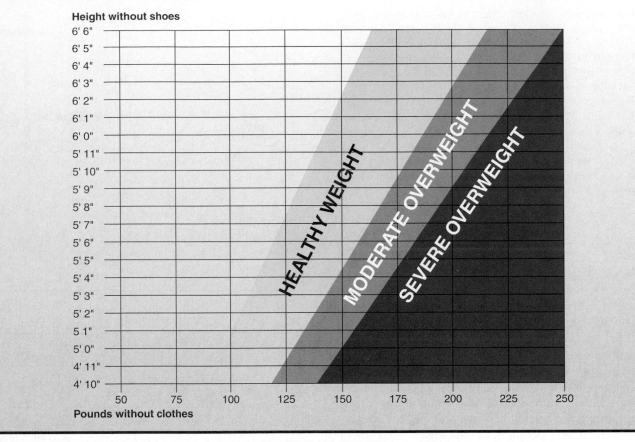

Source: *Report of the Dietary Guidelines Advisory Committee on the Dietary Guidelines for Americans, 2000.*

their weight in kilograms divided by their height in meters squared.

$$BMI = \frac{\text{weight in kilograms}}{(\text{height in meters})^2}$$

(Appendix 1 of this book gives conversions to the metric system of measurements.) For example, a person with a height of 5 feet 6 inches (1.68 meters) who weighs 165 pounds (75 kilograms) has a body mass index of 26.6 kg/m².

$$BMI = \frac{\text{kg}}{\text{m}^2} = \frac{75 \text{ kg}}{(1.68 \text{ m})^2} = \frac{75 \text{ kg}}{2.82 \text{ m}^2} = 26.6 \text{ kg/m}^2$$

The ideal body mass index for maintaining good health is between 18.5 and 25 kg/m². Therefore, the person described above would be slightly over the recommended weight. Those with a body mass index between 25 and 30 kg/m² are considered overweight, but there are no clear indications that there are significant health affects associated with this degree of overweight. Those with a body mass index of 30 kg/m² or more have a significantly increased risk of many different kinds of diseases. The higher the body mass index the more significant the risk.

Why is weight control a problem for such a large portion of the population? There are several metabolic pathways that convert carbohydrates (glucose) or proteins to fat. Stored body fat was very important for our prehistoric ancestors because it allowed them to survive periods of food scarcity. In periods of food scarcity the stored body fat can be used to supply energy. The glycerol portion of the fat molecule can be converted to a small amount of glucose, which can supply energy for red blood cells and nervous tissue that must have glucose. The fatty acid portion of the molecule can be metabolized by most other tissues directly to produce ATP. Today, however, for most of us food scarcity is not a problem, and even small amounts of excess food consumed daily tend to add to our fat stores.

Although energy doesn't weigh anything, the nutrients that contain the energy do. Weight control is a matter of balancing the kilocalories ingested as a result of dietary intake with the kilocalories of energy expended by normal daily activities and exercise. There is a limit to the rate at which a moderately active human body can use fat as an energy source. At most, 1 or 2 pounds (0.45–0.9 kilogram) of fat tissue per week are lost by an average person when dieting. Because 1 pound (0.45 kilogram) of fatty tissue contains about 3,500 kilocalories, decreasing your kilocalorie intake by 500 to 1,000 kilocalories per day while maintaining a balanced diet (including proteins, carbohydrates, and fats) will result in fat loss of 1 to 2 pounds (0.45–0.9 kilogram) per week. (A pound of pure fat contains about 4,100 kilocalories, but fat tissue contains other materials besides fat, such as water.)

Many diets promise large and rapid weight loss but in fact result only in temporary water loss. They may encourage eating and drinking foods that are diuretics, which increase the amount of urine produced and thus increase water loss. Or they may encourage exercise or other activities that cause people to lose water through sweating. Low carbohydrate diets deprive the body of glucose needed to sustain nervous tissue and red blood cells. If glucose is not available the body will begin to use protein from the liver and muscles to provide the glucose needed for these vital cells. This kind of weight loss is not healthy. Finally, just reducing the amount of food in the gut by fasting results in a temporary weight loss because the gut is empty.

For those who need to gain weight, increasing kilocalorie intake by 500 to 1,000 kilocalories per day will result in an increase of 1 or 2 pounds per week, provided the low weight is not the result of a health problem.

If you have calculated your body mass index and wish to modify your body weight, what are the steps you should take? First, you should check with your physician before making any drastic change in your eating habits. Second, you need to determine the number of kilocalories you are consuming. That means keeping an accurate diet record for at least a week. Record everything you eat and drink and determine the number of kilocalories in those nutrients. This can be done by estimating the amounts of protein, fat, and carbohydrate (including alcohol) in your foods. Roughly speaking, 1 gram of carbohydrate is the equivalent of 4 kilocalories, 1 gram of fat is the equivalent of 9 kilocalories, 1 gram of protein is the equivalent of 4 kilocalories, and 1 gram of alcohol is 7 kilocalories. Most nutrition books have food-composition tables that tell you how much protein, fat, and carbohydrate are in a particular food. Packaged foods also have serving sizes and nutritive content printed on the package. Do the arithmetic and determine your total kilocalorie intake for the week. If your intake (from your diet) in kilocalories equals your output (from basal metabolism plus voluntary activity plus SDA), you should not have gained any weight! You can double-check this by weighing yourself before and after your week of record keeping. If your weight is constant and you want to lose weight, reduce the amount of food in your diet. To lose 1 pound each week, reduce your kilocalorie intake by 500 kilocalories per day. Be careful not to eat less than 600 kilocalories of carbohydrates or reduce your total daily intake below 1,200 kilocalories unless you are under the care of a physician. It is important to have some carbohydrate in your diet because a lack of carbohydrate leads to a breakdown of the protein that provides the cells with the energy they need. Also you may not be getting all the vitamins required for efficient metabolism and you could cause yourself harm. To gain 1 pound, increase your intake by 500 kilocalories per day.

A second ingredient valuable in a weight loss plan is an increase in exercise while keeping food intake constant. This can involve organized exercise in sports or fitness programs. It can also include simple things like walking up the stairs rather than taking the elevator, parking at the back of the parking lot so you walk farther, riding a bike for short errands, or walking down the hall to someone's office rather than using the phone. Many people who initiate exercise plans as a way of reducing weight are frustrated because they may initially gain weight rather than lose it. This is because muscle weighs more than fat. Typically they are "out of shape" and have low muscle mass. If they gain a pound of muscle at the same time they lose a pound of fat they will not lose weight. However, if the fitness program continues they will eventually reach a point where they are not increasing muscle mass and weight loss will occur. Even so, weight as muscle is more healthy than weight as fat.

If, like millions of others, you believe that you are overweight, you have probably tried numerous diet plans. Not all of these plans are the same, and not all are suitable to your particular situation. If a diet plan is to be valuable in promoting good health, it must satisfy your needs in several ways. It must provide you with needed kilocalories, proteins, fats, and carbohydrates. It should also contain readily available foods from all the basic food groups, and it should provide enough variety to prevent you from becoming bored with the plan and going off the diet too soon. A diet should not be something you follow only for a while, then abandon and regain the lost weight.

19.3 The Chemical Composition of Your Diet

Nutritionists have divided nutrients into six major classes: carbohydrates, lipids, proteins, vitamins, minerals, and water. Chapters 2 and 3 presented the fundamental structures and examples of these types of molecules. A look at each of these classes from a nutritionist's point of view should help you better understand how your body works and how you might best meet its nutritional needs.

Carbohydrates

When the word *carbohydrate* is mentioned, many people think of things like table sugar, pasta, and potatoes. The term *sugar* is usually used to refer to mono- or disaccharides, but the carbohydrate group also includes more complex polysaccharides, such as starch, glycogen, and cellulose. Starch is the primary form in which we obtain carbohydrates. Each of

Figure 19.1

The Structure and Role of Various Carbohydrates

The diet includes a wide variety of carbohydrates. Some are monosaccharides (simple sugars); others are more complex disaccharides, trisaccharides, and polysaccharides. The complex carbohydrates differ from one another depending on the type of monosaccharides that are linked together by dehydration synthesis. Notice that the complex carbohydrates shown are primarily from plants. With the exception of milk, animal products are not a good nutritional source of carbohydrates because animals do not store them in great quantities or use large amounts of them as structural materials.

lized in cellular respiration to provide energy in the form of ATP molecules. Simple sugars are also used as building blocks in the manufacture of molecules such as nucleic acids. Complex carbohydrates can also be a source of **fiber** that slows the absorption of nutrients and stimulates peristalsis (rhythmic contractions) in the intestinal tract.

A diet deficient in carbohydrates results in fats being oxidized and converted to ATP. Unfortunately, protein is also metabolized to provide cells with the glucose they need for survival. In situations in which carbohydrates are absent, most of the fats are metabolized to keto acids. Large numbers of keto acids may be produced in extreme cases of fasting, resulting in a potentially dangerous change in the body's pH. If a person does not have stored fat to metabolize, a carbohydrate deficiency will result in an even greater use of the body's proteins as a source of energy. This is usually only encountered in starvation or extreme cases of fasting or in association with eating disorders. In extreme cases, this can be fatal because the oxidation of protein results in an increase in toxic, nitrogen-containing compounds.

these has a different structural formula, different chemical properties, and plays a different role in the body (figure 19.1). Many simple carbohydrates taste sweet and stimulate the appetite. When complex carbohydrates like starch or glycogen are broken down to monosaccharides, these may then be uti-

A more typical situation for us is the consumption of too much food. As with other nutrients, if there is an excess of carbohydrates in the diet, they are converted to lipids and stored by the body in fat cells—and you gain weight.

Lipids

The class of nutrients technically known as *lipids* is often called *fat* by many people. This is unfortunate and may lead to some confusion because fats are only one of three subclasses of lipids. Each subclass of lipids—phospholipids, steroids, and true fats—plays an important role in human nutrition. Phospholipids are essential components of all cell membranes. Although various kinds of phospholipids are sold as dietary supplements they are unnecessary because all food composed of cells contains phospholipids. Many steroids are hormones that help regulate a variety of body processes. With the exception of vitamin D, it is not necessary to specifically include steroids as a part of the diet. Cholesterol is a steroid commonly found in certain foods and may cause health problems in some people. The *true fats* (also called *triglycerides*) are an excellent source of energy. They are able to release 9 kilocalories of energy per gram compared to 4 kilocalories per gram of carbohydrate or protein. Some fats contain the **essential fatty acids,** linoleic acid and linolenic acid. Neither is synthesized by the human body and, therefore, must be a part of the diet. These essential fatty acids are required by the body for such things as normal growth, blood clotting, and maintaining a healthy skin. Most diets that incorporate a variety of foods including meats and vegetable oils have enough of these essential fatty acids. A diet high in linoleic acid has also been shown to help in reducing the amount of the steroid cholesterol in the blood. Some vitamins, such as vitamins A, D, E, and K, do not dissolve in water but dissolve in fat and, therefore, require fat for their absorption from the gut. • **omega fatty acids, p. 56**

Fat is an insulator against outside cold and internal heat loss and is an excellent shock absorber. Deposits in the back of the eyes serve as cushions when the head suffers a severe blow. During starvation, these deposits are lost, and the eyes become deep-set in the eye sockets, giving the person a ghostly appearance.

The pleasant taste and "mouth feel" of many foods is the result of fats. Their ingestion provides that full feeling after a meal because they leave the stomach more slowly than other nutrients. You may have heard people say, "When you eat Chinese food, you're hungry a half hour later." Because Chinese foods contain very little animal fat, it's understandable that after such a meal, the stomach will empty soon and people won't have that full feeling very long. Conversely, a buffet breakfast of sausages, bacon, eggs, fried potatoes, and pastries contains a great deal of fat and will remain in the stomach for four to five hours. Excess kilocalories obtained directly from fats are stored more efficiently than excess kilocalories obtained from carbohydrates and protein because the body does not need to expend energy to convert the molecules to fat. The fat molecules can simply be disassembled in the gut and reassembled in the cells.

Table 19.4

SOURCES OF ESSENTIAL AMINO ACIDS

Essential Amino Acids*	Food Sources
Threonine	Dairy products, nuts, soybeans, turkey
Lysine	Dairy products, nuts, soybeans, green peas, beef, turkey
Methionine	Dairy products, fish, oatmeal, wheat
Arginine (essential to infants only)	Dairy products, beef, peanuts, ham, shredded wheat, poultry
Valine	Dairy products, liverwurst, peanuts, oats
Phenylalanine	Dairy products, peanuts, calves' liver
Histidine (essential to infants only)	Human and cow's milk and standard infant formulas
Leucine	Dairy products, beef, poultry, fish, soybeans, peanuts
Tryptophan	Dairy products, sesame seeds, sunflower seeds, lamb, poultry, peanuts
Isoleucine	Dairy products, fish, peanuts, oats, lima beans

*The essential amino acids are required in the diet for protein building and, along with the nonessential amino acids, allow the body to metabolize all nutrients at an optimum rate. Combinations of different plant foods can provide essential amino acids even if complete protein foods (e.g., meat, fish, and milk) are not in the diet.

Proteins

Proteins are composed of amino acids linked together by peptide bonds; however, not all proteins contain the same amino acids. Proteins can be divided into two main groups: the **complete proteins** and the **incomplete proteins.** Complete proteins contain all the amino acids necessary for good health, whereas incomplete proteins lack certain amino acids that the body must have to function efficiently. Table 19.4 lists the **essential amino acids**—those that cannot be synthesized by the human body. Without adequate amounts of these amino acids in the diet, a person may develop a protein-deficiency disease. Proteins are essential components of hemoglobin and cell membranes, as well as antibodies, enzymes, some hormones, hair, muscle, and the connective tissue fiber, collagen. Plasma proteins are important because they can serve as buffers and help retain water in the bloodstream. Proteins also provide a last-ditch source of energy during starvation when carbohydrate and fat consumption falls below protective levels. • **three categories of protein, p. 48**

Unlike carbohydrates and fats, proteins cannot be stored for later use. Because they are not stored and because they serve many important functions in the body, it is important that adequate amounts of protein be present in the daily diet. However, a high-protein diet is not necessary. Only small amounts of protein are metabolized and lost from the body each day. This amounts to about 20 to 30 grams per day. Therefore, it is important to replace this with small amounts of protein in the diet. Any protein in excess of that needed to rebuild lost molecules is metabolized to provide the body with energy. Protein is the most expensive but least valuable energy source. Carbohydrate and fat are much better sources.

The body has several mechanisms that tend to protect protein from being metabolized to provide cells with energy. This relationship is called **protein-sparing**. During fasting or starvation many of the cells of the body can use fat as their primary source of energy, thus protecting the more valuable protein. However, red blood cells and nervous tissue must have glucose to supply their energy needs. Small amounts of carbohydrates can supply the glucose needed. Because very little glucose is stored, after a day or two of fasting the body begins to convert some of the amino acids from protein into glucose to supply these vital cells. Although fat can be used to supply energy for many cells during fasting or starvation it is not able to completely protect the proteins if there is no carbohydrate in the diet. With prolonged starvation the fat stores are eventually depleted and structural proteins are used for all the energy needs of the body.

Most people have a misconception with regard to the amount of protein necessary in their diets. The total amount necessary is actually quite small (about 50 grams/day) and can be easily met. The equivalent of ¼ pound of hamburger, a half chicken breast, or a fish sandwich contains the daily amount of protein needed for the majority of people.

Vitamins

Vitamins are the fourth class of nutrients. **Vitamins** are organic molecules needed in minute amounts to maintain essential metabolic activities. Like essential amino acids and fatty acids, vitamins cannot be manufactured by the body. Table 19.5 lists vitamins for which there are recommended daily intake data. Contrary to popular belief, vitamins do not serve as a source of energy, but play a role in assisting specific enzymes that bring about essential biochemical changes. In some cases the vitamin is actually incorporated into the structure of the enzyme. Such vitamins are called *coenzymes.* For example, a B-complex vitamin (niacin) helps enzymes in the respiration of carbohydrates • **cofactors, vitamins, and coenzymes, p. 94**

Most vitamins are acquired from food; however, vitamin D may be formed when ultraviolet light strikes a cholesterol molecule already in your skin, converting this cholesterol to vitamin D. This means that vitamin D is not really a vitamin at all. It came to be known as a vitamin because of the mistaken idea that it is only acquired through food rather than being formed in the skin on exposure to sunshine. It would be more correct to call vitamin D a hormone, but most people do not. Most people can get all the vitamins they need from a well-balanced diet. However, because vitamins are inexpensive, and people think they may not be getting the vitamins they need from their diets, many people take vitamin supplements.

Because many vitamins are inexpensive and their functions are poorly understood, there are many who advocate large doses of vitamins (megadoses) to prevent a wide range of diseases. Often the benefits advertised are based on fragmentary evidence and lack a clearly defined mechanism of action. Consumption of high doses of vitamins is unwise because high doses of many vitamins have been shown to be toxic. For example, fat-soluble vitamins such as vitamin A and vitamin D are stored in the fat of the body and the liver. Excess vitamin A is known to cause joint pain, loss of hair, and jaundice. Excess vitamin D results in calcium deposits in the kidneys, high amounts of calcium in the blood, and bone pain. Even high doses of some of the water-soluble vitamins may have toxic effects. Vitamin B_6 (pyridoxine) in high concentrations has been shown to cause nervous symptoms such as unsteady gait and numbness in the hands.

However, inexpensive multivitamins that provide 100% of the recommended daily allowance can prevent or correct deficiencies caused by poor diet without danger of toxic consequences. Most people have no need of vitamins *if they eat a well-balanced diet* (How Science Works 19.2).

Minerals

All **minerals** are inorganic elements found throughout nature and cannot be synthesized by the body. Table 19.6 lists the sources and functions of several common minerals. Because they are elements, they cannot be broken down or destroyed by metabolism or cooking. They commonly occur in many foods and in water. Minerals retain their characteristics whether they are in foods or in the body and each plays a different role in metabolism. Minerals can function as regulators, activators, transmitters, and controllers of various enzymatic reactions. For example, sodium ions (Na^+) and potassium ions (K^+) are important in the transmission of nerve impulses, whereas magnesium ions (Mg^{++}) facilitate energy release during reactions involving ATP. Without iron, not enough hemoglobin would be formed to transport oxygen, a condition called *anemia,* and a lack of calcium may result in *osteoporosis.* **Osteoporosis** is a condition that results from calcium loss leading to painful, weakened bones. There are many minerals that are important in your diet. In addition to those just mentioned, you need chlorine, cobalt, copper, iodine, phosphorus, potassium, sulfur, and zinc to remain healthy. With few exceptions, adequate amounts of minerals are obtained in a normal diet. Calcium and iron supplements may be necessary, particularly in women.

Table 19.5

VITAMINS: SOURCES AND FUNCTIONS

Name	Recommended Daily Intake for Adults*	Physiological Value	Readily Available Sources	Other Information
Vitamin A	800–1,000 µg	Important in vision; maintains skin and intestinal lining	Orange, red, and dark green vegetables	Fat soluble, stored in liver Children have little stored.
Vitamin B₁ (thiamin)	1.1–1.2 mg	Maintains nerves and heart; involved in carbohydrate metabolism	Whole grain foods, legumes, pork	Larger amounts needed during pregnancy and lactation
Vitamin B₂ (riboflavin)	1.1–1.3 mg	Central role in energy metabolism; maintains skin and mucous membranes	Dairy products, green vegetables, whole grain foods, meat	
Vitamin B₃ (niacin)	14–16 mg	Energy metabolism	Whole grain foods, meat	
Vitamin B₆ (pyridoxine pyridoxol, pyradoxamine)	1.3 mg	Form red blood cells; maintains nervous system	Whole grain foods, milk, green leafy vegetables, meats, legumes, nuts	
Vitamin B₁₂ (cobalamin)	2.4 µg	Protein and fat metabolism; form red blood cells; maintains nervous system	Animal products only: dairy products, meats, and seafood	Stored in liver Pregnant and lactating women and vegetarians need larger amounts.
Vitamin C	60 mg	Maintains connective tissue, bones, and skin	Citrus fruits, leafy green vegetables, tomatoes, potatoes	Toxic in high doses
Vitamin D	5 µg	Needed to absorb calcium for strong bones and teeth	Vitamin D-fortified milk; exposure of skin to sunlight	Fat soluble
Vitamin E	8–10 µg	Antioxidant; protects cell membranes	Whole-grain foods, seeds and nuts, vegetables, vegetable oils	Fat soluble Only two cases of deficiency ever recorded
Vitamin K	60–70 µg	Blood clotting	Dark green vegetables	Fat soluble; small amount of fat needed for absorption
Folate (folic acid)	400 µg	Coenzyme in metabolism	Most foods: beans, fortified cereals	Important in pregnancy
Pantothenic acid	5 mg	Involved in many metabolic reactions; formation of hormones, normal growth	All foods	Typical diet provides adequate amounts.
Biotin	30 µg	Maintains skin and nervous system	All foods	Typical diet provides adequate amounts. Added to intravenous feedings
Choline (lecithin)	425–550 mg	Component of cell membranes	All foods	Only important in people unable to consume food normally

*The recommended daily intake for various nutritional substances by age and gender.

HOW SCIENCE WORKS 19.2

Preventing Scurvy

Scurvy is a nutritional disease caused by the lack of vitamin C in the diet. This lack results in the general deterioration of health because vitamin C is essential to the formation of collagen, a protein important in most tissues. Disease symptoms include poor healing of wounds, fragile blood vessels resulting in bleeding, lack of bone growth, and loosening of the teeth.

Although this is not a common disease today, lack of fresh fruits and vegetables was a common experience for people who were on long sea voyages in previous centuries. This was such a common problem that the disease was often called sea scurvy. Many ship's captains and ship's doctors observed a connection between the lack of fresh fruits and vegetables and the increased incidence of scurvy. Excerpts from a letter by a Dr. Harness to the First Lord of the Admiralty of the British navy give a historical

background to the practice of using lemons to prevent scurvy on British ships.

"During the blockade of Toulon in the summer of 1793, many of the ships' companies were afflicted with symptoms of scurvy; . . . I was induced to propose . . . the sending a vessel into port for the express purpose of obtaining lemons for the fleet; . . . and the good effects of its use were so evident . . . that an order was soon obtained from the commander in chief, that no ship under his lordship's command should leave port without being previously furnished with an ample supply of lemons. And to this circumstance becoming generally known may the use of lemon juice, the effectual means of subduing scurvy, while at sea, be traced."

A common term applied to British seamen during this time was "limey."

Table 19.6

MINERALS: SOURCES AND FUNCTIONS

Name	Recommended Daily Intake for Adults (female; male)	Physiological Value	Readily Available Sources	Other Information
Calcium	1,000 mg	Builds and maintains bones and teeth	Dairy products	Children need 1,300 mg. Vitamin D needed for absorption.
Fluoride	3.1–3.8 mg	Maintains bones and teeth; reduces tooth decay	Fluoridated drinking water, seafood	
Iodine	150 µg	Necessary to make hormone thyroxine	Iodized table salt, seafood	The soils in some parts of the world are low in iodine so iodized salt is very important.
Iron	15–10 mg	Necessary to make hemoglobin	Grains, meat, seafood, poultry, legumes, dried fruits	Women need more than men; pregnant women need two times the normal dose.
Magnesium	320–420 mg	Bone mineralization; muscle and nerve function	Dark green vegetables, whole grains, legumes	
Phosphorus	700 mg	Acid/base balance; enzyme cofactor	Found in all foods	Children need 1,250 mg. Most people get more than recommended.
Selenium	55–70 µg	Involved in many enzymatic reactions	Meats, grains, seafood	
Zinc	12–15 mg	Wound healing; fetal development; involved in many enzymatic and hormonal activities	Meat, fish, poultry	

Water

Water is crucial to all life and plays many essential roles. You may be able to survive weeks without food, but you would die in a matter of days without water. It is known as the universal solvent because so many types of molecules are soluble in it. The human body is about 65% water. Even dense bone tissue consists of 33% water. All the chemical reactions in living things take place in water. It is the primary component of blood, lymph, and body tissue fluids. Inorganic and organic nutrients and waste molecules are also dissolved in water. Dissolved inorganic ions, such as sodium (Na^+), potassium (K^+), and chloride (Cl^-), are called **electrolytes** because they form a solution capable of conducting electricity. The concentration of these ions in the body's water must be regulated in order to prevent electrolyte imbalances. • electrolytes, p. 31

Excesses of many types of wastes are eliminated from the body dissolved in water; that is, they are excreted from the kidneys as urine or in small amounts from the lungs or skin through evaporation. In a similar manner, water acts as a conveyor of heat. Water molecules are also essential reactants in all the various hydrolytic reactions of metabolism. Without it, the breakdown of molecules such as starch, proteins, and lipids would be impossible. With all these important roles played by water, it's no wonder that nutritionists recommend that you drink the equivalent of at least eight glasses each day. This amount of water can be obtained from tap water, soft drinks, juices, and numerous food items, such as lettuce, cucumbers, tomatoes, and applesauce.

19.4 Amounts and Sources of Nutrients

In an attempt to give people some guidelines for planning a diet that provides adequate amounts of the six classes of nutrients, nutritional scientists and governmental officials in the United States and many other countries developed nutrient standards in the 1970s. In the United States, these guidelines are known as the **recommended dietary allowances, or RDAs.** RDAs are dietary recommendations, not requirements or minimum standards. They are based on what is thought to be the needs of a healthy person already eating an adequate diet. RDAs do not apply to persons with medical problems who are under stress or suffering from malnutrition. The amount of each nutrient specified by the RDAs has been set relatively high so that most of the population eating those quantities will be meeting their nutritional needs. Keep in mind that everybody is different and eating the RDA amounts may not meet your personal needs if you have an unusual metabolic condition.

General sets of RDAs have been developed for four groups of people: infants, children, adults, and pregnant and lactating women. The U.S. RDAs are used when preparing product labels. The federal government requires by law that labels list ingredients from the greatest to the least in quantity. The volume in the package must be stated along with the weight, and the name of the manufacturer or distributor. If any nutritional claim is made, it must be supported by factual information.

A product label that proclaims, for example, that a serving of cereal provides 25% of the RDA for vitamin A means that you are getting at least one-fourth of your RDA of vitamin A from a single serving of that cereal. To figure your total RDA of vitamin A, consult a published RDA table for adults. It tells you that an adult male requires 1,000 and a female 800 micrograms (μg) of vitamin A per day. Of this, 25% is 250 and 200 micrograms, respectively—the amount you are getting in a serving of that cereal. You will need to get the additional amounts (750 for men and 600 for women) by having more of that cereal or eating other foods that contain vitamin A. If a product claims to have 100% of the RDA of a particular nutrient, that amount must be present in the product. However, restricting yourself to that one product will surely deprive you of many of the other nutrients necessary for good health. Ideally, you should eat a variety of complex foods containing a variety of nutrients to ensure that all your health requirements are met.

In 2003, the World Health Organization (WHO) and the Food and Agriculture Organization (FAO) released an independent "*Expert Report*" on diet and chronic disease. The *Expert Report* contains the best currently available scientific evidence on the relationship of diet, nutrition, and physical activity to chronic diseases, based on the collective judgment of a group of experts from around the world. The report advises that the best ways of changing daily nutritional intake and increasing energy expenditure are to:

1. Reduce energy-rich foods high in saturated fat and sugar;
2. Cut the amount of salt in the diet;
3. Increase the amount of fresh fruit and vegetables in the diet; and
4. Be sure to do moderate-intensity physical activity for at least an hour a day.

19.5 The Food Guide Pyramid with Five Food Groups

Using RDAs and product labels is a pretty complicated way for a person to plan a diet. Planning a diet around basic food groups is generally easier. The basic food groups first developed and introduced in 1953 have been modified and updated several times to serve as guidelines in maintaining a balanced diet (figure 19.2). In May 1992, the U.S. Department of Agriculture released the results of its most recent study on how best to educate the public about daily nutrition. The federal government adopted the **Food Guide Pyramid** of the Department of Agriculture as one of its primary tools to help the general public plan for good nutrition. The Food Guide Pyramid contains five basic groups of foods

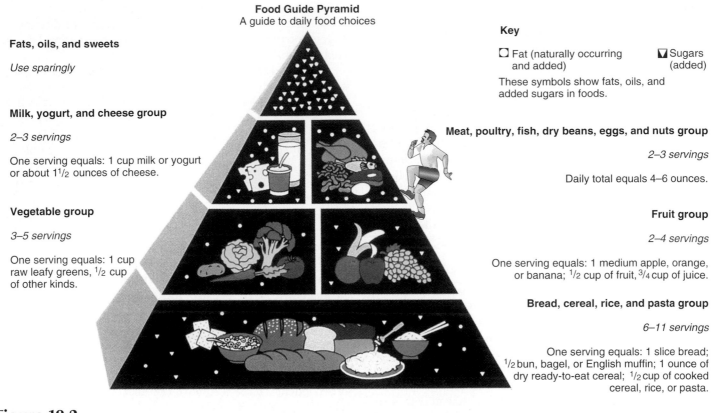

Figure 19.2

The Food Guide Pyramid
In May 1992, the Department of Agriculture released a new guide to good eating. This Food Guide Pyramid suggests that we eat particular amounts of five different food groups while decreasing our intake of fats and sugars. This guide should simplify our menu planning and help ensure that we get all the recommended amounts of basic nutrients. In order to be healthy, exercising on a regular basis is essential.

Source: U.S. Department of Agriculture, 1992.

with guidelines for the amounts one needs daily from each group for ideal nutritional planning. The Food Guide Pyramid differs from previous federal government information in that it encourages a reduction in the amount of fats and sugars in the diet and an increase in our daily servings of fruits and vegetables. In addition, the new guidelines suggest significantly increasing the amount of grain products we eat each day. Figure 19.3 shows typical serving sizes for each of the categories in the Food Guide Pyramid (Outlooks 19.1).
• the pyramid of numbers, p. 266

Grain Products Group

Grains include vitamin-enriched or whole-grain cereals and products such as breads, bagels, buns, crackers, dry and cooked cereals, pancakes, pasta, and tortillas. Items in this group are typically dry and seldom need refrigeration. They should provide most of your kilocalorie requirements in the form of complex carbohydrates like starch, which is the main ingredient in most grain products. You should have 6 to 11 servings from this group each day. This is a major change from previous recommendations of four servings each day. A serving is considered about ½ cup, or 1 ounce, or

the equivalent of about 100 kilocalories. Using product labels will help determine the appropriate serving size.*

In addition to their energy content, cereals and grains provide fiber and are rich sources of the B vitamins in your diet. Cereals are also important sources of minerals, iron, magnesium, and selenium. As you decrease the intake of proteins in the meat and poultry group, you should increase your intake of items from this group. These foods help you feel you have satisfied your appetite, and many of them are very low in fat.

Fruits Group

You probably can remember discussions of whether a tomato is a vegetable or a fruit. This controversy arises from the fact that the term *vegetable* is not scientifically precise but means a plant material eaten during the main part of the

---

*The measurement of ingredients during food preparation varies throughout the world. Some people measure by weight (e.g., grams), others by volume (e.g., cups); some use the metric system (grams), others the English system (pounds). Still others use units of measure that are even less uniform (e.g., "pinches"). This chapter describes quantities of nutrients using the units of measure most familiar to people in the United States.

(a) Meat Group

(b) Fruit Group

(c) Vegetable Group

(d) Cereal Group

(e) Dairy Group

(f) Fats, sugars, and alcohol

Figure 19.3

Serving Sizes of Basic Foods
These photos show typical serving sizes for many types of foods. Each item in each photo is equivalent to one serving. On a daily basis, an average adult should eat 6 to 11 servings of cereals, 2 to 3 servings from the meat group, 3 to 5 servings from the vegetable group, 2 to 4 servings from the fruit group, and 2 to 3 servings from the dairy group. Fats, sugars, and alcohol should be used sparingly.

OUTLOOKS 19.1

The Dietary Habits of Americans

The U.S. Department of Agriculture has provided Americans with information about diet and nutrition for many years. While there have been some changes in eating habits, there are still a great number of people who do not eat well, nor do they participate in other activities that have the potential to improve their health. The following statements about the dietary and exercise habits of Americans in 2000 illustrate that there are many changes that still need to be made.

- The quantity of fat consumed has increased in the past ten years.
- Fat consumption provides about 33% of Calories for the average American. It is recommended that less than 30% of Calories come from fat.
- Only 17% of the population eats the recommended amount of fruits.
- Only 31% of the population eats the recommended amount of vegetables.
- Only 26% of the population consumes the recommended amount of dairy products.
- 60% of the adult population is overweight or obese.
- 10% of children are overweight or obese.
- 25% of the adult population is obese.
- Less than 30% of the population gets the recommended amount of exercise.
- 25% of the population does not exercise at all.
- Less than 50% of meals eaten in the United States are cooked at home.

- Typical restaurant and fast-food outlets serve enormous portions compared to the serving sizes suggested by the Food Guide Pyramid.
- The proportion of overweight and obese persons has increased over the past 40 years.

In 2000 the U.S. Department of Agriculture responded by publishing a revised set of recommendations regarding food and exercise. The 2000 edition of *Dietary Guidelines for Americans* lists several criteria for maintaining a healthy lifestyle.

Aim for fitness

- Aim for a healthy weight.
- Be physically active each day—30 minutes five times a week.

Build a healthy base

- Let the Food Guide Pyramid guide your food choices.
- Choose a variety of grains daily, especially whole grains.
- Choose a variety of fruits and vegetables daily.
- Keep food safe to eat.

Choose sensibly

- Choose a diet that is low in saturated fat and cholesterol and moderate in total fat—less than 30% of Calories from fat and less than 10% from saturated fats.
- Choose beverages and foods to moderate your intake of sugars.
- Choose and prepare foods with less salt.
- If you drink alcoholic beverages, do so in moderation.

meal. *Fruit*, on the other hand, is a botanical term for the structure that is produced from the female part of the flower that surrounds the ripening seeds. Although, botanically, green beans, peas, and corn are all fruits, nutritionally speaking, they are placed in the vegetable category because they are generally eaten during the main part of the meal. Nutritionally speaking, fruits include such sweet plant products as melons, berries, apples, oranges, and bananas. The Food Guide Pyramid suggests two to four servings of fruit per day. However, because these foods tend to be high in natural sugars, consumption of large amounts of fruits can add significant amounts of kilocalories to your diet. A small apple, half a grapefruit, ½ cup of grapes, or 6 ounces of fruit juice is considered a serving. Fruits provide fiber, carbohydrate, water, and certain of the vitamins, particularly vitamin C.

Vegetables Group

The Food Guide Pyramid suggests three to five servings from this group each day. Items in this group include nonsweet plant materials, such as broccoli, carrots, cabbage, corn, green beans, tomatoes, potatoes, lettuce, and spinach. A serving is considered 1 cup of raw leafy vegetables or 1/2 cup of other types. It is wise to include as much variety as possible in this group. If you eat only carrots, several cups each day can become very boring. There is increasing evidence indicating that cabbage, broccoli, and cauliflower can provide some protection from certain types of cancers. This is a good reason to include these foods in your diet.

Foods in this group provide vitamins A and C as well as water and minerals. Leafy green vegetables are good sources of vitamin B_2 (riboflavin), vitamin B_6, vitamin K, and the mineral, magnesium. They also provide fiber, which assists in the proper functioning of the digestive tract.

Dairy Products Group

All of the cheeses, ice cream, yogurt, and milk are in this group. Two servings from this group are recommended each day. Each of these servings should be about 1 cup of ice cream, yogurt, or milk, or 2 ounces of cheese. Using product labels will help you determine the appropriate serving size of individual items. This group provides not only minerals, such as calcium, in your diet, but also water, vitamins, carbohydrates, and protein. Several of the B vitamins are present in milk and vitamin D and A often are added to the milk. You must remember that many cheeses contain large amounts of cholesterol and fat for each serving. Low-fat dairy products are increasingly common as manufacturers seek to match their products with the desire of the public for less fat in the diet.

Meat, Poultry, Fish, and Dry Beans Group

This group contains most of the things we eat as a source of protein; for example, nuts, peas, tofu, and eggs are considered members of this group. It is recommended that we include 5 to 7 ounces (140–200 grams) of these items in our daily diet. (A typical hamburger patty contains 60 grams of meat.) In general, people in the economically developed world eat at least this quantity and frequently much more. Because many sources of protein also include significant fat, and health recommendations suggest reduced fat, more attention is being paid to the quantity of protein-rich foods in the diet. We have not only decreased our intake of items from this group, but also shifted from the high-fat-content foods, such as beef and pork, to foods that are high in protein but lower in fat content, such as fish and poultry. Beans (except for the oil-rich soybean) are also excellent ways to get needed protein without unwanted fat. Modern food preparers tend to use smaller portions and cook foods in ways that reduce the fat content. Broiled fish, rather than fried, and baked, skinless portions of chicken or turkey (the fat is attached to the skin) are seen more and more often on restaurant menus and on dining room tables at home.

Remember that the recommended daily portion from this group is only 5 to 7 ounces (140–200 grams). This means that one double cheeseburger meets this recommendation for the daily intake. Actually the RDA for protein in the diet is about 60 grams for adults so the Food Guide Pyramid provides a generous amount of protein. Eating excessive amounts of protein can stress the kidneys by causing higher concentrations of calcium in the urine, increase the demand for water to remove toxic keto acids produced from the breakdown of amino acids, and lead to weight gain because of the fat normally associated with many sources of protein. It should be noted, however, that vegetarians must pay particular attention to acquiring adequate sources of protein because they have eliminated a major source from their diet. Although nuts and soybeans are high in protein they should not be consumed in large quantities because they are also high in fats.

19.6 Eating Disorders

The three most common eating disorders are obesity, bulimia, and anorexia nervosa. All three disorders are related to the prevailing perceptions and values of the culture in which we live. In many cases there is a strong psychological component as well (Outlooks 19.2).

Obesity

People who have a body mass index of 30 kg/m^2 or greater are considered **obese,** and suffer from a disease condition known as obesity. Approximately 34% of adults and 22% of preschool children in the U.S. population are overweight. Obesity is the condition of being overweight to the extent that a person's health and life span are adversely affected. The U.S. government blames 300,000 deaths a year on weight-related diseases. Because children of this generation are more physically inactive than their ancestors, they may be the first generation in American history to live a shorter life than their parents. The early death associated with this problem has recently been termed *sedentary death syndrome (SDS)*.

OUTLOOKS 19.2

Hungry?—The Molecular Basis of Obesity

Advances have recently been made in identifying genetic and biochemical bases that regulate body weight. While not yet fully understood, it is becoming increasingly clear that the ability to regulate food intake and body weight is controlled by how several genes function and how their proteins interact. This has turned out to be a very complex interaction, and simple solutions to obesity may not be so simple until we learn how body weight is regulated. The molecules involved are messengers that control the flow of information between the stomach and nerve cells in a portion of the brain called the *hypothalamus*. This flow of information controls the appetite—whether you are hungry or not depends on how these cells are stimulated. Several molecules are involved in this control:

1. *Leptin* is the product of the *ob* gene (formerly the *"obesity"* gene) and is a hormone produced by white fat tissue. Leptin acts on nerve cells in the brain to regulate food intake. The *absence* of leptin causes severe obesity; the presence of leptin acts to suppress the appetite center of the brain. As a person's weight increases, the additional fat cells increase the production of leptin. In animals, the increase in leptin signals the brain to eat less and the body to do more. In humans, however, this does not appear to be the case since many obese people have high levels of leptin and no suppression of appetite.

2. *Neuropeptide Y(NPY)* is a small protein that increases appetite. Leptin's appetite-suppressing effects may be due to the fact that leptin can inhibit NPY.

3. *Alpha-melanocyte-stimulating hormone (α-MSH)*, the brown pigmentation of the skin, acts to suppress the appetite by acting on the hypothalamus. Leptin is believed to stimulate production of α–MSH, which suppresses the appetite.

4. *Melanin-concentrating hormone (MCH)* is another neuropeptide. The gene for MCH is significantly more active in the brains of the obese. High levels of MCH increase food consumption and low levels decrease food consumption. Some believe that the smell of food may stimulate production of MCH and enhance appetite.

5. *Ghrelin* is a growth hormone peptide secreted by the stomach and has the opposite effect of leptin—it appears to be a powerful appetite stimulant.

Obesity occurs when people consistently take in more food energy than is necessary to meet their daily requirements. On the surface, it would appear that obesity is a simple problem to solve. To lose weight, all that people must do is consume fewer kilocalories, exercise more, or do both. On average, people in the United States gain 2 pounds a year, which equals 50 extra calories stored each day. Because the body can store half of the calories consumed, preventing a 2-pound weight gain might simply require eating 100 fewer calories a day.

Although all obese people have an imbalance between their need for kilocalories and the amount of food they eat, the reasons for this imbalance are complex and varied. It is clear that the prevailing culture has much to do with the incidence of obesity. For example, rates of obesity have increased over time, which strongly suggests that most cases of obesity are due to changes in lifestyle, not inherent biological factors. Furthermore, immigrants from countries with low rates of obesity show increased rates of obesity when they integrate into the American culture.

Many people attempt to cope with the problems they face by overeating. Overeating to solve problems is encouraged by our culture. Furthermore, food consumption is central to most kinds of celebrations. Social gatherings of almost every type are considered incomplete without some sort of food and drink. If snacks (usually high-calorie foods) are not made available by the host, many people feel uneasy or even unwelcome. It is also true that Americans and people of other cultures show love and friendship by sharing a meal. Many photographs in family albums have been taken at mealtime. In addition, less than half the meals consumed in the United States are prepared in the home. Under these conditions the consumer has reduced choices in the kind of food available, no control over the way foods are prepared, and little control over serving size. Meals prepared in restaurants and fast-food outlets emphasize meat and minimize the fruit, vegetable, and cereal portions of a person's diet, in direct contrast to the Food Guide Pyramid. The methods of preparation also typically involve cooking with oils and serving with dressings or fat-containing condiments. In addition, portion sizes are generally much larger than recommended.

Recent discoveries of genes in mice suggest that there are genetic components to obesity. Mice without a crucial gene gained an extraordinary amount of weight. There is some suggestion that there may be similar genes in humans. It is clear also that some people have much lower metabolic rates than the majority of the population and, therefore, need much less food than is typical. Still other obese individuals have a chemical imbalance of the nervous system that prevents them from feeling "full" until they have eaten an excessive amount of food. This imbalance prevents the brain from "turning off" the desire to eat after a reasonable amount of food has been eaten. Research into the nature and action of this brain chemical indicates that if obese people lacking this chemical receive it in pill form, they can feel "full" even when their food intake is decreased by 25%.

Health practitioners are changing their view of obesity from one of blaming the obese person for lack of self-control to one of treating the condition as a chronic disease that requires a varied approach to control. For the majority of people dietary counseling and increased exercise is all that is needed. But others need psychological counseling and some may need drug therapy or surgery. Regardless, controlling obesity can be very difficult because it requires basic changes in a person's eating habits, lifestyle, and value system.

Bulimia

Bulimia ("hunger of an ox" in Greek) is a disease condition in which the person has a cycle of eating binges followed by purging the body of the food by inducing vomiting or using laxatives. Many bulimics also use diuretics that cause the body to lose water and, therefore, reduce weight. It is often called the silent killer because it is difficult to detect. Bulimics are usually of normal body weight or are overweight. The cause is thought to be psychological, stemming from depression, low self-esteem, displaced anger, a need to be in control of one's body, or a personality disorder. Many bulimics have other compulsive behaviors such as drug abuse as well and are often involved in incidences of theft and suicide.

Vomiting may be induced physically or by the use of some nonprescription drugs. Case studies have shown that bulimics may take 40 to 60 laxatives a day to rid themselves of food. For some, the laxative becomes addictive. The binge-purge cycle and associated use of diuretics result in a variety of symptoms that can be deadly. The following is a list of the major symptoms observed in many bulimics:

> Excessive water loss
> Diminished blood volume
> Extreme potassium, calcium, and sodium deficiencies
> Kidney malfunction
> Increase in heart rate
> Loss of rhythmic heartbeat
> Lethargy
> Diarrhea
> Severe stomach cramps
> Damage to teeth and gums
> Loss of body proteins
> Migraine headaches
> Fainting spells
> Increased susceptibility to infections

Anorexia Nervosa

Anorexia nervosa (figure 19.4) is a nutritional deficiency disease characterized by severe, prolonged weight loss as a result of a voluntary severe restriction in food intake. An anorexic person's fear of becoming overweight is so intense

Figure 19.4

Anorexia Nervosa
Anorexia nervosa is a psychological eating disorder afflicting many Americans. These photographs were taken of an individual before and after treatment. Restoring a person with this disorder requires both medical and psychological efforts.

that even though weight loss occurs, it does not lessen the fear of obesity, and the person continues to diet, often even refusing to maintain the optimum body weight for his or her age, sex, and height. This nutritional deficiency disease is thought to stem from sociocultural factors. Our society's preoccupation with weight loss and the desire to be thin strongly influences this disorder.

Just turn on your television or radio, or look at newspapers, magazines, or billboards, and you can see how our culture encourages people to be thin. Male and female models are thin. Muscular bodies are considered healthy and any stored body fat unhealthy. Unless you are thin, so the advertisements imply, you will never be popular, get a date, or even marry. Our culture's constant emphasis on being thin has influenced many people to become anorexic and lose too much weight. Anorexic individuals frequently starve themselves to death. Individuals with anorexia are mostly adolescent and preadolescent females, although the disease does occur in males. Here are some of the symptoms of anorexia nervosa:

Thin, dry, brittle hair
Degradation of fingernails
Constipation
Amenorrhea (lack of menstrual periods)
Decreased heart rate
Loss of body proteins
Weaker-than-normal heartbeat
Calcium deficiency
Osteoporosis
Hypothermia (low body temperature)
Hypotension (low blood pressure)
Increased skin pigmentation
Reduction in size of uterus
Inflammatory bowel disease
Slowed reflexes
Fainting
Weakened muscles

19.7 Deficiency Diseases

Without minimal levels of the essential amino acids in the diet, a person may develop health problems that could ultimately lead to death. In many parts of the world, large populations of people live on diets that are very high in carbohydrates and fats but low in complete protein. This is easy to understand because carbohydrates and fats are inexpensive to grow and process in comparison to proteins. For example, corn, rice, wheat, and barley are all high-carbohydrate foods. Corn and its products (meal, flour) contain protein, but it is an incomplete protein that has very low amounts of the amino acids tryptophan and lysine. Without enough of these amino acids, many necessary enzymes cannot be made in sufficient amounts to keep a person healthy. One protein-deficiency disease is called **kwashiorkor,** and the symptoms

are easily seen (figure 19.5). A person with this deficiency has a distended belly, slow growth, slow movement, and is emotionally depressed. If the disease is caught in time, brain damage may be prevented and death averted. This requires a change in diet that includes expensive protein, such as poultry, fish, beef, shrimp, or milk. As the world food problem increases, these expensive foods will be in even shorter supply and will become more and more costly.

Starvation is also a common problem in many parts of the world. Very little carbohydrate is stored in the body. If you starve yourself, this small amount will last as a stored form of energy only for about 2 days. Even after a few hours of fasting the body begins to use its stored fat deposits as a source of energy; as soon as the carbohydrates are gone proteins will begin to be used to provide a source of glucose. Some of the keto acids produced during the breakdown of fats and amino acids are released in the breath and can be detected as an unusual odor. People who are fasting, anorexic, diabetic, or have other metabolic problems often have this "ketone breath." During the early stages of starvation, the amount of fat in the body steadily decreases, but the amount of protein drops only slightly (20–30 grams per day) (figure 19.6). This can continue only up to a certain point. After several weeks of fasting, so much fat has been

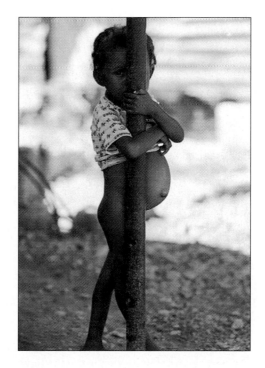

Figure 19.5

Kwashiorkor

This starving child shows the symptoms of kwashiorkor, a protein-deficiency disease. If treated with a proper diet containing all amino acids, the symptoms can be reduced.

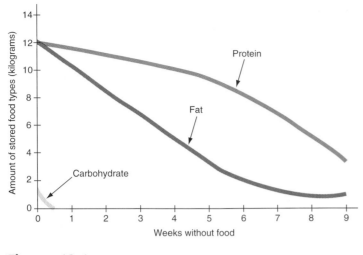

Figure 19.6

Starvation and Stored Foods

Starving yourself results in a very selective loss of the kinds of nutrients stored in the body. Notice how the protein level in the body has the slowest decrease of the three nutrients. This protein-conservation mechanism enables the body to preserve essential amounts of enzymes and other vital proteins.

lost from the body that proteins are no longer protected, and cells begin to use them as a primary source of energy (as much as 125 grams per day). This results in a loss of proteins from the cells that prevents them from carrying out their normal functions. When starvation gets to this point it is usually fatal. When not enough enzymes are available to do the necessary cellular jobs, the cells die. People who are chronically undernourished and lack fat do not have the protective effect of fat and experience the effects of starvation much more quickly than those who have stored fat. Children are particularly at risk because they also have a great need for nutrients to serve as building blocks for growth.

The lack of a particular vitamin in the diet can result in a **vitamin-deficiency disease.** A great deal has been said about the need for vitamin and mineral supplements in diets. Some people claim that supplements are essential; others claim that a well-balanced diet provides adequate amounts of vitamins and minerals. Supporters of vitamin supplements have even claimed that extremely high doses of certain vitamins can prevent poor health or even create "superhumans." It is very difficult to substantiate many of these claims, however. Because the function of vitamins and minerals and their regulation in the body is not completely understood, the RDAs are, at best, estimates by experts who have looked at the data from a variety of studies. In fact, the minimum daily requirement of a number of vitamins has not been determined. Vitamin-deficiency diseases that show recognizable symptoms are extremely rare except in cases of extremely poor nutrition.

19.8 Nutrition Through the Life Cycle

Nutritional needs vary throughout life and are related to many factors, including age, sex, reproductive status, and level of physical activity. Infants, children, adolescents, adults, and the elderly all require essentially the same types of nutrients but have special nutritional needs related to their stages of life, which may require slight adjustments in the kinds and amounts of nutrients consumed.

Infancy

A person's total energy requirements per kilogram are highest during the first 12 months of life: 100 kilocalories per kilogram of body weight per day. Fifty percent of this energy is required for an infant's basal metabolic rate. Infants (birth to 12 months) triple their weight and increase their length by 50% during that first year; this is their so-called first growth spurt. Because they are growing so rapidly they require food that contains adequate proteins, vitamins, minerals, and water. They also need food that is high in kilocalories. For many reasons, the food that most easily meets these needs is human breast milk (table 19.7). Even with breast milk's many nutrients, many physicians strongly recommend multivitamin supplements as part of an infant's diet.

Childhood

As infants reach childhood, their dietary needs change. The rate of growth generally slows between 1 year of age and puberty, and girls increase in height and weight slightly faster than boys. During childhood, the body becomes more lean, bones elongate, and the brain reaches 100% of its adult size between the ages of 6 and 10. To adequately meet growth and energy needs during childhood, protein intake should be high enough to take care of the development of new tissues. Minerals, such as calcium, zinc, and iron, as well as vitamins, are also necessary to support growth and prevent anemia. Although many parents continue to provide their children with multivitamin supplements, such supplements should be given only after a careful evaluation of their children's diets. There are four groups of children who are at particular risk and should receive such supplements:

1. Children from deprived families and those suffering from neglect or abuse
2. Children who have anorexia or poor eating habits, or who are obese
3. Pregnant teens
4. Children who are strict vegetarians

During childhood, eating habits are very erratic and often cause parental concern. Children often limit their intake of milk, meat, and vegetables while increasing their intake of sweets. To get around these problems, parents can provide calcium by serving cheeses, yogurt, and soups as al-

Table 19.7

A COMPARISON OF HUMAN BREAST MILK AND COW'S MILK*

Nutrient	Human Milk	Cow's Milk
Energy (kilocalories/1,000 grams)	690	660 (whole milk)
Protein (grams per liter)	9	35
Fat (grams per liter)	40	38
Lactose (grams per liter)	68	49
Vitamins		
A (international units)	1,898	1,025
C (micrograms)	44	17
D (activity units)	40	14
E (international units)	3.2	0.4
K (micrograms)	34	170
Thiamin (B_1) (micrograms)	150	370
Riboflavin (B_2) (micrograms)	380	1,700
Niacin (B_3) (milligrams)	1.7	0.9
Pyridoxine (B_6) (micrograms)	130	460
Cobalamin (B_{12}) (micrograms)	0.5	4
Folic acid (micrograms)	41–84.6	2.9–68
Minerals (all in milligrams)		
Calcium	241–340	1,200
Phosphorus	150	920
Sodium	160	560
Potassium	530	1,570
Iron	0.3–0.56	0.5
Iodine	200	80

*All milks are not alike. Each milk is unique to the species that produces it for its young, and each infant has its own special growth rate. Humans have one of the slowest infant growth rates, and human milk contains the least amount of protein. Because cow's milk is so different, many pediatricians recommend that human infants be fed either human breast milk or formulas developed to be comparable to breast milk during the first 12 months of life. The use of cow's milk is discouraged. This table lists the relative amounts of different nutrients in human breast milk and cow's milk.

ternatives to milk. Meats can be made more acceptable if they are in easy-to-chew, bite-sized pieces, and vegetables may be more readily accepted if smaller portions are offered on a more frequent basis. Steering children away from sucrose by offering sweets in the form of fruits can help reduce dental caries (*carrion* = rotten). You can better meet the dietary needs of children by making food available on a more frequent basis, such as every three to four hours. Obesity is an increasing problem among children. Parents sometimes encourage this by insisting that children eat everything served to them. Most children automatically regulate the food they eat to an appropriate amount; parents should be more concerned about the kinds of food children eat rather than the amounts.

Adolescence

The nutrition of an adolescent is extremely important because, during this period, the body changes from nonreproductive to reproductive. Puberty is usually considered to last between five and seven years. Before puberty, males and females tend to have similar proportions of body fat and muscle. Both body fat and muscle make up between 15% and 19% of the body's weight. Lean body mass, primarily muscle, is about equal in males and females. During puberty, female body fat increases to about 23%, and in males it decreases to about 12%. Males double their muscle mass in comparison to females.

The changes in body form that take place during puberty constitute the second growth spurt. Because of their more rapid rate of growth and unique growth patterns, males require more of certain nutrients than females (protein, vitamin A, magnesium, and zinc). During adolescence, youngsters will gain as much as 20% of their adult height and 50% of their adult weight, and many body organs will double in size. Nutritionists have taken these growth patterns and spurts into account by establishing RDAs for males and females 10 to 20 years old, including requirements at the peaks of growth spurts. RDAs at the peak of the growth spurt are much higher than they are for adults and children.

Adulthood

People who have completed the changes associated with adolescence are considered to have entered adulthood. Most of the information available to the public through the press, television, and radio focuses on this stage in the life cycle. During adulthood, the body enters a plateau phase, and diet and nutrition focus on maintenance and disease prevention. Nutrients are used primarily for tissue replacement and repair, and changes such as weight loss occur slowly. Because the BMR slows, as does physical activity, the need for food energy decreases from about 2,900 kilocalories in average young adult males (ages 20 to 40) to about 2,300 for elderly men. For women, the corresponding numbers decrease from 2,200 to 1,900 kilocalories. Protein intake for most U.S. citizens is usually in excess of the recommended amount. The RDA standard for protein is about 63 grams for men and 50 grams for women each day. About 25% to 50% should come from animal foods to ensure intake of the essential amino acids. The rest should be from plant-protein foods such as whole grains, legumes, nuts, and vegetables.

An adult who follows a well-balanced diet should have no need for vitamin supplements; however, improper diet, disease, or other conditions might require that supplements be added. The two minerals that demand special attention are calcium and iron, especially for women. A daily intake of 1,200 milligrams of calcium should prevent calcium loss from bones (osteoporosis; figure 19.7) and a daily intake of 15 mg of iron should allow adequate amounts of hemoglobin to be manufactured to prevent anemia in women over

Figure 19.7

Osteoporosis
These photographs are of a healthy bone (a) and a section of bone from a person with osteoporosis (b). This nutritional deficiency disease results in a change in the density of the bones as a result of the loss of bone mass. Bones that have undergone this change look "lacy" or like Swiss cheese, with larger than normal holes. A few risk factors found to be associated with this disease are being female and fair skinned; having a sedentary lifestyle; using alcohol, caffeine, and tobacco; and having reached menopause.

(a) Healthy

(b) Osteoporosis

50 and men over 60. In order to reduce the risk of chronic diseases such as heart attack and stroke, adults should definitely eat a balanced diet, participate in regular exercise programs, control their weight, avoid cigarettes and alcohol, and practice stress management.

Nutritional Needs Associated with Pregnancy and Lactation

Risk-management practices that help in avoiding chronic adult diseases become even more important when planning pregnancy. Studies have shown that an inadequate supply of the essential nutrients can result in infertility, spontaneous abortion, and abnormal fetal development. The period of pregnancy and milk production (lactation) requires that special attention be paid to the diet to ensure proper fetal development, a safe delivery, and a healthy milk supply.

The daily amount of essential nutrients must be increased, as should the kilocaloric intake. Kilocalories must be increased by 300 per day to meet the needs of an increased BMR; the development of the uterus, breasts, and placenta; and the work required for fetal growth. Some of these kilocalories can be obtained by drinking milk, which simultaneously supplies calcium needed for fetal bone development. In those individuals who cannot tolerate milk, supplementary sources of calcium should be used. In addition, the daily intake of protein should be at least 65 grams per day. As was mentioned earlier, most people consume much more than this amount of protein per day. Two essential nutrients, folic acid and iron, should be obtained through prenatal supplements because they are so essential to cell division and development of the fetal blood supply.

The mother's nutritional status affects the developing baby in several ways. If she is under 15 years of age or has had three or more pregnancies in a 2-year period, her nutritional stores are inadequate to support a successful pregnancy. The use of drugs such as alcohol, caffeine, nicotine, and "hard" drugs (e.g., heroin) can result in decreased nutrient exchange between the mother and fetus. In particular, heavy smoking can result in low birth weights and alcohol abuse is responsible for fetal alcohol syndrome (figure 19.8).

Old Age

As people move into their sixties and seventies, digestion and absorption of all nutrients through the intestinal tract is not impaired but does slow down. The number of cells undergoing mitosis is reduced, resulting in an overall loss in the number of body cells. With age, complex organs such as the kidneys and brain function less efficiently, and protein synthesis becomes inefficient. With regard to nutrition, energy requirements for the elderly decrease as the BMR slows, physical activity decreases, and eating habits also change.

The change in eating habits is particularly significant because it can result in dietary deficiencies. For example, linoleic acid, an essential fatty acid, may fall below required levels as an older person reduces the amount of food eaten. The same is true for some vitamins and minerals. Therefore, it may be necessary to supplement the diet daily with 1 tablespoon of vegetable oil. Vitamin E, multiple vitamins, or a mineral supplement may also be necessary. The loss of body protein means that people must be certain to meet their daily RDA for protein and participate in regular exercise to prevent muscle loss. As with all stages of the life cycle, regular exercise is important in maintaining a healthy, efficiently functioning body.

19.9 Nutrition for Fitness and Sports

In the past few years there has been a heightened interest in fitness and sports. Along with this, an interest has developed in the role nutrition plays in providing fuel for activities, controlling weight, and building muscle. The cellular respiration process described in chapter 6 is the source of the energy needed to take a leisurely walk or run a marathon. However,

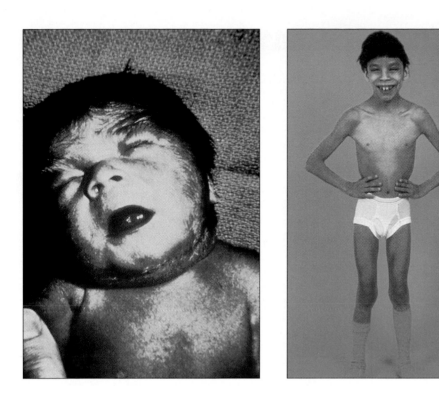

Figure 19.8

Fetal Alcohol Syndrome (FAS)

The 1-day-old infant on the left displays fetal alcohol syndrome; the child on the right with the syndrome is 8 years old. The 8-year-old was diagnosed at birth and has spent his entire life in a foster home where the quality of care has been excellent. His IQ has remained stable at 40 to 45.

the specific molecules used to get energy depend on the length of the period of exercise, whether or not you warm up before you exercise, and how much effort you exert during exercise. The molecules respired by muscle cells to produce ATP may be glucose, fatty acids, or amino acids. Glucose is stored as glycogen in the muscles, liver, and some other organs. Fatty acids are stored as triglycerides in fat cells. Amino acids are found as protein in muscles and other organs. Which molecules are respired depends on the duration and intensity of exercise. Glucose from glycogen and fatty acids from triglycerides are typically the primary fuels. Amino acids provide 10% or less of a person's energy needs even in highly trained athletes. • aerobic cellular respiration, p. 110

Aerobic exercise occurs when the level of exertion allows the heart and lungs to keep up with the oxygen needs of the muscles. **Anaerobic exercise** involves bouts of exercise that are so intense that the muscles cannot get oxygen as fast as they need it. Therefore, they must rely on anaerobic respiration of glucose to provide the energy needed. During a long, brisk walk, the heart and lungs of most people should be able to keep up with the muscle cells' requirement for oxygen. The oxygen is used by mitochondria to run the aerobic Krebs cycle and electron-transport system. (They are exercising aerobically.) Studies have shown that during the first 20 minutes of moderate exercise, glucose from the blood and glycogen from the muscles are used as the fuel for aerobic respiration. When the period of exercise is greater than 20 minutes, triglycerides stored in cells and fatty acids from the circulatory system are used as the prime sources of energy. Many exercise programs encourage a warm-up period before the exercise begins. During the warm-up period the body is using glucose and glycogen but will switch to the burning of triglycerides and fatty acids later in the exercise period. This is why moderate, longer periods of exercise are most beneficial in weight loss.

But what about extreme exertion or vigorous exercise for longer periods? Extreme bouts of activity that last only several seconds, such as a 100-meter dash or lifting weights, rely on the ATP already present in the muscle cells. Because the activity period is so short and the ATP is already in the muscles, oxygen is not needed. Intense exercise that lasts for a few minutes, such as a 400-meter run or gymnastic events, requires additional ATP—more than can be stored in the muscles. This is supplied primarily by anaerobic respiration of glycogen and glucose. As the period of exercise increases it is impossible to continue to rely on anaerobic respiration. Aerobic respiration of triglycerides and fatty acids becomes more important. Without a warm-up period, your muscle cells will begin to respire mostly muscle-stored glycogen. At first the heart and lungs are not able to supply all the oxygen needed by the mitochondria. Glycolysis will provide ATP for muscle contraction, but there will be an increase in the amount of pyruvic and lactic acids. About 20 minutes into vigorous exercise, 20% of muscle glycogen is gone and little triglyceride has been respired. By this time, the heart and lungs are "warmed up" and are able to provide the oxygen necessary to respire larger amounts of triglycerides and fatty acids along with the glycogen and glucose. Fatty acid levels in the blood increase greatly and ATP output increases dramatically. A person who has not warmed up experiences this metabolic shift as a "second wind."

From this point on, all sources of energy—glucose from glycogen, fatty acids from triglycerides, and even small amounts of amino acids from proteins—are utilized, but the

OUTLOOKS 19.3

Myths or Misunderstandings About Diet and Nutrition

Myth or Misunderstanding	Reality
1. Exercise burns calories.	You don't burn calories, but you do oxidize (burn) the fuels to provide yourself with the energy needed to perform various activities.
2. Active people who are increasing their fitness need more protein.	The amount of protein needed is very small. Most people get many times the amount of protein required.
3. Vitamins give you energy.	Vitamins are involved with enzymes in the release of energy from food items, but are not sources of energy.
4. Large amounts of protein are needed to build muscle.	A person can only build a few grams of new muscle per day. Therefore, consuming large amounts of protein will not increase the rate of muscle growth.
5. Large quantities (megadoses) of vitamins will fight disease, build strength, and increase length of life.	Quantities of vitamins that greatly exceed the recommended dietary allowance (RDA) have not been shown to be beneficial. Large doses of some vitamins are toxic (vitamins A, D, B_3).
6. Special protein supplements are more quickly absorbed and can build muscle faster.	There is adequate protein in nearly all diets. The supplements may be absorbed faster but that does not mean that they will be incorporated into muscle mass faster.

balance shifts during the period of exercise. The glycogen continues to be used but, at approximately 1 hour into exercise, glycogen stores have largely been depleted lowering blood glucose and the athlete experiences debilitating fatigue known as "hitting the wall."

There are also metabolic shifts that occur as one stops exercising. If exercise is suddenly stopped while there are high concentrations of fatty acids in the blood, they will be converted to keto acids, which could negatively affect kidney function; a loss of sodium, calcium, and other minerals; and change the pH of the blood. For this reason, many exercise physiologists suggest a cool-down period of light exercises that allow the body to slowly shift back to a more normal fuel balance of glucose and fatty acids.

Diet is an important aspect of an athlete's training program. The kind of diet needs to be tailored to the kind of performance expected from the athlete. To avoid or postpone hitting the wall, many marathon athletes practice **carbohydrate loading.** This should be done only by those engaged in periods of hard exercise or competition lasting 90 minutes or more. It requires a week-long diet and exercise program. On the first day, the muscles needed for the event are exercised for 90 minutes and carbohydrates, such as fruits, vegetables, or pasta provide a source of dietary kilocalories. On the second and third days of carbohydrate loading, the person continues the 50% carbohydrate diet, but the period of exercise is reduced to 40 minutes. On the fourth and fifth days, the workout period is reduced to 20 minutes and the carbohydrates are increased to 70% of total kilocalorie intake. On the sixth day, the day before competition, the person rests and continues the 70% carbohydrate diet. Following this program increases muscle glycogen levels and makes it possible to postpone "hitting the wall."

Conditioning includes many interrelated body adjustments in addition to energy considerations. Training increases the strength of muscles, including the heart, and increases the efficiency of operation. Practicing a movement allows for the development of a smooth action that is more energy efficient than a poorly trained motion. As the body is conditioned, there is an increase in the number of mitochondria per cell, the Krebs cycle and the ETS run more efficiently, the number of capillaries increases, fats are respired more efficiently and for longer periods, and weight control becomes easier.

The amount of protein in an athlete's diet has also been investigated. Understand that an increase in dietary protein does not automatically increase strength, endurance, or speed. In fact, most Americans eat the 10% additional protein that athletes require as a part of their normal diets. The additional percentage is used by the body for many things, including muscle growth. But increasing protein intake will not automatically increase muscle size. Only when there is a need will the protein be used to increase muscle mass. That means exercise. Your body will build the muscle it needs in order to meet the demands you place on it. Vitamins and minerals operate in much the same way. No supplements should be required as long as your diet is balanced and complex. Your meals should provide the vitamins and minerals needed to sustain your effort (Outlooks 19.3).

Athletes must monitor their water intake because dehydration can cripple an athlete very quickly. A water loss of only 5% of the body weight can decrease muscular activity by as much as 30%. One way to replace water is to drink 1 to 1½ cups of water 15 minutes before exercising and ½ cup during exercise. In addition, drinking 16 ounces of cool tap water for each pound (16 ounces) of body weight lost during exercise will prevent dehydration. Another method is to use diluted orange juice (1 part juice to 5 parts water). This is an excellent way to replace water and resupply a small amount of lost glucose and salt. Salt pills (so-called electrolyte pills) are not recommended because salt added to the digestive tract tends to reduce the rate at which water is absorbed from the gut. When the salt is absorbed into the bloodstream, additional water is needed to dilute the blood to the proper level.

SUMMARY

To maintain good health, people must receive nutrient molecules that can enter the cells and function in the metabolic processes. The proper quantity and quality of nutrients are essential to good health. Nutritionists have classified nutrients into six groups: carbohydrates, proteins, lipids, minerals, vitamins, and water. Energy for metabolic processes may be obtained from carbohydrates, lipids, and proteins, and is measured in kilocalories. An important measure of the amount of energy required to sustain a human at rest is the basal metabolic rate. To meet this and all additional requirements, the United States has established the RDAs, recommended dietary allowances, for each nutrient. Should there be metabolic or psychological problems associated with a person's normal metabolism, a variety of disorders may occur, including obesity, anorexia nervosa, bulimia, kwashiorkor, and vitamin-deficiency diseases. As people move through the life cycle, their nutritional needs change, requiring a reexamination of their eating habits in order to maintain good health.

THINKING CRITICALLY

For guidelines to answer this question, visit the Online Learning Center.

You're 21 years old, female, have never been involved in any kind of sports, and have suddenly become interested in rugby! This is a very demanding contact sport and many people are injured while playing. If you are to succeed and experience only minor injuries, you must get in condition. Describe changes you should make in your daily diet and exercise program that would prepare you for this new experience. Well, get busy, or you'll never make the team!

CONCEPT MAP TERMINOLOGY

Construct a concept map using these terms.

anorexia
basal metabolic rate
bulimia
calorie
carbohydrate
diet

essential amino acids
fat
nutrition
obesity
protein

Visit the Online Learning Center to use an interactive format to construct another concept map.

KEY TERMS

Use interactive flash cards, on the Online Learning Center, to help you learn the meanings of these terms.

absorption
aerobic exercise
anaerobic exercise
anorexia nervosa
assimilation
basal metabolic rate (BMR)
body mass index (BMI)
bulimia
calorie
carbohydrate loading
complete protein
diet
digestion
electrolytes
essential amino acids
essential fatty acid
fiber

Food Guide Pyramid
incomplete protein
ingestion
kilocalorie (kcal)
kwashiorkor
minerals
nutrients
nutrition
obese
osteoporosis
protein-sparing
recommended dietary
    allowances (RDAs)
specific dynamic action (SDA)
vitamin-deficiency disease
vitamins

LEARNING CONNECTIONS

Visit the Online Learning Center to find additional study resources including those highlighted below.

19.1   Living Things as Chemical Factories: Matter and Energy Manipulators

1. What is a nutrient?
2. What is the difference between digestion and assimilation?

19.2   Kilocalories, Basal Metabolism, and Weight Control

3. Define the terms *basal metabolism, specific dynamic action,* and *voluntary muscular activity.*
4. What is BMI? How is it calculated?

19.3   The Chemical Composition of Your Diet

Interactive Concept Map
• Nutrients

5. Why are some nutrients referred to as essential? Name them.
6. List the six main classes of nutrients as described by nutritionists.

19.4   Amounts and Sources of Nutrients

Animations and Review
• Nutrition
• Vitamins and minerals
Experience This!
• Comparison shopping

7. What do the initials RDA stand for?
8. List four of the dietary guidelines.
9. Americans are currently consuming 37% of their kilocalories in fat. According to the dietary goals, what should that percentage be?

19.5   The Food Guide Pyramid with Five Food Groups

Interactive Concept Map
• The Food Guide Pyramid

10. Name the five basic food groups and give two examples of each.
11. Why is there a person jogging up the pyramid?

19.6   Eating Disorders

Food for Thought
• Bulimia

12. What are the three primary eating disorders?
13. What is the role of leptin in controlling appetite?

19.7   Deficiency Diseases

14. What type of deficiency is associated with "ketone breath"?

15. How common are vitamin deficiency diseases in the United States?

19.8   Nutrition Through the Life Cycle

Interactive Concept Map
• Nutrition throughout life

16. During which phase of the life cycle is a person's demand for kilocalories per unit of body weight the highest?
17. As our population ages, what changes need to be made to the diets of the elderly?

19.9   Nutrition for Fitness and Sports

18. What is meant by "carbohydrate loading"?
19. How do you recognize when a person shifts from aerobic to anaerobic exercise?

Test yourself with multiple choice questions that provide immediate feedback
on the Online Learning Center with PowerWeb at

www.mhhe.com/enger11

The Body's Control Mechanisms

GOALS AND OBJECTIVES

Understand how information is carried by the nervous system.

❏ Describe the ionic events that take place at the nerve cell membrane.

❏ Explain how nerve cells carry information from one place to another.

❏ Describe the events that take place at the synapse.

❏ Explain why an impulse can go only in one direction across a synapse.

Recognize that specific functions are located in particular parts of the brain.

❏ Explain why damage to a particular part of the brain affects a specific sensory or motor ability.

❏ Describe how nervous disorders can be caused by incorrect amounts of specific neurotransmitters.

Understand how the endocrine system functions.

❏ Describe the role of receptor sites in the function of the endocrine system.

❏ Explain why a particular hormone affects only certain tissues.

❏ Explain how hormones can have long-term effects such as growth.

Appreciate how sense organs function.

❏ Describe the kinds of environmental changes to which the various sense organs respond.

❏ Explain why nerve damage can impair sensory ability.

Understand how muscles are stimulated and how they contract.

❏ Compare the functional abilities of cardiac, skeletal, and smooth muscle.

❏ Describe the molecular events that cause muscle contraction.

Physiological Processes

PART FIVE

20.1 Coordination in Multicellular Animals

A large, multicellular organism, which consists of many different kinds of systems, must have some way of integrating various functions so that it can survive. The various systems must be coordinated to maintain a reasonably constant internal environment. Recall from chapter 18 that this condition of maintaining a constant internal environment is called **homeostasis.** To allow for homeostasis there must be constant monitoring and modification of the way specific parts of the organism function. If the organism does not respond appropriately, it will die. There are many kinds of sense organs located within organs and on the surfaces that respond to specific kinds of stimuli. A **stimulus** is any change in the environment that the organism can detect. Some stimuli, like light or sound, are typically external to the organism; others, like the pain generated by an infection, are internal. The reaction of the organism to a stimulus is known as a **response** (figure 20.1). • homeostasis, p. 340.

The nervous and endocrine systems are the major systems of the body that integrate stimuli and generate appropriate responses necessary to maintain homeostasis. The **nervous system** consists of a network of cells with fibrous extensions that carry information along very specific pathways from one part of the body to another. The **endocrine system** consists of a number of glands that communicate with one another and with other tissues through chemicals distributed throughout the organism. **Glands** are organs that manufacture specific molecules that are either secreted into surrounding tissue, where they are picked up by the circulatory system, or are secreted through ducts into the cavity of an organ or to the body surface. **Endocrine glands** have no ducts and secrete their products into the circulatory system. The molecules produced by endocrine glands are called hormones. A **hormone** is a specific molecule released by one organ that is transported to another organ where it triggers a change in the other organ's activity. Other glands, such as the digestive glands and sweat glands, empty their contents through ducts. These kinds of glands are called **exocrine glands.**

Although the functions of the nervous and endocrine systems can overlap and be interrelated, these two systems have quite different methods of action. The nervous system functions very much like a computer. A message is sent along established pathways from a specific initiating point to a specific end point, and the transmission is very rapid. The endocrine system functions in a manner analogous to a radio broadcast system. Radio stations send their signals in all directions, but only those radio receivers that are tuned to the correct frequency can receive the message. Messenger molecules (hormones) are typically distributed throughout the body by the circulatory system, but only those cells that have the proper receptor sites can receive and respond to the molecules.

| Stimulus | Integration | Response |

Figure 20.1

Stimulus and Response
A stimulus is any detectable change in the surroundings of an organism. When an organism receives a stimulus, it processes the information and may ignore the stimulus or generate a response to it.

20.2 Nervous System Function

The nervous system is particularly well suited to manage adjustments that must take place in a short time period. The following sections describe the major structural and functional characteristics of the nervous system.

The Structure of the Nervous System

The basic unit of the nervous system is a specialized cell called a **neuron,** or **nerve cell.** A typical neuron consists of a central body called the **soma,** or *cell body,* which contains the nucleus and several long, protoplasmic extensions called nerve fibers. There are two kinds of fibers: **axons,** which carry information away from the cell body, and **dendrites,** which carry information toward the cell body (figure 20.2). Most nerve cells have one axon and several dendrites.

Neurons are arranged into two major systems. The **central nervous system,** which consists of the brain and spinal cord, is surrounded by the skull and the vertebrae of the spinal column. It receives input from sense organs, interprets information, and generates responses. The **peripheral nervous system** is located outside the skull and spinal column and consists of bundles of long axons and dendrites called **nerves.** There are two different sets of neurons in the peripheral nervous system. **Motor neurons** carry messages from the central nervous system to muscles and glands, and **sensory neurons** carry input from sense organs to the central nervous system. Motor neurons typically have one long axon that runs from the spinal cord to a muscle or gland; sensory neurons have long dendrites that carry input from the sense organs to the central nervous system.

The Nature of the Nerve Impulse

Because most nerve cells have long, fibrous extensions, it is possible for information to be passed along the nerve cell from one end to the other. The message that travels along a neuron is known as a **nerve impulse.** A nerve impulse is not like an electric current but involves a specific sequence of chemical events involving activities at the cell membrane.

Because all cell membranes are differentially permeable, it is difficult for some ions to pass through the membrane and the combination of ions inside the membrane is different from that on the outside. Cell membranes also contain proteins that actively transport specific ions from one side of the membrane to the other. Active transport involves the cell's use of adenosine triphosphate (ATP) to move materials from one side of the cell membrane to the other. Because ATP is required this is an ability that cells lose when they die. One of the ions that is actively transported from cells is the sodium ion (Na^+). At the same time sodium ions are being transported out of cells, potassium ions (K^+) are being transported into the normal resting cells. However, there are more sodium ions transported out than potassium ions transported in. • ions, p. 29

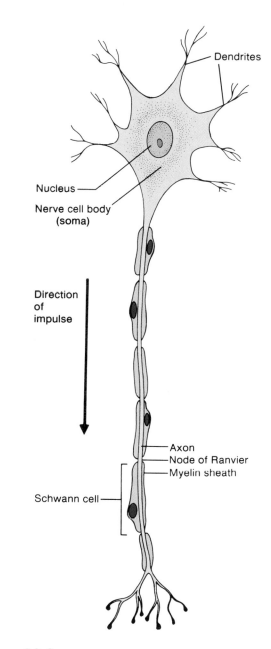

Figure 20.2

The Structure of a Nerve Cell
Nerve cells consist of a nerve cell body that contains the nucleus and several fibrous extensions. The shorter, more numerous fibers that carry impulses to the nerve cell body are dendrites. The long fiber that carries the impulse away from the cell body is the axon.

Because a normal resting cell has more positively charged Na^+ ions on the outside of the cell than on the inside, a small but measurable voltage exists across the membrane of the cell. (**Voltage** is a measure of the electrical charge difference that exists between two points or objects.) The voltage difference between the inside and outside of a cell membrane is about 70 millivolts (0.07 volt). The two sides of the cell membrane are, therefore, polarized in the same sense that a battery is polarized, with a positive and

Figure 20.3

The Polarization of Cell Membranes

All cells, including nerve cells, have an active transport mechanism that pumps Na+ out of cells and simultaneously pumps K+ into them. The end result is that there are more Na+ ions outside the cell and more K+ ions inside the cell. In addition, negative ions such as Cl− are more numerous inside the cell. Consequently, the outside of the cell is positive (+) compared to the inside, which is negative (−).

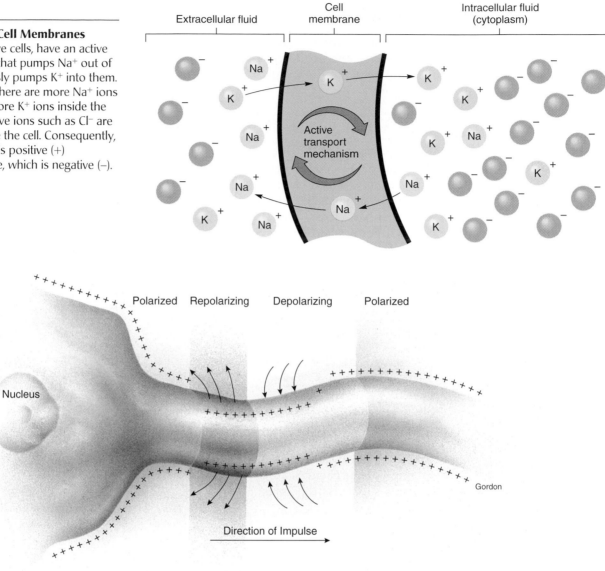

Figure 20.4

A Nerve Impulse

When a nerve cell is stimulated, a small portion of the cell membrane depolarizes as Na+ flows into the cell through the membrane. This encourages the depolarization of an adjacent portion of the membrane, and it depolarizes a short time later. In this way a wave of depolarization passes down the length of the nerve cell. Shortly after a portion of the membrane is depolarized, the ionic balance is re-established. It is repolarized and ready to be stimulated again.

negative pole. A resting neuron has its positive pole on the outside of the cell membrane and its negative pole on the inside of the membrane (figure 20.3).

When a cell is stimulated at a specific point on the cell membrane, the cell membrane changes its permeability and lets sodium ions (Na+) pass through it from the outside to the inside. The membrane is thus **depolarized;** it loses its difference in charge as sodium ions diffuse into the cell from the outside. Sodium ions diffuse into the cell because, initially, they are in greater concentration outside the cell than inside. When the membrane becomes more permeable, they are able

to diffuse into the cell, toward the area of lower concentration. The depolarization of one point on the cell membrane causes the adjacent portion of the cell membrane to change its permeability as well, and it also depolarizes. Thus, a wave of depolarization passes along the length of the neuron from one end to the other (figure 20.4). The depolarization and passage of an impulse along any portion of the neuron is a momentary event. As soon as a section of the membrane has been depolarized, potassium ions diffuse out of the cell. This re-establishes the original polarized state and the membrane is said to be *re-polarized*. Subsequently, the continuous active transport of

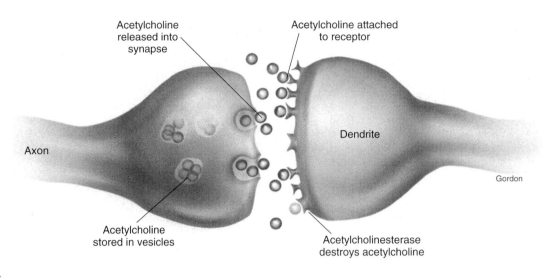

Acetylcholine
released into
synapse

Acetylcholine attached
to receptor

Axon

Dendrite

Gordon

Acetylcholine
stored in vesicles

Acetylcholinesterase
destroys acetylcholine

Figure 20.5

Events at the Synapse

When a nerve impulse reaches the end of an axon, it releases a neurotransmitter into the synapse. In this illustration, the neurotransmitter is acetylcholine. When acetylcholine is released into the synapse, acetylcholine molecules diffuse across the synapse and bind to receptors on the dendrite, initiating an impulse in the next neuron. Acetylcholinesterase is an enzyme that destroys acetylcholine, preventing continuous stimulation of the dendrite.

sodium ions out of the cell and potassium ions into the cell restores the original concentration of ions on both sides of the cell membrane. When the nerve impulse reaches the end of the axon, it stimulates the release of a molecule that stimulates depolarization of the next neuron in the chain.

Activities at the Synapse

Between the fibers of adjacent neurons in a chain is a space called the **synapse.** Many chemical events occur in the synapse that are important in the function of the nervous system. When a neuron is stimulated, an impulse passes along its length from one end to the other. When the impulse reaches a synapse, a molecule called a **neurotransmitter** is released into the synapse from the axon. It diffuses across the synapse and binds to specific receptor sites on the dendrite of the next neuron. When enough neurotransmitter molecules have been bound to the second neuron, an impulse is initiated in it as well. Several kinds of neurotransmitters are produced by specific neurons. These include dopamine, epinephrine, acetylcholine, and several other molecules. The first neurotransmitter identified was **acetylcholine.** Acetylcholine molecules are manufactured in the soma and migrate down the axon where they are stored until needed (figure 20.5).

As long as a neurotransmitter is bound to its receptor, it continues to stimulate the nerve cell. Thus, if acetylcholine continues to occupy receptors, the neuron continues to be stimulated again and again. An enzyme called **acetylcholinesterase** destroys acetylcholine and prevents this from happening. (The breakdown products of the acetylcholine can be used to remanufacture new acetylcholine molecules.) The destruction of acetylcholine allows the second neuron in

the chain to return to normal. Thus, it will be ready to accept another burst of acetylcholine from the first neuron a short time later. Neurons must also constantly manufacture new acetylcholine molecules or they will exhaust their supply and be unable to conduct an impulse across a synapse.

Certain drugs, such as curare and strychnine, interfere with activities at the synapse. Curare blocks the synapse and causes paralysis, whereas strychnine causes nerve cells to be continually stimulated. Many of the modern insecticides are also nerve poisons and are therefore quite hazardous.

Because of the way the synapse works, impulses can go in only one direction: Only axons secrete acetylcholine, and only dendrites have receptors. This explains why there are sensory and motor neurons to carry messages to and from the central nervous system.

As mentioned earlier, the nervous system is organized in a fashion similar to a computer. Information from various input devices (sense organs) is delivered to the central processing unit (brain) by way of wires (sensory nerves). The information is interpreted in the central processing unit. Eventually messages can be sent by way of cables (motor nerves) to drive external machinery (muscles and glands). This concept allows us to understand how the functions of various portions of the nervous system have been identified. It is possible to electrically stimulate specific portions of the nervous system or to damage certain parts of the nervous system in experimental animals and determine the functions of different parts of the brain and other parts of the nervous system. For example, because peripheral nerves carry bundles of both sensory and motor fibers, damage to a nerve may result in both a lack of feeling because sensory messages cannot get through and an inability to move because the motor nerves are damaged.

The Organization of the Central Nervous System

Major functions of specific portions of the brain have been identified. Certain parts of the brain are involved in controlling fundamental functions such as breathing and heart rate. Others are involved in generating emotions, whereas others decode sensory input, or coordinate motor activity. The human brain also has considerable capacity to store information and create new responses to environmental stimuli.

The brain consists of several regions, each of which has specific functions. The functions of the brain can be roughly divided into three major levels: automatic activities, basic decision making and emotions, and thinking and reasoning. If we begin with the spinal cord and work our way forward we will proceed from the more fundamental, automatic activities of the brain to the more complex thinking portions of the brain. The **spinal cord** is a collection of nerve fibers surrounded by the vertebrae that conveys information to and from the brain. At the base of the brain where the spinal cord enters the skull is a portion of the brain known as the *medulla oblongata*. The **medulla oblongata** is the region of the brain that controls fundamental activities such as blood pressure, breathing, and heart rate. Most of the fibers of the spinal cord cross from one side of the body to the other in the medulla oblongata. This is why the left side of the brain affects the right side of the body.

The **cerebellum** is a large bulge at the base of the brain that is connected to the medulla oblongata. The primary function of the cerebellum is coordination of muscle activity. It receives information from sense organs such as the portions of the ear that involve balance, the eyes, and pressure sensors in muscles and tendons. This information is used to make adjustments to the strength and order of contraction of muscles necessary to move in a coordinated fashion.

The **pons** is connected to the anterior end of the medulla oblongata. It also connects to the cerebellum and to higher levels of the brain. It is involved in controlling many sensory and motor functions of the sense organs of the head and face.

If we continue forward, the pons is connected to a portion of the brain that forms two bulblike structures that ultimately connect to the cerebrum. Although these portions of the brain still control many automatic activities, there are many activities that involve much more integration of information and some level of "decision making" occurs in this region. The primary regions are the **thalamus** and the **hypothalamus.** The thalamus relays information between the cerebrum and lower portions of the brain. It also provides some level of awareness in that it determines pleasant and unpleasant stimuli and is involved in sleep and arousal. The hypothalamus is also involved in sleep and arousal and is important in emotions such as anger, fear, pleasure, hunger, sexual response, and pain. Several other more automatic functions are regulated in this region, such as body temperature, blood pressure, and water balance. The hypothalamus also is connected to the pituitary gland and influences the manufacture and release of its hormones. Figure 20.6 shows the relationship of the various, more primitive parts of the brain.

Figure 20.6

Functions of the More Primitive Brain Regions

The brain is organized into several levels of function. The more primitive regions of the brain connected to the spinal cord monitor and manage many essential functions automatically. Some of the major functions are shown in this drawing.

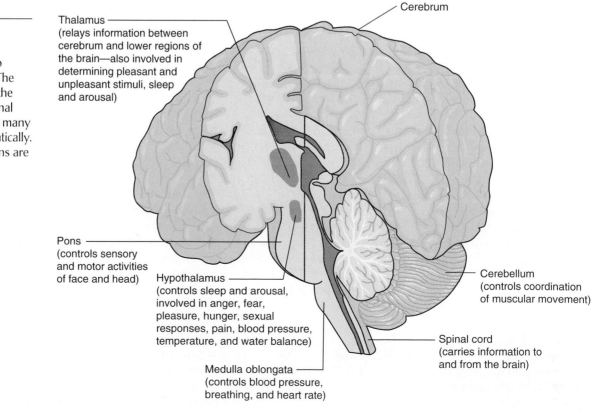

Thalamus (relays information between cerebrum and lower regions of the brain—also involved in determining pleasant and unpleasant stimuli, sleep and arousal)

Cerebrum

Pons (controls sensory and motor activities of face and head)

Hypothalamus (controls sleep and arousal, involved in anger, fear, pleasure, hunger, sexual responses, pain, blood pressure, temperature, and water balance)

Medulla oblongata (controls blood pressure, breathing, and heart rate)

Cerebellum (controls coordination of muscular movement)

Spinal cord (carries information to and from the brain)

The **cerebrum** is the thinking part of the brain. It is also the largest portion of the brain in humans. The two hemispheres of the cerebrum cover all other portions of the brain except the cerebellum. The surface of the cerebrum has been extensively mapped so that we know the location of many functions. Abilities such as memory, language, control of movement, interpretation of sensory input, and thought are associated with specific areas of the cerebrum. Figure 20.7 shows a diagram of the cerebrum and the locations of specific functions.

The function of the brain is not determined by structure alone. Many parts of the brain have specialized neurons that produce specific neurotransmitter molecules used only to stimulate particular nerve cells that have the proper receptor sites. As we learn more about the functioning of the brain, we are finding more kinds of specialized neurotransmitter molecules. Their discovery allows for the treatment of many types of mental and emotional diseases. Manipulating these neurotransmitter molecules can help correct inappropriate functioning of the brain. However, one should not assume that we understand the brain. We are still at an early stage in our search to comprehend this organ that sets us apart from other animals (How Science Works 20.1).

20.3 Endocrine System Function

As mentioned previously, the endocrine system is basically a broadcasting system in which glands secrete messenger molecules, called hormones, that are distributed throughout the body by the circulatory system (figure 20.8). However, each kind of hormone affects only certain cells. The specific cells that a particular hormone affects are often called **target cells.** The hormones target certain cells because the cells have specific receptor molecules on their surfaces to which particular hormones attach. The cells that receive the messages typically respond in one of three ways: (1) Some cells release products that have been previously manufactured; (2) other cells are stimulated to synthesize molecules or to begin metabolic activities; and (3) some are stimulated to divide and grow.

These different kinds of responses mean that some endocrine responses are relatively rapid, whereas others are very slow. For example, the release of the hormones **epinephrine** and **norepinephrine*** from the adrenal medulla, located near the kidney, causes a rapid change in the behavior

---

*Epinephrine and norepinephrine were formerly called adrenalin and noradrenalin.

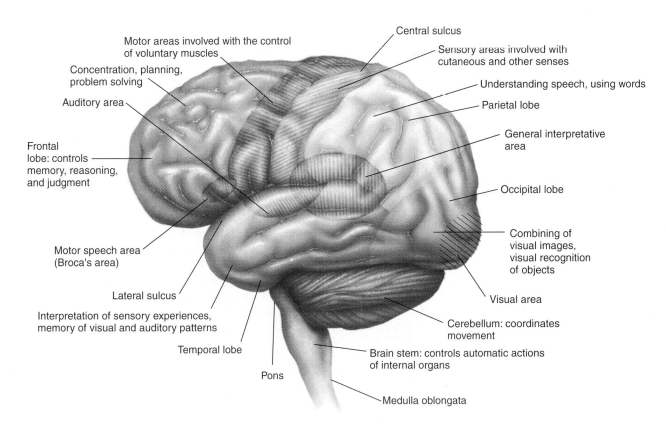

Figure 20.7

Specialized Areas of the Cerebrum
Each portion of the cerebrum has particular functions. This drawing identifies the general regions and their associated functions.

HOW SCIENCE WORKS 20.1

How Do We Know What the Brain Does?

Today we know a great deal about the function of the brain, although there is still much more to learn. Certain functions have been identified as residing in specific portions of the brain as a result of many different kinds of studies over the last century. For example, persons who have had specific portions of their brains altered by damage from accidents or strokes have been studied. Their changes in behavior or the way they perceive things can be directly correlated to the portion of the brain that was damaged. During surgeries that require that the brain be exposed, a local anesthetic can be given and the patient can be conscious while the surgery is taking place. (The brain perceives pain from pain receptors throughout the body, however, because the brain does not have many pain receptors within it, touching or manipulating the brain does not cause pain to be perceived.) Specific portions of the brain can be stimulated and the patient can be asked to describe the sensations they have or their motor functions can be observed.

Many kinds of experiments have also been done with animals in which specific portions of the brain are destroyed and the changes in the behavior of an animal are noted. Electrodes

have also been inserted into the brains of animals to stimulate certain portions of the brain.

More recently, techniques have been developed to make observations of changes in electrical activity of specific portions of the brain, without requiring electrodes or other invasive procedures. This allows researchers to present stimuli to human subjects and determine which parts of the brain alter their activity. In addition to localizing the part of the brain that responds, it is also possible to determine what parts of a complex stimulus are most important in changing brain activity. Some of the new bits of information gained by these techniques allow us to understand that languages learned by adults are processed in different places in the brain than the languages they learned as children and that the brain has a built-in mechanism for recognizing unexpected words or musical notes.

Although we have learned much about the brain, there is still much to learn. Current experiments are seeking ways to regenerate nerve cells that have been damaged. A better understanding of the chemical events that take place in the brain would enable us to cure many kinds of debilitating mental illnesses.

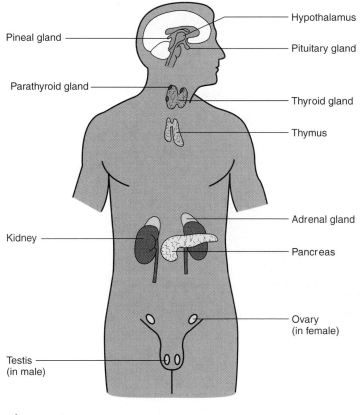

Figure 20.8

Endocrine Glands
The endocrine glands are located at various locations within the body and cause their effects by secreting hormones.

of an organism. The heart rate increases, blood pressure rises, blood is shunted to muscles, and the breathing rate increases. You have certainly experienced this reaction many times in your lifetime, such as when you nearly had an automobile accident or slipped and nearly fell.

Another hormone, called **antidiuretic hormone (ADH)**, acts more slowly. It is released from the posterior pituitary gland at the base of the brain and regulates the rate at which the body loses water through the kidneys. It does this by encouraging the reabsorption of water from the collecting ducts of the kidneys (see chapter 18). The effects of this hormone can be noticed in a matter of minutes to hours. Insulin is another hormone whose effects are quite rapid. Insulin is produced by the pancreas, located near the stomach, and stimulates cells—particularly muscle, liver, and fat cells—to take up glucose from the blood. After a high-carbohydrate meal, the level of glucose in the blood begins to rise, stimulating the pancreas to release insulin. The increased insulin causes glucose levels to fall as the sugar is taken up by cells. People with diabetes have insufficient or improperly acting insulin or lack the receptors to respond to the insulin, and therefore have difficulty regulating glucose levels in their blood.

The responses that result from the growth of cells may take weeks or years to occur. For example, **growth-stimulating hormone (GSH)** is produced by the anterior pituitary gland over a period of years and results in typical human growth. After sexual maturity, the amount of this hormone generally drops to very low levels, and body growth stops. Sexual development is also largely the result of the growth of specific tissues and organs. The male sex hormone **testosterone,** produced by the testes, causes the growth of male sex organs and a change to the adult body form. The female counterpart, **estrogen,** results in the development of female sex organs and body form. In all of these cases, it is the release of hormones over long periods, continually stimulating the growth of sensitive tissues, that results in a normal developmental pattern. The absence or inhibition of any of these hormones early in life changes the normal growth process.

Negative Feedback Control

Glands within the endocrine system typically interact with one another and control production of hormones. One common control mechanism is called *negative-feedback control.* In **negative-feedback control,** the increased amount of one hormone interferes with the production of a different hormone in the chain of events. The production of **thyroxine** and **triiodothyronine** by the thyroid gland exemplifies this kind of control. The production of these two hormones is stimulated by increased production of a hormone from the anterior pituitary called **thyroid-stimulating hormone (TSH).** The control lies in the quantity of the hormone produced. When the anterior pituitary produces high levels of thyroid-stimulating hormone, the thyroid is stimulated to grow and secrete more thyroxine and triiodothyronine. But when increased amounts of thyroxine and triiodothyronine are produced, these hormones have a negative effect on the pituitary so that it decreases its production of thyroid-stimulating hormone, leading to reduced production of thyroxine and triiodothyronine. If the amount of the thyroid hormones falls too low, the pituitary is no longer inhibited and releases additional thyroid-stimulating hormone. As a result of the interaction of these hormones, their concentrations are maintained within certain limits (figure 20.9).

20.4 Integration of Nervous and Endocrine Function

It is possible for the nervous and endocrine systems to interact (How Science Works 20.2). The pituitary gland is located at the base of the brain and is divided into two parts. The posterior pituitary is directly connected to the brain and it develops from nervous tissue during embryology. The other part, the anterior pituitary, develops from the lining of the roof of the mouth in early fetal development. Certain pituitary hormones are produced in the brain and transported down axons to the posterior pituitary where they are stored before being released. The anterior pituitary also receives a continuous input of messenger molecules from the brain, but these are delivered by way of a special set of blood vessels that pick up hormones produced by the hypothalamus of the brain and deliver them to the anterior pituitary.

The pituitary gland produces a variety of hormones that are responsible for causing other endocrine glands, such as the thyroid, ovaries and testes, and adrenals, to secrete

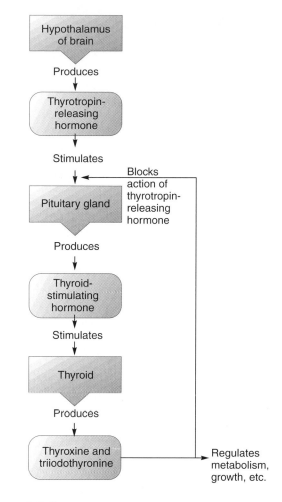

Figure 20.9

Negative-Feedback Control of Thyroid Hormone Levels
The levels of thyroxine and triiodothyronine increase and decrease in response to the amount of thyroid-stimulating hormone present. Increased levels of the thyroid hormones cause the pituitary to not respond to thyrotropin-releasing hormone, and the production of thyroid-stimulating hormone falls, so that eventually the thyroid hormone levels decrease. This is an example of negative-feedback control.

their hormones. Pituitary hormones also influence milk production, skin pigmentation, body growth, mineral regulation, and blood glucose levels (figure 20.10).

Because the pituitary is constantly receiving information from the hypothalamus of the brain, many kinds of sensory stimuli to the body can affect the functioning of the endocrine system. One example is the way in which the nervous system and endocrine system interact to influence the menstrual cycle. At least three different hormones are involved in the cycle of changes that affect the ovary and the lining of the uterus (see chapter 21 for details). It is well documented that stress caused by tension or worry can interfere with the normal cycle of hormones and delay or stop menstrual cycles. In addition, young women living in groups, such as in college dormitories, often find that their menstrual

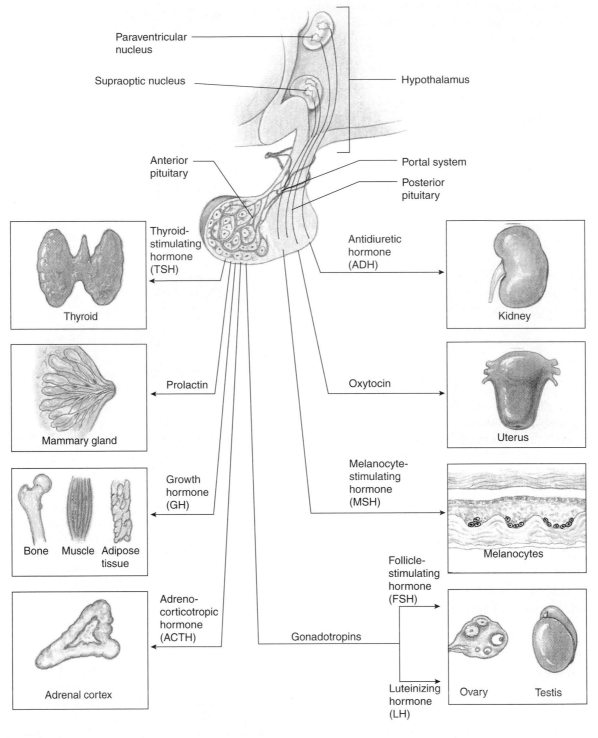

Paraventricular nucleus

Supraoptic nucleus

Hypothalamus

Anterior pituitary

Portal system

Posterior pituitary

Thyroid-stimulating hormone (TSH)

Thyroid

Antidiuretic hormone (ADH)

Kidney

Prolactin

Mammary gland

Oxytocin

Uterus

Growth hormone (GH)

Bone   Muscle   Adipose tissue

Melanocyte-stimulating hormone (MSH)

Melanocytes

Adreno-corticotropic hormone (ACTH)

Gonadotropins

Follicle-stimulating hormone (FSH)

Luteinizing hormone (LH)

Adrenal cortex

Ovary   Testis

Figure 20.10

Hormones of the Pituitary
The anterior pituitary gland produces several hormones that regulate growth and the secretions of target tissues. The posterior pituitary produces hormones that change the behavior of the kidney and uterus but do not influence the growth of these organs.

HOW SCIENCE WORKS 20.2

The Endorphins: Natural Pain Killers

The pituitary gland and brain produce a group of small molecules that act as pain suppressors. These are the *endorphins*. It is thought that these molecules are released when excessive pain or stress occurs in the body. They attach to the same receptor molecules of brain cells associated with the feeling of pain (see figure). The endorphins work on the brain in the same manner as morphine and opiate drugs. Once attached, the feeling of pain goes away, and a euphoric feeling takes over. Long-distance runners and other athletes talk about "feeling good" once they have "reached their stride," get their "second wind," or experience a "runner's high." These responses may be due to an increase in endorphin production. It is thought that endorphins are also released by mild electric stimulation or the use of acupuncture needles.

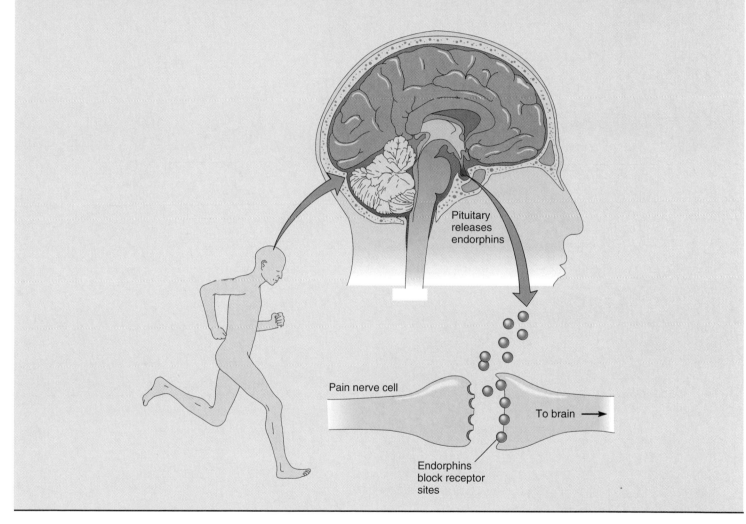

Pituitary releases endorphins

Pain nerve cell

To brain →

Endorphins block receptor sites

cycles become synchronized. Although the exact mechanism involved in this phenomenon is unknown, it is suspected that input from the nervous system causes this synchronization. (Odors and sympathetic feelings have been suggested as causes.)

In many animals, the changing length of the day causes hormonal changes related to reproduction. In the spring, birds respond to lengthening days and begin to produce hormones that gear up their reproductive systems for the summer breeding season. The pineal body, a portion of the brain, serves as the receiver of light stimuli and changes the amounts of hormones secreted by the pituitary, resulting in changes in the levels of reproductive hormones. These hormonal changes modify the behavior of birds. Courtship, mating, and nest-building behaviors increase in intensity. Therefore, it appears that a change in hormone level is affecting the behavior of the animal; the endocrine system is influencing the nervous system (figure 20.11).

It has been known for centuries that changes in the levels of sex hormones cause changes in the behavior of animals. Castration (removal of the testes) of male domesticated animals, such as cattle, horses, and pigs, is sometimes done

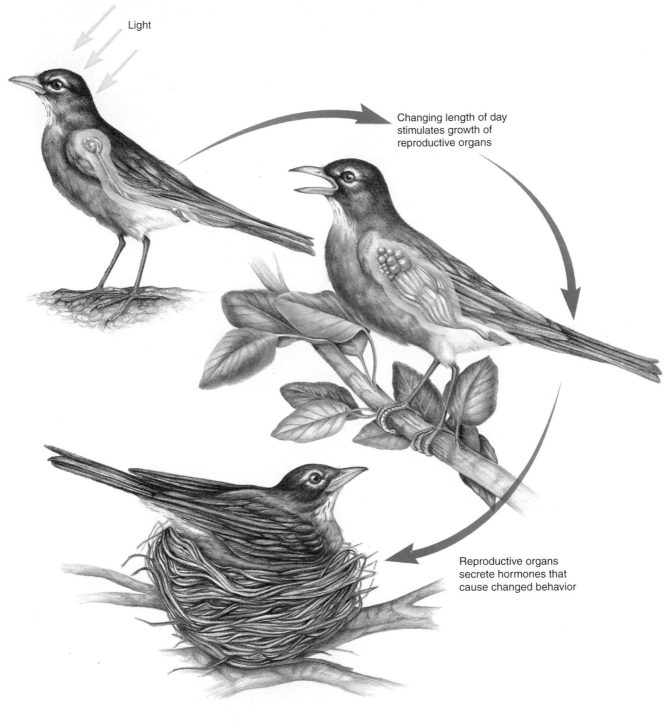

Light

Changing length of day
stimulates growth of
reproductive organs

Reproductive organs
secrete hormones that
cause changed behavior

Figure 20.11

Interaction Between the Nervous and Endocrine Systems
In birds and many other animals, the brain receives information about the changing length of day, which causes the pituitary to produce hormones that stimulate sexual development. The testes or ovaries grow and secrete their hormones in increased amounts. Increased levels of testosterone or estrogen result in changed behavior, with increased mating, aggression, and nest-building activity.

in part to reduce their aggressive behavior and make them easier to control. In humans, the use of anabolic steroids to increase muscle mass is known to cause behavioral changes and "moodiness."

Although we still tend to think of the nervous and endocrine systems as being separate and different, it is becoming clear that they are interconnected. As we learn more about the molecules produced in the brain, it is becoming clear that the brain produces many molecules that act as hormones. Some of these molecules affect adjacent parts of the brain, others affect the pituitary, and still others may have effects on more distant organs. In any case, these two systems cooperate to bring about appropriate responses to environmental challenges. The nervous system is specialized for receiving and sending short-term messages, whereas activities that require long-term, growth-related actions are handled by the endocrine system.

20.5 Sensory Input

The activities of the nervous and endocrine systems are often responses to some kind of input received from the sense organs. Sense organs of various types are located throughout the body. Many of them are located on the surface, where environmental changes can be easily detected. Hearing, sight, and touch are good examples of such senses. Other sense organs are located within the body and indicate to the organism how its various parts are changing. For example, pain and pressure are often used to monitor internal conditions. The sense organs detect changes, but the brain is responsible for **perception**—the recognition that a stimulus has been received. Sensory abilities involve many different kinds of mechanisms, including chemical recognition, the detection of energy changes, and the monitoring of physical forces.

Chemical Detection

All cells have receptors on their surfaces that can bind selectively to molecules they encounter. This binding process can cause changes in the cells in several ways. In some cells it causes depolarization. When this happens, the binding of molecules to the cell can stimulate neurons and cause messages to be sent to the central nervous system, informing it of some change in the surroundings. Many internal sense organs respond to specific molecules. For example, the brain and aorta contain cells that respond to concentrations of hydrogen ions, carbon dioxide, and oxygen in the blood. In other cases, a molecule binding to the cell surface may cause certain genes to be expressed, and the cell responds by changing the molecules it produces. This is typical of the way the endocrine system receives and delivers messages.

Taste

Most cells have specific binding sites for particular molecules. Others, such as the taste buds on the tongue, appear to respond to classes of molecules. Traditionally we have distinguished four kinds of tastes: sweet, sour, salt, and bitter. However, recently, a fifth kind of taste, *umami* (meaty), has been identified. *Umami* receptors respond to the amino acid glutamate, which is present in many kinds of foods and is added as a flavor enhancer (monosodium glutamate) to many kinds of foods.

The taste buds that give us the sour sensation respond to the presence of hydrogen ions (H^+). (Acidic foods taste sour.) The hydrogen ions stimulate the cells in two ways: they enter the cell directly or they alter the normal movement of sodium and potassium ions across the cell membrane. In either case, the cell depolarizes and stimulates a nerve cell. In similar fashion, sodium chloride stimulates the taste buds that give us the sensation of a salty taste by directly entering the cell, which causes the cell to depolarize.

However, the sensations of sweetness, bitterness, and *umami* occur when molecules bind to specific surface receptors on the cell. Sweetness can be stimulated by many kinds of organic molecules, including sugars and artificial sweeteners, and also by inorganic lead compounds. When a molecule binds to a sweetness receptor, a molecule is split and its splitting stimulates an enzyme that leads to the depolarization of the cell. The sweet taste of lead salts in old paints partly explains why children sometimes eat paint chips. Because the lead interferes with normal brain development, this behavior can have disastrous results. Many other kinds of compounds of diverse structures give the bitter sensation. The cells that respond to bitter sensations have a variety of receptor molecules on their surface. When a substance binds to one of the receptors, the cell depolarizes. In the case of *umami*, it is the glutamate molecule that binds to receptors on the cells of the taste buds.

Each of these tastes has a significance from an evolutionary point of view. Carbohydrates are a major food source and many carbohydrates taste sweet, therefore, this sense would be useful in identifying foods that have high food value. Similarly, proteins and salts are necessary in the diet. Therefore, being able to identify these items in potential foods would be extremely valuable. This is particularly true for salt, which must often be obtained from mineral sources. On the other hand, bitter and sour materials are often harmful. Many plants produce toxic materials that are bitter tasting and acids are often the result of bacterial decomposition (spoiling) of foods. Being able to identify bitter and sour would allow organisms to avoid foods that would be harmful.

It is also important to understand that much of what we often refer to as *taste* involves such inputs as temperature, texture, and smell. Cold coffee has a different taste than hot coffee even though they are chemically the same. Lumpy, cooked cereal and smooth cereal have different tastes. If you are unable to smell food, it doesn't taste as it should, which is why you sometimes lose your appetite when you have a stuffy nose. We still have much to learn about how the tongue detects chemicals and the role other associated senses play in modifying taste.

Figure 20.12

The Structure of the Eye

The eye contains a cornea and lens that focus the light on the retina of the eye. The light causes pigments in the rods and cones of the retina to decompose. This leads to the depolarization of these cells and the stimulation of neurons that send messages to the brain.

Smell

The other major chemical sense, the sense of smell, is much more versatile; it can detect thousands of different molecules at very low concentrations. The cells that make up the **olfactory epithelium,** the cells that line the nasal cavity and respond to smells, apparently bind molecules to receptors on their surfaces. Exactly how this can account for the large number of recognizably different odors is unknown, but the receptor cells are extremely sensitive. In some cases a single molecule of a substance is sufficient to cause a receptor cell to send a message to the brain, where the sensation of odor is perceived. These sensory cells also fatigue rapidly. You have probably noticed that when you first walk into a room, specific odors are readily detected, but after a few minutes you are unable to detect them. Most perfumes and aftershaves are undetectable after 15 minutes of continuous stimulation.

Vision

The eyes primarily respond to changes in the flow of light energy. The structure of the eye is designed to focus light on a light-sensitive layer of the back of the eye known as the **retina** (figure 20.12). There are two kinds of receptors in the retina of the eye. The cells called **rods** respond to a broad range of wavelengths of light and are responsible for black-and-white vision. Because rods are very sensitive to light, they are particularly useful in dim light. Rods are located over most of the retinal surface except for the area of most acute vision known as the **fovea centralis.** The other receptor cells, called **cones,** are not as sensitive to light, but they can detect different wavelengths of light. They are found throughout the retina but are particularly concentrated in the fovea centralis. This combination of receptors gives us

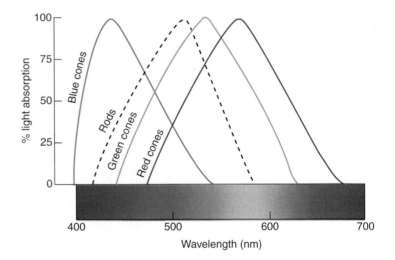

Figure 20.13

Light Reception by Cones

There are three different kinds of cones that respond differently to red, green, and blue wavelengths of light. Stimulation of combinations of these three kinds of cones gives us the ability to detect many different shades of color.

Sources: Data from W.B. Marks, W.H. Dobelle, and E.F. MacNichol, "Visual Pigments of Single Primate Cones," *Science* 143 (1964):45–52; and P.K. Brown and G. Wald, "Visual Pigments in Single Rods and Cones of the Human Retina," *Science* 144 (1964):45–52.

the ability to detect color when light levels are high, but we rely on black-and-white vision at night. There are three different varieties of cones: One type responds best to red light, another responds best to green light, and the third responds best to blue light. Stimulation of various combinations of these three kinds of cones allows us to detect different shades of color (figure 20.13).

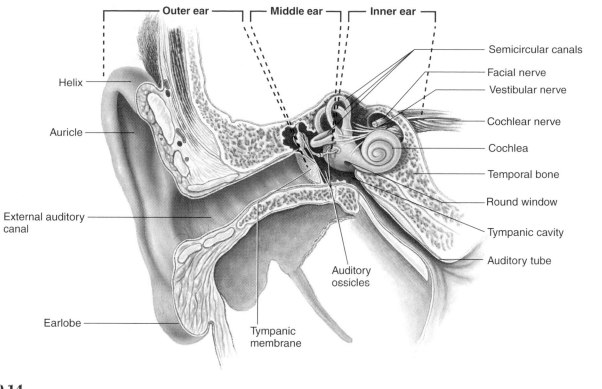

Figure 20.14

The Anatomy of the Ear
The ear consists of an external cone that directs sound waves to the tympanum. Vibrations of the tympanum move the ear bones and vibrate the oval window of the cochlea, where the sound is detected. The semicircular canals monitor changes in the position of the head, helping us maintain balance.

Rods and the three different kinds of cones each contain a pigment that decomposes when struck by light of the proper wavelength and sufficient strength. The pigment found in rods is called **rhodopsin.** This change in the structure of rhodopsin causes the rod to depolarize. Cone cells have a similar mechanism of action, and each of the three kinds of cones has a different pigment. Because rods and cones synapse with neurons, they stimulate a neuron when depolarized and cause a message to be sent to the brain. Thus the pattern of color and light intensity recorded on the retina is detected by rods and cones and converted into a series of nerve impulses that are received and interpreted by the brain.

Hearing and Balance

The ears respond to changes in sound waves. Sound is produced by the vibration of molecules. Consequently, the ears are detecting changes in the quantity of energy and the quality of sound waves. Sound has several characteristics. Loudness, or volume, is a measure of the intensity of sound energy that arrives at the ear. Very loud sounds will literally vibrate your body, and can cause hearing loss if they are too intense. Pitch is a quality of sound that is determined by the frequency of the sound vibrations. High-pitched sounds have short wavelengths; low-pitched sounds have long wavelengths.

Figure 20.14 shows the anatomy of the ear. The sound that arrives at the ear is first funneled by the external ear to the **tympanum,** also known as the *eardrum.* The cone-shaped nature of the external ear focuses sound on the tympanum and causes it to vibrate at the same frequency as the sound waves reaching it. Attached to the tympanum are three tiny bones known as the **malleus** (hammer), **incus** (anvil), and **stapes** (stirrup). The malleus is attached to the tympanum, the incus is attached to the malleus and stapes, and the stapes is attached to a small, membrane-covered opening called the **oval window** in a snail-shaped structure known as the *cochlea.* The vibration of the tympanum causes the tiny bones (malleus, incus, and stapes) to vibrate, and they in turn cause a corresponding vibration in the membrane of the oval window.

The **cochlea** of the ear is the structure that detects sound and consists of a snail-shaped set of fluid-filled tubes. When the oval window vibrates, the fluid in the cochlea begins to move, causing a membrane in the cochlea, called the **basilar membrane,** to vibrate. High-pitched, short-wavelength sounds cause the basilar membrane to vibrate at the base of the cochlea near the oval window. Low-pitched, long-wavelength sounds vibrate the basilar membrane far from the oval window. Loud sounds cause the basilar membrane to vibrate more vigorously than do faint sounds. Cells

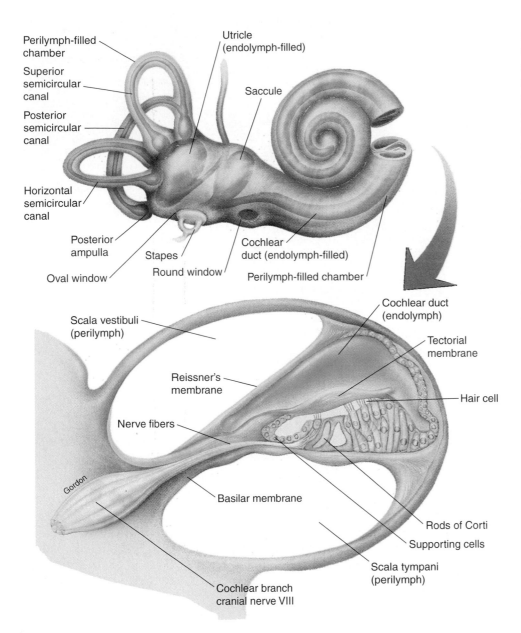

Perilymph-filled chamber

Superior semicircular canal

Posterior semicircular canal

Horizontal semicircular canal

Posterior ampulla

Oval window

Stapes

Round window

Utricle (endolymph-filled)

Saccule

Cochlear duct (endolymph-filled)

Perilymph-filled chamber

Scala vestibuli (perilymph)

Reissner's membrane

Nerve fibers

Gordon

Basilar membrane

Cochlear branch cranial nerve VIII

Cochlear duct (endolymph)

Tectorial membrane

Hair cell

Rods of Corti

Supporting cells

Scala tympani (perilymph)

Figure 20.15

The Basilar Membrane

The cells that respond to vibrations and stimulate neurons are located in the cochlea. Vibrations of the oval window cause the fluid in the cochlea to vibrate, and the basilar membrane moves also. This movement causes the receptor cells to depolarize and send a message to the brain.

Associated with the cochlea are two fluid-filled chambers and a set of fluid-filled tubes called the **semicircular canals**. These structures are not involved in hearing but are involved in maintaining balance and posture. In the walls of these canals and chambers are cells similar to those found on the basilar membrane. These cells are stimulated by movements of the head and by the position of the head with respect to the force of gravity. The constantly changing position of the head results in sensory input that is important in maintaining balance.

Touch

What we normally call the sense of *touch* consists of a variety of different kinds of input. Some receptors respond to pressure, others to temperature, and others, which we call *pain receptors,* usually respond to cell damage. When these receptors are appropriately stimulated, they send a message to the brain. Because receptors are stimulated in particular parts of the body, the brain is able to localize the sensation. However, not all parts of the body are equally supplied with these receptors. The tips of the fingers, lips, and external genitals have the highest density of these nerve endings, whereas the back, legs, and arms have far fewer receptors.

Some internal receptors, such as pain and pressure receptors, are important in allowing us to monitor our internal activities. Many pains generated by the internal organs are often perceived as if they were somewhere else. For example, the pain associated with heart attack is often perceived to be in the left arm. Pressure receptors in joints and muscles

on this membrane depolarize when they are stimulated by its vibrations. Because they synapse with neurons, messages can be sent to the brain (figure 20.15).

Because sounds of different wavelengths stimulate different portions of the cochlea, the brain is able to determine the pitch of a sound. Most sounds consist of a mixture of pitches. Louder sounds stimulate the membrane more forcefully, causing the sensory cells in the cochlea to send more nerve impulses per second. Thus the brain is able to perceive the loudness of various sounds as well as the pitch.

are important in providing information about the degree of stress being placed on a portion of the body. This is also important information to send back to the brain so that adjustments can be made in movements to maintain posture. If you have ever had your foot "go to sleep" because the nerve stopped functioning, you have experienced what it is like to lose this constant input of nerve messages from the pressure sensors that assist in guiding the movements you make. Your movements become uncoordinated until the nerve function returns to normal.

20.6 Output Coordination

The nervous system and endocrine system cause changes in several ways. Both systems can stimulate muscles to contract and glands to secrete. The endocrine system is also able to change the metabolism of cells and regulate the growth of tissues. The nervous system acts upon two kinds of organs: muscles and glands. The actions of muscles and glands are simple and direct: muscles contract and glands secrete.

Muscular Contraction

The ability to move is one of the fundamental characteristics of animals. Through the coordinated contraction of many muscles, the intricate, precise movements of a dancer, basketball player, or writer are accomplished. It is important to recognize that muscles can pull only by contracting; they are unable to push by lengthening. The work of any muscle is done during its contraction. Relaxation is the passive state of the muscle. There must always be some force available that will stretch a muscle after it has stopped contracting and relaxes. Therefore, the muscles that control the movements of the skeleton are present in antagonistic sets—for every muscle's action there is another muscle that has the opposite action. For example, the biceps muscle causes the arm to flex (bend) as the muscle shortens. The contraction of its antagonist, the triceps muscle, causes the arm to extend (straighten) and at the same time stretches the relaxed biceps muscle (figure 20.16).

What we recognize as a muscle is composed of many muscle cells, which are in turn made up of *myofibrils* that are composed of two kinds of *myofilaments* (figure 20.17). The mechanism by which muscle contracts is well understood and involves the movement of protein filaments past one another as ATP is utilized. ATP (adenosine triphosphate) is the primary molecule used by cells for their immediate energy needs. The filaments in muscle cells are of two types, arranged in a particular pattern. Thin filaments composed of the proteins **actin, tropomyosin,** and **troponin** alternate with thick filaments composed primarily of a protein known as **myosin** (figure 20.18).

The myosin molecules have a shape similar to a golf club. The head of the club-shaped molecule sticks out from the thick filament and can combine with the actin of the thin filament. However, the troponin and tropomyosin proteins associated with the actin cover the actin in such a way that myosin cannot bind with it. When actin is uncovered, myosin can bind to it and contraction of a muscle will occur when ATP is utilized.

The process of muscle-cell contraction involves several steps. When a nerve impulse arrives at a muscle cell, its arrival causes the muscle cell to depolarize. When muscle cells depolarize, calcium ions (Ca^{2+}) contained within membranes are released among the actin and myosin filaments. The calcium ions (Ca^{2+}) combine with the troponin molecules, caus-

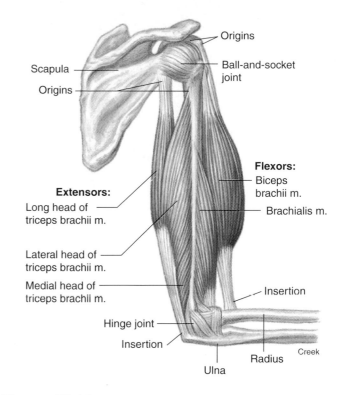

Figure 20.16

Antagonistic Muscles
Because muscles cannot actively lengthen, it is necessary to have sets of muscles that oppose one another. The contraction and shortening of one muscle cause the stretching of a relaxed muscle.

ing the troponin-tropomyosin complex to expose actin so that it can bind with myosin. While the actin and myosin molecules are attached, the head of the myosin molecule can flex as ATP is used and the actin molecule is pulled past the myosin molecule. Thus, a tiny section of the muscle cell shortens (figure 20.19). When one of our muscles contracts, thousands of such interactions take place within a tiny portion of a muscle cell, and many cells within a muscle all contract at the same time.

Types of Muscle

There are three major types of muscle: skeletal, smooth, and cardiac. These differ from one another in several ways. *Skeletal muscle* is voluntary muscle; it is under the control of the nervous system. The brain or spinal cord sends a message to skeletal muscles, and they contract to move the legs, fingers, and other parts of the body. This does not mean that you must make a conscious decision every time you want to move a muscle. Many of the movements we make are learned initially but become automatic as a result of practice. For example, walking, swimming, or riding a bicycle required a great

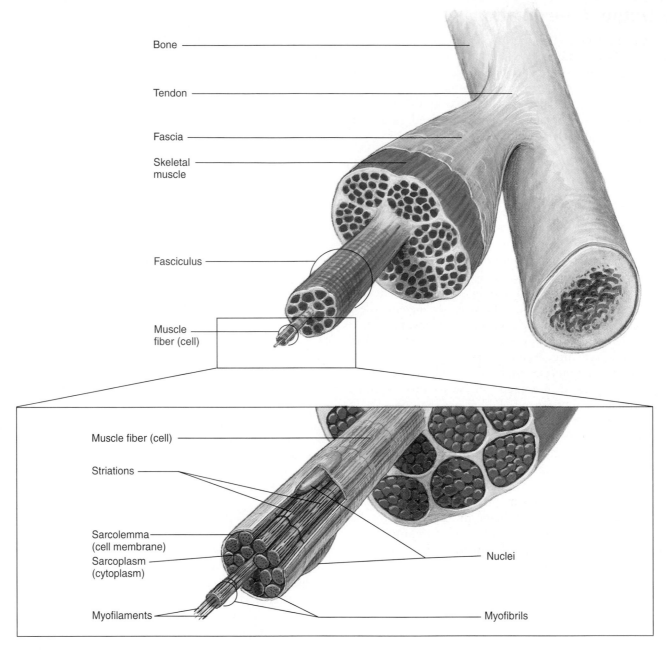

Figure 20.17

The Microanatomy of a Muscle
Muscles are made of cells that contain bundles known as myofibrils. The myofibrils are composed of myofilaments of two different kinds: thick myofilaments composed of myosin, and thin myofilaments containing actin, troponin, and tropomyosin.

amount of practice originally, but now you probably perform these movements without thinking about them. They are, however, still considered voluntary actions.

Skeletal muscles are constantly bombarded with nerve impulses that result in repeated contractions of differing strength. Many neurons end in each muscle, and each one stimulates a specific set of muscle cells called a **motor unit** (figure 20.20). Because each muscle consists of many motor units, it is possible to have a wide variety of intensities of contraction within one muscle organ. This allows a single set of muscles to serve a wide variety of functions. For example, the same muscles of the arms and shoulders that are used to play a piano can be used in other combinations to tightly grip and throw a baseball. If the nerves going to a muscle are destroyed, the muscle becomes paralyzed and begins to shrink. Regular nervous stimulation of skeletal muscle is nec-

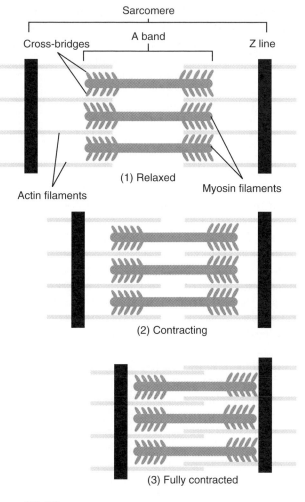

Figure 20.18

The Subcellular Structure of Muscle

The actin- and myosin-containing myofilaments are arranged in a regular fashion into units called sarcomeres. Each sarcomere consists of two sets of actin-containing myofilaments inserted into either end of bundles of myosin-containing myofilaments. The actin-containing myofilaments slide past the myosin-containing myofilaments, shortening the sarcomere.

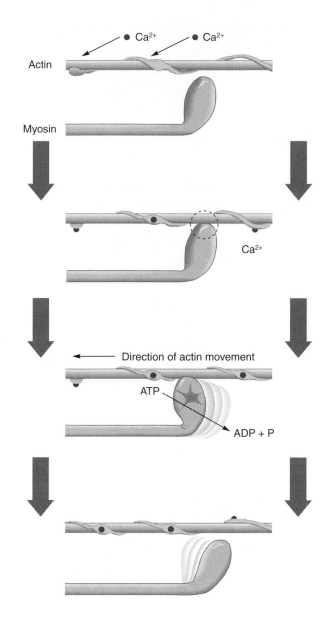

Figure 20.19

Interaction Between Actin and Myosin

When calcium ions (Ca^{2+}) enter the region of the muscle cell containing actin and myosin, they allow the actin and myosin to bind to each other. ATP is broken to ADP and P with the release of energy. This energy allows the club-shaped head of the myosin to flex and move the actin along, causing the two molecules to slide past each other.

essary for muscle to maintain size and strength. Any kind of prolonged inactivity leads to the degeneration of muscles known as atrophy. Muscle maintenance is one of the primary functions of physical therapy and a benefit of regular exercise.

Skeletal muscles are able to contract quickly, but they cannot remain contracted for long periods. Even when we contract a muscle for a minute or so, the muscle is constantly shifting the individual motor units within it that are in a state of contraction. A single skeletal muscle cell cannot stay in a contracted state.

Smooth muscles make up the walls of muscular internal organs, such as the gut, blood vessels, and reproductive organs. They have the property of contracting as a response to

being stretched. Because much of the digestive system is being stretched constantly, the responsive contractions contribute to the normal rhythmic movements associated with the digestive system. These are involuntary muscles; they can contract on their own without receiving direct messages from the nervous system. This can be demonstrated by removing portions

Axon 1   Axon 2

Motor unit   Motor unit

Figure 20.20

Motor Units
When a skeletal muscle contracts, individual groups of muscle cells are stimulated by specific nerve cells. They contract momentarily, then relax. The strength of contraction is determined by the number of motor units stimulated. The sequence of movements is determined by the order in which motor units are stimulated.

of the gut or uterus from experimental animals. When these muscular organs are kept moist with special solutions, they go through cycles of contraction without any possible stimulation from neurons. However, they do receive nervous stimulation, which can modify the rate and strength of their contraction. This kind of muscle also has the ability to stay contracted for long periods without becoming fatigued. Many kinds of smooth muscle, such as the muscle of the uterus, also respond to the presence of hormones. Specifically, the hormone **oxytocin,** which is released from the posterior pituitary, causes strong contractions of the uterus during labor and birth. Similarly, several hormones produced by the duodenum influence certain muscles of the digestive system to either contract or relax.

Cardiac muscle is the muscle that makes up the heart. It has the ability to contract rapidly like skeletal muscle, but does not require nervous stimulation to do so. Nervous stimulation can, however, cause the heart to speed or slow its rate of contraction. Hormones, such as epinephrine and norepinephrine, also influence the heart by increasing its rate

and strength of contraction. Cardiac muscle also has the characteristic of being unable to stay contracted. It will contract quickly but must have a short period of relaxation before it will be able to contract a second time. This makes sense in light of its continuous, rhythmic, pumping function. Table 20.1 summarizes the differences among skeletal, smooth, and cardiac muscles.

Activities of Glands

The glands of the body are of two different kinds. Those that secrete into the bloodstream are called endocrine glands. We have already talked about several of these: the pituitary, thyroid, ovary, and testis are examples. The exocrine glands are those that secrete to the surface of the body or into one of the tubular organs of the body, such as the gut or reproductive tract. Examples are the salivary glands, intestinal mucus glands, and sweat glands. Some of these glands, such as salivary glands and sweat glands, are under nervous control. When stimulated by the nervous system, they secrete their contents.

The Russian physiologist Ivan Petrovich Pavlov showed that salivary glands were under the control of the nervous system when he trained dogs to salivate in response to hearing a bell. You may recall from chapter 17 that, initially, the animals were presented with food at the same time the bell was rung. Eventually they would salivate when the bell was rung even if food was not present. This demonstrated that saliva release was under the control of the central nervous system.

Many other exocrine glands are under hormonal control. Many of the digestive enzymes of the stomach and intestine are secreted in response to local hormones produced in the gut. These are circulated through the blood to the digestive glands, which respond by secreting the appropriate digestive enzymes and other molecules.

Table 20.1

CHARACTERISTICS OF DIFFERENT KINDS OF MUSCLE

Kind of Muscle	Stimulus	Length of Contraction	Rapidity of Response
Skeletal	Nervous system	Short, tires quickly	Most rapid
Smooth	1. Self-stimulated 2. Also responds to nervous and endocrine systems	Long, doesn't tire quickly	Slow
Cardiac	1. Self-stimulated 2. Also responds to nervous and endocrine systems	Short, cannot stay contracted	Rapid

Table 20.2

COMPARISON OF THE NERVOUS AND ENDOCRINE SYSTEMS

System	Kind of Action	Effects
Nervous	1. Nerve impulse travels along established routes. 2. Neurotransmitters allow impulse to cross synapses. 3. Rapid action.	1. Causes skeletal muscle contraction. 2 Modifies contraction of smooth and cardiac muscles. 3. Causes gland secretion.
Endocrine	1. Hormones released into bloodstream. 2. Receptors bind hormones to their target organs. 3. Often slow to act.	1. Stimulates smooth muscle contraction. 2. Stimulates gland secretion. 3. Regulates growth.

Growth Responses

The hormones produced by the endocrine system can have a variety of effects. As mentioned earlier, hormones can stimulate smooth muscle to contract and can influence the contraction of cardiac muscle as well. Many kinds of glands, both endocrine and exocrine, are caused to secrete as a result of a hormonal stimulus. However, the endocrine system has one major effect that is not equaled by the nervous system: Hormones regulate growth. Several examples of the many kinds of long-term growth changes that are caused by the endocrine system were given earlier in the chapter. Growth-stimulating hormone (GSH) is produced over a period of years to bring about the increase in size of most of the structures of the body. A low level of this hormone results in a person with small body size. It is important to recognize that the amount of growth-stimulating hormone (GSH) present varies from time to time. It is present in fairly high amounts throughout childhood and results in steady growth. It also appears to be present at higher levels at certain times, resulting in growth spurts. Finally, as adulthood is reached, the level of this hormone falls, and growth stops.

Similarly, testosterone produced during adolescence influences the growth of bone and muscle to provide men with larger, more muscular bodies than those of women. In addition, there is growth of the penis, growth of the larynx, and increased growth of hair on the face and body. The primary female hormone, estrogen, causes growth of reproductive organs and development of breast tissue. It is also involved, along with other hormones, in the cyclic growth and sloughing of the wall of the uterus.

SUMMARY

Throughout this chapter we have been comparing the functions of the nervous and endocrine systems, the kinds of effects they have, and their characteristics. Table 20.2 summarizes these differences.

A nerve impulse is caused by sodium ions entering the cell as a result of a change in the permeability of the cell membrane. Thus, a wave of depolarization passes down the length of a neuron to the synapse. The axon of a neuron secretes a neurotransmitter, such as acetylcholine, into the synapse, where these molecules bind to the dendrite of the next cell in the chain, resulting in an impulse in it as well. The acetylcholinesterase present in the synapse destroys acetylcholine so that it does not repeatedly stimulate the dendrite. The brain is composed of several functional units. The lower portions of the brain control automatic activities, the middle portion of the brain controls basic categorizing of sensory input, and the higher levels of the brain are involved in thinking and self-awareness.

Several kinds of sensory inputs are possible. Many kinds of chemicals can bind to cell surfaces and be recognized. This is how the sense of taste and the sense of smell function. Light energy can be detected because light causes certain molecules in the retina of the eye to decompose and stimulate neurons. Sound can be detected because fluid in the cochlea of the ear is caused to vibrate, and special cells detect this movement and stimulate neurons. The sense of touch consists of a variety of receptors that respond to pressure, cell damage, and temperature.

Muscles shorten because of the ability of actin and myosin to bind to one another. A portion of the myosin molecule is caused to bend when ATP is used, resulting in the sliding of actin and myosin molecules past each other. Skeletal muscle responds to nervous stimulation to cause movements of the skeleton. Smooth muscle and cardiac muscle have internally generated contractions that may be modified by nervous stimulation or hormones.

Glands are of two types: exocrine glands, which secrete through ducts into the cavity of an organ or to the surface of the skin, and endocrine glands, which release their secretions into the circulatory system. Digestive glands and sweat glands are examples of exocrine glands. Endocrine glands such as the ovaries, testes, and pituitary gland change the activities of cells and often cause responses that result in growth over a period of time. It is becoming clear that the endocrine system and the nervous system are interrelated. Actions of the endocrine system can change how the nervous system functions, and the reverse is also true. Much of this interrelation takes place in the brain-pituitary gland association.

THINKING CRITICALLY

For guidelines to answer this question, visit the Online Learning Center.

Humans are considered to have a poor sense of smell. However, when parents are presented with baby clothing, they are able to identify the clothing with which their own infant had been in contact with a high degree of accuracy. Specially trained individuals, such as wine and perfume testers, are able to identify large numbers of different kinds of molecules that the average person cannot identify. Birds rely primarily on sound and sight for information about their environment; they have a poor sense of smell. Most mammals are known to have a very well-developed sense of smell. Is it possible that we have evolved into sound-and-sight-dependent organisms like birds and have lost the keen sense of smell of our ancestors? Or is it that we just don't use our sense of smell to its full potential? Can you devise an experiment that would help shed light on this question?

CONCEPT MAP TERMINOLOGY

Construct a concept map using these terms.

central nervous system	peripheral nervous system
motor neurons	retina
nerve impulse	sensory neurons
neurotransmitter	skeletal muscle
perception	synapse

Visit the Online Learning Center to use an interactive format to construct another concept map.

KEY TERMS

Use interactive flash cards, on the Online Learning Center, to help you learn the meanings of these terms.

acetylcholine	exocrine glands
acetylcholinesterase	fovea centralis
actin	gland
antidiuretic hormone (ADH)	growth-stimulating hormone
axon	(GSH)
basilar membrane	homeostasis
central nervous system	hormones
cerebellum	hypothalamus
cerebrum	incus
cochlea	malleus
cones	medulla oblongata
dendrites	motor neurons
depolarized	motor unit
endocrine glands	myosin
endocrine system	negative-feedback control
epinephrine	nerve cell
estrogen	nerve impulse

nerves	soma
nervous system	spinal cord
neuron	stapes
neurotransmitter	stimulus
norepinephrine	synapse
olfactory epithelium	target cells
oval window	testosterone
oxytocin	thalamus
perception	thyroid-stimulating hormone
peripheral nervous system	(TSH)
pons	thyroxine
response	triiodothyronine
retina	tropomyosin
rhodopsin	troponin
rods	tympanum
semicircular canals	voltage
sensory neurons	

LEARNING CONNECTIONS

Visit the Online Learning Center to find additional study resources including those highlighted below.

20.1   Coordination in Multicellular Animals

1. Describe two ways in which the function of the nervous system differs from that of the endocrine system.
2. How do exocrine and endocrine glands differ?

20.2   Nervous System Function

Animations and Review
- Nervous tissue
- Synapse

Labeling Exercise
- Synapse structure

3. Describe how the changing permeability of the cell membrane and the movement of sodium ions cause a nerve impulse.
4. What is the role of acetylcholine in a synapse? What is the role of acetylcholinesterase?
5. List the differences between the following:
   a. Central and peripheral nervous systems
   b. Motor and sensory nervous systems

20.3   Endocrine System Function

Labeling Exercise
- Endocrine system

Case Study
- Endocrine system

6. Give an example of negative-feedback control in the endocrine system.
7. List three hormones and give their functions.

20.4   Integration of Nervous and Endocrine Function

8. Give an example of the interaction between the endocrine system and the nervous system.
9. How do the anterior pituitary and the posterior pituitary differ in the way they function?

20.5 Sensory Input

Animations and Review
- Introduction
- Chemoreceptors
- Light receptors
- Mechanoreceptors

Labeling Exercises
- Eye anatomy I
- Eye anatomy II
- Ear anatomy

Interactive Concept Map
- Senses

Case Study
- Could your inner clocks make you a junkfood junkie?

Food for Thought
- Hearing loss and jobs

10. What is actually detected by the nasal epithelium, cochlea of the ear, and retina of the eye?

11. Name the five different taste receptors. What does each respond to?
12. List three different kinds of receptors associated with touch.

20.6 Output Coordination

Animations and Review
- Muscle
- Muscle cell function

Experience This!
- Muscle fatigue

13. How do skeletal, cardiac, and smooth muscles differ in (1) speed of contraction, (2) ability to stay contracted, and (3) cause of contraction?
14. What is the role of each of the following in muscle contraction: actin, myosin, ATP, troponin, and tropomyosin?
15. Why must muscles be in antagonistic pairs?

Test yourself with multiple choice questions that provide immediate feedback on the Online Learning Center with PowerWeb at

www.mhhe.com/enger11

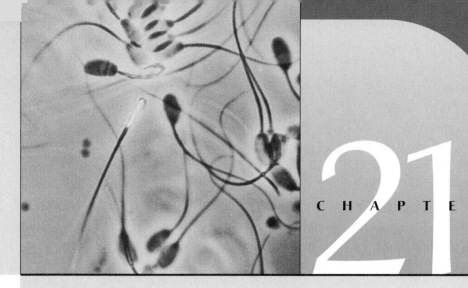

Human Reproduction, Sex, and Sexuality

GOALS AND OBJECTIVES

Understand that sexuality involves distinct hereditary, anatomical, and behavioral aspects.

❒ Explain how chromosomes determine the sex of a newborn child.

❒ Explain how gene of sexuality expression varies among individuals.

❒ Describe some aspects of sexual behavior that are strongly influenced by culture and others that may be hereditary.

Understand the normal structure and function of the male and female reproductive system.

❒ List the anatomical and behavioral differences between males and females.

❒ Describe the role played by each component of both the male and female reproductive system components.

❒ Describe the differences between spermatogenesis and oogenesis.

Recognize the importance of hormones in sexual development and function.

❒ Explain why abnormal development sometimes occurs.

❒ List the hormones that control and regulate the production of sperm.

❒ Understand the role of hormones in ovulation, in maintaining pregnancy, and in menopause.

❒ Recognize how various forms of contraception work.

Physiological Processes

PART FIVE

411

21.1 Sexuality from Different Points of View

Probably nothing interests people more than sex and sexuality. By **sexuality,** we mean all the factors that contribute to one's female or male nature. These include the structure and function of the sex organs, the behaviors that involve these structures, psychological components, and the role culture plays in manipulating our sexual behavior. Males and females have different behavior patterns for a variety of reasons. Some behavioral differences are learned (patterns of dress, use of facial makeup), whereas others appear to be less dependent on culture (degree of aggressiveness, frequency of sexual thoughts). We have an intense interest in the facts about our own sexual nature and the sexual behavior of members of the opposite sex and that of peoples of other cultures. • finding a mate, p. 330

There are several different ways to look at human sexuality. The behavioral sciences tend to focus on the behaviors associated with being male and female and what is considered appropriate and inappropriate sexual behavior. Sex is considered a strong drive, appetite, or urge by psychologists. They describe the sex drive as a basic impulse to satisfy a biological, social, or psychological need. Other social scientists (sociologists, cultural anthropologists) are interested in sexual behavior as it occurs in different cultures and subcultures. When a variety of cultures are examined, it becomes very difficult to classify various kinds of sexual behavior as normal or abnormal. What is considered abnormal in one culture may be normal in another. For example, public nudity is considered abnormal in many cultures but not in others.

Biologists have studied the sexual behavior of nonhuman animals for centuries. Biologists have long considered the function of sex and sexuality in light of its value to the population or species. Sexual reproduction results in new combinations of genes that are important in the process of natural selection. Many biologists today are attempting to look at human sexual behavior from an evolutionary perspective and speculate on why certain sexual behaviors are common in humans (How Science Works 21.1). The behaviors of courtship, mating, rearing of the young, and the division of labor between the sexes are complex in all social animals, including humans. These are demonstrated in the elaborate social behaviors surrounding mate selection and the establishment of families. It is difficult to draw the line between the biological development of sexuality and the social establishment of customs related to the sexual aspects of human life. However, the biological mechanism that determines whether an individual will develop into a female or male has been well documented.

21.2 Chromosomal Determination of Sex

When a human egg or sperm cell is produced, it contains 23 chromosomes. Twenty-two of these are **autosomes** that carry most of the genetic information used by the organism.

The other chromosome is a **sex-determining chromosome.** There are two kinds of sex-determining chromosomes: the **X chromosome** and the **Y chromosome** (see figure 9.23). The two sex-determining chromosomes, X and Y, do not carry equivalent amounts of information, nor do they have equal functions. X chromosomes carry typical genetic information about the production of specific proteins in addition to their function in determining sex. For example, the X chromosome carries information on blood clotting, color vision, and many other characteristics. The Y chromosome, however, appears to be primarily concerned with determining male sexual differentiation and has few other genes on it.

When a human sperm cell is produced, it carries 22 autosomes and a sex-determining chromosome. Unlike eggs, which always carry an X chromosome, half the sperm cells carry an X chromosome and the other half carries a Y chromosome. If an X-carrying sperm cell fertilizes an X-containing egg cell, the resultant embryo will develop into a female. A typical human female has an X chromosome from each parent. If a Y-carrying sperm cell fertilizes the egg, a male embryo develops. It is the presence or absence of the Y chromosome that determines the sex of the developing individual.

Evidence that the Y chromosome controls male development comes as a result of studying individuals who have an abnormal number of chromosomes. An abnormal meiotic division that results in sex cells with too many or too few chromosomes is called *nondisjunction* (nondisjunction is explained in chapter 9). If nondisjunction affects the X and Y chromosomes, a gamete might be produced that has only 22 chromosomes and lacks a sex-determining chromosome, or it might have 24, with 2 sex-determining chromosomes. If a cell with too few or too many sex chromosomes is fertilized, an abnormal embryo develops. If a normal egg cell is fertilized by a sperm cell with no sex chromosome, the offspring will have only 1 X chromosome. These people are designated as XO. They develop a collection of characteristics known as *Turner's syndrome* (figure 21.1). About 1 in 2,000 girls born is a Turner's syndrome person. An individual with this condition is female, is short for her age, and fails to mature sexually, resulting in sterility. In addition, she may have a thickened neck (termed webbing), hearing impairment, and some abnormalities in the cardiovascular system. When the condition is diagnosed, some of the physical conditions can be modified with treatment. Treatment involves the use of growth-stimulating hormone to increase growth rate and the use of female sex hormones to stimulate sexual development, although sterility is not corrected. • monosomy and trisomy, p. 184

An individual who has XXY chromosomes is basically male (figure 21.2). This genetic anomaly is termed *Klinefelter's syndrome,* and the symptoms include sterility because of small testes that do not usually produce viable sperm, lack of facial hair, and occasional breast tissue development. These persons are also more likely than most to experience difficulty with language development. Although they are sterile, men with this condition have normal sexual function. These characteristics vary greatly in degree and many men are diagnosed

Figure 21.1

Turner's Syndrome

Turner's syndrome individuals have 45 chromosomes. They have only one of the sex chromosomes and it is an X chromosome. Individuals with this condition are females and have delayed growth and fail to develop sexually. This woman is less than 150 cm (5 ft) tall and lacks typical secondary sexual development for her age. She also has a "webbed neck" that is common among Turner's syndrome individuals.

Figure 21.2

Klinefelter's Syndrome

Individuals with two X chromosomes and a Y chromosome are male, are sterile, and often show some degree of breast development and female body form. They are typically tall. The two photos show a Klinefelter's individual before and after receiving testosterone hormone therapy.

only after they undergo testing to determine why they are infertile. This condition is present in about 1 in 500 men. Treatment may involve breast-reduction surgery in males who have significant breast development and male hormone therapy.

Because both conditions involve abnormal numbers of X or Y chromosomes, they provide strong evidence that these chromosomes are involved in determining sexual development. The early embryo resulting from fertilization and cell division is neither male nor female but becomes female or male later in development—based on the sex-determining chromosomes that control the specialization of the cells of the undeveloped, embryonic gonads into female **ovaries,** or male **testes.** This specialization of embryonic cells is termed **differentiation.** The embryonic gonads begin to differentiate

into testes about seven weeks after **conception** (fertilization) if the Y chromosome is present. The Y chromosome seems to control this differentiation process in males because the gonads do not differentiate into female sex organs until later, and then only if 2 X chromosomes are present. It is the absence of the Y chromosome that determines female sexual differentiation.

Researchers were interested in how females, with 2 X chromosomes, handle the double dose of genetic material in comparison to males, who have only 1 X chromosome. M. L. Barr discovered that a darkly staining body was generally present in female cells but was not present in male cells. It was postulated, and has since been confirmed, that this structure is an X chromosome that is largely nonfunctional. Therefore, female cells have only one dose of X-chromosome genetic information that is functional; the other X chromosome coils up tightly and does not direct the manufacture of proteins. The 1 X chromosome of the male functions as expected, and the Y chromosome directs only male-determining activities. The tightly coiled structure in the cells of female mammals is called a **Barr body** after its discoverer (figure 21.3).

21.3 Male and Female Fetal Development

Development of embryonic gonads begins very early during fetal growth. First, a group of cells begins to differentiate into primitive gonads at about week 5. By week 6 or 7 if a Y chromosome is present, a gene product from the chromosome will begin the differentiation of these gonads into testes; they will develop into ovaries beginning about week 12 if 2 X chromosomes are present (Y chromosome is absent).

As soon as the gonad has differentiated into an embryonic testis at about week 8, it begins to produce testosterone.

HOW SCIENCE WORKS 21.1

Speculation on the Evolution of Human Sexual Behavior

There has been much speculation about how human sexual behavior evolved. It is important to recognize that this speculation is not fact, but an attempt to evaluate human sexual behaviors from an evolutionary perspective.

When we compare human sexuality with that of other mammals, there are several ways in which human sexuality is different from that of most other mammals. Whereas most mammals are sexually active during specific periods of the year, humans may engage in sexual intercourse at any time throughout the year. The sex act appears to be important as a pleasurable activity rather than a purely reproductive act. Associated with this difference is the fact that human females do not display changes that indicate they are releasing eggs (ovulating).

All other female mammals display changes in odor, appearance, or behavior that clearly indicate to the males of the species that the female is ovulating and sexually responsive. This is referred to as "being in heat." This is not true for humans. Human males are unable to differentiate ovulating females from those who are not ovulating.

In other mammals with few exceptions, infants grow to sexual maturity in a year or less. Although extremely long-lived mammals (elephants or whales) do not reach sexual maturity in a year, their young have well-developed muscles that allow them to move about with a high degree of independence. Although the young of these species still rely on their mothers for milk and protection, they are capable of obtaining other food for themselves as well. This is not true for human infants, which are extremely immature when born, develop walking skills slowly, and require several years of training before they are able to function independently.

Perhaps the extremely immature condition in which human infants are born is related to human brain size. The size of the head is very large and just fits through the birth canal in the mother's pelvis. One way to accommodate a large brain size and not need to redesign the basic anatomy of the female pelvis would be to have the young be born in a very immature condition while the brain is still small and in the process of growing. Having the young born in an immature condition can solve one problem but creates another. The immature condition of human infants is associated with a need to provide extensive care for them.

Raising young requires a considerable investment of time and resources. Females invest considerable resources in the pregnancy itself. Fat stores provide energy necessary to a successful pregnancy. Female mammals, including humans, that have little stored fat often have difficulty becoming pregnant in the first place and also are more likely to die of complications resulting from the pregnancy. Nutritional counseling is an important part of modern prenatal care because it protects the health of both mother and developing fetus. The long duration of pregnancy in humans requires good nutrition over an extended period. Once the child is born, the mother continues to require good nutrition because she provides the majority of food for the infant through her breast milk. As the child grows, other food items are added to its diet. Since the young child is unable to find

and prepare its own food, the mother or father or both must expend energy to feed the child.

With these ideas in mind we can speculate about how human sexual behavior may have evolved. Imagine a primitive stone age human culture. Females have a great deal invested in each child produced. They will only be able to produce a few children during their lifetimes, and many children will die because of malnutrition, disease, and accidents. Those females that have genes that will allow more of their offspring to survive will be selected for. Human males, on the other hand, have very little invested in each child and can impregnate many different females. Males that have many children that survive are selected for. How might these different male and female goals fit together to provide insight into the sexual behaviors we see in humans today?

The males of most mammals contribute little toward the raising of young. In many species, males meet with females only for mating (deer, cats, rabbits, mice). In some species, the male and female form short-term pair-bonds for one season and the males share the burden of raising the young (foxes). Only a few (wolves) form pair-bonds lasting for years in which males and females cooperate in the raising of young. However, pair-bonding in humans is usually a long-term relationship. The significance of this relationship can be evaluated from an evolutionary perspective. This long-term pair-bond can serve the interests of both males and females. When males form long-term relationships with females, the females gain an additional source of nutrition for their offspring, who will be completely dependent on their parents for food and protection for several years, thus increasing the likelihood that the young will survive. Human males benefit from the long-term pair-bond as well. Because human females do not display the fact that they are ovulating, the only way a male can be assured that the child he is raising is his is to have exclusive mating rights with a specific female. The establishment of bonding involves a great deal of sexual activity, much more than is necessary to just bring about reproduction. It is interesting to speculate that sexual behavior in humans is as much involved in maintaining pair-bonds as it is in creating new humans.

Much has been written about the differences in sexual behavior between men and women and that men and women look for different things when assessing individuals as potential mates. It is very difficult to distinguish behaviors that are truly biologically determined and those that are culturally determined. However, some differences may have biological roots. Females benefit from bonding with males who have access to resources that are shared in the raising of young. Do women look for financial security and a willingness to share in a mate? Because pregnancy and nursing young require a great deal of nutrition, it is in the male's interest to choose a mate who is healthy, young, and in good nutritional condition. Since the breasts and buttocks are places of fat storage in women, do men look for youth and appropriate amounts of nutritionally important fat stored in the breasts and buttocks? If these differences between men and women really exist, are they purely cultural, or is there an evolutionary input from our primitive ancestors?

Figure 21.3

Barr Body
In women, only one of the two X chromosomes functions. The extra dark body in the nucleus of this white blood cell from a woman is the nonfunctioning X chromosome.

The presence of testosterone results in the differentiation of male sexual anatomy and the absence of testosterone results in the differentiation into female sexual anatomy.

In normal males, at about the seventh month of gestation, the testes move from a position in the abdominal cavity to the external sac, called the scrotum, via an opening called the **inguinal canal** (figure 21.4). This canal closes off but continues to be a weakened area in the abdominal wall and may rupture later in life. This can happen when strain (e.g., from improperly lifting heavy objects) causes a portion of the intestine to push through the inguinal canal into the scrotum. This condition is known as an **inguinal hernia.**

Occasionally the testes do not descend and a condition known as **cryptorchidism** (*crypt* = hidden; *orchidos* = testes) develops. Sometimes the descent occurs during puberty; if not, there is an increased incidence of testicular cancer. Because of this increased risk, surgery is performed that allows the undescended testes to descend to their normal positions in the scrotum. The retention of the testes in the abdomen results in sterility because normal sperm cell development cannot occur in a very warm environment and the temperature in the abdomen is higher than the temperature in the scrotum. Normally the temperature of the testes is very carefully regulated by muscles that control their distance from the body. Physicians have even diagnosed cases of male infertility as being caused by tight-fitting pants that hold the testes so close to the body that the temperature increase interferes with normal sperm development. • **temperature effects on enzymes, p. 96**

21.4 Sexual Maturation of Young Adults

Following birth, sexuality plays only a small part in physical development for several years. Culture and environment shape the responses that the individual will come to recognize as normal behavior. During **puberty,** normally between 12 and 14 years of age, increased production of sex hor-

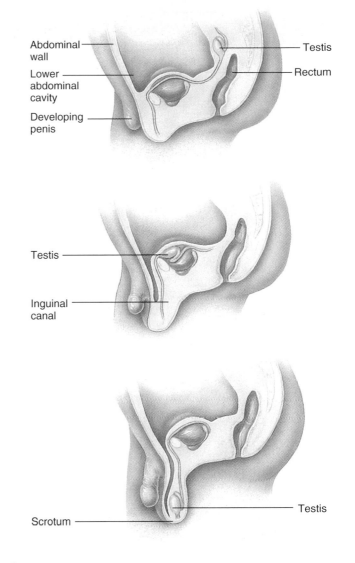

Figure 21.4

Descent of Testes
During the development of the male fetus, the testes originate in the abdomen and eventually descend through the inguinal canal to the scrotum. This usually happens prior to birth.

mones causes major changes as the individual reaches sexual maturity. Generally females reach puberty 6 months to 2 years before males. After puberty, humans are sexually mature and have the capacity to produce offspring.

The Maturation of Females

Female children typically begin to produce quantities of sex hormones from the hypothalamus, pituitary gland, ovaries, and adrenal glands at 8 to 13 years of age. This marks the onset of puberty. The **hypothalamus** controls the functioning of many other glands throughout the body, including the **pituitary gland.** At puberty the hypothalamus begins to release a hormone known as **gonadotropin-releasing hormone (GnRH),** which stimulates the pituitary to release **luteinizing hormone (LH)** and **follicle-stimulating hormone (FSH).**

Table 21.1

HUMAN REPRODUCTIVE HORMONES

Hormone	Production Site	Target Organ	Function
Prolactin, lactogenic, or luteotropic hormone	Anterior pituitary	Breast, ovary	Stimulates milk production; also helps maintain normal ovarian cycle
Follicle-stimulating hormone (FSH)	Anterior pituitary	Ovary, testes	Stimulates ovary and testis development; stimulates egg production in females and sperm production in males
Luteinizing hormone (LH) or interstitial cell-stimulating hormone (ICSH)	Anterior pituitary	Ovary, testes	Stimulates ovulation in females and sex-hormone (estrogens and testosterone) production in both males and females
Estrogens	Ovary	Entire body	Stimulates development of female reproductive tract and secondary sexual characteristics
Testosterone	Testes	Entire body	Stimulates development of male reproductive tract and secondary sexual characteristics
Progesterone	Corpus luteum of ovary	Uterus, breasts	Causes uterine thickening and maturation; maintains pregnancy
Oxytocin	Posterior pituitary	Uterus, breasts	Causes uterus to contract and breasts to release milk
Androgens	Testes, adrenal glands	Entire body	Stimulates development of male reproductive tract and secondary sexual characteristics in males and females
Gonadotropin-releasing hormone (GnRH)	Hypothalamus	Anterior pituitary	Stimulates the release of FSH and LH from anterior pituitary
Human chorionic gonadotropin	Placenta	Corpus luteum	Maintains corpus luteum so that it continues to secrete progesterone and maintain pregnancy

Increased levels of FSH stimulate the development of **follicles,** saclike structures that produce oocytes in the ovary, and the increased luteinizing hormone stimulates the ovary to produce larger quantities of **estrogens.** The increasing supply of estrogen is responsible for the many changes in sexual development that can be noted at this time. These changes include breast growth, changes in the walls of the uterus and vagina, increased blood supply to the clitoris, and changes in the pelvic bone structure.

Estrogen also stimulates the female adrenal gland to produce **androgens,** male sex hormones. The androgens are responsible for the production of pubic hair and they seem to have an influence on the female sex drive. The adrenal gland secretions may also be involved in the development of acne. Those features that are not primarily involved in sexual reproduction but are characteristic of a sex are called **secondary sexual characteristics.** In women, the distribution of body hair, patterns of fat deposits, and a higher voice are examples.

A major development during this time is the establishment of the **menstrual cycle.** This involves the periodic growth and shedding of the lining of the uterus. These changes are under the control of a number of hormones produced by the pituitary and ovaries. The ovaries are stimulated to release their hormones by the pituitary gland, which is in turn influenced by the ovarian hormones. Both follicle-stimulating hormone (FSH) and luteinizing hormone (LH) are produced by the pituitary gland. FSH causes the maturation and development of the ovaries. LH is important in causing ovulation and converting the ruptured follicle into a structure known as the **corpus luteum.** The corpus luteum produces the hormone **progesterone,** which is important in maintaining the lining of the uterus. Changes in the levels of progesterone result in a periodic buildup and shedding of the lining of the uterus known as the menstrual cycle. Table 21.1 summarizes the activities of these various hormones. Associated with the menstrual cycle is the periodic

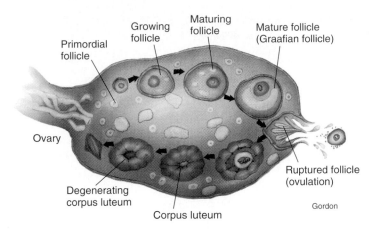

Figure 21.5

Ovulation

In the ovary, the egg begins development inside a sac of cells known as a follicle. Each month, one of these follicles develops and releases its product. This release through the wall of the ovary is known as ovulation.

release of sex cells from the surface of the ovary, called **ovulation** (figure 21.5). Initially, these two cycles, menstruation and ovulation, may be irregular, which is normal during puberty. Eventually, hormone production becomes regulated so that ovulation and menstruation take place on a regular monthly basis in most women, although normal cycles may vary from 21 to 45 days.

As girls progress through puberty, curiosity about the changing female body form and new feelings leads to self-investigation. Studies have shown that sexual activity such as manipulation of the clitoris, which causes pleasurable sensations, is performed by a large percentage of young women. Self-stimulation, frequently to orgasm, is a common result. This stimulation is termed **masturbation,** and it should be stressed that it is considered a normal part of sexual development. **Orgasm** is a complex response to mental and physical stimulation that causes rhythmic contractions of the muscles of the reproductive organs and an intense frenzy of excitement.

The Maturation of Males

Males typically reach puberty about 2 years later (ages 10 to 15) than females, but puberty in males also begins with a change in hormone levels. At puberty the hypothalamus releases increased amounts of gonadotropin-releasing hormone (GnRH), resulting in increased levels of follicle-stimulating hormone (FSH) and luteinizing hormone. These are the same changes that occur in female development. Luteinizing hormone is often called **interstitial cell-stimulating hormone (ICSH)** in males. ICSH stimulates the testes to produce **testosterone,** the primary sex hormone in males. The testosterone produced by the embryonic testes caused the differentiation of internal and external genital anatomy in the male

embryo. At puberty the increased amount of testosterone is responsible for the development of male secondary sexual characteristics and is also important in the maturation and production of sperm.

The major changes during puberty include growth of the testes and scrotum, pubic-hair development, and increased size of the penis. Secondary sex characteristics begin to become apparent at age 13 or 14. Facial hair, underarm hair, and chest hair are some of the most obvious. The male voice changes as the larynx (voice box) begins to change shape. Body contours also change, and a growth spurt increases height. In addition, the proportion of the body that is muscle increases and the proportion of body fat decreases. At this time, a boy's body begins to take on the characteristic adult male shape, with broader shoulders and heavier muscles.

In addition to these external changes, increased testosterone causes the production of seminal fluid by the **seminal vesicles,** prostate gland, and the bulbourethral glands. FSH stimulates the production of sperm cells. The release of sperm cells and seminal fluid begins during puberty and is termed **ejaculation.** This release is generally accompanied by the pleasurable sensations of orgasm. The sensations associated with ejaculation may lead to self-stimulation, or masturbation. Masturbation is a common and normal activity as a boy goes through puberty. Studies of sexual behavior have shown that nearly all men masturbate at some time during their lives.

21.5 Spermatogenesis

One of the biological reasons for sexual activity is the production of offspring. The process of producing gametes includes meiosis and is called **gametogenesis** (gamete formation) (figure 21.6). The term **spermatogenesis** is used to describe gametogenesis that takes place in the testes of males. The two bean-shaped testes are composed of many small sperm-producing tubes, or **seminiferous tubules,** and collecting ducts that store sperm. These are held together by a thin covering membrane. The seminiferous tubules join together and eventually become the epididymis, a long, narrow convoluted tube in which sperm cells are stored and mature before ejaculation (figure 21.7).

Leading from the epididymis is the vas deferens, or sperm duct; this empties into the urethra, which conducts the sperm out of the body through the **penis** (figure 21.8). Before puberty, the seminiferous tubules are packed solid with diploid cells called spermatogonia. These cells, which are found just inside the tubule wall, undergo *mitosis* and produce more spermatogonia. Beginning about age 11, some of the spermatogonia specialize and begin the process of *meiosis,* whereas others continue to divide by mitosis, assuring a constant and continuous supply of spermatogonia. Once spermatogenesis begins, the seminiferous tubules become hollow and can transport the mature sperm.

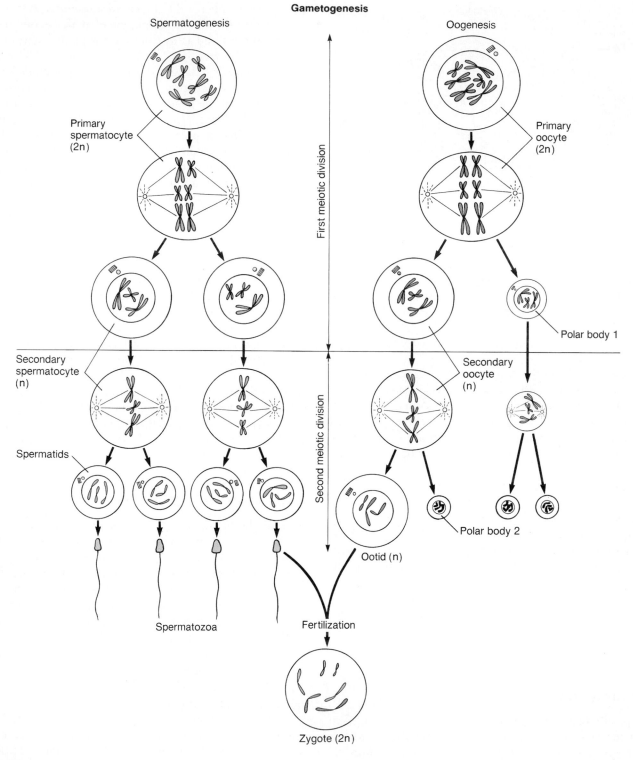

Gametogenesis

Figure 21.6

Gametogenesis
This diagram illustrates the process of gametogenesis in human males and females. Not all of the 46 chromosomes are shown. Carefully follow the chromosomes as they segregate, recalling the details of the process of meiosis explained previously.

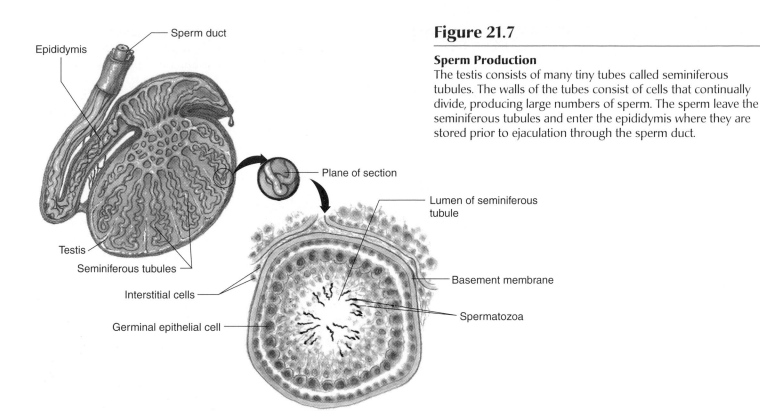

Figure 21.7

Sperm Production

The testis consists of many tiny tubes called seminiferous tubules. The walls of the tubes consist of cells that continually divide, producing large numbers of sperm. The sperm leave the seminiferous tubules and enter the epididymis where they are stored prior to ejaculation through the sperm duct.

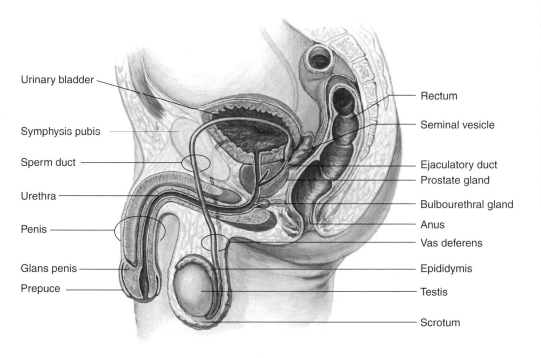

Figure 21.8

The Human Male Reproductive System

The male reproductive system consists of two testes that produce sperm, ducts that carry the sperm, and various glands. Muscular contractions propel the sperm through the vas deferens past the seminal vesicles, prostate gland, and bulbourethral gland, where most of the liquid of the semen is added. The semen passes through the urethra of the penis to the outside of the body.

Spermatogenesis involves several steps. It begins when some of the spermatogonia in the walls of the seminiferous tubules differentiate and enlarge to become **primary spermatocytes.** These diploid cells undergo the first meiotic division, which produces two haploid **secondary spermatocytes.** The secondary spermatocytes go through the second meiotic division, resulting in four haploid **spermatids,** which lose much of their cytoplasm and develop long tails. These cells are then known as **sperm** (figure 21.9). The sperm have only a small amount of food reserves. Therefore, once they are released and become active swimmers, they live no more than 72 hours. However, if the sperm are placed in a special protective solution, the temperature can be lowered drastically to −196°C. Under these conditions the sperm become deactivated,

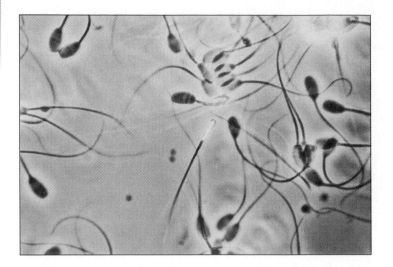

Figure 21.9

Human Sperm Cells
These cells are primarily DNA-containing packages produced by the male.

freeze, and can survive live for years outside the testes. This has led to the development of sperm banks. Artificial insemination of cattle, horses, and other domesticated animals with sperm from sperm banks is common.

Spermatogenesis in human males takes place continuously throughout a male's reproductive life, although the number of sperm produced decreases as a man ages. Sperm counts can be taken and used to determine the probability of successful fertilization. For reasons not totally understood, a man must be able to release at least 100 million sperm at one insemination to be fertile. It appears that enzymes in the head of sperm are needed to digest through mucus and protein found in the female reproductive tract. Millions of sperm contribute in this way to the process of fertilization, but only one is involved in fertilizing the egg. A healthy male probably releases about 300 million sperm with each ejaculation (although the numbers of sperm per ejaculate may be reduced with frequent ejaculation) during **sexual intercourse**, also known as **coitus** or **copulation**.

21.6 Oogenesis

The term **oogenesis** refers to the production of egg cells. This process starts during prenatal development of the ovary, when diploid oogonia cease dividing by *mitosis* and enlarge to become **primary oocytes.** All of the primary oocytes that a woman will ever have are already formed prior to her birth. At this time they number approximately 2 million, but that number is reduced by cell death to between 300,000 to 400,000 cells by the time of puberty. Oogenesis halts at this point and all the primary oocytes remain just under the surface of the ovary.

Primary oocytes begin to undergo *meiosis* in the normal manner at puberty. At puberty and on a regular basis thereafter, the sex hormones stimulate a primary oocyte to continue its maturation process, and it goes through the first meiotic division. But in telophase I, the two cells that form receive unequal portions of cytoplasm. You might think of it as a lopsided division (figure 21.6). The smaller of the two cells is called a **polar body,** and the larger haploid cell is the **secondary oocyte** (people commonly refer to this cell as an *egg* or *ovum* although technically it is not). The other primary oocytes remain in the ovary. Ovulation begins when the soon-to-be-released secondary oocyte, encased in a saclike structure known as a follicle, grows and moves near the surface of the ovary. When this maturation is complete, the follicle erupts and the secondary oocyte is released. It is swept into the **oviduct** (fallopian tube) by ciliated cells and travels toward the **uterus** (figure 21.10). Because of the action of the luteinizing hormone, the follicle from which the oocyte ovulated develops into a glandlike structure, the corpus luteum, which produces hormones (progesterone and estrogen) that prevent the release of other secondary oocytes.

If the secondary oocyte is fertilized, it completes meiosis with the sperm DNA inside and becomes the zygote. If the secondary oocyte is not fertilized, it passes through the **vagina** to the outside during menstruation. During her lifetime, a female releases about 300 to 500 secondary oocytes. Obviously, few of these cells are fertilized.

One of the characteristics to note here is the relative age of the sex cells. In males, sperm production is continuous throughout life. Sperm do not remain in the tubes of the male reproductive system for very long. They are either released shortly after they form or die and are harmlessly absorbed. In females, meiosis begins before birth, but the oogenesis process is not completed, and the cell is not released for many years. A secondary oocyte released when a woman is 37 years old began meiosis 37 years before! During that time, the cell was exposed to many influences, a number of which may have damaged the DNA or interfered with the meiotic process. The increased risk of abnormal births in older mothers may be related to the age of their eggs. Such alterations are less likely to occur in males because new gametes are being produced continuously. Also, defective sperm appear to be much less likely to be involved in fertilization. • **mutations, p. 141**

Hormones control the cycle of changes in breast tissue, in the ovaries, and in the uterus. In particular, estrogen and progesterone stimulate milk production by the breasts and cause the lining of the uterus to become thicker and filled with blood vessels prior to ovulation. This ensures that if fertilization occurs, the resultant embryo will be able to attach itself to the wall of the uterus and receive nourishment. If the cell is not fertilized, the lining of the uterus is shed. This is known as *menstruation, menstrual flow,* the *menses,* or a *period.* Once the wall of the uterus has been shed, it begins to build up again. As noted previously, this continual building up and shedding of the wall of the uterus is known as the *menstrual cycle.*

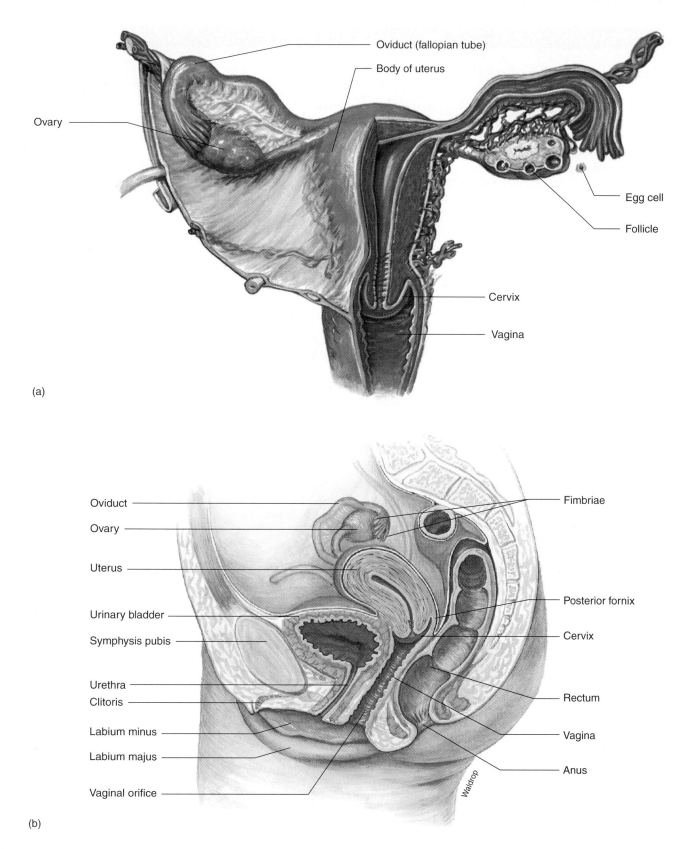

Oviduct (fallopian tube)

Body of uterus

Ovary

Egg cell

Follicle

Cervix

Vagina

(a)

Oviduct

Fimbriae

Ovary

Uterus

Urinary bladder

Posterior fornix

Symphysis pubis

Cervix

Urethra

Clitoris

Rectum

Labium minus

Labium majus

Vagina

Vaginal orifice

Anus

Waldrop

(b)

Figure 21.10

The Human Female Reproductive System
(a) After ovulation, the cell travels down the oviduct to the uterus. If it is not fertilized, it is shed when the uterine lining is lost during menstruation. (b) The human female reproductive system, side view.

HOW SCIENCE WORKS 21.2

Can Humans Be Cloned?

Recent advances in the understanding of the reproductive biology of humans and other mammals have led to the possibility of cloning in humans. An understanding of the function of hormones allows manipulation of the reproductive cycles of women so that developing oocytes can be harvested from the ovary. These oocytes can be fertilized in dishes and the early development of the embryos can be observed. These embryos can be implanted into the uterus of a woman who does not need to be the woman who donated the oocyte. In nonhuman mammals, oocytes have been manipulated such that the nucleus of the oocyte was removed and the nucleus from a mature cell was inserted. The manipulated cell begins

embryological development and can be implanted in the uterus of a female. In several species (sheep, mouse, monkey) the process has successfully resulted in a cloned offspring that is genetically identical to the individual that donated the nucleus. The technology for cloning humans is present. The oocytes can be harvested. The nuclei can be transferred and the techniques for introducing them into a uterus are well known. Can humans be cloned? The answer is yes. In 2001 a group of researchers cloned a human embryo. However, development stopped at an early stage and the embryo died. Should humans be cloned? That is a question for ethicists and social thinkers to answer.

The activities of the ovulatory cycle and the menstrual cycle are coordinated. During the first part of the menstrual cycle, increased amounts of FSH cause the follicle to increase in size. Simultaneously, the follicle secretes increased amounts of estrogen that cause the lining of the uterus to increase in thickness. When ovulation occurs, the remains of the follicle is converted into a corpus luteum by the action of LH. The corpus luteum begins to secrete progesterone and the nature of the uterine lining changes by becoming more vascularized. This is choreographed so that if an embryo arrives in the uterus shortly after ovulation, it meets with a uterine lining prepared to accept it. If pregnancy does not occur, the corpus luteum degenerates, resulting in a reduction in the amount of progesterone needed to maintain the lining of the uterus, and the lining is shed.

At the same time that hormones are regulating ovulation and the menstrual cycle, changes are taking place in the breasts. The same hormones that prepare the uterus to receive the embryo also prepare the breasts to produce milk. These changes in the breasts, however, are relatively minor unless pregnancy occurs.

21.7 Hormonal Control of Fertility

An understanding of how various hormones regulate the menstrual cycle, ovulation, milk production, and sexual behavior has led to the medical use of certain hormones. Some women are unable to have children because they do not release oocytes from their ovaries or they release them at the wrong time. Physicians can now regulate the release of oocytes from the ovary using certain hormones, commonly called *fertility drugs*. These hormones can be used to stimulate the release of oocytes for capture and use in what is called *in vitro* fertilization (*test-tube* fertilization) or to increase the probability of natural conception; that is, *in vivo* fertilization (*in-life* fertilization).

Unfortunately, the use of these drugs often results in multiple implantations because they may cause too many secondary oocytes to be released at one time. The implantation of multiple embryos makes it difficult for one embryo to develop properly and be carried through the entire nine-month gestation period. When we understand the action of hormones better, we may be able to control the effects of fertility drugs and eliminate the problem of multiple implantations.

A second medical use of hormones is in the control of conception by the use of birth-control pills—oral contraceptives. Birth-control pills have the opposite effect of fertility drugs. They raise the levels of estrogen and progesterone, which suppresses the production of FSH and LH, preventing the release of secondary oocytes from the ovary. Hormonal control of fertility is not as easy to achieve in men because there is no comparable cycle of gamete release. The use of drugs and laboratory procedures to help infertile couples have children has also raised the technical possibility of cloning in humans (How Science Works 21.2).

21.8 Fertilization and Pregnancy

In most women, a secondary oocyte is released from the ovary about 14 days before the next menstrual period. The menstrual cycle is usually said to begin on the first day of menstruation. Therefore, if a woman has a regular 28-day cycle, the cell is released approximately on day 14 (figure 21.11). If a woman normally has a regular 21-day menstrual cycle, ovulation would occur about day 7 in the cycle. If a woman has a regular 40-day cycle, ovulation would occur about day 26 of her menstrual cycle. Some women, however, have very irregular menstrual cycles, and it is difficult to determine just when the oocyte will be released to become available for fertilization. Once the cell is released, it is swept into the oviduct and moved toward the uterus. If sperm are present, they swarm around the secondary oocyte as it passes down the oviduct,

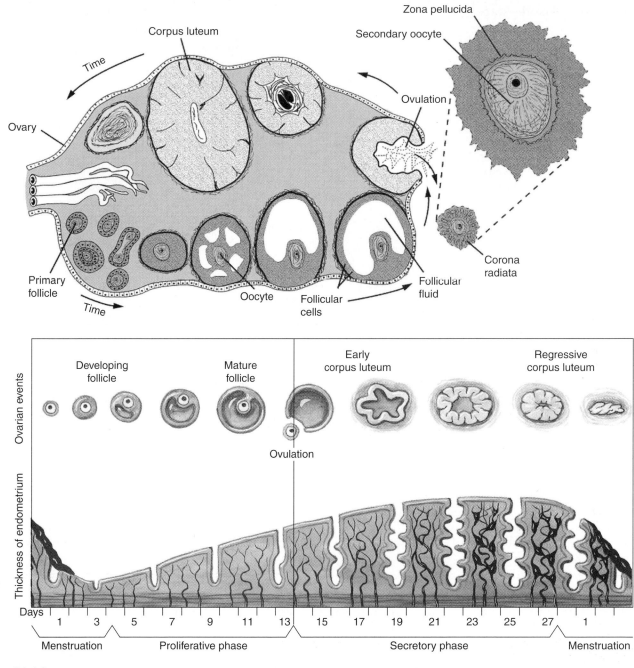

Figure 21.11

The Ovarian and Uterine Cycles in Human Females

The release of a secondary oocyte (ovulation) is timed to coincide with the thickening of the lining of the uterus. The uterine cycle in humans involves the preparation of the uterine wall to receive the embryo if fertilization occurs. Knowing how these two cycles compare, it is possible to determine when pregnancy is most likely to occur.

but only one sperm penetrates the outer layer to fertilize it and cause it to complete meiosis II. The other sperm contribute enzymes that digest away the protein and mucus barrier between the egg and the successful sperm.

During this second meiotic division, the second polar body is pinched off and the *ovum* (egg) is formed. Because chromosomes from the sperm are already inside, they simply intermingle with those of the ovum, forming a diploid **zygote** or fertilized egg. As the zygote continues to travel down the oviduct, it begins to divide by mitosis into smaller and smaller cells without having the mass of cells increase in size (figure 21.12). This division process is called *cleavage*. Eventually, a solid ball of cells is produced, known as the morula stage of embryological development, and attaches to the uterine wall. Following the morula stage, the solid ball of cells becomes hollow and begins to increase in size and is then known as the blastula stage. During this stage, when the embryo is about 6 days old, it becomes embedded, or implanted, in the lining of the uterus. In mammals, the blastula has a region of cells, called the *inner-cell mass,* that develops into the embryo proper. The outer cells become membranes associated with the embryo. • **stages of mitosis, p. 159**

The next stage in the development is known as the gastrula stage because the gut is formed during this time (*gastro* = stomach). In many kinds of animals, the gastrula is formed by an infolding of one side of the blastula, a process similar to poking a finger into a balloon. Gastrula formation in mammals is more difficult to visualize, but the result is the same. The embryo develops a tube that eventually becomes the gut. The formation of the primitive gut is just one of a series of changes that eventually result in an embryo that is recognizable as a miniature human being (figure 21.12). Most of the time during its development, the embryo is enclosed in a water-filled membrane, the amnion, which protects it from blows and keeps it moist. Two other membranes, the chorion and allantois, fuse with the lining of the uterus to form the **placenta** (figure 21.13). A fourth sac, the yolk sac, is well developed in birds, fish, amphibians, and reptiles. The yolk sac in these animals contains a large amount of food used by the developing embryo. Although a yolk sac is present in mammals, it is small and does not contain yolk. The nutritional needs of the embryo are met through the placenta. The placenta also produces the hormone chorionic gonadotropin that stimulates the corpus luteum to continue producing progesterone and thus prevents menstruation and ovulation during gestation.

As the embryo's cells divide and grow, some of them become differentiated into nerve cells, bone cells, blood cells, or other specialized cells. In order to divide, grow, and differentiate, cells must receive nourishment. This is provided by the

Figure 21.12

Human Embryonic Development
During the period of time between fertilization and birth, many changes take place in the embryo. Here we see some of the changes that take place during the first 8 weeks.

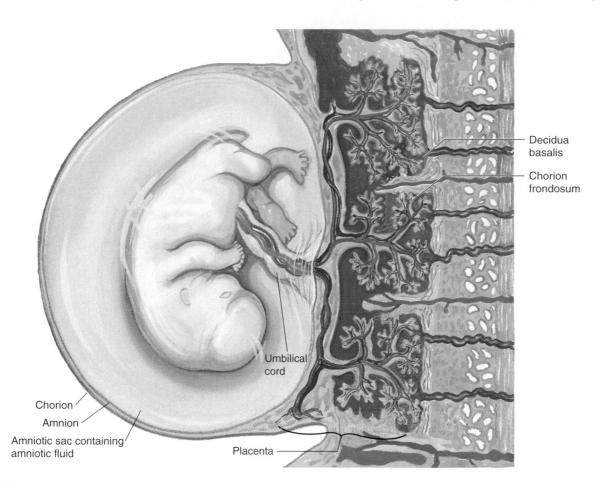

Decidua
basalis

Chorion
frondosum

Umbilical
cord

Chorion

Amnion

Amniotic sac containing
amniotic fluid

Placenta

Figure 21.13

Placental Structure
The embryonic blood vessels that supply the developing child with nutrients and remove the metabolic wastes are separate from the blood vessels of the mother. Because of this separation, the placenta can selectively filter many types of incoming materials and microorganisms.

mother through the placenta, in which both fetal and maternal blood vessels are abundant, allowing for the exchange of substances between the mother and embryo. The materials diffusing across the placenta include oxygen, carbon dioxide, nutrients, and a variety of waste products. The materials entering the embryo travel through blood vessels in the umbilical cord. The diet and behavior of the mother are extremely important. Any molecules consumed by the mother can affect the embryo. Cocaine, alcohol, heroin, and chemicals in cigarette smoke can all cross the placenta and affect the development of the embryo. The growth of the embryo results in the development of major parts of the body by the 10th week of pregnancy. After this time, the embryo continues to increase in size, and the structure of the body is refined.

Twins

Approximately 1 in 70 pregnancies in the United States results in a multiple birth. The vast majority of these are twin births. Twins happens in two ways. In the case of identical twins (approximately one-third of twins), during cleavage the embryo splits into two separate groups of cells. Each de-

velops into an independent embryo. Because they come from the same single fertilized ovum, they have the same genes and are of the same sex.

Fraternal twins do not contain the same genetic information and may be of different sexes. They result from the fertilization of two separate oocytes by different sperm. Therefore, they no more resemble each other than do ordinary brothers and sisters.

Birth

At the end of about 9 months, hormone changes in the mother's body stimulate contractions of the muscles of the uterus during a period prior to birth called *labor*. These contractions are stimulated by the hormone oxytocin, which is released from the posterior pituitary. The contractions normally move the baby headfirst through the vagina, or birth canal. One of the first effects of these contractions may be bursting of the amnion (bag of water) surrounding the baby. Following this, the uterine contractions become stronger, and shortly thereafter the baby is born. In some cases, the baby becomes turned in the uterus before labor. If this occurs, the feet or buttocks appear first. Such a birth is called a *breech birth*. This can be a dangerous situation because the baby's source of oxygen is being cut off as the placenta begins to separate from the mother's body.

If for any reason the baby does not begin to breathe on its own, it will not receive enough oxygen to prevent the death of nerve cells; thus brain damage or death can result.

Occasionally, a baby may not be able to be born normally because of the position of the baby in the uterus, the location of the placenta on the uterine wall, the size of the birth canal, the number of babies in the uterus, or many other reasons. A common procedure to resolve this problem is the surgical removal of the baby through the mother's abdomen. This procedure is known as a cesarean, or C-section. Currently, over 20% of births in the United States are by cesarean section. This rate reflects the fact that many women who are prone to problem pregnancies are having children rather than forgoing pregnancy. In addition, changes in surgical techniques have made the procedure much safer. Finally, many physicians who are faced with liability issues related to problem pregnancy may encourage cesarean section rather than normal birth. • endocrine glands, p. 394

Following the birth of the baby, the placenta, also called the *afterbirth*, is expelled. Once born, the baby begins to function on its own. The umbilical cord collapses and the baby's lungs, kidneys, and digestive system must now support all bodily needs. This change is quite a shock, but the baby's loud protests fill the lungs with air and stimulate breathing.

Over the next few weeks, the mother's body returns to normal, with one major exception. The breasts, which have undergone changes during the period of pregnancy, are ready to produce milk to feed the baby. Following birth, prolactin, a hormone from the pituitary gland, stimulates the production of milk, and oxytocin stimulates its release. If the baby is breast-fed, the stimulus of the baby's sucking will prolong the time during which milk is produced. This response involves both the nervous and endocrine systems. The sucking stimulates nerves in the nipple and breast that results in the release of prolactin and oxytocin from the pituitary.

In some cultures, breast-feeding continues for 2 to 3 years, and the continued production of milk often delays the reestablishment of the normal cycles of ovulation and menstruation. Many people believe that a woman cannot become pregnant while she is nursing a baby. However, because there is so much variation among women, relying on this as a natural conception-control method is not a good choice. Many women have been surprised to find themselves pregnant again a few months after delivery.

21.9 Contraception

Throughout history people have tried various methods of conception control (figure 21.14). In ancient times, conception control was encouraged during times of food shortage or when tribes were on the move from one area to another in search of a new home. Writings as early as 1500 B.C. indicate that the Egyptians used a form of tampon medicated with the ground powder of a shrub to prevent fertilization. This may sound primitive, but we use the same basic principle today to destroy sperm in the vagina.

Contraceptive jellies and foams make the environment of the vagina more acidic, which diminishes the sperm's chances of survival. The spermicidal (sperm-killing) foam or jelly is placed in the vagina before intercourse. When the sperm make contact with the acidic environment, they stop swimming and soon die. Aerosol foams are an effective method of conception control, but interfering with the hormonal regulation of ovulation is more effective.

The first successful method of hormonal control was "the pill," or "birth control pill." However, the way the pill works is by preventing ovulation, not preventing a birth. The quantity and balance of hormones (estrogen and progesterone) in the pill serves to fool the ovary into functioning as if the woman was already pregnant. Therefore, conception is highly unlikely. The emergency contraceptive pill (ECP), or "morning-after pill," uses a high dose of the same hormones found in oral contraceptives, which prevents the woman from becoming pregnant in the first place. ECPs prevent conception in several ways: (1) delay or inhibit ovulation; (2) inhibit transport of the egg or sperm, (3) prevent fertilization; (4) inhibit implantation of a fertilized egg, or (5) stimulate an autoimmune response that kills the sperm. This method is effective if used within 72 hours after unprotected intercourse.

Another method of conception control also involves hormones. The hormones are contained within small rods or capsules, which are placed under a woman's skin. These rods, when properly implanted, slowly release hormones and prevent the maturation and release of oocytes from the follicle. The major advantage of the implant is its convenience. Once the implant has been inserted, the woman can forget about contraceptive protection for several years. If she wants to become pregnant, the implants are removed and her normal menstrual and ovulation cycles return over a period of weeks.

The vaginal ring is still another method that utilizes hormones. The ring is inserted into the vagina and releases a continuous low dose of estrogen and progestin for 21 days. At the end of the 21 days, the ring is removed for 7 days to allow the menstrual period to occur. A new ring is then inserted monthly.

Research has also shown that men may also have gamete production controlled by altering hormone levels. While still in development, the male contraceptive pill contains the hormones desogestrel and testosterone. This combination prevents the production of sperm while maintaining male characteristics and sex drive. In clinical trials, all sperm counts dropped to zero, and therefore the male pill is more effective than the condom or the female pill. In order to be effective, it must be taken daily. As with the female pill, the male pill does not protect against AIDS and other sexually transmitted diseases (STDs).

Killing sperm and preventing ovulation are not the only methods of preventing conception. Any method that pre-

(a)

(b)

(c)

(d)

(e)

(f)

(g)

(h)

(i)

Figure 21.14

Contraceptive Methods

These are the primary methods of conception control used today: (*a*) oral contraception (pills), (*b*) contraceptive implants, (*c*) diaphragm and spermicidal jelly, (*d*) intrauterine device, (*e*) spermicidal vaginal foam, (*f*) Depo-Provera injection, (*g*) male condom, (*h*) female condom, and (*i*) the patch.

vents the sperm from reaching the oocyte prevents conception. One method is to avoid intercourse during those times of the month when a secondary oocyte may be present. This is known as the *rhythm method* of conception control. Although at first glance it appears to be the simplest and least expensive, but determining just when a secondary oocyte is likely to be present can be very difficult. A woman with a regular 28-day menstrual cycle will typically ovulate about 14 days before the onset of the next menstrual flow. In order to avoid pregnancy, couples need to abstain from intercourse a few days before and after this date. However, if a woman has an irregular menstrual cycle, there may be only a few days each month for intercourse without the possibility of pregnancy. In addition to calculating safe days based on the length of the menstrual cycle, a woman can better estimate the time of ovulation by keeping a record of changes in her body temperature and vaginal pH. Both changes are tied to the menstrual cycle and can therefore help a woman predict ovulation. In particular, at about the time of ovulation, a woman has a slight rise in body temperature—less than 1°C. Thus, one should use an extremely sensitive thermometer. A digital-readout thermometer on the market spells out the word "yes" or "no." • **pH scale, p. 37**

Other methods of conception control that prevent the sperm from reaching the secondary oocyte include the diaphragm, cap, sponge, and condom. The diaphragm is a specially fitted membranous shield that is inserted into the vagina before intercourse and positioned so that it covers the cervix, which contains the opening of the uterus. Because of anatomical differences among females, diaphragms must be fitted by a physician. The effectiveness of the diaphragm is increased if spermicidal foam or jelly is also used. The vaginal cap functions in a similar way. The contraceptive sponge, as the name indicates, is a small amount of absorbent material that is soaked in a spermicide. The sponge is placed within the vagina, and chemically and physically prevents the sperm cells from reaching the oocyte. The contraceptive sponge is no longer available for use in the United States, but is still available in many other parts of the world.

The male condom is probably the most popular contraceptive device. It is a thin sheath that is placed over the erect penis before intercourse. In addition to preventing sperm from reaching the secondary oocyte, this physical barrier also helps prevent the transmission of the microbes that cause STDs, such as syphilis, gonorrhea, and AIDS, from being passed from one person to another during sexual intercourse (Outlooks 21.1). The most desirable condoms are made of a thin layer of latex that does not reduce the sensitivity of the penis. Latex condoms have also been determined to be the most effective in preventing transmission of the AIDS virus. The condom is most effective if it is prelubricated with a spermicidal material such as nonoxynol-9. This lubricant also has the advantage of providing some protection against the spread of the HIV virus.

Recently developed condoms for women are now available for use. One called the Femidom is a polyurethane sheath that, once inserted, lines the contours of the woman's vagina. It has an inner ring that sits over the cervix and an outer ring that lies flat against the labia. Research shows that this device protects against STDs and is as effective a contraceptive as the condom used by men.

The intrauterine device (IUD) is not a physical barrier that prevents the gametes from uniting. How this device works is not completely known. It may in some way interfere with the implantation of the embryo. The IUD must be fitted and inserted into the uterus by a physician, who can also remove it if pregnancy is desired. IUDs are used successfully in many countries. Current research with new and different intrauterine implants indicates that they are able to prevent pregnancy, and one is currently available in the United States. The IUD can also be used for "emergency contraception"—in cases of unprotected sex (forced sex) or failure of a conception control method (a condom slips or breaks). The IUD must be inserted within 7 days of unprotected sex.

Two contraceptive methods that require surgeries are tubal ligation and vasectomy (figure 21.15). Tubal ligation involves the cutting and tying off of the oviducts and can be done on an outpatient basis in most cases. An alternative to sterilization by tubal ligation (Essure) involves the insertion of small, flexible devices called Micro-inserts into each fallopian tube. Once inserted, tissue grows into the inserts, blocking the tubes. Ovulation continues as usual, but the sperm and egg cannot unite. Vasectomy can be performed in a physician's office and does not require hospitalization. A small opening is made above the scrotum, and the spermatic cord (vas deferens) is cut and tied. This prevents sperm from moving through the ducts to the outside. Because most of the

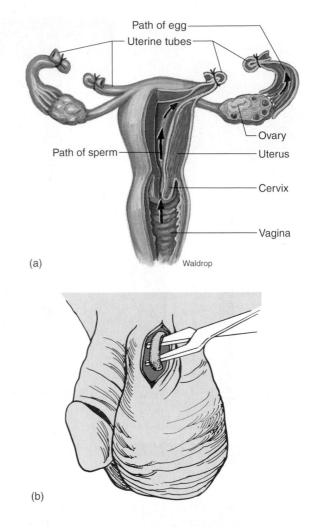

(a)

(b)

Figure 21.15

Tubal Ligation and Vasectomy
Two very effective contraceptive methods require surgery. Tubal ligation (a) involves severing the oviducts and suturing or sealing the cut ends. This prevents the sperm cell and the secondary oocyte from meeting. This procedure is generally considered ambulatory surgery, or at most requires a short hospitalization period. Vasectomy (b) requires minor surgery, usually in a clinic under local anesthesia. Following the procedure, minor discomfort may be experienced for several days. The severing and sealing of the vas deferens prevents the release of sperm cells from the body by ejaculation.

sperm-carrying fluid, called **semen,** is produced by the seminal vesicles, prostate gland, and bulbourethral glands, a vasectomy does not interfere with normal ejaculation. The sperm that are still being produced die and are reabsorbed in the testes. Neither tubal ligation nor vasectomy interferes with normal sex drives. However, these medical procedures are generally not reversible and should not be considered by those who may want to have children at a future date. A description of various contraceptive methods is summarized in table 21.2.

Sexually Transmitted Diseases

Diseases currently referred to as *sexually transmitted diseases (STDs)* were formerly called *venereal diseases (VDs)*. The term *venereal* is derived from the name of the Roman goddess for love, Venus. Although these kinds of illnesses are most frequently transmitted by sexual activity, many can also be spread by other methods of direct contact such as: hypodermic needles, blood transfusions, and blood-contaminated materials. Currently, the Centers for Disease Control and Prevention (CDC) in Atlanta, Georgia, recognizes over 20 diseases as being sexually transmitted (table 21.A).

The United States has the highest rate of sexually transmitted disease among industrially developed countries—65 million people (nearly one-fifth of the population) in the United States have incurable sexually transmitted diseases. The CDC estimates there are 15 million new cases of sexually transmitted diseases each year, nearly 4 million among teenagers. Table 21.B lists the most common STDs and estimates of the number of new cases each year. The portions of the public that are most at risk are teenagers, minorities, and women. Some of the most important STDs are described here because of their high incidence in the population and our inability to bring some of them under control. For example, there is no known cure for the HIV virus that is responsible for AIDS. There has also been a sharp rise in the number of gonorrhea cases in the United States caused by a form of the bacterium *Neisseria gonorrhoeae* that has become resistant to the drug penicillin by producing an enzyme that actually destroys the antibiotic. However, most of the infectious agents can be controlled if diagnosis occurs early and treatment programs are carefully followed by the patient.

The spread of STDs during sexual intercourse is significantly diminished by the use of condoms. Other types of sexual contact (i.e., hand, oral, anal) and congenital transmission (i.e., from the mother to the fetus during pregnancy) help maintain some of these diseases in the population at high enough levels to warrant attention by public health officials, the U.S. Public Health Service, the CDC, and state and local public health agencies. All of these agencies are involved in attempts to raise the general public health to a higher level. Their investigations have resulted in the successful control of many diseases and the identification of special problems, such as those associated with the STDs. Because the United States has an incidence rate of STDs that is 50 to 100 times higher than other industrially developed countries, there is still much that needs to be done.

All public health agencies are responsible for warning members of the public about things that may be dangerous to them. In order to meet these obligations when dealing with sexually transmitted diseases, such as AIDS and syphilis, they encourage the use of one of their most potent weapons, sex education. Individuals must know about their own sexuality if they are to understand the transmission and nature of STDs. Then it will be possible to alter their behavior in ways that will prevent the spread of these diseases. The intent is to present people with biological facts, not scare them. Public health officials do not have the luxury of advancing their personal opinions when it comes to their jobs. The biological nature of sexual behavior is not a moral issue, but biological facts are needed if people are to make intelligent decisions relating to their sexual behavior. It is hoped that, through education, people will alter their high-risk sexual behaviors and avoid situations where they could become infected with one of the STDs.

High-risk behaviors associated with contracting STDs include: sex with multiple partners and failing to use condoms. While some STDs are simply inconvenient or annoying, others severely compromise health and can result in death. As one health official stated, "We should be knowledgeable enough about our own sexuality and the STDs to answer the question, Is what I'm about to do worth dying for?"

Table 21.A

SEXUALLY TRANSMITTED DISEASES

Disease	Agent
Genital herpes	Virus
Gonorrhea	Bacterium
Syphilis	Bacterium
Acquired immunodeficiency syndrome (AIDS)	Virus
Candidiasis	Yeast
Chancroid	Bacterium
Genital warts	Virus
Gardnerella vaginalis	Bacterium
Genital *Chlamydia* infection	Bacterium
Genital cytomegalovirus infection	Virus
Genital *Mycoplasma* infection	Bacterium
Group B *Streptococcus* infection	Bacterium
Nongonococcal urethritis	Bacterium
Pelvic inflammatory disease (PID)	Bacterium
Molluscum contagiosum	Virus
Crabs	Body lice
Scabies	Mite
Trichomoniasis	Protozoan
Hepatitis B	Virus

Table 21.B

YEARLY ESTIMATES OF THE NUMBER OF NEW CASES OF SEXUALLY TRANSMITTED DISEASES*

Sexually Transmitted Disease	New Cases Each Year (Estimate)
Genital warts (Human papillomavirus)	5.5 million
Trichomoniasis	5 million
Chlamydia	3 million
Genital herpes	1 million
Gonorrhea	650,000
Hepatitis B	120,000
Syphilis	70,000
Human immunodeficiency virus (HIV) (AIDS)	40,000

*Data from the Centers for Disease Control and Prevention, *Tracking the Hidden Epidemics: Trends in STDs in the United States* 2000.

Table 21.2

EFFECTIVENESS OF VARIOUS METHODS OF BIRTH CONTROL

Birth Control Guide

Failure rates in this chart are based on information from clinical trials submitted to the Food and Drug Administration during product reviews. This number represents the percentage of women who become pregnant during the first year of use of a birth control method. For methods that the FDA does not review, such as periodic abstinence, numbers are estimated from published literature. For comparison, about 85 out of 100 sexually active women who wish to become pregnant would be expected to become pregnant in a year. The most effective way to avoid both pregnancy and sexually transmitted disease is to practice total abstinence (refrain from sexual contact).

This chart is a summary; it is not intended to be used alone. All product labeling should be followed carefully, and a health care professional should be consulted for some methods.

Type of Contraceptive	Description	Failure Rate (number of pregnancies expected per 100 women per year)	Some Risks (serious medical risks from contraceptives are rare)	Protection from Sexually Transmitted Diseases (STDs)
Male condom, latex/polyurethane	A sheath placed over the erect penis blocking the passage of sperm.	11 (a, b)	Irritation and allergic reactions (less likely with polyurethane).	Except for abstinence, latex condoms are the best protection against STDs, including herpes and AIDS.
Female condom	A lubricated, polyurethane sheath shaped similarly to the male condom. The closed end has a flexible ring that is inserted into the vagina.	21	Irritation and allergic reactions.	May give some STD protection; not as effective as latex condom.
Diaphragm with spermicide	A dome-shaped, rubber disk with a flexible rim that covers the cervix so that sperm cannot reach the uterus. A spermicide is applied to the diaphragm before insertion.	17 (b, d, e)	Irritation and allergic reactions, urinary tract infection. (c) Risk of toxic shock syndrome, a rare but serious infection, when kept in place longer than recommended.	None
Cervical cap with spermicide	A soft, rubber cup with a round rim that fits snugly around the cervix.	17 (b, d, e)	Irritation and allergic reactions, abnormal Pap test. (c) Risk of toxic shock syndrome, a rare but serious infection, when kept in place longer than recommended.	None
Sponge with spermicide	A disk-shaped polyurethane device containing the spermicide nonoxynol-9.	14–28 (d, e)	Irritation and allergic reactions, difficulty in removal. (c) Risk of toxic shock syndrome, a rare but serious infection, when kept in place longer than recommended.	None
Spermicide alone	A foam, cream, jelly, film, suppository, or tablet that contains nonoxynol-9, a sperm-killing chemical.	20–50 (studies have shown varying failure rates)	Irritation and allergic reactions, urinary tract infections. (c)	None
Oral contraceptives—combined pill	A pill that suppresses ovulation by the combined actions of the hormones estrogen and progestin.	1	Dizziness; nausea; changes in menstruation, mood, and weight; rarely, cardiovascular disease, including high blood pressure, blood clots, heart attack, and strokes.	None, except some protection against pelvic inflammatory disease.
Oral contraceptives—progestin-only minipill	A pill containing only the hormone progestin that reduces and thickens cervical mucus to prevent the sperm from reaching the egg.	2	Irregular bleeding, weight gain, breast tenderness, less protection against ectopic pregnancy.	None
Patch (Ortho Evra)	Skin patch worn on the lower abdomen, buttocks, or upper body that releases the hormones progestin and estrogen into the bloodstream.	1 (appears to be less effective in women weighing more than 198 pounds)	Similar to oral contraceptives—combined pill.	None

Table 21.2   *continued*

Type of Contraceptive	Description	Failure Rate (number of pregnancies expected per 100 women per year)	Some Risks (serious medical risks from contraceptives are rare)	Protection from Sexually Transmitted Diseases (STDs)
Vaginal contraceptive ring (NuvaRing)	A flexible ring about 2 inches in diameter that is inserted into the vagina and releases the hormones progestin and estrogen.	1	Vaginal discharge, vaginitis, irritation. Similar to oral contraceptives—combined pill.	None
Post-coital contraceptives (Preven and Plan B)	Pills containing either progestin alone or progestin plus estrogen.	Almost 80% reduction in risk of pregnancy for a single act of unprotected sex.	Nausea, vomiting, abdominal pain, fatigue, headache.	None
Injection (Depo-Provera)	An injectable progestin that inhibits ovulation, prevents sperm from reaching the egg, and prevents the fertilized egg from implanting in the uterus.	Less than 1	Irregular bleeding, weight gain, breast tenderness, headaches.	None
Injection (Lunelle)	An injectable form of progestin and estrogen.	Less than 1	Changes in menstrual cycle, weight gain. Similar to oral contraceptives—combined.	None
Implant (Norplant)	Six matchstick-sized rubber rods that are surgically implanted under the skin of the upper arm, where they steadily release the contraceptive steroid levonorgestrel.	Less than 1	Irregular bleeding, weight gain, breast tenderness, headaches, difficulty in removal.	None
IUD (intrauterine device)	A T-shaped device inserted into the uterus by a health professional.	Less than 1	Cramps, bleeding, pelvic inflammatory disease, infertility, perforation of uterus.	None
Periodic abstinence	To deliberately refrain from having sexual intercourse during times when pregnancy is more likely.	20	None	None
Surgical sterilization—female	The woman's fallopian tubes are blocked so the egg and sperm can't meet in the fallopian tube, preventing conception. (g)	Less than 1	Pain, bleeding, infection, other post-surgical complications.	None
Surgical sterilization—male	Sealing, tying, or cutting a man's vas deferens so that the sperm can't travel from the testicles to the penis. (g)	Less than 1	Pain, bleeding, infection, other minor postsurgical complications.	None

(a) Projected from 6-month study and adjusted for use of emergency contraception.

(b) If spermicides are used with barrier methods, be sure that the spermicide is compatible with the condom or diaphragm (won't cause it to weaken or break). Oil-based lubricants (such as petroleum jelly or baby oil) will cause latex to weaken and should not be used with these methods.

(c) Spermicides should not be used during pregnancy.

(d) Medications for vaginal yeast infections may decrease effectiveness of spermicides.

(e) Less effective for women who have had a baby because the birth process stretches the vagina and cervix, making it more difficult to achieve a proper fit.

(f) First approval date of currently marketed IUDs. Some IUDs were sold before premarket approval was required. Those products are no longer on the market.

(g) A contraceptive option for people who don't want children. Considered permanent because reversal is typically unsuccessful.

Source: Food and Drug Administration, August 2002.

21.10 Abortion

Another medical procedure often associated with birth control is abortion, which has been used throughout history. Abortion involves various medical procedures that cause the death and removal of the developing embryo. Abortion is obviously not a method of conception control; rather, it prevents the normal development of the embryo and causes its death. Abortion is a highly charged subject. Some people believe that abortion should be prohibited by law in all cases. Others think that abortion should be allowed in certain situations, such as in pregnancies that endanger the mother's life or in pregnancies that are the result of rape or incest. Still others think that abortion should be available to any woman under any circumstances. Regardless of the moral and ethical issues that surround abortion, it is still a common method of terminating unwanted pregnancies.

The abortion techniques used in the United States today all involve the possibility of infections, particularly if done by poorly trained personnel. The three most common techniques are scraping the inside of the uterus with special instruments (called a *D and C* or *dilation and curettage*), injecting a saline solution into the uterine cavity, or using a suction device to remove the embryo from the uterus. In the future, abortion may be accomplished by a medication prescribed by a physician. One drug, RU-486, is currently used in about 15% or more of the elective abortions in France. It has received approval for use in the United States. The medication is administered orally under the direction of a physician, and several days later, a hormone is administered. This usually results in the onset of contractions that expel the fetus. A follow-up examination of the woman is made after several weeks to ensure that there are no serious side effects of the medication.

21.11 Changes in Sexual Function with Age

Although there is a great deal of variation, somewhere around the age of 50, a woman's hormonal balance begins to change because of changes in the production of hormones by the ovaries. At this time, the menstrual cycle becomes less regular and ovulation is often unpredictable. The changes in hormone levels cause many women to experience mood swings and physical symptoms, including cramps and hot flashes. This period when the ovaries stop producing viable secondary oocytes and the body becomes nonreproductive is known as the **menopause.** Occasionally, the physical impairment becomes so severe that it interferes with normal life and the enjoyment of sexual activity, and a physician might recommend *hormone replacement therapy (HRT)* to augment the natural production of the hormones estrogen and progestin. Normally, the sexual enjoyment of a healthy woman continues during the time of menopause and for many years thereafter.

The U.S. Preventive Services Task Force (USPSTF) was set up by the United States Public Health Service to determine the merits of hormone replacement therapy. In 1996, the USPSTF recommended that physicians advise all women about the potential benefits and problems of HRT, while stating that the evidence was insufficient to recommend for or against HRT for all women after menopause. The Task Force concluded in 1996 that HRT might have long-term benefits for women with osteoporosis and heart disease, but that it might increase the risk for breast cancer.

Although recent evidence confirmed that HRT reduces the risk for bone fractures and may reduce the likelihood of developing colorectal cancer, the evidence has also confirmed that HRT increases the risk for blood clots, breast cancer, and stroke. Most importantly, new evidence indicates that HRT does not reduce the risk for coronary heart disease and that combined estrogen and progestin may actually increase the risk. The USPSTF now (2003) recommends against the routine use of combined estrogen and progestin for the prevention of chronic conditions in postmenopausal women because the problems are likely to outweigh the benefits for most women. So far the evidence isn't strong enough to recommend for or against the use of estrogen alone for the prevention of chronic conditions in postmenopausal women who have had a hysterectomy.

Human males do not experience a relatively abrupt change in their reproductive or sexual lives. Rather, their sexual desires tend to wane slowly as they age. They produce fewer sperm cells and less seminal fluid. Nevertheless, healthy individuals can experience a satisfying sex life during aging. Recent evidence also indicates that men also experience hormonal and emotional changes similar to those seen as women go through menopause. Human sexual behavior is quite variable. The same is true of older persons. The whole range of responses to sexual partners continues but generally in a diminished form. People who were very active sexually when young continue to be active, but are less active as they reach middle age. Those who were less active tend to decrease their sexual activity also. It is reasonable to state that one's sexuality continues from before birth until death.

SUMMARY

The human sex drive is a powerful motivator for many activities in our lives. Although it provides for reproduction and improvement of the gene pool, it also has a nonbiological, sociocultural dimension. Sexuality begins before birth, as sexual anatomy is determined by the sex-determining chromosome complement that we receive at fertilization. Females receive 2 X sex-determining chromosomes. Only one of these is functional; the other remains tightly coiled as a Barr body. A male receives 1 X and 1 Y sex-determining chromosomes. It is the presence of the Y chromosome that causes male development and the absence of a Y chromosome that allows female development.

At puberty, hormones influence the development of secondary sex characteristics and the functioning of gonads. As the ovaries and testes begin to produce gametes, fertilization becomes possible.

Sexual reproduction involves the production of gametes by meiosis in the ovaries and testes. The production and release of these gametes is controlled by the interaction of hormones. In males, each cell that undergoes spermatogenesis results in four sperm; in females, each cell that undergoes oogenesis results in one oocyte and two polar bodies. Humans have specialized structures for the support of the developing embryo, and many factors influence its development in the uterus. Successful sexual reproduction depends on proper hormone balance, proper meiotic division, fertilization, placenta formation, proper diet of the mother, and birth. Hormones regulate ovulation and menstruation and may also be used to encourage or discourage ovulation. Fertility drugs and birth-control pills, for example, involve hormonal control. In addition to the pill, a number of contraceptive methods have been developed, including the diaphragm, condom, IUD, spermicidal jellies and foams, contraceptive implants, the sponge, tubal ligation, and vasectomy.

Hormones continue to direct our sexuality throughout our lives. Even after menopause, when fertilization and pregnancy are no longer possible for a female, normal sexual activity can continue in both men and women.

THINKING CRITICALLY

For guidelines to answer this question, visit the Online Learning Center.

A great world adventurer discovered a tribe of women in the jungles of Brazil. After many years of very close study and experimentation, he found that sexual reproduction was not possible, yet women in the tribe were getting pregnant and having children. He also noticed that the female children resembled their mothers to a great degree and found that all the women had a gene that prevented meiosis. Ovulation occurred as usual, and pregnancy lasted 9 months. The mothers nursed their children for 3 months after birth and became pregnant the next month. This cycle was repeated in all the women of the tribe.

Consider the topics of meiosis, mitosis, sexual reproduction, and regular hormonal cycles in women, and explain in detail what may be happening in this tribe.

CONCEPT MAP TERMINOLOGY

Construct a concept map using these terms.

estrogen	sexual intercourse
hypothalamus	testosterone
placenta	X chromosome
puberty	Y chromosome
secondary sexual characteristics	zygote

Visit the Online Learning Center to use an interactive format to construct another concept map.

KEY TERMS

Use interactive flash cards, on the Online Learning Center, to help you learn the meanings of these terms.

androgens	oviduct
autosomes	ovulation
Barr bodies	penis
coitus	pituitary gland
conception	placenta
copulation	polar body
corpus luteum	primary oocyte
cryptorchidism	primary spermatocyte
differentiation	progesterone
ejaculation	puberty
estrogens	secondary oocyte
follicle	secondary sexual characteristics
follicle-stimulating hormone (FSH)	secondary spermatocyte
gametogenesis	semen
gonadotropin-releasing hormone (GnRH)	seminal vesicle
	seminiferous tubules
hypothalamus	sex-determining chromosome
inguinal canal	sexual intercourse
inguinal hernia	sexuality
interstitial cell-stimulating hormone (ICSH)	sperm
	spermatids
luteinizing hormone (LH)	spermatogenesis
masturbation	testes
menopause	testosterone
menstrual cycle	uterus
oogenesis	vagina
orgasm	X chromosome
ovary	Y chromosome
	zygote

LEARNING CONNECTIONS

Visit the Online Learning Center to find additional study resources including those highlighted below.

21.1   Sexuality from Different Points of View

Interactive Concept Map
• Components of sexuality

Experience This!
• Sex education in the school systems

Case Study
• Space sex?

1. Define the term *sexuality*.
2. Name one advantage of a sexually reproducing species.

21.2   Chromosomal Determination of Sex

Interactive Concept Map
• Variations on sex chromosome numbers

3. Describe the processes that cause about 50% of babies to be born male and 50% to be born female.
4. Name two genetic abnormalities associated with nondisjunction of chromosomes.

21.3   Male and Female Fetal Development

Case Study
• Embryonic stem cells: The future of organ transplants

Food for Thought
• Development

5. List the events that occur as an embryo matures.
6. What is cryptorchidism?

21.4   Sexual Maturation of Young Adults

7. What are the effects of secretions of the pituitary, the gonads, and adrenal glands at puberty in females?
8. What role does testosterone play in male sexual maturation?

21.5   Spermatogenesis

Animations and Review
• Introduction

Labeling Exercises
• Human male reproductive system
• Male reproductive system

Interactive Concept Map
• Sperm development

9. What structures are associated with the human male reproductive system? What are their functions?
10. Explain the path taken by sperm from their source until they exit the body.

21.6   Oogenesis

Animations and Review
• Female reproductive system

Labeling Exercise
• Sperm and egg anatomy

Interactive Concept Map
• Egg development

11. What structures are associated with the human female reproductive system? What are their functions?

12. What are the differences between oogenesis and spermatogenesis in humans?

21.7   Hormonal Control of Fertility

Food for Thought
• Hormones and multiple births

13. What changes occur in ovulation and menstruation during pregnancy?
14. How are ovulation and menses related to each other?

21.8   Fertilization and Pregnancy

Case Study
• The case of the embryos without parents
• Is Dr. Melissa Walker too old to have a baby?

Food for Thought
• Frozen embryos
• Treatment of critically ill newborns
• Artificial womb

15. What are the functions of the placenta?
16. What causes the genetic differences between identical and fraternal twins?

21.9   Contraception

Food for Thought
• AIDS vaccine

17. Describe hormonal methods of conception control.
18. Describe barrier methods of conception control.

21.10   Abortion

19. What is a D & C?
20. How does RU-486 work?

21.11   Changes in Sexual Function with Age

21. What is menopause?
22. How does sexual function change with age?

Test yourself with multiple choice questions that provide immediate feedback on the Online Learning Center with PowerWeb at

www.mhhe.com/enger11

The Origin of Life and Evolution of Cells

GOALS AND OBJECTIVES

Know the history of the scientific interest in the origin of life.

❏ Describe the differences between the concepts of spontaneous generation and biogenesis.

❏ Explain how scientists have conducted experiments that test the ideas of spontaneous generation and biogenesis.

Recognize that the nature of the Earth has changed considerably over the past 4.5 billion years.

❏ Describe the most probable physical conditions on early Earth and the changes thought to have happened before life could exist.

❏ Explain the experimental evidence for the origin of life from inorganic material.

❏ Describe different hypotheses for what the first living thing might have been like.

❏ Explain the chemical and physical events that must have occurred to have life originate on Earth.

❏ Describe the way in which organisms have caused the atmosphere of the Earth to change

Recognize the evolutionary changes necessary to give rise to the variety of living things we see on Earth today.

❏ Describe the relationships among Eubacteria, Archaea, and Eucarya.

❏ Explain the endosymbiotic theory

❏ State the order and approximate times for major evolutionary events

22.1 Spontaneous Generation Versus Biogenesis

For centuries humans have studied the basic nature of their environment. The vast amount of information presented in previous chapters is evidence of our ability to gather and analyze information. These efforts have resulted in solutions to many problems and have simultaneously revealed new and more challenging topics to study. Despite the growth in scientific knowledge, two questions have continued to be subjects of speculation: What is the nature of life? How did life originate?

In earlier times, no one ever doubted that life originated from nonliving things. The Greeks, Romans, Chinese, and many other ancient peoples believed that maggots arose from decaying meat; mice developed from wheat stored in dark, damp places; lice formed from sweat; and frogs originated from damp mud. The concept of **spontaneous generation**—the idea that living organisms arise from nonliving material—was proposed by Aristotle (384–322 B.C.) and became widely accepted until the seventeenth century. However, there were some who doubted this belief. These people subscribed to an opposing idea, called *biogenesis*. **Biogenesis** is the concept that life originates only from preexisting life.

One of the earliest challenges to the concept of spontaneous generation came in 1668. Francesco Redi, an Italian physician, set up a controlled experiment designed to disprove spontaneous generation (figure 22.1). He used two sets of jars that were identical except for one aspect. Both sets of jars contained decaying meat, and both were exposed to the atmosphere; however, one set of jars was covered by gauze, and the other was uncovered. Redi observed that flies settled on the meat in the open jar, but the gauze blocked their access to the covered jars. When maggots appeared on the meat in the uncovered jars but not on the meat in the covered ones, Redi concluded that the maggots arose from the eggs of the flies and not from spontaneous generation in the meat. • experiments, p. 6

Even after Redi's experiment, some people still supported the idea of spontaneous generation. After all, a belief that has been prevalent for over 2,000 years does not die a quick death. In 1748, John T. Needham, an English priest, placed a solution of boiled mutton broth in containers that he sealed with corks. Needham reasoned that boiling would kill any organisms in the broth and that the corks would prevent any microorganisms from entering the broth. When, after several days, the broth became cloudy with a large population of microorganisms, he concluded that life in the broth was the result of spontaneous generation.

In 1767, another Italian scientist, Abbe Lazzaro Spallanzani, challenged Needham's findings. Spallanzani boiled a meat and vegetable broth, placed this medium in clean glass containers, and sealed the openings by melting the glass over a flame. He placed the sealed containers in boiling water to make certain all microorganisms were destroyed. As a control, he set up the same conditions but did not seal the necks, allowing air to enter the flasks (figure 22.2). Two days later, the open containers had a large population of microorganisms, but there were none in the sealed containers.

Figure 22.1

Redi's Experiment
The two sets of jars here are identical in every way except one—the gauze covering. The set on the right had nothing special done to it. It is called the control group. The set on the left has been manipulated by placing a gauze covering over the jars. It is the experimental group. Any differences seen between the control and the experimental groups are the result of this single variable (the gauze coverings). In this manner, Redi concluded that the presence of maggots in meat was due to flies laying their eggs on the meat and not spontaneous generation.

Spallanzani's experiment did not completely disprove spontaneous generation to everyone's satisfaction. The supporters of spontaneous generation attacked Spallanzani by stating that he excluded air, a factor believed necessary for spontaneous generation. Supporters also argued that boiling had destroyed a "vital element." When Joseph Priestly discovered oxygen in 1774, the proponents of spontaneous generation claimed that oxygen was the "vital element" that Spallanzani had excluded in his sealed containers.

In 1861, the French chemist Louis Pasteur convinced most scientists that spontaneous generation could not occur. He placed a fermentable sugar solution and yeast mixture in a flask that had a long swan neck. The mixture and the flask were boiled for a long time. The flask was left open to allow oxygen, the "vital element," to enter, but no organisms developed in the mixture. The organisms that did enter the flask settled on the bottom of the curved portion of the neck and could not reach the sugar-water mixture. As a control, he cut off the swan neck (figure 22.3). This allowed microorganisms from the air to fall into the flask, and within two days the fermentable solution was supporting a population of microorganisms. In his address to the French Academy, Pasteur stated, "Never will the doctrine of spontaneous generation arise from this mortal blow."

22.2 Current Thinking About the Origin of Life

Although Pasteur thought that he had defeated those who believed in spontaneous generation and strongly supported biogenesis, today we still have modified forms of these two major contrasting concepts regarding the origin of life. One holds that

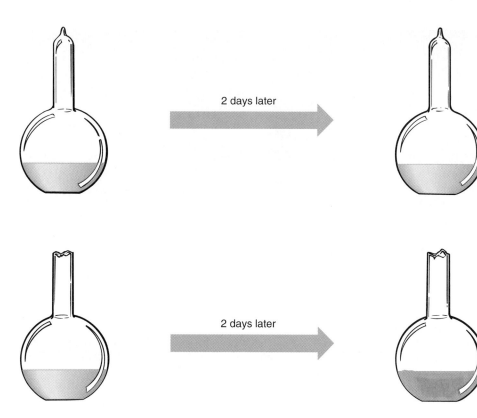

Figure 22.2

Spallanzani's Experiment
Spallanzani carried the experimental method of Redi one step further. He boiled a meat and vegetable broth and placed this medium into clean flasks. He sealed one and put it in boiling water. As a control, he subjected another flask to the same conditions, except he left it open. Within 2 days, the open flask had a population of microorganisms. Spallanzani maintained that this demonstrated that spontaneous generation did not occur and that something from the air was responsible for the growth in the broth.

life arrived on Earth from some extraterrestrial source (biogenesis) and the other maintains that life was created on Earth from nonliving material (spontaneous generation).

Extraterrestrial Origin for Life on Earth

Early in the 1900s, Svante Arrhenius proposed a different twist on biogenesis. His concept, called **panspermia**, hypothesized that life arose outside the Earth and that living things were transported to Earth serving to seed the planet with life. While his ideas had little scientific support at that time, his basic idea has since been revived and modified as a result of new evidence gained from space explorations, as you will see later in the chapter. However, panspermia does not explain *how* life arose originally. Explanations of how life might have originated are now focused on chemical evolution, suggesting that life arose from natural chemical processes and that these processes can be observed and eval-

uated by scientific experimentation. These hypotheses proposed that inorganic matter changed into organic matter composed of complex carbon-containing molecules and that these in turn combined to form the first living cell. It is important to recognize that we will probably never know for sure how life on Earth came to be, but it is interesting to speculate and examine the evidence related to this fundamental question.

The biogenesis concept (referred to as "directed panspermia") received renewed support when in 1969 in Murchison, Australia, a meteorite was found to contain amino acids and other complex organic molecules. In 1996 a meteorite from Antarctica was also analyzed. It has been known for many years that meteorites often contain organic molecules and this suggested that life may have existed elsewhere in the solar system. The chemical makeup of the Antarctic meteorite suggests that it was a portion of the planet Mars, which was ejected from Mars as a result of a collision between the planet and an

Figure 22.3

Pasteur's Experiment
Pasteur used swan–neck flasks that allowed oxygen, but not airborne organisms, to enter the flask. He broke the neck off another flask. Within 2 days, there was growth in this second flask but none in the intact flask. Pasteur demonstrated that it was not oxygen in the air that caused growth in the flasks but living things, which were prevented from entering the flask with the swan neck. This provided additional evidence against spontaneous generation.

asteroid. Analysis of the meteorite shows the presence of complex organic molecules and small globules that resemble those found on Earth that are thought to be the result of the activity of ancient micro-organisms. While scientists no longer think these structures are from micro-organisms, many still think it is quite possible Mars may have supported living things in the past. Since Mars currently has some water as ice and shows features that resemble dried up river systems, Mars may have had much more water in the past. For these reasons many believe it is reasonable to consider that life of a nature similar to that presently found on Earth could have existed on Mars.

Although previous exploration of Mars's surface has failed to find evidence for life, many researchers are not satisfied with the results. In June and July 2003, two spacecrafts were launched by the U.S. space agency to explore the surface of Mars. One of the important goals of these missions is to search for signs of present or past life. The spacecrafts will arrive at Mars in January 2004 and the rovers they contain will land on opposite sides of Mars. The rovers will spend several months gathering information about the surface of Mars. Perhaps they will collect information that answers the question "Did life exist on Mars?"

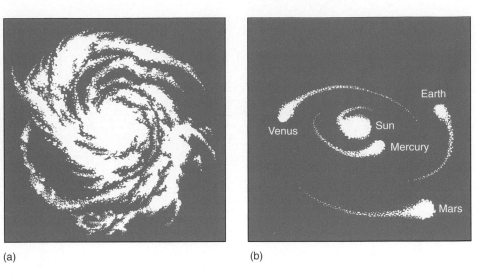

(a)   (b)

Figure 22.4

Formation of Our Solar System
As gravity pulled the gases and other elements toward the center, the accumulating matter resulted in the formation of the Sun (*a*). In other regions, smaller gravitational forces caused the formation of the Sun's planets (*b*).

Earth Origin for Life on Earth

The alternative view that life originated on the Earth has also received support. Let us look at several lines of evidence.

1. The Earth is the only planet in our solar system with a temperature range that allows for water to exist as a liquid on its surface, and water is the most common compound in most kinds of living things.

2. Analysis of the atmospheres of other planets shows that they all lack oxygen. The oxygen in the Earth's atmosphere is the result of current biological activity. Therefore, before life on Earth the atmosphere probably lacked oxygen.

3. Experiments demonstrate that organic molecules can be generated in an atmosphere that lacks oxygen.

4. Because it is assumed that all of the planets have been cooling off as they age, it is very likely that the Earth was much hotter in the past. The large portions of the Earth's surface that are of volcanic origin strongly suggest a hotter past. There is also the likelihood that various large bodies collided with the Earth early in its history and that they could have led to increased temperatures at least in the site of the collision.

5. Recognition that there are distinct prokaryotic organisms that live in extreme environments of high temperature, high salinity, low pH, or the absence of oxygen suggests that they may have been adapted to life in a world that is very different from today's Earth. These kinds of organisms are found today in unusual locations such as hot springs and around thermal vents in the ocean floor and may be descendants of the first organisms formed on the primitive Earth.

22.3 The "Big Bang" and the Origin of the Earth

As astronomers and others look at the current stars and galaxies, it can be observed that they are moving apart from one another. This and other evidence has led to the concept that our current universe began as a very dense mass of matter that had a great deal of energy. This dense mass of matter exploded in a "big bang" about 13 billion years ago that resulted in the formation of atoms. According to this scientific theory, the original universe consisted primarily of atoms of hydrogen and helium. The *protoplanet nebula model* proposes that the solar system was formed from a large cloud of gases and elements formed from previously existing stars (figure 22.4). The simplest and most abundant gases would have been hydrogen and helium but other heavier elements were present as well. A gravitational force was created by the collection of particles within this cloud that caused other particles to be pulled from the outer edges to the center. As particles collected into larger bodies, gravity increased and more particles were attracted to the bodies. Ultimately a central body (the Sun) was formed and several other bodies (planets) formed that moved around it (How Science Works 22.1). The Sun consists primarily of

HOW SCIENCE WORKS 22.1

Gathering Information About the Planets

Our exploration of our solar system has become quite sophisticated with the development of space travel. Only the planet Pluto has not been visited by a spacecraft. Spacecraft that took close-up pictures of the planets as they traveled past them on their way to other solar systems have visited the distant planets of Jupiter, Saturn, and Neptune. The *Galileo* spacecraft released a probe that entered the atmosphere of Jupiter in 1996. In addition to the Earth's Moon, Mars and Venus have had spacecraft land on their surfaces and send back pictures and data about temperature, atmospheric composition, and the geologic nature of their surfaces.

Several characteristics of planets affect how likely it is that life could be found or might have been present on a planet. The distance from the Sun determines the amount of energy received from the Sun. Planets that are near the Sun receive more solar energy and distant planets receive less. The larger the mass of a planet the greater its force of gravity. The force of gravity will influence how many other bodies it can capture (moons) and how much atmosphere it can hold. Some of the planets are gases, whereas others have solid surfaces. In order for life as we know it on Earth to exist a planet must have liquid water, an appropriate atmosphere, and a solid surface. The only planet that may have had these conditions at one time is Mars.

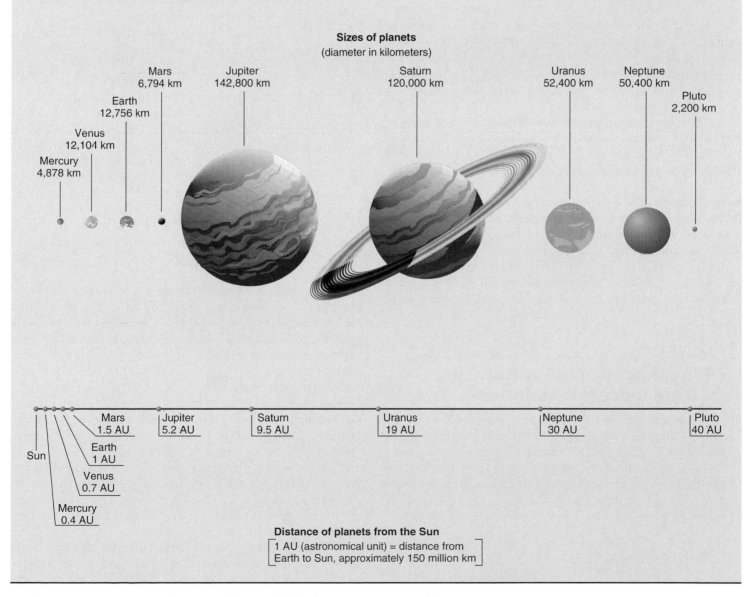

Sizes of planets
(diameter in kilometers)

Mercury 4,878 km
Venus 12,104 km
Earth 12,756 km
Mars 6,794 km
Jupiter 142,800 km
Saturn 120,000 km
Uranus 52,400 km
Neptune 50,400 km
Pluto 2,200 km

Sun
Mercury 0.4 AU
Venus 0.7 AU
Earth 1 AU
Mars 1.5 AU
Jupiter 5.2 AU
Saturn 9.5 AU
Uranus 19 AU
Neptune 30 AU
Pluto 40 AU

Distance of planets from the Sun
1 AU (astronomical unit) = distance from Earth to Sun, approximately 150 million km

hydrogen and helium atoms, which are being fused together to form larger atoms with the release of large amounts of thermonuclear energy. Many scientists believe that Earth—along with other planets, meteors, asteroids, and comets—was formed at least 4.6 billion years ago. A large amount of heat was generated as the particles became concentrated to form Earth. Geologically, this is called the "Hadean Eon." The term Hadean means "hellish." Although not as hot as the Sun, the material of Earth formed a molten core that became encased by a thin outer crust as it cooled. In its early stages of formation, about 4 billion years ago, there may have been a considerable amount of volcanic activity on Earth.

Physically, Earth was probably much different than it is today. Because the surface was hot, there was no water on the surface or in the atmosphere. In fact, the tremendous amount of heat probably prevented any atmosphere from forming. The gases associated with our present atmosphere (nitrogen, oxygen, carbon dioxide, and water vapor) were contained in the planet's molten core. These hostile conditions (high temperature, lack of water, lack of atmosphere) on early Earth could not have supported any form of life similar to what we see today.

Over hundreds of millions of years, Earth is thought to have slowly changed. As it cooled, volcanic activity probably caused the release of water vapor (H_2O), carbon dioxide (CO_2), methane (CH_4), ammonia (NH_3), and hydrogen (H_2), and the early atmosphere was formed. These gases formed a **reducing atmosphere**—an atmosphere that did not contain molecules of oxygen (O_2). Any oxygen would have quickly combined with other atoms to form compounds, so a significant quantity of molecular oxygen would have been highly unlikely. Further cooling enabled the water vapor in the atmosphere to condense into droplets of rain. The water ran over the land and collected to form the oceans we see today.

22.4 Steps Needed to Produce Life from Inorganic Materials

When we consider the nature of the simplest forms of life today, we find that living things consist of an outer membrane that separates the cell from its surroundings, genetic material in the form of nucleic acids, and many kinds of enzymes that control the activities of the cell. Therefore, when we speculate about the origin of life from inorganic material it seems logical that several events or steps were necessary:

1. Organic molecules must first be formed from inorganic molecules.

2. Simple organic molecules must combine to form larger organic molecules like RNA, proteins, carbohydrates, and lipids.

3. A molecule must serve as genetic material.

4. Some molecules must function as enzymes.

5. Genetic material must become self-replicating.

6. The molecules that serve as genetic material and other large organic molecules must be collected together and segregated from their surroundings by a membrane.

7. The first life-forms would need a way to obtain energy from their surroundings in order to maintain their complex structure.

Formation of the First Organic Molecules

In the 1920s, a Russian biochemist, Alexander I. Oparin, and a British biologist, J. B. S. Haldane, working independently, proposed that the first organic molecules were formed spontaneously in the reducing atmosphere thought to be present on the early Earth. The molecules of water vapor, ammonia, methane, carbon dioxide, and hydrogen supplied the atoms of carbon, hydrogen, oxygen, and nitrogen, and lightning, heat from volcanoes, and ultraviolet radiation furnished the energy needed for the synthesis of simple organic molecules (figure 22.5). It is important to understand the significance of a reducing atmosphere to this proposed mechanism for the origin of life. The absence of oxygen in the atmosphere would have allowed these organic molecules to remain and combine with one another. This does not happen today because organic molecules are either consumed by organisms or oxidized to simpler inorganic compounds because of the oxygen present in our current atmosphere. For example, today many kinds of organic air pollutants (hydrocarbons) eventually degrade into smaller molecules in the atmosphere. Unfortunately they participate in the formation of photochemical smog as they are broken down. • **organic molecules, p. 42**

After simple organic molecules were formed in the atmosphere, they probably would have been washed from the air and carried into the newly formed oceans by rain and runoff from the land. Here, the molecules could have reacted with one another to form the more complex molecules of simple sugars, amino acids, and nucleic acids. This accumulation is thought to have occurred over half a billion years, resulting in oceans that were a dilute organic soup. These simple organic molecules in the ocean served as the building materials for more complex organic macromolecules, such as complex carbohydrates, proteins, lipids, and nucleic acids. Recognize that all the ideas presented so far cannot be confirmed by direct observation because we cannot go back in time. However, several assumptions central to this model for the origin of life have been laboratory tested.

In 1953 Stanley L. Miller conducted an experiment to test the idea that organic molecules could be synthesized in a reducing environment. Miller constructed a simple model of the early Earth's atmosphere (figure 22.6). In a glass apparatus he placed distilled water to represent the early oceans. Adding hydrogen, methane, and ammonia to the water simulated the reducing atmosphere. Electrical sparks provided the energy needed to produce organic

- Carbon
- Nitrogen
- Oxygen
- Hydrogen

Figure 22.5

Formation of Organic Molecules in the Atmosphere

The environment of the primitive Earth was harsh and lifeless. But many scientists believe that it contained the necessary molecules and sources of energy to fashion the first living cell. The energy furnished by volcanoes, lightning, and ultraviolet light could have broken the bonds in the simple inorganic molecules in the atmosphere. New bonds could have formed as the atoms from the smaller molecules were rearranged and bonded to form simple organic compounds in the atmosphere. Rain and runoff from the land would have carried these chemicals into the oceans. Here they could have reacted with each other to form more complex organic molecules.

From electricity source

Water
Ammonia
Methane
Hydrogen

Gases

Spark discharging

To vacuum pump

Boiling water

Water containing organic molecules

Figure 22.6

Miller's Apparatus

Stanley Miller developed this apparatus to demonstrate that the spontaneous formation of complex organic molecules could take place in a reducing atmosphere.

compounds. By heating parts of the apparatus and cooling others, he simulated the rains that are thought to have fallen into the early oceans. After a week of operation, he removed some of the water from the apparatus. When this water was analyzed, it was found to contain many simple organic compounds. Although Miller demonstrated nonbiological synthesis of simple organic molecules like amino acids and simple sugars, his results did not account for complex organic molecules like proteins and nucleic acids (e.g., DNA). However, other researchers produced some of the components of nucleic acid under similar primitive conditions.

Formation of Macromolecules

Several ideas have been proposed for the concentration of simple organic molecules and their combination into macromolecules. The first hypothesis suggests that a portion of the early ocean could have been separated from the main ocean by geologic changes. The evaporation of water from this pool could have concentrated the molecules, which might have led to the formation of macromolecules by dehydration synthesis. Second, it has been proposed that freezing may have been the means of concentration. When a mixture of alcohol and water is placed in a freezer, the water freezes solid and the alcohol becomes concentrated into a small portion of liquid. A similar process could have occurred on Earth's early surface, resulting in the concentration of simple organic molecules. In this concentrated solution, dehydration synthesis in a reducing atmosphere could have occurred, resulting in the formation of macromolecules. A third theory proposes that clay particles may have been a factor in concentrating simple organic molecules. Small particles of clay have electrical charges that can attract and concentrate organic molecules like protein from a watery solution. Once the molecules became concentrated, it would have been easier for them to interact to form larger macromolecules.

The Origin of Genetic Material

The reproduction of most current organisms involves the replication of DNA and the distribution of the copied DNA to subsequent cells (those that do not use DNA use RNA as their genetic material). DNA is responsible for the manufacture of RNA, which subsequently leads to the manufacture of proteins. Some of the proteins produced serve as enzymes that control chemical reactions. This is the central dogma of modern molecular biology. However, it is difficult to see how this complicated sequence of events, which involves many steps and the assistance of several enzymes, could have been generated spontaneously, so scientists have looked for simpler systems that could have led to the DNA system we see today.

Science works simultaneously on several fronts. Scientists involved in studying the structure and function of viruses discovered that many viruses do not contain DNA but store their genetic information in the structure of RNA. In order for these RNA-viruses to reproduce, they must enter a cell. The cell makes copies of the RNA and proteins necessary to make more of the virus. Certain plant diseases are caused by viroids that consist of naked RNA, which enters cells and causes the cells to make copies of the RNA. In other words, in both of these cases, RNA serves as genetic material.

Other scientists who study viral diseases find that it is difficult to develop vaccines for many viral diseases because their genetic material easily mutates. Because of this, researchers have been studying the nature of viral DNA or RNA to see what causes the high rate of mutation. This has led others to explore the RNA viruses and ask the question: Can RNA replicate itself without DNA? This is an important question, because if RNA can replicate itself it would have all of the properties necessary to serve as genetic material. It could store information, mutate, and make copies of itself.

Other research about the nature of RNA provides interesting food for thought. RNA can be assembled from simpler subunits that could have been present on the early Earth. Scientists have also shown that RNA molecules are able to make copies of themselves without the need for enzymes, and they can do so without being inside cells. Furthermore, some RNA molecules serve as enzymes for specific reactions. This new evidence suggests that RNA may have been the first genetic material and helps solve one of the

Figure 22.7

Stromatolites in Australia
This photo of stromatolites was taken at Hamelin Pool, Western Australia, a marine nature reserve. The dome-shaped structures shown in the photograph are composed of cyanobacteria and materials they secrete, and grow up to 60 centimeters tall. By taking samples from fossils of similar structures and cutting them into slices or sections, microscopic images can be produced that display some of the world's oldest cells.

problems associated with the origin of life: how genetic information was stored. Because RNA is a much simpler molecule than DNA and can make copies of itself without the aid of enzymes, perhaps it was the first genetic material. Once a primitive life-form had the ability to copy its genetic material it would be able to reproduce. Reproduction is one of the most fundamental characteristics of living things.

Isolating Organic Molecules

It has been estimated that the formation of simple organic molecules in the atmosphere began about 4 billion years ago and lasted approximately 1.5 billion years. The oldest known fossils of living cells are thought to have formed 3.5 billion years ago. Fossilized, photosynthetic bacteria have been found in geological formations called *stromatolites* on the coasts of South Africa and Western Australia (figure 22.7). The question is, How do you get from the spontaneous formation of macromolecules to primitive cells in half a billion years?

Three hypotheses are proposed for the formation of **prebionts,** nonliving structures that led to the formation of the first living cells. Oparin speculated that a prebiont consisted of carbohydrates, proteins, lipids, and nucleic acids that accumulated to form a **coacervate.** Such a structure could have consisted of a collection of organic macromolecules surrounded by a film of water molecules. This arrangement of water molecules, although not a membrane, could have functioned as a physical barrier between the organic

Figure 22.8

Coacervates
One hypothesis proposes that a film of water that acted as a primitive cell membrane could have surrounded organic molecules forming a structure that resembles a living cell. Such a structure can easily be produced in the lab.

molecules and their surroundings. They could selectively take in materials from their surroundings and incorporate them into their structure.

Coacervates have been synthesized in the laboratory (figure 22.8). They can selectively absorb molecules from the surrounding water and incorporate them into their structure. Also, the chemicals within coacervates have a specific arrangement—they are not random collections of molecules. Some coacervates contain proteins (enzymes) that direct a specific type of chemical reaction. Because they lack a definite membrane, no one claims coacervates are alive, but they do exhibit some lifelike traits: They are able to increase in size and will split into smaller particles if the right conditions exist.

An alternative hypothesis is that this early prebiotic cell structure could have been a *microsphere*. A **microsphere** is a nonliving collection of organic macromolecules with a double-layered outer boundary. Sidney Fox demonstrated the ability to build microspheres from *proteinoids*. **Proteinoids** are proteinlike structures consisting of branched chains of amino acids. Proteinoids are formed by reacting amino acids at a temperature of 180°C. Fox, from the University of Miami, showed that it is feasible to combine single amino acids into polymers of proteinoids. He also demonstrated the ability to build microspheres from these proteinoids.

Microspheres can be formed when proteinoids are placed in boiling water and slowly allowed to cool. Some of the proteinoid material produces a double-boundary structure that encloses the microsphere. Although these walls do not contain lipids, they do exhibit some membranelike characteristics and suggest the structure of a cellular membrane. Microspheres swell or shrink depending on the osmotic potential in the surrounding solution. They also display a type of internal movement (streaming) similar to that exhibited by cells and contain some proteinoids that function as enzymes. Using ATP as a source of energy, microspheres can direct the formation of polypeptides and nucleic acids. They can absorb material from the surrounding medium and form buds, which results in a second generation of microspheres.

A third possibility is that a membrane could form as a result of lipid materials interacting with water. Lipids do not mix with water and whenever lipids are mixed with water, spherical structures form. You have seen this in vinegar and oil salad dressing.

There is no way to know if any of these models represents what really happened in the origin of living things. However, some kind of structure was necessary to separate the complex, organized molecules from the watery environment in which they were dissolved. Once this separation occurred, the structures and their contents could be called **protocells.**

Toward the Central Dogma

The laboratory synthesis of coacervates and microspheres and the recognition that RNA can serve as enzymes and genetic material help us understand how the first primitive living cells might have developed. However, it leaves a large gap in our understanding because it does not explain how these first cells might have become the highly complex, living cells we see today. Once the first primitive cell existed, several additional steps were required to reach our current state of complexity:

1. Proteins must become the catalysts (enzymes) of the cells.

2. The control of protein synthesis must be taken over by RNA.

3. DNA must replace RNA as the self-replicating genetic material of the cell.

4. Ultimately these first cellular units must be able to reproduce more of themselves.

5. These first cellular units needed a way to obtain energy from their surroundings

Research into these questions will continue. However, there has been a great deal of thought about the last of these questions: What kind of metabolism did the first living things utilize?

Meeting Metabolic Needs— Heterotrophs or Autotrophs

Fossil evidence indicates that there were primitive forms of life on Earth about 3.5 billion years ago. Regardless of how they developed, these first primitive cells would have needed a way to add new organic molecules to their structures as previously existing molecules were lost or destroyed. There are two ways to accomplish this. **Heterotrophs** capture organic molecules such as sugars, amino acids, or organic acids from their surroundings, which they use to make new molecules and provide themselves with a source of energy. **Autotrophs** use some external energy source such as sunlight or the energy from inorganic chemical reactions to allow them to combine simple inorganic molecules like water and carbon dioxide to make new organic molecules. These new organic molecules can then be used as building materials for new cells or can be broken down at a later date to provide a source of energy.

Many scientists support the idea that the first living things produced on Earth were heterotrophs that lived off the organic molecules that would have been found in the oceans. Because the early heterotrophs are thought to have developed in a reducing atmosphere that lacked oxygen, they would have been of necessity anaerobic organisms; therefore they did not obtain the maximum amount of energy from the organic molecules they obtained from their environment. At first, this would not have been a problem. The organic molecules that had been accumulating in the ocean for millions of years served as an ample source of organic material for the heterotrophs. However, as the population of these primitive heterotrophs increased through reproduction, the supply of organic material would have been consumed faster than it was being spontaneously produced in the atmosphere. If there was no other source of organic compounds, they would have eventually exhausted their nutrient supply, and would have become extinct.

Even though the early heterotrophs probably contained nucleic acids and were capable of producing enzymes that could regulate chemical reactions, they probably carried out a minimum of biochemical activity. There is evidence to suggest that a wide variety of compounds were present in the early oceans, some of which could have been used unchanged by primitive cells. There was no need to modify the compounds to meet their needs.

Those compounds that could be easily used by these cells would have been the first to become depleted from the early environment. However, some of the heterotrophs may have contained a mutated form of nucleic acid, which allowed them to convert material that was not directly usable into a compound that could be used. Mutations may have been common because the amount of ultraviolet light, one cause of mutations, would have been high. The absence of ozone in the upper atmosphere of the early Earth would have allowed high amounts of ultraviolet light to reach the Earth's surface. Heterotrophic cells with such mutations could have survived, whereas those without it would have become extinct as the compounds they used for food became

scarce. It has been suggested that through a series of mutations in the early heterotrophic cells, a more complex series of biochemical reactions originated within some of the cells. Such cells could use chemical reactions to convert unusable chemicals into usable organic compounds.

As with many areas of science there are often differences of opinion. Although this heterotroph hypothesis for the origin of living things was the prevailing theory for many years, recent discoveries have caused many scientists to consider an alternative—that the first organism was an autotroph. Several kinds of information support this theory. Many members of the Domain Archaea, are autotrophic and live in extremely hostile environments. For this reason, they are referred to as "extremophiles" (lovers of extremes).

Today these organisms are found in hot springs like those found in Yellowstone National Park; Kamchatka, Russia (Siberia); or near hot thermal vents—areas where hot, mineral-rich water enters seawater from the deep ocean floor. They use inorganic chemical reactions as a source of energy to allow them to synthesize organic molecules from inorganic components. Because of this they are called *chemoautotrophs*. If their ancient ancestors had similar characteristics, it is possible that the first organisms may have been autotrophs. The nutrients they could have utilized most likely were CO_2, CO, H_2, H_2S, N_2, and S. The end products of their metabolism were probably such compounds as H_2SO_4, CH_4, and H_2O.

The fact that today many of these organisms live in very hot environments suggests that they may have originated on an Earth that was much hotter than it is currently. There is much evidence that the Earth was a much hotter place in the past. If the first organisms were autotrophs there could have been subsequent evolution of a variety of kinds of cells, both autotrophic and heterotrophic, that could have led to the diversity of different prokaryotic cells seen today in the Domains Eubacteria and Archaea (see chapter 4).

As a result of this discussion you should understand that we do not know how life on Earth originated. Scientists look at many kinds of evidence and continue to explore new avenues of research. So we currently have three competing theories for the origin of life on Earth:

1. Life arrived from some extraterrestrial source (directed panspermia/biogenesis).

2. Life originated on Earth as a heterotroph (spontaneous generation).

3. Life originated on Earth as an autotroph (spontaneous generation).

22.5 Major Evolutionary Changes in the Nature of Living Things

Once living things existed and had a genetic material that stored information but was changeable (mutational), living things could have proliferated into a variety of kinds that were adapted to specific environmental conditions.

Remember that the Earth has not been static but has been changing as a result of its cooling, volcanic activity, and encounters with asteroids. In addition, the organisms have had an impact on the way in which the Earth has developed. Regardless of the way in which life originated on Earth, there have been several major events in the subsequent evolution of living things.

The Development of an Oxidizing Atmosphere

Ever since its formation, Earth has undergone constant change. In the beginning, it was too hot to support an atmosphere. Later, as it cooled and as gases escaped from volcanoes, a reducing atmosphere (lacking oxygen) was likely to have been formed. The early life-forms would have lived in this reducing atmosphere. However, today we have an oxidizing atmosphere and most organisms use this oxygen as a way to extract energy from organic molecules through a process of aerobic respiration. But what caused the atmosphere to change? Today it is clear that the oxygen in our atmosphere is the result of the process of photosynthesis.

There are several kinds of eubacteria that perform photosynthesis. Several of these perform a type of photosynthesis that does not result in the release of oxygen. However, one major group, the cyanobacteria, utilize a form of photosynthesis that results in the release of oxygen. Therefore, it seems plausible that the first organisms, regardless of whether they were heterotrophs or chemoautotrophs, may have accumulated many mutations over time that could have resulted in photosynthetic autotrophs. Since oxygen is released from the most common form of photosynthesis, this would have resulted in the development of an oxygen-containing **oxidizing atmosphere.**

The development of an oxidizing atmosphere created an environment unsuitable for the continued formation of organic molecules. Organic molecules tend to break down (oxidize) when oxygen is present. The presence of oxygen in the atmosphere would make it impossible for life to spontaneously originate in the manner described earlier in this chapter because an oxidizing atmosphere would not allow the accumulation of organic molecules in the seas. However, new life is generated through reproduction, and new *kinds* of life are generated through mutation and evolution. The presence of oxygen in the atmosphere had one other important outcome: It opened the door for the evolution of aerobic organisms.

It appears that an oxidizing atmosphere began to develop about 2.3 billion years ago. Although various chemical reactions released small amounts of molecular oxygen into the atmosphere, it was photosynthesis that generated most of the oxygen. The oxygen molecules also reacted with one another to form ozone (O_3). Ozone collected in the upper atmosphere and acted as a screen to prevent most of the ultraviolet light from reaching Earth's surface. The reduction of ultraviolet light diminished the spontaneous formation of complex organic molecules. It also reduced the number of mutations in cells. In an oxidizing atmosphere, it was no longer possible for organic molecules to accumulate over millions of years to be later incorporated into living material.

The appearance of oxygen in the atmosphere also allowed for the evolution of aerobic respiration. Because the first heterotrophs were of necessity anaerobic organisms, they did not derive large amounts of energy from the organic materials available as food. With the evolution of aerobic heterotrophs, there could be a much more efficient conversion of food into usable energy. Aerobic organisms would have a significant advantage over anaerobic organisms: They could use the newly generated oxygen as a final hydrogen acceptor and, therefore, generate many more ATPs (adenosine triphosphates) from the food molecules they consumed.

The Establishment of Three Major Domains of Life

In 1977 Carl Woese published the idea that the "bacteria" (organisms that lack a nucleus), which had been considered a group of similar organisms, were really made up of two very different kinds of organisms: the Eubacteria and Archaea. Furthermore, the Archaea shared some characteristics with eukaryotic organisms. Subsequent investigations have supported these ideas and led to an entirely different way of looking at the classification and evolution of living things. Although biologists have traditionally divided organisms into kingdoms based on their structure and function, it was very difficult to do this with microscopic organisms. With the newly developed ability to decode the sequence of nucleic acids, it became possible to look at the genetic nature of organisms without being confused by their external structures. Woese studied the sequences of ribosomal RNA and compared similarities and differences. As a result of his studies and those of many others a new concept of the relationships between various kinds of organisms has emerged.

The three main kinds of living things, Eubacteria, Archaea, and Eucarya, have been labeled "domains." Within each domain there are several kingdoms. In the Eucarya there are four kingdoms that we already recognize: Animalia, Plantae, Fungi, and Protista. However, previously all of the Eubacteria and Archaea have been lumped into the same kingdom: Prokaryote. It has become clear that there are great differences between the Eubacteria and Archaea, and within each of these groups there are greater differences than are found among the other four kingdoms (Animalia, Plantae, Fungi, and Protista).

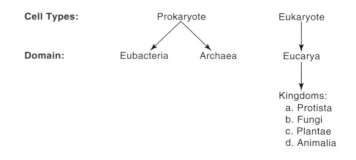

Table 22.1

MAJOR DOMAINS OF LIFE

Eubacteria	Archaea	Eucarya
No nuclear membrane.	No nuclear membrane.	Nuclear membrane present.
Chlorophyll-based photosynthesis is a bacterial invention but most can function anaerobically.	None have been identified as pathogenic.	Many kinds of membranous organelles present in cells.
Oxygen-generating photosynthesis was an invention of the cyanobacteria.	Probably have a common ancestor with Eucarya.	Chloroplasts are probably derived from cyanobacteria.
Large number of heterotrophs oxidize organic molecules for energy.	Many obtain energy from inorganic reactions to make organic matter.	Mitochondria probably derived from certain aerobic bacteria.
Some use energy from inorganic chemical reactions to produce organic molecules.	Typically found in extreme environments.	Probably have a common ancestor with Archaea.
Some live at high temperatures and may be ancestral to Archaea.	Few heterotrophs.	Common evolutionary theme is the development of complex cells through endosymbiosis with other organisms.
Much metabolic diversity among closely related organisms.	Example: *Pyrolobus fumarii,* deep-sea hydrothermal vents, hot springs, volcanic areas, growth to 113°C	Example: *Homo sapiens,* all cells of the human body
Example: *Streptococcus pneumoniae,* one cause of pneumonia		

This new picture of living things requires us to reorganize our thinking. It appears that the oldest organisms gave rise to two major types of prokaryotic organisms (Eubacteria and Archaea). Perhaps most startling is the idea that the Archaea and Eucarya share many characteristics, suggesting that they are more closely related to each other than either is to the Eubacteria.

It appears that each domain developed specific abilities. The Archaea are primarily organisms that use inorganic chemical reactions to generate the energy they need to make organic matter. Often these reactions result in the production of methane (CH_4). These organisms are known as methanogens. Others use sulfur and produce hydrogen sulfide (H_2S). Most of these organisms are found in extreme environments such as hot springs or in extremely salty or acid environments.

The Eubacteria developed many different metabolic abilities. Today many are able to use organic molecules as a source of energy, some are able to carry on photosynthesis, and still others are able to get energy from inorganic chemical reactions similar to Archaea.

The Eucarya are the most familiar and appear to have exploited the metabolic abilities of other organisms by incorporating them into their own structure. Chloroplasts and mitochondria are both bacterialike structures found inside eukaryotic cells. Table 22.1 summarizes the major characteristics of these three domains.

The Origin of Eukaryotic Cells

The earliest fossils appear to be similar in structure to that of present-day bacteria. Therefore it is likely that the early heterotrophs and autotrophs were probably simple one-celled organisms like bacteria. They were **prokaryotes** that lacked nuclear membranes and other membranous organelles, such as mitochondria, an endoplasmic reticulum, chloroplasts, and a Golgi apparatus. Present-day bacteria and archaea are prokaryotes. All other forms of life are **eukaryotes,** which possess a nuclear membrane and other membranous organelles.

Biologists generally believe that the eukaryotes evolved from the prokaryotes. The **endosymbiotic theory** attempts to explain this evolution. This theory suggests that present-day eukaryotic cells evolved from the combining of several different types of primitive prokaryotic cells. It is thought that some organelles found in eukaryotic cells may have originated as free-living prokaryotes. For example, because mitochondria and chloroplasts contain DNA and ribosomes that resemble those of bacteria, control their own reproduction, and synthesize their own enzymes, it has been suggested that they were once free-living prokaryotes. These

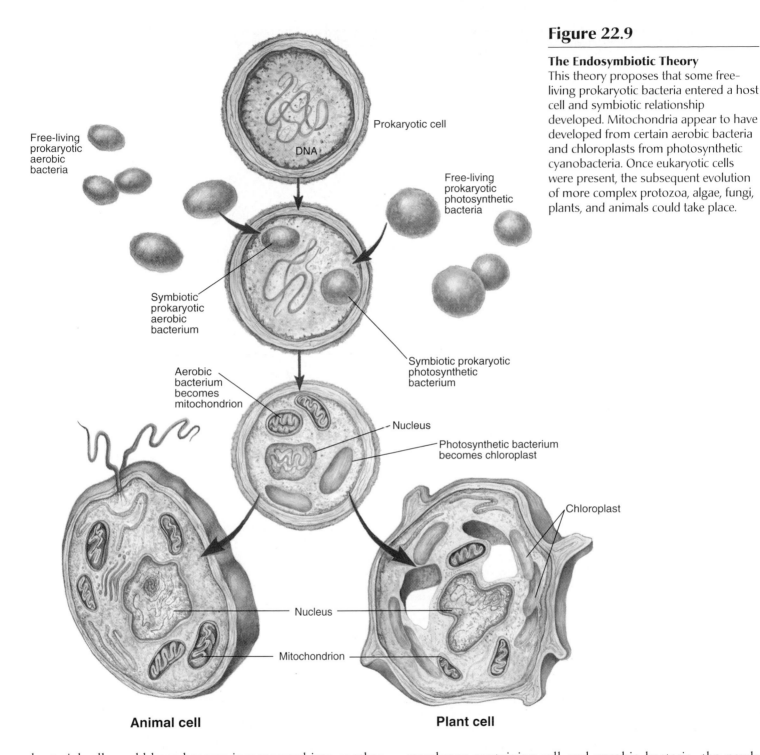

Free-living prokaryotic aerobic bacteria

Prokaryotic cell

DNA

Free-living prokaryotic photosynthetic bacteria

Symbiotic prokaryotic aerobic bacterium

Symbiotic prokaryotic photosynthetic bacterium

Aerobic bacterium becomes mitochondrion

Nucleus

Photosynthetic bacterium becomes chloroplast

Chloroplast

Nucleus

Mitochondrion

Animal cell

Plant cell

Figure 22.9

The Endosymbiotic Theory
This theory proposes that some free-living prokaryotic bacteria entered a host cell and symbiotic relationship developed. Mitochondria appear to have developed from certain aerobic bacteria and chloroplasts from photosynthetic cyanobacteria. Once eukaryotic cells were present, the subsequent evolution of more complex protozoa, algae, fungi, plants, and animals could take place.

bacterial cells could have become incorporated into another primitive nuclear membrane-containing cell type in a symbiotic relationship (figure 22.9). When this theory was first suggested it met with a great deal of criticism. However, continuing research has uncovered several other instances of the probable joining of two different prokaryotic cells to form one.

   If these cells adapted to one another and were able to survive and reproduce better as a team, it is possible that this relationship may have evolved into present-day eukaryotic cells. If this relationship had included only a nuclear membrane-containing cell and aerobic bacteria, the newly evolved cell would have been similar to present-day heterotrophic protozoa, fungi, and animal cells. If this relationship had included both aerobic bacteria and photosynthetic bacteria, the newly formed cell would have been similar to present-day autotrophic algae and plant cells. In addition it is likely that endosymbiosis occurred among eukaryotic organisms as well. Several kinds of eukaryotic red and brown algae contain chloroplastlike structures that appear to have originated as free-living eukaryotic cells (figure 22.10).

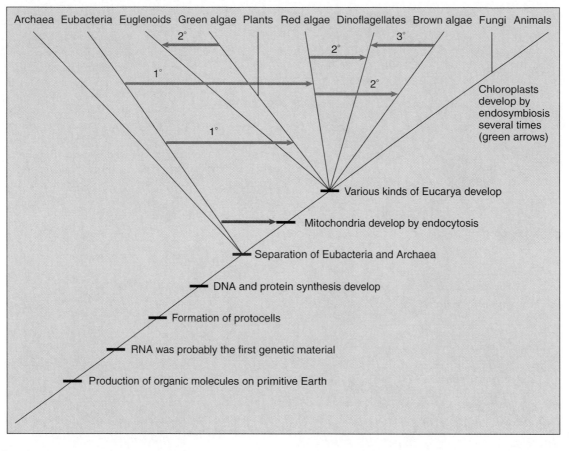

Figure 22.10

Major Steps in the Origin and Evolution of Life
This graphic illustrates some of the major events in the origin and evolution of living things. The horizontal colored arrows indicate possible cases of endosymbiosis. It appears that chloroplasts have been passed around among many different photosynthetic organisms by endosymbiosis. Primary endosymbiosis involving prokaryotic cells is designated 1°. Cases of secondary or tertiary endosymbiosis of one eukaryote appropriating the chloroplast of another are designated 2° or 3°.

Regardless of the type of cell (prokaryotic or eukaryotic), or whether the organisms are heterotrophic or autotrophic, all organisms have a common basis. DNA is the universal genetic material; protein serves as structural material and enzymes; and ATP is the source of energy. Although there is a wide variety of organisms, they all are built from the same basic molecular building blocks. Therefore, it is probable that all life derived from a single origin and that the variety of living things seen today evolved from the first protocells.

Let us return for a moment to the question that perplexed early scientists and caused the controversies surrounding the opposing concepts of spontaneous generation and biogenesis. From our modern perspective we can see that all life we experience comes into being as a result of reproduction. Life is generated from other living things, the process of biogenesis. However, reproduction does not answer the question: Where did life come from in the first place? We can speculate, test hypotheses, and discuss various possibilities, but we will probably never know for sure. Life either always was or it started at some point in the past. If it started, then spontaneous generation of some type had to occur at least once, but it is not happening today.

Figure 22.10 illustrates several of the major events thought to have been involved in the evolution of living things.

22.6 Evolutionary Time Line

A geological time chart shows a chronological history of living organisms based on the fossil record. The largest geological time units are called *eons*. From earliest to most recent, the geological eons are the Hadean, Archaean, Proterozoic, and Phanerozoic. The Phanerozoic Eon is divided into the eras Paleozoic, Mesozoic, and Cenozoic. Each of these *eras* is subdivided into smaller time units called *periods*. For example, Jurassic is a period of the Mesozoic Era that began 208 million years ago (figure 22.11).

Recent evidence suggests that the first living thing most likely came into existence approximately 3.8 billion years ago during the Archaean Eon. By about 3.5 billion years

Eon	Era	Period	Epoch	Million Years Ago	Major Events and Characteristics
Phanerozoic	Cenozoic	Quaternary	Holocene (Recent)	0.01 to present	Dominance of modern humans Much of Earth modified by humans
			Pleistocene	1.8–0.01	Extinction of many large mammals Most recent period of glaciation
		Tertiary	Pliocene	5–1.8	Cool, dry period Grasslands and grazers widespread Origin of ancestors of humans
			Miocene	23–5	Warm, moist period Grasslands and grazers common
			Oligocene	38–23	Cool period—tropics diminish Woodlands and grasslands expand
			Eocene	54–38	All modern forms of flowering plants present All major groups of mammals present
			Paleocene	65–54	Warm period Many new kinds of birds and mammals
	Mesozoic	Cretaceous		144–65	Meteorite impact causes mass extinction Dinosaurs dominant Many new kinds of flowering plants and insects
		Jurassic		208–144	Giant continent begins to split up Dinosaurs dominant First flowering plants and birds
		Triassic		245–208	Giant continent—warm and dry Explosion in reptile and cone-bearing plant diversity First dinosaurs and first mammals
	Paleozoic	Permian		286–245	90% of species go extinct at end of Permian New giant continent forms Modern levels of oxygen in atmosphere Gymnosperms, insects, amphibians, and reptiles dominant
		Carboniferous		360–286	Gymnosperms and reptiles present by end Vast swamps of primitive plants—formed coal Amphibians and insects common
		Devonian		408–360	Glaciation the probable cause of extinction of many warm-water marine organisms Abundant fish, insects, forests, coral reefs
		Silurian		436–408	Melting of glaciers caused rise in sea level Numerous coral reefs First land animals (arthropods), jawed fish, and vascular plants
		Ordovician		505–436	Sea level drops causing major extinction of marine animals at end of Ordovician Primitive plants present—jawless fish common
		Cambrian		540–505	Major extinction of organisms at end of Cambrian All major phyla of animals present Continent splits into several parts and drift apart
Proterozoic	This period of time is also known as the Precambrian			2,500–540	First multicellular organisms about 1 billion years ago First eukaryotes about 1.8 billion years ago Increasing oxygen in atmosphere Single large continent about 1.1 billion years ago
Archaean				3,800–2,500	Continents form—no oxygen in atmosphere Fossil prokaryotes—about 3.5 billion years ago
Hadean				4,500–3,800	Crust of Earth in process of solidifying Origin of the Earth

Figure 22.11

Geologic Time Chart

This chart shows the various geologic time designations, their time periods, and major events and characteristics of each time period.

ago, prokaryotic cell types (Domain Eubacteria) are present in the fossil record. Also about this time it is thought that the Archaea diverged from the Eubacteria. For approximately 2 billion years the only organisms on Earth were Eubacteria and Archaea. Since there has been life on Earth for about 3.8 billion years, that means that for the majority of its existence the Earth was dominated by prokaryotic organisms. The photosynthetic eubacterial cyanobacteria are thought to have been responsible for the production of molecular oxygen (O_2) that began to accumulate in the atmosphere, making conditions favorable for the evolution of other types of cells. The first members of the Domain Eucarya, the eukaryotic organisms, appeared approximately 1.8 billion years ago.

There is fossil evidence of multicellular algae at about 1 billion years ago and multicellular animals at about 0.6 billion years ago. During the Cambrian Period of the Paleozoic Era, an evolutionary explosion of multicellular animals occurred. Examples of most of the present-day phyla of marine invertebrate animals (echinoderms, arthropods, mollusks, etc.) are found at this time.

Several other "evolutionary explosions," or *adaptive radiations,* followed. Note on the geological time chart that its divisions are associated with the major events in the evolutionary history of living things. For example, the Paleozoic Era was dominated by nonvascular and primitive vascular plants; the Mesozoic Era by cone-bearing evergreens; and the Cenozoic by the flowering plants with which we are most familiar. Likewise, many periods of time are associated with specific animal groups. Among vertebrates, the Devonian Period is considered the Age of Fishes and the Carboniferous Period the Age of Amphibians. The Mesozoic Era is considered the Age of Reptiles and the Cenozoic Era is considered the Age of Mammals. In each instance, dominance of a particular animal group resulted from adaptive radiation events. It is also important to note that the ends of many geologic time designations are associated with major extinctions. These extinctions appear to be related to major changes in climate, sea level, or asteroid impacts. Following each major extinction there was an adaptive radiation that led to a new group taking a dominant position.

For about 90% of Earth's history, life was confined to the sea. Primitive land plants probably arose about 430 million years ago and the first land animals (millipedes) at about 420 million years ago. In order to live on land, organisms needed several characteristics: an ability to breathe air, a way to prevent dehydration, some sort of skeleton for support, and an ability to reproduce out of water. The first major group of land animals to become abundant was various kind of insects.

Among the vertebrates, the first land animals most likely evolved from a lobe-finned fish of the Devonian Period. They possessed two important adaptations: lungs and paired, lobed fins that allowed the organisms to pull themselves onto land and travel to new water holes during times of drought. Selective pressures resulted in fins evolving into legs, and the first amphibian came into being. The first vertebrates to spend part of their lives on land found a variety of unexploited niches, resulting in the rapid evolution of new amphibian species and their dominance during the Carboniferous Period. During this time there were abundant swamps that allowed amphibians to return to water to reproduce. However, to totally exploit the land environment, a modification of the method of reproduction was required.

During this time, mutations continued to occur, and valuable modifications were passed on to future generations that eventually led to the development of reptiles. One change allowed the male to deposit sperm directly within the female. Because the sperm could directly enter the female and remain in a moist interior, it was no longer necessary for the animals to return to the water to mate, as amphibians still must do. However, developing young still required a moist environment for early growth. A second modification, the amniotic egg, solved this problem. An amniotic egg, like a chicken egg, protects the developing young from injury and dehydration while allowing for the exchange of gases with the external environment. A third adaptation, the development of protective scales and relatively impermeable skin, protected reptiles from dehydration. With these adaptations, reptiles were able to outcompete amphibians in most terrestrial environments. Amphibians that did survive were the ancestors of present-day frogs, toads, and salamanders. With extensive adaptive radiation, reptiles took to the land, sea, and air. A particularly successful group of reptiles was the dinosaurs, which were the dominant terrestrial vertebrates for more than 100 million years (figure 22.12)

As reptiles diversified, some developed characteristics common to other classes of vertebrates found today, such as warm-bloodedness, feathers, and hair. Warm-blooded reptiles with scales modified as feathers for insulation eventually evolved into organisms capable of flight. Through natural selection, reptilian characteristics were slowly eliminated and characteristics typical of today's modern birds (multiple adaptations to flight, keen senses, and complex behavioral instincts) developed. *Archaeopteryx,* the first bird, had characteristics typical of both birds and reptiles.

Also evolving from reptiles were the mammals. The first reptiles with mammalian characteristics appeared in the Permian Period, although the first true mammals did not appear until the Triassic Period. These organisms remained relatively small in number and size until after the mass extinction of the reptiles. Extinctions opened many niches and allowed for the subsequent adaptive radiation of mammals. As with the other adaptive radiations, mammals possessed unique characteristics that made them better adapted to the changing environment; the characteristics include insulating hair, constant body temperature, and internal development of young. Figure 22.13 summarizes some of the major events in the evolutionary history of the Earth.

Figure 22.12

The Age of the Dinosaurs
Dinosaurs were the dominant vertebrates for about 100 million years.

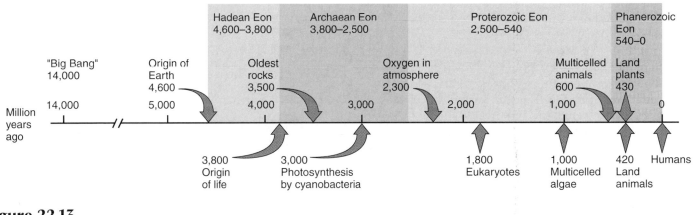

Figure 22.13

Evolutionary Time Line
This chart displays how science sees the order of major, probable events in the origin and evolution of life from the "Big Bang" to the present day.

SUMMARY

The centuries of research outlined in this chapter illustrate the development of our attempts to understand the origin of life. Current theories speculate that either the primitive Earth's environment led to the spontaneous organization of organic chemicals into primitive cells or primitive forms of life arrived on Earth from space. Regardless of how the first living things came to be on Earth, these basic units of life were probably similar to present-day prokaryotes. These primitive cells could have changed through time as a result of mutation and in response to a changing environment. The recognition that many prokaryotic organisms have characteristics that clearly differentiate them from the rest of the bacteria has led to the development of the concept that there are three major domains of life: the Eubacteria, the Archaea, and the Eucarya. The Eubacteria and Archaea are similar in structure but the Archaea have distinctly different metabolic processes from the Eubacteria. Some people consider the Archaea, many of which can live in very extreme environments, good candidates for the first organisms to inhabit Earth. The origin of the Eucarya is less contentious. Similarities between cyanobacteria and chloroplasts and between aerobic bacteria and mitochondria suggest that eukaryotic cells may really be a combination of ancient cell ancestors that lived together symbiotically. The likelihood of these occurrences is supported by experiments that have simulated primitive Earth environments and investigations of the cellular structure of simple organisms. Despite volumes of information, the question of how life began remains unanswered.

THINKING CRITICALLY

For guidelines to answer this question, visit the Online Learning Center.

It has been postulated that there is "life" on another planet in our galaxy. The following data concerning the nature of this life have been obtained from "reliable" sources. Using these data, what additional information is necessary, and how would you go about verifying these data in developing a theory of the origin of life on planet X?

1. The age of the planet is 5 billion years.
2. Water is present in the atmosphere.
3. The planet is farther from the Sun than our Earth is from our Sun.
4. The molecules of various gases in the atmosphere are constantly being removed.
5. Chemical reactions on this planet occur at approximately half the rate at which they occur on Earth.

CONCEPT MAP TERMINOLOGY

Construct a concept map using these terms.

autotroph	oxidizing atmosphere
biogenesis	prokaryotes
endosymbiotic theory	reducing atmosphere
eukaryote	spontaneous generation
heterotroph	

Visit the Online Learning Center to use an interactive format to construct another concept map.

KEY TERMS

Use interactive flash cards, on the Online Learning Center, to help you learn the meanings of these terms.

autotrophs	panspermia
biogenesis	prebionts
coacervate	prokaryote
endosymbiotic theory	proteinoid
eukaryote	protocell
heterotroph	reducing atmosphere
microsphere	spontaneous generation
oxidizing atmosphere	

LEARNING CONNECTIONS

Visit the Online Learning Center to find additional study resources including those highlighted below.

22.1 Spontaneous Generation Versus Biogenesis

Interactive Concept Map
• Origin of life

Experience This!
• Spontaneous generation or biogenesis

1. What is meant by the term *spontaneous generation*?
2. What is meant by the term *biogenesis*?
3. Describe the contribution of each of the following scientists to the biogenesis/spontaneous generation debate: Spallanzani, Needham, Pasteur, and Redi.

22.2 Current Thinking About the Origin of Life

Animations and Review
• Fossils
• Origin of life

4. Provide two kinds of evidence that support the idea that life could have originated on Earth.
5. Provide two kinds of evidence that support the idea that life could have arrived on Earth from space.

22.3 The "Big Bang" and the Origin of the Earth

6. How did the atmosphere, the temperature, and the surface of the newly formed Earth differ from what exists today?
7. What is the approximate age of the Earth?

22.4 Steps Needed to Produce Life from Inorganic Materials

Animations and Review
• Key events

Interactive Concept Map
• Formation of organic molecules

8. The current theory of the origin of life as a result of nonbiological manufacture of organic molecules depends on our knowing something of Earth's history. Why is this so?
9. List two kinds of evidence that suggest RNA was the first genetic material.
10. Describe two models that suggest how collections of organic molecules could have been segregated from other molecules.
11. Why must the first organism on Earth have been anaerobic?

22.5 Major Evolutionary Changes in the Nature of Living Things

Animations and Review
- Continental drift
- Extinction
- Evolutionary trends

12. What evidence supports the theory that eukaryotic cells arose from the development of an endosymbiotic relationship between primitive prokaryotic cells?
13. Can spontaneous generation occur today? Explain.
14. List two important effects caused by the increase of oxygen in Earth's atmosphere.

15. In what sequence did the following things happen: living cell, oxidizing atmosphere, respiration, photosynthesis, reducing atmosphere, first organic molecule?
16. Why do scientists believe life originated in the seas?

22.6 Evolutionary Time Line

17. For each of the following pairs of terms, select the one that is the earliest in geologic time.
    a. eukaryote—prokaryote
    b. marine—terrestrial
    c. vertebrate—invertebrate
    d. flowering plant—cone-bearing plant
    e. aerobic respiration—photosynthesis
    f. plants—animals

Test yourself with multiple choice questions that provide immediate feedback on the Online Learning Center with PowerWeb at

www.mhhe.com/enger11

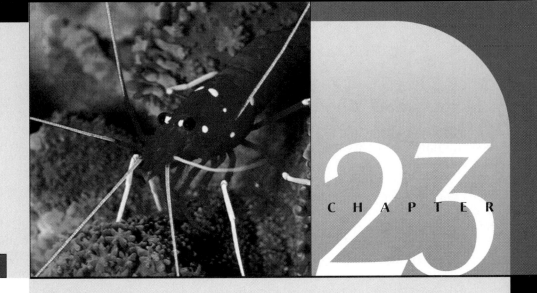

The Classification and Evolution of Organisms

GOALS AND OBJECTIVES

Understand why and how scientists categorize organisms.

❒ Describe the scientific method for naming organisms.

❒ Distinguish between taxonomy and phylogeny.

❒ Describe the kinds of tools used to establish phylogenetic relationships.

❒ List the domains of organisms.

Know the criteria used to classify organisms into different kingdoms.

❒ List and give distinguishing characteristics of the kingdoms within the Domain Eucarya.

❒ Distinguish between Eubacteria and Archaea.

❒ Distinguish among viruses, viroids, and prions.

The Origin and Classification of Life

PART SIX

455

23.1 The Classification of Organisms

Every day you see a great variety of living things. Just think of how many different species of plants and animals you have observed. Biologists at the Smithsonian Institution estimated that there are over 30 million species in the world; over 1.5 million of these have been named. What names do you assign to each? Is the name you use the same as that used in other sections of the country or regions of the world? In much of the United States and Canada, the fish pictured in figure 23.1*a* is known as a largemouth black bass, but in sections of the southern United States it is called a trout. This use of local names can lead to confusion. If a student in Mississippi writes to a friend in Wisconsin about catching a 6-pound trout, the person in Wisconsin thinks that the friend caught the kind of fish pictured in figure 23.1*b*. In the scientific community, accuracy is essential; local names cannot be used. When a biologist is writing about a species, all biologists in the world who read that article must know exactly what that species is.

Taxonomy

Taxonomy is the science of naming organisms and grouping them into logical categories. Various approaches have been

(a)

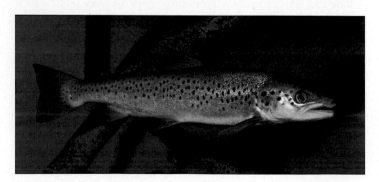

(b)

Figure 23.1

Scientific Names Reduce Confusion
Although both of the fish pictured here are called trout in certain parts of the world, they are very different organisms. The use of the scientific names *Micropterus salmoides* for largemouth black bass (*a*) and *Salmo trutta* for brown trout (*b*) prevent confusion because the scientific names are accepted by scientists throughout the world, regardless of the local common names used.

used to classify organisms. The Greek philosopher Aristotle (384–322 B.C.) had an interest in nature and was the first person to attempt a logical classification system. The root word for *taxonomy* is the Greek word *taxis*, which means *arrangement*. Aristotle used the size of plants to divide them into the categories of trees, shrubs, and herbs.

During the Middle Ages, Latin was widely used as the scientific language. As new species were identified, they were given Latin names, often using as many as 15 words. Although using Latin meant that most biologists, regardless of their native language, could understand a species name, it did not completely do away with duplicate names. Because many of the organisms could be found over wide geographic areas and communication was slow, there could be two or more Latin names for a species. To make the situation even more confusing, ordinary people still called organisms by their common local names.

The modern system of classification began in 1758 when Carolus Linnaeus (1707–1778), a Swedish doctor and botanist, published his tenth edition of *Systema Naturae* (figure 23.2). (Linnaeus's original name was Carl von Linné, which he "latinized" to Carolus Linnaeus.) In the previous editions, Linnaeus had used a polynomial (many-name) Latin system. However, in the tenth edition he introduced the **binomial** (two-name) **system of nomenclature**. This system used two Latin names—genus and specific epithet (*epithet* = descriptive word) for each species of organism.

Recall that a species is a population of organisms capable of interbreeding and producing fertile offspring. Individual organisms are members of a species. A **genus** (plural, *genera*) is a group of closely related organisms; the **specific epithet** is a word added to the genus name to identify which one of several species within the genus we are discussing. It is similar to the naming system we use with people. When you look in the phone book you look for the last name (surname), which gets you in the correct general category. Then

Figure 23.2

Carolus Linnaeus (1707–1778)
Carolus Linnaeus, a Swedish doctor and botanist, originated the modern system of taxonomy known as binomial nomenclature.

you look for the first name (given name) to identify the individual you wish to call. The unique name given to a particular type of organism is called its species name or scientific name. In order to clearly identify the scientific name, binomial names are either *italicized* or underlined. The first letter of the genus name is capitalized. The specific epithet is always written in lowercase. *Micropterus salmoides* is the binomial name for the largemouth black bass. • species, p. 206

When biologists adopted Linnaeus's binomial method, they eliminated the confusion that was the result of using common local names. For example, with the binomial system the white water lily is known as *Nymphaea odorata*. Regardless of which of the 245 common names is used in a botanist's local area, when botanists read *Nymphaea odorata*, they know exactly which plant is being referred to. The binomial name cannot be changed unless there is compelling evidence to justify doing so. The rules that govern the world-wide classification and naming of species are expressed in the International Rules for Botanical Nomenclature, the International Rules for Zoological Nomenclature, and the International Bacteriological Code of Nomenclature.

In addition to assigning a specific name to each species, Linnaeus recognized a need for placing organisms into groups. This system divides all forms of life into **kingdoms,** the largest grouping used in the classification of organisms. Originally there were two kingdoms—Plantae and Animalia—but the number of kingdoms was increased as better knowledge of the nature of organisms accumulated. Recently biologists recognized a category above that of kingdoms, known as *domains*. There are three domains: Eubacteria, Archaea, and Eucarya. Each domain is subdivided into kingdoms. There are four kingdoms of life in the Domain Eucarya: Plantae, Animalia, Fungi, and Protista (protozoa and algae) (figure 23.3). Each of these kingdoms is

(a)

Figure 23.3

Representatives of the Domains of Life
(*a*) The Domain Eubacteria is represented by the bacterium *Streptococcus pyogenes* (the cause of strep throat). The Domain Eucarya is represented by: (*b*) *Morchella esculenta,* kingdom Fungi; (*c*) *Amoeba proteus,* kingdom Protista; (*d*) *Homo sapiens,* kingdom Animalia; and (*e*) *Acer saccharum,* the kingdom Plantae.

(b)

(c)

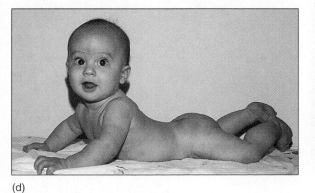
(d)

(e)

CLASSIFICATION OF HUMANS

Table 23.1

Taxonomic Category	Human Classification	Other Representative Organisms in the Same Category
Domain	Eucarya	Plants, animals, fungi, protozoans, and algae
Kingdom	Animalia	Heterotrophic organisms with specialized tissues that are usually mobile: insects, snails, starfish, worms, snakes, fish, dogs
Phylum	Chordata	Animals with stiffening rod in the back: reptiles, amphibians, birds, fish
Class	Mammalia	Animals with hair and mammary glands: dogs, whales, mice

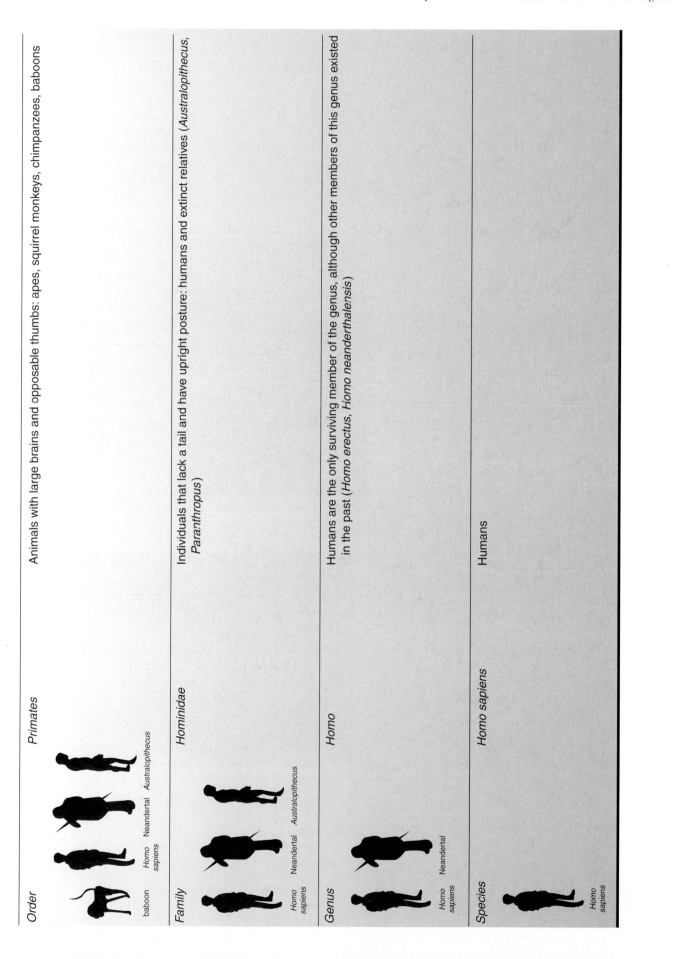

divided into smaller units and given specific names. The taxonomic subdivision under each kingdom is usually called a **phylum**, although microbiologists and botanists replace this term with the word *division*. All kingdoms have more than one phylum. For example, the kingdom Plantae contains several phyla: flowering plants, conifer trees, mosses, ferns, and several other groups. Organisms are placed in phyla based on careful investigation of the specific nature of their structure, metabolism, and biochemistry. An attempt is made to identify natural groups rather than artificial or haphazard arrangements. For example, although nearly all plants are green and carry on photosynthesis, only flowering plants have flowers and produce seeds; conifers lack flowers but have seeds in cones; ferns lack flowers, cones, and seeds; and mosses lack tissues for transporting water.

A **class** is a subdivision within a phylum. For example, within the phylum Chordata within the kingdom Animalia, there are seven classes: mammals, birds, reptiles, amphibians, and three classes of fishes. An **order** is a category within a class. The order Carnivora is an order of meat-eating animals within the class Mammalia. There are several other orders of mammals including horses and their relatives, cattle and their relatives, rodents, rabbits, bats, seals, whales, and many others. A **family** subdivision of an order consists of a group of closely related genera, which in turn are composed of groups of closely related species. The cat family, Felidae, is a subgrouping of the order Carnivora and includes many species in several genera, including the Canada lynx and bobcat (genus *Lynx*), the cougar (genus *Puma*) the leopard, tiger, jaguar, and lion (genus *Panthera*), the house cat (genus *Felis*), and several other genera. Thus, in the present-day science of taxonomy, each organism that has been classified has its own unique, binomial name. In turn, it is assigned to larger groupings that are thought to have a common evolutionary history. Table 23.1 on pages 458 and 459 uses the classification of humans to show how the various categories are used.

Phylogeny

Phylogeny is the science that explores the evolutionary relationships among organisms and seeks to reconstruct evolutionary history. Taxonomists and phylogenists work together so that the products of their work are compatible. A taxonomic ranking should reflect the evolutionary relationships among the organisms being classified. Although taxonomy and phylogeny are sciences, there is no complete agreement as to how organisms are classified or how they are related. Just as there was dissension 200 years ago when biologists disagreed on the theories of spontaneous generation and biogenesis, there are still differing opinions about the evolutionary relationships of organisms. People arrive at different conclusions because they use different kinds of evidence or interpret this evidence differently. Phylogenists use several lines of evidence to develop evolutionary histories: fossils, comparative anatomy, life cycle information, and biochemical and molecular evidence.

Fossils are physical evidence of previously existing life and are found in several different forms. Some fossils may be preserved whole and relatively undamaged. For example, mammoths and humans have been found frozen in glaciers, and bacteria and insects have been preserved after becoming embedded in plant resins. Other fossils are only parts of once-living organisms. The outlines or shapes of extinct plant leaves are often found in coal deposits, and individual animal bones that have been chemically altered over time are often dug up (figure 23.4). Animal tracks have also been discovered in the dried mud of ancient riverbeds.

(a)   (b)

(c)

Figure 23.4

Fossil Evidence

A fossil is any evidence of previously existing life. Fossils may be the intact, preserved remains of organisms as in (a) the remains of an ancient fly preserved in amber, or preserved parts of an organism as in (b), the fossilized skeleton and body outline of a bony fish. It is even possible to have evidence of previously existing living things that are not the remains of organisms as in (c), a case of a dinosaur footprint.

It is important to understand that some organisms are more easily fossilized than others. Those that have hard parts like cell walls, skeletons, and shells are more likely to be preserved than are tiny, soft-bodied organisms. Aquatic organisms are much more likely to be buried in the sediments at the bottom of the oceans or lakes than are their terrestrial counterparts. Later, when these sediments are pushed up by geologic forces, aquatic fossils are found in layers of sediments on dry land.

Evidence obtained from the discovery and study of fossils allows biologists to place organisms in a time sequence. This can be accomplished by comparing one type of fossil with another. As geologic time passes and new layers of sediment are laid down, the older organisms should be in deeper layers, providing the sequence of layers has not been disturbed (figure 23.5). In addition, it is possible to age-date rocks by comparing the amounts of certain radioactive iso-

topes they contain. The older sediment layers have less of these specific radioactive isotopes than do younger layers. A comparison of the layers gives an indication of the relative age of the fossils found in the rocks. Therefore, fossils found in the same layer must have been alive during the same geologic period. • evolutionary time line, p. 451

It is also possible to compare subtle changes in particular kinds of fossils over time. For example, the size of the leaf of a specific fossil plant has been found to change extensively through long geologic periods. A comparison of the extremes, the oldest with the newest, would lead to their classification into different categories. However, the fossil links between the extremes clearly show that the younger plant is a descendant of the older.

Comparative anatomy studies of fossils or currently living organisms can be very useful in developing a phylogeny. Because the structures of an organism are determined by its genes, those organisms having similar structures are thought to be related. Plants can be divided into several categories: All plants that have flowers are thought to be more closely related to one another than to plants like ferns, which do not have flowers. In the animal kingdom, all organisms that have hair and mammary glands are grouped together, and all animals in the bird category have feathers, wings, and beaks.

Life cycle information is another line of evidence useful to phylogenists and taxonomists. Many organisms have complex life cycles that include many completely different stages. After fertilization, some organisms grow into free-living developmental stages that do not resemble the adults of their species. These are called *larvae* (singular, *larva*). Larval stages often provide clues to the relatedness of organisms. For example, adult barnacles live attached to rocks and other solid marine objects and look like small, hard cones. Their outward appearance does not suggest that they are related to shrimp; however, the larval stages of barnacles and shrimp are very similar. Detailed anatomical studies of barnacles confirm that they share many structures with shrimp; their outward appearance tends to be misleading (figure 23.6). Both birds and reptiles lay eggs with shells. However, reptiles lack feathers and have scales covering their bodies. The fact that these two groups share this fundamental eggshell characteristic implies that they are more closely related to each other than they are to other groups, but they can be divided into two groups based on their anatomical differences.

This same kind of evidence is available in the plant kingdom. Many kinds of plants, such as peas, peanuts, and lima beans, produce large, two-parted seeds in pods (you can easily split the seeds into two parts). Even though peas grow as vines, lima beans grow as bushes, and peanuts have their seeds underground, all these plants are considered to be related.

Biochemical and molecular studies are recent additions to the toolbox of phylogenists. Like all aspects of biology, the science of phylogeny is constantly changing as new techniques develop. Recent advances in DNA analysis are being used to determine genetic similarities among species

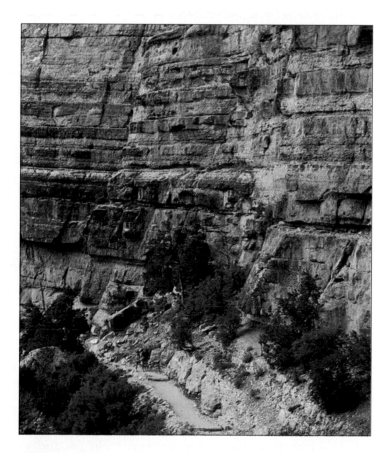

Figure 23.5

Determining the Age of Fossils

Because new layers of sedimentary rock are formed on top of older layers of sedimentary rock, it is possible to determine the relative ages of fossils found in various layers. The layers of rock shown here represent on the order of hundreds of millions of years of formation. The fossils of the lower layers are millions of years older than the fossils in the upper layers.

(a)

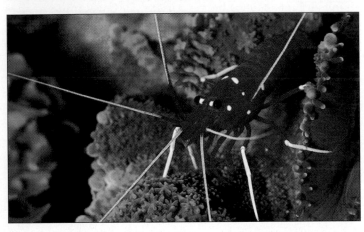

(b)

Figure 23.6

Developmental Biology
The adult barnacle (a) and shrimp (b) are very different from each other, but the early larval stages look very much alike.

(How Science Works 23.1). In the field of ornithology, which deals with the study of birds, there are those who believe that storks and flamingos are closely related; others believe that flamingos are more closely related to geese. An analysis of the DNA points to a higher degree of affinity between flamingos and storks than between flamingos and geese. This is interpreted to mean that the closest relationship is between flamingos and storks.

Algae and plants have several different kinds of chlorophyll: chlorophyll *a, b, c, d,* and *e.* Most photosynthetic organisms contain a combination of two of these chlorophyll molecules. Members of the kingdom Plantae have chlorophyll *a* and *b.* The large seaweeds, like kelp, superficially resemble terrestrial plants like trees and shrubs. However, a comparison of the chlorophylls present shows that kelp has chlorophyll *a* and *d.* When another group of algae, called the *green algae,* are examined, they are found to have chlorophyll *a* and *b.* Along with other anatomical and developmental evidence, this biochemical information has helped establish an evolutionary link between the green

algae and plants. All of these kinds of evidence (fossils, comparative anatomy, life cycle information, and biochemical evidence) have been used to develop phylogenetic relationships and taxonomic categories.

Given all these sources of evidence, biologists have developed a hypothetical picture of how all organisms are related (figure 23.7). At the base of this evolutionary scheme is the biochemical evolution of cells first postulated by Oparin (see chapter 22). These first cells are thought to be the origin of all the categories of living things. Although protocells no longer exist, their descendants have diversified over millions of years. Of these groups, Eubacteria and Archaea have the simplest structure and are probably most similar to some of the first cellular organisms on Earth.

23.2 Domains Eubacteria and Archaea

Members of the Domains Eubacteria and Archaea are commonly known as *bacteria.* They are all tiny, prokaryotic cells that are difficult to distinguish from one another. Because of this, in the past it was assumed that the members of these groups were closely related. However, they show a wide variety of metabolic abilities. Many are heterotrophs that utilize organic molecules as food. A few are disease-causing, such as *Streptococcus pneumoniae,* but most are not. In addition, many are photosynthetic and a few are chemosynthetic. Individual species are distinguished from one another primarily by differences in their metabolic activities and details of their chemical structure. Evidence gained from studying DNA and RNA nucleotide sequences and a comparison of the amino acid sequences of proteins indicate that there are major differences between the Eubacteria and Archaea. The Eubacteria evolved first and their ancestors probably used DNA as their genetic material. The Eubacteria gave rise to the Archaea and, finally, the Eucarya evolved. • **domains, p. 66**

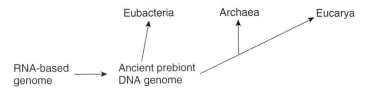

Eubacteria

The "true bacteria" (*eu* = true) are small, prokaryotic, single-celled organisms ranging from 1 to 10 micrometers (μm). Their cell walls typically contain complex organic molecules, such as peptidoglycan, polymers of unique sugars, and unusual amino acids not found in other kinds of organisms. The most common shapes of the cells are spheres, rods, and spirals.

Since they are prokaryotic, Eubacteria have no nucleus, and the genetic material is a single loop of DNA. Some have as few as 5,000 genes. The cells reproduce by binary fission. This is a type of asexual cell division that does not involve the more complex structures used by eukaryotes in mitosis or meiosis. As a result, the daughter cells produced have a single

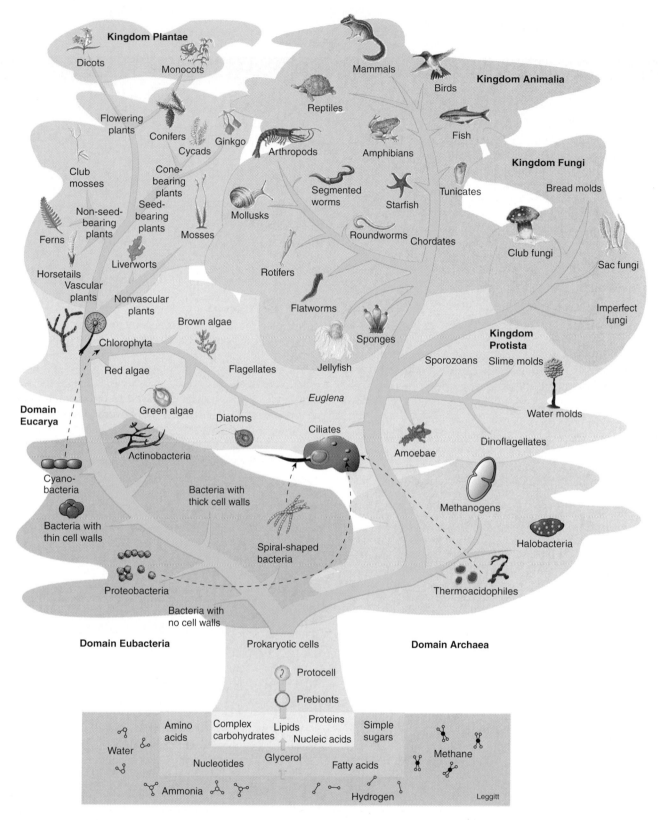

Figure 23.7

Molecules to Organisms

The theory of chemical evolution proposes that the molecules in the early atmosphere and early oceans accumulated to form prebionts—nonliving structures composed of carbohydrates, proteins, lipids, and nucleic acids. The prebionts are believed to have been the forerunners of the protocells—the first living cells. These protocells probably evolved into primitive kinds of prokaryotic cells, which gave rise to members of the domains Eubacteria and Archaea. Some prokaryotic cells probably gave rise to eukaryotic cells through the process of endosymbiosis. The organisms formed from these early eukaryotic cells were probably similar to single-celled members of the kingdom Protista. Members of this kingdom are thought to have given rise to the kingdoms Animalia, Plantae, and Fungi. Thus, all present-day organisms evolved from the protocells.

HOW SCIENCE WORKS 23.1

Cladistics: A Tool for Taxonomy and Phylogeny

*C*lassification, or *taxonomy,* is one part of the much larger field of phylogenetic systematics. Classification involves placing organisms into logical categories and assigning names to those categories. *Phylogeny,* or *systematics,* is an effort to understand the evolutionary relationships of living things in order to interpret the way in which life has diversified and changed over billions of years of biological history. Phylogeny attempts to understand how organisms have changed over time.

Cladistics (*klados* = branch) is a method biologists use to present their knowledge of the degree of relatedness among different organisms within a group based on how similar they are genetically. The basic assumptions behind cladistics are that:

1. Groups of organisms are related by descent from a common ancestor.
2. The relationships among groups can be represented by a branching pattern with new evolutionary groups arising from a common ancestor.
3. Changes in characteristics occur in organisms over time.

To be considered "close relatives," groups of organisms must share a large number of *genetically derived* characteristics, the characteristics must be derived from the same origin, and all members of the group must be considered to be derived from the same common ancestor. More distantly related subgroups will share fewer characteristics. The more characteristics two groups share, the more closely related they are likely to be. In a simplified example, we could compare jellyfish, starfish, and humans. To do so we must select several characteristics to compare, such as: organism is made of many cells; organism contains three layers of cells; organ systems are present; and a circulatory system is present. Taking all of these characteristics into account, we would conclude that the starfish and the human are more closely related than either is to the jellyfish. Starfish and humans share the characteristics of being made of many cells, having three layers of cells, and organ systems. The jellyfish and the other two groups only share the characteristic of being made of many cells. While humans are different from the other two organisms, they differ only in genetic characteristics that arose after their ancestors had diverged from the other two organisms. The presence of a circulatory system is a recently developed characteristic found in humans but not in jellyfish or starfish.

A *cladogram* is a chart designed to show how closely groups of organisms are related genetically. A cladogram for the previous example would be:

The choice of characteristics with which to make comparisons is important. Two organisms may share many characteristics and still not be members of the same evolutionary group if the characteristics being compared are not from the same genetic background. For example, if you were to compare butterflies, birds, and squirrels and you were to use the presence or absence of wings and the presence or absence of bright colors as your characteristics, you would conclude that butterflies and birds are more closely related than birds and squirrels. This is not a valid comparison, since the wings of birds and butterflies are not of the same evolutionary origin.

A cladogram for fishes, amphibians, mammals, and reptiles would appear as:

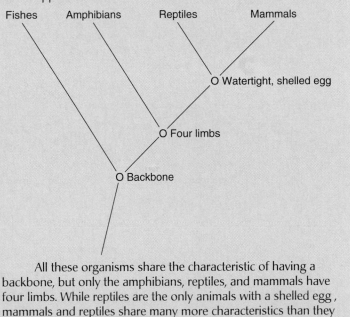

All these organisms share the characteristic of having a backbone, but only the amphibians, reptiles, and mammals have four limbs. While reptiles are the only animals with a shelled egg, mammals and reptiles share many more characteristics than they do with amphibians and fishes.

---

copy of the parental DNA loop (figure 23.8). Some cells move by secreting a slime that glides over the cell's surface, causing it to move through the environment. Others move by means of a kind of flagellum. The structure of the flagellum is different from the flagellum found in eukaryotic organisms.

Because the early atmosphere is thought to have been a reducing atmosphere, the first Eubacteria were probably anaerobic organisms. Today there are both anaerobic and aerobic

Eubacteria. Many heterotrophic Eubacteria are **saprophytes,** organisms that obtain energy by the decomposition of dead organic material; others are parasites that obtain energy and nutrients from living hosts and cause disease; still others are mutualistic and have a mutually beneficial relationship with their host; finally, some are commensalistic and derive benefit from a host without harming it. Several kinds of Eubacteria are autotrophic. Many are called cyanobacteria because they

Figure 23.8

Binary Fission in Bacteria
Two cells of the bacterium *Bacillus megaterium* formed by binary fission (*a*). (*b*) Binary fission consists of DNA replication and cytoplasmic division.

(a)

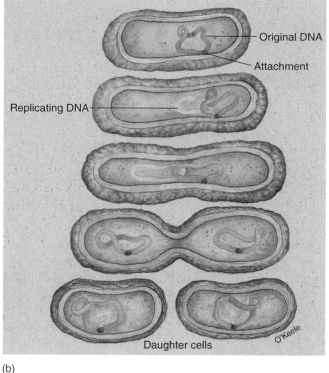

Original DNA

Attachment

Replicating DNA

Daughter cells

O'Keefe

(b)

contain a blue-green pigment, which allows them to capture sunlight and carry on photosynthesis. They can become extremely numerous in some polluted waters where nutrients are abundant. Others use inorganic chemical reactions for their energy sources and are called chemosynthetic.

Archaea

The term *archaea* comes from the Greek term *archaios*, meaning "ancient." This is a little misleading since the Eubacteria preceded them and the Archaea are thought to have branched off from the Eubacteria somewhere between 2 to 3 billion years ago. Although the Archaea are prokaryotic like the Eubacteria, they have many chemical similarities to the Eucarya. For example, the Archaea and the Eucarya both lack peptidoglycan as their cell wall building material and they both have introns as components of their DNA, while the bacteria have peptidoglycan and lack introns. Members of the Archaea exist in many shapes including rods, spheres, spirals, filaments, and flat plates. Because they are found in many kinds of extreme environments, they have become known as *extremophiles*. Based on the particular extreme habitats they occupy, the Archaea are divided into three groups: (1) *Methanogens* are methane-producing bacteria that are anaerobic. They can be found in the intestinal tracts of humans, sewage, and swamps. (2) *Halobacteria* (*halo* = salt) are found growing in very salty environments such as the Great Salt Lake (Utah), salt ponds, and brine solutions. Some contain a special kind of chlorophyll and are therefore capable of generating their ATP by photosynthesis. (3) The *thermophilic* Archaea live in environments that normally have very high temperatures and high concentrations of sulfur (e.g., hot sulfur springs or around deep-sea hy-

drothermal vents). Over 500 species of thermophiles have been identified at the openings of hydrothermal vents in the open oceans. One such thermophile, *Pyrolobus fumarii*, grows in a hot spring in Yellowstone National Park. Its maximum growth temperature is 113°C, its optimum is 106°C, and its minimum is 90°C. • **Eubacteria and Archaea, p. 445**

23.3 Domain Eucarya

Some biologists hypothesize that eukaryotic cells evolved from prokaryotic cells by a process of endosymbiosis. This hypothesis proposes that structures like mitochondria, chloroplasts, and other membranous organelles originated from separate cells that were ingested by larger, more primitive cells. Once inside, these structures and their functions became integrated with the host cell and ultimately became essential to its survival. This new type of cell was the forerunner of present-day eukaryotic cells.

Eukaryotic cells are usually much larger than those of the prokaryotes, typically having more than a thousand times the volume of prokaryotic cells. Their larger size was made possible by the presence of specialized membranous organelles, such as mitochondria, the endoplasmic reticulum, chloroplasts, and nuclei. • **endosymbiosis, p. 447**

Kingdom Protista

The changes in cell structure that led to eukaryotic organisms most probably gave rise to single-celled organisms similar to those currently grouped in the kingdom Protista (see figure 23.7). Most members of this kingdom are one-celled organisms, although there are some colonial forms.

There is a great deal of diversity among the approximately 60,000 known species of Protista. Many species live in freshwater; others are found in marine or terrestrial habitats, and some are parasitic, commensalistic, or mutualistic. All species can undergo mitosis, resulting in asexual reproduction. Some species can also undergo meiosis and reproduce sexually. Many contain chlorophyll in chloroplasts and are autotrophic; others require organic molecules as sources of energy and are heterotrophic. Both autotrophs and heterotrophs have mitochondria and respire aerobically.

Because members of this kingdom are so diverse with respect to form, metabolism, and reproductive methods, most biologists do not think that the Protista form a valid phylogenetic unit. However, it is still a convenient taxonomic grouping. By placing these organisms together in this group it is possible to gain a useful perspective on how they relate to other kinds of organisms. After the origin of eukaryotic organisms, evolution proceeded along several different pathways. Three major lines of evolution can be seen today in the plantlike autotrophs (algae), animal-like heterotrophs (protozoa), and the funguslike heterotrophs (slime molds). *Amoeba* and *Paramecium* are commonly encountered examples of protozoa. Many seaweeds and pond scums are collections of large numbers of algal cells. Slime molds are less frequently seen because they live in and on the soil in moist habitats; they are most often encountered as slimy masses on decaying logs.

Through the process of evolution, the plantlike autotrophs probably gave rise to the kingdom Plantae, the animal-like heterotrophs probably gave rise to the kingdom Animalia, and the funguslike heterotrophs were probably the forerunners of the kingdom Fungi (see figure 23.7).

Kingdom Fungi

Fungus is the common name for members of the kingdom Fungi. The majority of fungi are nonmotile. They have a rigid, thin cell wall, which in most species is composed of chitin, a complex carbohydrate containing nitrogen. Members of the kingdom Fungi are nonphotosynthetic, eukaryotic organisms. The majority of the approximately 70,000 species (mushrooms and molds) are multicellular, but a few, like yeasts, are single-celled. In the multicellular fungi the basic structural unit is a network of multicellular filaments. Because all of these organisms are heterotrophs, they must obtain nutrients from organic sources. Most are saprophytes and secrete enzymes that digest large molecules into smaller units that are absorbed. They are very important as decomposers in all ecosystems. They feed on a variety of nutrients ranging from dead organisms to such products as shoes, food-stuffs, and clothing. Most synthetic organic molecules are not attacked as readily by fungi; this is why plastic bags, foam cups, and organic pesticides are slow to decompose.

Some fungi are parasitic, whereas others are mutualistic. Many of the parasitic fungi are important plant pests. Some attack and kill plants (chestnut blight, Dutch elm disease); others injure the fruit, leaves, roots, or stems and reduce yields. The fungi that are human parasites are responsible for athlete's foot, vaginal yeast infections, valley fever, "ringworm," and other diseases. Mutualistic fungi are important in lichens and in combination with the roots of certain kinds of plants.

Kingdom Plantae

Another major group thought to be derived from the kingdom Protista are the green, photosynthetic plants. The ancestors of plants were most likely specific kinds of algae commonly called *green algae*. Members of the kingdom Plantae are nonmotile, terrestrial, multicellular organisms that contain chlorophyll and produce their own organic compounds. All plant cells have a cellulose cell wall. Over 300,000 species of plants have been classified; about 85% are flowering plants, 14% are mosses and ferns, and the remaining 1% are cone-bearers and several other small groups within the kingdom.

A wide variety of plants exist on Earth today. Members of the kingdom Plantae range from simple mosses to vascular plants with stems, roots, leaves, and flowers. Most biologists believe that the evolution of this kingdom began nearly 500 million years ago when the green algae of the kingdom Protista gave rise to two lines: The nonvascular plants like the mosses evolved as one type of plant and the vascular plants like the ferns evolved as a second type (figure 23.9). Some of the vascular plants evolved into seed-producing plants, which today are the cone-bearing and flowering plants, whereas the ferns lack seeds. The development of vascular plants was a major step in the evolution of plants from an aquatic to a terrestrial environment.

Plants have a unique life cycle. There is a haploid **gametophyte stage** that produces a haploid sex cell by mitosis. There is also a diploid **sporophyte stage** that produces haploid spores by meiosis. This **alternation of generations** is a unifying theme that ties together all members of this kingdom. In addition to sexual reproduction, plants are able to reproduce asexually.

Kingdom Animalia

Like the fungi and plants, the animals are thought to have evolved from the Protista. Over a million species of animals have been classified. These range from microscopic types, like mites or aquatic larvae of marine animals, to huge animals like elephants or whales. Regardless of their types, all animals have some common traits. All are composed of eukaryotic cells and all species are heterotrophic and multicellular. All animals are motile, at least during some portion of their lives; some, like the sponges, barnacles, mussels, and corals, are sessile (nonmotile, i.e., not able to move) when they are most easily recognized—the adult portion of their lives. All animals are capable of sexual reproduction, but many of the less complex animals are also able to reproduce asexually.

It is thought that animals originated from certain kinds of Protista that had flagella (see figure 23.7). This idea proposes that colonies of flagellated Protista gave rise to simple multicellular forms of animals like the ancestors of present-

Flowering plants

Mosses

Conifers

Seed plants

Nonseed plants

Vascular | Nonvascular

Ferns

Algae

Figure 23.9

Plant Evolution
Two lines of plants are thought to have evolved from the green algae of the kingdom Protista. The nonvascular mosses evolved as one type of plant. The second type, the vascular plants, evolved into more complex plants. The early vascular plants did not produce seeds as is the case with present-day ferns. Eventually the evolution of the ability to produce seeds resulted in two distinct lines: the cone-bearing plants and the flowering plants.

day sponges. These first animals lacked specialized tissues and organs. As cells became more specialized, organisms developed special organs and systems of organs and the variety of kinds of animals increased.

Although taxonomists have grouped organisms into six kingdoms, some organisms do not easily fit into these categories. Viruses, which lack all cellular structures, still show some characteristics of life. In fact, some scientists consider them to be highly specialized parasites that have lost their complexity as they developed as parasites. Others consider them to be the simplest of living organisms. Some even consider them to be nonliving. For these reasons, viruses and certain other infectious particles are considered separate from the three domains.

23.4 Acellular Infectious Particles

All of the groups discussed so far fall under the category of cellular forms of life. They all have at least the following features in common. They have (a) cell membranes, (b) nucleic acids as their genetic material, (c) cytoplasm, (d) enzymes and

Host cell recognition proteins

Nucleic acid

Protein coat

Nucleic acid

Envelope

Protein coat

Figure 23.10

Typical Viruses
Viruses consist of a core of nucleic acid, either DNA or RNA, surrounded by a protein coat. Some have an additional layer, called an envelope, that surrounds the protein coat.

coenzymes, (e) ribosomes, and (f) use ATP as their source of chemical-bond energy. Since the three groups to follow lack this cellular organization, they are referred to as *acellular* (the prefix *a* = lacking) or are known as infectious particles. In order for these particles to make more of their own kind, they must make their way into true cells where they become parasites eventually causing harm or death to their host cells. The only infectious particles that are considered beneficial are a few that have been modified through bioengineering to help in the genetic transformation of cells. One example of an infectious agent that has been "domesticated" is HIV, human immunodeficiency virus. Many bioengineers use this "tame" form of the virus to carry laboratory-attached genes into host animal cells in an attempt to change their genetic makeup. Since evolutionary biologists can only speculate on the origin of acellular infectious particles, they are not classified using the same methods outlined previously. Therefore, the names given to various kinds of *viruses, viroids,* and *prions* are not as strictly regulated as the scientific names for species of the Eubacteria, Archaea, and Eucarya.

Viruses

A **virus** consists of a nucleic acid core surrounded by a coat of protein (figure 23.10). Viruses are **obligate intracellular parasites,** which means they are infectious particles that can function only when inside a living cell. Because of their unusual characteristics, viruses are not members of any of the three domains. Biologists do not consider them to be living because they are not capable of living and reproducing by themselves and show the characteristics of life only when inside living cells.

Soon after viruses were discovered in the late part of the nineteenth century, biologists began to speculate on how they originated. One early hypothesis was that they were either prebionts or parts of prebionts that did not evolve into cells. This idea was discarded as biologists learned more about the complex relationship between viruses and host cells. A second hypothesis was that viruses developed from intracellular parasites that became so specialized that they needed only the nucleic acid to continue their existence.

Once inside a cell, this nucleic acid can take over and direct the host cell to provide for all of the virus's needs. A third hypothesis is that viruses are runaway genes that have escaped from cells and must return to a host cell to replicate. Regardless of how the viruses came into being, today they are important as parasites in all forms of life.

Viruses are typically host-specific, which means that they usually attack only one kind of cell. The **host** is a specific kind of cell that provides what the virus needs to function. Viruses can infect only those cells that have the proper receptor sites to which the virus can attach. This site is usually a glycoprotein molecule on the surface of the cell membrane. For example, the virus responsible for measles attaches to the membranes of skin cells, hepatitis viruses attach to liver cells, and mumps viruses attach to cells in the salivary glands. Host cells for the HIV virus include some types of human brain cells and several types belonging to the immune system (Outlooks 23.1).

Once it has attached to the host cell, the virus either enters the cell intact or it injects its nucleic acid into the cell. If it enters the cell, the virus loses its protein coat, releasing the nucleic acid. Once released into the cell, the nucleic acid of the virus may remain free in the cytoplasm or it may link with the host's genetic material. Some viruses contain as few as 3 genes, others contain as many as 500. A typical eukaryotic cell contains tens of thousands of genes. Most viruses need only a small number of genes because they rely on the host to perform most of the activities necessary for viral multiplication. Viruses do not "reproduce" as do true cells—that is, mitosis or meiosis. In those processes, the contents of the cell are doubled prior to splitting the cell into daughter cells. However, viruses are not reproduced, they are "*replicated,*" which means copies are made but the one virus does not become two. Virus particles are replicated by using a set of genetic instructions from the virus and new building materials and enzymes from the host cell.

Some viruses have DNA as their genetic material but many have RNA. Many RNA viruses can be replicated directly but others must have their RNA reverse-transcribed to DNA before the virus can reproduce. Reverse transcriptase, the enzyme that accomplishes this has become very important in the new field of molecular genetics because its use allows scientists to make large numbers of copies of a specific molecule of DNA.

Viral genes are able to take command of the host's metabolic pathways and direct it to carry out the work of making new copies of the original virus. The virus makes use of the host's available enzymes and ATP for this purpose. When enough new viral nucleic acid and protein coat are produced, complete virus particles are assembled and released from the host (figure 23.11). The number of viruses released ranges from 10 to thousands. The virus that causes polio releases about 10,000 new virus particles from each human host cell. Some viruses remain in cells and are only occasionally triggered to reproduce, causing symptoms of disease. Herpes viruses, which cause cold sores, genital herpes, and shingles reside in nerve cells.

Viruses vary in size and shape, which helps in classifying them. Some are rod-shaped, others are round, and still others are in the shape of a coil or helix. Viruses are some of the

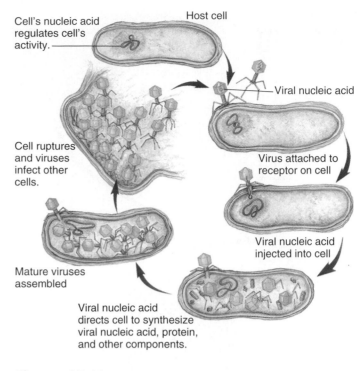

Figure 23.11

Viral Invasion of a Bacterial Cell
The viral nucleic acid takes control of the activities of the host cell. Because the virus has no functional organelles of its own, it can become metabolically active only while it is within a host cell. The viral genetic material causes the host cell to make copies of the virus, ultimately resulting in the destruction of the host cell.

Table 23.2	
VIRAL DISEASES	
Type of Virus	**Disease**
Papovaviruses	Warts in humans
Paramyxoviruses	Mumps and measles in humans; distemper in dogs
Adenoviruses	Respiratory infections in most mammals
Poxviruses	Smallpox
Wound-tumor viruses	Diseases in corn and rice
Potexviruses	Potato diseases
Bacteriophage	Infections in many types of bacteria

smallest infecting agents known to humans. Only a few can be seen with a standard laboratory microscope; most require an electron microscope to make them visible. A great deal of work is necessary to isolate viruses from the environment and prepare them for observation with an electron microscope. For this reason, most viruses are more quickly identified by their activities in host cells. It is very likely that all species on Earth serve as hosts to some form of virus (table 23.2).

OUTLOOKS 23.1

The AIDS Pandemic

Epidemiology is the study of the transmission of diseases through a population. Diseases that occur throughout the world population at extremely high rates are called *pandemics*. Influenza, the first great pandemic of the first part of the twentieth century, killed millions of people. AIDS is an acronym for *a*cquired *i*mmuno*d*eficiency *s*yndrome and is caused by the *h*uman *i*mmunodeficiency *v*iruses (HIV-1 and HIV-2), shown in figure 1.

The Magnitude of the Pandemic

HIV/AIDS has become the greatest pandemic of recent times, claiming about 22 million victims so far. This viral disease has been reported in all countries of the world. UNAIDS (the United Nations Joint HIV/AIDS Program) reported that 3.1 million adults died of AIDS in 2002, and about 5 million became infected with HIV. Of the 3.1 million deaths during 2002, about 40% were adult females, 40% were adult males, and about 20% were children under age 15.

Figure 2 shows World Health Organization (WHO) estimates of the numbers of HIV-infected people in various parts of the world. The incidence of the virus is lowest in the economically developed countries and highest in the developing countries. In economically developed countries, people can be diagnosed and treated. In the less-developed world, there is little medical care to treat AIDS and a lack of resources to identify those who have HIV. Many people do not know they are infected and continue to pass the disease to others. AIDS has become the fourth-leading cause of death in the world. In some countries in southern Africa, AIDS is the leading cause of death resulting from disease. Over 2 million African children lose one parent to AIDS in a typical year. Most of these deaths have occurred in southern Africa.

The Nature and Life Cycle of the Virus

Evidence strongly supports the belief that HIV originated through many mutations of an African monkey virus sometime during the late 1950s or early 1960s. The virus probably moved from its original monkey host to humans as a result of an accidental scratch or bite. Not until the late 1970s was the virus identified in human populations. The first reported case of AIDS was diagnosed in the United States in 1981 at the UCLA Medical Center. The virus had been present in Africa, however, for some time prior to 1981 without being recognized and diagnosed.

HIV is a spherical virus containing an RNA genome, a protein shell, and a lipid-protein envelope. Before the virus can be replicated by its host cell, its RNA must be used to make DNA. This is the reverse of the normal transcription process, in which a DNA template is used to manufacture an RNA molecule. Therefore, HIV and similar viruses are called *retroviruses* (*retro* = reverse). To accomplish reverse transcription, HIV has a gene for an enzyme called *reverse transcriptase*. The virus gains entry into a suitable host cell through a very complex series of events involving the virus envelope and the host-cell membrane. Certain types of human cells can serve as hosts because they have a specific viral receptor site on their surface identified as *CD4* (*CD* stands for "clusters of differentiation," a group of molecules on the surface of cells. *CD4* refers to a specific group of CD molecules). There are several kinds of cells that have CD4 receptors on their surface. These include some types of brain cells and several types of cells belonging to the immune system—namely, monocytes, macrophages, and T4-helper/inducer lymphocytes.

Once inside the host cell, the RNA of the HIV is used to make a DNA copy with the help of the enzyme, reverse transcriptase. When reverse transcriptase has completed its job, this piece of viral DNA is spliced into the host cell's DNA. In this integrated form, the virus is called a *provirus*. As a provirus, it can remain inside some host cells for an extended period without causing any harm. Some researchers estimate that this dormant period can last more than 30 years. Eventually, something triggers the host cell to replicate the virus. Then, multiple copies of the virus are produced, the host cell dies, and new viruses are released into surrounding body fluids where they can be transmitted to other cells in the body or to other individuals. Because of the dormant period, it is possible to differentiate two stages of an HIV infection. Persons can be HIV positive (HIV+) without showing the symptoms of AIDS. However, once replication of the virus begins and CD4 cells begin to die, the symptoms of AIDS become evident.

Figure 1

The Nature of the Disease

A decrease in one type of CD4 cell—the T4 lymphocytes—is an important diagnostic indicator of HIV infection and an indicator of the onset of AIDS symptoms. The initial symptoms of the disease have been referred to as *ARC,* or *AIDS-related complex,* or *pre-AIDS.* The immune system cells killed by HIV are responsible for several important defense mechanisms against disease:

1. They assist in the production of antibodies.
2. They encourage the killing of tumor cells, microorganisms, and cells infected by microorganisms.
3. They encourage disease-fighting cells to reproduce and increase their number.

When T4 lymphocytes are destroyed by HIV, all of these defensive efforts of the immune system are depressed. This leaves the body vulnerable to invasion by many types of infecting microbes or to being overtaken by body cells that have changed into tumor cells. This means that HIV infections normally do not directly cause the death of the infected individual. Instead, interference with the immune system allows the development of a variety of secondary infections and cancers that produce the symptoms known as AIDS. Most of the disease conditions associated with AIDS are rare in people with healthy immune systems. Some of the more common diseases associated with AIDS include:

1. Kaposi's sarcoma, a form of skin cancer that shows up as purple-red bruises. This is one of the most common forms of cancer found among AIDS patients.

2. A rare lung infection, *Pneumocystis carinii* pneumonia (PCP), caused by a fungus.
3. Gastroenteritis (severe diarrhea) caused by the protozoan *Isospora.*
4. Cytomegalovirus (CMV) infections of the retina of the eye.

Prevention and Treatment

Because lymphocyte host cells are found in the blood and other body fluids, it is logical that these fluids serve as carriers for the transmission of HIV. The virus is transmitted through direct contact with contaminated blood, semen, mucus secretions, serum, breast milk, or blood-contaminated hypodermic needles. If these body fluids contain the free viruses or infected cells (monocytes, macrophages, T4-helper/inducer lymphocytes) in sufficient quantity, they can be a source of infection.

To control the spread of the virus, there must be wide public awareness of the nature of the disease and how it is transmitted. People must be able to recognize high-risk behavior and take action to avoid or change it. The most important risk factor is promiscuous sexual behavior (i.e., sex with many partners). This behavior increases the probability that eventually a person will be exposed to a partner who is HIV+. Other high-risk behaviors include intravenous drug use with shared needles, contact with blood-contaminated articles, and failure to use a condom during intercourse (vaginal, anal, oral) with persons whose HIV status is unknown. Babies born to women known to be HIV+ are at high risk of becoming infected.

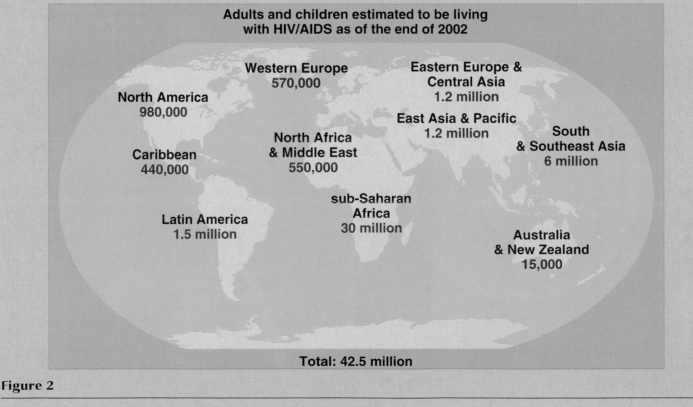

Adults and children estimated to be living with HIV/AIDS as of the end of 2002

Western Europe 570,000

Eastern Europe & Central Asia 1.2 million

North America 980,000

East Asia & Pacific 1.2 million

South & Southeast Asia 6 million

Caribbean 440,000

North Africa & Middle East 550,000

Latin America 1.5 million

sub-Saharan Africa 30 million

Australia & New Zealand 15,000

Total: 42.5 million

Figure 2

Source: World Health Organization of the United Nations

Although it appears that the virus first entered the United States through the homosexual population, AIDS is not a disease unique to homosexuals. Transmission of HIV can occur in homosexual and heterosexual individuals. Today in all parts of the world, HIV is being spread primarily through heterosexual contact, transmission from infected mothers to their infants, and intravenous drug use.

There is no evidence indicating that the virus can be transmitted through the air; on toilet seats; by mosquitoes; by casual contact such as shaking hands, hugging, touching, or closed-mouth kissing; by utensils such as silverware or glasses; or by caring for AIDS patients. The virus is too fragile to survive transmission by these routes.

At the present time, the progress of the infection can be slowed (but not cured) by using drugs that can kill infected cells, improve the body's immune system, or selectively interfere with the life cycle of the virus. The life cycle may be disrupted when the virus enters the cell and the reverse transcriptase converts the RNA to DNA. If this enzyme does not operate, the virus is unable to function. Several drugs have been developed that block this reverse transcriptase step that is necessary for the replication of the virus. In addition, protease inhibitors block the protease enzyme that the virus needs. Usually two or more drugs are given simultaneously or in sequence. This reduces the chance that the virus will develop resistance to the drugs being used.

What about the development of a vaccine to prevent the virus from infecting the body? Experimental vaccines have been developed based on the body's ability to produce antibodies against the virus. Vaccines, however, have been shown to be effective only in monkeys. One of the problems associated with developing a vaccine is the high mutation rate of HIV. Studies have indicated that over 100 mutant strains of the virus have developed from a single infection in one individual. Such an astronomical mutation rate makes it very difficult to develop a vaccine against HIV because any vaccine would have to stimulate an immune response against all possible mutant forms.

Blood tests (the ELISA and Western blot) can indicate whether a person has been exposed to the virus. The tests should be taken on a voluntary basis, absolutely anonymously, and with intensive counseling before and after. People who test positively (HIV+) should make lifestyle changes to prevent others from being exposed. They should also take measures to prevent re-infection, since new varieties of the virus may be more dangerous than those to which they were previously exposed. HIV+ individuals also should do everything they can to maintain good health—exercise regularly, eat a balanced diet, get plenty of rest, and reduce stress. We cannot stop this pandemic in its tracks, but it can be slowed.

Viroids: Infectious RNA

The term **viroid** refers to infectious particles that are only like (-*oid* = similar to) viruses. Viroids are composed solely of a small, single strand of RNA. To date no viroids have been found to parasitize animals. The hosts in which they have been found are cultivated crop plants such as potatoes, tomatoes, and cucumbers. Viroid infections result in stunted or distorted growth and may sometimes cause the plant to die. Pollen, seeds, or farm machinery can transmit viroids from one plant to another. Some scientists believe that viroids may be parts of normal RNA that have gone wrong.

Prions: Infectious Proteins

Several kinds of brain diseases appear to be caused by proteins that can be passed from one individual to another. These infectious agents are called **prions**. All the diseases of this type currently known cause changes in the brain that result in a spongy appearance to the brain called spongiform encephalopathies. The symptoms typically involve abnormal behavior and eventually death. In animals the most common examples are scrapie in sheep and goats and mad cow disease in cattle. Scrapie got its name because one of the symptoms of the disease is an itching of the skin associated with nerve damage that causes the animals to rub against objects and scrape their hair off.

The occurrence of mad cow disease (BSE-bovine spongiform encephalitis) in Great Britain was apparently caused by the spread of prions from sheep to cattle. This occurred because of the practice of processing unusable parts of sheep carcasses into a protein supplement that was fed to cattle. Other similar diseases are known from mink, cats, dogs, elk, and deer. It now appears that the original form of BSE has changed to a variety that is able to infect humans. This new form is called vCJD, which makes scientists believe that BSE and CJD (Creutzfeldt-Jakob disease) are in fact the same prion.

In humans there are several similar diseases. Kuru is a disease known to have occurred in the Fore people of the highlands of Papua New Guinea. The disease was apparently spread because the people ate small amounts of brain tissue of dead relatives. (This ritual is performed as an act of love and respect for the relative.) When the Fore people were encouraged to discontinue this ritual, the incidence of the disease declined. Creutzfeldt-Jakob disease (CJD) is found throughout the world. Its spread is associated with medical treatment. Contaminated surgical instruments and tissue transplants such as corneal transplants are the most likely causes of transfer from affected to uninfected persons.

It now appears to be well-established that these proteins can be spread from one animal to another and they do cause disease, but how are they formed and how do they multiply? The multiplication of the prion appears to result from the disease-causing prion protein coming in contact with a normal body protein and converting it into the disease-causing form, a process called *conversion*. Since this normal protein is produced as a result of translating a DNA message, scientists looked for the genes that make the protein and have found it in a wide variety of mammals. The normal allelle produces a protein that does not cause disease, but is able to be changed by the invading prion protein into the prion form. Prions do not "reproduce" or "replicate" as do viruses or viroids. A prion protein (pathogen) presses up against a normal (not harmful) body protein and may cause it to change shape to that of the dangerous protein. When this conversion happens to a number of proteins, they stack up and interlock, as do the individual pieces of a Lego toy. When enough link together they have a damaging effect—they form plaques (patches) of protein on the surface of nerve cells that disrupt the flow of the nerve impulses and eventually cause nerve cell death. Brain tissues taken from animals that have died of such diseases appear to be full of holes, thus the name spongiform (spongelike in appearance) encephalitis (inflammation of the brain). Because infected organisms lose muscle mass and weight as a result of prion infection, these diseases are often called *chronic wasting diseases (CWDs)*. Chronic wasting disease is present in elk and deer in parts of the United States and Canada.

A person's susceptibility to acquiring a prion disease such as CJD depends on many factors, such as their genetic makeup. If a person produces a functional protein with a particular amino acid sequence, the prion may not be able to convert it to its dangerous form. Other people may produce a protein with a slightly different amino acid sequence that can be converted to the prion form. Once formed, these abnormal proteins resist being destroyed by enzymes and most other agents used to control infectious diseases. Therefore, individuals with the disease-causing form of the protein can serve as the source of the infectious prions.

There is still much to learn about the function of the prion protein and how the abnormal, infectious protein can cause copies of itself to be made. A better understanding of the alleles that produce proteins that can be transformed by prions will eventually lead to effective treatment and prevention of these serious diseases in humans and other animals.

SUMMARY

To facilitate accurate communication, biologists assign a specific name to each species that is cataloged. The various species are cataloged into larger groups on the basis of similar traits.

Taxonomy is the science of classifying and naming organisms. Phylogeny is the science of trying to figure out the evolutionary history of a particular organism. The taxonomic ranking of organisms reflects their evolutionary relationships. Fossil evidence, comparative anatomy, developmental stages, and biochemical evidence are employed in the sciences of taxonomy and phylogeny.

The first organisms thought to have evolved were prokaryotic, single-celled organisms of the domain Eubacteria. Current thinking is that the domain Archaea developed from the Eubacteria and, ultimately, more complex, eukaryotic, many-celled organisms

evolved. These organisms have been classified into the kingdoms Protista, Fungi, Plantae, and Animalia within the domain Eucarya.

Three kinds of acellular infectious particles are currently recognized: viruses, viroids, and prions. Viruses consists of pieces of genetic material surrounded by a protein coat. Viroids consist of naked pieces of RNA. Prions are proteins. Viruses and viroids can be replicated within the cells they invade. Prions cause already existing proteins within an organism to deform, resulting in disease.

THINKING CRITICALLY

For guidelines to answer this question, visit the Online Learning Center.

A minimum estimate of the number of species of insects in the world is 750,000. Perhaps then it would not surprise you to see a fly with eyes on stalks as long as its wings, a dragonfly with a wingspread greater than 1 meter, an insect that can revive after being frozen at −35°C, and a wasp that can push its long, hairlike, egg-laying tool directly into a tree. Only the dragonfly is not presently living, but it once was!

What other curious features of this fascinating group can you discover? Have you looked at a common beetle under magnification? It will hold still if you chill it.

CONCEPT MAP TERMINOLGY

Construct a concept map using these terms.

binomial system of nomenclature	order
	phylogeny
class	phylum
family	species
genus	specific epithet
kingdom	taxonomy

Visit the Online Learning Center to use an interactive format to construct another concept map.

KEY TERMS

Use interactive flash cards, on the Online Learning Center, to help you learn the meanings of these terms.

alternation of generations	order
binomial system of nomenclature	phylogeny
class	phylum
family	prion
fungus	saprophyte
gametophyte stage	specific epithet
genus	sporophyte stage
host	taxonomy
kingdom	viroid
obligate intracellular parasites	virus

LEARNING CONNECTIONS

Visit the Online Learning Center to find additional study resources including those highlighted below.

23.1   The Classification of Organisms

Animations and Review
- Hierarchies
- Phylogeny
- Three domains

Experience This!
- Developing a classification key

1. List two ways that scientific names are different from common names for organisms.
2. Who designed the present-day system of classification? How does this system differ from previous systems?
3. What is the goal of phylogeny?
4. Name the different categories of the classification system.

23.2   Domains Eubacteria and Archaea

Animations and Review
- Characteristics—bacteria
- Diversity—bacteria

5. List two ways that Eubacteria and Archaea differ.
6. Describe typical modes of nutrition for Eubacteria.
7. Describe typical environmental conditions under which Archaea live.

23.3   Domain Eucarya

Animations and Review
- Characteristics—protozoa
- Characteristics—fungi
- Protozoa
- Photosynthetic protists
- Fungus-like protists
- Diversity—fungi

Interactive Concept Map
- The kingdoms

8. List the four kingdoms of the domain Eucarya and give distinguishing characteristics for each.
9. Which of the following groups of organisms contain members that are autotrophic: Eubacteria, Archaea, Protista, Fungi, Plantae, Animalia?

23.4   Acellular Infectious Particles

Animations and Review
- Characteristics
- Life cycles

10. Why do viruses invade only specific types of cells?
11. How do viruses reproduce?
12. Describe how viruses and viroids differ in structure.

Test yourself with multiple choice questions that provide immediate feedback
on the Online Learning Center with PowerWeb at

www.mhhe.com/enger11

Microorganisms:
Bacteria, Protista, and Fungi

GOALS AND OBJECTIVES

Know the characteristics of the microbes.

❏ List the basic characteristics of the bacteria.

❏ List the basic characteristics of members of the Protista.

❏ List the basic characteristics of members of the Fungi.

❏ Explain the features that differentiate organisms as microbes.

❏ List the types of environments in which microorganisms live.

❏ Describe microbes that live in and on all humans.

Understand how microbes interact with other organisms.

❏ Give examples of microbes that are harmful and describe the diseases they cause.

❏ Give examples of microbes that are helpful and describe the benefits they provide.

❏ Diagram the life cycle of a pathogenic protozoan such as *Plasmodium vivax*.

The Origin and Classification of Life

PART SIX

24.1 Microorganisms

Members of the bacteria, Protista, and Fungi share several characteristics that set them apart from plants and animals. These are organisms that rely primarily on asexual reproduction. Some microbes are autotrophic, whereas many others are heterotrophic. Because the majority of organisms in these kingdoms are small and cannot be seen without some type of magnification, they are called **microorganisms,** or **microbes.**

There are only the most basic forms of cooperation among the different cells of microorganisms. Some microbes are free-living, single-celled organisms; others are collections of cells that cooperate to a limited extent. The latter types are called **colonial microbes.** The limited cooperation of individual cells within a colony may take several forms. Some cells within a colony may specialize for reproduction and others do not. Some colonial microbes coordinate their activities so that the colony moves as a unit. Some cells are specialized to produce chemicals that are nutritionally valuable to other cells in the colony.

Microbes are typically found in aquatic or very moist environments; most lack the specialization required to withstand drying. Because they are small, the moist habitat does not need to be large. Microbes can maintain huge populations in very small moist places like the skin of your armpits, temporary puddles, and tiled bathroom walls. Others have the special ability to become dormant and survive long periods without water. When moistened, they become actively growing cells again. The simplest of microbes are the bacteria.

24.2 Bacteria

The Domains Archaea and Eubacteria contain microorganisms that are commonly referred to as **bacteria.** Another common name for them is *germs.* Some unusual bacteria (the Archaea) have the genetic ability to function in extreme environments such as sulfur hot springs, on glaciers, and at the openings of submarine volcanic vents. They are single-celled prokaryotes that lack an organized nucleus and other complex organelles (figure 24.1). *Bergey's Manual of Determinative Bacteriology* first published in 1923 now lists in its latest edition over 2,000 species of bacteria and describes the subtle differences among them. As investigators have discovered more bacteria, they have come to suspect that the known species may represent only 1% of all the bacteria on Earth. For general purposes, bacteria are divided into the three groups based on such features as their staining properties, ability to form endospores, shape (morphology), motility, metabolism, and reproduction (How Science Works 24.1).

Many forms of bacteria are beneficial to humans. Some forms of bacteria decompose dead material, sewage, and other wastes into simpler molecules that can be recycled. Organisms that function in this manner are called saprophytic.

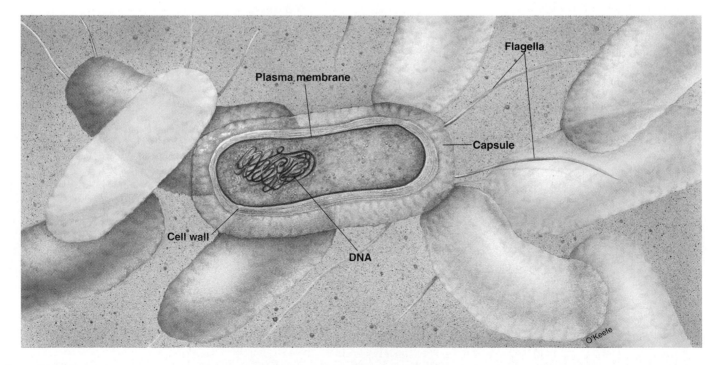

Figure 24.1

Bacteria Cell

The plasma membrane regulates the movement of material between the cell and its environment. A rigid cell wall protects the cell and determines its shape. Some bacteria, usually pathogens, have a capsule to protect them from the host's immune system. The genetic material consists of numerous replicated strands of DNA resembling an unraveled piece of twine.

HOW SCIENCE WORKS 24.1

Gram Staining

Gram staining was first developed in 1843 by the Danish bacteriologist Christian Gram, who discovered that most bacteria could be divided into two main groups based on their staining reactions. Such a technique is called differential staining because it allows the microbiologist to highlight the differences between cell types. Bacteria not easily decolorized with 95% ethyl alcohol after staining with crystal violet and iodine are said to be Gram-positive. Those bacteria decolorized are Gram-negative, and thus very difficult to see through the microscope. Another stain,

called a counterstain, is added to make Gram-negative cells more visible. A number of different stains can be used as a counterstain, but red is preferred because it provides the greatest contrast.

Knowing how some pathogenic bacteria react to Gram staining is of great value in determining how to handle those microbes in cases of infection. The Gram stain is probably the most widely performed diagnostic test in microbiology and can provide guidance in such matters as selecting the right antibiotic for treatment and predicting the kinds of symptoms a patient will show.

The food industry uses bacteria to produce cheeses, yogurt, sauerkraut, and many other foods. Alcohols, acetones, acids, and other chemicals are produced by bacterial cultures. The pharmaceutical industry employs bacteria to produce antibiotics and vitamins. Some bacteria can even metabolize oil and are used to clean up oil spills.

There are also mutualistic relationships between bacteria and other organisms. Some intestinal bacteria benefit humans by producing antibiotics that inhibit the development of pathogenic bacteria. They also compete with disease-causing bacteria for nutrients, thereby helping keep the pathogens in check. They aid digestion by releasing various nutrients. They produce and release vitamin K. Mutualistic bacteria establish this symbiotic relationship when they are ingested along with food or drink. When people travel, they consume local bacteria along with their food and drink and may have problems establishing a new symbiotic relationship with these foreign bacteria. Both the host and the symbionts have to make adjustments to their new environment, which can result in a very uncomfortable situation for both. Some people develop traveler's diarrhea as a result (Outlooks 24.1).

Animals do not produce the enzymes needed for the digestion of cellulose. Methanogens, bacteria that obtain metabolic energy by reducing carbon dioxide (CO_2) to methane (CH_4), digest the cellulose consumed by herbivorous animals, such as cows, thereby permitting the cow to obtain simple sugars from the otherwise useless cellulose. There is a mutualistic relationship between the cow and the methanogens. Some methanogens are also found in the human gut and are among the organisms responsible for the production of intestinal gas. In some regions of the world methanogens are used to digest organic waste, and the methane is used as a source of fuel. • decomposers, p. 262

The Romans knew that bean plants somehow enriched the soil, but it was not until the 1800s that bacteria were recognized as the enriching agents. Certain types of bacteria have a symbiotic relationship with the roots of bean plants and other legumes. These bacteria are capable of converting atmospheric nitrogen into a form that is usable to the plants.

Early forms of life consisted of prokaryotic cells living in a reducing atmosphere. Photosynthetic bacteria released oxygen, and the Earth's atmosphere began to change to an

---

OUTLOOKS 24.1

Bismuth, Bowels, and Bugs

In its pure state, bismuth (German = *bisemutum*) is a solid, lustrous, reddish-white material that has been known since ancient times. Today, the element bismuth is commonly found in over-the-counter medications such as Pepto-Bismol, Bismatral, and Pink Bismuth, in the form of the ingredient bismuth subsalicylate. It is used to control traveler's diarrhea (TD)—the diarrhea that affects people who visit countries or regions with poor public sanitation and hygiene. TD may be caused by a number of microbes including the bacterium *E. coli*, protozoans such as *Crytosporidium* and *Cyclospora,* and some viruses. These microbes can be found in contaminated water and food. Taking eight pills per day has been shown to reduce the symptoms of TD by as much as 60%. However, there can be some undesirable side effects. Nausea and

vomiting, along with the temporary blackening of the stool, teeth, and tongue, may occur.

How does bismuth do its job? The hypothesis is that it promotes the formation of a protective layer of fluids on the surface of the intestinal tract that prevents the microbes from doing their dirty work. However, children and teens have been advised by the FDA (Food and Drug Administration) to avoid bismuth subsalicylate-containing medications when they have the flu or chicken pox, since they are at increased risk of developing Reye syndrome. More recently, bismuth subsalicylate is being used to treat patients with peptic ulcers since it also inhibits the growth of *Helicobacter pylori*, the bacterium that causes many peptic ulcers.

From F. Ross, *Chemistry and Cell Biology for Allied Health Science,* 5th ed., Kendall/Hunt Publishers, 1999.

Table 24.1

COMMON BACTERIA FOUND IN OR ON YOUR BODY

Skin	*Corynebacterium* sp., *Staphylococcus* sp., *Streptococcus* sp., *E. coli*, *Mycobacterium* sp.
Eye	*Corynebacterium* sp., *Neisseria* sp., *Bacillus* sp., *Staphylococcus* sp., *Streptococcus* sp.
Ear	*Staphylococcus* sp., *Streptococcus* sp., *Corynebacterium* sp., *Bacillus* sp.
Mouth	*Streptococcus* sp., *Staphylococcus* sp., *Lactobacillus* sp., *Corynebacterium* sp., *Fusobacterium* sp., *Vibrio* sp., *Haemophilus* sp.
Nose	*Corynebacterium* sp., *Staphylococcus* sp., *Streptococcus* sp.
Intestinal tract	*Lactobacillus* sp., *E. coli*, *Bacillus* sp., *Clostridium* sp., *Pseudomonas* sp., *Bacteroides* sp., *Streptococcus* sp.
Genital tract	*Lactobacillus* sp., *Staphylococcus* sp., *Streptococcus* sp., *Clostridium* sp., *Peptostreptococcus* sp., *E. coli*

Streptococcus   E. coli   Lactobacillus   Corynebacterium

oxidizing atmosphere. Photosynthetic, colonial blue-green bacteria are still present in large numbers on Earth and continue to release significant quantities of oxygen. Colonies of blue-green bacteria are found in aquatic environments, where they form long, filamentous strands commonly called *pond scum*. Some of the larger cells in the colony are capable of nitrogen fixation and convert atmospheric nitrogen, N_2, to ammonia, NH_3. This provides a form of nitrogen usable to other cells in the colony—an example of division of labor.

The word *bacteria* usually brings to mind visions of tiny things that cause diseases or are used as dangerous biological weapons; however, the majority are free living and not harmful. Their roles in the ecosystem include those of decomposers, nitrogen fixers, and other symbionts. It is true that some diseases are caused by bacteria, but only a minority of bacteria are **pathogens**, microbes that cause infectious diseases. It is normal for all organisms to have symbiotic relationships with bacteria. Most organisms are lined and covered by populations of bacteria called *normal flora* (table 24.1). In fact, if an organism lacks bacteria it is considered abnormal. Some pathogenic bacteria may be associated with an organism yet do not cause disease. For example, *Streptococcus pneumoniae* may grow in the throats of healthy people without any pathogenic effects. But if a person's resistance is lowered, as after a bout with viral flu, *Streptococcus pneumoniae* may reproduce rapidly in the lungs and cause pneumonia; the relationship has changed from commensalistic to parasitic.

Bacteria may invade the healthy tissue of the host and cause disease by altering the tissue's normal physiology. Bacteria living in the host release a variety of enzymes that cause the destruction of tissue. The disease ends when the pathogens are killed by the body's defenses or some outside agent, such as an antibiotic. Examples are the infectious diseases strep throat, syphilis, pneumonia, tuberculosis, and leprosy (figure 24.2).

Many other bacterial illnesses are caused by toxins or poisons produced by bacteria that may be consumed with food or drink (Outlooks 24.2). In this case, disease can be caused even though the pathogens may never enter the host. For example, botulism is an extremely deadly disease caused by the presence of bacterial toxins in food or drink. Some other bacterial diseases are the result of toxins released from bacteria growing inside the host tissue; tetanus and diphtheria are examples. In general, toxins may cause tissue damage, fever, and aches and pains.

Bacterial pathogens are also important factors in certain plant diseases. Bacteria are the causative agents in many types of plant blights, wilts, and soft rots. Apples and other fruit trees are susceptible to fire blight, a disease that lowers the fruit yield because it kills the tree's branches. Citrus canker, a disease of citrus fruits that causes cancerlike growths, can generate widespread damage. In a three-year period, Florida citrus growers lost $2.5 billion because of this disease (figure 24.3).

Figure 24.2

Leprosy

More than 20 million people worldwide are infected with
Mycobacterium leprae and have leprosy (Hansen's disease). This
disease alters the host's physiology, resulting in these open sores.
Another species of *Mycobacterium, M. tuberculosis,* is again
becoming a public health concern because it is becoming
increasingly resistant to the controlling effects of antibiotics. New
standards of control have been issued by the Centers for Disease
Control and Prevention in Atlanta, Georgia.

Despite large investments of time and money, scientists
have found it difficult to control bacterial populations. Two
factors operate in favor of the bacteria: their reproductive
rate and their ability to form spores. Under ideal conditions
some bacteria can grow and divide every 20 minutes. If one
bacterial cell and all its offspring were to reproduce at this
ideal rate, in 48 hours there would be 2.2×10^{43} cells. In reality, bacteria cannot achieve such incredibly large populations because they would eventually run out of food and be
unable to dispose of their wastes.

Because bacteria reproduce so rapidly, a few antibiotic-
resistant cells in a population can increase to dangerous levels in a very short time. This requires the use of stronger
doses of antibiotics or new types of antibiotics to bring the
bacteria under control. Furthermore, these resistant strains
can be transferred from one host to another. For example,
sulfa drugs and penicillin, once widely used to fight infections, are now ineffective against many strains of pathogenic
bacteria. As new antibiotics are developed, natural selection
encourages the development of resistant bacterial strains.
Therefore humans are constantly waging battles against new
strains of resistant bacteria.

Another factor that enables some bacteria to survive a
hostile environment is their ability to form *endospores*. An
endospore is a unique bacterial structure with a low meta-

Figure 24.3

Plant Disease

Citrus canker growth on an orange tree promotes rotting of the
infected part of the tree.

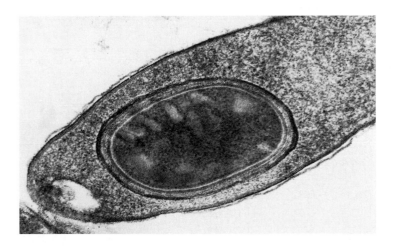

Figure 24.4

Bacterial Endospore

The darker area in the cell is the endospore. It contains the bacterial
DNA as well as a concentration of cytoplasmic material that is
surrounded and protected by a thick wall (approx. 63,000×).
Endospores thought to be *Bacillus sphaericus* and estimated to be
25 million to 40 million years old have been isolated from the
intestinal tract of a bee fossilized in amber. When placed in an
optimum growth environment, they germinated and grew into
numerous colonies.

bolic rate that can germinate under favorable conditions to
form a new, actively growing cell (figure 24.4). For example, people who preserve food by canning often boil the
food in the canning jars to kill the bacteria. But not all bacteria are killed by boiling; some of them form endospores.

OUTLOOKS 24.2

A Look at Microbes and Biological Warfare

Because bacteria and other microbes can spread through a population and cause disease, they have been used as biological warfare agents. However, there are only a limited number that can be used effectively as agents of biological warfare. To be an effective biological warfare agent, the agents should:

1. Be human pathogens;
2. Be able to be easily cultured in a lab;
3. Survive outside the lab long enough to reach the target;
4. Enter a person in large enough numbers and in a way that causes disease;
5. Cause a large number of soldiers and civilians to die or become severely ill;
6. Spread from person to person easily;
7. Cause panic and disruption in the infected population;
8. Be difficult to diagnose;
9. Have few, if any, antidotes; and
10. Require special action for public health preparedness.

A serious drawback to the use of biological warfare agents is the problem of protecting the soldiers and civilian population of the country using the agent. Unlike explosives or chemical warfare agents, biological agents have the ability to reproduce, spread through the population, and cause epidemics, even after a war has ended. This is why only a few kinds of microbes or their toxins could be used as agents of warfare. As the historical chronology suggests (Table 24.A), deaths due to biological warfare agents are very few. Other mechanisms (bombs, guns, automobile crashes, smoking, and obesity) cause many more deaths per day than the use of biological weapons ever has. Nevertheless, the history of biological warfare is a lone one and is becoming an increasing concern with the development of genetic engineering and technology. Equally if not more important, the potential and actual release of biological agents can be used to frighten people. Since one of the main goals of terrorism is to create panic and disrupt normal life, even a small number of infected persons and limited numbers of deaths due to biological agents can be effective.

Table 24.A

History of Biological Warfare

Date	Event
Throughout history	Spear tips, swords, and knives were jabbed into the soil to inoculate them with pathogens before battle. Wounded soldiers later became incapacitated or died from their infections. In many conflicts, sharpened sticks were dipped into human and animal excrement and used as booby trap weapons.
6th century B.C.	Wells were poisoned with fungal rye ergot toxins. The purgative herb, black hellebore, was used to poison the well water during the siege of the Greek city, Krissa.
1346	Corpses of persons who died of plague were hurled by catapult over the city walls during the siege of the Genoese city of Kaffa, Italy.
1754–1767	The Englishman Sir Jeffery Amherst gave smallpox-contaminated blankets to Indians fighting with the French against the English during the French and Indian war at Fort Carillon, renamed Fort Ticonderoga (New York).
World War I	German secret agents inoculated horses and mules awaiting shipment from the U.S. to Europe with the bacterium responsible for Glanders disease.
1937–1945	Japan began a biological warfare program known as Unit 731 in which prisoners of war were used as research subjects to investigate anthrax, plague, and other agents.
1943–1969	The U.S. began biological warfare agent research for offensive purposes at Fort Detrick, Maryland. All stockpiles of agents were destroyed between 1971 and 1972. Russia produced and stored anthrax endospores for use as biological weapons (BW).
1953–present	The U.S. began a medical defensive program called the U.S. Army Medical Research Institute of Infectious Diseases (USAMRIID).
1972	The U.S., U.K., and USSR signed the *Convention on the Prohibition of the Development, Production and Stockpiling of Bacteriological (Biological) and Toxic Weapons and on Their Destruction*, commonly called the Biological Weapons Convention.
1978	A Bulgarian exile was assassinated in London by a Bulgarian secret service agent using a modified umbrella to inject lethal biotoxin.
1979	Sixty-six fatalities resulted from a Soviet military experiment in Sverdlovsk (now Yekaterinburg) in which *Bacillus anthracis* endospores were released into the air from the Soviet military microbiology facility known as Compound 19. The nature of the incident was denied until 1992.
1980	Members of the German Red Army Brigade were arrested in Paris, France for growing *Clostridium botulinum* for use as biological weapons.
1984	The religious cult, Bhagwan Shree Rajneesh, sprayed the salad bars of four restaurants in The Dalles, Oregon, with a solution containing *Salmonella* sp.; 751 people were sickened, no fatalities.
1991	The United Nations carried out its first inspection of Iraq's biological warfare capabilities.
1992	A biological weapons program manager who defected from Russia in 1992 revealed the existence of a biological warfare program.
1995	The religious cult Aum Shinriky released *Botulinum*-containing aerosols in downtown Tokyo, Japan and near U.S. military installations in Japan.
1996	Twelve workers at a medical center in Texas contracted a rare form of dysentery. The bacterium in their stool was identical to one that was missing from the lab's stock of samples. No arrests were made, but the infections were clearly the result of an intentional act.
2001	Letters containing anthrax endospores were mailed from New Jersey. Five persons died and 22 were infected.

For example, botulism poison is usually found in foods that are improperly canned. The endospores of *Clostridium botulinum,* the bacterium that causes botulism, can withstand boiling and remain for years in the endospore state. However, endospores do not germinate and produce botulism toxin if the pH of the canned goods is in the acid range; in that case, the food remains preserved and edible. If conditions become favorable for endospores to germinate, they become actively growing cells and produce toxin. Home canning is the major source of botulism. Using a pressure cooker and heating the food to temperatures higher than 121°C for 15 to 20 minutes destroys both botulism toxin and the endospores.

24.3 Kingdom Protista

The first protists evolved about 1.5 billion years ago. Like the prokaryotes, most of the protists are one celled organisms. However, there is a significant difference between the two kingdoms: All the protists are eukaryotic cells and all

the prokaryotes are prokaryotic cells. Prokaryotic cells usually have a volume of 1 to 5 cubic micrometers. Most eukaryotic cells have a volume greater than 5,000 cubic micrometers. This means that eukaryotic cells usually have a volume at least 1,000 times greater than prokaryotic cells. The presence of membranous organelles such as the nucleus, endoplasmic reticulum, mitochondria, and chloroplasts allows protists to be larger than prokaryotes. These organelles provide a much greater surface area within the cell upon which specialized reactions may occur. This allows for more efficient cell metabolism than is found in prokaryotic cells.

Because of the great diversity within the more than 60,000 species, it is a constant challenge to separate the kingdom Protista into subgroupings as research reveals new evidence about members of this group. Usually the species are divided into three groups: **algae,** autotrophic unicellular organisms; **protozoa,** heterotrophic unicellular organisms; and funguslike protists. However, emerging evidence suggests a much more complex evolutionary pattern as noted in the cladogram seen in table 24.2.

Table 24.2

A CLADOGRAM SUGGESTING EVOLUTIONARY RELATIONSHIP AMONG THE VARIOUS GROUPS OF PROTISTS

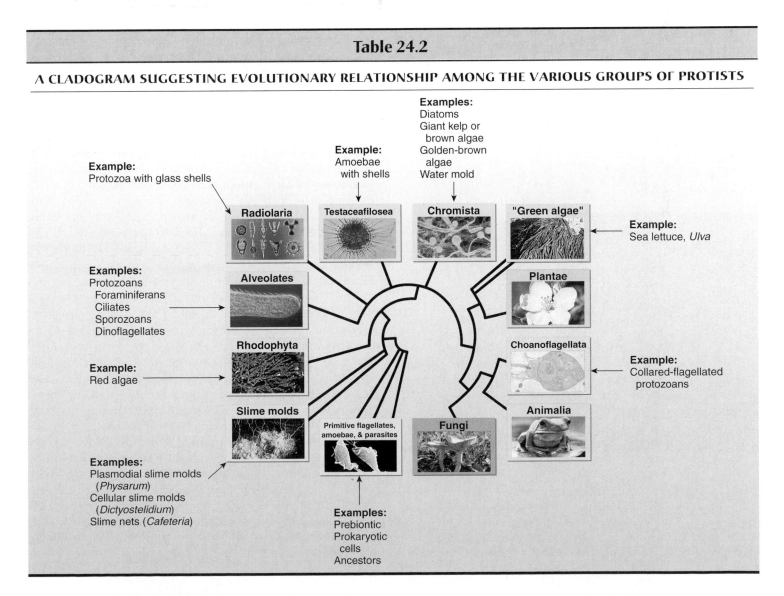

Plantlike Protists

Algae are protists that have a cellulose cell wall. They contain chlorophyll and can therefore carry on photosynthesis. Unicellular and colonial types occur in a variety of habitats. There are two major forms of algae in a variety of marine and freshwater habitats: planktonic and benthic. **Plankton** consists of small floating or weakly swimming organisms. **Benthic** organisms live attached to the bottom or to objects in the water. **Phytoplankton** consists of photosynthetic plankton that forms the basis for most aquatic food chains (figure 24.5). The large number of benthic and planktonic algae makes them an important source of atmospheric oxygen (O_2). It is estimated that 30% to 50% of atmospheric oxygen is produced by algae. • cell walls, p. 64

Because algae require light, phytoplankton is found only near the surface of the water. Even in the clearest water, photosynthesis does not usually occur any deeper than 100 meters. To remain near the surface, some of the phytoplankton are capable of locomotion. Others maintain their position by storing food as oil, which is less dense than water and enables the cells to float near the surface.

Three common forms of single-celled algae typically found as phytoplankton are the Euglenophyta (euglenas), and Chrysophyta (golden-brown algae = diatoms, yellow-green algae), and Pyrrophyta (dinoflagellates). *Euglena* is found mainly in freshwater. They are widely studied because they are easy to culture. Under low levels of light, these photosynthetic species can ingest food. *Euglena* can be either autotrophic or heterotrophic.

There are over 10,000 species of diatoms. Diatoms are commonly found in freshwater, marine, and soil environments. They can reproduce both sexually and asexually. When conditions are favorable, asexual reproduction can result in what is called an algal **bloom**—a rapid increase in the population of microorganisms in a body of water. The population can become so large that the water looks murky. These algae are unique because their cell walls contain silicon dioxide (silica). The algal walls fit together like the lid and bottom of a shoe box; the lid overlaps the bottom. Because their cell walls contain silicon dioxide, they readily form fossils. The fossil cell walls have large, abrasive surface areas with many tiny holes and can be used in a number of commercial processes. They are used as filters for liquids and as abrasives in specialty soaps, toothpastes, and scouring powders.

Along with diatoms, dinoflagellates are the most important food producers in the ocean's ecosystem. All members of this group of algae have two flagella, which is the reason for their name (*dino* = two). Many marine forms are bioluminescent; they are responsible for the twinkling lights seen at night in ocean waves or in a boat's wake.

Some species of dinoflagellates have symbiotic relationships with marine animals, such as the reef corals; the dinoflagellates provide a source of nutrients for the reef-building coral. Corals that live in the light and contain dinoflagellates grow 10 times faster than corals without this symbiont. Thus, in coral reef ecosystems, dinoflagellates form the foundation of the food chain. Some forms of dinoflagellates produce toxins that can be accumulated by such filter-feeding marine animals as clams and oysters. Filter-feeding shellfish ingest large amounts of the toxins, which has no effect on the shellfish but can cause sickness or death in animals that feed on them, such as fish, birds, and mammals. Many of the toxin-producing dinoflagellates contain red pigment. Blooms of this kind are responsible for *red tides*. Red tides usually occur in the warm months, during which people should refrain from collecting and eating oysters. The expression "Oysters 'R' in season" comes from the fact that most of the months with an R in their spelling are cold-weather months, during which oysters are safer to eat. Commercially available shellfish are tested for toxin content; if they are toxic, they are not marketed. Red tides not only have occurred off the coast of Florida in North America, but also have more recently developed off the coast of China. Hundreds of thousands of fish and other marine life have been killed as a result of toxin release, thus having a significant impact on the economy and food supply.

In recent years a new problem has surfaced caused by the dinoflagellate, *Pfiesteria piscidia*. In combination with certain fungi, these algae have been responsible for the death of millions of fish in estuaries of the eastern United States. These dinoflagellates release toxins that paralyze fish and feed on the fish. They have also been responsible for human and wildlife poisoning.

(a)   (b)

Figure 24.5

Algae

Algae can be found in a variety of types and colors: (a) a single-celled green alga, *Micrasterias*; (b) a colonial red alga, *Antithamnium*.

Multicellular algae, commonly known as *seaweed,* are large colonial forms usually found attached to objects in shallow water. Two types, red algae (Rhodophyta) and brown algae (Phaeophyta), are mainly marine forms. The green algae (Chlorophyta) are a third kind of seaweed; they are primarily freshwater species.

Red algae live in warm oceans and attach to the ocean floor by means of a holdfast structure. They may be found from the splash zone, the area where waves are breaking, to depths of 100 meters. Some red algae become encrusted with calcium carbonate and are important in reef building; other species are of commercial importance because they produce agar and carrageenin. *Agar* is widely used as a jelling agent for growth media in microbiology. *Carrageenin* is a gelatinous material used in paints, cosmetics, and baking. It is also used to make gelatin desserts harden faster and to make ice cream smoother. In Asia and Europe some red algae are harvested and used as food.

Brown algae are found in cooler marine environments than are the red algae. Most species of brown algae have a holdfast organ. Colonies of these algae can reach 100 meters in length (figure 24.6). Brown algae produce *alginates,* which are widely used as stabilizers in frozen desserts, emulsifiers in salad dressings, and as thickeners that give body to foods such as chocolate milk and cream cheeses; they are also used to form gels in such products as fruit jellies.

The Sargasso Sea is a large mat of free-floating brown algae between the Bahamas and the Azores. It is thought that this huge mass (as large as the European continent) is the result of brown algae that have become detached from the ocean bottom, have been carried by ocean currents, and accumulate in this calm region of the Atlantic Ocean. This large mass of floating algae provides a habitat for a large number of marine animals, such as marine turtles, eels, jellyfish, and innumerable crustaceans. • **pelagic organisms, p. 528**

Green algae are found primarily in freshwater ecosystems, where they may attach to a variety of objects. Members of this group can also be found growing on trees, in the soil, and even on snowfields in the mountains. Like land plants, green algae have cellulose cell walls and store food as starch. Green algae also have the same types of chlorophyll as do plants. Biologists believe that land plants evolved from the green algae.

Animal-like Protists

A second major group of organisms in the kingdom Protista, the protozoa, lacks all types of chlorophyll. The word *protozoa* literally means "first animal." It is a descriptive term that includes all eukaryotic, heterotrophic, unicellular organisms that lack cell walls. The protozoa are classified into subgroups according to their method of locomotion.

Most members of the Zoomastigina have flagella and live in freshwater. They have no cell walls and no chloroplasts, and they can be parasitic or free living (Outlooks 24.3).

Figure 24.6

A Kelp Grove
These multicellular brown algae are attached to the ocean floor by holdfasts. Their blades may reach a length of 100 meters and float upward because they have a bladderlike sac filled with air.

There is a mutualistic relationship between some flagellates and their termite hosts. Certain protozoa live in termite guts and are capable of digesting cellulose into simple sugars that serve as food for the termite. Of the parasitic protozoa, two different species produce sleeping sickness in humans and domestic cattle. In both cases the protozoan enters the host as the result of an insect bite. The parasite develops in the circulatory system and moves to the cerebrospinal fluid surrounding the brain. When this occurs, the infected person develops the "sleeping" condition, which, if untreated, is eventually fatal. Many biologists believe that all other types of protozoa, and even the multicellular animals, evolved from primitive flagellated microorganisms similar to the Zoomastigina. • **cilia and flagella, p. 82**

Members of the group Sarcodina range from the most well-known *Amoeba*, with its constantly changing shape, to species having a rigid outer cover (figure 24.7). *Amoeba* uses pseudopods to move about and to engulf food. A pseudopod is a protoplasmic extension of the cell that contains moving cytoplasm. Many pseudopods are temporary extensions that

OUTLOOKS 24.3

Don't Drink the Water!

Giardia lamblia is a protozoan in streams and lakes throughout the world, found even in "pure" mountain water in U.S. wilderness areas. Over 40 species of animals harbor this organism in their small intestines. Its presence may cause diarrhea, vomiting, cramps, or nausea. *Giardia* may be found even if good human sanitation is practiced. No matter how inviting it may be to drink directly from that cold mountain stream, don't. Deer, beaver, or other animals could have contaminated the water with *Giardia*. Treat the water before drinking. The most effective way to eliminate the spores formed by this protozoan is to use special filters that can filter out particles as small as 1 micrometer; otherwise, boil the water for at least five minutes before drinking.

The species called *Entamoeba histolytica* (*ent* = inside; *amoeba* = amoeba; *histo* = tissue; *lytica* = destroying) is responsible for the diarrheal disease known as dysentery. People become infected with this protozoan when they travel to a foreign country and drink contaminated water. If you plan on such a trip, be sure to see your physician *several weeks* before you go! The infection can be prevented by taking an antiprotozoal antibiotic, but you must start treatment ahead of time.

Giardia lamblia

Entamoeba histolytica

(a)

(b)

Figure 24.7

Sarcodina

These protozoa range from (a) the *Amoeba,* which changes shape to move and feed, to (b) organisms that are enclosed in a shell (shelled amoeba). The extensions from the cell are called pseudopods (*pseudo* = false; *pod* = foot).

form and disappear as the cell moves. Most amoeboid protists are free living and feed on bacteria, algae, or even small multicellular organisms. Some forms are parasitic, such as the one that causes amoebic dysentery in humans.

Members of another group of the Sarcodina, the foraminiferans, live in warm oceans and are enclosed in a

shell. As these cells die, the shells collect on the ocean floor, and their remains form limestone. The cliffs of Dover, England, were formed from such shells. Oil companies have a vested interest in foraminiferans because they are often found where oil deposits are located.

All members of the group Sporozoa are nonmotile parasites that have a sporelike stage in their life cycles. Malaria, one of the leading causes of disability and death in the world, results from a type of sporozoan. Two billion people live in malaria-prone regions of the world. There are an estimated 150 to 300 million new cases of malaria each year, and the disease kills 2 to 4 million people annually.

Like most sporozoans, the one that causes malaria has a complex life cycle involving a mosquito vector for transmission (figure 24.8). Recall from chapter 15 that a *vector* is an organism capable of transmitting a parasite from one organism to another. While in the mosquito vector, the parasite goes through the sexual stages of its life cycle. One of the best ways to control this disease is to eliminate the vector, which usually involves using some sort of pesticide. Many of us are concerned about the harmful effects of pesticides in the environment. However, in parts of the world where malaria is common, the harmful effects of pesticides are of less concern than the harm generated by the disease. Many

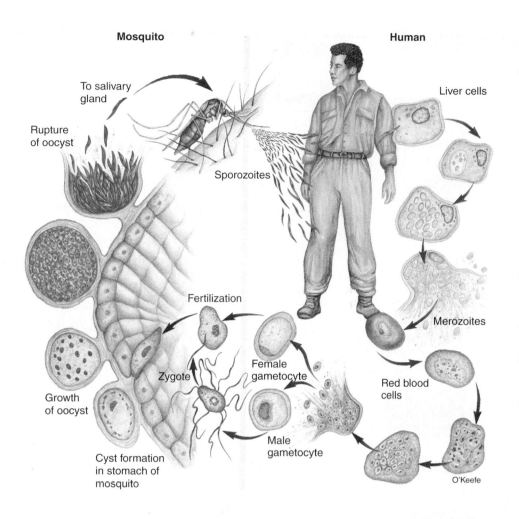

Mosquito

Human

To salivary gland

Rupture of oocyst

Sporozoites

Liver cells

Fertilization

Zygote

Growth of oocyst

Female gametocyte

Male gametocyte

Merozoites

Red blood cells

Cyst formation in stomach of mosquito

O'Keefe

Figure 24.8

The Life Cycle of *Plasmodium vivax*

The complex life cycle of the member of the Protista that causes malaria requires two hosts, the *Anopheles* mosquito and the human. Humans get malaria when they are bitten by a mosquito carrying the larval stage of *Plasmodium*. The larva undergoes asexual reproduction and releases thousands of individuals that invade the red blood cells. Their release from massive numbers of infected red blood cells causes the chills, fever, and headache associated with malaria. Inside the red blood cell, more reproduction occurs to form male gametocytes and female gametocytes. When the mosquito bites a person with malaria, it ingests some gametocytes. Fertilization occurs and zygotes develop in the stomach of the mosquito. The resulting larvae are housed in the mosquito's salivary gland. Then, when the mosquito bites someone, some saliva containing the larvae is released into the person's blood and the cycle begins again.

diseases of domestic and wild animals are also caused by members of this group.

The group Ciliophora contains the most structurally complex protozoans. They are commonly known as *ciliates* and derive their name from the fact that they have numerous short, flexible filaments called *cilia* (figure 24.9). These move in an organized, rhythmic manner and propel the cell through the water. Some types of ciliates, such as *Paramecium*, have nearly 15,000 cilia per cell and move at a rapid speed of 1 millimeter per second. Most ciliates are free-living cells found in fresh and salt water, where they feed on bacteria and other small organisms.

Funguslike Protists

Funguslike protists have a motile, amoeboid reproductive stage, which differentiates them from true fungi. There are two kinds of funguslike protists: slime molds and water molds. Some slime molds, members of Myxomycota, can be found growing on rotting damp logs, leaves, and soil. They look like giant amoebae whose nucleus and other organelles have divided repeatedly within a single large cell (figure 24.10). No cell membranes partition this mass into separate segments.

Figure 24.9

Ciliated Protozoa

The many hairlike cilia on the surface of this cell are used to propel the protozoan through the water.

They vary in color from white to bright red or yellow, and may reach relatively large sizes (45 centimeters in length) when in an optimum environment.

Other kinds of slime molds, members of Acrasiomycota, exist as large numbers of individual, amoebalike cells. These haploid cells get food by engulfing microorganisms.

Figure 24.10

Slime Mold

Slime molds grow in moist conditions and are important decomposers. As the slime mold grows, additional nuclei are produced by mitosis, but there is no cytoplasmic division. Thus, at this stage, it is a single mass of cytoplasm with many nuclei.

Figure 24.11

Water Mold

Rapidly reproducing water molds quickly produce a large mass of filamentous hyphae. These hyphae are the cause of fuzzy growth often seen on dead fish or other dead material in the water.

They reproduce by mitosis. When their environment becomes dry or otherwise unfavorable, the cells come together into an irregular mass. This mass glides along rather like an ordinary garden slug and is labeled the sluglike stage. This sluglike form may flow about for hours before it forms spores. When the mass gets ready to form spores, it forms a stalk with cells that have cell walls. At the top of this specialized structure, cells are modified to become haploid spores. When released, these spores may be carried by the wind and, if they land in a favorable place, may develop into new amoebalike cells.

Another group of funguslike protists includes the water molds (figure 24.11). This group, the Oomycota, has repro-ductive cells with two flagella. A wide variety of water molds are saprophytes, which are usually found growing in a moist environment. They differ in structure from the true fungi in that some filaments have no cross walls, thus allowing the cell contents to flow from cell to cell.

Water molds are important saprophytes and parasites in aquatic ecosystems. They are often seen as fluffy growths on dead fish or other organic matter floating in water. A parasitic form of this fungus is well known to people who rear tropical fish; it causes a cottonlike growth on the fish. Although these organisms are usually found in aquatic habitats, they are not limited to this environment. Some species cause downy mildew on plants such as grapes. In the 1880s this mildew almost ruined the French wine industry when it spread throughout the vineyards. A copper-based fungicide called *Bordeaux mixture*—the first chemical used against plant diseases—was used to save the vineyards. A water mold was also responsible for the Irish potato blight. In the nineteenth century, potatoes were the staple of the Irish diet. Cool, wet weather in 1845 and 1847 damaged much of the potato crop, and more than a million people died of starvation. Nearly one-third of the survivors left Ireland and moved to Canada or the United States.

24.4 Multicellularity in the Protista

The three major types of the kingdom Protista (algae, protozoa, and funguslike protists) include both single-celled and multicellular forms. Biologists believe that there has been a similar type of evolution in all three of these groups. The most primitive organisms in each group are thought to have been single-celled, and to have given rise to the more advanced multicellular forms. Most protozoan organisms are single-celled; however, there is a group that contains numerous colonial forms. The multicellular forms of funguslike protists are the slime molds, which have both single-celled and multicellular stages. Perhaps the most widely known example of this trend from a single-celled to a multicellular condition is found in the green algae. A very common single-celled green alga is *Chlamydomonas*, which has a cell wall and two flagella. It looks just like the individual cells of the colonial green algae *Volvox*. *Volvox* can be composed of more than half a million cells (figure 24.12). All the flagella of each cell in the colony move in unison, allowing the colony to move in one direction. Many of the cells cannot reproduce sexually; other cells assume this function for the colony. In some *Volvox* species, certain cells have even specialized to produce sperm or eggs. Biologists believe that the division of labor seen in colonial protists represents the beginning of specialization that led to the development of true multicellular organisms with many different kinds of specialized cells. Three types of multicellular organisms—fungi, plants, and animals— eventually developed.

(a)

View of *Gonium* colony

Matrix

Flagellum

Individual
cell

(b)

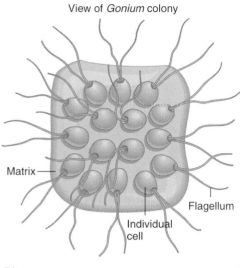

(c)

Figure 24.12

Algae

(a) *Chlamydomonas* is a green, single-celled alga containing the same type of chlorophyll as that found in green plants. (b) *Gonium*, a green alga similar to *Chlamydomonas*, forms colonies composed of 4 to 32 cells. (c) *Volvox*, another green alga, is a more complex form in the evolution of colonial green algae.

(a)

(b)

(c)

(d)

Figure 24.13

Spore Production

Some fungi, like the puffball (*a*), produce spores on the inside. The puffball must be broken (*b*) to release the spores. Other forms, like the club fungus (*c*), have exposed gills with spore-producing basidia (*d*).

24.5 Kingdom Fungi

Members of the kingdom Fungi are nonphotosynthetic eukaryotic organisms with rigid cell walls. The majority are multicellular, but a few, like yeasts, are single-celled. The majority also do not move. All of these organisms are heterotrophs; that is, they must obtain nutrients from organic sources. Most secrete chemicals that digest large molecules into smaller units that are absorbed. Fungi can be either free living or parasitic. Fungi that are free living, like mushrooms, decompose dead organisms as they absorb nutrients. Fungi that are parasitic are responsible for athlete's foot, vaginal yeast infections, ringworm, as well as many plant diseases. There is no unanimity regarding the divisions within the kingdom Fungi. Originally, fungi were thought to be members of the Plantae kingdom. In fact, the term *division* is used with this kingdom because this is the term used by botanists in place of *phylum*.

Even though fungi are nonmotile, they successfully survive and disperse because of their ability to form spores, which some produce sexually and others produce asexually. Spores may be produced internally or externally (figure 24.13). An average-sized mushroom can produce over 20 billion spores; a good-sized puffball can produce as many as 8 trillion spores. When released, the spores can be transported by wind or water. Because of their small size, spores can remain in the

HOW SCIENCE WORKS 24.2

Penicillin

The discovery of the antibiotic penicillin is an interesting story. In 1928 Dr. Alexander Fleming was working at St. Mary's Hospital in London. As he sorted through some old petri dishes on his bench, he noticed something unusual. The mold *Penicillium notatum* was growing on some of the petri dishes. Apparently, the mold had found its way through an open window and onto a bacterial culture of *Staphylococcus aureus*. The bacterial colonies that were growing at a distance from the fungus were typical, but there was no growth close to the mold. Fleming isolated the agent responsible for this destruction of the bacteria and named it *penicillin*.

Through Fleming's research efforts and those of several colleagues, the chemical was identified and used for about 10 years in microbiological work in the laboratory. Many suspected that penicillin might be used as a drug, but the fungus could not produce enough of the chemical to make it worthwhile. When World War II began, and England was being firebombed, there was an urgent need for a drug that would control bacterial infections in burn wounds. Two scientists from England were sent to the United States to begin research into the mass production of penicillin.

Their research in isolating new forms of *Penicillium* and purifying the drug were so successful that cultures of the mold now produce over 100 times more of the drug than the original mold discovered by Fleming. In addition, the price of the drug dropped considerably—from a 1944 price of $20,000 per kilogram to a current price of less than $250.00. The species of *Penicillium* used to produce penicillin today is *P. chrysogenum*, which was first isolated in Peoria, Illinois, from a mixture of molds found growing on a cantaloupe. The species name, *chrysogenum*, means "golden" and refers to the golden-yellow droplets of antibiotic that the mold produces on the surface of its hyphae. The spores of this mold were isolated and irradiated with high dosages of ultraviolet light, which caused mutations to occur in the genes. When some of these mutant spores were germinated, the new hyphae were found to produce much greater amounts of the antibiotic.

---

atmosphere a long time and travel thousands of kilometers. Fungal spores have been collected as high as 50 kilometers above the Earth.

In a favorable environment, a fungus produces dispersal spores, which are short-lived and germinate quickly under suitable conditions. If the environment becomes unfavorable—too cold or hot, or too dry—the fungus produces survival spores. These may live for years before germinating. Fungi are so prolific that their spores are almost always present in the air; as soon as something dies, fungal spores settle on it, and decomposition usually begins.

Fungi play a variety of roles. They are used in the processing of food and are vital in the recycling processes within ecosystems. As decomposers, they destroy billions of dollars worth of material each year; as pathogens, they are responsible for certain diseases. They are beneficial in the production of antibiotics and other chemicals used in the treatment of diseases. *Penicillium chrysogenum* is a mold that produces the antibiotic penicillin, which was the first commercially available antibiotic and is still widely used (How Science Works 24.2). • fermentation, p. 117

There are over 100 species of *Penicillium*, and each characteristically produces spores in a brushlike border; the word *penicillus* means "little brush." Members of this group do more than produce antibiotics; they are also widely used in processing food. Many people are familiar with the blue, cottony growth that sometimes occurs on citrus fruits. The *P. italicum* growing on the fruit appears to be blue because of the pigment produced in the spores. The blue cheeses, such as Danish, American, and the original Roquefort, all have this color. Each has been aged with *P. roquefortii* to produce the color, texture, and flavor. Differences in the cheeses are determined by the kinds of milk used and the conditions under which the aging occurs. Roquefort cheese is made from sheep's milk and aged in Roquefort, France, in particular caves. American blue cheese is made from cow's milk and aged in many places around the United States. The blue color has become a very important feature of these cheeses. The same research laboratory that first isolated *P. chrysogenum* also found a mutant species of *P. roquefortii* that would produce spores having no blue color. The cheese made from this mold is "white" blue cheese. The flavor is exactly the same as "blue" blue cheese, but commercially it is worthless: People want the blue color.

Fungi and their by-products have been used as sources of food for centuries. When we think of fungi and food, mushrooms usually come to mind. The common mushroom found in the grocer's vegetable section is grown in many countries and has an annual market value in the billions of dollars. But there are other uses for fungi as food. *Shoyu* (soy sauce) was originally made by fermenting a mixture of wheat, soybeans, and an ascomycote fungus for a year. Most of the soy sauce used today is made by a cheaper method of processing soybeans with hydrochloric acid. True connoisseurs still prefer soy sauce made the original way. Another mold is important to the soft-drink industry. The citric acid that gives a soft drink its sharp taste was originally produced by squeezing juice from lemons and purifying the acid. Today, however, a mold is grown on a nutrient medium with table sugar (sucrose) to produce great quantities of citric acid at a low cost.

All fungi are capable of breaking down organic matter to provide themselves with the energy and building materials they need. This may be either beneficial or harmful, depending on what is being broken down. In order for any ecosystem to survive, it must have a source of carbon, nitrogen, phosphorus, and other elements that can be incorporated into new carbohydrates, fats, proteins, and other molecules necessary for growth. The fungi, along with bacteria, are the primary recycling agents for these elements in ecosystems. Spores are an efficient method of dispersal, and when they land in a favorable environment with moist conditions, they germinate and begin the process of decomposition. As decomposers, fungi cause billions of dollars worth of damage each year. Clothing, wood, leather, and all types of food are susceptible to damage by fungi. One of the best ways to protect against such damage is to keep the material dry because fungi grow best in a moist environment. More recently, a fungal disease known as *sudden oak death* (SOD) was identified in Germany, the Netherlands, California, and Oregon. Caused by the fungus *Phytophthora ramorum*, this disease is misnamed because it also infects rhododendrons, redwoods, and fir trees. Once a tree or plant is infected, its leaves turn yellow to brown within a few weeks after an oozing canker develops on the lower trunk. Control of the disease will require spending millions of dollars for fungicides (called *phosphides*). It will also be expensive and difficult to dispose of the millions of cubic tons of infected plant materials—a process that is necessary to prevent the spread of the fungus.
• ecosystems, p. 261

Some fungi have a symbiotic relationship with plant roots; **mycorrhiza** usually grow inside a plant's root-hair cells—the cells through which plants absorb water and nutrients. The hyphae from the fungus grow out of the root-hair cells and greatly increase the amount of absorptive area (figure 24.14). Plants with mycorrhizal fungi can absorb as much as 10 times more minerals than those without the fungi. Some types of fungi also supply plants with growth hormones, while the plants supply carbohydrates and other organic compounds to the fungi. Mycorrhizal fungi are found in 80% to 90% of all plants.

In some situations, mycorrhizae may be essential to the life of a plant. Botanists are investigating a correlation between mycorrhizae and acid-rain damage to trees. Acid-rain conditions can leach certain necessary plant minerals from the soil, making them less accessible to plants. The increased soil acidity also makes certain toxic chemicals, such as copper, more accessible to plants. When the roots of trees suspected of being killed by acid rain are examined, there is often no evidence of the presence of mycorrhizal fungi, whereas a healthy tree growing next to a dead one has the root fungus.

Figure 24.14

Mycorrhiza

The symbiotic relationship between fungi and the roots of the two plants on the right increases the intake of water and nutrients into the plant. As a result, these plants have more growth than the control plant on the left.

Figure 24.15

Fairy Ring

Legend tells us that fairies danced in a circle in the moonlight and rested on the mushrooms. Mycologists tell us that the mushrooms began to grow in the center; as the organic material was consumed, the mushrooms grew in an ever-widening circle and formed this "fairy ring."

One of the most interesting formations caused by mushroom growth can be seen in soil that is rich in mushroom hyphae, such as in lawns, fields, and forests. These formations, known as *fairy rings,* result from the expanding growth of the mushrooms (figure 24.15). The inner circle is normal grass and vegetation. The mushroom population originally began to grow at the center, but grew out from there because it exhausted the soil nutrients necessary for fungal growth. As the microscopic hyphae grow outward from the center, they stunt the growth of grass, forming a ring of short, inhibited grass. Just to the outside of this

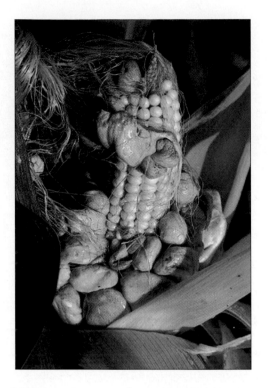

Figure 24.16

Corn Smut

Most people who raise corn have seen corn smut. Besides being unsightly, it decreases the corn yield.

growth ring, the grass is luxuriant because the hyphae excrete enzymes that decompose soil material into rich nutrients for growth. The name *fairy ring* comes from an old superstition that such rings were formed by fairies tramping down the grass while dancing in a circle.

There are also pathogenic fungi that feed on living organisms; those that cause ringworm and athlete's foot are two examples. A number of diseases are caused by fungi that grow on human mucus membranes, such as those of the vagina, lungs, and mouth. Plants are also susceptible to fungal attacks. Chestnut blight and Dutch elm disease almost caused these two species of trees to become extinct. The fungus that causes Dutch elm disease is a parasite that kills the tree; then it functions as a saprophyte and feeds on the dead tree. Fungi also damage certain domestic crops. Wheat rust gets its common name because infected plants look as if they are covered with rust. Corn smut is also due to a fungal pathogen of plants (figure 24.16).

A number of fungi produce deadly poisons called **mycotoxins.** There is no easy way to distinguish those that are poisonous from those that are safe to eat. The poisonous forms are sometimes called *toadstools* and the nonpoisonous ones, *mushrooms.* However, they are all properly called mushrooms. The origin of the name toadstools is unclear. One idea is that toadstools are mushrooms on which toads sit; another is that the word is derived from the German *todstuhl,* "seat of death." The most deadly of these,

Amanita verna, is known as "the destroying angel" and can be found in woodlands during the summer. Mushroom hunters must learn to recognize this deadly, pure white species. This mushroom is believed to be so dangerous that food accidentally contaminated by its spores can cause illness and possible death. Another mushroom, *Psilocybe mexicana,* has been used for centuries in religious ceremonies by certain Mexican tribes because of the hallucinogenic chemical that it produces. These mushrooms have been grown in culture, and the drug psilocybin has been isolated. In the past, it was used experimentally to study schizophrenia. *Claviceps purpurea,* a sac fungus, is a parasite on rye and other grains. The metabolic activity of *C. purpurea* produces a toxin that can cause hallucinations, muscle spasms, insanity, or even death. However it is also used to treat high blood pressure, to stop bleeding after childbirth, and to treat migraine headaches.

Lichens

Lichens are usually classified with the Fungi, but they actually represent a very close mutualistic relationship between a fungus and an algal protist or a cyanobacterium. Algae and cyanobacteria require a moist environment. Certain species of these photosynthetic organisms grow surrounded by fungus. The fungal covering maintains a moist area, and the photosynthesizers in turn provide nourishment for the fungus. These two species growing together are what we call a **lichen** (figure 24.17). Lichens grow slowly; a patch of lichen may grow only 1 centimeter per year in diameter. • symbiosis, p. 289

Because the fungus provides a damp environment and the algae produce the food, lichens require no soil for growth. For this reason, they are commonly found growing on bare rock, and are the pioneer organisms in the process of succession. Lichens are important in the process of soil formation. They secrete an acid that weathers the rock and makes minerals available for use by plants. When lichens die, they provide a source of humus—dead organic material—that mixes with the rock particles to form soil.

Lichens are found in a wide variety of environments, ranging from the frigid arctic to the scorching desert. One reason for this success is their ability to withstand drought conditions. Some lichens can survive with only 2% water by weight. In this condition they stop photosynthesis and go into a dormant stage, remaining so until water becomes available and photosynthesis begins again.

Another factor in the success of lichens is their ability to absorb minerals. However, because air pollution increases the amounts of minerals they absorb, many lichens are damaged. Some forms of lichens absorb concentrations of sulfur 1,000 times greater than those found in the atmosphere. This increases the amount of sulfuric acid in the lichen, resulting in damage or death. For this reason, areas with heavy air pollution are "lichen deserts." Because they can absorb minerals, certain forms of lichens have been used to monitor the amount of various pollutants in the atmosphere, including ra-

(a)

(b)

Figure 24.17

Lichens

Lichens grow in a variety of habitats: (*a*) the shrubby British soldier lichen is growing on soil; (*b*) the crustlike lichen is growing on rock. The different coloring is due to the different species of algae or cyanobacteria in the lichens.

Table 24.3

A CLADOGRAM SUGGESTING EVOLUTIONARY RELATIONSHIP AMONG THE VARIOUS GROUPS OF FUNGI

dioactivity. The absorption of radioactive fallout from Chernobyl by arctic lichens made the meat of the reindeer that fed on them unsafe for human consumption (table 24.3).

SUMMARY

Organisms in the Domains Archaea and Eubacteria, and the kingdoms Protista and Fungi rely mainly on asexual reproduction, and each cell usually satisfies its own nutritional needs. In some species, there is minimal cooperation between cells. The bacteria have the genetic ability to function in various environments. Most species of bacteria are beneficial, although some are pathogenic.

Members of the kingdom Protista are one-celled organisms. They differ from the prokaryotes in that they are eukaryotic cells, whereas the prokaryotes are prokaryotic cells. Protists include algae, autotrophic cells that have a cell wall and carry on photosynthesis; protozoa, which lack cell walls and cannot carry on photosynthesis; and funguslike protists, whose motile, amoeboid reproductive stage distinguishes them from true fungi. Some species of Protista developed a primitive type of specialization, and from these evolved the multicellular fungi, plants, and animals.

The kingdom Fungi consists of nonphotosynthetic, eukaryotic organisms with cell walls. Most species are multicellular. Fungi are nonmotile organisms that disperse by producing spores. Lichens are a combination of organisms involving a mutualistic relationship between a fungus and an algal protist or cyanobacterium.

THINKING CRITICALLY

For guidelines to answer this question, visit the Online Learning Center.

Throughout much of Europe there has been a severe decline in the mushroom population. On study plots in Holland, data collected since 1912 indicate that the number of mushroom species has dropped from 37 to 12 per plot in recent years. Along with the reduction in the number of species there is a parallel decline in the number of individual plants; moreover, the surviving plants are smaller.

The phenomenon of the disappearing mushrooms is also evident in England. One study noted that in 60 fungus species, 20 exhibited declining populations. Mycologists are also concerned about a decline in the United States; however there are no long-term studies, such as those in Europe, to provide evidence for such a decline.

Consider the niche of fungi in the ecosystem. How would an ecosystem be affected by a decline in their numbers?

CONCEPT MAP TERMINOLOGY

Construct a concept map using these terms.

algae	eukaryotic
archaea	microorganism
bacteria	prokaryotic
colony	protozoa
endospore	

Visit the Online Learning Center to use an interactive format to construct another concept map.

KEY TERMS

Use interactive flash cards, on the Online Learning Center, to help you learn the meanings of these terms.

algae	microorganisms (microbes)
bacteria	mycorrhiza
benthic	mycotoxin
bloom	pathogen
colonial microbes	phytoplankton
endospore	plankton
lichen	protozoa

LEARNING CONNECTIONS

Visit the Online Learning Center to find additional study resources including those highlighted below.

24.1   Microorganisms

Experience This!
• Useful microbes!

1. What is meant by the term *bloom*?
2. What is a pathogen? Give two examples.
3. Name a disease caused by each of the following: bacteria, fungi, protozoa.
4. Name two beneficial results of fungal growth and activity.
5. Define the term *saprophytic*.
6. Give an example of a symbiotic relationship.

24.2   Bacteria

7. What is a bacterial endospore?
8. What is meant by the term *pathogen*?

24.3   Kingdom Protista

9. Why are the protozoa and the algae in different subgroups of the kingdom Protista?
10. What is phytoplankton?
11. Name three commercial uses of algae.
12. What is the best method to prevent the spread of malaria?

24.4   Multicellularity in the Protista

13. What is the difference between a tissue and the multicellularity found in some protists?
14. Give an example of a multicellular protozoan.

24.5   Kingdom Fungi

Interactive Concept Maps
• Beneficial microbes
• Problem microbes

15. What types of spores do fungi produce?
16. What is SOD and why is it significant?

Test yourself with multiple choice questions that provide immediate feedback on the Online Learning Center with PowerWeb at

www.mhhe.com/enger11

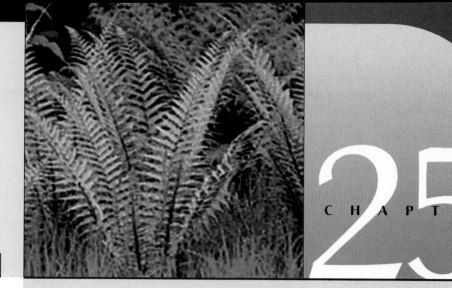

CHAPTER

25

Plantae

GOALS AND OBJECTIVES

Understand the nature and characteristics common to most plants.

❏ List the feature of an organism that is needed to qualify it as a plant.

❏ List examples of plants.

❏ Define the term *alternation of generations*.

❏ Diagram the life cycle of a fern.

Recognize the basic characteristics of the major plant groups.

❏ Know the differences among the plant groups.

❏ Describe the evolutionary relationships among plants.

❏ Explain how plants adapted to terrestrial habitats.

❏ Describe the advantage of vascular tissue to plants.

Understand the basic structure of seed plants.

❏ Differentiate among roots, stems, and leaves.

❏ Distinguish between vascular and nonvascular tissue.

❏ Compare gymnosperms and angiosperms.

The Origin and Classification of Life

PART SIX

25.1 What Is a Plant?

Because plants quietly go about their business of feeding and helping maintain life on Earth, they often are not noticed or appreciated by the casual observer. Yet we are aware enough about plants to associate them with the color green. Grass, garden plants, and trees are all predominantly green, the color associated with photosynthesis. It is the green pigment chlorophyll that captures light and allows the process of photosynthesis to store the energy as the chemical-bond energy needed by all other organisms. Yet in this quiet sea of green, there is incredible variety and complexity. Plants range in size from tiny floating duckweed the size of your pencil eraser to giant sequoia trees as tall as the length of a football field. A wide range of colors (e.g., red, yellow, orange, purple, white, pink, violet) stand out against the basic green we associate with plants. Bright spots of color are often flowers and fruit, where the colors may serve as attractants for animals.

Plants are adapted to live in just about any environment (figure 25.1). They live on the shores of oceans, in shallow freshwater, the bitter cold of the polar regions, the dryness of the desert, and the driving rains of tropical forests. There are plants that we eat, plants that eat animals, plants that are parasites, plants that don't carry on photosynthesis, and plants that strangle other plants (table 25.1). They show a remarkable variety of form, function, and activity.

If we were asked to decide what all plants have in common, the list might include:

1. They are anchored to soil, rocks, bark, and other solid objects.
2. They have hard, woody tissues that support the plants and allow them to stand upright.
3. They are green and carry on photosynthesis.

Although there are exceptions to these criteria, they are good starting points to explore what it is to be a plant.

The scientific study of plants is called *botany* and the first classification of plants was devised in the fourth century by one of Aristotle's students, Theophrastus of Eresus. His system of classification was based on the shapes of leaves and whether they were trees, shrubs, or herbs. In the first century A.D., a Greek physician, Dioscorides, classified plants according to their medicinal value. In 1623 a Swiss botanist, Gaspard Bauhin, was the first to begin naming plants using two-part Latin names. About 100 years later, Carolus Linnaeus (1707–1778) categorized plants according to the number and position of male parts in flowers. Although almost everyone has looked closely at a flower, few people recognize it as a structure associated with sexual reproduction in plants. It wasn't until more recently (early eighteenth century) that botanists such as John Ray began to base plant classification on a more detailed examination of plant parts and their hypothetical evolutionary relationships. Today, botanists classify plants based on the following assumptions:

(a)   (b)

Figure 25.1

Variety of Flowering Plants
Flowering plants are adapted to living in many different kinds of environments and show much variety in size, color, and structure: (*a*) Cacti at the Saguaro National Monument, Arizona and (*b*) coconut palms along a tropical shore, Belize, Central America.

1. Plants display various similarities and differences.
2. Plants that are similar in nearly all respects are members of the same species.
3. Species that share some of their features comprise a genus.
4. On the basis of their similar features, similar genera can be grouped into a family; and families can be organized into successively higher levels of a taxonomic hierarchy, the most broad category being the division.
5. The greater the number of shared features among plants, the closer their relationship.

The cladogram seen in table 25.2 located at the end of this chapter illustrates botanists' current thinking on the evolutionary relationships of the major subgroups of the kingdom Plantae.

25.2 Alternation of Generations

Plants live alternatively between two different forms during their **life cycle,** the series of stages in their life. One stage in the life cycle of a plant is a diploid ($2n$) stage called the **sporophyte** because special cells of this stage undergo meiosis and form numerous haploid, *n*, **spores.** The release of spores allows the plant to be dispersed through the environment and explore new areas. In plants, spores are reproductive cells

Table 25.1

PLANTS AND THEIR PRODUCTS

Common Name	Scientific Name	Product/Use	Biogeographical Location
Brazil nuts	*Bertholletia excelsa*	Edible nut	Amazonian forest, South American tropics
Papaya	*Carica papaya*	Edible fruit	Tropics and subtropics
Cocoa plant	*Theobroma cacao*	Cocoa bean products and chocolate	Tropical America and West Africa (Ghana)
Rice, arroz, riz	*Oryza sativa*	Edible starch	Worldwide but native to Asia
Cassava, manioc, manihot, or yuca	*Manihot esculenta*	Edible root	Tropical America
Sweet potato	*Ipomoea batatas*	Edible root	Native to South American Andes but grows throughout the world
Taro	*Colocasia esculenta*	Edible root	Tropics and subtropics
Water chestnuts	*Eleocharis dulcis*	Edible root	China, Asia
Yams	*Dioscorea* spp.	Edible root	Tropics
Cashew nuts	*Anacardium occidentale*	Edible nut	Tropics
Hot pepper, chili pepper, sweet bell pepper	*Capsicum annum*	Edible vegetable	North America and tropics
Tomato	*Lycopersicon esculentum*	Edible vegetable	Worldwide
Bamboo	*Bambusa* spp.	Edible vegetable; also used as construction material	Tropics and subtropics
Coffee	*Coffea arabica*	Flavoring bean for beverage	Central and South America, Africa, India, and Indonesia
Kola or cola	*Cola nitida*	Beverage flavor seed	West Africa, Brazil, West Indies
Vanilla	*Vanilla planifolia*	Flavoring bean	Mexico and Central America
Black pepper	*Piper nigrum*	Flavoring fruit	India and Cambodia
Tea	*Camellia sinensis*	Short tips and young leaves; fermented	Worldwide
Capers	*Capparis spinosa*	Flower bud	Mediterranean and parts of Asia

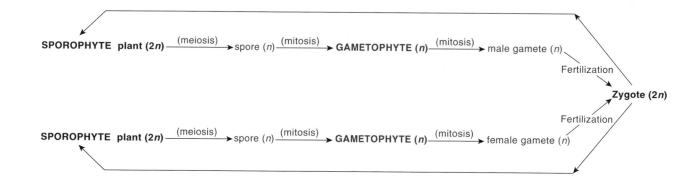

that are capable of developing into a haploid, multicellular adult without fusion with another cell. In land plants, a hard shell covers the spore. • reduction division, p. 173

When spores land in suitable areas they **germinate** (begin to grow) into their alternate stage in the life cycle, the **gametophyte**. Gametophytes are composed of haploid cells and look very different from the sporophyte plant. They are involved in the sexual reproduction of the plant since it is the gametophytes that produce male or female gametes. When the gametes unite at the time of fertilization, the newly formed embryo undergoes mitosis and grows into the sporophyte form of the next generation (see diagram above).

The term **alternation of generations** is used to describe the fact that plants cycle between two different stages in their life—the diploid sporophyte and haploid gametophyte. To understand plants, and how and why they may have evolved, we need to go back in time and examine how plant structures and functions were modified over time.

25.3 Ancestral Plants: The Bryophytes

One group of plants that shows primitive or ancestral characteristics is the bryophytes. All bryophytes have several things in common:

1. They are small, compact, slow-growing, green plants with motile sperm that swim to eggs.
2. There are no well-developed vascular tissues (tubes for conducting water through the plant body) and no mechanism that would provide support to large, upright plant parts such as stems.
3. Bryophytes do not have true leaves or roots as are found in more highly evolved plants.
4. Nutrients are obtained from the surfaces upon which they grow or from rainwater.
5. Their life cycle consists of two stages. The gametophyte (gamete-producing plant) dominates the sporophyte (spore-producing plant).

There are three types of bryophytes: Bryophyta (mosses), Hepatophyta (liverworts), and Anthocerotophyta (hornworts). Mosses grow as a carpet composed of many parts. Each individual moss plant is composed of a central stalk less than 5 centimeters tall with short, leaflike structures that are sites of photosynthesis. If you look at the individual cells in the leafy portion of a moss, you can distinguish the cytoplasm, cell wall, and chloroplasts (figure 25.2). You may also distinguish the nucleus of the cell. This nucleus is haploid (n), meaning that it has only one set of chromosomes. In fact, every cell in the moss plant body is haploid.

Although all the cells have the haploid number of chromosomes (the same as gametes), not all of them function as gametes. Because this plant produces cells that are capable of acting as gametes, it is called the gametophyte, or gamete-producing stage in the plant life cycle. Special structures in the moss, called **antheridia,** produce mobile sperm cells capable of swimming to a female egg cell (figure 25.3). The sperm cells are enclosed within a jacket of cells (the antheridium) that opens when the sperm are mature. The sperm swim by the undulating motion of flagella through a film of dew from splashing rainwater, carrying their packages of genetic information. Their destination is the egg cell of another moss plant with a different package of genetic information. The egg is produced within a jacket called the **archegonium** (figure 25.3). There is usually only one egg cell in each archegonium. The sperm and egg nuclei fuse, resulting in a diploid cell, the zygote. The zygote grows, divides, and differentiates into an embryo. The gametophyte generation is dominant over the sporophyte generation in mosses. This means that the gametophyte generation is more likely to be seen.

The casual observer usually overlooks liverworts and hornworts (figure 25.4) because they are rather small, low-growing plants composed of a green ribbon of cells. The name *liverwort* comes from the fact that these plants resemble the moist surface of a liver. Although they do not have

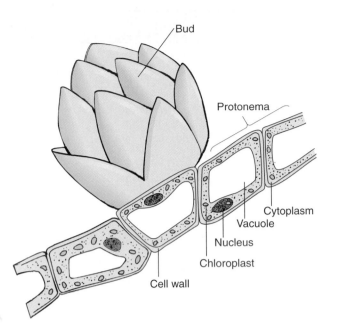

Figure 25.2

The Cells of a Moss Plant
The chain of cells seen here are typical plant cells. Note the large central vacuole and cell wall. The leaflike structure, the bud, grows from this filament.

well-developed roots or stems, the leaflike ribbons of tissue are well suited to absorb light for photosynthesis.

25.4 Adaptations to Land

Botanists consider mosses and the other bryophytes the lowest step of the evolutionary ladder in the plant kingdom. They are considered "primitive" (ancestral) because they have not developed an efficient network of tubes or vessels that can be used to transport water throughout their bodies; they must rely on the physical processes of diffusion and osmosis to move dissolved materials through their bodies. The fact that mosses do not have a complex vascular system to move water limits their size to a few centimeters and their location to moist environments. Another characteristic of mosses points out how closely related they are to their aquatic ancestors, the algae: They require water for fertilization. The sperm cells must "swim" from the antheridia to the archegonia. Small size, moist habitat, and swimming sperm are considered characteristics of ancestral plants. In a primitive way, mosses have adapted to a terrestrial niche.
• geologic time line, p. 449

Vascular Tissue: What It Takes to Live on Land

A small number of currently existing plants show some of the more ancient directions of evolution. You might think of these evolutionary groups as experimental models or transitional

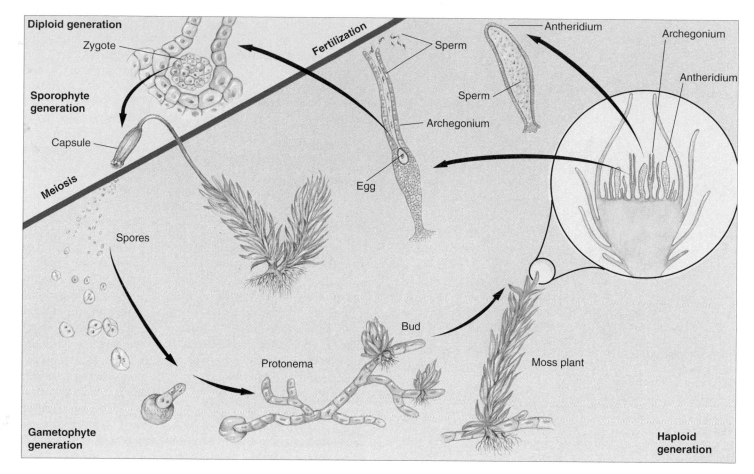

Figure 25.3

The Life Cycle of a Moss

In this illustration the portion with the darker-colored background represents cells that have the haploid (*n*) nuclei (gametophyte generation). The portion with a lighter-colored background represents cells that have the diploid (2*n*) nuclei (sporophyte generation). Notice that the haploid and diploid portions of the life cycle alternate.

(a)

(b)

Figure 25.4

Bryophytes: Liverworts and Hornworts

These ribbon-shaped plants are related to the mosses (*a*). The name liverwort comes from the fact that they resemble thin layers of green-colored animal liver. Liverworts feel like a moist rubber material. The gametophyte is the stage of the life cycle that is most easily recognized. Similar to liverworts in many ways, there are about 100 species of hornworts. (*b*) *Anthoceros* sporophytes, a hornwort.

(*a*) Courtesy of Carolina Biological Supply Co.

H₂O

Figure 25.5

Vascular Tissue
The vascular tissues of plants are used to carry materials between the roots and leaves. One kind of tissue, called xylem (shown in red), carries dissolved raw materials from roots to leaves. Another kind, called phloem (shown in green), carries manufactured food from leaves to stems and roots. Notice the hollow nature of these tissues.

plants. They successfully filled certain early terrestrial niches, but did not evolve into other niches. However, their features were important in the evolution of more successful land plants. The advances all have to do with cell specializations enabling a plant to do a better job of acquiring, moving, and retaining water while living out of an aquatic environment. Groups of closely associated cells that work together to perform a specialized or particular function are called tissues. The tissue important in moving water within a plant is called **vascular tissue.**

When a plant with vascular tissue is wounded it usually drips liquid, called *sap,* from the cut surface. This is because some of the thick-walled cells that serve as "pipes" or "tubes" for transporting liquids throughout the plant are broken and their contents leak out. Vascular tissues are used to transport water and minerals to the leaves where photosynthesis takes place. They also transport manufactured food from the leaves to storage sites found in the roots or the stems. Vascular tissues have cells connected end to end forming many long tubes, similar to a series of pieces of pipe hooked together (figure 25.5). The long celery strands that get stuck between your teeth are bundles of vascular tissue.

There are two kinds of vascular tissue: **xylem** and **phloem.** *Xylem* consists of a series of hollow cells arranged end to end so that they form a tube. These tubes carry water absorbed from the soil into the roots and transport it to the above-ground parts of the plant. Associated with these tube-like cells are cells with thickened cell walls that provide strength and support for the plant. *Phloem* carries the organic molecules produced in one part (e.g., leaves) of the plant to storage areas in other parts (e.g., roots). The specialization of cells into vascular tissues has allowed for the development of **roots, stems,** and **leaves.**

Roots

When you attempt to pull some plants from the ground, you quickly recognize that the underground parts, the roots, anchor them firmly in place. Roots have a variety of functions in addition to serving as anchors. Foremost among them is taking up water and other nutrients from the soil. The primary nutrients plants obtain from the soil are inorganic molecules, which are incorporated into the organic molecules they produce. By constantly growing out from the main plant body, roots explore new territory for available nutrients. As a plant becomes larger it needs more root surface to absorb nutrients and hold the plant in place. The actively growing portions of the root near the tips have large numbers of small, fuzzy, hairlike cell extensions called **root hairs** that provide a large surface area for the absorption of nutrients.

We eat many kinds of roots such as carrots, turnips, and radishes. The food value they contain is an indication of another function of roots. Most roots are important storage places for the food produced by the above-ground parts of the plant. Many kinds of plants store food in their roots during the growing season and use this food to stay alive during the winter. The food also provides the raw materials necessary for growth for the next growing season. Although we do not eat the roots of plants such as maple trees, rhubarb, or grasses, their roots are as important to them in food storage as those of carrots, turnips, and radishes (figure 25.6).

Stems

Stems are, in most cases, the above-ground structures of plants that support the light-catching leaves in which photosynthesis occurs. Many kinds of plants also have buds on their stems that may grow to produce leaves, flowers, or new branches. Trees have stems that support large numbers of branches; vines have stems that require support; and some plants, like dandelions, have very short stems with all their leaves flat against the ground. Stems have two main functions:

1. They serve as supports for the leaves.
2. They transport raw materials from the roots to the leaves and manufactured food from the leaves to the roots.

Fibrous root   Tap root

Figure 25.6

Kinds of Roots
The roots of grasses are often in the upper layers of the soil and form a dense network that traps water in the dry environment. The roots of trees and many other plants typically extend deep into the soil where they obtain moisture and serve as anchors that hold large plants upright.

When you chew on toothpicks, which are stems or made from stems, you recognize that they contain hard, tough materials. These are the cell walls of the plant cells (refer to figure 25.5). All plant cells are surrounded by a cell wall made of the carbohydrate *cellulose.* Cellulose fibers are interwoven to form a box within which the plant cell is contained. Because the cell wall consists of fibers, it can be compared to a wicker basket. It has spaces between these cellulose fibers through which materials pass relatively easily. However, the cell wall does not stretch very much, and if the cell is full of water and other cellular materials it will become quite rigid. Because of these forces, the many cells that make up a plant stem are able to keep a large nonwoody plant upright. The word **herb** (L. *herba* = grass) actually refers to nonwoody plants such as the grasses and many annual flowers like petunias and marigolds. You might think of a plant body as being similar to the bubble wrap used to protect fragile objects during shipping. Each little bubble contains air and can be easily popped. However, in combination, they will support considerable weight.

In addition to cellulose, some plants deposit other compounds in the cell walls that strengthen them, make them more rigid, and bind them to other neighboring cell walls. **Woody vascular plants** deposit a material called *lignin* while the grasses deposit *silicon dioxide,* the same kind of material of which sand is made. Stems and roots of plants tend to have large numbers of cells with strengthened cell walls. This is such an effective support mechanism that large trees and bushes are supported against the pull of gravity and can

withstand strong winds for centuries without being broken or blown over. Some of the oldest trees on Earth have been growing for several thousand years.

Stems not only provide support and nutrient transport, but also may store food. This is true of sugar cane, yams, and potatoes. Many plant stems are green and, therefore, involved in the process of photosynthesis.

Leaves

Green leaves are the major sites of photosynthesis for most plants. Photosynthesis involves trapping light energy and converting it into the chemical-bond energy of complex organic molecules like sugar (refer to chapter 6). Thus, there is a flow of energy from the Sun into the organic matter of plants. Light energy is needed to enable the smaller inorganic molecules (water and carbon dioxide) to be combined to make the organic compounds. In the process, oxygen is released for use in other biochemical processes such as aerobic cellular respiration. • **photosystems, p. 125**

Leaves have vascular tissue to allow for the transport of materials, but they also have cells containing chloroplasts. Chloroplasts are the cellular structures responsible for photosynthesis. They contain the green pigment chlorophyll. The organic molecules (e.g., glucose) produced by plants as a result of photosynthesis can be used by the plant to make the other kinds of molecules (e.g., cellulose and starch) needed for its structure and function. In addition, these molecules can satisfy the energy needs of the plant. Organisms that eat plants also use the energy captured by photosynthesis.

To carry out photosynthesis, leaves must have certain characteristics (figure 25.7). Because it is a solar collector, a leaf should have a large surface area. In addition, most plants have their leaves arranged so they do not shade one another. This assures that the maximum number of cells in the leaf will be exposed to sunlight. Most leaves are relatively thin in comparison to other plant parts. Thick leaves would not allow penetration of light to the maximum number of photosynthetic cells.

A drawback to having large, flat, thin leaves is an increase in water loss because of evaporation. To help slow water loss, the epidermal (skin) layer produces a waxy coat on the outside surface of the leaf. However, water loss is not always a disadvantage to the plant. The loss of water helps power the flow of more water and nutrients from the roots to the leaves. Water lost from the leaf is in effect pulled through the xylem into the leaf, a process called *transpiration.* Because too much water loss can be deadly, it is necessary to regulate transpiration. The amount of water, carbon dioxide, and oxygen moving into and out of the leaves of many plants is regulated by many tiny openings in the epidermis, called *stomates* (figure 25.8). The stomates can close or open to control the rate at which water is lost and gases are exchanged. Often during periods of drought or during the hottest, driest part of the day, the stomates are closed, thus reducing the rate at which the plant loses water.

Figure 25.7

The Structure of a Leaf

Though a leaf is thin, it consists of several specialized layers. An outer layer (epidermis) has openings called stomates that can open and close to regulate the movement of gases into and out of the leaf. The internal layers have many cells with chloroplasts, air spaces, and bundles of vascular tissue all organized so that photosynthetic cells can acquire necessary nutrients and transport metabolic products to other locations in the plant.

Closed   Open

Figure 25.8

Stomates

The stomates are located in the covering layer (epidermis) on the outside of leaves. When these two elongated guard cells are swollen, the space between them is open and leaves lose water and readily exchange oxygen and carbon dioxide. In their less rigid and relaxed state, the two stomatal cells close. In this condition the leaf conserves water but is not better able to exchange oxygen and carbon dioxide with the outside air.

25.5 Transitional Plants: Non–Seed–Producing Vascular Plants

The transitional groups of plants [e.g., Psilotophyta (whisk ferns), Equisetophyta (horsetails), Lycopodophyta (club mosses), and Pteridophyta (ferns)] have vascular tissue (figure 25.9). Members of these divisions are evolutionary links between the nonvascular bryophytes and the highly successful land plants, the gymnosperms and angiosperms. These plants display many common features:

1. Their diploid sporophytes produce haploid spores by meiosis, which develop into gametophytes. The gametophytes produce sperm and egg in antheridia and archegonia. Sperms require water through which they swim in order to reach eggs.
2. Fertilization results in a multicellular embryo that gets its nutrients from the gametophyte. The embryo eventually grows into the sporophyte.
3. The sporophyte generation is more dominant in the life cycle than the gametophyte and is usually highly branched.
4. All have well-developed vascular tissue to transport water and nutrients.
5. Many have the ability to support upright, above-ground plant parts, for example, leaves.

(a)   (b)

Figure 25.9

Club Mosses

These plants are sometimes called ground pines because of their slight resemblance to the evergreen trees. Most club mosses grow only a few centimeters in height. The sporophyte is the stage of the life cycle that is most easily recognized. Club mosses are a group of low-growing plants that are somewhat more successful than bryophytes in adapting to life on land. They have a stemlike structure that holds the leafy parts above other low-growing plants, enabling them to compete better for available sunlight. Thus, they are larger than mosses and not as closely tied to wet areas. Although not as efficient in transporting water and nutrients as the stems of higher plants, the stem of the club moss, with its vascular tissue, is a hint of what was to come.

With fully developed vascular tissues, these non-seed-producing plants are no longer limited to wet areas. They can absorb water and distribute it to leaves many meters above the surface of the soil. The ferns are the most primitive vascular plants truly successful at terrestrial living. They have not only a wider range and greater size than mosses and club mosses, but also an additional advantage: The sporophyte generation has assumed more importance and the gametophyte generation has decreased in size and complexity. Figure 25.10 illustrates the life cycle of a fern. The diploid condition of the sporophyte is an advantage because a recessive gene can be masked until it is combined with another identical recessive gene. In other words, the plant does not suffer because it has one bad allele. On the other hand, a mutation may be a good change, but time is lost by having it hidden in the heterozygous condition. In a haploid plant, any change, whether recessive or not, shows up. Not only is a diploid condition beneficial to an individual, but the population benefits when many alleles are available for selection (refer to chapter 11). As in most terrestrial plants, the sporophyte generation of ferns is the dominant generation. The green, leafy structure with which most people are familiar is the sporophyte generation. • **advantage of sexual reproduction, p. 224**

Figure 25.10

The Life Cycle of a Fern

In this illustration the dark green color represents cells that have diploid (2n) nuclei (sporophyte generation). On the back of some fern leaves there are small dots—clusters of sporangia. The sporangia produce spores. These spores develop into cells that have haploid (n) nuclei (gametophyte generation). This stage is shown in a light green color. Notice that the gametophyte and sporophyte generations alternate. Compare this life cycle with that of the moss (figure 25.3). In the moss the gametophyte generation is considered dominant, whereas in the fern the sporophyte is dominant. The sporophyte is the part of the fern most people recognize.

Ferns take many forms, including the delicate, cloverlike maidenhair fern of northern wooded areas; the bushy bracken fern (figure 25.11); and the tree fern, known primarily from the fossil record but seen today in some tropical areas. In spite of all this variety, however, they still lack one tiny but very important structure—the seed. Without seeds, ferns must rely on fragile spores to spread the species from place to place.

25.6 Advanced Plants: Seed-Producing Vascular Plants

Gymnosperms

The next advance made in the plant kingdom was the evolution of the **seed**. A seed is a specialized structure that contains an embryo enclosed in a protective covering called the *seed coat*. It also has some stored food for the embryo. The first attempt at seed production is exhibited in the conifers, which are cone-bearing plants such as pine trees. **Cones** are

Figure 25.11

A Typical Fern

Most ferns live in shaded areas of the forest. The most recognized part of the life cycle of a fern is the sporophyte generation (a). As a fern leaf grows, it "unrolls" from a coiled structure known as a "fiddlehead" (b) because it resembles the coiled end of a violin, i.e., the head of the fiddle. Fiddleheads are often used in gourmet cooking. Some ferns reach tree-size (c) and can be used in the construction of basic dwellings.

reproductive structures. The male cone produces **pollen.** Grains of pollen are actually the miniaturized male gametophyte plants. Each of these small dustlike particles contains a sperm nucleus. The female cone is usually larger than the male cone and produces the female gametophyte. Pollen is produced in smaller, male cones and released in such large quantities that clouds of pollen can be seen in the air when sudden gusts of wind shake the branches of the trees. The archegonia in the female gametophyte contain eggs. Pollen is carried by wind to the female cone, which holds the archegonium in a position to gather the airborne pollen. The process of getting the pollen from the male cone to the female cone is called **pollination.** Fertilization occurs when the sperm cell from the pollen unites with the egg cell in the archegonium. This may occur months or even years following pollination. The fertilized egg develops into an embryo within the seed (also called a mature *ovule*). The production of seeds and

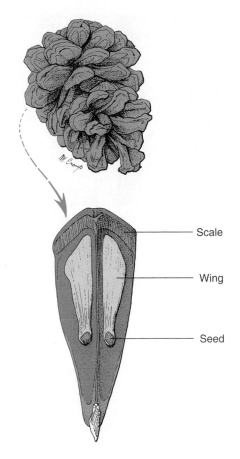

Figure 25.12

A Pine Cone with Seeds

On the scaly, leaflike portions of the cone are the seeds. Because these cones produce seeds "out in the open," they are aptly named the naked seed plants—gymnosperms. Pine tree seeds can be harvested by simply shaking a dried, open female cone. The seeds appear as small, brown, and papery winglike structures.

pollination are features of conifers that place them higher on the evolutionary ladder than ferns.

Because conifer seeds with their embryos inside are produced on the bare surface of woody, leaflike structures (the female cone), they are said to be *naked*, or out in the open (figure 25.12). The cone-producing plants such as conifers are called **gymnosperms,** which means "naked seed" plants. Producing seeds out in the open makes this very important part of the life cycle vulnerable to adverse environmental influences, such as attack by insects, birds, and other organisms. • **sudden oak death, p. 489**

Many gymnosperms generally produce needle-shaped leaves, which do not all fall off at once. Such trees are said to be **nondeciduous.** (A few gymnosperms do lose their leaves all at once—for example, *Larix* (tamarack) and *Taxodium* (bald cypress)—like most angiosperms.) This term may be misleading because it suggests that the needles do not fall off at all. Actually, they are constantly being shed a few at a time. Perhaps you have seen the mat of needles under a conifer. Because these trees retain some green leaves year-round they are

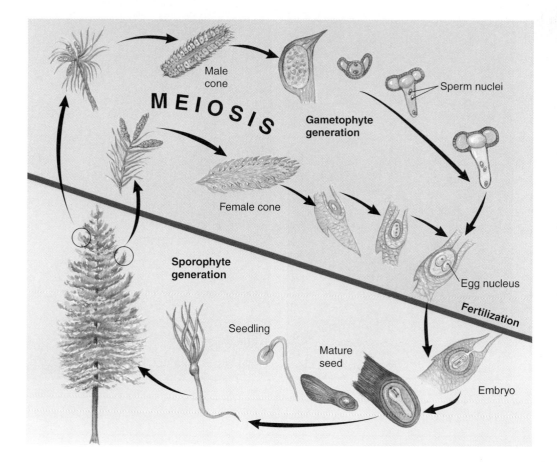

Figure 25.13

The Life Cycle of a Pine
In this illustration, the portion with the darker-colored background represents cells that have haploid (*n*) nuclei (gametophyte generation). The lightly colored background represents cells that have diploid (*2n*) nuclei (sporophyte generation). Notice that the gametophyte and sporophyte generations alternate, and that the sporophyte is dominant in the gymnosperms. Compare this life cycle of the pine with the life cycle of the moss (figure 25.3) and that of the fern (figure 25.10). Notice the ever-increasing dominance in the sporophyte generation.

called *evergreens*. The portion of the evergreen with which you are familiar is the sporophyte generation; the gametophyte, or haploid, stages have been reduced to only a few cells, the pollen grains. Look closely at figure 25.13, which shows the life cycle of a pine with its alternation of haploid and diploid generations.

Gymnosperms are also called **perennials;** that is, they live year after year. Unlike **annuals,** which complete their life cycle in 1 year, gymnosperms take many years to grow from seeds to reproducing adults. The trees get taller and larger in diameter each year, continually adding layers of strengthening cells and vascular tissues. As a tree becomes larger, the strengthening tissue in the stem becomes more and more important.

A layer of cells in the stem, called the **cambium,** is responsible for this increase in size. Xylem tissue is the innermost part of the tree trunk or limb, and phloem is outside the cambium. The cambium layer of cells is positioned between the xylem and the phloem. Cambium cells go through a mitotic cell division, and two cells form. One cell remains cambium tissue, and the other specializes to form vascular tissue. If the cell is on the inside of the cambium ring, it becomes xylem; if it is on the outside of the cambium ring, it becomes phloem. As cambium cells divide again and again, one cell always remains cambium, and the other becomes vascular tissue. Thus, the tree constantly increases in diameter (figure 25.14).

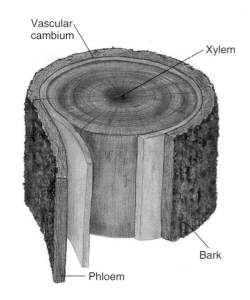

Figure 25.14

A Cross Section of Woody Stem
Notice that the xylem makes up most of what we call wood. The approximate age of a tree can be determined by counting the growth rings seen on the cut surface. It is also possible to learn something about the environment from these rings. Wide rings indicate good growth years with high rainfall, whereas narrow rings indicate poor growth and low rainfall. Can you picture the relative positions of the labeled structures 20 years from now?

(a)   (b)   (c)

Figure 25.15

Several Gymnosperms
Do you recognize these gymnosperms? They are (*a*) redwood, (*b*) Torrey pine, and (*c*) cedar. One cedar known as the cedar of Lebanon (*Cedrus libani*) is displayed on the national flag of Lebanon and is regarded as a symbol of strength, prosperity, and long life. These trees were used in ancient times as a source of perfumes, and their wood was used to make coffins. In fact, coffins made from cedar and found in the pyramids have been found intact and still smell of cedar fragrance. In many parts of the world, cedar wood is used in "cedar chests" because its aromatic molecules are able to inhibit the destructive effects of wool–eating moths.

The accumulation of the xylem in the trunk of gymnosperms is called **wood.** Wood is one of the most valuable biological resources of the world. We get lumber, paper products, turpentine, and many other valuable materials from the wood of gymnosperms. You are already familiar with many examples of gymnosperms, three of which are pictured in figure 25.15.

Angiosperms

The group of plants considered most highly evolved is known as the **angiosperms.** This name means that the seeds, rather than being produced naked, are enclosed within the surrounding tissues of the **ovary.** The ovary and other tissues mature into a protective structure known as the **fruit.** Many of the foods we eat are the seed-containing fruits of angiosperms: green beans, melons, tomatoes, and apples are only a few of the many edible fruits (figure 25.16; How Science Works 25.1).

Angiosperm trees generally produce broad, flat leaves. In colder parts of the world, most angiosperms lose all their leaves during the fall. Such trees are said to be **deciduous** (figure 25.17). (However, there are exceptions. Some angiosperms are nondeciduous, keeping their leaves and staying green throughout the winter; for example, American holly—*Ilex opaca.*) However, the majority of angiosperms are not trees; they are small plants like grasses, "weeds," vines, houseplants, garden plants, wildflowers, and green houseplants. Look closely at figure 25.18, which shows the life cycle of an angiosperm with its alternation of haploid and diploid generations.

The **flower** of an angiosperm is the structure that produces sex cells and other structures that enable the sperm cells to get to egg cells. The important parts of the flower are the female **pistil** (composed of the *stigma, style,* and *ovary*) and the male **stamen** (composed of the *anther* and *filament*). In figure 25.19, notice that the egg cell is located inside the ovary. Any flower that has both male and female parts is called a **perfect flower;** a flower containing just female or just male parts is called an **imperfect flower.** Any additional parts of the flower are called **accessory structures** because fertilization can occur without them. **Sepals,** which form the outermost whorl of the flower, are accessory structures that serve a protective function. **Petals,** also accessory structures, increase the probability of fertilization. Before the sperm cell (contained in the pollen) can join with the egg cell, it must somehow get to the egg. This is the process called *pollination.* Some flowers with showy petals are adapted to attracting insects, which unintentionally

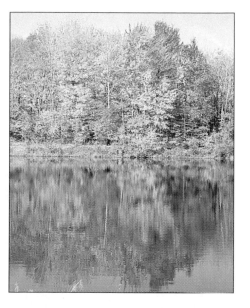

Figure 25.17

Fall Colors
The color change you see in leaves in the fall of the year in certain parts of the world is the result of the breakdown of the green chlorophyll. Other pigments (red, yellow, orange, brown) are always present but are masked by the presence of the green chlorophyll pigments. In the fall, a layer of waterproof tissue forms at the base of the leaf and cuts off the flow of water and other nutrients. The cells of the leaf die and their chlorophyll disintegrates, revealing the reds, oranges, yellows, and browns that make a trip through the countryside a colorful experience.

Figure 25.16

Types of Edible and Inedible Fruits
Fruits are the structures that contain seeds. The seed containers of the peach, apple, and tomato are used by humans as food. The other fruits are not usually used by humans as food. Although these are familiar foods, it is becoming increasingly common to find "unusual" fruits and vegetables in our food markets as the time needed to transport foods from around the world decreases. Still, it has been estimated that a full one-third of our foods is lost to spoilage.

HOW SCIENCE WORKS 25.1

Banana Extinction—A Real Slip-Up?

There may seem to be a lot of bananas on supermarket shelves, but it seems that bananas could be in danger of becoming extinct. A new fungal disease is plaguing banana plants, and because bananas lack the genetic diversity to fight it, some scientists speculate that they may disappear within 10 years. The banana appears to have been around as long ago as 10,000 years, but because of human-controlled asexual reproduction, bananas have very low genetic variety. Their gene pool is becoming very stagnant. Since the common yellow bananas found in U.S. markets (many of which are imported from Honduras) produce no seeds, there is no way for them to diversify their own genetic makeup. As a result, they do not have the genetic capability to defend themselves against the fungal pathogen. It is hoped that through genetic engineering, fungal-resistant genes might be introduced before extinction occurs. It has been estimated that 500 million people depend on bananas for their income.

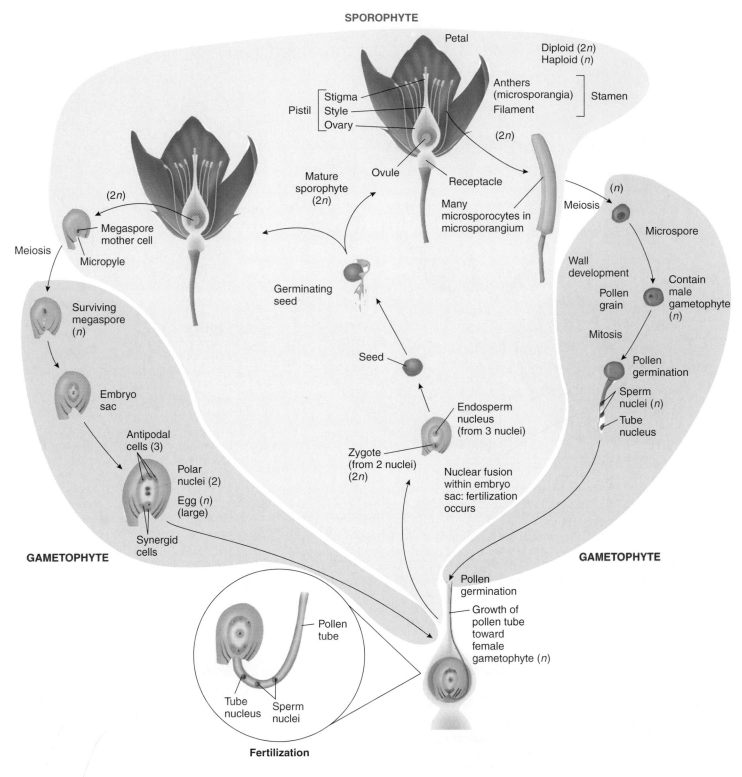

Figure 25.18

Life Cycle of Angiosperm
Compare the life cycle of a conifer with that of an angiosperm. Although there are significant differences, the sporophyte dominates the gametophyte, which has been reduced to a small portion of the life cycle.

(a)

(b)

(c)

Figure 25.19

The Flower
The flower (*a*) is the structure in angiosperms that produces sex cells. Notice that the egg is produced within a structure called the ovule. The seeds, therefore, will not be naked, as in the gymnosperms, but will be enclosed in a fruit. The dried, fragrant stigmas of the crocus flower (*Crocus sativus*) (*b*) are used as the cooking spice, saffron. The people of Iran, Crete, and Turkey have harvested and used saffron throughout their histories. Today saffron commonly comes from Spain. (*c*) Saffron's small size and difficulty in harvesting the stigmas make it the most expensive spice on the planet.

OUTLOOKS 25.1

Spices and Flavorings

Think about all the plant materials we use to season our foods. Black pepper comes from the hard, dried berries of a tropical plant, *Piper nigrum*. Cayenne pepper is made from the ground-up fruits of *Capsicum annum*, and the hot, spicy chemical in the fruit and seeds is known as capsaicin. The seeds of the dill plant, *Anethum graveolens*, are used to flavor pickles and many other foods. The dried or fresh leaves of many herbs such as thyme, rosemary, chives, and parsley are also used as flavorings. If you examine your kitchen cabinet, you may also find the following: cinnamon from the bark of a tree found in India; cloves, which are the dried flower buds of a tropical tree; ginger from the root of a tropical plant of Africa and China; and nutmeg from the seed of a tropical tree of Asia.

Centuries ago, spices like these were so highly prized that fortunes were made in the "spice trade." Beginning in the early 1600s, ships from Europe regularly visited the tropical regions of Asia and Africa, returning with cargoes of spices and other rare commodities that could be sold at great profit. Consequently, India has been greatly influenced by Britain, Indonesia has been greatly influenced by the Netherlands, and Britain and France have influenced the development of different portions of Africa.

---

carry the pollen to the pistil. Others have become adapted for wind pollination. The important thing is to get the genetic information from one parent to the other.

All the flowering plants have retained the evolutionary advances of previous groups. That is, they have well-developed vascular tissue with true roots, stems, and leaves. They have pollen and produce seeds within the protective structure of the ovary.

There are over 300,000 kinds of plants that produce flowers, fruits, and seeds (Outlooks 25.1). Almost any

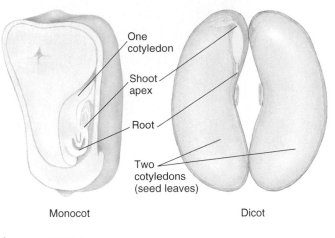

Figure 25.20

Embryos in Dicots and Monocots
The number of seed leaves (cotyledons) attached to an embryo is one of the characteristics botanists use to classify flowering plants. It has been estimated that about 80% of all angiosperms are dicots.

plant you can think of is an angiosperm. If you made a list of these familiar plants, you would quickly see that they vary a great deal in structure and habitat. The mighty oak, the delicate rose, the pesky dandelion, and the expensive orchid are all flowering plants. How do we organize this diversity into some sensible and useful arrangement? Botanists classify all angiosperms into one of two groups: **dicots** or **monocots**. The names *dicot* and *monocot* refer to a structure in the seeds of these plants. If the embryo has two **seed leaves** (*cotyledons*), the plant is a dicot; those with only one seed leaf are the monocots (figure 25.20). A peanut is a dicot; lima beans and apples are also dicots; grass, lilies, and orchids are all monocots. Even with this separation, the diversity is staggering. The characteristics used to classify and name plants are listed in figure 25.21, which includes a comparison of the extremes of these characteristics.

25.7 Response to the Environment— Tropisms

Our casual impression of plants is that they are unchanging objects. However, on closer examination we recognize that plants change over time. They grow new leaves in the spring, produce flowers and fruits at certain times of the year, and grow toward a source of light. Furthermore, they will respond to organisms that harm them, and may even mount an attack against competitors. Any action resulting from a particular stimulus is referred to as a **tropism.**

One of the first responses studied in plants is their ability to grow toward a source of sunlight. This action is known as *phototropic* motion. The value of this response is obvious because plants need light to survive. The mechanism that allows this response involves a hormone. In the case of plants growing toward light, the growing tip of the stem produces a hormone, **auxin,** that is transported down the stems. The hormone stimulates cells to elongate, divide, and grow. If the growing tip of a plant is shaded on one side, the shaded side produces more of the hormone than the lighted side. The larger amount of auxin on the shaded side causes greater growth in that area and the tip of the stem bends toward the light. When all sides of the stem are equally illuminated the stem will grow equally on all sides and will grow straight. If you have house plants near a window it is important to turn them regularly or they will grow more on one side than the other (figure 25.22).

Plants also respond to changes in exposure to daylight. They are able to measure day length and manufacture hormones that cause changes in the growth and development of specific parts of the plant. Some plants produce flowers only when the days are getting longer, some only when the days are getting shorter, and some only after the days have reached a specific length. Other activities are triggered by changing day length. Probably the most obvious is the mechanism that leads to the dropping of leaves in the fall.

Many kinds of climbing vines are able to wrap rapidly growing, stringlike *tendrils* around sturdy objects in a matter of minutes. As the tendrils grow, they slowly wave about. When they encounter an object, their tropic response is to wrap around it and anchor the vine. Once attached, the tendrils change into hard, tough structures that bind the vine to its attachment. Sweet peas, grape vines, and the ivy on old buildings spread in this manner. Ivy can cause great damage as it grows and its tendrils loosen siding and serve as a haven for the growth of other destructive organisms (figure 25.23).

Plants may even have the ability to communicate with one another. When the leaves of plants are eaten by animals, the new leaves produced to replace those lost often contain higher amounts of toxic materials than the original leaves. An experiment carried out in a greenhouse produced some interesting results. Some of the plants had their leaves mechanically "eaten" by an experimenter, whereas nearby plants were not harmed. Not only did the cut plants produce new leaves with more toxins, but the new growth on neighboring, nonmutilated plants had increased toxin levels as well. This raises the possibility that plants communicate in some way, perhaps by the release of molecules that cause changes in the receiving plant.

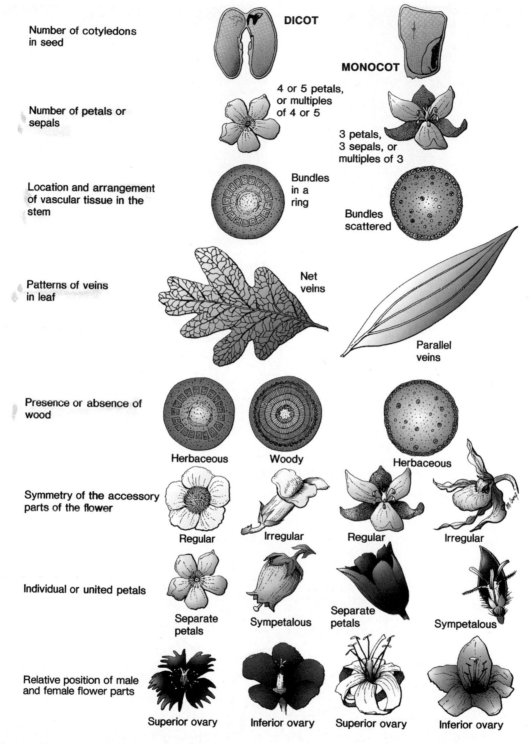

Number of cotyledons in seed

DICOT

MONOCOT

Number of petals or sepals

4 or 5 petals, or multiples of 4 or 5

3 petals, 3 sepals, or multiples of 3

Location and arrangement of vascular tissue in the stem

Bundles in a ring

Bundles scattered

Patterns of veins in leaf

Net veins

Parallel veins

Presence or absence of wood

Herbaceous   Woody   Herbaceous

Symmetry of the accessory parts of the flower

Regular   Irregular   Regular   Irregular

Individual or united petals

Separate petals   Sympetalous   Separate petals   Sympetalous

Relative position of male and female flower parts

Superior ovary   Inferior ovary   Superior ovary   Inferior ovary

Figure 25.21

A Comparison of Structures in Dicots and Monocots
Botanists classify all angiosperms into these two groups.

Figure 25.22

Phototropism

The above-ground portions of plants grow toward a source of light. One of the first studies done on phototropism was done by Charles Darwin and his son Francis in the late 1870s. They studied canary grass (*Phalaris canariensis*) and oats (*Avena sativa*). They concluded that the process was controlled in some way by the tip of the plants. Later this was substantiated with the discovery of the plant hormone, auxin. Auxin proved to be a plant hormone that controls cell elongation and stimulates plant growth toward light.

Figure 25.23

Clinging Stems

Some stems are modified to wrap themselves around objects and give support. The *tendrils* of this grapevine are a good example. The tendrils of the Virginia creeper (*Parthenocissus* sp.) have adhesive pads that help them stick to objects.

Table 25.2

A CLADOGRAM OF PLANTS

SPERMATOPHYTES

Examples:
Gymnosperms
Conifers
Pine tree

Examples:
Angiosperms
Flowering plants
Rose

Example:
Fern

Example:
Probably first to dominate land, *Rhynia*

"rhyniophytes"   "seed plants"   Pteridopsida   Cladoxylopsida

Example:
Fossil ancestors of ferns and horsetails

Example:
Moss

Bryophyta

Sphenopsida

Example:
Horsetails

Example:
Hornworts

Anthocerotophyta

Trimero-phytophyta

Example:
Fossil plants similar to rhyniophytes

Example:
Liverworts

Hepaticophyta

"charophytes"   Zosterophyllophyta   Lycophyta

Examples:
Club moss
Scale trees

B R Y O P H Y T E S

P T E R I D O P H Y T E S

Ancestoral origin:
Green algae

Extinct plants:
Devonian Period

SUMMARY

The plant kingdom is composed of organisms that are able to manufacture their own food by the process of photosynthesis. They have specialized structures for producing the male sex cell (the sperm) and the female sex cell (the egg). The relative importance of the haploid gametophyte and the diploid sporophyte that alternate in plant life cycles is a major characteristic used to determine an evolutionary sequence. The extent and complexity of the vascular tissue and the degree to which plants rely on water for fertilization are also used to classify plants as primitive (ancestral) or complex. Among the gymnosperms and the angiosperms, the methods of production, protection, and dispersal of pollen are used to name and classify the organisms into an evolutionary sequence. Based on the information available, mosses are the most primitive plants. Liverworts and club mosses are experimental models. Ferns, seed-producing gymnosperms, and angiosperms are the most advanced and show the development of roots, stems, and leaves.

The kingdom Plantae is summarized in table 25.2.

THINKING CRITICALLY

For guidelines to answer this question, visit the Online Learning Center.

Some people say the ordinary "Irish" potato is poisonous when the skin is green, and they are at least partly correct. A potato develops a green skin if the potato tuber grows so close to the surface of the soil that it is exposed to light. An alkaloid called *solanine* develops under this condition and may be present in toxic amounts. Eating such a potato raw may be dangerous. However, cooking breaks down the solanine molecules and makes the potato as edible and tasty as any other.

The so-called Irish potato is of interest to us historically. Its real country of origin is only part of the story. Check your local library to find out about this potato and its relatives. Are all related organisms edible? Where did this group of plants develop? Why is it called the Irish potato?

CONCEPT MAP TERMINOLOGY

Construct a concept map using these terms.

accessory structures	monocot
angiosperm	root hairs
dicot	roots
flower	stems
leaves	

Visit the Online Learning Center to use an interactive format to construct another concept map.

KEY TERMS

Use interactive flash cards, on the Online Learning Center, to help you learn the meanings of these terms.

accessory structures	ovary
alternation of generations	perennial
angiosperms	perfect flowers
annual	petals
antheridia (singular, antheridium)	phloem
archegonium	pistil
auxin	pollen
cambium	pollination
cone	root
deciduous	root hairs
dicot	seed
flower	seed leaves
fruit	sepals
gametophyte	spores
germinate	sporophyte
gymnosperms	stamen
herb	stem
imperfect flowers	tropism
leaves	vascular tissues
life cycle	wood
monocot	woody vascular plants
nondeciduous	xylem

LEARNING CONNECTIONS

Visit the Online Learning Center to find additional study resources including those highlighted below.

25.1 What Is a Plant?

Animations and Review
• Introduction and characteristics

1. What do all plants appear to have in common?
2. On what five bases do botanists classify plants?

25.2 Alternation of Generations

Animations and Review
• Asexual reproduction
• Life cycles

3. What is the ploidy of a sporophyte? Of a gametophyte?
4. What are the dominant generations in mosses, ferns, gymnosperms, and angiosperms?

25.3 Ancestral Plants: The Bryophytes

Animations and Review
• Nonvascular Plants

5. What sex cells are associated with antheridia? Archegonia?
6. Give two examples of bryophytes.

25.4 Adaptations to Land

Animations and Review
• Introduction to plant organs: stems, leaves, roots

7. Give two examples of vascular tissue.
8. What is the significance of the cambium tissue in woody perennials?

25.5 Transitional Plants: Non-Seed-Producing Vascular Plants

Animations and Review
• Vascular-seedless I
• Vascular-seedless II

Interactive Concept Map
• Spore producers

Experience This!
• Rooting plants

9. What are the differences between xylem and phloem?
10. Ferns have not been as successful as gymnosperms and angiosperms. Why?

25.6 Advanced Plants: Seed-Producing Vascular Plants

Labeling Exercises
• Plant body

• Dicot root tip
• Stem tip
• Stem anatomy

Animations and Review
• Gymnosperms
• Angiosperms
• Fruits

Interactive Concept Map
• Seed producers

11. What is the difference between an annual and a perennial?
12. What is meant by the term *perfect flower*?
13. How is a seed different from pollen, and how do both of these differ from a spore?
14. How are cones and flowers different?
15. How are cones and flowers similar?

25.7 Response to the Environment—Tropisms

Interactive Concept Map
• Text concept map

16. What is an auxin? What role does it play in plant metabolism?
17. What advantage does a plant with tendrils have?

Test yourself with multiple choice questions that provide immediate feedback
on the Online Learning Center with PowerWeb at

www.mhhe.com/enger11

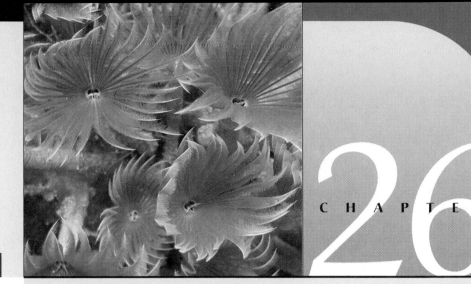

CHAPTER OUTLINE

Animalia

GOALS AND OBJECTIVES

List the characteristics that are typical of animals.

❏ Describe the seven features that distinguish animals from other kinds of
   organisms.

❏ Give examples that illustrate that there are a wide variety of kinds of animals.

Understand the nature of the different structural traits of animals.

❏ Explain the difference between homeotherms and poikilotherms.

❏ List the three body forms typical of animals.

❏ Distinguish between diploblastic and triploblastic animals.

❏ Explain what is meant by the terms *protostome* and *deuterostome*.

❏ Understand the role of a skeleton.

Understand the evolutionary history of the animal kingdom.

❏ Describe why primitive animals are primarily marine.

❏ Explain why the body plans of some animals are only suitable for aquatic
   habitats.

❏ Recognize how aquatic organisms can reproduce by external fertilization.

❏ Describe the adaptations of terrestrial animals that allow them to be successful on
   land.

❏ Explain why the parasitic way of life is beneficial to parasites.

❏ Define the terms *pelagic* and *benthic*.

❏ List specific adaptations that allow animals to succeed in their habitats and
   niches.

The Origin and Classification of Life

PART SIX

26.1 What Is an Animal?

Animals are adapted to live in just about any environment. They live on the shores of oceans, in shallow freshwater, the bitter cold of the arctic, the dryness of the desert, and the driving rains of tropical forests. There are animals that eat only other animals, animals that are parasites, animals that eat both plants and other animals, and animals that feed only on plants. They show a remarkable variety of form, function, and activity. They can be found in an amazing variety of sizes, colors, and body shapes (figure 26.1).

Because we humans display the characteristics of many other animals, most people find it easy to distinguish animals from plants. In comparison to plants, many animals move great distances, attack and eat their prey, and have soft bodies. Their energy is supplied from the food they eat by the biochemical process of cellular respiration, not photosynthesis. Traits that allow people to classify an unknown organism in the kingdom Animalia might include:

1. *Animals are multicellular organisms.* Multicellular organisms have some advantages over one-celled organisms. In multicellular organisms, individual cells are able to specialize. Thus, multicellular organisms can perform some tasks better than can single-celled organisms. For example, some unicellular organisms

Figure 26.1

Variety of Animals
Today there are at least 4 million known species of animals, ranging in size from microscopic rotifers, 40 micrometers in length, to giant blue whales, 30 meters long. Animals not only vary in size, but also inhabit widely diverse habitats—from the frigid Arctic to the scorching desert, from the dry African savanna to the South American tropical rainforest, and from violent tidal zones to the ocean's depths: (*a*) bumblebee, (*b*) mayfly, (*c*) giant green tree frog, (*d*) turtle, (*e*) brain coral, (*f*) stingray, (*g*) anole lizard, (*h*) mallard duck.

(a)

(b)

(c)

(d)

(e)

(f)

(g)

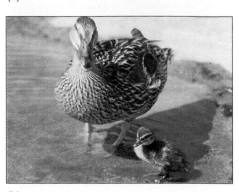
(h)

can move by using cilia, flagella, or amoeboid motion. However, such movement is relatively slow, weak, and allows the organism to move only a short distance. Many animals have muscle cells that are specialized for movement and allow for a greater variety of more efficient kinds of movement. For example, some animals use muscle power to migrate thousands of kilometers—a feat beyond the capability of unicellular organisms. Multicellular organisms can also be larger than single-celled organisms. Many animals are large but the vast majority are small. Many animals are actually smaller than some of the larger protozoa. Biologists classify animals into two categories based on their cellular organization. *Parazoa* is a group of animals comprised of individual cells that do not work together in a coordinated manner. Members of the *Eumetazoa*, on the other hand, have true tissues.
• **tissue, p. 15**

2. *The bodies of animals are usually composed of groups of cells organized into tissues, organs, and organ systems.* Functions such as ingesting food, exchanging gases, and removing wastes are more complicated in animals than in unicellular organisms. One-celled organisms are in direct contact with the environment; any exchange between the organism and the external environment occurs through the plasma membrane. However, in animals the majority of cells are not on the body surface and are therefore not in direct contact with the external environment. Animals must have a specialized means of exchange between the internal environment and the external environment. This often takes the form of gills, lungs, and digestive systems. They must also have developed a method of transporting materials between the body surface and internal cells. As the size of an animal increases, the amount of body surface increases more slowly than the volume, and the body systems for exchanging and transporting material become more complex. In many larger animals this is known as a circulatory system. There are many different kinds of systems that accomplish this distribution function. • **human circulatory system, p. 342**

The majority of cells in an animal are internal and do not directly perceive changes in the external environment, such as changes in light and temperature. Some of the surface cells have developed specialized sensory sites that do perceive environmental changes. These external changes can then be communicated to the internal regions of the body through a network of sensory neurons or chemical messengers. Again, as an animal gets larger, the body systems become more complex. Chapters 18 through 21 describe the functions of some of the systems in the human body. Most animals have systems that function in basically the same ways. Naturally there are some modifications—fish have gills, not lungs—but the purpose of the respiratory system is the same in fish and in animals with lungs. • **surface/volume ratio, p. 340**

3. *All animal cells have a nucleus but lack cell walls.* All organisms except the bacteria and archaea are eukaryotic; that is, they have a nucleus and other cellular organelles. These include the protists, fungi, plants, and animals. However, only the fungi, plants, and animals are multicellular. Of these three the cells of animals differ in that they do not have cell walls like plants and fungi.

4. *Animals are heterotrophic. They do not carry on photosynthesis.* Animals must eat other organisms to obtain molecules that are used as building materials for new cells, and for energy to drive life's processes. Most animals specialize in certain kinds of food. Sharks eat other aquatic animals, mosquitos suck blood, sparrows are seedeaters, jellyfish eat small aquatic organisms. Each style of eating involves specialized structures like teeth, stinging cells, or digestive tracts to process the food that is eaten. • **symbiotic relationships, p. 289**

5. *During all or part of their lives, animals are able to move from place to place or move one part of their body with respect to other parts.* Movement is typical of animals. Most animals move freely from place to place as they seek food and other necessities of life. However, some become permanently attached to surfaces after a period of dispersal early in their lives. For example, barnacles are related to lobsters; oysters and mussels are related to other mollusks that move freely; and many kinds of marine worms live in tubes they have constructed.

In addition animals can move one part of their body with respect to other parts. Many animals have appendages (legs, wings, tentacles, spines, or soft muscular organs) that bend or change shape. Often these structures are involved in moving the animal from place to place but they are also used to capture food, clean their surfaces, or move things in their environment.

Nearly all of these different kinds of movements are the result of specialized cells that shorten to move one part of the body with respect to another. In most animals these cells are known as muscle cells.

6. *Animals respond quickly and appropriately to changes in their environment.* Movement is an important characteristic of animals, but the movement is usually in response to some stimulus in their surroundings. This activity involves coordinating sensory input and transmitting this information from a sense organ to muscle, which actually brings about the movement.

7. *Sexual reproduction is a characteristic of animals, although many animals reproduce asexually as well as sexually.* Reproduction is essential to the survival of any species. Most animals are involved in some variety of sexual reproduction. The methods vary greatly—from the simultaneous release of sperm and eggs into

Figure 26.2

Regulating Body Temperature
The body temperature of a homeotherm remains constant regardless of changes in environmental temperature. The body temperature of a poikilotherm is dependent upon environmental temperature.

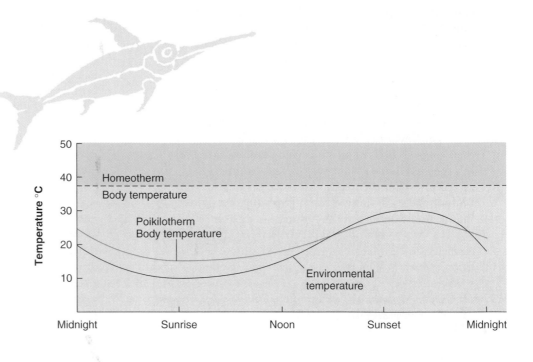

the water for many marine organisms, to fertilization inside the body for most terrestrial organisms, and high degrees of parental care in birds and mammals.

Although there are exceptions to these taxonomic criteria, they are a good starting point to explore what it is to be an animal. In addition to the generalities just listed, there are a few other concepts that need to be introduced before we begin a more detailed discussion of the nature of animals.

26.2 Temperature Regulation

Unicellular organisms are all **poikilotherms**—organisms whose body temperature varies. Most animals, including insects, worms, and reptiles, are poikilotherms. The body temperature of an ant changes as the external temperature changes, which means that at colder temperatures, poikilotherms also have lower metabolic rates. Animals such as deer, however, are **homeotherms**, which means that they maintain a constant body temperature that is generally higher than the environmental temperature, regardless of the external temperature (figure 26.2). These animals, birds and mammals, all have high metabolic rates.

Ectothermic describes animals that regulate body temperature by moving to places where they can be most comfortable. A good example of this occurs in snakes and other reptiles. **Endothermic** describes animals that have internal temperature-regulating mechanisms and can maintain a relatively constant body temperature in spite of wide variations in the temperature of their environment as occurs in humans.

• heat and temperature, p. 27

26.3 Body Plans

The form of animal bodies is diverse. Some animals have no regular body form, a condition called **asymmetry.** Asymmetrical body forms are rare and occur only in certain species of sponges, which are the simplest kinds of animals (figure 26.3a). **Radial symmetry** (figure 26.3b) occurs when a body is constructed around a central axis. Any division of the body along this axis results in two similar halves. Although many animals with radial symmetry are capable of movement, they do not always lead with the same portion of the body; that is, there is no anterior, or head, end.

Animals with **bilateral symmetry** (figure 26.3c) are constructed along a plane running from a head to a tail region. There is only one way to divide bilateral animals into two mirrored halves. Animals with bilateral symmetry move head first, and the head typically has sense organs and a mouth. When these animals develop as embryos, the first opening formed may become the mouth or the anus. If the first opening becomes the mouth, the animals are called *protostomia.* If the first opening becomes the anus, they are called *deuterostomia.* This difference in developmental pattern marked a major evolutionary separation among certain animals. The vast majority of animals display bilateral symmetry (figure 26.4).

Animals also differ in the number of layers of cells of which they are composed. The simplest animals (sponges and jellyfishes) are **diploblastic,** which means that they only have two layers of cells—an outer layer and an inner hollow layer that is involved in processing food. All the other major groups of animals are **triploblastic.** Their bodies are made up of three layers and have many structures in the middle layer. In most of

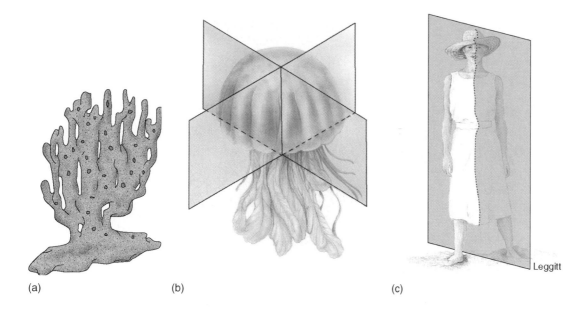

Figure 26.3

Radial and Bilateral Symmetry
(*a*) This sponge has a body that cannot be divided into symmetrical parts and is therefore asymmetrical. (*b*) In animals such as this jellyfish with radial symmetry, any cut along the central body axis results in similar halves. (*c*) In animals with bilateral symmetry, only one cut along one plane results in similar halves.

(a)   (b)   (c)

Leggitt

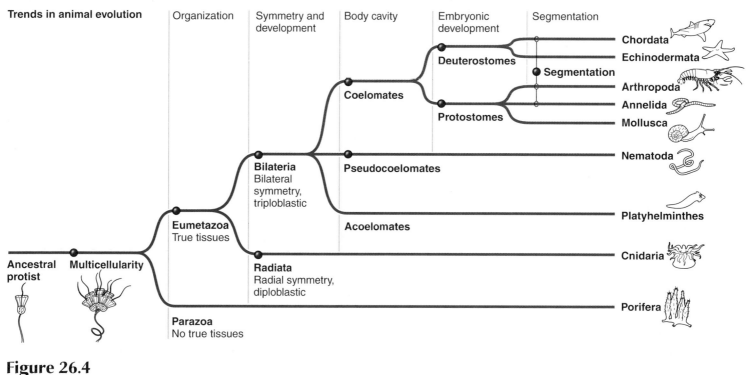

Figure 26.4

A Cladogram of Animalia
The classification of animals reflects complexity of body form and developmental features.

them, this resembles a tube within a tube (figure 26.5). The outer layer contains muscles and nerves, and is exposed to the environment. In many animals the outer layer is not protected by specialized structures, but other animals have such things as shells, scales, feathers, or hair protecting the outer layer of skin.

The inner tube layer constitutes the digestive system, with a mouth at one end and anus at the other. Many por-

tions of this food tube are specialized for the digestion, absorption, and reabsorption of nutrients. Other organs associated with the food tube secrete digestive enzymes into it, and are located between the digestive tube and the outside body wall. Also located in this area are other organs that are involved in excretion of waste, circulation of material, exchange of gases, and body support.

Figure 26.5

Body Structure

All animals with bilateral symmetry have a body consisting of a digestive tube running through an outer tube. This is often called a tube-within-a-tube body plan. The outer tube consists of two layers. The outermost thin layer forms the surface of the skin and other structures such as a cuticle (covering material), hair, scales, or feathers. This is shown in blue in the drawing. Attached to the outer layer of the skin is a layer of muscle and connective tissue (shown in red). The gut is lined with a thin layer of cells called the epithelial lining. This is shown in yellow in the drawing. Surrounding the gut lining is a layer of muscle and connective tissue (shown in red). Most of the internal organs of the body have the same origin as the muscle and connective tissue.

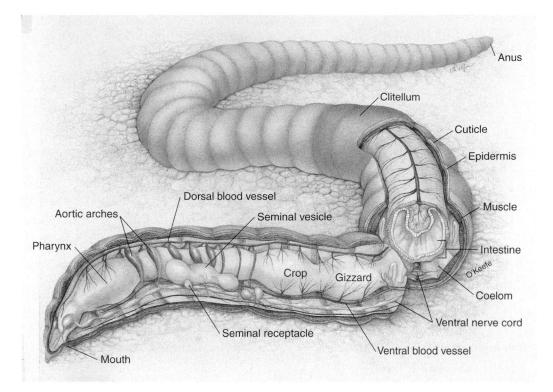

Simple animals, such as jellyfish and flatworms, have no space between the inner and outer tubes (figure 26.6a). More advanced animals, such as earthworms, insects, reptiles, birds, and mammals, have a **coelom,** or body cavity, between these two tubes (figure 26.6b). The coelom in a turkey is the cavity where you stuff the dressing. In the living bird this cavity contains a number of organs, including those of the digestive, excretory, and circulatory systems. The development of the coelom was significant in the evolution of animals. There are also animals with **pseudocoeloms,** or a false body cavity. The pseudocoelom of roundworms is a space but it is not completely filled with cells. In **acoelomates,** animals without a coelom, the internal organs are packed closely together. In coelomates there is less crowding of organs and less interference among them. Organs such as the heart, lungs, stomach, liver, and intestines have ample room to grow, move, and function. The coelom allows for separation of the inner tube and the body-wall musculature; thus, the inner tube functions freely, independent of the outer wall. This results in organ systems that are more highly specialized than acoelomate systems. Organs are not loose in the coelom but are held in place by sheets of connective tissue called **mesenteries.** Mesenteries also serve as support for blood vessels connecting the various organs (figure 26.6c).

Segmentation is another anatomical plan. There are two types: superficial and metameric. Superficial segmentation is seen in more primitive animals (ex. rotifer) and appears as a series of subdivisions of the animal's body from head to tail. This kind of segmentation has little evo-

lutionary importance. Superficial segmentation might help with the animal's ability to move by allowing bending or telescoping. Metameric segmentation (ex. human) does have evolutionary significance. The evolution of this type of segmentation served as a springboard to increased complexity and specialization as seen in the more highly evolved animals because the coelom is divided into segments that can become specialized.

26.4 Skeletons

Most animals have some sort of structural support we call a skeleton. A skeleton is important for several reasons. First of all, it serves as strong scaffolding to which other organs can be attached. The skeleton also provides places for muscle attachment and if the skeleton has joints, the muscles can move one part of the skeleton with respect to others. Aquatic organisms are generally supported by the dense medium in which they live and many marine animals lack well-developed skeletons. In terrestrial animals, however, a strong supportive structure is necessary to support the animal in the thin medium of the atmosphere.

There are two major types of skeletons: internal skeletons (**endoskeleton**) and external skeletons (**exoskeleton**). The vertebrates (fish, amphibians, reptiles, birds, mammals) have internal skeletons. The various organs are attached to and surround the skeleton, which grows in size as the animal grows. Arthropods (crustaceans, spiders, insects, millipedes, centipedes) have an external

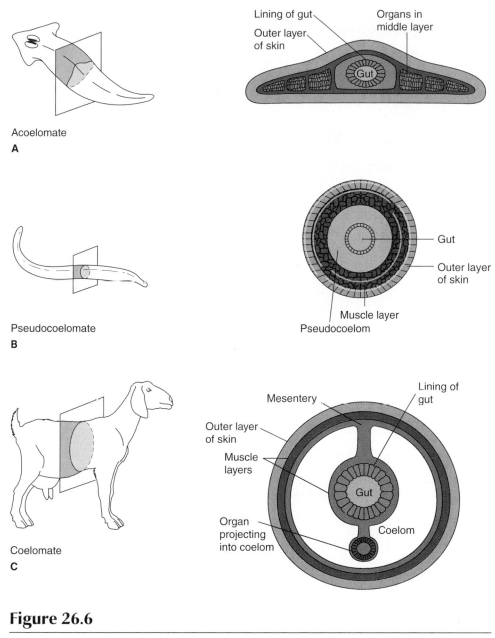

Figure 26.6

Body Cavities

(a) Some animals, like flatworms, have no open space between the gut and outer body layer. (b) Roundworms commonly found in soil have body cavities that have some cells and are not completely lined with thin sheets of connective tissue. (c) Other animals, including all vertebrates, have a coelom, an open area within the middle layer. Organs form from this middle layer, projecting into the coelom. The mesenteries are thin sheets of connective tissue that hold the organs in place within the coelom.

skeleton that surrounds all other organs. It is generally hard and has joints. Growth in these animals is accommodated by shedding the old skeleton and producing a new larger one. This period in the life of an arthropod is dangerous because for a short period it is without its hard, protective outer layer.

Many other animals have structures that have a supportive or protective function (such as clams, snails, and corals) and these are sometimes called skeletons but they do not have joints.

26.5 Animal Evolution

Scientists estimate that the Earth is at least 4.5 billion years old and that life originated in the ocean about 3.8 to 3.7 billion years ago (chapter 22). For approximately 2 billion years, one-celled, prokaryotic organisms were the only forms of life present in the ocean, and there were no life-forms on land. These early life-forms probably evolved into unicellular, eukaryotic, plantlike and animal-like organisms that were the forerunners of present-day plants and animals about 1.8 billion years ago. The earliest animal-like fossils date to about 600 million years ago (figure 26.7).

All groups of animals appear to have started in the sea and many groups have remained aquatic to the present. In the ocean, animals did not have a problem with dehydration. Also, the ion content of the early ocean approximated that of the animals' cells, so little energy was required to keep the cell in osmotic balance. Finally, the temperature range in the ocean is not as great as that on land and the rate of temperature change is lower. Therefore, animals in the ocean did not require mechanisms to deal with rapid or extreme changes in the environment.

Most of the earliest animals were probably small planktonic organisms that floated in the water column or were wormlike and crawled on the bottom or through the sediment on the bottom of the ocean. All of the different kinds of animals showed much adaptive radiation and produced many new forms based on their original body plans. For example, clams, snails, and octopus in mollusks; starfish, sea urchins, and sea cucumbers in echinoderms; and sharks, bony fishes, and lamprey among the vertebrates.

The backbone is a recent development in the evolution of animals. Animals with backbones made of vertebrae are called **vertebrates;** those without backbones are called **invertebrates.** All early animals lacked backbones and still constitute 99.9% of all animal species in existence today.

Members of the arthropods (notably the insects) and vertebrates (particularly the reptiles, birds, and mammals) made the transition to a terrestrial existence about 400 million years ago, with the arthropods preceding the vertebrates by a few million years.

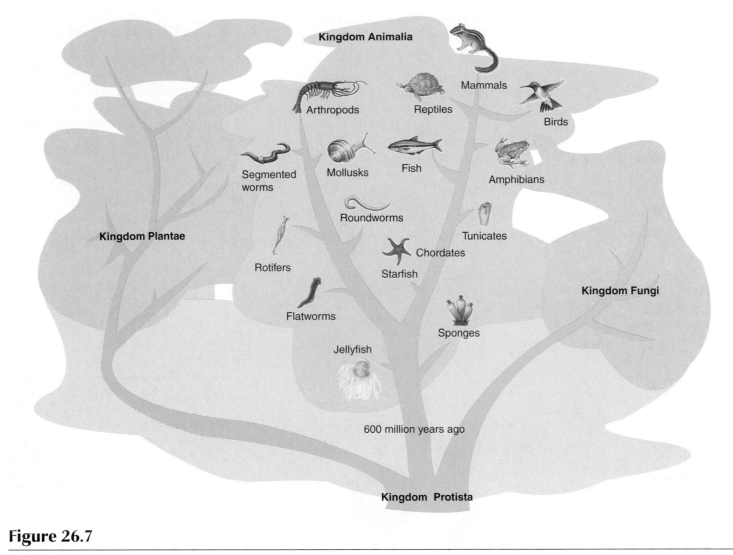

Figure 26.7

Animal Evolution
The first recognizable animals show up in the fossil record about 600 million years ago and have evolved into the variety of creatures we see today.

26.6 Primitive Marine Animals

Sponges, jellyfish, and corals are the simplest multicellular animals. They evolved about 600 million years ago and are usually found in saltwater environments. Even though sponges are classified as multicellular, in many ways they are colonial. Most cells are in direct contact with the environment. All adult sponges are sessile (permanently attached) filter feeders with ciliated cells that cause a current of water to circulate within the organism. The individual cells obtain their nutrients directly from the water (figure 26.8).

Reproduction in sponges can occur by fragmentation. Wave action may tear off a part of a sponge, which eventually settles down, attaches itself, and begins to grow. Sponges also reproduce by **budding,** a type of asexual reproduction in which the new organism is an outgrowth of the parent. They also reproduce sexually, and external fertiliza-

tion results in a free-swimming, ciliated larval stage. The larva swims in the plankton and eventually settles to the bottom, attaches, and grows into an adult sponge.

Cnidarians include the jellyfish, corals, and sea anemones. Like many sponges, they have radial symmetry. Many species of Cnidaria exhibit alternation of generations and have both sexual and asexual stages of reproduction. The **medusa** is a free-swimming adult stage that reproduces sexually. The **polyp** is a sessile larval stage that reproduces asexually (figure 26.9). All species have a single opening that leads into a saclike interior. Surrounding the opening is a series of tentacles (figure 26.10). These long, flexible, armlike tentacles have specialized cells called *nematocysts* that can sting and paralyze small organisms. Nematocysts are unique to the Cnidaria. Even though they are primitive organisms, cnidarians are carnivorous.
• trophic levels, p. 261

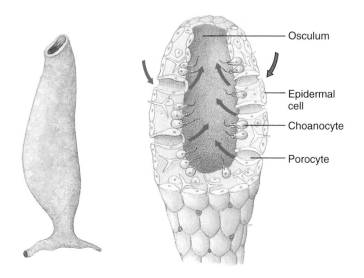

Figure 26.8

A Sponge

Verongia is an example of a tube sponge. The cells are in direct contact with the environment. The choanocytes form an inner layer of flagellated cells. These flagella create a current that brings water in through the openings formed by the porocyte cells, and it flows out through the osculum. The current brings food and oxygen to the inner layer of cells. The food is filtered from the water as it passes through the animal. Although some sponges grow this "vase shape," most others have no distinct form. Sponges can be found growing in both salt water and freshwater adhering to dock pilings, rocks, and other fixed objects.

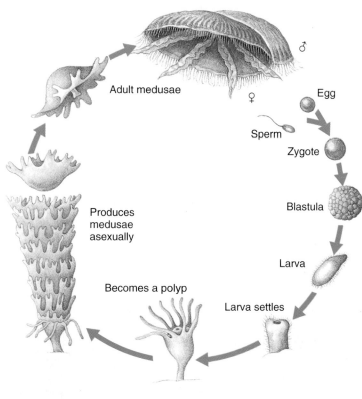

Figure 26.9

The Life Cycle of Cnidaria

The life cycle of *Aurelia* is typical of the alternation of generations seen in most species of Cnidaria. The free-swimming adult medusae (jellyfish) reproduce sexually, and the resulting larva develops into a polyp. The polyp undergoes asexual reproduction, which produces the free-swimming medusa stage. The polyp stage is microscopic in many cases and is not well known to most people.

(a)   (b)

Figure 26.10

Phylum Cnidaria

The Portuguese man-of-war, or *Physalia physalis* (a), is commonly called the blue-bottle in Australia. The name "blue-bottle" comes from the body, which really is a large, gas-filled, blue float, that can be up to 30 centimeters in length and rise above the water as much as 15 centimeters. The float has a crest that is used much as a sail to propel the colony across the water when the wind blows. It is widely distributed throughout the warmer seas of the world. Various species of the Portuguese man-of-war have been found in the tropical Atlantic, sometimes reaching as far north as the Bay of Fundy (Canada), the Mediterranean Sea, the Indo-Pacific region, the ocean around Hawaii, and up to southern Japan. Although the sting from the Portuguese man-of-war is rarely fatal, a person stung by a Portuguese man-of-war will still experience severe pain. Single or multiple welts will appear on the skin. The sting can cause fever, shock, and circulatory and respiratory problems. The severe pain from the sting may last about two hours, and depending upon treatment, the pain will usually subside and go away in seven or eight hours. (b) The sessile sea anemone is also a typical cnidarian. Each has a saclike body structure with a single opening into the gut.

HOW SCIENCE WORKS 26.1

Emerald Ash Borer (*Agrilus planipennis*)— An Exotic Pest from Asia that Affects Ash Trees

The Emerald Ash Borer (*Agrilus planipennis*) is a parasite belonging to a group of insects known as metallic wood-boring beetles. It is native to Asia and is found in China, Korea, Japan, Mongolia, the Russian Far East, and Taiwan. Even though the Emerald is not native to Michigan or anywhere in the United States, it has already killed many ash trees (*Fraxinus* sp.) in Michigan. Researchers speculate that if not brought under control, this insect parasite may cause the death of all ash trees in Michigan. It has the potential to destroy ash trees in much the same way that Dutch elm disease killed American elms throughout the United States and Canada.

The adults are active during the day and can be found in protected locations in bark crevices or on foliage. Chinese literature indicates that these beetles usually fly tree-to-tree within 2 meters of the ground. Whether these insects can fly long distances is yet to be discovered. The adults consume up to 0.45 cm² of foliage per day, leaving irregularly shaped patches of leaf tissue with jagged edges. Females lay 65 to 90 eggs during their lifetime, about 21 to 22 days. After hatching, the larvae chew (bore) through the bark into the cambium and feed on the tree's phloem. As a result, branches in the canopy of the tree die.

In some areas, woodpeckers feed on larvae. Control and management techniques are being developed. Recent experiments suggest that two insecticides, Imidacloprid and Bidrin, may be helpful.

(Photos courtesy of United States Department of Agriculture Forest Service)

(a) Adult insect

(b) Larvae

26.7 A Parasitic Way of Life

Many types of animals have become successful by adopting a parasitic way of life (How Science Works 26.1). One such group is the flatworms. There are three basic types of flatworms (the Platyhelminthes): free-living flatworms (often called *planarians*), flukes, and tapeworms (figure 26.11). The majority of free-living flatworms are nonparasitic bottom dwellers in marine or freshwater. A few species are found in moist terrestrial habitats. Free-living flatworms have muscular, nervous, and excretory systems.

All flukes and tapeworms are parasites. All parasites are extremely well adapted to their way of life but must have solved several kinds of problems. Although some parasites live on the outside of their host, many others live within the body of the organisms on which they are feeding. There are several specializations parasites must have in order to use a living host for food:

1. They must be able to find a suitable host.
2. They must be resistant to the efforts of the host to rid itself of the parasite.

(a)   (b)

Figure 26.11

Flatworms
(*a*) Planarians are free-swimming, nonparasitic flatworms that inhabit freshwater. (*b*) Adult tapeworms are parasites found in the intestines of many carnivores.

Figure 26.12

The Life History of *Schistosoma mansoni*
(1) Cercaria larvae in water penetrate human skin and are carried through the circulatory system to the veins of the intestine. They develop into (2) adult worms, which live in the blood vessels of the intestine. (3) Copulating worms are shown. The female produces eggs, which enter the intestine and leave with the feces. (4) A miracidium larva within an eggshell is shown. (5) The miracidium larva hatches in water and burrows into a snail (6) where it develops into a mother sporocyst (7). The mother sporocyst produces many daughter sporocysts (8), each of which produces many cercaria larvae. These leave the snail's body and enter the water, thus completing the life cycle.

3. They must have a method of anchoring themselves to the host.
4. They need to keep their host alive as long as possible.

We may consider this form of nutrition rather unusual, but if you count up all the kinds of animals in the world, there are more that are parasites than are not. Some of the flukes are external parasites on the gills and scales of fish, but the majority are internal parasites. Most flukes have a complex life cycle involving more than one host. Usually, the larval stage infects an invertebrate host, whereas the adult parasite infects a vertebrate host. Roundworms, flatworms, segmented worms, and insects contain many examples of parasites.

Schistosomiasis, which causes diarrhea, liver damage, anemia, and a lowering of the body's resistance, is caused by

adult *Schistosoma mansoni* flukes that live in the blood vessels of the human digestive system. Fertilized eggs pass out with the feces. Eggs released into the water hatch into free-swimming larvae. If a larva infects a snail, it undergoes additional reproduction and produces a second larval stage. A single infected snail may be the source of thousands of larvae. These new larvae swim freely in the water. Should they encounter a human, the larvae bore through the skin and enter the circulatory system, which carries them to the blood vessels of the intestine (figure 26.12).

Two hosts are also involved in the tapeworm's life cycle, but both hosts are usually vertebrate animals. A herbivore eats tapeworm eggs that have been passed from another infected host through its feces. The eggs are eaten along with the vegetation the herbivore uses for food. An egg develops

Figure 26.13

The Life Cycle of a Tapeworm
The adult beef tapeworm lives in the human small intestine. Proglottids, individual segments containing male and female sex organs, are the site of egg production. When the eggs are ripe, the proglottids drop off the tapeworm and pass out in the feces. If a cow eats an egg, the egg will develop into a cyst in the cow's muscles. When humans eat the cyst in the meat, the cyst develops into an adult tapeworm.

Adult tapeworm

Infected meat eaten

Mature bladder worm

Gravid (ripe) proglottid

Eggs

Mature cysts in striated muscle

Tapeworm eggs

Eggs eaten

O'Keefe

into a larval stage that encysts in the muscle of the herbivore. When the herbivore is eaten by a carnivore, the tapeworm cyst develops into the adult form in the intestine of the carnivore. When the worms reproduce, the eggs can be easily dispersed in the feces (figure 26.13).

Another major group of animals that has many parasitic species is the Nematoda, commonly called roundworms. Few animals are found in as many diverse habitats or in such numbers as the roundworms. Most are free living, but many are economically important parasites. Some are parasitic on plants, whereas others infect animals, and collectively they do untold billions of dollars worth of damage to our crops and livestock (figure 26.14).

Roundworm parasites range from the relatively harmless human intestinal pinworms *Enterobius*, which may cause irritation but no serious harm, to *Dirofilaria*, which can cause heartworm disease in dogs. If untreated, this infection may be fatal. Often the amount of damage inflicted by roundworms is directly proportional to their number. For example, hookworms (figure 26.15) feed on the host's blood. A slight infestation often results in anemia, but a heavy infestation of hookworms may result in mental or physical retardation.

(a)

(b)

Figure 26.14

Roundworms

When meat infected with trichina cysts is eaten by a host, the cysts develop into adults. The adult worms reproduce, and the resulting larvae encyst in the muscles of the host. A heavy infection can result in the death of the host. (*a*) A cross section of an infected muscle. Within each circle is an encysted larva. (*b*) Some roundworms are parasitic on plants and may cause extensive damage.

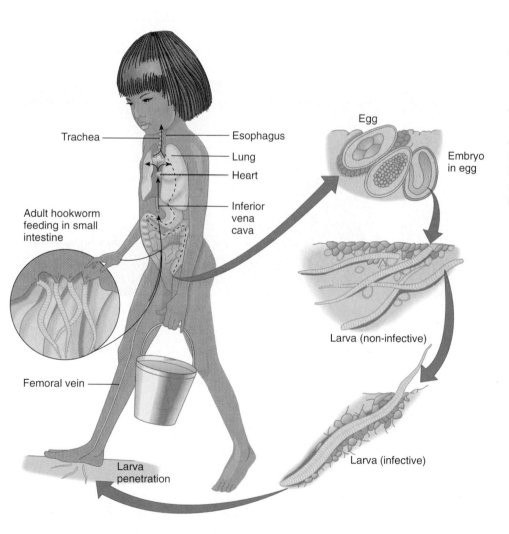

Trachea

Esophagus

Lung

Heart

Adult hookworm feeding in small intestine

Inferior vena cava

Femoral vein

Larva penetration

Egg

Embryo in egg

Larva (non-infective)

Larva (infective)

Figure 26.15

The Life Cycle of a Hookworm

Fertilized hookworm eggs pass out of the body in the feces and develop into larvae. In the soil, the larvae feed on bacteria. The larvae bore through the skin of humans and develop into adults. The adult hookworms live in the digestive system, where they suck blood from the host. The loss of blood can result in anemia, mental and physical retardation, and a loss of energy.

26.8 Advanced Benthic Marine Animals

A major ecological niche in the oceans includes large numbers of organisms that live on the bottom, called **benthic** organisms. Among benthic organisms are such animals as the segmented worms, the Annelida; clams and snails, the Molluska; and lobsters, crabs, and shrimp, the Arthropoda. When people think of annelids, they commonly think of the terrestrial earthworm. However, most annelids are not terrestrial but are found in marine benthic habitats, where most burrow into the ocean floor (figure 26.16).

The bilaterally symmetrical annelids have a well-developed musculature and circulatory, digestive, excretory, and nervous systems that are organized into repeating segments. For this reason the annelids are called *segmented worms.* Annelida are the first evolutionary group to display

(a)   (b)

Figure 26.16

Annelids

Annelids include (*a*) the sandworm (a polychaete), which is common in marine environments, and (*b*) sessile forms that are filter feeders.

(a)   (b)

Figure 26.17

Mollusks

Mollusks may range in complexity from (*a*) a small, slow-moving, grazing animal like a chiton to (*b*) intelligent, rapidly moving carnivores like an octopus.

this feature—the linear repetition of body parts, *segments* or *somites*. In annelids the segments are essentially alike; in the more highly evolved arthropods they are specialized to perform certain functions. Depending on the species, the individual may be male, female, or hermaphroditic (contain both male and female reproductive organs). Because most marine annelids live on the bottom and do not travel great distances, a free-swimming larval stage is important in their distribution. Like many other marine animals, many annelids are filter feeders, straining small organic materials from their surroundings. Others are primarily scavengers, and a few are predators of other small animals.

Another major group of benthic animals is the mollusks (figure 26.17). Like most other forms of animal life, the mollusks originated in the ocean, and even though some forms have made the move to freshwater and terrestrial envi-

ronments, the majority still live in the oceans. They range from microscopic organisms to the giant squid, which is up to 18 meters long.

A primary characteristic of mollusks is the presence of a soft body enclosed by a hard shell. Clams and oysters have two shells, whereas snails have a single shell. Some forms, such as the slugs, have no shell; they are unprotected. In the squids and octopuses, there is no external shell that serves as a form of support structure. Members of this phylum display a true body cavity, the coelom. Reproduction is generally sexual; some species have separate sexes and others are hermaphroditic.

Except for the squids and octopuses, mollusks are slow-moving benthic animals. Some are herbivores and feed on marine algae; others are scavengers and feed on dead organic matter. A few are even predators of other slow-moving or ses-

(a)   (b)   (c)

Figure 26.18

Echinoderms

(a) Starfish move by means of a water vascular system. Water enters the system through an opening, travels to the radial canals, and is forced into the tube feet. Other echinoderms include (b) the sea urchin and (c) the sea cucumber.

sile neighbors. As with most other marine animals, the mollusks produce free-swimming larval stages that aid in dispersal.

Echinoderms such as starfish are strictly marine benthic animals and are found in all regions, from the shoreline to the deep portions of the ocean. Echinoderms are often the most common type of animal on much of the ocean floor. Most species are free moving and are either carnivores or feed on detritus. They are unique among more advanced invertebrates in that they display radial symmetry. However, the larval stage has bilateral symmetry, leading many biologists to believe that the echinoderm ancestors were bilaterally symmetrical. Another unique characteristic of this group is the water vascular system (figure 26.18). In this system, water is taken in through a structure on the top side of the animal and then moves through a series of canals. The passage of water through this water vascular system is involved in the organism's locomotion.

Animals that live in shallow coastal areas must withstand tidal changes and the forces of wave action. Some are free moving and migrate with the tidal changes. Others are firmly attached to objects; these are said to be **sessile.** Most sessile animals are **filter feeders** that use cilia or other appendages to create water currents to filter food out of the water. Mussels, oysters, and barnacles are sessile marine animals.

Reproduction presents special problems for sessile organisms because they cannot move to find mates. However, because they are in an aquatic environment, it is possible for the sperm to swim to the egg and fertilize it. The fertilized egg develops into a larval stage—the juvenile stage of the organism (figure 26.19). The larvae are usually ciliated or have appendages that enable them to move, even though the adults are sessile. The free-swimming, ciliated larval stages allow the animal to disperse through its environment.

The larva differs from the sessile adult not only because it is free swimming, but also because it usually uses a different source of food and often becomes part of the

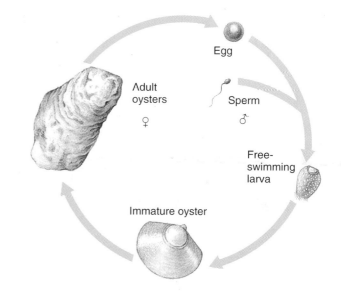

Figure 26.19

The Life Cycle of an Oyster

Each individual oyster can be either male or female at different times in its life. Sperm are released and swim to the egg. Fertilization results in the formation of a free-swimming larva. This larva undergoes several changes during the first 12 to 14 days and eventually develops into an immature oyster that becomes attached and develops into the adult oyster.

plankton community. The larval stages of most organisms are subjected to predation, and the mortality rate is high. The larvae move to new locations, settle down, and develop into adults. Even marine animals that do not have sessile stages typically produce free-swimming larvae. For example, crabs, starfish, and eels move freely about and produce free-swimming larvae.

(a)   (b)

Figure 26.20

The Lamprey
The lamprey uses its round mouth (a) to attach to a fish and then (b) suck blood from the fish.

(a)   (b)

Figure 26.21

Cartilaginous Fish
Although sharks (a) and rays (b) are large animals, they do not have bones; their skeletal system is made entirely of cartilage.

26.9 Pelagic Marine Animals: Fish

Animals that swim freely as adults are called **pelagic.** Many kinds of animals belong in this ecological niche, including squids, swimming crabs, sea snakes, and whales. However, the major kinds of pelagic animals are commonly called *fish*. There are several different kinds of fish that are as different from one another as reptiles are from birds, or birds from mammals.

Hagfish and lampreys lack jaws and are the most primitive of the fish. Hagfish are strictly marine forms and are scavengers; lampreys are mainly marine but may also be found in freshwater (figure 26.20). Adult lampreys suck blood from their larger fish hosts. Lampreys reproduce in freshwater streams, where the eggs develop into filter-feeding larvae. After several years, the larvae change to adults and migrate to open water.

Sharks and rays are marine animals that have an internal skeleton made entirely of cartilage (figure 26.21). These animals have no swim bladder to adjust their body density in order to maintain their position in the water; therefore, they must constantly swim or they will sink. Many of us have developed some misconceptions about sharks and rays as a result of movies or TV. Rays feed by gliding along the bottom

(a) (b) (c)

Figure 26.22

Bony Fish

Fish have a variety of body shapes. The animals pictured here have skeletons made of bones and, like most bony fish, they have a swim bladder: (a) sea perch, (b) moray eel, and (c) sea horse.

and dredging up food, usually invertebrates. Sharks are predatory and feed primarily on other fish. They travel great distances in search of food. Of the 40 species of sharks, only 7 are known to attack people. Most sharks grow no longer than a meter. The whale shark, the largest shark, grows to 16 meters, but it is strictly a filter feeder.

The bony fish are the class most familiar to us (figure 26.22). The skeleton is composed of bone. Most species have a swim bladder and can regulate the amount of gas in the bladder to control their density. Thus, the fish can remain at a given level in the water without expending large amounts of energy. Bony fish are found in marine and freshwater habitats, and some, like the salmon, can live in both. Bony fish feed on a wide variety of materials, including algae, detritus, and other animals. Like the sharks, many range widely in search of food. However, many fish are highly territorial and remain in a small area their entire lives.

26.10 The Movement to Land

Plants began to colonize land over 400 million years ago, during the Silurian period, and they were well established on land before the animals. Thus, they served as a source of food and shelter for the animals. When the first terrestrial animals evolved, there were many unfilled niches; therefore much adaptive radiation occurred, resulting in a large number of different animal species. Of all the many phyla of animals in the ocean, only a few made the transition from the ocean to the extremely variable environments found on the land. The annelids and the mollusks evolved onto the land but were confined to moist habitats. Many of the arthropods (insects and spiders) and vertebrates (reptiles, birds, and mammals) adapted to a wide variety of drier terrestrial habitats.

Regardless of their type, all animals that live on land must overcome certain common problems. Terrestrial animals must have (1) a moist membrane that allows for adequate gas exchange between the atmosphere and the organism, (2) a means of support and locomotion suitable for land travel, (3) methods to conserve internal water, (4) a means of reproduction and early embryonic development in which large amounts of water are not required, and (5) methods to survive the rapid and extreme climatic changes that characterize many terrestrial habitats. When we consider the transition of animals from an aquatic to a terrestrial environment, it is important to understand that this process required millions of years. There had to be countless mutations resulting in altered structures, functions, and behavioral characteristics that enabled animals to successfully adapt.

One large group of animals, the arthropods, has been incredibly successful in all kinds of habitats. They can be found in the plankton, as benthic inhabitants, and as pelagic organisms. This phylum includes nearly three-quarters of all known animal species. No other phylum lives in such a wide range of habitats. Although they include carnivores and omnivores, the majority of arthropods are herbivores.

The crustaceans are the best-known class of aquatic arthropods (figure 26.23). Copepods are common in the plankton of the oceans, crabs and their relatives are found as benthic organisms, and shrimp and krill are pelagic. However, the major success of this group is seen in the huge variety of terrestrial insects. Other terrestrial arthropod groups include the millipedes, centipedes, spiders, and scorpions. Half of all life forms that have been identified and named are insects.

Insects and other arthropods probably developed the adaptations for success on land at about the same time as the plants. They developed an internal tracheal system of thin-walled tubes extending into all regions of the body,

(a)   (b)   (c)

Figure 26.23

Crustaceans
Crustaceans include marine forms such as the king crab (*a*). The microscopic water flea (*b*) is a freshwater organism, and the pill (sow) bug (*c*) is terrestrial.

Figure 26.24

Insect Respiratory and Waste-Removal Systems
(*a*) Spiracles are openings in the exoskeleton of an insect. These openings connect to a series of tubes (tracheae) that allow for the transportation of gases in the insect's body. (*b*) Malpighian tubules are used in the elimination of waste materials and the reabsorption of water into the insect's body. Both systems are means of conserving body water.

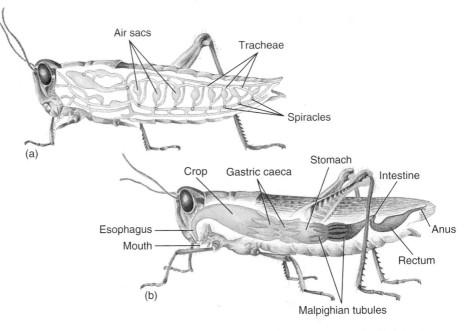

thus providing a large surface area for gas exchange (figure 26.24*a*). These tubes have small openings to the outside, which reduce the amount of water lost to the environment. They also developed a rigid outer layer that provides body support and an area for muscle attachment that permits rapid muscular movement. Because it is waterproof, this outer layer also reduces water loss. Another important method of conserving water in insects and spiders is the presence of Malpighian tubules, thin-walled tubes that surround the gut (figure 26.24*b*). If the insect is living in a dry environment, most of the water in waste materials is reabsorbed into the body by the Malpighian tubules and conserved.

Insects have separate male and female individuals and fertilization is internal, which means that the insects do not require water to reproduce. Insects have evolved a number of means of survival under hostile environmental conditions. Their rapid reproductive rate is one means. Most of a population may be lost because of an unsuitable environmental change, but when fa-

vorable conditions return, the remaining individuals can quickly increase in number. Other insects survive unfavorable conditions in the egg or larval form and develop into adults when conditions become suitable (figure 26.25). Some insects survive because of a lower metabolic rate during unfavorable conditions.

The terrestrial arthropods occupy an incredible variety of niches. Many are herbivores that compete directly with humans for food. They are capable of completely decimating plant populations that serve as food for human consumption. Many farming practices, including the use of pesticides, are directed at controlling insect populations. Other kinds of insects are carnivores that feed primarily on herbivorous insects. These insects are beneficial in controlling herbivore populations. Wasps and ladybird beetles have been used to reduce the devastating effects of insects that feed on agricultural crops. Insects have evolved in concert with the flowering plants; their role in pollination is well understood. Bees, butterflies, and beetles transfer pollen from one flower to an-

(a)

(b)

(c)

(d)

Figure 26.25

The Life History of a Moth

The fertilized egg (a) of the moth hatches into the larval stage (b). This wormlike stage then covers itself with a case and becomes a pupa (c). After emerging from the pupa (d), the adult moth's wings expand to their full size.

other as they visit the flowers in search of food. Many kinds of crops rely on bees for pollination, and farmers even rent beehives to ensure adequate seed for fruit production.

The first vertebrates on land were probably the ancestors of present-day amphibians (frogs, toads, and salamanders). Certain of the bony fishes have lobe-fins that can serve as primitive legs. It is likely that the amphibians evolved from a common ancestor of present-day lobe-finned fishes. The first amphibians made the transition to land some 360 million years ago during the Devonian period. This was 50 million years after plants and arthropods had become established on land. Thus, when the first vertebrates developed the ability to live on land, shelter and food for herbivorous as well as carnivorous animals were available. But vertebrates faced the same five problems that the insects and spiders faced in their transition ashore.

In amphibians (figure 26.26), the development of lungs was an adaptation that provided a means for land animals to exchange oxygen and carbon dioxide with the atmosphere. However, amphibians do not have an efficient method of breathing; they swallow air to fill the lungs, and most gas exchange between amphibians and the atmosphere must occur through the skin. In addition to needing water to keep their skin moist, amphibians must reproduce in water. When they mate, the female releases eggs into the water, and the male releases sperm amid the eggs. External fertilization occurs in the water, and the fertilized eggs must remain in water or they will dehydrate. Thus, with the appearance of amphibians, vertebrate animals moved onto "dry" land, but the processes of gas exchange and reproduction still limit the range of movement of amphibians from water.

Their buoyancy in water helps support the bodies of aquatic animals. This form of support is lost when animals move ashore; thus the amphibians developed a skeletal structure that prevented the collapse of their bodies on land. Even though they have an appropriate skeletal structure, amphibians must always be near water because they dehydrate and require water for reproduction. The extreme climatic changes were a minor problem: When conditions on land became too hostile, the amphibians retreated to an aquatic environment.

For 40 million years amphibians were the only vertebrate animals on land. During this time, mutations continued to occur, and valuable modifications were passed on to future generations. One change allowed the male to deposit sperm directly within the female. Because the sperm could directly enter the female and remain within a moist interior,

(a)   (b)

Figure 26.26

Amphibians

Amphibian larvae (*a*) are aquatic organisms that have external gills and feed on vegetation. The adults (*b*), such as the salamander, are terrestrial and feed on insects, worms, and other small animals.

(a)   (b)

Figure 26.27

Reptiles

Present–day reptiles include (*a*) turtles and (*b*) lizards. The green sea turtle, *Chelonia mydas,* is between 78 to 112 centimeters (31 to 44 inches) in length and inhabits tropical and subtropical waters near continental coasts and around islands. This endangered species feeds primarily on seagrasses and algae. Some populations of green sea turtles migrate over 2,000 kilometers (1,242 miles) across the ocean from nesting grounds to feeding grounds.

it was no longer necessary for the animals to return to water to reproduce. The reptiles had evolved (figure 26.27). • **geologic time table, p. 449**

Internal fertilization was not enough to completely free the reptiles from returning to water, however. The developing young still required a moist environment for their early growth. Reptiles became completely independent of an aquatic environment with the development of the amniotic egg, which protects the developing young from injury and dehydration (figure 26.28). The covering on the egg retains moisture and protects the developing young from dehydration while allowing for the exchange of gases. The reptiles were the first animals to develop such an egg. Some even use this egg in a form of asexual reproduction known as parthenogenesis (Outlooks 26.1.)

The development of a means of internal fertilization and the amniotic egg allowed the reptiles to spread over much of the Earth and occupy a large number of previously unfilled niches. For about 200 million years they were the only large vertebrate animals on land. The evolution of reptiles increased competition with the amphibians for food and space. The amphibians generally lost in this competition; consequently, most became extinct. Some, however, were able to evolve into the present-day frogs, toads, and salamanders.

There have been several periods of mass extinction on the Earth. One such period occurred about 65 million years ago, when many kinds of reptiles became extinct. Before that period of mass extinction, about 150 million years ago, birds evolved (figure 26.29). Although the amniotic egg remained the method of protecting the young, a series of changes in the reptiles pro-

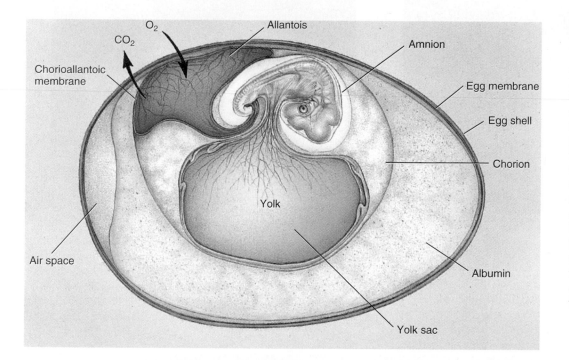

Figure 26.28

The Amniotic Egg
An amniotic egg has a shell and a membrane that prevent the egg from dehydrating and allow for the exchange of gases between the egg and the environment. The egg yolk provides a source of nourishment for the developing young. The embryo grows three extraembryonic membranes: the amnion is a fluid-filled sac that allows the embryo to develop in a liquid medium, the allantois collects the embryo's metabolic waste material and exchanges gases, and the chorion is a membrane that encloses the embryo and the other two membranes.

OUTLOOKS 26.1

Parthenogenesis

Parthenogenesis is an unusual method of reproduction used by some insects, crustaceans, and rotifers. In these cases an "egg" cell is produced but it is not fertilized. These cells develop into exact copies of the female parent. In most species, parthenogenesis may be a major method of reproduction, but at other times these species also engage in sexual reproduction, often when environmental conditions become challenging. In some species, however, true sexual reproduction is not known to occur. Because the offspring have only one parent, they are genetically identical to that parent. Certain fishes, amphibians, and reptiles also reproduce by parthenogenesis. Populations of certain whip-tail lizards are entirely female (see figure).

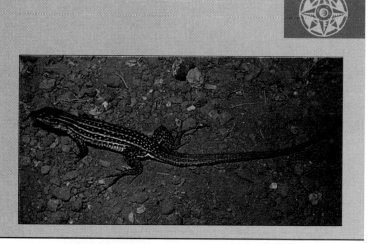

duced animals with a more rapid metabolism, feathers, and other adaptations for flight. There are several values to flight. Animals that fly are able to travel long distances in a short time and use less energy than animals that must walk or run. They are able to cross barriers like streams, lakes, oceans, bogs, ravines, or mountains that other animals cannot cross. They can also escape many kinds of predators by quickly taking flight. These were the first birds. They also possessed behavioral instincts, such as nest building, defense of their young, and feeding of the young. Because of these adaptations and their invasion of the air, a previously unoccupied niche, birds became one of the very successful groups of animals.

Even though the reptiles and birds had mastered the problems of coming ashore, mutations and natural selection continued, and so did evolution. As good as the amniotic egg is, it does have drawbacks: It lacks sufficient protection from sudden environmental changes and from predators that use eggs as

food. Other mutations in the reptile line of evolution resulted in animals that overcame the disadvantages of the external egg by providing for internal development of the young. Such development allowed for a higher survival rate. The internal development of the young, along with milk-gland development, a constant body temperature, a body covered with hair, and care of young by parents marked the emergence of mammals.

The first mammals to evolve were egg-laying mammals (figure 26.30), whose young still developed in an external egg. The marsupials (pouched mammals) have internal development of the young. However, the young are all born prematurely and must be reared in a pouch (figure 26.31). In the pouch, the young attach to a nipple and remain there until they are able to forage for themselves. The young of placental mammals remain within the female much longer, and they are born in a more advanced stage of development than is typical for marsupials (figure 26.32).

(a)

(b)

Figure 26.29

Birds

Birds range in size from (a) the small hummingbird to (b) the large ostrich. The rapid wing beat, seen as a blur, may be as much as 90 or more beats per second. The humming sound is the result of these rapid wing movements. The ostrich is flightless but can go from 0 to 45 mph in 2½ seconds!

(a)

(b)

Figure 26.31

Marsupials

All marsupials are born prematurely. The young of the opossum (a) and other marsupials crawl into a pouch and complete development there. (b) Even after the young are fully developed, some marsupials still carry the young in a pouch.

Figure 26.30

The Duck-Billed Platypus

The duck-billed platypus is a primitive type of mammal. It has the mammalian characteristics of fur and milk production, but the young are hatched from eggs. The platypus, *Ornithorhynchus anatinus*, lives in the streams, rivers, and lakes of eastern Australia. Its body grows up to 0.6 meters (2 feet) long, and has short, dense fur. Its bill is about 5 centimeters (2 inches) wide and it has hairless, webbed feet, and a flat, furry tail.

(a)

(b)

Figure 26.32

Placental Mammals

Placental mammals are not born until the development of the young is complete. Included in this group are animals ranging from (a) small terrestrial animals like the least shrew (5 centimeters) to (b) large marine animals like the humpback whale (approximately 12 meters).

Table 26.1

SUMMARY OF KINGDOM ANIMALIA

Phylum (Common Name)	Class (Common Name)	General Features	Example
Porifera (sponges)		Pores for water circulation Sessile as adults Aquatic	Marine and freshwater sponges
Cnidaria (cnidarians)		Radial symmetry Tentacles and stinging cells Aquatic habitat Some are sessile	Jellyfish, coral
Platyhelminthes (flatworms)		Flattened body Not segmented Sac-type digestive system	Planaria, liver fluke
Nematoda (roundworms)		Cylindrical body Not segmented Digestive tract Many parasitic forms Many in soil habitat	Pinworm, soil nematode
Annelida (segmented worms)		Segmented body Oxygen uptake through skin Circulatory, muscular, digestive, and nervous systems	Fanworm, earthworm, leech
Arthropoda (arthropods)	Insecta (insects)	Head, thorax, and abdomen Appendages: antennae, legs, mouthparts, wings, etc. External skeleton Metamorphosis common: from egg, larva, pupa, to adult	Ant, wasp, beetle, grasshopper
	Arachnida (spiders)	Head, thorax, and abdomen Appendages: four pairs of legs External skeleton	Wolf spider, tick, scorpion, mite
	Diplopoda (millipedes); Chilopoda (centipedes)	Segmented body Two, or one, pair of walking legs per segment	Centipedes, millipedes
	Crustacea (crustaceans)	External skeleton Walking legs and other appendages Mostly marine habitat	Lobster, crab, crayfish, isopods, and barnacles
Molluska (mollusks)		Soft body, frequently protected with shell Bilateral symmetry Anterior head region Internal organs in visceral region Ventral foot	Clam, squid, chitin
Echinodermata (spiny-skinned animals)		Spines on surface Radial symmetry Marine habitats only Tube feet and water vascular system	Starfish, sand dollars

Table 26.1 *(continued)*

Phylum (Common Name)	Class (Common Name)	General Features	Example
Chordata Vertebrata (vertebrates with backbone)	Mammalia (mammals)	Hair covers body Internal fertilization Mammary glands to feed young Placenta Four-chambered heart	Monotremes: egg layers—platypus Marsupials: pouched—kangaroo, opossum Artiodactyla: hoofed, even number of toes—sheep, deer, giraffe Carnivores: meat eaters—wolves, seals, weasels Cetaceans: marine, fishlike—whales, dolphins Chiroptera: flying—bats Edentata: no teeth—sloth, anteater Insectivora: insect eaters—mole, shrew Lagomorphs: chisel-like teeth—rabbit, hare Perissodactyla: hoofed, odd number of toes—horse, rhinoceros, zebra Primate: large brain, eyes front—human, lemur, ape Proboscidia: trunk and tusks—elephants Rodents: incisor teeth with continual growth—squirrel, mice, beaver Sirenia: aquatic, only forelimbs—sea cow, manatee
	Aves (birds)	Feathers cover body Internal fertilization Eggs with calcified shell Forelimbs adapted for flight Four-chambered heart	Chicken, eagle, sparrow
	Reptilia (reptiles)	Scales cover body Internal fertilization Membrane-enclosed egg Lungs Poikilotherm	Turtle, snake
	Amphibia (amphibians)	Moist skin, no scales External fertilization typical Metamorphosis common Lungs in adult form Three-chambered heart	Toad, frog, salamander
	Osteichthyes (bony fishes)	Scales Gills to acquire oxygen from water External fertilization usual Limbs are fins Two-chambered heart	Perch, trout, carp
	Chondrichthyes (cartilaginous fishes)	No bone Limbs are fins	Sharks, rays

SUMMARY

The 4 million known species of animals, which inhabit widely diverse habitats, are all multicellular and heterotrophic. Animal body shape is asymmetrical, radial, or bilateral. All animals with bilateral symmetry have a body structure composed of three layers.

Animal life originated in the ocean about 600 million years ago, and for the first 200 million years, all animal life remained in the ocean. Many simple marine animals have life cycles that involve alternation of generations.

Parasitism is still a very successful way of life for many animals. A major ecological niche for many marine animals is the ocean bottom—the benthic zone. Large, free-swimming marine animals dominate the pelagic ocean zone.

Animals that adapted to a terrestrial environment had to have (1) a moist membrane for gas exchange, (2) support and locomotion suitable for land, (3) a means of conserving body water, (4) a means of reproducing and providing for early embryonic development out of water, and (5) a means of surviving in rapid and extreme climatic changes. Table 26.1 starting on page 535 provides a summary of the kingdom Animalia.

THINKING CRITICALLY

For guidelines to answer this question, visit the Online Learning Center.

Animals have been used routinely as models for the development of medical techniques and strategies. They have also been used in the development of pharmaceuticals and other biomedical products such as heart valves, artificial joints, and monitors. The techniques necessary to perform heart, kidney, and other organ transplants were first refined using chimpanzees, rats, and calves. Antibiotics, hormones, and chemotherapeutic drugs have been tested for their effectiveness and for possible side effects using laboratory animals that are very sensitive and responsive to such agents. Biologists throughout the world have bred research animals that readily produce certain types of cancers that resemble cancers found in humans. By using these animals instead of humans to screen potential drugs, the risk to humans is greatly reduced. The emerging field of biotechnology is producing techniques that enable researchers to manipulate the genetic makeup of organisms. Research animals are used to perfect these techniques and highlight possible problems.

Animal-rights activists are very concerned about using animals for these purposes. They are concerned about research that seems to have little value in relation to the suffering these animals are forced to endure. Members of the American Liberation Front (ALF), an animal-rights organization, vandalized a laboratory at Michigan State University where mink were used in research to assess the toxicity of certain chemicals. Members of this group poured acid on tables and in drawers containing data, smashed equipment, and set fires in the laboratory. This attack destroyed 32 years of research records, including data used for developing water-quality standards. In one year, 80 similar actions were carried out by groups advocating animal rights.

What type of restrictions or controls should be put on such research? Where do you draw the line between "essential" and "nonessential" studies? Do you support the use of live animals in experiments that may alleviate human suffering?

CONCEPT MAP TERMINOLOGY

Construct a concept map using these terms.

bats	heart rate
bears	hibernation
birds	homeotherm
coelomate	metabolism
endotherm	torpor

Visit the Online Learning Center to use an interactive format to construct another concept map.

KEY TERMS

Use interactive flash cards, on the Online Learning Center, to help you learn the meanings of these terms.

acoelomates	homeotherms
asymmetry	invertebrates
benthic	medusa
bilateral symmetry	mesenteries
budding	pelagic
coelom	poikilotherms
diploblastic	polyp
ectothermic	pseudocoelom
endoskeleton	radial symmetry
endothermic	sessile
exoskeleton	triploblastic
filter feeders	vertebrates

LEARNING CONNECTIONS

Visit the Online Learning Center to find additional study resources including those highlighted below.

26.1 What Is an Animal?

1. List the six features used to identify an organism as an animal.
2. What makes an animal a heterotroph?

26.2 Temperature Regulation

3. What is the value in being able to maintain a constant body temperature?
4. Define the term *ectotherm.*

26.3 Body Plans

Animations and Review
- Body organization
- Tissues
- Homeostasis

Interactive Concept Map
- Body plans

5. Describe body forms that show asymmetry, radial symmetry, and bilateral symmetry.
6. Give an example of an animal that has a coelom.

26.4 Skeletons

Interactive Concept Map
- Skeleton

7. Give an example of an animal with an exoskeleton; an endoskeleton.
8. How does an animal with an exoskeleton grow?

26.5 Animal Evolution

Human Explorations
- Evolution of the heart

9. Which is speculated to have been most recently evolved: a backbone or exoskeleton?
10. Which is probably the most recently evolved animal: jellyfish or segmented worms?

26.6 Primitive Marine Animals

11. Give an example of an animal that is a cnidarian.
12. What are the differences between a polyp and a medusa?

26.7 A Parasitic Way of Life

13. Explain the tapeworm's life cycle.
14. What animal causes the disease schistosomiasis?

26.8 Advanced Benthic Marine Animals

Animations and Review
- Invertebrate characteristics

15. What is a sessile filter feeder?

16. How does the medusa stage of an animal differ from the polyp stage?
17. Describe a benthic environment.

26.9 Pelagic Marine Animals: Fish

Animations and Review
- Vertebrate introduction
- Fish

Case Study
- Sharks given a bum deal by rumors

18. How does a shark differ from most freshwater fish?
19. Define the term *pelagic*.

26.10 The Movement to Land

Animations and Review
- Amphibians
- Reptiles
- Birds
- Mammals

Experience This!
- Insect behavior

20. List the problems animals had to overcome to adapt to a terrestrial environment.
21. Why can't amphibians live in all types of terrestrial habitats?
22. What is the importance of the amniotic egg?
23. How does a marsupial differ from a placental mammal?

**Test yourself with multiple choice questions that provide immediate feedback
on the Online Learning Center with PowerWeb at**

www.mhhe.com/enger11

Appendix 1

The Metric System

Standard Metric Units

Standard Metric Units		Abbreviation
Standard unit of mass	gram	g
Standard unit of length	meter	m
Standard unit of volume	liter	L

Common Prefixes   **Examples**

Common Prefixes		Examples
kilo	1,000	a kilogram is 1,000 grams
centi	0.01	a centimeter is 0.01 meter
milli	0.001	a milliliter is 0.001 liter
micro (µ)	one-millionth	a micrometer is 0.000001 (one-millionth) of a meter
nano (n)	one-billionth	a nanogram is 10^{-9} (one-billionth) of a gram
pico (p)	one-trillionth	a picogram is 10^{-12} (one-trillionth) of a gram

Units of Length

Unit	Abbreviation	Equivalent
meter	m	approximately 39 in
centimeter	cm	10^{-2} m
millimeter	mm	10^{-3} m
micrometer	µm	10^{-6} m
nanometer	nm	10^{-9} m
angstrom	Å	10^{-10} m

Length Conversions

1 in = 2.5 cm	1 mm = 0.039 in
1 ft = 30 cm	1 cm = 0.39 in
1 yd = 0.9 m	1m = 39 in
1 mi = 1.6 km	1m = 1.094 yd
	1 km = 0.6 mi

To Convert	Multiply by	To Obtain
inches	2.54	centimeters
feet	30	centimeters
centimeters	0.39	inches
millimeters	0.039	inches

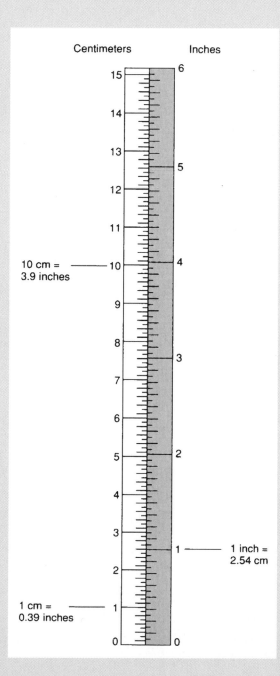

Centimeters   Inches

10 cm = 3.9 inches

1 inch = 2.54 cm

1 cm = 0.39 inches

Units of Volume

Unit	Abbreviation	Equivalent
liter	L	approximately 1.06 qt
milliliter	mL	10^{-3} L (1 mL = 1 cm^3 = 1 cc)
microliter	μL	10^{-6} L

Volume Conversions

1 tsp	=	5mL	1 mL	=	0.03 fl oz
1 tbsp	=	15mL	1L	=	2.1 pt
1 fl oz	=	30 mL	1L	=	1.06 qt
1 cup	=	0.24 L	1L	=	0.26 gal
1 pt	=	0.47 L			
1 qt	=	0.95 L			
1 gal	=	3.8 L			

To Convert	Multiply by	To Obtain
fluid ounces	30	milliliters
quarts	0.95	liters
milliliters	0.03	fluid ounces
liters	1.06	quarts

Units of Weight

Unit	Abbreviation	Equivalent
kilogram	kg	10^3 g (approximately 2.2 lb)
gram	g	approximately 0.035 oz
milligram	mg	10^{-3} g
microgram	μg	10^{-6} g
nanogram	ng	10^{-9} g
picogram	pg	10^{-12} g

Weight Conversions

1 oz	=	28.3 g	1 g	=	0.035 oz
1 lb	=	453.6 g	1 kg	=	2.2 lb
1 lb	=	0.45 kg			

To Convert	Multiply by	To Obtain
ounces	28.3	grams
pounds	453.6	grams
pounds	0.45	kilograms
grams	0.035	ounces
kilograms	2.2	pounds

Apothecary System of Weight and Volume*

Metric Weight		Apothecary Weight	Metric Volume		Apothecary Volume
30 g	=	1 ounce	1,000 ml	=	1 quart
15 g	=	4 drams	500 ml	=	1 pint
10 g	=	2.5 drams	250 ml	=	8 fl ounces
4 g	=	60 grains	90 ml	=	3 fl ounces
		(= 1 dram)	30 ml	=	1 fl ounce
2 g	=	30 grains			
1 g	=	15 grains			

*Used by pharmacists in preparing medications.

Temperature

Boiling point of water
100° C (212° F)

Human body temperature
37° C (98.6° F)

Freezing point of water
0° C (32° F)

Temperature Conversions

$$°C = \frac{(°F - 32) \times 5}{9}$$

$$°F = \frac{(°C \times 9)}{5} + 32$$

Some Equivalents

0° C	=	32° F
37° C	=	98.6° F
100° C	=	212° F

Appendix 2

Acronyms

ADH	antidiuretic hormone		LH	luteinizing hormone
AIDS	acquired immunodeficiency syndrome		Lp(a)	lipoprotein a
AMU	atomic mass unit		NAD	nicotinamide adenine dinucleotide
ATP	adenosine triphosphate		NADP	nicotinamide adenine dinucleotide phosphate
BMI	body mass index		pH	hydrogen ion concentration
BMR	basal metabolic rate		PCB	polychlorinated biphenyl
CAM	crassulacean acid metabolism		PKU	phenylketone urea
DDT	dichlorodiphenyltrichloroethane		RBC	red blood cell
DNA	deoxyribonucleic acid		RDA	recommended dietary allowance
EGF	epidermal growth factor		RNA	ribonucleic acid
ER	endoplasmic reticulum		RuBisCo	ribulose bisphosphate carboxylase
ETS	electron-transport system		SARS	servere acute respiratory syndrome
FAS	fetal alcohol system		SDA	specific dynamic action
FAD	flavin adenine dinucleotide		SOD	sudden oak death
FSH	follicle-stimulating hormone		STD	sexually transmitted disease
GI	glycemic index		TCA	tricarboxylic acid cycle
GnRH	gonadotropin-releasing hormone		TSH	thyroid-stimulating hormone
GSH	growth-stimulating hormone		UK	United Kingdom
HDL	high-density lipoprotein		UNAIDS	United Nations Joint HIV/AIDS Program
HIV	human immunodeficiency virus		USAMRIID	U.S. Army Medical Research Institute of Infectious Diseases
HLA	human leukocyte antigens			
ICSH	interstitial cell-stimulating hormone		USSR	Union of Soviet Socialist Republics
IPCC	Intergovernmental Panel on Climate Change		VLDL	very-low-density lipoproteins
LDL	low-density lipoprotein		WBC	white blood cell

Glossary

A

abiotic factors (a-bi-ot′ik fak′tōrz) Nonliving parts of an organism's environment. 260

absorption (ab-sorp′shun) The movement of simple molecules from the digestive system to the circulatory system for dispersal throughout the body. 364

accessory pigments (ak-ses′o-re pig′ment) Photosynthetic pigments other than the chlorophylls that enable an organism to utilize more colors of the visible light spectrum for photosynthesis; e.g., *carotinoids* (yellow, red, and orange); *phycoerythrins* (red); and *phycocyanin* (blue). 122

accessory structures (ak-ses′o-re struk′churz) The parts of some flowers that are not directly involved in gamete production. 504

acetyl (ă-sēt′l) The 2-carbon remainder of the carbon skeleton of pyruvic acid that is able to enter the mitochondrion. 112

acetylcholine (ă-sēt″l-kō′lēn) A neurotransmitter secreted into the synapse by many axons and received by dendrites. 391

acetylcholinesterase (ă-sēt″l-kō′lĭ-nes′tĕ-rās) An enzyme present in the synapse that destroys acetylcholine. 391

acid (ăs′id) Any compound that releases a hydrogen ion in a solution. 36

acid-base reactions (ăs′id bās re-ak′ shun) When the ions of one compound (acid) interact with the ions of another compound (base), forming a salt and water. 36

acoelomates (a-se′lă-mats) Animals without a coelom. The internal organs have no spaces between them. 518

acquired characteristics (ă-kwīrd′ kar″ak-ter-iss′tiks) A characteristic of an organism gained during its lifetime, not determined genetically, and therefore not transmitted to the offspring. 227

actin (ak′tin) A protein found in the thin filaments of muscle fibers that binds to myosin. 403

activation energy (ak″tĭ-va′shun en′ur-je) Energy required to start a reaction. 92

active site (ak″tiv sīt) The place on the enzyme that causes the substrate to change. 93

active transport (ak′tive trans′port) Use of a carrier molecule to move molecules across a cell membrane in a direction opposite that of the concentration gradient. The carrier requires an input of energy other than the kinetic energy of the molecules. 73

adaptive radiation (uh-dap′tiv ra-de-a′shun) A specific evolutionary pattern in which there is a rapid increase in the number of kinds of closely related species. 249

adenine (ad′ĕ-nēn) A double-ring nitrogenous-base molecule in DNA and RNA. It is the complementary base of thymine or uracil. 134

adenosine triphosphate (ATP) (uh-den′o-sēn tri-fos′fāt) A molecule formed from the building blocks of adenine, ribose, and phosphates. It functions as the primary energy carrier in the cell. 106

aerobic cellular respiration (aer-o′-bik sel′yu-lar res″pi-ra′shun) The biochemical pathway that requires oxygen and converts food, such as carbohydrates, to carbon dioxide and water. During this conversion, it releases the chemical-bond energy as ATP molecules. 79

aerobic exercise (aer-o′-bik ek-ser-sīz) A level of exercise where the level of exertion allows the heart and lungs to keep up with the oxygen needs of the muscles. 383

age distribution (āj dis″trĭ-biu′shun) The number of organisms of each age in the population. 304

alcoholic fermentation (al-ko-hol′ik fur″men-ta′shun) The anaerobic respiration pathway in yeast cells. During this process, pyruvic acid from glycolysis is converted to ethanol and carbon dioxide. 117

algae (al′je) Protists that have cell walls and chlorophyll and can therefore carry on photosynthesis. 481

allele frequency (a-lēl′ fre′kwen-sē) A term used to describe how common a specific allele is compared to other alleles for the same characteristic. See **gene frequency.** 210

alleles (a-lēlz′) Alternative forms of a gene for a particular characteristic (e.g., attached earlobe and free earlobe are alternative alleles for ear shape). 190

alternation of generations (awl″tur-na′shun uv gen″uh-ra′shunz) A term used to describe that aspect of the life cycle in which there are two distinctly different forms of an organism. Each form is involved in the production of the other and only one form is involved in producing gametes. The cycling of a diploid sporophyte generation and a haploid gametophyte generation in plants. 466

alveoli (al-ve′o-li″) Tiny sacs found in the lungs where gas exchange takes place. 349

amino acid (ah-mēn′o ă′sid) A basic subunit of protein consisting of a short carbon skeleton that contains an amino group, a carboxylic acid group, and one of various side groups. 49

anabolism (ah-nab′o-lizm″) Metabolic pathways that result in the synthesis of new, larger compounds; e.g., protein synthesis. 106

anaerobic cellular respiration (an′uh-ro″bik sel′yu-lar res″pi-ra′shun) A biochemical pathway that does not require oxygen for the production of ATP and does not use O_2 as its ultimate hydrogen ion acceptor. 104

anaerobic exercise (an′uh-ro″bik ek-ser-sīz) Involves bouts of exercise that are so intense that the muscles cannot get oxygen as fast as they need it. 383

anaphase (an'ă-fāz) The third stage of mitosis, characterized by dividing of the centromeres and movement of the chromosomes to the poles.   160

anatomy (a'-na-ta-mē) The study of the structure of living things.   24

androgens (an'dro-jenz) Male sex hormones produced by the testes that cause the differentiation of typical internal and external genital male anatomy.   416

anemia (uh-nēm'e-ah) A disease condition in which the oxygen-carrying capacity of the blood is reduced.   342

angiosperms (an'je-o"spurmz) Plants that produce flowers, seeds, and fruits.   504

annual (an'yu-uhl) A plant that completes its life cycle in one year.   503

anorexia nervosa (an"o-rek'se-ah ner-vo'sah) A nutritional deficiency disease characterized by severe, prolonged weight loss for fear of becoming obese. This eating disorder is thought to stem from sociocultural factors.   378

anther (an-ther) The sex organ in plants that produces the pollen that contains the sperm.   170

antheridia (singular, *antheridium*) (an"thur-id'e-ah) The structures in lower plants that bear sperm.   496

anthropomorphism (an-thro-po-mōr'fizm) The ascribing of human feelings, emotions, or meanings to the behavior of animals.   321

antibiotics (an-tc-bi-ot'iks) Drugs that selectively kill or inhibit the growth of a particular cell type.   86

anticodon (an"te-ko'don) A sequence of three nitrogenous bases on a tRNA molecule capable of forming hydrogen bonds with three complementary bases on an mRNA codon during translation.   139

antidiuretic hormone (ADH) (an-tĭ-di"u-rĕtik hōr'mōn) A hormone produced by the pituitary gland that stimulates the kidney to reabsorb water.   394

aorta (a-or'tah) The large blood vessel that carries blood from the left ventricle to the majority of the body.   344

apoptosis (ap"op-to'sis) Also known as "programmed cell death"; death that has a genetic basis and not the result of injury.   167

archegonium (ar"ke-go'ne-um) The structure in lower plants that bears eggs.   496

arteries (ar'-tĕ-rēz) The blood vessels that carry blood away from the heart.   342

arterioles (ar-ter'e-ōlz) Small arteries located just before capillaries that can expand and contract to regulate the flow of blood to parts of the body.   345

assimilation (ă-sĭ"mĭ-la'shun) The physiological process that takes place in a living cell as it converts nutrients in food into specific molecules required by the organism.   364

association (ă-sō"sē-ă'-shun) An animal learns that a particular outcome is associated with a particular stimulus.   324

asymmetry (a-sĭm'ĭ-tre) The characteristic of animals with no particular body shape.   516

atom (ă'tom) The fundamental unit of matter. The smallest part of an element that still acts like that element.   25

atomic mass unit (AMμ) (ă-tom'ik mas yu'nit) A unit of measure used to describe the mass of atoms and is equal to 1.67×10^{-24} grams, approximately the mass of one proton.   28

atomic nucleus (ă-tom'ik nu'kle-us) The central region of the atom.   28

atomic number (ă-tom'ik num'bur) The number of protons in an atom.   28

atomic weight (mass number) (ă-tom'ik wāt) The weight of an atomic nucleus expressed in atomic mass units (the sum of the protons and neutrons).   28

atria (singular, *atrium*) (a'trē-uh) Thin-walled sacs of the heart that receive blood from the veins of the body and empty into the ventricles.   344

atrioventricular valves (a"trē-o-ven-trĭk'u-lar valvz) Valves located between the atria and ventricles of the heart that prevent the blood from flowing backward from the ventricles into the atria.   344

attachment site (uh-tatch'munt sīt) A specific point on the surface of the enzyme where it can physically attach itself to the substrate; also called **binding site.**   93

autosomes (aw'to-sōmz) Chromosomes that typically carry genetic information used by the organism for characteristics other than the primary determination of sex.   186

autotrophs (aw'to-trōfs) Organisms that are able to make their food molecules from inorganic raw materials by using basic energy sources such as sunlight.   104

auxin (awk'sin) A plant hormone that stimulates cell elongation, cell division, and growth.   508

axon (ak'sahn) A neuronal fiber that carries information away from the nerve cell body.   389

B

bacteria (bak-tĭr'ē-ah) Prokaryotic, unicellular organisms of the Domains Archaea and Eubacteria.   476

Barr bodies (bar bod'ēz) Tightly coiled X chromosomes in the cells of female mammals, described by Dr. M. L. Barr.   413

basal metabolic rate (BMR) (ba'sal mĕ-ta-bah'lik rāt) The amount of energy required to maintain normal body activity while at rest.   364

base (bās) Any compound that releases hydroxide ions or accepts hydrogen ions in a solution.   36

basilar membrane (ba'sĭ-lar mem'brān) A membrane in the cochlea containing sensory cells that are stimulated by the vibrations caused by sound waves.   401

behavior (be-hāv'yur) How an organism acts, what it does, and how it does it.   320

behavioral isolating mechanism (be-hāv'yu-ral eye'-so-lā'-ting mek'-ă-nizm) A genetic isolating mechanism that prevents interbreeding between species because of differences in behavior.   244

benign tumor (be-nīn' too'mor) A cell mass that does not fragment and spread beyond its original area of growth.   165

benthic (ben'thik) A term used to describe organisms that live in bodies of water, attached to the bottom or to objects in the water.   482

bilateral symmetry (bi-lat'er-al sĭm'ĭ-tre) The characteristic of animals that are constructed along a plane running from a head to a tail region, so that only a cut along one plane of this axis results in two mirror halves.   516

bile (bīl) A product of the liver, stored in the gallbladder, which is responsible for the emulsification of fats.   354

binding site (bin'ding sīt) See **attachment site.**   93

binomial system of nomenclature (bi-no' mi-al sis'tem uv no'men-kla-ture) A naming system that uses two Latin names, genus and specific epithet, for each species of organism.   456

biochemical isolating mechanism (bi'o-kem" i-kal eye'-so-lā'-ting mek'-ă-nizm) Differences in biochemical activities that prevent mating between individuals of different species.   245

biochemical pathway (bi'o-kem"ĭ-kal path'wā) A major series of enzyme-controlled reactions linked together.   104

biochemistry (bi-o-kem'iss-tre) The chemistry of living things, often called biological chemistry.   42

bioengineering (bi'o-en"jĭn-ee-ring) The process of purposely altering the genetic makeup of organisms.   150

biogenesis (bi-o-jen'uh-sis) The concept that life originates only from preexisting life.   436

biological species concept (bi'o-loj'ĭ-cal spe'shez kon'sept) The concept that species are distinguished from one another by their inability to interbreed.   206

biology (bi-ol′o-je) The science that deals with the study of living things and how living entities interact with things around them.   2

biomagnification (bi″o-mag′nĭf-fĭ-ka′shun) The accumulation of a compound in increasing concentrations in organisms at successively higher trophic levels.   297

biomass (bi′o-mas) The dry weight of a collection of designated organisms.   265

biomes (bi′ōmz) Large regional communities primarily determined by climate.   270

biotechnology (bi-o-tek-nol′uh-je) The science of gene manipulation.   150

biotic factors (bi-ah′tik fak′tōrz) Living parts of an organism's environment.   260

biotic potential (bi-ah′tik po-ten′shul) See **reproductive capacity.**   306

blood (blud) The fluid medium consisting of cells and plasma that assists in the transport of materials and heat.   342

bloom (bloom) A rapid increase in the number of microorganisms in a body of water.   482

body mass index (BMI) (bah′-dee măs in′-deks) The weight of a person in kilograms divided by the person's height in meters squared.   365

Bowman's capsule (bo′manz kap′sl) A saclike structure at the end of a nephron that surrounds the glomerulus.   357

breathing (bre′thing) The process of pumping air in and out of the lungs.   349

bronchi (brŏng′ki) Major branches of the trachea that ultimately deliver air to bronchioles in the lungs.   349

bronchioles (brŏng′ke-ōlz) Small tubes that deliver air to the alveoli in the lung.   349

budding (bud′ing) A type of asexual reproduction in which the new organism is an outgrowth of the parent.   520

bulimia (bu-lēm′e-ah) A nutritional deficiency disease characterized by a binge-and-purge cycle of eating. It is thought to stem from psychological disorders.   378

C

calorie (kal′or-e) The amount of heat energy necessary to raise the temperature of 1 gram of water 1°C.   27

Calvin cycle (kal′vin sĭ′kl) A cyclic sequence of reactions that make up the light-independent reaction stage of photosynthesis.   126

cambium (kam′be-um) A tissue in higher plants that produces new xylem and phloem.   503

cancer (kan′sur) A tumor that is malignant.   165

capillaries (cap′ĭ-lair-ez) Tiny blood vessels through which exchange between surrounding tissue and the blood takes place.   342

carbohydrate (kar-bo-hi′drāt) One class of organic molecules composed of carbon, hydrogen, and oxygen in a ratio of 1:2:1. The basic building block of carbohydrates is a simple sugar (monosaccharide).   47

carbohydrate loading (kar-bo-hi′drāt lo′ding) A week-long program of diet and exercise that results in an increase in muscle glycogen stores.   384

carbonic anhydrase (car-bon′ik an-hī′drās) An enzyme present in red blood cells that assists in converting carbon dioxide to bicarbonate ions.   342

carbon skeleton (kar′bon skel′uh-ton) The central portion of an organic molecule composed of rings or chains of carbon atoms.   44

carcinogens (kar-sin′o-jen) Agents responsible for causing cancer.   165

carnivores (kar′nĭ-vōrz) Animals that eat other animals.   262

carrier (ka′re-er) Any individual having a hidden, recessive allele.   191

carrier proteins (ka′re-er prō′tēns) A category of proteins that pick up molecules at one place and transport them to another.   49

carrying capacity (ka′re-ing kuh-pas′i-te) The maximum sustainable population for an area.   308

catabolism (kah-tab′o-lizm) Metabolic pathways that result in the breakdown of compounds; e.g., glycolysis.   106

catalyst (cat′uh-list) A chemical that speeds up a reaction but is not used up in the reaction.   92

cell (sel) The basic structural unit that makes up all living things. The smallest unit that displays the characteristics of life.   13

cell membrane (sel mem′brān) The outer boundary membrane of the cell; also known as the **plasma membrane.**   76

cell plate (sel plāt) A plant-cell structure that begins to form in the center of the cell and proceeds to the cell membrane, resulting in cytokinesis.   163

cellular membranes (sel′yu-lar mem′brāns) Thin sheets of material composed of phospholipids and proteins. Some of the proteins have attached carbohydrates or fats.   66

cellular respiration (sel′yu-lar res″ pĭ-ra′shun) A major biochemical pathway by which cells release the chemical-

bond energy from food and convert it into a usable form (ATP).   104

cell wall (sel wawl) An outer covering on some cells; may be composed of cellulose, chitin, or peptidoglycan depending on the kind of organism.   64

central dogma (sen′-tral dog′-mah) A phrase that refers to the process by which a cell uses and passes on genetic information. It includes the processes of DNA replication, transcription, and translation.   134

central nervous system (sen′trul ner′vus sis′tem) The portion of the nervous system consisting of the brain and spinal cord.   389

centriole (sen′tre-ōl) Two sets of nine short microtubules, each set of tubules is arranged in a cylinder.   81

centromere (sen′tro-mēr) The unreplicated region of a chromosome where two chromatids are joined.   159

cerebellum (ser″ĕ-bel′um) A region of the brain connected to the medulla oblongata that receives many kinds of sensory stimuli and coordinates muscle movement.   392

cerebrum (ser′ĕ-brum) A region of the brain that surrounds most of the other parts of the brain and is involved in consciousness and thought.   393

chemical bonds (kem′ĭ-kal bonds) Forces that combine atoms or ions and hold them together.   29

chemical equation (kem′i-kal e-kwa′-zhun) A way of describing what happens in a chemical reaction.   35

chemical reaction (kem′ĭ-kal re-ak′shun) The formation or rearrangement of chemical bonds, usually indicated in an equation by an arrow from the reactants to the products.   33

chlorophyll (klo′ro-fil) The green pigment located in the chloroplasts of plant cells associated with trapping light energy.   79

chloroplast (klo′ro-plast) An energy-converting, membranous, saclike organelle in plant cells containing the green pigment chlorophyll.   79

chromatid (kro′mah-tid) One of two component parts of a chromosome formed by replication and attached at the centromere.   159

chromatin (kro′mah-tin) Areas or structures within the nucleus of a cell composed of long molecules of deoxyribonucleic acid (DNA) in association with proteins.   83

chromatin fibers (kro′mah-tin fi′bers) See **nucleoproteins.**   148

chromosomal aberrations (kro-mo-sōm′al ab-e-rā-shens) Changes in the structure of chromosomes, which can affect the expression of genes; e.g., translocation, duplication mutations.   150

chromosomes (kro′mo-sōmz) Double-stranded DNA molecules with attached protein (nucleoprotein) coiled into a short, compact unit.   53

cilia (sil′e-ah) Numerous short, hairlike structures projecting from the cell surface that enable locomotion.   82

class (class) A group of closely related families found within a phylum.   460

classical conditioning (klas′ĭ-kul kon-dĭ′shuning) Learning that occurs when an involuntary, natural, reflexive response to a natural stimulus is transferred from the natural stimulus to a new stimulus.   324

cleavage furrow (kle′vaj fuh′ro) An indentation of the cell membrane of an animal cell that pinches the cytoplasm into two parts during cell division.   162

climax community (klī′maks ko-miu′nĭ-te) A relatively stable, long-lasting community.   275

clone (klōn) A population made up of genetically identical individuals, that were reproduced asexually.   183

coacervate (ko-as′ur-vāt) A spherical body, which is a collection of organic macromolecules surrounded by water molecules.   442

cochlea (kŏk′lē-ah) The part of the ear that converts sound into nerve impulses.   401

coding strand (kō-ding strand) One of the two DNA strands that serves as a template, or pattern, for the synthesis of RNA.   138

codominance (ko-dom′ĭ-nans) A situation in which both alleles in a heterozygous organism express themselves.   192

codon (ko′don) A sequence of three nucleotides of an mRNA molecule that directs the placement of a particular amino acid during translation.   136

coelom (se′lōm) A body cavity in which internal organs are suspended.   518

coenzyme (ko-en′zīm) A molecule that works with an enzyme to enable the enzyme to function as a catalyst.   94

coitus (ko′ĕ-tus) See **sexual intercourse.**   420

colonial microbes (ko-lo′ne-al mī′krōbs) A term used to describe a collection of cells that cooperate to a small extent.   476

commensalism (ko-men′sal-izm) A relationship between two organisms in which one organism is helped and the other is not affected.   288

community (ko-miu′nĭ-te) Populations of different kinds of organisms that interact with one another in a particular place.   261

competition (com-pe-tĭ′shun) A relationship between two organisms in which both organisms are harmed.   289

competitive exclusion principle (com-pĕ′-tĭ-tiv eks-klu′zhun prin′sĭ-pul) No two species can occupy the same niche at the same time.   289

competitive inhibition (com-pĕ′-tĭ-tiv in″hĭ-bĭ′shun) The formation of a temporary enzyme-inhibitor complex that interferes with the normal formation of enzyme-substrate complexes, resulting in a decreased turnover.   100

complementary base (kom″plĕ-men′tah-re bās) A base that can form hydrogen bonds with another base of a specific nucleotide.   135

complete protein (kom-plēt′ prō′tēn) Protein molecules that provide all the essential amino acids.   369

complex carbohydrates (kom′pleks kar-bo-hi′drāts) Macromolecules composed of simple sugars combined by dehydration synthesis to form a polymer.   47

compound (kom′pound) A kind of matter that consists of a specific number of atoms (or ions) joined to each other in a particular way and held together by chemical bonds.   25

concentration gradient (kon″sen-tra′shun gra′de-ent) The gradual change in the number of molecules per unit of volume over distance.   70

conception (kon-sep′shun) Fertilization.   413

conditioned response (kon-dĭ′shund re-spons′) The modified behavior displayed in which a new response is associated with a natural stimulus.   324

cone (kōn) A reproductive structure of gymnosperms that produces pollen in males or eggs in females.   501

cones (kōnz) Light-sensitive cells in the retina of the eye that respond to different colors of light.   400

consumers (kon-soom′urs) Organisms that must obtain energy in the form of organic matter.   261

control group (con-trōl′ grŭp) The situation used as the basis for comparison in a controlled experiment. The group in which there are no manipulated variables.   6

controlled experiment (con-trold′ ek-sper′ĭ-ment) An experiment that includes two groups; one in which the variable is manipulated in a particular way and one in which there is no manipulation.   6

control processes (con-trōl′ pro′ses-es) Mechanisms that ensure an organism will carry out all life activities in the proper sequence (coordination) and at the proper rate (regulation).   12

convergent evolution (kon-vur′jent ĕv-o-lu′shun) An evolutionary pattern in which widely different organisms show similar characteristics.   250

copulation (kop-yu-la′shun) See **sexual intercourse.**   420

corpus luteum (kŏr′pus lu′te-um) Remainder of the follicle after the release of the secondary oocyte. It develops into a gland-like structure that produces hormones (progesterone and estrogen) that prevent the release of other eggs.   416

covalent bond (ko-va′lent bond) The attractive force formed between two atoms that share a pair of electrons.   31

cristae (krĭs′te) Folded surfaces of the inner membranes of mitochondria.   78

critical period (krit′ĭ-kl pir-ē-ed) Period of time during the life of an animal when imprinting can take place.   326

crossing-over (kro′sing o′ver) The exchange of a part of a chromatid from one chromosome with an equivalent part of a chromatid from a homologous chromosome.   173

cryptorchidism (krip-tōr′kĭ-dizm) A developmental condition in which the testes do not migrate from the abdomen through the inguinal canal to the scrotum.   415

cytokinesis (si-to-kĭ-ne′sis) Division of the cytoplasm of one cell into two new cells.   158

cytoplasm (si′to-plazm) The portion of the protoplasm that surrounds the nucleus.   64

cytosine (si′to-sēn) A single-ring nitrogenous-base molecule in DNA and RNA. It is complementary to guanine.   134

cytoskeleton (si″to-skel′ĕ-ton) The internal framework of eukaryotic cells composed of intermediate filaments, microtubules, and microfilaments; provides the cell with a flexible shape, the ability to move through the environment, move molecules internally, and respond to environmental changes.   81

D

daughter cells (daw′tur sels) Two cells formed by cell division.   161

daughter chromosomes (daw′tur kro′mo-somz) Chromosomes produced by DNA replication that contain identical genetic information; separated during chromosome division in anaphase.   160

daughter nuclei (daw′tur nu′kle-i) Two nuclei formed by mitosis.   161

death phase (deth fāz) The portion of some population growth curves in which the size of the population declines.   310

deciduous (de-sid′yu-us) A term used to describe trees that lose their leaves at the end of the growing season.   504

decomposers (de-kom-po′zurs) Organisms that use dead organic matter as a source of energy.   262

deductive reasoning (deduction) (de′-duk-tiv re′son-ing) (de′duk-shun) The mental process of using accepted generalizations to predict the outcome of specific events. From the general to the specific.   7

dehydration synthesis reaction (de-hi-dra′shun sin′thuh-sis re-ak′shun) A reaction that results in the formation of a macromolecule when water is removed from between the two smaller component parts.   35

denature (de-nā′chur) To alter the structure of a protein so that some of its original properties are diminished or eliminated.   52

dendrites (den′drīts) Neuronal fibers that receive information from axons and carry it toward the nerve-cell body.   389

denitrifying bacteria (de-ni′trĭ-fi-ing bak-te′re-ah) Several kinds of bacteria capable of converting nitrite to nitrogen gas.   293

density-dependent factors (den′sĭ-te de-pen′dent fak′torz) Population-limiting factors that become more effective as the size of the population increases.   311

density-independent factors (den′sĭ-te in-de-pen′dent fak′torz) Population-controlling factors that are not related to the size of the population.   311

deoxyribonucleic acid (DNA) (de-ok″se-ri-bo-nu-kle′ik ă′sid) A polymer of nucleotides that serves as genetic information. In prokaryotic cells, it is a double-stranded loop and contains attached HU proteins. In eukaryotic cells, it is found in strands with attached histone proteins. When tightly coiled, the DNA and histone structure is known as a chromosome.   53

deoxyribose (de-ok″se-ri′bōs) A 5-carbon sugar molecule; a component of DNA.   134

dependent variable (de-pen′dent var′ē-a-bul) Variable that changes in direct response to (depends on) how another variable (independent variable) is manipulated.   6

depolarized (de-po′lă-rīzd) Having lost the electrical difference existing between two points or objects.   390

diaphragm (di′uh-fram) A muscle separating the lung cavity from the abdominal cavity that is involved in exchanging the air in the lungs.   349

diastolic blood pressure (di″uh-stol′ik blud presh′yur) The pressure present in a large artery when the heart is not contracting.   345

dicot (di′kot) An angiosperm whose embryo has two seed leaves.   508

diet (di′et) The food and drink consumed by a person from day to day.   364

differentiation (dĭf-ĕ-ren″she-a′shun) The process of forming specialized cells within a multicellular organism.   163

diffusion (dĭ-fiu′zhun) Net movement of a kind of molecule from an area of higher concentration to an area of lesser concentration.   69

diffusion gradient (dĭ-fiu′zhun gra′de-ent) The difference in the concentration of diffusing molecules over distance.   70

digestion (di-jest′yun) The breakdown of complex food molecules to simpler molecules; the chemical reaction of hydrolysis.   364

diploblastic (dip-lo-blas′tik) A condition in which some simple animals only consist of two layers of cells.   516

diploid (dip′loid) Having two sets of chromosomes: one set from the maternal parent and one set from the paternal parent.   170

directional selection (dye-rek′-shun-l sĕ-lek′shun) A kind of selection that occurs when individuals at one extreme of the range of a characteristic are consistently selected for.   230

disruptive selection (dis-rup′tive sĕ-lek′shun) A kind of selection that occurs when both extremes of a range for a charcteristic are selected for and the intermediate condition is selected against.   230

distal convoluted tubule (dis′tul kon′vo-lu-ted tūb′yūl) The downstream end of the nephron of the kidney, that is primarily responsible for regulating the amount of hydrogen and potassium ions in the blood.   357

divergent evolution (di-vur′jent ĕv-o-lu′shun) A basic evolutionary pattern in which individual speciation events cause many branches in the evolution of a group of organisms.   247

DNA polymerase (po-lim′er-ās) An enzyme that bonds DNA nucleotides together when they base-pair with an existing DNA strand.   152

DNA replication (rep″lĭ-ka′shun) The process by which the genetic material (DNA) of the cell reproduces itself prior to its distribution to the next generation of cells.   147

domain (dō′mān) The first (broadest) classification unit of organisms. There are three domains: Eubacteria, Archaea, and Eucarya.   66

dominance hierarchy (dom′in-ants hi′ur-ar-ke) A relatively stable, mutually understood order of priority within a group.   333

dominant allele (dom′in-ant a-lēl′) An allele that expresses itself and masks the effects of other alleles for the trait.   191

double bond (dub′l bond) A pair of covalent bonds formed between two atoms when they share two pairs of electrons.   43

double-factor cross (dub′l fak′tur kros) A genetic study in which two pairs of alleles

are followed from the parental generation to the offspring.   198

Down syndrome (down sin′drom) A genetic disorder resulting from the presence of an extra chromosome number 21. Symptoms include slightly slanted eyes, flattened facial features, a large tongue, and a tendency toward short stature and fingers. Individuals usually display mental retardation.   184

duodenum (dew″o-de′num) The first part of the small intestine, which receives food from the stomach and secretions from the liver and pancreas.   353

dynamic equilibrium (di-nam′ik e-kwĭ-lib′re-um) The condition in which molecules are equally dispersed, therefore movement is equal in all directions.   69

E

ecological isolating mechanism (e-kŏ-loj′ĭ-kal eye′-sō-lā-ting mek′-ă-nizm) A genetic isolating mechanism that prevents interbreeding between species because they live in different areas; also called **habitat preference.**   244

ecology (e-kol′o-je) The branch of biology that studies the relationships between organisms and their environment.   260

ecosystem (e′ko-sis″tum) A unit consisting of a community of organisms (populations) and its interactions with the physical surroundings.   261

ectothermic (ek″to-therm′ik) Animals that are unable to regulate their body temperature by automatic physiological processes, but can regulate their temperature by moving to places where they can be most comfortable.   516

egg (eg) The haploid sex cell produced by sexually mature females.   170

ejaculation (e-jak″u-lā′shun) The release of sperm cells and seminal fluid through the penis of a male.   417

electrolytes (ĕ-lek′tro-līts) Ionic compounds dissolved in water. Their proper balance is essential to life.   31

electrons (e-lek′trons) The negatively charged particles moving at a distance from the nucleus of an atom that balance the positive charges of the protons.   28

electron-transport system (ETS) (e-lek′tron trans′port sis′tem) The series of oxidation-reduction reactions in aerobic cellular respiration in which the energy is removed from hydrogens and transferred to ATP.   110

elements (el′ĕ-ments) Matter consisting of only one kind of atom.   25

empirical evidence (em-pir′ĭ-cal ev′i-dens) The information gained by observing an event.   3

endocrine glands (en′do-krĭn glandz) Glands that secrete into the circulatory system. 388

endocrine system (en′do-krĭn sis′tem) A number of glands that communicate with one another and other tissues through chemical messengers transported throughout the body by the circulatory system. 388

endocytosis (en″ dō-sī-tō′-sis) Also phagocytosis; the process cells use to wrap membrane around a particle (usually food) and engulf it. 73

endoplasmic reticulum (ER) (en″do-plaz′mik re-tĭk′yu-lum) Folded membranes and tubes throughout the eukaryotic cell that provide a large surface upon which chemical activities take place. 76

endoskeleton (en″do-skel′ĕ-ton) A skeleton typical of vertebrates in which the skeleton is surrounded by muscles and other organs. 518

endospore (en′do-spōr″) A unique bacterial structure with a low metabolic rate that germinates under favorable conditions to grow into a new cell. 479

endosymbiotic theory (en″do-sim-be-ot′ik the′o-re) A theory suggesting that some organelles found in eukaryotic cells may have originated as frcc-living cells. 446

endothermic (en″do-therm′ik) Animals that have internal temperature-regulating mechanisms and can maintain a relatively constant body temperature in spite of wide variations in the temperature of their environment. 516

energy (en′er-je) The ability to do work or cause things to move. 11, 24

energy level (en′er-je lev′el) Regions surrounding an atomic nucleus that contain electrons moving at approximately the same speed and having approximately the same amount of kinetic energy. 28

environment (en-vi′ron-ment) Anything that affects an organism during its lifetime. 260

environmental resistance (en-vi-ron-men′tal rezis′tans) The collective set of factors that limit population growth. 308

enzymatic competition (en-zi-mătik com-pĕ-tĭ′shun) Competition among several different available enzymes to combine with a given substrate material. 98

enzyme (en′zīm) A specific protein that acts as a catalyst to speed the rate of a reaction. 92

enzymes (en′zīms) Molecules, produced by organisms, that are able to control the rate at which chemical reactions occur. 12

enzyme-substrate complex (en′zīm sub′strāt kom′pleks) A temporary molecule formed when an enzyme attaches itself to a substrate molecule. 93

epinephrine (ĕ″pĭ-nef′rin) A hormone produced by the adrenal medulla and certain nerve cells that increases heart rate, blood pressure, and breathing rate. 393

epiphyte (ep′e-fīt) A plant that lives on the surface of another plant without doing harm. 288

essential amino acids (ĕ-sen′shul ah-me′no ă′sids) Those amino acids that cannot be synthesized by the human body and must be part of the diet (e.g., lysine, tryptophan, and valine). 369

essential fatty acid (ĕ-sen′shul fă′te ă′sid) The fatty acids linoleic and linolenic, which cannot be synthesized by the human body and must be part of the diet. 369

estrogen (es′tro-jen) One of the female sex hormones responsible for the growth and development of female sexual anatomy. 395

estrogens (es′tro-jens) Female sex hormones that cause the differentiation of typical female internal and external genital anatomy; responsible for the changes in breasts, vagina, uterus, clitoris, and pelvic bone structure at puberty. 416

ethology (e thol′uh-je) The scientific study of the nature of behavior and its ecological and evolutionary significance in its natural setting. 321

eugenics laws (yu-jen′iks laws) Laws designed to eliminate "bad" genes from the human gene pool and encourage "good" gene combinations. 217

eukaryote (yu-kār′e-ōt) An organism composed of cells possessing a nucleus and other membranous organelles. 446

eukaryotic cells (yu′ka-rc-ah″tik sels) One of the two major types of cells; characterized by cells that have a true nucleus, as in plants, fungi, protists, and animals. 66

evolution (ĕv-o-lu′shun) A change in the frequency of genetically determined characteristics within a population over time. 222

exocrine glands (ek′so-krĭn glandz) Glands that secrete through ducts to the surface of the body or into hollow organs of the body. 388

exocytosis (ek′ sō-sī-to′-sis) The process cells use to wrap membrane around a particle (usually cell products or wastes) and eliminate it from a cell. 73

exoskeleton (ek″sō-skel′ĕ-ton) A skeleton typical of many invertebrates in which the skeleton is on the outside of the animal. 518

experiment (ek-sper′ĭ-ment) A re-creation of an event that enables a scientist to gain valid and reliable empirical evidence. 6

experimental group (ek-sper′ĭ-men′tal grūp) The group in a controlled experiment that has a variable manipulated. 6

exploratory learning (ek-splor′a-to-re lur′ning) Learning about the surroundings that occur as a result of wandering through the environment. 325

exponential growth phase (ek-spo-nen′shul grōth fāz) A period of time during population growth when the population increases at an accelerating rate. 308

expressivity (ek-spre′si″vi-te) A term used to describe situations in which the gene expresses itself but not equally in all individuals that have it. 225

external parasite (ek-stur′nal pĕr′uh-sīt) A parasite that lives on the outside of its host. 286

extinction (eks-tingt′-shun) The loss of a species. 248

extrinsic factors (eks-trin′sik fak′tōrz) Population-controlling factors that arise outside the population. 311

F

facilitated diffusion (fah-sil′ĭ-ta″ted dĭ fiu′zhun) Diffusion assisted by carrier-molecules. 72

FAD (flavin adenine dinucleotide) (F-A-D) A hydrogen carrier used in respiration. 108

family (fam′ĭ-le) A group of closely related genera within an order. 460

fat (fat) A class of water-insoluble macromolecules composed of a glycerol and fatty acids. 56

fatty acid (fat′ē ă′sid) One of the building blocks of a fat, composed of a long-chain carbon skeleton with a carboxylic acid functional group. 56

fermentation (fer-men-ta′shun) Pathways that oxidize glucose to generate ATP energy using something other than O_2 as the ultimate hydrogen acceptor. 117

fertilization (fer″tĭ-lĭ-za′shun) The joining of haploid nuclei, usually from an egg and a sperm cell, resulting in a diploid cell called the zygote. 170

fiber (fi′ber) Natural (plant) or industrially produced polysaccharides that are resistant to hydrolysis by human digestive enzymes. 368

filter feeders (fil′ter fēd′erz) Animals that use cilia or other appendages to create water currents and filter food out of the water. 527

fitness (fit′nes) The concept that those who are best adapted to their environment produce the most offspring. 222

flagella (flah-jel′luh) Long, hairlike structures projecting from the cell surface that enable locomotion.   82

flower (flow′er) A complex plant reproductive structure made from modified stems and leaves that produces pollen and eggs.   504

fluid-mosaic model (flu′id mo-za′ik mod′l) The concept that the cell membrane is composed primarily of protein and phospholipid molecules that are able to shift and flow past one another.   66

follicle (fol′i-kul) The saclike structure near the surface of the ovary that encases the soon-to-be-released secondary oocyte.   416

follicle-stimulating hormone (FSH) (fol′i-kul stim′yu-lā-ting hōr′mōn) The pituitary secretion that causes the ovaries to begin to produce larger quantities of estrogen and to develop the follicle and prepare the egg for ovulation.   415

food chain (food chān) A sequence of organisms that feed on one another, resulting in a flow of energy from a producer through a series of consumers.   262

Food Guide Pyramid (food gīd pī′ra-mĭd) A tool developed by the U.S. Department of Agriculture to help the general public plan for good nutrition. It contains guidelines for required daily intake from each of the five food groups.   373

food web (food web) A system of interlocking food chains.   267

fossils (fŏ′-silz) Physical evidences of former life.   241

founder effect (faùn′der e′fekt) The concept that small, newly established populations are likely to have reduced genetic diversity because of the small number of individuals in the founding population.   211

fovea centralis (fo′ve-ah sen-tral′is) The area of sharpest vision on the retina, containing only cones, where light is sharply focused.   400

free-living nitrogen-fixing bacteria (ni′tro-jen fik′sing bak-te′re-ah) Soil bacteria that convert nitrogen gas molecules into nitrogen compounds plants can use.   292

fruit (froot) The structure (mature ovary) in angiosperms that contains seeds.   504

functional groups (fung′shun-al grūps) Specific combinations of atoms attached to the carbon skeleton that determine specific chemical properties.   94

fungus (fung′us) The common name for members of the kingdom Fungi.   466

G

gallbladder (gol′blad″er) An organ attached to the liver that stores bile.   354

gamete (gam′ēt) A haploid sex cell.   170

gametogenesis (gă-me″to-jen′ĕ-sis) The generating of gametes; the meiotic cell-division process that produces sex cells; oogenesis and spermatogenesis.   417

gametophyte (gă-me′to-fīt) A haploid plant that produces gametes; it alternates with the sporophyte through the life cycle.   495

gametophyte stage (gă-me′to-fīt stāj) A life cycle stage in plants in which a haploid sex cell is produced by mitosis.   406

gas (gas) The state of matter in which the molecules are more energetic than the molecules of a liquid, resulting in only slight attraction for each other.   27

gastric juice (gas′trik jūs) The secretions of the stomach that contain enzymes and hydrochloric acid.   353

gene (jēn) Any molecule, usually a segment of DNA, that is able to (1) replicate by directing the manufacture of copies of itself; (2) mutate, or chemically change, and transmit these changes to future generations; (3) store information that determines the characteristics of cells and organisms; and (4) use this information to direct the synthesis of structural and regulatory proteins.   134

gene expression (jēn ek-spresh′un) The cellular process of transcribing and translating genetic information.   145

gene flow (jēn flo) The movement of genes within a population because of migration or the movement of genes from one generation to the next by gene replication and reproduction.   240

gene frequency (jēn fre′kwen-see) A measure of how often a specific gene shows up in the gametes of a population.   53

gene pool (jēn pool) All the genes of all the individuals of the same species.   208

generative processes (jen′uh-ra″tiv pros′es-es) Actions that increase the size of an individual organism (growth) or increase the number of individuals in a population (reproduction).   12

gene-regulator proteins (jēn reg′yu-la-tor prō′tēns) Chemical messengers within a cell that inform the genes as to whether protein-producing genes should be turned on or off or whether they should have their protein-producing activities in-

creased or decreased; for example, gene-repressor proteins and gene-activator proteins.   98

genetic bottleneck (jĕ-ne′tik ba′tul-nek) The concept that when populations are severely reduced in size they may lose some of their genetic diversity.   212

genetic counselor (jĕ-ne′tik kown′sel-or) A professional with specific training in human genetics who can advise on the likelihood of genetic defects being passed to children.   216

genetic diversity (jĕ-ne′tik dĭ-ver′sĭ-te) The degree to which individuals in a population possess alternative alleles for characteristics.   210

genetic drift (jĕ-ne′tik drĭft) A change in gene frequency that is not the result of natural selection. This typically occurs in a small population.   231

genetic finger printing (jĕ-ne′-tik fing′-er-prin′ting) A technique that uniquely identifies individuals on the basis of how their DNA migrates in an electrophoresis gel.   152

genetic heterogeneity (jĕ-net′ik het″er-o-jĕ-ne′ĭ-te) The fact that there are many different alleles for characteristics and that individuals will differ from one another.   200

genetic isolating mechanism (jĕ-net′ik i-so-la′ting mek′an-izm) See **reproductive isolating mechanism.**   244

genetic recombination (jĕ-net′ik re-kom-bĭ-na′shun) The gene mixing that occurs during sexual reproduction.   224

genetically modified (GM) organisms (jĕ-ne′tik-le ma′di-fĭd or′gun-izm) Organisms or their offspring that have been engineered to contain genes from at least one other species; also called **transgenic organisms.**   150

genetics (jĕ-net′iks) The study of genes, how genes produce characteristics, and how the characteristics are inherited.   190

genome (jĕ′nōm) A set of all the genes necessary to specify an organism's complete list of characteristics.   144

genomics (je-nō′-miks) A new field of science that has developed since the sequencing of the human genome. The field looks at how genomes are organized and compares them to genomes of other organims   154

genotype (je′no-tīp) The catalog of genes of an organism, whether or not these genes are expressed.   190

genus (je′nus) (plural, *genera*) A group of closely related species within a family.   456

geographic barriers (je-o-graf′ik băr′yurz) Geographic features that keep different portions of a species from exchanging genes.   241

geographic isolation (je-o-graf′ik i-so-la′shun) A condition in which part of the gene pool is separated by geographic barriers from the rest of the population.   241

germinate (jur′min-āt) To begin to grow (as from a seed).   495

gland (gland) An organ that manufactures and secretes a material either through ducts or directly into the circulatory system.   388

glomerulus (glo-mer′u-lus) A cluster of blood vessels, surrounded by Bowman's capsule in the kidney.   357

glyceraldehyde-3-phosphate (glĭ-ser-al′-dĕ-hīd 3 fos′-fāt) A 3-carbon compound produced during glycolysis and photosynthesis, that can be converted to other organic molecules.   123

glycerol (glis′er-ol) One of the building blocks of a fat, composed of a carbon skeleton that has three alcohol groups (OH) attached to it.   56

glycolysis (gli-kol′ĭ-sis) The anaerobic first stage of cellular respiration, consisting of the enzymatic breakdown of a sugar into two molecules of pyruvic acid.   110

Golgi apparatus (gōl′je ap″pah-rat′us) A stack of flattened, smooth, membranous sacs; the site of synthesis and packaging of certain molecules in eukaryotic cells.   77

gonad (go′nad) A generalized term for organs in which meiosis occurs to produce gametes; ovary or testis.   170

gonadotropin-releasing hormone (GnRH) (go″nad-o-tro′-pin re-le′sing hōr′mōn) A hormone released from the hypothalamus that stimulates the release of follicle-stimulating hormone (FSH) and luteinizing hormone.   415

gradualism (grad′u-al-izm) The theory stating that evolution occurred gradually with an accumulated series of changes over a long period of time.   252

grana (gra′nuh) Areas of the chloroplast membrane where chlorophyll molecules are concentrated.   79

granules (gran′yūls) Materials whose structure is not as well defined as that of other organelles.   82

growth-stimulating hormone (GSH) (grōth stĭ′mu-la-ting hōr′ mōn) A hormone produced by the anterior pituitary gland that stimulates tissues to grow.   394

guanine (gwah′nēn) A double-ring nitrogenous-base molecule in DNA and RNA. It is the complementary base of cytosine.   134

gymnosperms (jim′no-spurmz) Plants that produce their seeds in cones.   502

H

habitat (hab′-ĭ-tat) The place or part of an ecosystem occupied by an organism.   284

habitat preference (hab′i-tat pref′ur-ents) See ecological isolating mechanism.   244

habituation (hah-bit′u-a′shun) A change in behavior in which an animal ignores a stimulus after repeated exposure to it.   324

haploid (hap′loid) Having a single set of chromosomes resulting from the reduction division of meiosis.   170

Hardy-Weinberg concept (har′de wīn′burg kon′sept) Populations of organisms will maintain constant gene frequencies from generation to generation as long as mating is random, the population is large, mutation does not occur, migration does not occur, and no genes provide more advantageous characteristics than others.   232

heart (hart) The muscular pump that forces the blood through the blood vessels of the body.   342

heat (hēt) The total internal kinetic energy of molecules.   27

hemoglobin (hēm′o-glo-bin) An iron-containing molecule found in red blood cells, to which oxygen molecules bind.   342

hepatic portal vein (hĕ-pat′ik pōr′tul vān) A blood vessel that collects blood from capillaries in the intestine and delivers it to a second set of capillaries in the liver.   356

herb (herb) Nonwoody plants; e.g., grasses, petunias, and marigolds.   499

herbivores (her′bĭ-vōrz) Animals that feed directly on plants.   262

heterotrophs (hĕ′tur-o-trōfs) Organisms that require a source of organic material from their environment. They cannot produce food on their own.   104

heterozygous (hĕ″ter-o-zi′gus) Describes a diploid organism that has two different alleles for a particular characteristic.   191

high-energy phosphate bond (hi en′ur-je fos′fāt bond) The bond between two phosphates in an ADP or ATP molecule that readily releases its energy for cellular processes.   107

homeostasis (ho″me-o-sta′sis) The maintenance of a constant internal environment.   12

homeotherms (ho′me-o-thermz) Animals that maintain a constant body temperature.   516

homologous chromosomes (ho-mol′o-gus kro′mo-sōmz) A pair of chromosomes in a diploid cell that contain similar genes at corresponding loci throughout their length.   170

homozygous (ho″mo-zi′gus) Describes a diploid organism that has two identical alleles for a particular characteristic.   191

hormone (hōr′mōn) A chemical messenger secreted by an endocrine gland to regulate other parts of the body.   388

host (host) An organism that a parasite lives in or on.   286

hydrogen bond (hi′dro-jen bond) Weak attractive forces between molecules. Important in determining how groups of molecules are arranged.   32

hydrolysis (hi-drol′ĭ-sis) A process that occurs when a large molecule is broken down into smaller parts by the addition of water.   46

hydrophilic (hi-dro-fil′ik) Readily absorbing or dissolving in water.   66

hydrophobic (hi′dro-fo′bik) Tending not to combine with, or incapable of dissolving in, water.   66

hydroxide ion (hi-drok′sīd i′on) A negatively charged particle (OH^-) composed of oxygen and hydrogen atoms released from a base when dissolved in water.   36

hypertonic (hi′pur-tŏn′ik) A comparative term describing one of two solutions. The hypertonic solution is the one with the higher amount of dissolved material.   71

hypothalamus (hi″po-thal′ah-mus) A region of the brain located in the floor of the thalamus and connected to the pituitary gland. Is involved in sleep and arousal; emotions such as anger, fear, pleasure, hunger, sexual response, and pain; and automatic functions such as temperature, blood pressure, and water balance.   392

hypothesis (hi-poth′e-sis) A possible answer to or explanation for a question that accounts for all the observed facts and is testable.   4

hypotonic (hi′po-tŏn′ik) A comparative term describing one of two solutions. The hypotonic solution is the one with the lower amount of dissolved material.   71

I

imitation (i′mi-tā′shun) A form of learning gained while watching another animal being rewarded or punished after performing a particular behavior.   325

immune system (ĭ-myun′ sis′tem) A system of white blood cells specialized to provide the body with resistance to disease. There are two types: antibody-mediated immunity and cell-mediated immunity.   343

imperfect flowers (im″pur′fekt flow′erz) Flowers that contain either male (stamens) or female (pistil) reproductive structures, but not both.   504

imprinting (im′prin-ting) A form of learning, that occurs in a very young animal that is genetically primed to learn a specific behavior in a very short period.   326

inclusions (in-klu′zhuns) A general term referring to materials inside a cell that are usually not readily identifiable; stored materials.   82

incomplete protein (in-kom-plēt′ prō′tēn) Protein molecules that do not provide all the essential amino acids.   369

incus (in′kus) The ear bone that is located between the malleus and the stapes.   401

independent assortment (in″de-pen′dent ă-sort′ment) The segregation, or assortment, of one pair of homologous chromosomes independently of the segregation, or assortment, of any other pair of chromosomes.   174

independent variable (in″de-pen′dent var′ē-a-bul) A variable that is purposely manipulated to determine how it will affect the outcome of an event.   6

inductive reasoning (induction) (in-duk′tiv re′son-ing) (in′duk-shun) The mental process of examining many sets of facts and developing generalizations. From the specific to the general.   7

ingestion (in-jes′chun) The process of taking food into the body through eating.   364

inguinal canal (ing′gwĭ-nal că-nal′) An opening in the floor of the abdominal cavity through which the testes in a human male fetus descend into the scrotum.   415

inguinal hernia (ing′gwĭ-nal her′ne-ah) An opening in the abdominal wall in the area of the inguinal canal that allows a portion of the intestine to push through the abdominal wall.   415

inhibitor (in-hib′ĭ-tōr) A molecule that temporarily attaches itself to an enzyme, thereby interfering with the enzyme's ability to form an enzyme-substrate complex.   99

inorganic molecules (in-or-gan′ik mol′uh-kiuls) Molecules that do not contain carbon atoms in rings or chains.   42

insecticide (in-sek′tĭ-sīd) A poison used to kill insects.   296

insight (in′sīt) Learning in which past experiences are reorganized to solve new problems.   326

instinctive behavior (in-stink′tiv be-hāv′yur) Automatic, preprogrammed, genetically determined behavior.   320

intermediate filaments (in″ter-me′de-it fil′ah-ments) Protein fibers that connect microtubules and microfilaments as part of the cytoskeleton.   80

internal parasite (in-tur′nal per′uh-sīt) A parasite that lives inside its host.   286

interphase (in′tur-fāz) The stage between cell divisions in which the cell is engaged in metabolic activities.   158

interspecific hybrids (in-ter-spĕ-sĭ′-fik hī′-bridz) Hybrids between two different species.   215

interstitial cell-stimulating hormone (ICSH) (in″ter-stĭ′shal sel stim′yu-lā-ting hōr′mōn) The chemical messenger molecule released from the pituitary that causes the testes to produce testosterone, the primary male sex hormone. Same as **follicle-stimulating hormone.**   417

intraspecific hybrids (in-tră-spĕ-sĭ′-fik hī′-bridz) Organisms that are produced by the controlled breeding of separate varieties of the same species.   215

intrinsic factors (in-trin′sik fak′tōrz) Population-controlling factors that arise from within the population.   311

invertebrates (in-vur′tuh-brāts) Animals without backbones.   519

ionic bond (i-on′ik bond) The attractive force between ions of opposite charge.   29

ions (i′ons) Electrically unbalanced or charged atoms.   29

isomers (i′so-meers) Molecules that have the same empirical formula but different structural formulas.   44

isotonic (i′so-tŏn′ik) A term used to describe two solutions that have the same concentration of dissolved material.   71

isotopes (i′so-tōps) Atoms of the same element that differ only in the number of neutrons.   28

K

kidneys (kid′nēz) The primary organs involved in regulating blood levels of water, hydrogen ions, salts, and urea.   357

kilocalorie (kcal) (kil″o-kal′o-re) A measure of heat energy 1,000 times larger than a calorie. Food calories are kilocalories.   364

kinetic energy (ki-net′ik en′er-je) Energy of motion.   24

kingdom (king′dom) A classification category larger than a phylum and smaller than a domain.   457

Krebs cycle (krebs si′kl) The series of reactions in aerobic cellular respiration that results in the production of two carbon dioxides, the release of four pairs of hydrogens, and the formation of an ATP molecule.   110

kwashiorkor (kwa″she-or′kōr) A protein-deficiency disease common in malnourished children caused by prolonged protein starvation leading to reduced body size, lethargy, and low mental ability.   379

L

lacteals (lak′tēlz) Tiny lymphatic vessels located in the villi.   355

lactic acid fermentation (lak′-tik ă′-sid fer-men-tā′-shun) A process during which the pyruvic acid ($CH_3COCOOH$) that results from glycolysis is converted to lactic acid ($CH_3CHOHCOOH$) by the transfer of electrons that had been removed from the original glucose.   118

lag phase (lag fāz) A period of time following colonization when the population remains small or increases slowly.   307

large intestine (larj in-tes′tin) The last portion of the food tube. It is primarily involved in reabsorbing water.   354

law of conservation of energy States that energy is never created or destroyed.   24

law of dominance (law uv dom′in-ans) When an organism has two different alleles for a trait, the allele that is expressed and overshadows the expression of the other allele is said to be dominant. The allele whose expression is overshadowed is said to be recessive.   194

law of independent assortment (law uv in″de-pen′dent ă-sort′ment) Members of one allelic pair will separate from each other independently of the members of other allele pairs.   194

law of segregation (law uv seg″rĕ-ga′shun) When haploid gametes are formed by a diploid organism, the two alleles that control a trait separate from one another into different gametes, retaining their individuality.   194

learned behavior (lur′ned be-hāv′yur) A change in behavior as a result of experience.   320

learning (lur′ning) A change in behavior as a result of experience. 323

leaves (lēvz) Specialized portions of higher plants that are the sites of photosynthesis. 498

lichen (li′kĕn) A mutualistic relation between fungi and algal protists or cyanobacteria. 490

life cycle (līf sīkl) The series of stages in the life of any organism. 494

light-capturing events (līt kap′chu-ring e-vents) The first stage in photosynthesis that involves photosynthetic pigments capturing light energy in the form of "excited" electrons. 120

light-dependent reactions (līt de-pen′dent re-ak′shuns) Also known as the light reaction. The second stage in photosynthesis during which "excited" electrons from the light-capturing events are used to make ATP, and water is broken down to hydrogen and oxygen. The hydrogens are transferred to electron carrier coenzymes, NADP⁺. 120

light-independent reactions (līt in′de-pen″dent re-ak′shuns) Also known as the dark reaction, this third stage of photosynthesis involves cells using ATP and NADPH from the light-dependent reactions to attach CO_2 to 5-carbon starter molecules to manufacture organic molecules; for example, glucose ($C_6H_{12}O_6$). 121

limiting factors (lim′ĭ-ting fak′tōrz) Environmental influences that limit population growth. 308

linkage group (lingk′ij grūp) Genes located on the same chromosome that tend to be inherited together. 193

lipids (lī′pids) Large organic molecules that do not easily dissolve in water; classes include true or neutral fats, phospholipids, and steroids. 56

liquid (lik′wid) The state of matter in which the molecules are strongly attracted to each other, but because they have more energy and are farther apart than in a solid, they move past each other more freely. 27

liver (lī′vur) An organ of the body responsible for secreting bile, filtering the blood, detoxifying molecules, and modifying molecules absorbed from the gut. 354

locus (*loci*) (lo′kus) (lo′si) The spot on a chromosome where an allele is located. 190

loop of Henle (loop uv hen′le) The middle portion of the nephron, which is primarily involved in regulating the amount of water lost from the kidney. 357

lung (lung) A respiratory organ in which air and blood are brought close to one another and gas exchange occurs. 349

luteinizing hormone (loo′te-in-i″zing hōr′mōn) A hormone produced by the anterior pituitary gland that stimulates ovulation. 415

lymph (limf) Liquid material that leaves the circulatory system to surround cells. 347

lymphatic system (lim-fă′tik sis′tem) A collection of thin-walled tubes that collects, filters, and returns lymph from the body to the circulatory system. 347

lysosome (li′so-sōm) A specialized, submicroscopic organelle that holds a mixture of hydrolytic enzymes. 77

M

macromolecules (mă-krō-mol′-ĕ-kyūlz) Very large molecules, many of which are composed of many smaller, similar monomers that are chemically bonded together. 46

malignant tumor (mah-lig′nant too′mor) Nonencapsulated growths of tumor cells that are harmful; they may spread or invade other parts of the body. 165

malleus (mă′le-us) The ear bone that is attached to the tympanum. 401

mass number (mas num′ber) The weight of an atomic nucleus expressed in atomic mass units (the sum of the protons and neutrons). 28

masturbation (măs″tur-ba′shun) Stimulation of one's own sex organs. 417

matter (mat′er) Anything that has weight (mass) and also takes up space (volume). 11

mechanical (morphological) isolating mechanism (mĕ-kan′i-kal [mor″fo-loj-i-kal] i-so-la′ting me′kan-izm) Structural differences that prevent mating between members of different species. 245

medulla oblongata (mĕ-dul′ah ob″long-ga′tah) A region of the more primitive portion of the brain connected to the spinal cord that controls such automatic functions as blood pressure, breathing, and heart rate. 392

medusa (muh-du′sah) A free-swimming adult stage in the phylum Cnidaria that reproduces sexually. 520

meiosis (mi-o′sis) The specialized pair of cell divisions that reduces the chromosome number from diploid ($2n$) to haploid (n). 170

Mendelian genetics (men-dē′le-an jĕ-net′iks) The pattern of inheriting characteristics that follows the laws formulated by Gregor Mendel. 190

menopause (mĕn′o-pawz) The period beginning at about age 50 when the ovaries stop producing viable secondary oocytes and ovarian hormones. 432

menstrual cycle (men′stru-al si′kul) (**menses, menstrual flow, period**) The repeated building up and shedding of the lining of the uterus. 416

mesenteries (mes′en-ter″ēz) Connective tissues that hold the organs in place and also serve as support for blood vessels connecting the various organs. 518

messenger RNA (mRNA) (mes′en-jer) A molecule composed of ribonucleotides that functions as a copy of the gene and is used in the cytoplasm of the cell during protein synthesis. 55

metabolic pathway (me-tah-bol′ik path′-wā) A series of enzyme-controlled reactions. 104

metabolic processes (me-tah-bol′ik pros′es-es) The total of all chemical reactions within an organism; for example, nutrient uptake and processing, and waste elimination. 12

metabolism (mĕ-tab′o-lizm) The total of all the chemical reactions and energy changes that take place in an organism. 12

metaphase (me′tah-fāz) The second stage in mitosis, characterized by alignment of the chromosomes at the equatorial plane. 159

metastasize (me-tas′tah-sīz) The process of cells of tumors moving from the original site and establishing new colonies in other regions of the body. 165

microfilaments (mi″kro-fil′ah-ments) Long, fiberlike, submicroscopic structures made of protein and found in cells, often in close association with the microtubules; provide structural support and enable movement. 80

microorganisms (microbes) (mi″kro-or′guh-niz′mz) Small organisms that cannot be seen without some type of magnification. 476

microscope (mi′kro-skōp) An instrument used to produce an enlarged image of a small object. 64

microsphere (mi′kro-sfēr) A collection of organic macromolecules in a structure with a double-layered outer boundary. 443

microtubules (mi′kro-tū″byuls) Submicroscopic, hollow tubes of protein that function throughout the cytoplasm to provide structural support and enable movement. 80

microvilli (mi″kro-vil′ē) Tiny projections from the surfaces of cells that line the intestine. 340

minerals (mĭn′er-alz) Inorganic elements that cannot be manufactured by the body but are required in low concentrations; essential to metabolism. 370

mitochondrion (mi-to-kahn′dre-on) A membranous organelle resembling a small bag with a larger bag inside that is folded back on itself; serves as the site of aerobic cellular respiration. 78

mitosis (mi-to′sis) A process that results in equal and identical distribution of replicated chromosomes into two newly formed nuclei.   158

mixture (miks′chur) Matter that contains two or more substances *not* in set proportions.   33

molecule (mol′ĕ-kūl) The smallest particle of a chemical compound; also the smallest naturally occurring part of an element or compound.   25

monocot (mon′o-kot) An angiosperm whose embryo has one seed leaf (cotyledon).   508

monoculture (mon″o-kul′chur) The agricultural practice of planting the same varieties of a species over large expanses of land.   215

monosomy (mon′o-so″me) A cell with only one of the two chromosomes of a homologous pair.   184

morphological species concept (mor″fo-loj′i-kal spe′shēz kon′sept) The concept that different species can be distinguished from one another by structural, chemical, or behavioral differences.   206

mortality (mor-tal′ĭ-te) The number of individuals leaving the population by death per thousand individuals in the population.   307

motor neurons (mo′tur noor′onz″) Those neurons that carry information from the central nervous system to muscles or glands.   389

motor unit (mo′tur yoo′nit) All the muscle cells stimulated by a single neuron.   404

multiple alleles (mul′tĭ-pul a-lēlz′) A term used to refer to conditions in which there are several different alleles for a particular characteristic within a population, not just two.   199

multiregional hypothesis (mul″ti-re′jun-al hi-poth′e-sis) The concept that *Homo erectus* migrated to Europe and Asia from Africa and evolved into *Homo sapiens.*   255

mutagenic agent (miu-tah-jen′ik a-jent) Anything that causes permanent change in DNA.   141

mutation (miu-ta′shun) Any change in the genetic information of a cell.   136

mutualism (miu′chu-al-izm) A relationship between two organisms in which both organisms benefit.   288

mycorrhiza (my″ko-rye′zah) A symbiotic relation between fungi and plant roots.   489

mycotoxin (mi″ko-tok′sin) A deadly poison produced by fungi.   490

myosin (mi′o-sin) A protein molecule found in the thick filaments of muscle fibers that attaches to actin, bends, and moves actin molecules along its length, causing the muscle fiber to shorten.   403

N

NAD⁺ (**n**icotinamide **a**denine **d**inucleotide) (N-A-D) An electron acceptor and hydrogen carrier used in respiration.   108

NADP⁺ (**n**icotinamide **a**denine **d**inucleotide **p**hosphate) (N-A-D-P) An electron acceptor and hydrogen carrier used in photosynthesis.   108

natality (na-tal′ĭ-te) The number of individuals entering the population by reproduction per thousand individuals in the population.   307

natural selection (nat′ chu-ral se-lek′shun) A broad term used in reference to the various mechanisms that encourage the passage of beneficial genes to future generations and discourage harmful or less valuable genes from being passed to future generations.   222

negative-feedback control (nĕg′ăh-tiv fēd′bak con-trōl′) A kind of control mechanism in which the product of one activity inhibits an earlier step in the chain of events.   395

negative-feedback inhibition (neg′ăh-tiv fēd′băk in-hib′ĭ-shun) A metabolic control process that operates at the surfaces of enzymes. This process occurs when one of the end products of the pathway alters the three-dimensional shape of an essential enzyme in the pathway and interferes with its operation long enough to slow its action.   98

nephrons (nef′ronz) Tiny tubules that are the functional units of kidneys.   357

nerve cell (nerv sel) See **neuron.**   389

nerve impulse (nerv im′puls) A series of changes that take place in the neuron, resulting in a wave of depolarization that passes from one end of the neuron to the other.   389

nerves (nervz) Bundles of neuronal fibers.   389

nervous system (ner′vus sis′tem) A network of neurons that carry information from sense organs to the central nervous system and from the central nervous system to muscles and glands.   388

net movement (net mūv′ment) Movement in one direction minus the movement in the other.   69

neuron (noor′on″) The cellular unit consisting of a cell body and fibers that makes up the nervous system; also called **nerve cell.**   389

neurotransmitter (noor″o-trans′mĭt-er) A molecule released by the axons of neurons that stimulates other cells.   391

neutrons (nu′trons) Particles in the nucleus of an atom that have no electrical charge; they were named *neutrons* to reflect this lack of electrical charge.   28

niche (nitch) The functional role of an organism.   284

nitrifying bacteria (ni′trĭ-fi-ing bak-te′re-ah) Several kinds of bacteria capable of converting ammonia to nitrite, or nitrite to nitrate.   292

nitrogenous base (ni-trah′jen-us bās) A category of organic molecules found as components of the nucleic acids. There are five common types: thymine, guanine, cytosine, adenine, and uracil.   136

nondeciduous (non″de-sid′yu-us) A term used to describe trees that do not lose their leaves all at once.   502

nondisjunction (non″dis-junk′shun) An abnormal meiotic division that results in sex cells with too many or too few chromosomes.   183

norepinephrine (nor-ĕ″pĭ-nef′rin) A hormone produced by the adrenal medulla and certain nerve cells that increases heart rate, blood pressure, and breathing rate.   393

nuclear membrane (nu′kle-ar mem′brăn) The structure surrounding the nucleus that separates the nucleoplasm from the cytoplasm.   78

nucleic acids (nu-kle′ik ă′sids) Complex molecules that store and transfer information within a cell. They are constructed of fundamental monomers known as nucleotides.   53

nucleoli (singular, *nucleolus*) (nu-kle′o-li) Nuclear structures composed of completed or partially completed ribosomes and the specific parts of chromosomes that contain the information for their construction.   83

nucleoplasm (nu′kle-o-plazm) The liquid matrix of the nucleus composed of a mixture of water and the molecules used in the construction of the rest of the nuclear structures.   83

nucleosomes (nu′kle-o-somz) Histone clusters with their encircling DNA.   148

nucleotide (nu″kle-o-tīd′) A fundamental subunit of nucleic acid constructed of a phosphate group, a sugar, and an organic nitrogenous base.   53

nucleus (nu′kle-us) The central body that contains the information system for the cell. Also, the central part of an atom containing protons and neutrons.   64

nutrients (nu′tre-ents) Molecules required by organisms for growth, reproduction, or repair.   92

nutrition (nu-trĭ′shun) Collectively, the processes involved in taking in, assimilating, and utilizing nutrients.   364

O

obese (o-bēs) A term describing a person who gains a great deal of unnecessary weight. A person with a body mass index of 30 kg/m² or greater.   376

obligate intracellular parasites (ob′lĭ-gāt in″trah-sel′yu-lar pĕr′uh-sīts) Infectious particles (viruses) that can function only when inside a living cell.   467

observation (ob-sir-vā′shun) The process of using the senses or extensions of the senses to record events.   3

observational learning (imitation) (ob-sir-vā′ shun-uhl lur′ning) A form of associative learning that involves a complex set of associations involved in watching another animal being rewarded for performing a particular behavior and then performing that same behavior oneself.   325

offspring (of′spring) Descendants of a set of parents.   190

olfactory epithelium (ōl-fak′to-re ĕ″pĭ-thē′le-um) The cells of the nasal cavity that respond to chemicals.   400

omnivores (om′nĭ-vōrz) Animals that are carnivores at some times and herbivores at others.   262

oogenesis (oh″ŏ-jen′ĕ-sis) The specific name given to the gametogenesis process that leads to the formation of eggs.   420

operant (instrumental) conditioning (op′ĕ-rant kon′ dĭ′shun-ing) A change in behavior that results from associating a stimulus with a response by either rewarding or punishing the behavior after it has occurred.   324

order (or′der) A group of closely related classes within a phylum.   460

organ (or′gun) A structure composed of groups of tissues that performs particular functions.   15

organelles (or-gan-elz′) Cellular structures that perform specific functions in the cell. The function of an organelle is directly related to its structure.   66

organic molecules (or-gan′ik mol′uh-kiuls) Complex molecules whose basic building blocks are carbon atoms in chains or rings.   42

organism (or′gun-izm) An independent living unit.   13

organ system (or′gun sis′tem) A structure composed of groups of organs that perform particular functions.   15

orgasm (or′gaz-um) A complex series of responses to sexual stimulation that results in an intense frenzy of sexual excitement.   417

osmosis (os-mo′sis) The net movement of water molecules through a selectively permeable membrane.   71

osteoporosis (os″te-o-po-ro′sis) A disease condition resulting from the demineralization of the bone, resulting in pain, deformities, and fractures; related to a loss of calcium.   370

out-of-Africa hypothesis The concept that modern humans (*Homo sapiens*) originated in Africa and migrated from Africa to Europe and Asia and displaced existing hominids.   255

oval window (o′val win′do) The membrane-covered opening of the cochlea, to which the stapes is attached.   401

ovary (o′vah-re) The female sex organ that produces haploid sex cells, called eggs.   413

oviduct (o′vĭ-dukt) The tube (*fallopian tube*) that carries the oocyte to the uterus.   420

ovulation (ov-yu-lā′shun) The release of a secondary oocyte from the surface of the ovary.   417

oxidation-reduction (redox) reactions (ok′sĭ-day-shun re-duk′shun re-ak′shunz) Electron-transport reactions in which the molecules losing electrons become oxidized and those gaining electrons become reduced.   35

oxidizing atmosphere (ok′sĭ-di-zing at′mos-fĕr) An atmosphere that contains molecular oxygen.   445

oxytocin (ok″sĭ-to′sin) A hormone released from the posterior pituitary that causes contraction of the uterus.   406

P

pancreas (pan′kre-as) An organ of the body that secretes many kinds of digestive enzymes into the duodenum.   353

panspermia (pan-sper′me-a) A hypothesis by Svante Arrhenius in the early 1900s that life arose outside the earth and that living things were transported to Earth serving to seed the planet with life.   437

parasite (pĕr′uh-sīt) An organism that lives in or on another organism and derives nourishment from it.   286

parasitism (pĕr′uh-sĭt-izm) A relationship between two organisms that involves one organism living in or on another organism and deriving nourishment from it.   286

pathogen (path′uh-jen) An agent that causes a specific disease.   478

pelagic (pĕ-lăj′ĭk) A term used to describe animals that swim freely as adults.   528

penetrance (pen′ĕ-trens) A term used to describe how often an allele expresses itself when present.   225

penis (pe′nis) The portion of the male reproductive system that deposits sperm in the female reproductive tract.   417

pepsin (pep′sin) An enzyme produced by the stomach that is responsible for beginning the digestion of proteins.   353

peptide bond (pep′tīd bond) A covalent bond between amino acids in a protein.   49

perception (per-sep′shun) Recognition by the brain that a stimulus has been received.   399

perennial (pur-en′e-uhl) A plant that lives for many years.   503

perfect flowers (pur′fekt flow′erz) Flowers that contain both male (stamen) and female (pistil) reproductive structures.   504

periodic table of the elements (pĭr-ē-od′ik tā′bul uv the el′ĕ-ments) A list of all the elements in order of increasing atomic number (number of protons).   25

peripheral nervous system (pŭ-rĭ′fĕ-ral ner′vus sis′tem) The fibers that communicate between the central nervous system and other parts of the body.   389

peroxisome (pĕ-roks′ĭ-sōm) A single membrane-bound, submicroscopic organelle that holds enzymes capable of producing hydrogen peroxide that aids in the control of infections and other dangerous compounds.   78

pesticide (pes'tĭ-sīd) A poison used to kill pests. This term is often used interchangeably with *insecticide*.   296

petals (pě'tuls) Modified leaves of angiosperms; accessory structures of a flower.   504

pH A scale used to indicate the concentration of an acid or base.   37

phagocytosis (fa"go-si-to'sis) The process by which the cell wraps around a particle and engulfs it.   73

pharynx (far'inks) The region at the back of the mouth cavity; the throat.   353

phenotype (fēn'o-tīp) The physical, chemical, and behavioral expression of the genes possessed by an organism.   191

pheromone (fĕr'uh-mōn) A chemical produced by an animal and released into the environment to trigger behavioral or developmental processes in some other animal of the same species.   330

phloem (flo'em) One kind of vascular tissue found in higher plants. It transports food materials from the leaves to other parts of the plant.   498

phospholipid (fos"fo-lĭ'pid) A class of water-insoluble molecules that resembles fats but contains a phosphate group (PO_4) in its structure.   58

phosphorylation reaction (fos"for lā shun re-ak shun) Takes place when a custer of atoms known as a phosphate group is added to another molecule.   36

photoperiod (fo"to-pĭr'e-ud) The length of the light part of the day.   334

photosynthesis (fo-to-sin'thuh-sis) A series of reactions that take place in chloroplasts and result in the storage of sunlight energy in the form of chemical-bond energy.   79

photosystem (fo'to-sis"tem) Clusters of photosynthetic pigments (e.g., chlorophyll) that serve as energy gathering or concentrating mechanisms. They are utilized during the light-capturing events of photosynthesis.   123

phylogeny (fi-laj'uh-ne) The science that explores the evolutionary relationships among organisms and seeks to reconstruct evolutionary history.   460

phylum (fi'lum) A subdivision of a kingdom.   460

physiology Scientific study of living processes, activities, and functions.   24

phytoplankton (fye-tuh-plank'tun) Microscopic, photosynthetic species that form the basis for most aquatic food chains.   482

pinocytosis (pǐ'no-si-to'sis) The process by which a cell engulfs some molecules dissolved in water.   73

pioneer community (pi"o-nēr' ko-miu'nǐ-te) The first community of organisms in the successional process established in a previously uninhabited area.   275

pioneer organisms (pi"o-nēr' or'gu-nizms) The first organisms in the successional process.   275

pistil (pis'til) The female reproductive structure in flowers that contains the ovary that produces eggs.   170

pituitary gland (pǐ-tu'ǐ-tě-re gland) The gland at the base of the brain that controls the functioning of other glands throughout the organism.   415

placenta (plah-sen'tah) An organ made up of tissues from the embryo and the uterus of the mother that allows for the exchange of materials between the mother's bloodstream and the embryo's bloodstream. It also produces hormones.   424

plankton (plank'tun) Small floating or weakly swimming organisms.   482

plasma (plaz'muh) The watery matrix that contains the molecules and cells of the blood.   342

plasma membrane (plaz'muh mem'brān) The outer boundary membrane of the cell; see **cell membrane.**   76

pleiotropy (ple'o-tro'-pe) The multiple effects that a gene may have on the phenotype of an organism.   206

poikilotherms (poy-ki'luh-thermz) Animals with a variable body temperature that changes with the external environment.   516

point mutation (point miu-ta'shun) A change in the DNA of a cell as a result of a loss or change in a nitrogenous-base sequence.   143

polar body (po'lar bod'ē) The smaller of two cells formed by unequal meiotic division during oogenesis.   420

pollen (pol'en) The male gametophyte in gymnosperms and angiosperms.   502

pollination (pol"ǐ-na'shun) The transfer of pollen in gymnosperms and angiosperms.   502

polygenic inheritance (pol"e-jen'ik in-her'ǐ-tans) The concept that a number of different pairs of alleles may combine their efforts to determine a characteristic.   200

polyp (pol'ǐp) A sessile larval stage in the phylum Cnidaria that reproduces asexually.   520

polypeptide chain (pǒ'le-pep'tīd chān) A macromolecule composed of a specific sequence of amino acids.   50

polyploidy (pah'lǐ-ploy'de) A condition in which cells contain multiple sets of chromosomes.   244

pons (ponz) A region of the brain immediately anterior to the medulla oblongata that connects to the cerebellum and higher regions of the brain and controls several sensory and motor functions of the head and face.   392

population (pop"u-la'shun) A group of organisms of the same species located in the same place at the same time.   206

population density (pop"u-la'shun den'sǐ-te) The number of organisms of a species per unit area.   306

population growth curve (pop"u-la'shun grōth kurv) A graph of the change in population size over time.   307

population pressure (pop"u-la'shun presh'yur) Intense competition as a result of high population density that leads to changes in the environment and dispersal of organisms.   306

potential energy (po-ten'shul en'er-je) The energy an object has because of its position.   24

prebionts (pre"bi'onts) Nonliving structures thought to have led to the formation of the first living cells.   442

predation (prě-da'shun) A relationship between two organisms that involves the capturing, killing, and eating of one by the other.   284

predator (pred'uh-tor) An organism that captures, kills, and eats another animal.   284

prey (prā) An organism captured, killed, and eaten by a predator.   284

primary carnivores (pri'mar-e kar'nǐ-vōrz) Carnivores that eat herbivores and are therefore on the third trophic level.   262

primary consumers (pri'mar-e kon-su'merz) Organisms that feed directly on plants—herbivores.   262

primary oocyte (pri'mar-e o'o-sīt) The diploid cell of the ovary that begins to undergo the first meiotic division in the process of oogenesis.   420

primary spermatocyte (pri'mar-e spur-mat'o-sīt) The diploid cell in the testes that undergoes the first meiotic division in the process of spermatogenesis.   419

primary succession (pri'mar-e suk-sě'shun) The orderly series of changes that begins in a previously uninhabited area and leads to a climax community.   275

prion (pri'on) Infectious protein particles responsible for diseases such as Creutzfeldt-Jakob disease and bovine spongiform encephalitis.   471

probability (prob"a-bil'ǐ-te) The chance that an event will happen, expressed as a percentage or fraction.   195

producers (pro-du'surz) Organisms that produce new organic material from inorganic material with the aid of sunlight.   261

productivity (pro-duk-tiv´ĭ-te) The rate at which an ecosystem can accumulate new organic matter.   278

products (prō´dukts) New molecules resulting from a chemical reaction.   35

progesterone (pro-jes´ter-ōn) A hormone produced by the corpus luteum that maintains the lining of the uterus.   416

prokaryote (pro-kār´e-ōt) An organism composed of cells that lack a nuclear membrane and other membranous organelles.   446

prokaryotic cells (pro´kār´e-ot˝ik sels) One of the two major types of cells. They do not have a typical nucleus bound by a nuclear membrane and lack many of the other membranous cellular organelles; for example, members of the Eubacteria and Archaea.   66

promoter (pro-mo´ter) A region of DNA at the beginning of each gene, just ahead of an initiator code.   145

prophase (pro´fāz) The first phase of mitosis during which individual chromosomes become visible.   159

protein (prō´tēn) Macromolecules made up of one or more polypeptides attached to each other by bonds.   48

proteinoid (prō´tēn oid) The proteinlike structure of branched amino acid chains that is the basic structure of a microsphere.   443

protein-sparing (prō´tēn spē´ring) The conservation of proteins by first oxidizing carbohydrates and fats as a source of ATP energy.   370

protein synthesis (prō´tēn sin´thĕ-sis) The process whereby the tRNA utilizes the mRNA as a guide to arrange the amino acids in their proper sequence according to the genetic information in the chemical code of DNA.   134

proteomics (prō-tee-o´-miks) A new field of science that has developed since the sequencing of the human genome. It groups proteins by similarities to help explain their function and how they may have evolved.   154

protocell (pro´to-sel) The first living cell.   443

protons (pro´tons) Particles in the nucleus of an atom that have a positive electrical charge.   28

protoplasm (pro´to-plazm) The living portion of a cell as distinguished from the nonliving cell wall.   64

protozoa (pro˝to-zo´ah) Heterotrophic, eukaryotic, unicellular organisms.   481

proximal convoluted tubule (prok´sĭ-mal kon´vōl-lu-ted tŭb´yŭl) The upstream end of the nephron of the kidney, which is responsible for reabsorbing most of the valuable molecules filtered from the glomerulus into Bowman's capsule.   357

pseudoscience (su-dō-si´ens) An activity that uses the appearance or language of science to convince or mislead people into thinking that something has scientific validity.   10

puberty (pu´ber-te) A time in the life of a developing individual characterized by the increasing production of sex hormones, which cause it to reach sexual maturity.   415

pulmonary artery (pul´muh-nă-rē ar´tuh-rē) The major blood vessel that carries blood from the right ventricle to the lungs.   344

pulmonary circulation (pul´muh-nă-rē ser-kyu-lă´shun) The flow of blood through certain chambers of the heart and blood vessels to the lungs and back to the heart.   344

punctuated equilibrium (pung´chu-a-ted e-kwĭ-lib´re-um) The theory stating that evolution occurs in spurts, between which there are long periods with little evolutionary change.   252

Punnett square (pun´net sqwār) A method used to determine the probabilities of allele combinations in a zygote.   196

pyloric sphincter (pi-lor´ik sfingk´ter) A valve located at the end of the stomach that regulates the flow of food from the stomach to the duodenum.   353

pyruvic acid (pi-ru´vik ăs´id) A 3-carbon carbohydrate that is the end product of the process of glycolysis.   117

R

radial symmetry (ra´de-ul sĭm´ĭ-tre) The characteristic of an animal with a body constructed around a central axis. Any division of the body along this axis results in two similar halves.   516

range (rānj) The geographical distribution of a species.   241

reactants (re-ak´tants) Materials that will be changed in a chemical reaction.   35

recessive allele (re-sĕ´siv a-lēl´) An allele that, when present with its homolog, does not express itself and is masked by the effect of the other allele.   191

recombinant DNA (re-kom´bĭ-nant) DNA that has been constructed by inserting new pieces of DNA into the DNA of an organism.   150

recommended dietary allowances (RDAs) (rĕ-ko-men´ded di´ĕ-tĕ-rē ă-lao´an-ses) U.S. dietary guidelines for a healthy person that focus on the amounts of foods desired from six classes of nutrients.   373

redirected aggression (re-di-rek´ted a-grĕ´shun) A behavior in which the aggression of an animal is directed away from an opponent to some other animal or object.   333

reducing atmosphere (re-du´sing at´mos-fēr) An atmosphere that does not contain molecular oxygen (O_2).   440

reduction division (also **meiosis**) (re-duk´shun dĭ-vĭ´zhun) A type of cell division in which daughter cells get only half the chromosomes from the parent cell.   173

regulator proteins (reg´yu-la-tōr prō´tēns) Proteins that influence the activities that occur in an organism, for example, enzymes and some hormones.   48

reliable (re-li´a-bul) A term used to describe results that remain consistent over successive trials.   6

reproductive capacity (re-pro-duk´tiv kuh-pas´ĭ-te) The theoretical maximum rate of reproduction; also called **biotic potential.**   306

reproductive isolating mechanism (re-pro-duk´tiv i-so-la´ting me´kan-izm) A mechanism that prevents interbreeding between species; also called **genetic isolating mechanism.**   244

response (re-spons´) The reaction of an organism to a stimulus.   322

responsive processes (re-spon´siv pros´es-es) Those abilities to react to external and internal changes in the environment; for example, irritability, individual adaptation, and evolution.   12

retina (rĕ´tĭ-nah) The light-sensitive region of the eye.   400

rhodopsin (ro-dop´sin) A light-sensitive pigment found in the rods of the retina.   401

ribonucleic acid (RNA) (ri-bo-nu-kle´ik ă´sid) A polymer of nucleotides formed on the template surface of DNA by transcription. Three forms that have been identified are mRNA, rRNA, and tRNA.   55

ribose (ri´bōs) A 5-carbon sugar molecule that is a component of RNA.   134

ribosomal RNA (rRNA) (ri-bo-sōm´al) A globular form of RNA; a part of ribosomes.   55

ribosomes (ri´bo-sōmz) Small structures composed of two protein and ribonucleic acid subunits, involved in the assembly of proteins from amino acids.   80

ribulose (ri´bu-lōs) A 5-carbon sugar molecule used in photosynthesis.   123

ribulose biphosphate carboxylase (RuBisCo) (ri′bu-lōs bi-fos′fāt kar-bak′se-lāz) An enzyme found in the stroma of chloroplast that speeds the combining of the CO_2 with an already-present, 5-carbon carbohydrate, ribulose.   123

RNA See **ribonucleic acid.**   55

RNA polymerase (po-lim′er-ās) An enzyme that bonds RNA nucleotides together during transcription after they have aligned on the DNA.   138

rods (rahdz) Light-sensitive cells in the retina of the eye that respond to low-intensity light but do not respond to different colors of light.   400

root (root) A specialized organ that functions in the absorption of water and minerals in higher plants.   498

root hairs (root hārs) Tiny cellular outgrowths of roots that improve the ability of plants to absorb water and minerals.   498

S

salivary amylase (să′lĭ-vĕ-rē ă′mĭ-lās) An enzyme present in saliva that breaks starch molecules into smaller molecules.   353

salivary glands (să′lĭ-vĕ′rē glanz) Glands that produce saliva.   353

salts (salts) Ionic compounds formed from a reaction between an acid and a base.   37

saprophyte (sap′ruh-fīt) An organism that obtains energy by the decomposition of dead organic material.   464

saturated (sat′yu-ra-ted) A term used to describe the carbon skeleton of a fatty acid that contains no double bonds between carbons.   56

science (si′ens) A process used to solve problems or develop an understanding of natural events.   3

scientific law (si-en-tif′ik law) A uniform or constant feature of nature that describes what happens in nature.   7

scientific method (si-en-tif′ik meth′ud) A way of gaining information (facts) about the world around you that involves observation, hypothesis formation, testing of hypotheses, theory formation, and law formation.   3

seasonal isolating mechanism (se′zun-al eye-sō-lā′ting mek-ă-nizm) A genetic isolating mechanism that prevents interbreeding between species because they reproduce at different times of the year.   244

secondary carnivores (sĕk′on-dĕr-e kar′nĭ-vōrz) Carnivores that feed on primary carnivores and are therefore at the fourth trophic level.   262

secondary consumers (sek′on-dĕr-e kon-su′merz) Animals that eat other animals—carnivores.   262

secondary oocyte (sek′on-dĕr-e o′o-sīt) The larger of the two cells resulting from the unequal cytoplasmic division of a primary oocyte in meiosis I of oogenesis.   420

secondary sexual characteristics (sek′on-dĕr-e sek′shoo-al kăr-ak-tĕ-ris′tiks) Characteristics of the adult male or female, including the typical shape that develops at puberty: broader shoulders, heavier long-bone muscles, development of facial hair, axillary hair, and chest hair, and changes in the shape of the larynx in the male; rounding of the pelvis and breasts and changes in deposition of fat in the female.   416

secondary spermatocyte (sek′on-dĕr-e spur-mat′o-sīt) Cells in the seminiferous tubules that go through the second meiotic division, resulting in four haploid spermatids.   419

secondary succession (sek′on-dĕr-e suk-sĕ′shun) The orderly series of changes that begins with the disturbance of an existing community and leads to a climax community.   275

seed (sēd) A specialized structure produced by gymnosperms and angiosperms that contains the embryonic sporophyte.   501

seed leaves (sēd lēvz) Cotyledons; embryonic leaves in seeds.   508

segregation (seg″rĕ-ga′shun) The separation and movement of homologous chromosomes to the opposite poles of the cell.   174

selecting agent (se-lek′ting a′jent) Any factor that affects the probability that a gene will be passed to the next generation.   227

selectively permeable (se-lek′tiv-le per′me-uh-bul) The property of a membrane that allows certain molecules to pass through it but interferes with the passage of others.   70

semen (se′men) The sperm-carrying fluid produced by the seminal vesicles, prostate gland, and bulbourethral glands of males.   428

semicircular canals (sĕ-mi-ser′ku-lar că-nalz′) A set of tubular organs associated with the cochlea that sense changes in the movement or position of the head.   402

semilunar valves (sĕ-me-lu′ner valvz) Valves located in the pulmonary artery and aorta that prevent the flow of blood backward into the ventricles.   344

seminal vesicle (sĕm′ĭ-nal ves′ĭ-kul) A part of the male reproductive system that produces a portion of the semen.   417

seminiferous tubules (sem″ĭ-nif′ur-us tub′yūlz) Sperm-producing tubes in the testes.   417

sensory neurons (sen′so-re noor′onz″) Those neurons that send information from sense organs to the central nervous system.   389

sepals (se′pals) Accessory structures of flowers.   504

sessile (ses′il) Firmly attached.   527

sex chromosomes (seks kro′mo-sōmz) Chromosomes that carry genes that determine the sex of the individual (X and Y in humans).   186

sex-determining chromosome (seks de-ter′mĭ-ning kro′mo-sōm) The chromosomes X and Y that are primarily responsible for determining if an individual will develop as a male or female.   412

sex ratio (seks ra′sho) The number of males in a population compared to the number of females.   305

sexual intercourse (sek′shoo-al in′ter-kors) The mating of male and female; the deposition of the male sex cells, or sperm cells, in the reproductive tract of the female; also known as **coitus** or **copulation.**   420

sexual reproduction (sek′shu-al re″pro-duk′shun) The propagation of organisms involving the union of gametes from two parents.   170

sexual selection (sek′shu-al se-lek′shun) Selection that is the result of specific individuals being chosen by members of the opposite sex for mating purposes.   230

sexuality (sek″shu-al′ĭ-te) A term used in reference to the totality of the aspects—physical, psychological, and cultural—of our sexual nature.   412

sickle-cell anemia (sĭ-kul sel ah-ne′me-ah) A disease caused by a point mutation. This malfunction produces sickle-shaped red blood cells.   143

signal transduction (sig′-nal trans-duk′-shun) The term used to describe how cells detect specific signals and how they transmit these signals to the cell's interior.   74

single-factor cross (sing′ul fak-tur kros) A genetic study in which a single characteristic is followed from the parental generation to the offspring.   196

small intestine (smahl in-tes'ten) The portion of the digestive system immediately following the stomach; responsible for digestion and absorption. 353

society (so-si'uh-te) Interacting groups of animals of the same species that show division of labor. 335

sociobiology (so-se-o-bi-ol'o-je) The systematic study of all forms of social behavior, both human and nonhuman. 336

solid (sol'id) The state of matter in which the molecules are packed tightly together; they vibrate in place. 27

solute (sol'lūt) Is the component that dissolves in the solvent. 33

solution (sol'ū shun) Is a homogeneous mixture of ions or molecules of two or more substances. 3

solvent (sol'vent) Is the component present in the larger amount. 33

soma (so'mah) The cell body of a neuron, which contains the nucleus. 389

speciation (spe-she-a'shun) The process of generating new species. 241

species (spe'shēz) A population of organisms potentially capable of breeding naturally among themselves and having offspring that also interbreed. 206

specific dynamic action (SDA) (spě-si'fik di-nă'mik ak'shun) The amount of energy required to digest and assimilate food. SDA is equal to approximately 10% of your total daily kilocalorie intake. 365

specific epithet (spě-si'fik ep''i-the't) A word added to the genus name to identify which one of several species within the genus is being identified (i.e., *Homo sapiens*: *Homo* is the genus name and *sapiens* is the specific epithet). 456

sperm (spurm) The haploid sex cells produced by sexually mature males. 170

spermatids (spurm'ah-tids) Haploid cells produced by spermatogenesis that change into sperm. 419

spermatogenesis (spur-mat-o-jen'uh-sis) The specific name given to the gametogenesis process that leads to the formation of sperm. 417

spinal cord (spi'nal 'kōrd) The portion of the central nervous system located within the vertebral column, which carries both sensory and motor information between the brain and the periphery of the body. 392

spindle (spin'dul) An array of microtubules extending from pole to pole; used in the movement of chromosomes. 159

spontaneous generation (spon-ta'ne-us jen-uh-ra'shun) The idea that living organisms arose from nonliving material. 436

spontaneous mutation (spon-ta'ne-us miu-ta'shun) Natural changes in the DNA caused by unidentified environmental factors. 224

spores (spōrz) Haploid specialized cells produced by sporophytes. 494

sporophyte (spōr'o-fīt) A diploid plant that produces haploid spores by meiosis; it alternates with the gametophyte through the life cycle. 494

sporophyte stage (spōr'o-fīt stāj) A life cycle stage in plants in which a haploid spore is produced by meiosis. 466

stabilizing selection (stā-bill-ī-zing se-lěk'-shun) A kind of selection that occurs when individuals at the extremes of the range of a characteristic are consistently selected against. 230

stable equilibrium phase (stā'bul e-kwi-lib're-um fāz) A period of time during population growth when the number of individuals entering the population and the number leaving the population are equal, resulting in a stable population size. 308

stamen (sta'men) The male reproductive structure of a flower. 504

stapes (sta'pēz) The ear bone that is attached to the oval window. 401

states of matter (stātes uv mat'er) Physical conditions of matter (solid, liquid, and gas) determined by the relative amounts of energy of the molecules. 27

stem (stem) The upright portion of a higher plant. 498

stem cells (stem selz) Cells that have the ability to differentiate into any type of cell, including liver cells, skin cells, and brain cells; embryonic and hematopoietic cells. 164

steroid (ster'oid) One of the three kinds of lipid molecules characterized by their arrangement of interlocking rings of carbon. 58

stimulus (stim'yu-lus) Any change in the internal or external environment of an organism that it can detect. 322

stroma (stro'muh) The region within a chloroplast that has no chlorophyll. 79

structural proteins (struk'chu-ral prō'tēns) Proteins that are important for holding cells and organisms together, such as the proteins that make up the cell membrane, muscles, tendons, and blood. 48

subspecies (races, breeds, strains, or varieties) (sub'spe-shēz) Regional groups within a species that are significantly different structurally, physiologically, or behaviorally, yet are capable of exchanging genes by interbreeding. 209

substrate (sub'strāt) A reactant molecule with which the enzyme combines. 93

succession (suk-sě'shun) The process of changing one type of community to another. 275

successional community (stage) (suk-sě'shun-al ko-miu'nĭ-te) An intermediate stage in succession. 275

surface area-to-volume ratio (SA/V) (ser'fas a're-uh to vol'yūm ra'sho) The relationship between the surface area of an object and its volume. As objects increase in size, their volume increases more rapidly than their surface area. 340

symbiosis (sim-bi-o'sis) A close physical relationship between two kinds of organisms. Parasitism, commensalism, and mutualism may all be examples of symbiosis. 289

symbiotic nitrogen-fixing bacteria (sim-bi-ah'tik ni'tro-jen fik'sing bak-te're-ah) Bacteria that live in the roots of certain kinds of plants, where they convert nitrogen gas molecules into compounds that plants can use. 292

synapse (sin-aps) The space between the axon of one neuron and the dendrite of the next, where chemicals are secreted to cause an impulse to be initiated in the second neuron. 391

synapsis (sin-ap'sis) The condition in which the two members of a pair of homologous chromosomes come to lie close to one another. 173

systemic circulation (sis-tě'mik ser-kyu-la'shun) The flow of blood through certain chambers of the heart and blood vessels to the general body and back to the heart. 345

systolic blood pressure (sis-tah'lik blud presh'yur) The pressure generated in a large artery when the ventricles of the heart are in the process of contracting. 345

T

target cells (tar'get sels) The specific cells to which a hormone binds. The cells respond in a predetermined manner. 393

taxonomy (tak-son'uh-me) The science of classifying and naming organisms. 456

telomeres (tel'o-mērs) Chromosome caps composed of repeated, specific sequences of nucleotide pairs; their activity or inactivity is associated with cell aging and cancer. 148

telophase (tel'uh-fāz) The last phase in mitosis characterized by the formation of daughter nuclei. 161

temperature (tem'per-ă-chiur) A measure of molecular energy of motion. 27

territorial behavior (ter″ĭ-tōr′e-al be-hāv′yur) Behavior involved in establishing, defending, and maintaining a territory for food, mating, or other purposes.   332

territory (ter′ĭ-tōr-e) A space that an animal defends against others of the same species.   332

testes (tes′tēz) The male sex organs that produce haploid cells—the sperm.   170

testosterone (tes-tos-tur-ōn) The male sex hormone produced in the testes that controls male sexual development.   394

thalamus (thal′ah-mus) A region of the brain that relays information between the cerebrum and lower portions of the brain. It also provides some level of awareness in that it determines pleasant and unpleasant stimuli and is involved in sleep and arousal.   392

theory (the′o-re) A widely accepted, plausible generalization about fundamental concepts in science that is supported by many experiments and explains why things happen in nature.   7

theory of natural selection (the′o-re uv nat′chural se-lek′shun) In a species of genetically differing organisms, the organisms with the genes that enable them to survive better in the environment and thus reproduce more offspring than others will transmit more of their genes to the next generation.   222

thinking (thingk′ing) A mental process that involves memory, a concept of self, and an ability to reorganize information.   330

thylakoids (thī′la-koids) Thin, flat disks found in the chloroplast of plant cells that are the site of the light-capturing events and light-dependent reactions of photosynthesis.   79

thymine (thī′mēn) A single-ring nitrogenous-base molecule in DNA but not in RNA. It is complementary to adenine.   134

thyroid-stimulating hormone (TSH) (thi′roid stĭ′mu-la-ting hōr′mōn) A hormone secreted by the pituitary gland that stimulates the thyroid to secrete thyroxine.   395

thyroxine (thi-rok′sin) A hormone produced by the thyroid gland that speeds up the metabolic rate.   395

tissue (tish′yu) A group of specialized cells that work together to perform a specific function.   15

trachea (trā′ke-uh) A major tube supported by cartilage rings that carries air to the bronchi; also known as the windpipe.   349

transcription (tran-skrip′shun) The process of manufacturing RNA from the template surface of DNA. Three forms of RNA that can be produced are mRNA, rRNA, and tRNA.   134

transcription factors (tran-skrip′shun fak′-torz) Proteins that help to control the transcription process by binding DNA or other transcription factors and regulating when RNA polymerase begins transcription.   145

transcriptomics (tran-skrip-tō′-miks) A new field of science developed since the sequencing of the human genome. It looks at when and how much of a particular transcript is made by an organism.   154

transfer RNA (tRNA) (trans′fur) A molecule composed of ribonucleic acid. It is responsible for transporting a specific amino acid into a ribosome for assembly into a protein.   55

transgenic organisms (trans-jen′ik or′gunizms) See **genetically modified (GM) organisms.**   150

translation (trans-la′shun) (protein synthesis) The process whereby the tRNA utilizes the mRNA as a guide to arrange the amino acids in their proper sequence according to the genetic information in the chemical code of DNA.   134

translocation (tran″-slo-ká-shun) A chromosomal mutation in which a portion of one chromosome breaks off and becomes attached to another chromosome.   186

transpiration (tran″spĭ-ra′shun) In plants, the process of water being transported from the soil by way of the roots to the leaves where it evaporates.   290

triiodothyronine (tri″i-ō″dō-thī′row-nēn) A hormone produced by the thyroid gland that speeds up the metabolic rate; similar to thyroxine but more potent.   395

triploblastic (trĭp″low-bla′stik) A condition typical of most animals in which their bodies consist of three layers of cells.   516

trisomy (trīs′oh-me) The presence of three chromosomes instead of the normal number of two resulting from the nondisjunction of homologous chromosomes during meiosis; for example, as in Down syndrome.   184

trophic level (tro′fik lĕ′vel) A step in the flow of energy through an ecosystem.   261

tropism (tro′pizm) Any action to a particular stimulus in which the organism orients toward or away from the stimulus.   508

tropomyosin (tro″po-mi′o-sin) A molecule found in thin filaments of muscle that helps regulate when muscle cells contract.   403

troponin (tro′po-nin) A molecule found in thin filaments of muscle that helps regulate when muscle cells contract.   403

true (neutral) fats (troo fats) Important organic molecules composed of glycerol and fatty acids that are used to provide energy.   56

tumor (too′mor) A mass of undifferentiated cells not normally found in a certain portion of the body.   165

turnover number (turn′o-ver num′ber) The number of molecules of substrate that a single molecule of enzyme can react with in a given time.   96

tympanum (tim′pă-num) The eardrum.   401

U

unique structural organization The organization of living things can be seen at both the organism and cellular levels.   13

unsaturated (un-sat′yu-ra-ted) A term used to describe the carbon skeleton of a fatty acid containing carbons that are double-bonded to each other at one or more points.   56

uracil (yu′rah-sil) A single-ring nitrogenous-base molecule in RNA but not in DNA. It is complementary to adenine.   134

uterus (yu′tur-us) The organ in female mammals in which the embryo develops.   420

V

vacuole (vak′yu-ōl) A large sac within the cytoplasm of a cell, composed of a single membrane.   73

vagina (vuh-ji′nah) The passageway between the uterus and outside of the body; the birth canal.   420

valid (val′id) A term used to describe meaningful data that fit into the framework of scientific knowledge.   6

variable (var′e-ă-bul) Factors in an experimental situation or other circumstance that are changeable.   6

vascular tissues (vas′kyu-lar tish′yus) Specialized tissues that transport fluids in higher plants; xylem and phloem.   498

vector (vek′tor) An organism that carries a disease or parasite from one host to the next.   286

veins (vānz) The blood vessels that return blood to the heart.   342

ventricles (ven′tri-klz) The powerful muscular chambers of the heart whose contractions force blood to flow through the arteries to all parts of the body.   344

vertebrates (vur′tuh-brāts) Animals with backbones.   519

vesicles (ve̅′si̅-kuls) Small, intracellular, membrane-bound sacs in which various substances are stored.   74

villi (vil′e) Tiny fingerlike projections in the lining of the intestine that increase the surface area for absorption.   355

viroid (vi′roid) Infectious particles that are composed solely of single-stranded RNA.   471

virus (vi′rus) A nucleic acid particle coated with protein that functions as an obligate intracellular parasite.   467

vitamin-deficiency disease (vi′tah-min de-fish′en-se di-zēz′) Poor health caused by the lack of a certain vitamin in the diet; for example, scurvy from lack of vitamin C.   380

vitamins (vi′tah-minz) Organic molecules that cannot be manufactured by the body but are required in very low concentrations for good health.   370

voltage (vōl′tij) A measure of the electrical difference that exists between two different points or objects.   389

W

wood (wood) The xylem of gymnosperms and angiosperms.   504

woody vascular plants (wood-e vas′kul-ar plantz) Plants with deposits of cellulose and other compounds (e.g., lignin) in the cell walls that strengthen them, make them more rigid, and bind them

to other neighboring cell walls (e.g., maple and pine trees.)   499

X

X chromosome (eks kro′mo-sōm) The chromosome in a human female egg (and in one-half of sperm cells) that is associated with the determination of sexual characteristics.   412

X-linked gene (eks-lingt jēn) A gene located on the sex-determining X chromosome.   193

xylem (zi′lem) A kind of vascular tissue that transports water from the roots to other parts of the plant.   498

Y

Y chromosome (wi kro′mo-sōm) The sex-determining chromosome in one-half the sperm cells of human males responsible for determining maleness.   412

Z

zygote (zi′gōt) A diploid cell that results from the union of an egg and a sperm.   170

Credits

Photographs

Chapter 1

Opener: © Science VU/Visuals Unlimited; 1.1: © Eldon Enger & Fred Ross/Bob Coyle, photographer; 1.3: © Barrington Brown/Photo Researchers; 1.4a: © David M. Phillips/Visuals Unlimited; 1.4b: © Hank Morgan/Photo Researchers; 1.5(left): © Bettman/CORBIS; 1.5(right): © Bob Coyle/Red Diamond Photos; 1.6(left&right): © Mark E. Gibson/Visuals Unlimited; 1.7: © McGraw-Hill Higher Education/Jim Shaffer, photographer; 1.9(tl): © A. M. Siegelman/Visuals Unlimited; 1.9(tr): © Science VU/Visuals Unlimited; 1.9(bl): © Ray Coleman/Photo Researchers; 1.9(br): © CORBIS website; 1.12: Courtesy EOSAT; 1.12(inset): © Gary Braasch/Stone/Getty Images

Chapter 2

Figure 2.1: Courtesy of Dr. Harold Schott, Equine Exercise Physiology Laboratory at Michigan State University; 2.2a, b: © 2002 Fred Ross

Chapter 3

Opener: From C. E. Creutz, "The annexins and exocytosis" from *Science* 258, Nov. 6, 1992, p. 927 fig 3b (photo provided by A. Burger & R. Huber). Reprinted by permission of the author; 3.1(all): © Eldon Enger & Fred Ross/Jim Shaffer, photographer; 3.2(left): © Susan McCartney/Photo Researchers; 3.2(right): © Will & Deni McIntyre/Photo Researchers; p. 50: © Vol. 12/PhotoDisc; 3.15b: From C. E. Creutz, "The annexins and exocytosis" from *Science* 258, Nov. 6, 1992, p. 927 fig 3b (photo provided by A. Burger & R. Huber). Reprinted by permission of the author

Chapter 4

Opener: M. Sameni and B. F. Sloane, Wayne State University School of Medicine, Detroit, MI; 4.1a: © Stock Montage; 4.1b: © Dr. Jeremy Burgess/Science Photo Library/Photo Researchers; p. 65a: © Kathy Talaro/Visuals Unlimited; p. 65b: © 2003 Fred Ross; p. 65c: © William Ormerod/Visuals Unlimited; 4.6a-c: © David M. Phillips/Visuals Unlimited; 4.11(top): © K. G. Murti/Visuals Unlimited; 4.11(mr): © Warren Rosenberg/Biological Photo Service; 4.11(bl): © David M. Phillips/Visuals Unlimited; 4.11(br): © Richard Rodewald/Biological Photo Service; 4.12: © Fred Ross; 4.15: © Don Fawcett/Photo Researchers; 4.17a: M. Sameni and B. F. Sloane, Wayne State University School of Medicine, Detroit, MI; 4.18: © 2002 Warren Rosenberg/Biological Photo

Service; 4.20: From William Jensen and R. B. Park, *Cell Ultrastructure*, 1967, p. 57. © Brooks/Cole Publishing, a division of International Thompson Publishing, Inc.

Chapter 5

Opener: 5.3a(top & bottom): © McGraw-Hill Higher Education/Jim Shaffer, photographer

Chapter 6

Opener: © Brian Parker/Tom Stack & Associates; 6.16(right): © Brian Parker/Tom Stack & Associates; 6.18 (left): © Manfred Kage/Peter Arnold, Inc.; 6.18 (right): Courtesy of Kenneth Miller/Brown University

Chapter 7

7.10a,b: © Stanley Flegler/Visuals Unlimited; 7.14b: © Biophoto Associates/Photo Researchers; 7.16(tl): Courtesy of USDA/Agriculture Research Service; 7.16(tm): © Vol. 275/CORBIS; 7.16(tr): © Vol. 40/CORBIS; 7.16(bl): © Swiss Federal Institute of Technology; 7.16(bm): © PhotoDisc Website; 7.16(br): © Vol. 16/PhotoDisc

Chapter 8

Opener: © Biophoto Associates/Photo Researchers; 8.10b (1-7 whitefish): © McGraw-Hill Higher Education/Kingsley Stern, photographer; 8.10b (8-daughter cells): © Ed Reschke; 8.10d(all-onion root): © McGraw-Hill Higher Education/Kingsley Stern, photographer; 8.12(left): © Biophoto Associates/Photo Researchers; 8.13: © James Stevenson/Photo Researchers

Chapter 9

Page 183(top & bottom): Photo Courtesy: College of Veterinary Medicine, Texas A&M University 9.24: © M. Coleman/Visuals Unlimited

Chapter 10

Opener: © Renee Lynn/Photo Researchers; 10.1a,b: © McGraw-Hill Higher Education/Bob Coyle, photographer; 10.3(left): © Renee Lynn/Photo Researchers; 10.3(middle, right): Courtesy of Mary Drapeau; 10.4: © John D. Cunningham/Visuals Unlimited; p. 195: Courtesy of Advanced Respiratory, St. Paul, MN; 10.8a-c: Courtesy of Jeanette Navia; 10.9: Courtesy of Neurofibromatosis, Inc., Lanham, MD; 10.10: © McGraw-Hill Higher Education/Bob Coyle, photographer

Chapter 11

Opener: © Earl Roberge/Photo Researchers; 11.1a-c: © Fred Ross; 11.1d: © Reynolds Photography; 11.3(top): © John Cancalosi/Tom Stack & Associates; 11.3(bottom):

© John Serrao/Visuals Unlimited; 11.4(African): © Linda Bartlett/Photo Researchers; 11.4(Europe): © Henry Bradshaw/Photo Researchers; 11.4(Asia): © Lawrence Migdale/Photo Researchers; 11.5: © Gary Milburn/Tom Stack & Associates; 11.7(left & right): Courtesy of Ball Seed Company; 11.8: © Earl Roberge/Photo Researchers; 11.10a,b: © Stanley L. Flegler/Visuals Unlimited

Chapter 12

Opener: © T. Kitchin/Tom Stack & Associates; p. 223: © Walt Anderson/Visuals Unlimited; 12.1: © Gary Connor/Photo Edit; 12.2: © Joe McDonald/Visuals Unlimited; 12.3(left): © Eldon Enger; 12.3(middle): © John Colwell/Grant Heilman Photography; 12.3(right): © Kees VanDen Berg/ Photo Researchers; 12.4: © Ted Levin/Animals Animals/Earth Scenes; 12.5: © Mark Phillips/Photo Researchers; 12.8: © T. Kitchin/ Tom Stack & Associates; 12.10: © John Shaw/ Tom Stack & Associates; 12.11: © Mitch Reardon/Photo Researchers

Chapter 13

Opener: © Mero/Jacana/Photo Researchers, Inc.; 13.1a: © Walt Anderson/Visuals Unlimited; 13.1b: © Eldon Enger; 13.1c: © William J. Weber/ Visuals Unlimited; 13.2(left&right): © Tom & Pat Leeson; 13.5a: Courtesy of Dr. Tim Spira. from *Nature,* Vol. 352 No. 6338, August 6, 29, 1991, cover; 13.5b: Courtesy of Sakata Seed America, Inc.; 13.6: © Mike Blair; 13.7a: © Joel Arrington/Visuals Unlimited; 13.7b: © Mero/ Jacana/Photo Researchers

Chapter 14

Opener: © Eldon Enger; 14.1a: © Gregory K. Scott/ Photo Researchers; 14.1b: © Eldon Enger; 14.12: © Eldon Enger; 14.13: © Pat & Tom Leeson/Photo Researchers; 14.14: © C. P. Hickman/Visuals Unlimited; 14.15, 14.16: © Eldon Enger; 14.17: © Michael Giannenchini/Photo Researchers; 14.18: © Eldon Enger

Chapter 15

Opener: © Eldon Enger; 15.2a: © John D. Cunningham/Visuals Unlimited; 15.2b: © Eldon Enger; 15.3a: © David Walters/Envision; 15.3b: © Stephen Dalton/Photo Researchers; 15.4a: © J. H. Robinson/Photo Researchers; 15.4b: © Gary Milburn/Tom Stack & Associates; 15.4c: © Stephen J. Kraseman/Peter Arnold; 15.6a: © Douglas Faulkner/Sally Faulkner Collection; 15.6b: © Kjell Sanved/Visuals Unlimited; 15.7: © Eldon Enger; 15.8: © Susanna Pashko/Envision

Chapter 16

Opener: © G. R. Higbee/Photo Researchers; 16.3: © G. R. Higbee/Photo Researchers; 16.4a: © Stephen J. Krasemann/Photo Researchers; 16.4b: © Walt Anderson/Visuals Unlimited

Chapter 17

Opener: © Eldon Enger; 17.4: © Eldon Enger; 17.5a,b: © Lincoln P. Brower Nature Photography; 17.6: Courtesy of Sybille Kalas; 17.10: © Eldon Enger; 17.11: © Harry Rogers/Photo Researchers; 17.13: © Eldon Enger; 17.14: © David Cayless/ OSF/Animals, Animals/Earth Scenes

Chapter 18

Figure 18.2a: © Dr. Keith R. Porter

Chapter 19

Opener: Courtesy of David Dempster; p. 365: Copyright © LECO Corporation, St. Louis, Michigan. Reprinted by permission.; 19.3(all): © McGraw-Hill Higher Education/Bob Coyle, photographer; 19.4(all): Courtesy of Dr. Randy Sansone; 19.5: © Paul A. Souders; 19.7a,b: Courtesy of David Dempster; 19.8(left& right): From A. P. Streissguth, S. K. Clarren, and K. L. Jones, "Natural History of the Fetal Alcohol Syndrome; a ten-year follow-up of eleven patients" in *Lancet*, 2:85-91, July 1985. Courtesy of Ann Streissguth, Fetal Alcohol Research Studies, School of Medicine, University of Washington, Seattle, WA

Chapter 21

Opener: © SIU/Peter Arnold, Inc.; 21.1: From M. Bartalos and T. A. Baramski, *Medical Cytogentetics*, Williams & Wilkins, 1967, p. 133, f. 9-2; 21.2(left, right): Courtesy of Dr. Kenneth L. Becker from *Fertility and Sterility*, 23:5668-78, Williams & Wilkins, 1972.; 21.3: Courtesy of Thomas G. Brewster, Foundation for Blood Research, Scarborough, ME. From T. G. Brewster and S. Gerald Park, "Chromosome Disorders Associated with Mental Retardation" in *Pediatric Annals*, 7(2), 1978.; 21.9: © SIU/Peter Arnold, Inc.; 21.14a,c-h: © Mcgraw-Hill Higher Education/ Bob Coyle, photographer; 21.14b: © Hank Morgan/Photo Researchers, Inc.; 21.14i: © Ortho-McNeil Pharmaceutical, Inc.

Chapter 22

Figure 22.7: © Fred Ross; 22.8: Courtesy of Abe Tetsuya; 22.12: "The Age of Reptiles" a mural by Rudolph F. Zallinger, © 1966, 1975, 1985, 1989, Peabody Museum of Natural History, Yale University, New Haven, CT

Chapter 23

Opener: © Don Valenti/Tom Stack & Associates; 23.1a: © Tom McHugh/Photo Researchers; 23.1b: © Russ Kinne/Comstock; 23.2: © Stock Montage; 23.3a: © A. M. Siegelman/Visuals Unlimited; 23.3b: © John Gerlach/Tom Stack & Associates; 23.3c: © Paul W. Johnson/Biological Photo Service; 23.3d: © Eldon Enger; 23.3e: © Glenn Oliver/Visuals Unlimited; 23.4a: © W. B. Saunders/Biological Photo Service; 23.4b: © Eldon Enger; 23.4c: US Geological Survey; 23.5: © Frank T. Aubrey/ Visuals Unlimited; 23.6a: © Tom Stack/Tom Stack & Associates; 23.6b: © Don Valenti/Tom Stack & Associates; 23.8a: © Alfred Pasieka/SPL/Photo Researchers; 23.9(flower): © Eldon Enger; 23.9(conifer): © Eldon Enger; 23.9(moss): © Claudia E. Mills/Biological Photo Service; 23.9(algae): © P. Nuridsany/ Photo Researchers; 23.9(fern): © Eldon Enger

Chapter 24

Opener: © CNRI/SPL/Photo Researchers; Table 24.2(left): © David Phillips/Visuals Unlimited; Table 24.2(ml): © R. Kessel/G. Shih/ Visuals Unlimited; Table 24.2(mr): © Moredum Animal Health/SPL/Photo Researchers; Table 24.2(right): © CNRI/SPL/Photo Researchers; 24.2: © Y. Arthus-Bertrand/Peter Arnold; 24.3: USDA; 24.4: © T. J. Beveridge/Biological Photo Service; Table 24.3(chromista): © Cabisco/ Visuals Unlimited; Table 24.3(green algae): © John D. Cunningham/Visuals Unlimited; Table 24.3(plantae): © Eldon Enger; Table 24.3(choanoflagellata): Courtesy of Serguel Karpov; Table 24.3(animalia): © Fred Ross; Table 24.3(fungi): © Fred Ross; Table 24.3(flagallates): © David M. Phillips/Visuals Unlimited; Table 24.3(slime mold): © B. Beatty/ Visuals Unlimited; Table 24.3(Rhodophyta): © Brian Parker/Tom Stack & Associates; Table 24.3(alveolates): © M. Abbey/Visuals Unlimited; Table 24.3(radiolaria): © E. R. Degginger/Photo Researchers; Table 24.3 (testaceafilosea): © A. M. Siegelman/Visuals Unlimited; 24.6: © David J. Wrobel/Biological Photo Service; 24.5a: Courtesy of Leland Johnson; 24.5b: © Carolina Biological Supply/Phototake; p. 484(top box): © Richard J. Green/Photo Researchers; p. 484(bottom box): © Eric Grave/Photo Researchers; 24.7a: © Michael Abbey/Photo Researchers; 24.7b: © Science VU/Visuals Unlimited; 24.9: © M. Abbey/Visuals Unlimited; 24.10: © Edward S. Ross; 24.11: © Cabisco/Visuals Unlimited; 24.12a: © Cabisco/Visuals Unlimited; 24.12c: © Manfred Kage/Peter Arnold; 24.13a: © Barbara J. Miller/Biological Photo Service; 24.13b: © Bill Keogh/Visuals Unlimited; 24.13c: © S. Flegler/Visuals Unlimited; 24.13d: © Fred Ross; p. 488: © Richard Humbert/Biological Photo Service; 24.14: © R. Roncardori/Visuals Unlimited; 24.15: © Nancy M. Wells/Visuals Unlimited; 24.16: © David M. Dennis/Tom Stack & Associates; 24.17a,b: © Edward S. Ross; Table 24.4(1): Courtesy of Tom Volk, University of Wisconsin-LaCrosse; Table 24.4(2): © D. Yeske/ Visuals Unlimited; Table 24.4(3): © M. Abbey/ Visuals Unlimited; Table 24.4(4): © Edward S. Ross; Table 24.4(5): © Fred Ross

Chapter 25

Opener: © Eldon Enger; 25.1a,b: Fred Ross; 25.4b: © Robert & Linda Mitchell Photography; 25.9a: © John Sohlden/Visuals Unlimited; 25.9b: © McGraw-Hill Higher Education/Carlyn Iverson, photographer; 25.11a: © Eldon Enger; 25.11b,c: © Fred Ross; 25.15a: © John Gerlach/ Tom Stack & Associates; 25.15b: © Bruce Wilcox/ Biological Photo Service; 25.15c: © L. West/ Photo Researchers; 25.17: © Fred Ross; 25.19b: © Steve Callahan/Visuals Unlimited; 25.19c: © Fred Ross; 25.22: © Cathlyn Melloan/Stone/Getty Images; 25.23: © Fred Ross; Table 25.1(sphenopsida): © Biophoto Associates/ Photo Researchers; Table 25.1 (lycophyta): © David Sieren/Visuals Unlimited; Table 25.1(zostero): Courtesy of Michael E. Kotyk, Patrick C. Gensel and James F. Basinger from Kotyk, M. E. and J. F. Bassinger, "The Early Devonian" (Pragian zosterophyll) Bathurstia denticulata. Hueber, *Canadian Journal of Botany*, 2000; Table 25.1(charophytes): © John D. Cunningham/ Visuals Unlimited; Table 25.1(hepaticophyta): © David Sieren/Visuals Unlimited; Table 25.1 (anthocerotophyta): © Henry Robison/Visuals Unlimited; Table 25.1 (bryophyta): © David Wrobel/Visuals Unlimited; Table 25.1 (rhyniophytes): © Sinclair Stammers/Science Source Photo Library/Photo Researchers; Table 25.1(seed plants): © Frcd Ross; Table 25.1(pteriodopsida): © Eldon Enger

Chapter 26

Opener: © Paul Janosi/Valan Photos; 26.1a-h: © Fred Ross; 26.10a: © Peter Parks/ Oxford Scientific Films Ltd; 26.10b: © Robert Evans/ Peter Arnold, Inc.; p. 522a(top & bottom): © Howard Russell, Michigan State University; 26.11a,b: © Bruce Russell/ BioMedia Associates; 26.14a: © Ed Reschke/Peter Arnold, Inc.; 26.14b: © Norm Thomas/Photo Researchers; 26.16a: © Michael DiSpezio; 26.16b: © Paul Janos/Valan Photos; 26.17a: © Robert A. Ross/RARE Photography; 26.17b: © David D. Fleetham/Tom Stack & Associates; 26.18b,c: © Michael DiSpezio; 26.20a: © Russ Kinne/Photo Researchers; 26.20b: © Tom Stack/Tom Stack & Associates; 26.21a: © Marty Snyderman; 26.21b: © Ed Robinson/ Tom Stack & Associates; 26.22a: © Tom McHugh/Photo Researchers, Inc.; 26.22b: © David Doubilet; 26.22c: © Michael Dispezio; 26.23a: © Kent Dannen/Photo Researchers; 26.23b: © Bruce Russell/BioMedia Associates; 26.23c: © Michael DiSpezio; 26.25a-d: © Peter J. Bryant/Biological Photo Service; 26.26a,b: © Dwight Kuhn; 26.27a: © Darryl Torckler/ Stone/Getty Images; 26.27b: © Don & Esther Phillips/Tom Stack & Associates; p. 533: © Charles Cole; 26.29a: © Tom Ulrich/Visuals Unlimited; 26.29b: Joe McDonald/Visuals Unlimited; 26.30 © Tom McHugh/Photo Researchers; 26.31a: © T. J. Cawley/Tom Stack & Associates; 26.31b: © Fred Ross; 26.32a: © John Gerlach/Tom Stack & Associates; 26.32b: © Kim Westerkov/ Stone/Getty Images

Index